Brian J. Huntley • Vladimir Russo
Fernanda Lages • Nuno Ferrand
Editors

Biodiversity of Angola

Science & Conservation: A Modern Synthesis

Editors
Brian J. Huntley
CIBIO-InBIO, Centro de Investigação em Biodiversidade e Recursos Genéticos
Universidade do Porto
Vairão, Portugal

Fernanda Lages
ISCED – Instituto Superior de Ciências da Educação da Huíla
Lubango, Angola

Vladimir Russo
Fundação Kissama
Luanda, Angola

Nuno Ferrand
CIBIO-InBIO, Centro de Investigação em Biodiversidade e Recursos Genéticos
Laboratório Associado, Campus de Vairão
Universidade do Porto
Vairão, Portugal

Departamento de Biologia, Faculdade de Ciências
Universidade do Porto
Porto, Portugal

Department of Zoology, Auckland Park
University of Johannesburg
Johannesburg, South Africa

https://doi.org/10.1007/978-3-030-03083-4

To the new generation of Angolan students of biodiversity. May they stand on the shoulders of giants: the founders of Angola's biodiversity science

Friedrich Martin Josef Welwitsch (1806–1872)
José Vicente Barbosa du Bocage (1823–1907)
José Alberto de Oliveira Anchieta (1832–1897)
Johannes (John) Gossweiler (1873–1952)

Contributors to this volume record with sadness the passing of William Roy (Bill) Branch (1946–2018) - herpetologist, indefatigable field researcher and mentor of young Angolan scientists

Foreword

His Excellency the President of the Republic of Angola, João Manuel Gonçalves Lourenço

Angola occupies only four percent of the terrestrial area of the African continent, yet it possesses the highest number of biomes of any African country. It is second in terms of the number of ecoregions represented within its borders. It has ecosystems as diverse as the rainforests of Maiombe of Cabinda to the vegetation-less dunes of Namibe and the endless savannas and woodlands of the Cuando Cubango to the tiny remnant forests of the highest valleys of Mount Moco in Huambo. It is the only home to the most magnificent mammal in the world – the Giant Sable Antelope.

It was in Angola that one of the most extraordinary plant species *Welwitschia mirabilis* was discovered and described – the enigmatic 'living fossil' of the desert. It even puzzled Charles Darwin, who compared its evolutionary importance in the plant kingdom to that of the Duck-billed Platypus in the animal kingdom. Considering real fossils, Angola's history goes back hundreds of millions of years, to the earliest known living organisms, the bacterial stromatolites of the limestones of Bembe and Humpata. Angola's fossils range in size from microorganisms to the gigantic dinosaur – *Angolatitan adamastor* – recently discovered in the sediments of the Bengo coastline. Yet, despite this globally significant natural wealth, Angola remains one of the least well-documented countries in the world in terms of its biodiversity. This situation is about to change.

Angolan scientists have collaborated with over 40 colleagues from 7 countries to produce a new synthesis of knowledge of Angola's remarkable biodiversity. They have produced a magnificent volume that seeks to review all what is known about Angola's biodiversity, especially that which has been revealed through studies undertaken in the twenty-first century. For several decades, field studies were rendered nearly impossible because of the disruptions of war. But since peace was achieved in 2002, a new generation of research has been made possible, bringing many foreign specialists into partnerships with Angolan scientists and institutions and introducing new technologies that have helped stimulate an unprecedented wave of research activity.

Angola's indigenous knowledge acquired over millennia provided the foundations to the information documented and materials collected by visiting researchers from the eighteenth to twenty-first centuries. Many detailed accounts have been published over the past century and more, but often in scientific journals and official reports that have been lost with the passage of time. More important, the knowledge that does exist is fragmentary and largely inaccessible to Angolan students and researchers. Much is published in foreign languages, and thus the inexistence of a comprehensive synthesis of studies on Angola's fauna, flora and ecosystems is a challenge for young researchers. Angolan students, researchers and government officials have very limited access to data sources where they can find a reliable, science-based and up-to-date summary of that which has been recorded on the country's biodiversity.

It was the need for an integrated 'state-of-knowledge' summary, recognised during the past decade by many Angolan university and government colleagues, that was the catalyst that stimulated this important project. From humble beginnings by a few Angolan and foreign partners, the effort expanded into the present volume of over 500 pages of authoritative accounts on our landscapes, seascapes, vegetation, flora and fauna, its past and future. Most importantly, this work identifies the exciting opportunities for research and conservation that Angolan scientists, conservationists, government officials and the general public can embrace as the country moves forward to an ever greater and more prosperous and environmentally sustainable future.

It is a pleasure to endorse this valuable contribution to the new wave of Angolan scientific and conservation literature, a source of inspiration to our students and a reminder to all our leaders, young and old, of our responsibility to treasure and safeguard Angola's unsurpassed, but vulnerable, biodiversity and natural resources.

President of the Republic of Angola
Luanda, Angola

João Manuel Gonçalves Lourenço

Acknowledgements

This book was conceived as a collaborative and voluntary endeavour shared by students of Angola's biodiversity. Contributors were drawn from Angola, Britain, Germany, Namibia, Portugal, Swaziland, South Africa, the Netherlands and the United States. To all contributors to this synthesis, the Editors offer their thanks for the unstinting efforts of the synthesis team to meet the strict demands of quality and of timelines.

Over the past decades, biodiversity research in Angola has been encouraged by successive leaders within the government, from academics and from the general public of Angola. On behalf of all the contributors to this volume, the Editors wish to thank the Ministers, past and present, of Science and Technology, Dr Cândida Teixeira and Dr Maria do Rosário Bragança Sambo, and of Environment, Dr Fátima Jardim and Dr Paula Francisco, for the support given to students of Angola's biodiversity. Similarly, the encouragement and logistical support of Prof Liz Matos, Prof Serôdio d'Almeida (Universidade Agostinho Neto), Dr Charles Skinner (De Beers, Angola) and General João Traguedo (Lubango) are gratefully acknowledged. Without their strong support, the scientific results described in this synthesis would not have been possible.

Special thanks are due to Martim Melo, who provided his superb copy-editing, proofreading and technical support during the final preparation of the manuscript for submission to the publishers. Similarly, Pedro Tarroso and John Mendelsohn offered their graphic design skills to greatly improve many figures and maps.

Photos were also generously offered by the chapter contributors and by Maans Booysen, Merle Huntley, Tassos Leventis, Lars Petersson, Fiona Tweedie and Alexandre Vaz.

The financial and logistic support of CIBIO and the UNESCO Chair *Life on Land* (University of Porto, in association with *Twinlab* partners in Angola, Namibia, Mozambique, South Africa and Zimbabwe) in this project was fundamental to its success.

Finally, our thanks go to Margaret Deignan, Esther Rentmeester, Malini Arumugam and Maadhuri Kandrakota of Springer (English Edition) and to Jorge Reis-Sa of *Arte e Ciência* (Portuguese Edition), for their consistent and professional support throughout the project.

Contents

Contributors

Ninda Baptista Instituto Superior de Ciências da Educação da Huíla, Rua Sarmento Rodrigues, Lubango, Angola

National Geographic Okavango Wilderness Project, Wild Bird Trust, Parktown, Gauteng, South Africa

CIBIO-InBIO, Centro de Investigação em Biodiversidade e Recursos Genéticos, Laboratório Associado, Universidade do Porto, Vairão, Portugal

Pedro Beja CIBIO-InBIO, Centro de Investigação em Biodiversidade e Recursos Genéticos, Universidade do Porto, Vairão, Portugal

CEABN-InBio, Centro de Ecologia Aplicada "Professor Baeta Neves", Instituto Superior de Agronomia, Universidade de Lisboa, Lisboa, Portugal

Elena Bersacola Nocturnal Primate Research Group, Faculty of Humanities and Social Sciences, Oxford Brookes University, Oxford, UK

A. Bivar-de-Sousa Museu Nacional de História Natural e da Ciência, Universidade de Lisboa, Lisboa, Portugal

Sociedade Portuguesa de Entomologia, Lisboa, Portugal

William R. Branch (deceased) National Geographic Okavango Wilderness Project, Wild Bird Trust, Parktown, Gauteng, South Africa

Department of Zoology, Nelson Mandela University, Port Elizabeth, South Africa

Pedro M. Callapez CITEUC; Departamento de Ciências da Terra, Faculdade de Ciências e Tecnologia, Universidade de Coimbra, Coimbra, Portugal

Viola Clausnitzer Senckenberg Museum for Natural History, Görlitz, Görlitz, Germany

Werner Conradie National Geographic Okavango Wilderness Project, Wild Bird Trust, Hogsback, South Africa

School of Natural Resource Management, Nelson Mandela University, George, South Africa

Port Elizabeth Museum (Bayworld), Humewood, South Africa

W. Richard J. Dean DST-NRF Centre of Excellence at the FitzPatrick Institute, University of Cape Town, Rondebosch, South Africa

Klaas-Douwe B. Dijkstra Naturalis Biodiversity Center, Leiden, The Netherlands

Sara R. F. Fernandes Elizalde SASSCAL – BID GBIF, Instituto de Investigação Agronómica, Huambo, Angola

Ezequiel Fabiano Department of Wildlife Management and Ecotourism, Katima Mulilo Campus, Faculty of Agriculture and Natural Resources, University of Namibia, Katima Mulilo, Namibia

Nuno Ferrand CIBIO-InBIO, Centro de Investigação em Biodiversidade e Recursos Genéticos, Laboratório Associado, Campus de Vairão, Universidade do Porto, Vairão, Portugal

Departamento de Biologia, Faculdade de Ciências, Universidade do Porto, Porto, Portugal

Department of Zoology, Auckland Park, University of Johannesburg, Johannesburg, South Africa

Rui Figueira CIBIO-InBIO, Centro de Investigação em Biodiversidade e Recursos Genéticos, Universidade do Porto, Vairão, Portugal

CEABN-InBio, Centro de Ecologia Aplicada "Professor Baeta Neves", Instituto Superior de Agronomia, Universidade de Lisboa, Lisboa, Portugal

Manfred Finckh Institute for Plant Science and Microbiology, University of Hamburg, Hamburg, Germany

Amândio Gomes Institute for Plant Science and Microbiology, University of Hamburg, Hamburg, Germany

Faculty of Sciences, Agostinho Neto University, Luanda, Angola

António Olímpio Gonçalves Departamento de Geologia, Faculdade de Ciências, Universidade Agostinho Neto, Luanda, Angola

Francisco Maiato P. Gonçalves National Geographic Okavango Wilderness Project, Wild Bird Trust, Parktown, Gauteng, South Africa

Instituto Superior de Ciências da Educação da Huíla, Lubango, Angola

University of Hamburg, Institute for Plant Science and Microbiology, Hamburg, Germany

David J. Goyder Herbarium, Royal Botanic Gardens, Kew, Richmond, Surrey, UK

National Geographic Okavango Wilderness Project, Wild Bird Trust, Parktown, Gauteng, South Africa

Brian J. Huntley CIBIO-InBIO, Centro de Investigação em Biodiversidade e Recursos Genéticos, Universidade do Porto, Vairão, Portugal

Louis L. Jacobs Roy M. Huffington Department of Earth Sciences, Southern Methodist University, Dallas, TX, USA

Jens Kipping BioCart Ökologische Gutachten, Taucha/Leipzig, Germany

Stephen P. Kirkman Department of Environmental Affairs, Oceans and Coasts Research, Cape Town, South Africa

Fernanda Lages ISCED – Instituto Superior de Ciências da Educação da Huíla, Lubango, Angola

Octávio Mateus GeoBioTec, Faculdade de Ciências e Tecnologia, Universidade Nova de Lisboa, Lisbon, Portugal

Museu da Lourinhã, Rua João Luis de Moura, Lourinhã, Portugal

Martim Melo DST-NRF Centre of Excellence at the FitzPatrick Institute, University of Cape Town, Rondebosch, South Africa

CIBIO-InBIO, Centro de Investigação em Biodiversidade e Recursos Genéticos, Laboratório Associado, Universidade do Porto, Vairão, Portugal

Instituto Superior de Ciências da Educação de Huíla, Lubango, Angola

John M. Mendelsohn RAISON (Research & Information Services of Namibia), Windhoek, Namibia

Luís F. Mendes Museu Nacional de História Natural e da Ciência, Universidade de Lisboa, Lisboa, Portugal

CIBIO, Centro de Investigação em Biodiversidade e Recursos Genéticos, Vairão, Portugal

Michael S. L. Mills Instituto Superior de Ciências da Educação de Huíla, Lubango, Angola

A. P. Leventis Ornithological Research Institute, University of Jos, Jos, Plateau State, Nigeria

Ara Monadjem Department of Biological Sciences, University of Swaziland, Kwaluseni, Swaziland

Mammal Research Institute, Department of Zoology and Entomology, University of Pretoria, Pretoria, South Africa

Pedro Monterroso CIBIO-InBIO, Centro de Investigação em Biodiversidade e Recursos Genéticos, Universidade do Porto, Vairão, Portugal

Miguel Morais Faculdade de Ciências, Universidade Agostinho Neto, Luanda, Angola

Kumbi Kilongo Nsingi Benguela Current Convention, Swakopmund, Namibia

Jorge M. Palmeirim Departamento de Biologia Animal, Faculdade de Ciências, cE3c – Centre for Ecology, Evolution and Environmental Changes, Universidade de Lisboa, Lisboa, Portugal

Pedro Vaz Pinto Fundação Kissama, Luanda, Angola

CIBIO-InBIO, Centro de Investigação em Biodiversidade e Recursos Genéticos, Universidade do Porto, Campus de Vairão, Vairão, Portugal

Michael J. Polcyn Roy M. Huffington Department of Earth Sciences, Southern Methodist University, Dallas, TX, USA

Rasmus Revermann Institute for Plant Science and Microbiology, University of Hamburg, Hamburg, Germany

Vladimir Russo Fundação Kissama, Luanda, Angola

Anne S. Schulp Naturalis Biodiversity Center, Leiden, The Netherlands

Faculty of Earth and Life Sciences, VU University Amsterdam, Amsterdam, The Netherlands

Paul H. Skelton South African Institute for Aquatic Biodiversity (SAIAB), Grahamstown, South Africa

Wild Bird Trust, National Geographic Okavango Wilderness Project, Hogsback, South Africa

Magdalena S. Svensson Nocturnal Primate Research Group, Faculty of Humanities and Social Sciences, Oxford Brookes University, Oxford, UK

Peter John Taylor School of Mathematical & Natural Sciences, University of Venda, Thohoyandou, South Africa

Luís Veríssimo Fundação Kissama, Luanda, Angola

Caroline R. Weir Ketos Ecology, Kingsbridge, Devon, UK

Mark C. Williams Pretoria University, Pretoria, South Africa

Paulina Zigelski Institute for Plant Science and Microbiology, University of Hamburg, Hamburg, Germany

Part I
Introduction: Setting the Scene

Chapter 1
Angolan Biodiversity: Towards a Modern Synthesis

Brian J. Huntley and Nuno Ferrand

Abstract Angola possesses an unusually rich diversity of ecosystems and species, but this natural wealth is poorly documented when compared with other countries in the region. Both colonial history and extended wars challenged progress in biodiversity research and conservation, but since peace was achieved in 2002 a rapidly increasing level of collaboration between Angolan and visiting scientists and institutions has seen a blossoming of biodiversity research. The absence of comprehensive reviews and syntheses of existing knowledge, often published in extinct journals and inaccessible official reports, necessitates a modern synthesis. This volume brings together the existing body of scientific results from studies on Angola's landscapes, ecosystems, flora and fauna, and presents an outline of opportunities for biodiversity discovery, understanding and conservation as well as collaborative research.

Keywords Africa · Biomes · Collaborative research · Conservation · Ecoregions

Background and Context

Angola is a country of unusually rich physiographic, climatic and biological diversity. It occupies only 4% of the terrestrial area of Africa, yet it possesses the highest diversity of biomes and is second only to mega-diverse South Africa in terms of the number of ecoregions found within its borders. However, scientific literature on its

B. J. Huntley (✉)
CIBIO-InBIO, Centro de Investigação em Biodiversidade e Recursos Genéticos, Universidade do Porto, Vairão, Portugal
e-mail: brianjhuntley@gmail.com

N. Ferrand
CIBIO-InBIO, Centro de Investigação em Biodiversidade e Recursos Genéticos, Laboratório Associado, Campus de Vairão, Universidade do Porto, Vairão, Portugal

Departamento de Biologia, Faculdade de Ciências, Universidade do Porto, Porto, Portugal

Department of Zoology, Auckland Park, University of Johannesburg, Johannesburg, South Africa
e-mail: nferrand@cibio.up.pt

B. J. Huntley et al. (eds.), *Biodiversity of Angola*,
https://doi.org/10.1007/978-3-030-03083-4_1

biodiversity is extremely limited when compared with most African countries. Much of that which has been published is difficult to access or out of print. This volume seeks to redress this situation.

Here we present a review of what is known about Angola's biodiversity. Much of the existing literature dates from the nineteenth and early to mid-twentieth centuries. Following independence in 1975, field studies were curtailed by the instabilities of an extended civil war. It was not until after the peace settlement of 2002 that a new wave of research has been possible. Initial attempts to establish collaborative field expeditions were frustrated by visa and permit restrictions, but these challenges were gradually overcome and by the 2010s a vibrant programme of joint projects has evolved. Today many foreign specialists work in partnership with Angolan researchers and institutions, producing a new flow of scientific results of which many are presented in this volume.

For any comprehensive synthesis, both temporal depth and spatial breadth is necessary. An historical perspective is presented in each chapter. Angolan indigenous knowledge has contributed to the insights and materials that have informed visiting researchers from the eighteenth century to the present day. The pioneering studies and exhaustive botanical collections of the Austrian botanist, Friedrich Welwitsch (1806–1872), the zoological collections of the indefatigable Portuguese naturalist José Anchieta (1832–1897) and the Swiss botanist John Gossweiler (1873–1952) set benchmarks for later work (Swinscow 1972; de Andrade 1985). Each succeeding student of Angola has added to the description of its biological diversity. While botanists such as Romero Monteiro (1970) and zoologists such as Crawford-Cabral (1983) have summarised available biogeographic information within a national context, no comprehensive synthesis of studies on Angola's fauna, flora and ecosystems has yet been undertaken. The need for an integrated account has become evident in the past decade, as increasing numbers of expeditions and collaborative projects have evolved as part of the country's 'peace dividend'.

Approach and Purpose of This Synthesis

A modern synthesis is not easily achieved. Much of the early literature on Angola's biodiversity resides in publications and reports that are difficult to source. This review attempts to reference these important but sometimes elusive accounts, in order to provide students with access to what information is available. While focusing on papers in peer-reviewed journals, some topics need to draw on unpublished reports filed in government departments. It also seeks to bring together the findings of recent, post-independence studies, many of which are still in progress or in press. It is intended to serve the new generation of Angolan students by providing a comprehensive but focused synopsis of what is known on the biomes, landscapes, flora and fauna of Angola. It should also bring Angola to the attention of researchers across Africa and beyond, revealing the great diversity of life, and the multiple questions on the structure and functioning of Angola's biodiversity that await exploration, examination and explanation.

In structuring this book, this introduction leads through synopses on the country's terrestrial and marine biogeography, paleontological record, recent landscape evolution and land transformation, to chapters on its flora and vegetation. The main body of this volume is devoted to accounts of its fauna – selected invertebrate groups that have promise as indicators of environmental stress, and all vertebrate groups. In each treatment, the need for increased conservation measures for threatened taxa and habitats is a recurrent theme, while research opportunities are highlighted. While general inventories and checklists are progressing well, the state of ecological knowledge remains rudimentary. Topics as fundamental as ecological processes such as the flows of energy, water and nutrients; the ecological impacts of phenomena such as fire, invasive species, herbivory, droughts and frosts; community structure, plant-animal interactions and the impacts of land-transformation and of climate change are yet to be researched in Angola. This volume's content is limited by the availability of information. It is therefore opportunistic, covering those taxonomic groups and those features and processes for which a critical mass of information is available. The focus is primarily on the terrestrial ecosystems and biota of Angola, but the importance of the marine environment is described in accounts on marine biodiversity and ocean dynamics, and on the richness of the whale, dolphin and marine turtle faunas of Angolan waters.

In comparison with similar reviews for other African countries with long and strong traditions of research into their biodiversity and ecology, and for which comprehensive syntheses of the state of knowledge are available (e.g. Namibia: Barnard 1998; Southern Africa: Davis 1964; Werger and van Bruggen 1978; Huntley 1989; Tanzania: Sinclair 2012), this account reveals both the strengths and weaknesses of the research agenda of the colonial era, and the challenges of the recent past. While institutions such as the *Instituto de Investigação Científica de Angola* and the *Instituto de Investigação Agronómica de Angola* undertook very important studies on many taxa, and on vegetation, soils and agronomy, and the *Museu do Dundo* amassed and distributed a vast series of collections of the animal species of the Lundas, the coverage of disciplines and of the remote regions of Angola was weak.

Biodiversity Surveys: Historical Synopsis

The history of scientific exploration and biological collection in Angola is relatively modest. Whereas South Africa, by 1975, had over three million herbarium specimens collected by some 2500 botanists since the late eighteenth century (Gunn and Codd 1981), Angola had less than 300,000 specimens collected by just 300 botanists during the same period (Figueiredo and Smith 2008). Despite the relatively limited coverage of Angolan collections, the great botanist Francisco Mendonça was occasioned to state in his preface to Gossweiler and Mendonça (1939):

> We are happily able to confirm that the flora of Angola is the best known in tropical Africa, due to the attention given by the state towards the botanical exploration of the colony, and the great interest and zeal of scientists in its study.

The Swiss zoologist, Monard (1935), had been less sanguine:

> A regrettable fact about the Natural History of Angola is the scarcity of concrete information about the nature, distribution, and habits of the large game. The Boers … never communicated their observations. The Portuguese hunters … did not write reports on their hunts, or if they did, did so in newspapers or magazines that never enter the scientific literature. The observations remain, in this way, lost to the naturalist who is not able to locate such work.

In truth, during the colonial era, investment in research on the country's biodiversity was limited. The achievements of early pioneers such as Friedrich Welwitsch, José Anchieta and John Gossweiler were quite remarkable, and those of more recent agronomists, botanists and zoologists such as Castanheira Diniz, Romero Monteiro, Grandvaux Barbosa, Brito Teixeira, Crawford-Cabral, Rosa Pinto, Barros Machado, etc., were equally laudable, indeed amazing.

The war years from 1975 to 2002 saw very few researchers venturing into the field. Most activities were limited to brief searches for remnant populations of giant sable (Estes 1982), marine turtles (Carr and Carr 1991), birds (Günther and Feiler 1986a, b; Hawkins 1993), and a countrywide assessment of the state of wildlife populations (Huntley and Matos 1992). The Southern African Botanical Diversity Network (SABONET) project attempted to stimulate botanical studies in Angola from the mid-1990s (Huntley et al. 2006), while the Kissama Foundation funded a vegetation survey of the northern extreme of Quiçama (Jeffrey 1996) and introduced a mixed assemblage of antelope and ostriches into the park in 2000 (Walker 2004). The last decades of the twentieth century were aptly described as a period of *confusão* – confusion (Maier 2007). In brief, from Angola's independence in 1975, until the twenty-first century, cooperative field research over most of the country was challenged by the impacts of war. But dramatic and positive change came with the dawn of the new millennium.

Research Collaboration in the Twenty-First Century

From 2000, especially after the peace agreement of April 2002, field activities expanded rapidly. Most notably, Vaz Pinto has focused on a long-term study on giant sable in Cangandala (Walker 2004; Vaz Pinto 2018), Morais (2017) has led surveys of marine turtles along the Angolan coast, and Mills (2010, 2018) has undertaken field studies on birds across the country.

International support for environmental conservation and research strengthened from 2001, when the Global Environment Facility, through the United Nations Development Programme, initiated a multidisciplinary project to develop a transboundary diagnostic analysis of hydro-environment threats within the Okavango River Basin, known as the Environmental Protection and Sustainable Management of the Okavango River Basin Project (EPSMO). The project's aim was to facilitate the protection of the Basin's aquatic ecosystems and biological diversity (OKACOM 2009, 2011). The project included participation from Angola, Botswana and

Namibia and provided a strong impetus to future multi-national projects in the Basin. A further initiative, the Integrated River Management Project, was funded by USAID/Southern Africa between 2004 and 2009 and provided both institutional and management planning support to the national partners (OKACOM 2009). The EPSMO project was succeeded by the SAREP project described below.

The OKACOM projects were focused on major water management needs and did not embrace detailed biodiversity surveys. Indeed, until 2009, biodiversity research activities in Angola had been essentially individual efforts, with limited funding. Difficulties experienced in obtaining visas to visit, and permits to collect specimens in Angola were a continuing challenge faced by foreign scientists. With the signing of an agreement between the South African National Biodiversity Institute (SANBI), the Angolan Ministry of Environment and the *Instituto Superior de Ciências da Educação* (ISCED), Lubango, in 2009, more ambitious cooperative biodiversity projects became possible. Initially designed as training exercises, the series of Rapid Biodiversity Assessments, in Huíla/Namibe (Huntley 2009), Lunda-Norte (Huntley 2011; Huntley and Francisco 2015) and across western Angola (Rejmánek et al. 2017), brought over 40 scientists from 14 countries to Angola to work with local students and researchers.

By the early 2010s, a wide diversity of major cooperative programmes had developed, including those of the Southern African Regional Environmental Programme (SAREP), the Southern African Science Service Centre for Climate Change and Adaptive Land Management (SASSCAL) (Revermann et al. 2018), the National Geographic Okavango Wilderness Project (NGOWP 2018), and conservation initiatives of NGOs such as *Elephants without Borders*, *Panthera*, Peace Parks Foundation, the Kavango-Zambezi Transfrontier Conservation Area (KAZA) project and several others. Collaboration between foreign museums and universities and Angolan counterparts stimulated additional specialist interests, collectively gaining momentum until the present. In October 2012, CIBIO (Research Centre in Biodiversity and Genetic Resources) at the University of Porto, Portugal, and ISCED-Huíla (Lubango) established a collaborative research, capacity building and advanced training project – the ISCED/CIBIO TwinLab initiative. The initiative was soon replicated in South Africa, Mozambique, Namibia and Zimbabwe, and the whole network of TwinLabs now forms a UNESCO Chair *Life on Land*, awarded at the end of 2017.

For much of the post-independence period, biodiversity research efforts had been un-coordinated and opportunistic. With the establishment of the *Instituto Nacional de Biodiversidade e Áreas de Conservação* (INBAC) in 2011, the opportunity arose for a greater level of coordination and priority setting. The *Plano Estratégico da Rede Nacional da Áreas de Conservação de Angola* (GoA 2011), provided a stimulus to studies in key biodiversity hotspots such as Mount Moco, Mount Namba, Serra da Neve, Serra Pingano, Cumbira, Lagoa Carumbo, and to the vast and very poorly researched catchments of the Cuando Cubango. While a greater level of inter-institutional collaboration is still possible, the momentum developed over the past decade has been unprecedented since 1975. The successes of the recent past are presented in this volume, often drawing on work that is still in progress, is unpublished, or is in press.

Chapter Outlines

Angola is a large country, and as emphasised throughout this volume, it has a rich diversity of landscapes, seascapes and associated biomes and ecoregions. The history of biodiversity research in Angola stretches over 200 years. The spatial, temporal and taxonomic scales embraced in this book results in it being structured in five parts. Part I, Chap. 1 (Huntley and Ferrand this chapter) provides an introduction to the book and its content. Chap. 2 (Huntley 2019) outlines the country's biogeography, drawing on the long history of geomorphological and landscape analysis in Angola, and describes the diversity of seven terrestrial biomes, 15 ecoregions and 32 vegetation types. In Chap. 3 Kirkman and Nsingi (2019) synthesise the findings of recent multi-national research activities on the Benguela Current Large Marine Ecosystem project and other studies on Angola's coastal and marine systems. The long history of the evolution of Angola's biota is introduced by Mateus et al. (2019) in Chap. 4, where the exciting recent discoveries in Angola's fossil record, most especially that of the Cretaceous, is described. A highlight was the discovery of the sauropod dinosaur *Angolatitan adamastor*, the first dinosaur to be found in Angola (Mateus et al. 2011). These authors emphasise the fact that for very long periods – hundreds of millions of years – the absence of fossiliferous rocks in Angola excludes the possibility of tracking animal and plant evolution in Angola over such long periods.

Part II presents an historical and contemporary analysis of our understanding of the country's flora and vegetation and on curious patterns and evolutionary processes in some typical Angolan plant communities. In Chap. 5 Goyder and Gonçalves (2019) note that the vascular flora now totals 6850 species, with 14.8% of these being endemic. The two early vegetation maps of Angola, prepared by the pioneers Gossweiler and Mendonça (1939) and Barbosa (1970), having served the country for many decades, now deserve renewed mapping efforts at a finer scale, using modern remote sensing and numerical analysis approaches, as recommended by Revermann and Finckh (2019) in Chap. 6. Of the many intriguing features of Angolan vegetation, the patterns of plant community/soil/animal associations, such as the 'fairy circles' of the Namib (Juergens 2013; Cramer and Barger 2013), 'fairy forests' of the miombo, and the influence of coastal fog on desert vegetation and fauna; are of special interest to ecologists. Few of these phenomena have been adequately interpreted, but Zigelski et al. (2019) in Chap. 7 presents recent studies on the 'underground forests' of the *chanas de ongote* of the Angolan plateau. The landscapes of Angola are not static, being subject to multiple processes of transformation. In Chap. 8, Mendelsohn (2019) uses results from satellite technologies and ground surveys to describe the dramatic impacts of deforestation, fires, mining and agricultural activities on vegetation, soils and water quality at landscape scales.

Part III details the results of surveys that have advanced rapidly over the past two decades on two invertebrate groups – dragonflies and butterflies. These colourful and taxonomically distinctive insects are known to be sensitive to subtle changes in environmental conditions, such as forest cover and water quality, and serve as effective indicators of change in environmental health. The chapters (9 and 10) on butterflies (Mendes et al. 2019) and on dragonflies (Kipping et al. 2019) have enriched Angola's knowledge of these important ecological groups. Prior to 2009, for example, only 158 species of dragonflies and damselflies were known from Angola. By 2018 this number had increased to 260. The butterfly checklist now stands at 792 species and subspecies, up by over 220 species and subspecies since the turn of the millennium. In contrast to the encouraging progress in these taxa, key environmental engineers – ants and termites – remain poorly documented and await study.

A major component of this volume has been devoted to the vertebrate taxa that have enjoyed the attention of scientists active in Angola since the mid-nineteenth century. Part IV presents detailed accounts of the pioneering work of such luminaries as Anchieta, Bocage, Boulenger, Machado, Rosa Pinto and Crawford-Cabral, but also of the very many other contributors to the inventory of Angola's vertebrate fauna. Skelton (2019), Chap. 11, provides a concise summary of what is known of Angola's ca. 358 species of freshwater fishes (of which 22% are endemic), and also presents a model of post-Cretaceous biogeography of Angola and the roles of regional tectonics and river capture on the speciation and distribution of the fish fauna. Baptista et al. (2019), Chap. 12, cover the amphibian fauna, noting that the group clearly deserves further survey given that thus far only 111 species have been recorded (compared to 128 species for similar-sized but much drier and cooler South Africa). In Chap. 13, Branch et al. (2019) offer a comprehensive account on the 278 species of Angolan reptiles and their patterns of diversity and endemism, documenting key reptile hotspots deserving further exploration. They predict that as many as 75 new species of lizards await discovery in Angola. Both Branch and Baptista demonstrate the value of molecular phylogenetics in clarifying taxonomic complexes in reptiles and frogs.

Angola, with ca. 940 bird species recorded, has in recent years become a favoured destination of ecotourists searching for the country's 29 endemic bird species, and Dean et al. (2019) provide in Chap. 14, a chronology of ornithological surveys, a listing of endemics and near-endemics, and sites of special interest to bird enthusiasts, both professional and amateur. They emphasise, as highlighted by Hall (Hall 1960), the faunistic importance of the Angolan Escarpment, and also of the relict Afromontane forests of the highlands (Vaz Silva 2015), as areas of critical importance to understanding the evolution of Africa's avifauna. These isolated, fragmented and rapidly declining forests merit the highest level of protection to secure their futures as evolutionary fingerprints of the past.

A team of ten mammal specialists, coordinated by Beja et al. (2019), present a major synthesis (Chap. 15) on Angola's 291 mammal species. This chapter fills a need felt since Hill and Carter's (1941) benchmark study, and the more recent coverage of ungulates by Crawford-Cabral and Veríssimo (2005). With 73 species of bats (one third of the bat species known for Africa) Angola has the highest number of species for southern Africa, despite the comparatively limited intensity of surveys undertaken to date. The most diverse mammal group, rodents, has 85 species listed for Angola, of which 13 are endemic or near-endemic. While the number of endemic mammal species is modest, the vulnerability of many species to extinction within Angola is high and deserves urgent conservation measures. Less well known to Angolans than the terrestrial mammals is the country's unusually rich marine mammal fauna. The 28 species of cetaceans (whales and dolphins) found off Angola's coast have been the subject of surveys and research undertaken by Weir (2019) since 2003. As noted in Chap. 16, the possible presence of a further seven species of cetaceans in Angolan waters makes the country globally important for marine mammal conservation.

The Angolan mammal that has enjoyed national and international attention is the Giant Sable Antelope, which has been the subject of an intense research and conservation project since 2002 (Chap. 17), led by Vaz Pinto (2019). The successful rescue and rehabilitation of this national icon, from the brink of extinction, is a conservation model for which Angola can justifiably be proud. The success of the Giant Sable Project needs replication for the many mammal species that are known to have been reduced to very low numbers, or have been hunted to extinction in Angola. These include most large carnivores – Cheetah, Lion, Wild Dog, plus many herbivores – Black-faced Impala, Red Hartebeest, Lichtenstein's Hartebeest, Tsessebe, Southern Lechwe, Puku, Forest Buffalo, Giraffe, Black Rhino, Western Gorilla, Chimpanzee, Forest Elephant and Manatee.

The final section of this volume (Part V) presents an overview of the country's conservation history and current opportunities for action, Chap. 18 (Huntley et al. 2019); and an introduction to the importance of natural history museums and herbaria in the country's biodiversity science and conservation agenda, Chap. 19 (Figueira and Lages 2019). What is abundantly clear, as expressed in the concluding chapter, (Russo et al. 2019), is that Angola is alive with research and conservation opportunities, stimulated by recent initiatives led by the Angolan government and supported by the international community.

This volume provides a first synthesis of what is known and published about Angola's diverse landscapes, biomes and ecosystems and the species that inhabit them. It is a humble attempt by its 46 contributors to place this knowledge before researchers and conservationists in Angola and beyond, especially those who might be stimulated to strengthen the imperfect understanding and vulnerable state of Angola's biodiversity. It is the fervent hope of this book's editors that this volume will provide an entry point for many young Angolan students to study the literature, be inspired by the dedication, tenacity and wisdom of the early pioneers and contemporary explorers, and enter careers in field-based biodiversity research and conservation in Angola.

References

Baptista N, Conradie W, Vaz Pinto P et al (2019) The amphibians of Angola: early studies and the current state of knowledge. In: Huntley BJ, Russo V, Lages F, Ferrand N (eds) Biodiversity of Angola. Science & conservation: a modern synthesis. Springer Nature, Cham

Barbosa LAG (1970) *Carta Fitogeográfica da Angola.* Instituto de Investigação Científica de Angola, Luanda, 343 pp

Barnard P (1998) Biological diversity in Namibia: a country study. Namibian National Biodiversity Task Force, Windhoek, 325 pp

Beja P, Vaz Pinto P, Veríssimo L et al (2019) The mammals of Angola. In: Huntley BJ, Russo V, Lages F, Ferrand N (eds) Biodiversity of Angola. Science & conservation: a modern synthesis. Springer Nature, Cham

Branch WR, Vaz Pinto P, Baptista N et al (2019) The reptiles of Angola: history, diversity, endemism and hotspots. In: Huntley BJ, Russo V, Lages F, Ferrand N (eds) Biodiversity of Angola. Science & conservation: a modern synthesis. Springer Nature, Cham

Carr T, Carr P (1991) Surveys of the sea turtles of Angola. Biol Conserv 58(1):19–29

Cramer MD, Barger NN (2013) Are Namibian "fairy circles" the consequence of self-organizing spatial vegetation patterning? PLoS One 8(8):e70876

Crawford-Cabral J (1983) Esboço zoogeográfico de Angola. Unpublished manuscript. Instituto de Investigação Científica Tropical, Lisboa, 50 pp + 13 maps

Crawford-Cabral J, Veríssimo LN (2005) The ungulate fauna of Angola: systematic list, distribution maps, database report. Estudos, Ensaios e Documentos do Instituto de Investigação Científica Tropical 163:1–277

Davis DHS (ed) (1964) Ecological studies in Southern Africa. Junk, The Hague, 415 pp

de Andrade AAB (1985) O Naturalista José de Anchieta. Instituto de Investigação Científica Tropical, Lisboa, 187 pp

Dean WRJ, Melo M, Mills MSL (2019) The avifauna of Angola: richness, endemism and rarity. In: Huntley BJ, Russo V, Lages F, Ferrand N (eds) Biodiversity of Angola. Science & conservation: a modern synthesis. Springer Nature, Cham

Estes RD (1982) The giant sable and wildlife conservation in Angola. Report to IUCN Species Survival Commission. Gland, Switzerland

Figueira R, Lages F (2019) Museum and herbarium collections for biodiversity research in Angola. In: Huntley BJ, Russo V, Lages F, Ferrand N (eds) Biodiversity of Angola. Science & conservation: a modern synthesis. Springer Nature, Cham

Figueiredo E, Smith GF (eds) (2008) Plants of Angola / Plantas de Angola. Strelitzia 22:1–279

GoA (Government of Angola) (2011) Plano Estratégico da Rede Nacional de Áreas de Conservação de Angola (PLENARCA). Ministry of Environment, Luanda

Gossweiler J, Mendonça FA (1939) Carta Fitogeográfica de Angola. Ministério das Colónias, Lisboa, 242 pp

Goyder DJ, Gonçalves FMP (2019) The Flora of Angola: collectors, richness and endemism. In: Huntley BJ, Russo V, Lages F, Ferrand N (eds) Biodiversity of Angola. Science & conservation: a modern synthesis. Springer Nature, Cham

Gunn M, Codd LE (1981) Botanical exploration of Southern Africa. AA Balkema, Cape Town, 400 pp

Günther R, Feiler A (1986a) Zur phänologie, ökologie und morphologie angolanischer Vögel (Aves). Teil I: Non-Passeriformes Faunistische Abhandlungen aus dem Staatlichen Museum für Tierkunde in Dresden 13:189–227

Günther R, Feiler A (1986b) Zur phänologie, ökologie und morphologie angolanischer Vögel (Aves). Teil II: Passeriformes Faunistische Abhandlungen aus dem Staatlichen Museum für Tierkunde in Dresden 14:1–29

Hall BP (1960) The faunistic importance of the scarp of Angola. Ibis 102:420–442

Hawkins F (1993) An integrated biodiversity conservation project under development: the ICBP Angola Scarp Project. Proceedings of the VIII Pan-African Ornithological Congress: 279–284. Kigali, Rwanda, 1992. Koninklijk Museum voor Midden-Afrika, Tervuren

Hill JE, Carter TD (1941) The mammals of Angola, Africa. *Bulletin of the American Museum of Natural History* 78, 211 pp

Huntley BJ (ed) (1989) Biotic diversity in southern Africa: concepts and conservation. Oxford University Press, Oxford, 380 pp

Huntley BJ (2009) SANBI/ISCED/UAN Angolan biodiversity assessment capacity building project. Report on pilot project. Unpublished Report to Ministry of Environment, Luanda 97 pp, 27 figures

Huntley BJ (2011) Biodiversity rapid assessment of the Lagoa Carumbo area, Lunda-Norte, Angola. Expedition report. Ministry of Environment, Luanda

Huntley BJ (2019) Angola in outline: physiography, climate and patterns of biodiversity. In: Huntley BJ, Russo V, Lages F, Ferrand N (eds) Biodiversity of Angola. Science & conservation: a modern synthesis. Springer Nature, Cham

Huntley BJ, Francisco P (eds) (2015) Avaliação Rápida da Biodiversidade de Região da Lagoa Carumbo, Lunda-Norte – Angola / Rapid Biodiversity Assessment of the Carumbo Lagoon Area, Lunda-Norte – Angola. Ministério do Ambiente, Luanda, 219 pp

Huntley BJ, Matos L (1992) Biodiversity: Angolan environmental status quo assessment report. IUCN Regional Office for Southern Africa, Harare, 55 pp

Huntley BJ, Siebert SJ, Steenkamp Y, et al (2006) The achievements of the southern African botanical diversity network (SABONET) – a southern African botanical capacity building project. In: Ghazanfar SA, Beentje H (eds.) Taxonomy and ecology of African plants, their conservation and sustainable use: Proceedings of the 17th AETFAT Congress. Addis Ababa, Ethiopia, 2003. Royal Botanic Gardens, Kew, pp 531–543

Huntley BJ, Beja P, Vaz Pinto P et al (2019) Biodiversity conservation: history, protected areas and hotspots. In: Huntley BJ, Russo V, Lages F, Ferrand N (eds) Biodiversity of Angola. Science & conservation: a modern synthesis. Springer Nature, Cham

Jeffrey R (1996) A Phytosociological Survey of the Northern Sector of the Quicama National Park in Angola. B.Sc. (Hons.) Dissertation. Faculty of Biological and Agricultural Sciences, University of Pretoria, Pretoria

Juergens N (2013) The biological underpinnings of Namib Desert fairy circles. Science 339:1618–1621

Kipping J, Clausnitzer V, Fernandes Elizalde SRF et al (2019) The dragonflies and damselflies of Angola. In: Huntley BJ, Russo V, Lages F, Ferrand N (eds) Biodiversity of Angola. Science & conservation: a modern synthesis. Springer Nature, Cham

Kirkman SP, Nsingi KK (2019) Marine biodiversity of Angola: biogeography and conservation. In: Huntley BJ, Russo V, Lages F, Ferrand N (eds) Biodiversity of Angola. Science & conservation: a modern synthesis. Springer Nature, Cham

Maier K (2007) Angola: promises and lies. Serif, London, 224 pp

Mateus O, Jacobs LL, Schulp AS et al (2011) *Angolatitan adamastor*, a new sauropod dinosaur and the first record from Angola. An Acad Bras Cienc 83(1):221–233

Mateus O, Callapez P, Polcyn M et al (2019) Biodiversity in Angola through time: a paleontological perspective. In: Huntley BJ, Russo V, Lages F, Ferrand N (eds) Biodiversity of Angola. Science & conservation: a modern synthesis. Springer Nature, Cham

Mendelsohn JM (2019) Landscape changes in Angola. In: Huntley BJ, Russo V, Lages F, Ferrand N (eds) Biodiversity of Angola. Science & conservation: a modern synthesis. Springer Nature, Cham

Mendes L, Bivar-de-Sousa A, Williams M (2019) The butterflies and skippers of Angola. In: Huntley BJ, Russo V, Lages F, Ferrand N (eds) Biodiversity of Angola. Science & conservation: a modern synthesis. Springer Nature, Cham

Mills MSL (2010) Angola's central scarp forests: patterns of bird diversity and conservation threats. Biodivers Conserv 19:1883–1903

Mills MSL (2018) The special birds of Angola / as Aves Especiais de Angola. Go-away-birding/ Kissama Foundation, Cape Town/Luanda

Monard A (1935) Contribution à la mammologie d'Angola et prodrome d'une faune d'Angola. Arquivos do Museu Bocage 6:1–314

Monteiro RFR (1970) Estudo da Flora e da Vegetacão das Florestas Abertas do Planalto do Bié. Instituto de Investigacão Cientifíca de Angola, Luanda, 352 pp

Morais M (2017) Projecto Kitabanga – Conservação de tartarugas marinhas. Relatório final da temporada 2016/2017. Universidade Agostinho Neto / Faculdade de Ciências, Luanda

NGOWP (2018) National Geographic Okavango Wilderness Project (2018) Initial findings from exploration of the upper catchments of the Cuito, Cuanavale and Cuando rivers in Central and South-Eastern Angola (May 2015 to December 2016). National Geographic Okavango Wilderness Project, 352 pp

OKACOM (2009) Final report: Okavango integrated river management project. The Permanent Okavango River Basin Water Commission, Maun

OKACOM (2011) Cubango-Okavango river basin transboundary diagnostic analysis. The Permanent Okavango River Basin Water Commission, Maun

Rejmánek M, Huntley BJ, le Roux JJ et al (2017) A rapid survey of the invasive plant species in western Angola. Afr J Ecol 55:56–69

Revermann R, Finckh M (2019) Vegetation survey, classification and mapping in Angola. In: Huntley BJ, Russo V, Lages F, Ferrand N (eds) Biodiversity of Angola. Science & conservation: a modern synthesis. Springer Nature, Cham

Revermann R, Krewenka KM, Schmeidel U et al (eds) (2018) Climate change and adaptive land management in Southern Africa – assessments, changes, challenges, and solutions. Biodivers Ecol 6:1–497

Russo V, Huntley BJ, Ferrand N (2019) Biodiversity research and conservation opportunities. In: Huntley BJ, Russo V, Lages F, Ferrand N (eds) Biodiversity of Angola. Science & conservation: a modern synthesis. Springer Nature, Cham

Sinclair ARE (2012) Serengeti story: life and science in the world's greatest wildlife region. Oxford University Press, Oxford, 270 pp

Skelton PH (2019) The freshwater fishes of Angola. In: Huntley BJ, Russo V, Lages F, Ferrand N (eds) Biodiversity of Angola. Science & conservation: a modern synthesis. Springer Nature, Cham

Swinscow TDV (1972) Friedrich Welwitsch, 1806–72: a centennial memoir. Biol J Linn Soc 4:269–289

Vaz da Silva B (2015) Evolutionary history of the birds of the Angolan highlands – the missing piece to understand the biogeography of the Afromontane forests. MSc Thesis. University of Porto, Porto

Vaz Pinto P (2018) Evolutionary history of the critically endangered giant Sable Antelope (Hippotragus niger variani): insights into its phylogeography, population genetics, demography and conservation. PhD Thesis. University of Porto, Porto

Vaz Pinto P (2019) The Giant sable Antelope: Angola's National Icon. In: Huntley BJ, Russo V, Lages F, Ferrand N (eds) Biodiversity of Angola. Science & conservation: a modern synthesis. Springer Nature, Cham

Walker JF (2004) A certain curve of horn. The hundred-year quest for the giant sable antelope of Angola. Grove/Atlantic Inc., New York, 514 pp

Weir CR (2019) The Cetaceans (Whales and Dolphins) of Angola. In: Huntley BJ, Russo V, Lages F, Ferrand N (eds) Biodiversity of Angola. Science & conservation: a modern synthesis. Springer Nature, Cham

Werger MJA, van Bruggen AC (eds) (1978) Biogeography and ecology of Southern Africa. The Junk, Hague, 1444 pp

Zigelski P, Gomes A, Finckh M (2019) Suffrutex dominated ecosystems in Angola. In: Huntley BJ, Russo V, Lages F, Ferrand N (eds) Biodiversity of Angola. Science & conservation: a modern synthesis. Springer Nature, Cham

Chapter 2
Angola in Outline: Physiography, Climate and Patterns of Biodiversity

Brian J. Huntley

Abstract Angola is a large country of 1,246,700 km^2 on the southwest coast of Africa. The key features of the country's diverse geomorphological, geological, pedological, climatic and biotic characteristics are presented. These range from the ultra-desert of the Namib, through arid savannas of the coastal plains to a biologically diverse transition up the steep western Angolan Escarpment. Congolian rainforests are found in Cabinda and along the northern border with the Democratic Republic of Congo, with outliers penetrating southwards along the Angolan Escarpment, or up the tributaries of the Congo Basin. Above the escarpment, high mountains rise to 2620 m above sea level, with isolated remnants of Afromontane forests and grasslands. Extensive *Brachystegia/Julbernardia* miombo moist woodlands dominate the plateaus and peneplains of the Congo and Zambezi basins, and dry woodlands of *Colophospermum/Acacia* occur in the southeast towards the Cunene River, with *Baikiaea/Burkea/Guibourtia* woodlands dominating the Kalahari sands of the endorheic basins of the Cubango and Cuvelai rivers. Rainfall varies from lower than 20 mm per year in the southwest to over 1600 mm in the northwest and northeast. At a regional scale, Angola is notable for having representatives of seven of Africa's nine biomes, and 15 of the continent's ecoregions, placing Angola second only after South Africa for its diversity of African ecoregions.

Keywords Afromontane forest · Biogeography · Biomes · Climate change · Congolian forest · Ecoregions · Namib · Kalahari Basin · Zambezian savannas

Introduction

This chapter presents a general outline of the physical geography and biodiversity characteristics of Angola, as background to the chapters that follow. It draws on the work of the great Portuguese agro-ecologist Alberto Castanheira Diniz, who

B. J. Huntley (✉)
CIBIO-InBIO, Centro de Investigação em Biodiversidade e Recursos Genéticos, Universidade do Porto, Vairão, Portugal
e-mail: brianjhuntley@gmail.com

B. J. Huntley et al. (eds.), *Biodiversity of Angola*,
https://doi.org/10.1007/978-3-030-03083-4_2

synthesised the diverse drivers of Angola's ecological systems and agricultural potential, based on his many decades of fieldwork in the country (Diniz and Aguiar 1966; Diniz 1973, 1991, 2006). Colonial records of climatic variables (Silveira 1967) are used in the absence of recent time series. The pioneer studies on Angolan vegetation by Gossweiler and Mendonça (1939) and Barbosa (1970) are fundamental to any account on Angola biodiversity. Surveys of Angola's protected areas and biodiversity 'hotspots' (Huntley 1974a, b, 2010, 2015, 2017) provide conservation context. This outline also draws on the recent regional geographies of Angola by Mendelsohn and co-workers (Mendelsohn et al. 2013; Mendelsohn and Weber 2013, 2015; Mendelsohn and Mendelsohn 2018). The chapter is also strengthened by material detailed in the specialist papers that form the core of this volume.

Location and Extent

As a large country of 1,246,700 km^2 on the southwest coast of Africa, Angola is roughly square in outline, lying between 4° 22′ and 18° 02′ south latitude, and 11° 41′ and 24° 05′ east longitude. It is bounded to the west by an arid 1600 km coastline along the Atlantic Ocean; to the north by the moist forest and savanna ecosystems of the Republic of Congo and the Democratic Republic of Congo (DRC); to the east by the moist savanna and woodland ecosystems of the DRC and Zambia; and by arid woodlands, savannas and desert along its 1200 km southern border with Namibia.

Geomorphology and Landscape Evolution

The general topography of Angola is illustrated in Fig. 2.1. In summary, coastal lowlands lying below 200 m altitude and of 10–150 km breadth occupy 5% of the country's land surface area, leading to a stepped and mountainous escarpment rising to 1000 m (23%), and an extensive interior plateau of 1000–1500 m (65%). Seven percent of the country lies above 1500 m, reaching its highest point at 2620 m on Mount Moco.

The ecological importance of the major physiographic divisions in Angola was recognised as early as the 1850s by the pioneer Austrian botanist Friedrich Welwitsch who categorised the 5000 plant species that he collected in Angola within three regions: *região litoral*, *região montanhosa*, and *região alto-plano* (Welwitsch 1859). Besides his remarkable contribution to the founding of Angolan botany, Welwitsch prepared detailed geological profiles across the landscapes inland of Luanda and Moçâmedes (Albuquerque and Figueirôa 2018), probably the first such analyses in western Africa (Fig. 2.2). His understanding of the patterns and relationships of geology, physiography and vegetation set a strong ecological tradition that has been followed by successive students of Angola's biodiversity.

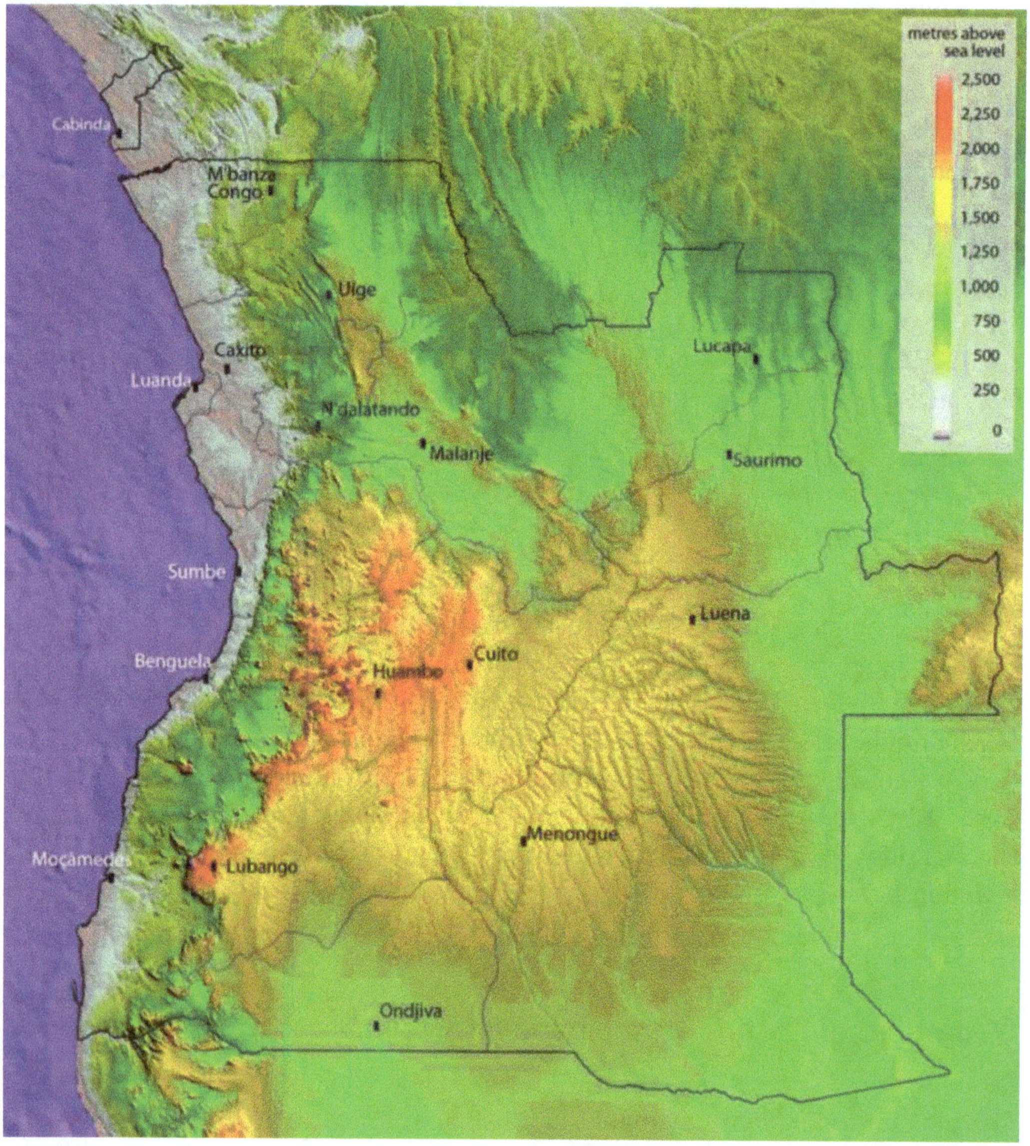

Fig. 2.1 Topography of Angola, indicating provincial boundaries and capitals. The coastal lowlands, western escarpments, central highlands and plateaus, and the major drainage basins of the Cuanza, Congo and Zambezi rivers are clearly revealed

A further detailed and indeed classic study of Angola's geomorphology and local ecology was that of the German geographer Otto Jessen (1936). Jessen undertook a series of 11 transects from the coast inland, traversing the escarpment to the interior plateau from Moçâmedes and thereafter at intervals northwards to Luanda. Describing, illustrating and mapping selected vegetation communities, geological exposures, landscapes, landuse and ethnological features of the country, Jessen's Angolan work remains unique in its diversity of interest and originality. He recognised five major erosional planation surfaces in western Angola at a time when geomorphology was evolving as a discipline, and he was recognised by King (1962) as one of the founders of peneplanation theory. Geomorphological studies in Angola

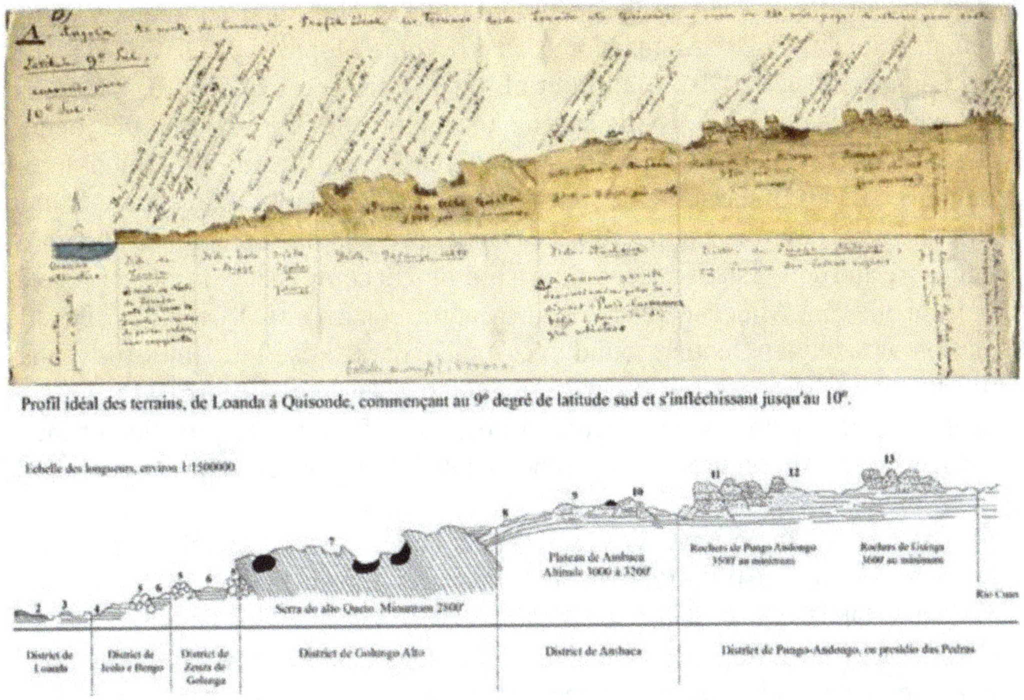

Fig. 2.2 Geological profile from Luanda to Quisonde, scanned from the original manuscript produced by Friedrich Welwitsch during his expeditions between 1853 and 1860. The lower profile is a redrafted version of the upper profile adapted from Choffat (1888), and reproduced with permission from Albuquerque and Figueirôa (2018) and of the Museums of the University of Lisbon Historical Archives

continued from the 1950s to 1970s by Portuguese researchers, including Marques (1963), Feio (1964) and Amaral (1969), whose work is summarised by Costa (2006).

More recent research, in particular that on the evolution and of the major tectonic and erosional patterns across southern Africa (Cotterill 2010, 2015; Cotterill and De Wit 2011) and on the biogeography of the freshwater fishes of Angola (Skelton 2019) provide a picture of a very dynamic landscape since the breakup of Gondwana in the late Cretaceous. These and other authors are providing an improved understanding of the processes of uplift, back-tilting, down-warping, deposition, erosion and river capture on the evolution of the Kalahari Basin. The impacts of sea-level fluctuations and of the flow of the Congo River on coastal waters and on the erosional forces of the Congo Basin as it impacts on the Zambezi Basin are guiding our interpretation of the dramatic events shaping the faunal and floral patterns of today. Cotterill (2015) presents a synthesis of hypotheses on the evolution of the Kalahari from the late Mesozoic into the early Cenozoic, events which were followed by the later overlying suite of younger Kalahari sediments – the world's largest sandsea. The interplay of geological and paleoclimatic drivers described by Cotterill (2010, 2015), through the pulsing of hot wet and cool dry episodes during the Plio-Pleistocene, was accompanied by the expansion and contraction of forest and savanna habitats responding to climatic and fire regimes.

The role of fire in shaping the landscapes of Angola – and particularly of the dominant miombo moist savanna biome – has become a topic of discussion in recent years (Zigelski et al. 2019). Maurin et al. (2014) provide evidence based on the dated phylogenies of 1400 woody species to support the proposal that the 'underground forests' (White 1976) that are so prominent across the moist miombo savannas and woodlands of the south-central African plateau, evolved in response to high fire frequency. They suggest that moist savannas pre-date the emergence of anthropogenic fire and deforestation, becoming a prominent component of tropical vegetation from the late Miocene (ca. 8 Ma). Maurin et al. (2014) conclude that the evolution of geoxyles ('underground trees') that characterise these moist savannas define the timing of the transition to fire-maintained savannas occurring in climates suitable for and previously occupied by forests. The further interpretation of these key drivers of evolution processes is fundamental to an improved understanding of the biogeography of Angola.

A major contribution towards an ecological understanding Angola's contemporary landscapes and natural regions, and their agro-forestry potential, was that of Castanheira Diniz. Diniz (Diniz and Aguiar 1966, Diniz 1973, 1991) provides a series of maps illustrating the key features of Angola's topography, geomorphology, geology, climate, soils, and phyto-geographic and bio-climatic zones. Diniz's 11 'mesological' units (Fig. 2.3) provide a useful framework for discussions on Angola's ecology and biodiversity. Indeed his mesological concept closely corresponds with current perceptions of ecoregions. He also delineated and described 32 agro-ecological zones (Diniz 2006). Although some of his 11 mesological units need more rigorous and objective definition and delineation, they have become widely adopted within Angola. Important aspects of these 11 broad units will be summarised here, integrating these with insights from other sources.

1. Coastal Belt (*Faixa litoranea* sensu *Diniz*). This is a mostly continuous platform at 10–200 m above sea level, broken occasionally by broad river valleys. In contrast to the situation on the east coast of Africa at similar latitudes, the Angolan coastline is notable for the absence of coral reefs and coastal dune forests. Long sandbars stretch northwards from rivers such as the Cunene and Cuanza. Mudflats and mangroves occur at most river mouths from Lobito northwards, increasing in dimension and diversity towards the Congo. Much of the coast is uplifted, resulting in sharp sea-cliffs of 10–100 m. In places as narrow as 10 km, the coastal belt is mostly of about 40 km width, broadening to 150 km northwards of Sumbe and up the lower Cuanza. The coastal plains are composed mostly of fossiliferous marine sediments of the Cabinda, Cuanza, Benguela and Namibe geological basins. The northern coastal platforms are covered by deep red Pleistocene sands (*terras de musseque*) of former beaches. Lying below the sands, and exposed over large areas, are Cretaceous to Miocene clays, gypsipherous marls, dolomitic limestones and sandstones. Important beds of Cretaceous fossils occur at Bentiaba and Iembe, the latter including the sauropod dinosaur *Angolatitan adamastor* (Mateus et al. 2011, 2019). The

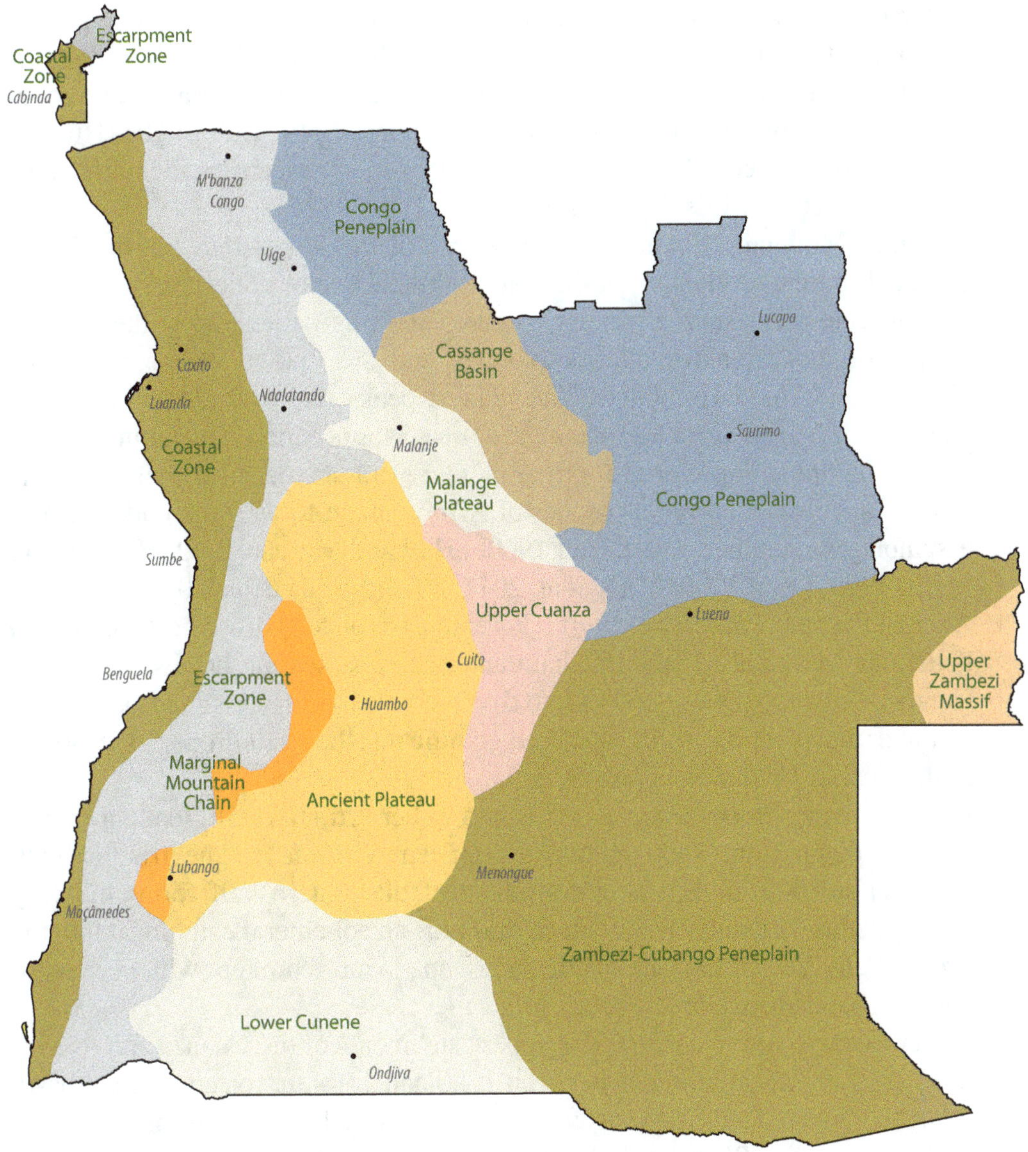

Fig. 2.3 Main geomorphological and landscape units of Angola. (After Diniz 1973)

southernmost segment of the Coastal Belt includes the mobile and mostly vegetation-less dunes of the Namib Desert.

2. Escarpment Zone (*Faixa subplanaltica* sensu *Diniz; região montanhosa* sensu *Welwitsch*). A broad transition belt lies between the coastal plains and the interior plateaus – variable in breadth and gradient. Over much of the zone, the transition advances up several steep steps of between 400 and 600 m. In the south, between Moçâmedes and Lubango, the escarpment of the Serra da Chela is very sharp, rising 1000 m at Tundavala and Bimbe. The geology of the Escarpment Zone is complex, comprising crystalline rocks of the Precambrian: granites, gneisses, schists, quartzites and amphibolites. The Escarpment Zone (also referred to as the Western Angolan Scarp) includes very hilly country,

with mountainous belts in the north, and some major inselbergs in the south, the most important of which is Serra de Neve, which rises to 2489 m from the surrounding plains and low hills. The Angolan Escarpment has long been recognised for its biogeographic importance (Humbert 1940, Hall 1960a, Huntley 1974a) and has been the centre of interest of many recent studies (Hawkins 1993, Dean 2000, Mills 2010, Cáceres et al. 2015, 2017).

3. Marginal Mountain Chain (*Cadeia Marginal de Montanhas* sensu *Diniz*). Residual mountain lands, mostly at 1800–2200 m, underlain mostly by Precambrian rocks such as gneiss, granites and migmatites, lie at the western margin of the extensive interior plateau, and are known as the Benguela, Huambo and Huíla Highlands. The highest peaks rise to 2420 m on Mount Namba, 2582 m on Serra Mepo and 2620 m at Mount Moco. The mountains are of biogeographic importance for their montane grasslands, with some elements of the Cape flora, and relict patches of Afromontane forests and endemic bird assemblages (Humbert 1940; Hall 1960b; Hall and Moreau 1962; Huntley and Matos 1994; Dean 2000; Mills et al. 2011, 2013; Vaz da Silva 2015).
4. Ancient Plateau (*Planalto Antigo*). This extensive plateau drops eastwards from below the Marginal Mountain Chain and encompasses the headwaters of the Cunene, Cubango, Queve and Cutato rivers, comprising rolling landscapes with wetlands and low ridges with scattered granitic inselbergs. It drops from 1800 m in the west to 1400 m in central Angola.
5. Lower Cunene (*Baixo Cunene*). This is a rather artificial unit, leading imperceptibly down from 1400 m on the 'Ancient Plateau' to the frontier with Namibia at 1000 m. The gentle gradient of the eastern half forms the very clearly defined Cuvelai Basin, which drains as an ephemeral catchment into the Etosha Pan. West of the Cunene the landscape is more broken, with pockets of Kalahari sands between low rocky hills.
6. Upper Cuanza (*Alto Cuanza*). The upper catchments of the Cuanza and its tributary the Luando, at altitudes between 1200 and 1500 m, form a distinct basin of slow drainage feeding extensive wetlands during the rain season.
7. Malange Plateau (*Planalto de Malange*). A gently undulating plateau at 1000–1250 m, dropping abruptly, on its northeastern margin, some several hundred metres to the *Baixa de Cassange* and the Cuango drainage. The escarpment ravines hold important moist forest outliers (such as at Tala Mungongo) that deserve investigation. To the west, the plateau is drained by the rivers flowing to the Atlantic, most spectacularly by a tributary of the Cuanza, the Lucala, that drops over 100 m at the famous Calandula Falls (formerly Duque da Bragança Falls).
8. Congo Peneplain (*Peneplanície do Zaire*). This is a vast sandy peneplain, drained by the northward flowing tributaries of the Cassai/Congo Basin, and stretching eastwards from the margins of the mountainous northern end of the Escarpment Zone in Uíge, to the extensive *Chanas da Borracha* of the Lundas. These gently dipping plains, mostly at 1100–800 m, are being aggressively dissected by the many northward flowing, parallel tributaries of the Congo Basin. The Cuango River, draining the *Baixa de Cassange*, drops to 500 m at the

frontier with the Democratic Republic of Congo. The southern boundary of this Congo Peneplain is defined imperceptibly by the watershed between the Zambezi and Congo basins, lying at ca. 1200 m.

9. Cassange Basin (*Baixa de Cassange*). A wide depression, several hundred metres below the surrounding plateaus, is demarcated by abrupt escarpments to the west and the densely dendritic catchment of the Cuango to the northeast. The geological substrate comprises Triassic Karoo Supergroup sediments of limestone, sandstone and conglomerates. Within the Basin, several large tablelands – remnants of the old planation surface – rise above the depression as extensive plateaus flanked by sheer, 300 m escarpments, exemplified by Serra Mbango, which awaits biological survey.
10. Zambezi-Cubango Peneplain (*Peneplanície do Zambeze-Cubango*). This is the vast peneplain draining deep Kalahari sands, with slow-flowing rivers that meander across the gently dipping plateau from 1200 m at the watershed with the Congo Basin to 1000 m at the frontier with Namibia. Within this extensive peneplain, the Bulozi Floodplain occupies an area in excess of 150,000 km^2 in Angola and Zambia.
11. Upper Zambezi Massif (*Maciço do Alto Zambeze*). The Calunda Mountains of eastern Moxico, composed of Precambrian schists and norites, dolorites, sandstones and limestones, rise to 1628 m above the Zambezi peneplain which lies at 1150 m. The mountains form a striking contrast to the almost featureless landscape that stretches some 800 km eastwards from Huambo to Calunda.

Rivers and Hydrology

Angolan river systems fall into two categories. First, coastal rivers drain the central and western highlands and flow rapidly westwards where they have penetrated the steep escarpment to the Atlantic Ocean. Most of these coastal rivers are relatively short, are highly erosive and carry high sediment loads. Backward erosion by some of these has produced minor basins, such as the amphitheatres of the upper Queve and Catumbela. The biogeographic importance of the river captures associated with these systems, especially the Congo, Cuanza, and Cunene, have been profound, as described by Skelton (2019). Most of the coastal rivers south of Benguela are ephemeral.

The second major category of river systems is that of the vast interior plateaus. Drained by nine large hydrographic basins, seven of which are transnational, Angola serves as the 'water tower' for much of southern and central Africa. Many of these rivers arise in close proximity on either side of the gently undulating watershed between the Cuanza, Cassai (Congo), Lungue-Bungo (Zambezi), Cunene, and endorheic Cubango (Okavango) basins. These rivers drain the vast and deep Kalahari sands, are slow moving and due to the filtering action of the sands, are crystal clear and nutrient poor. A separate ephemeral, endorheic system, the Cuvelai Basin, drains southwards into the Etosha Pan.

The conservation importance of the Angolan river systems is of great significance, feeding as they do two wetlands (Okavango and Etosha) of global importance, and the still under-researched Bulozi Floodplain of Moxico. This is possibly the largest ephemeral floodplain in Africa – 800 km from north to south and 200 km from east to west – straddling the Angola/Zambia frontier (Mendelsohn and Weber 2015).

Geology and Soils

The geological history and soil genesis of Angola is complex and interrelated, and influenced by rainfall, drainage, evaporation and wind. Mateus et al. (2019) provide a map and stratigraphic profile of the geology of Angola which summarises the major geological features of the country. The predominance of a broad belt of Precambrian systems along the western margin of the country, with Cenozoic systems occupying most of the eastern half, is striking. Over three-quarters of the country (Fig. 2.4) is covered by two main soil groups arenosols and ferralsols – an understanding of which provides an essential introduction to Angolan pedology. For simplicity, soils will be described with reference to their geological substrate.

First, Angola's main soil groups are the sandy arenosols (*solos psamíticos*) that cover more than 53% of the country. These sands are dominant features of three major landscapes: the dunes of the Namib Desert; the red *'terras de musseque'* of the coastal belt northwards from Sumbe; and the vast Kalahari Basin. The great majority of the arenosols lie to the east of approximately 18° longitude – the aeolian sands of the Kalahari Basin which cover nearly 50% of Angola and hides nearly all of the underlying geological formations. The Kalahari Basin, extending across 2500 km from the Cape in the south to the Congo Basin in the north, and up to 1500 km in breadth, is reputedly the largest body of sand in the world. The sands have been deposited by wind and water over the past 65 million years. Composed of quartz grains that hold no mineral nutrients, and with very little accumulated organic matter, they are thus of very low fertility and water-holding capacity. Waters passing through the vast catchments of the Congo, Cubango and Zambezi basins that drain the Kalahari are therefore extremely pure.

Second, the higher ground of the western half of Angola (the Ancient Massif) is dominated by ferralsols (*solos ferralíticos*) derived from underlying rocks (gneisses, granites, metamorphosed sediments of the Precambrian Basement Complex; and schists, limestones and quartzites of the West Congo System). Ferralsols cover approximately 23% of Angola. The soils are mostly of low water-holding capacity. Because they are heavily leached in higher rainfall areas, the loss of mineral nutrients and organic matter results in low fertility. They are characteristically reddish due to the oxidation of their high iron and aluminium content, which also accounts for the presence in many areas of ferricrete hardpan horizons a metre or two below the surface, impeding root and water penetration and resulting in the formation of extensive areas of laterite.

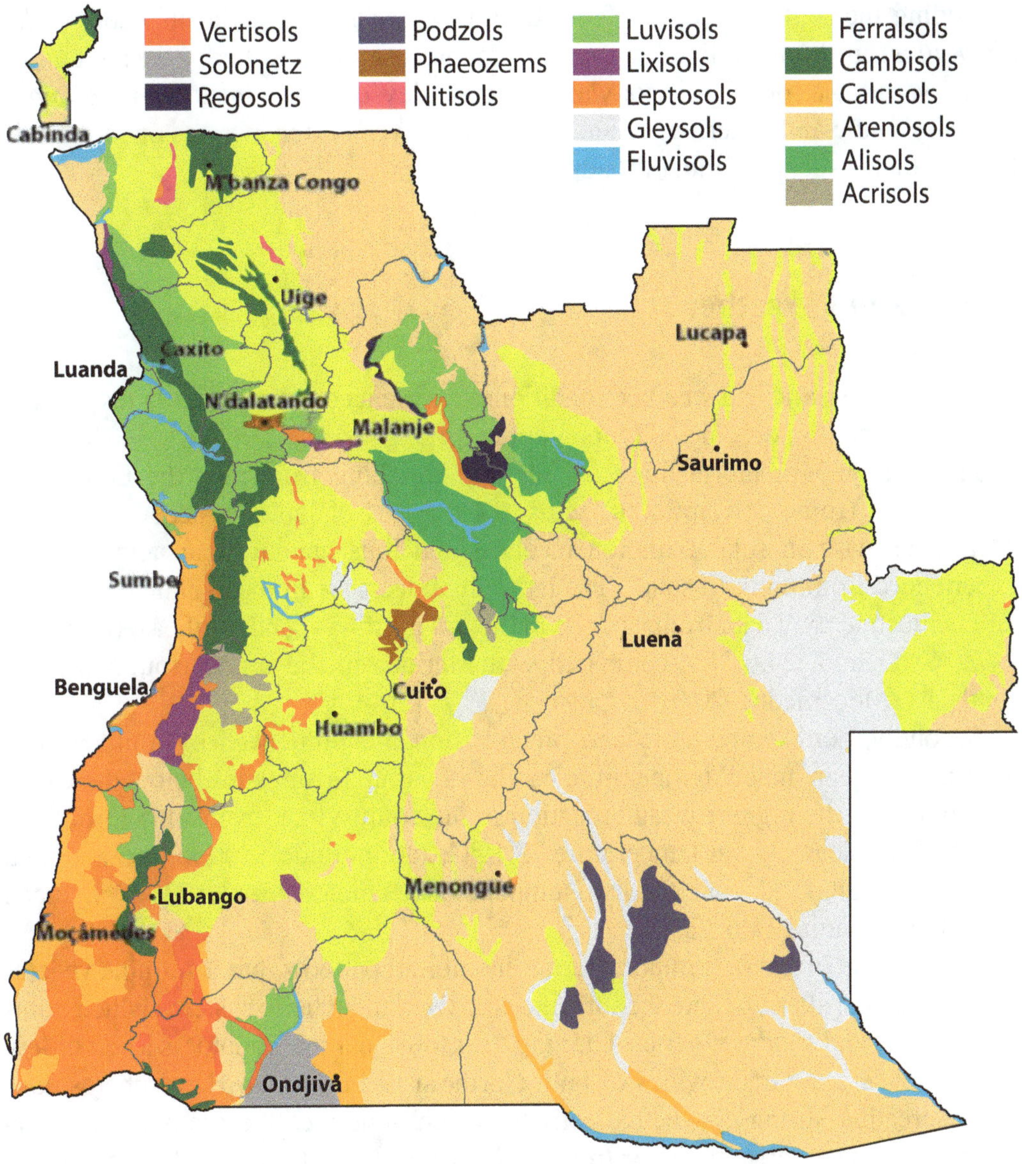

Fig. 2.4 Outline of the main soil types of Angola (from Jones et al. 2013), illustrating the predominance of arenosols in the eastern half of the country, and ferralsols across the western and central plateaus

These two low-fertility soil groups (arenosols and ferralsols) cover over 76% of the country, thus despite the adequacy of rainfall over most of Angola, agricultural production faces the challenges of low soil fertility (Neto et al. 2006; Ucuassapi and Dias 2006). The natural vegetation types that cover both arenosols and ferralsols – predominantly miombo woodlands – are well adapted to these soil conditions and the untransformed landscape gives the appearance of great vitality and luxuriance.

The next soil grouping in terms of landcover, occupying 6% of Angola, are the shallow regosols (*litosolos*) of rocky hills and gravel plains, most extensive in the arid southwest. Other important soil types include luvisols, calcisols and cambisols (*solos calcários, solos calcialíticos*), which provide fertile loam soils for crops (including the 'coffee forests' of the Escarpment Zone); alluvial fluvisols (*solos aluvionais*) in drainage lines with high organic content and high water retaining capacity, suitable for crops if not water-logged; gleysol clays (*solos hidromórficos*), typically acidic and waterlogged and occasionally very extensive – as on seasonally flooded plains such as Bulozi Floodplains.

Climate and Weather

The diverse climatic and weather conditions experienced across Angola result from many atmospheric, oceanic and topographic driving forces.

First, the geographic position of Angola, stretching from near the Equator to close to the Tropic of Capricorn, across 14 degrees of latitude, accounts for the overall decrease in solar radiation received and thus annual mean temperatures experienced from north to south. The latitudinal decrease in mean annual temperature is illustrated by data from stations in the hot northwest and northeast (Cabinda: 24.7 °C; Dundo: 24.6 °C), compared with stations in the milder southwest and southeast (Moçâmedes: 20.0 °C; Cuangar: 20.7 °C).

Secondly, both temperature and precipitation are influenced by altitude. The decrease in mean annual temperature can be illustrated from sites below the Chela escarpment to the highest weather station in the country: i.e. from Chingoroi: altitude 818 m, mean annual temperature 23.1 °C; Jau: altitude 1700 m, mean annual temperature 18.0 °C; and finally Humpata-Zootécnica: altitude 2300 m, mean annual temperature 14.6 °C.

Thirdly, and of greatest importance to the rainfall patterns that determine vegetation and habitat structure, are the influences of the atmospheric systems which dominate central and southern Africa. Circling the globe near the Equator is a belt of low pressure where the trade winds of both Northern and Southern Hemispheres converge, creating strong convective activity which generates the dramatic thunderstorms that characterise the inter-tropics. Known as the Inter-tropical Convergence Zone (ITCZ), the belt moves southwards over Angola during summer, and then returns northwards to the Equator as winter approaches. The rainfall season that is triggered by the ITCZ passes across northern Angola from early summer, reaching southern Angola in late summer. The climate is strongly seasonal, with hot wet summers (October to May) and mild to cool dry winters (June to September). Some stations in northern Angola receive two peaks of rainfall, early summer and late summer, often with a short drier period in mid-summer.

Moving in tandem with the ITCZ are two high-pressure systems – over the Atlantic and over southern Africa – the South Atlantic Anticyclone and the Botswana Anticyclone. In simple terms, these two anticyclones block the southward movement of moist air from the ITCZ during winter (preventing cloud formation) and as the high-pressure cells move southwards in summer, the conditions required for cloud formation return. This pulsing of rainfall systems is clearly illustrated in the series of rainfall maps prepared by Mendelsohn et al. (2013) from weather satellite imagery (Fig. 2.5).

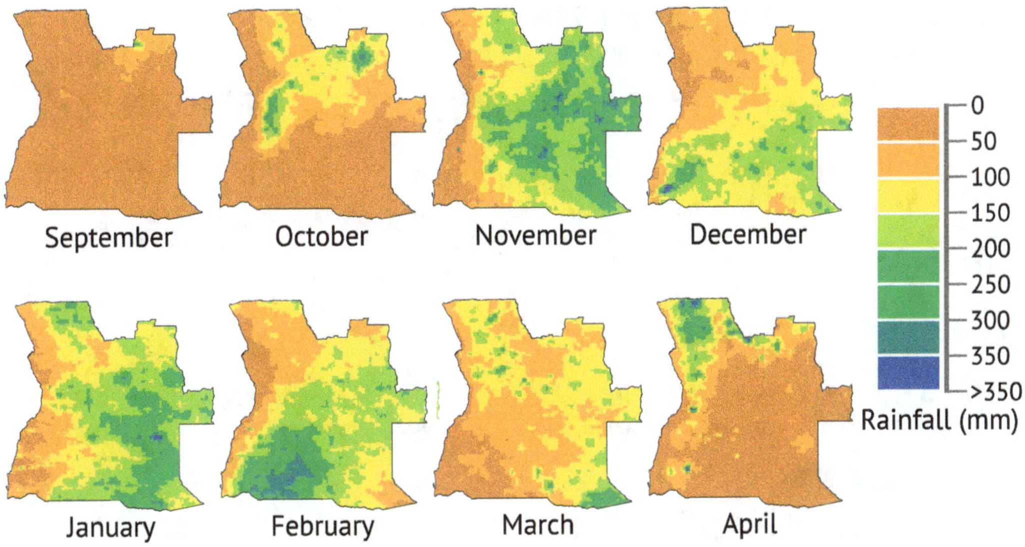

Fig. 2.5 The impact of the southwards and northwards pulsing of the Inter-tropical Convergence Zone on rainfall seasonality across Angola during 2009/2010. (From Mendelsohn et al. 2013)

During winter and early summer, the Botswana Anticyclone generates strong winds that blow across Angola from east to west, with impacts on micro-relief over much of the country. In the southwest, the winds pick up dust from the arid lands and create hot, choking dust storms that feed the sand dunes of the Namib. The winds are also notorious in the north, where they desiccate the grasslands of the Lundas. In the east, the winds and their sand deposits account for dune formation across the Bulozi Floodplain (Mendelsohn and Weber 2015).

Rainfall and temperature seasonality and other climatic parameters are illustrated by the climate diagrams in Fig. 2.6. The distribution of mean annual rainfall across Angola is summarised in Fig. 2.7.

Fourthly, as noted above, altitude and seasonality determine temperature conditions. However, an anomaly to this general rule occurs in the coastal belt of Angola, especially in the far south, through the influence of the temperature inversion created by the cold, upwelling Benguela Current. The Benguela Current has a stabilizing effect on the lower atmosphere and prevents the upward movement of moist, cloud-forming air off the ocean, accounting for the evolution of the Namib Desert. Its impact also extends as far north as Cabinda, where a narrow belt of arid savanna woodland and dry forest, of acacias, sterculias and baobabs, flanks the rainforests of the Maiombe.

Despite the aridity of the coastal zone, the cooling effect of the Benguela Current results in low stratus cloud and fog (*cacimbo*) through much of winter, with heavy dew condensing on vegetation along the coast even during the driest months of winter. The fog belt is most pronounced between Moçâmedes and Benguela, where epiphytic lichens reach great abundance in an otherwise desertic environment. The Benguela Current also results in a gradient of increasing precipitation from south to north and from west to east. The rainfall gradients are locally accentuated by the orographic influence of the escarpment and the highland mountain massifs. The

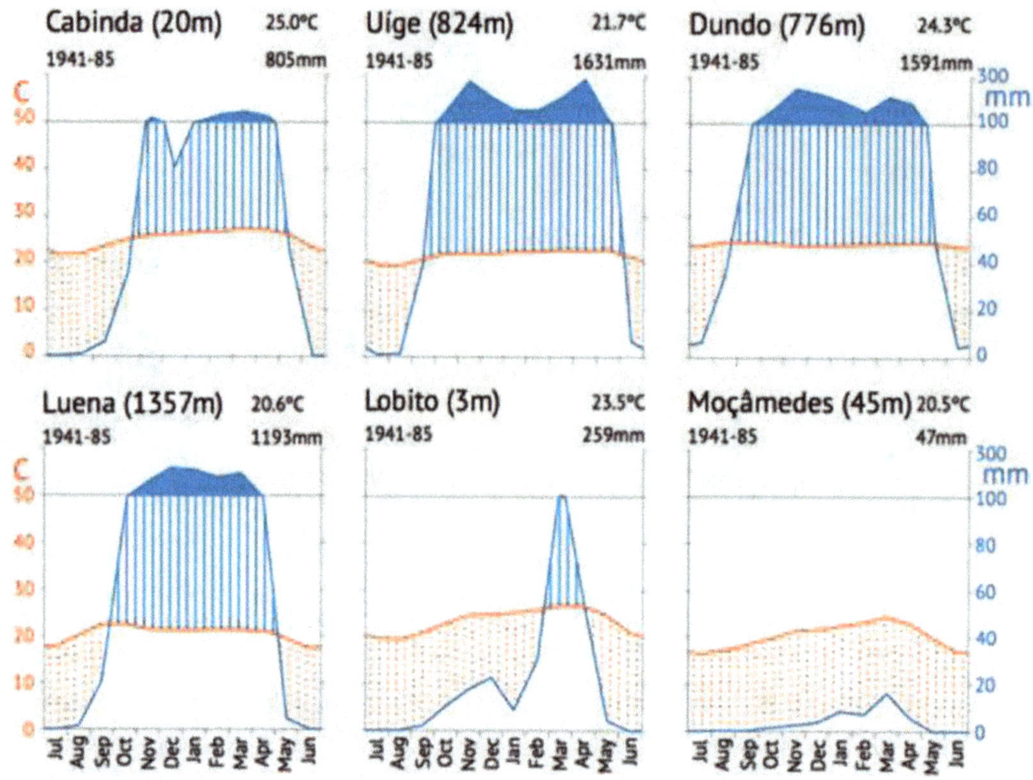

Fig. 2.6 Climate diagrams illustrating rainfall and temperature seasonality and other climatic parameters. Note weak bimodal rainfall maxima for stations in northern Angola and unimodal maxima in central and southern Angola

sharp relief of the escarpment creates conditions for orographic rainfall along most of this zone, supporting the 'coffee forests' of Seles, Gabela, Cuanza-Norte and Uíge.

Attempts to synthesise climatic characteristics into simple formulae or graphics have resulted in a wide range of classification systems. A synthesis of climatic data provided by the widely used Köppen and Thornthwaite classification systems was undertaken by Azevedo (1972) to map and quantify, at a national scale, the climatic regions of Angola, based on the substantial data set available at that time. Interestingly, despite some of its shortcomings, the Azevedo map provides a closer fit with general features of Angola's bio-climatic patterns than a much more recent map (Peel et al. 2007). The latter map is based on a global synthesis and review of the Köppen system, and draws on a very limited data set for Angola (5 stations for temperature; 16 stations for precipitation). The northern region of Angola is typical of Köppen's Tropical Wet Savanna (Aw) group, the plateau of the Temperate Mesothermal (Cw) group, and the southwest and coastal plain the Dry Desert and Semi-desert (Bsh, Bwh) group.

Mean annual rainfall and mean monthly temperatures for hottest and coldest months illustrate a few climatic characteristics of the Köppen regions (Table 2.1). The absence of data on extreme minimum temperatures and of frost occurrence is regrettable, as these factors, in tandem with fires and herbivory, play significant roles in the floristic composition and physiognomic structure of Angolan vegetation (Zigelski et al. 2019).

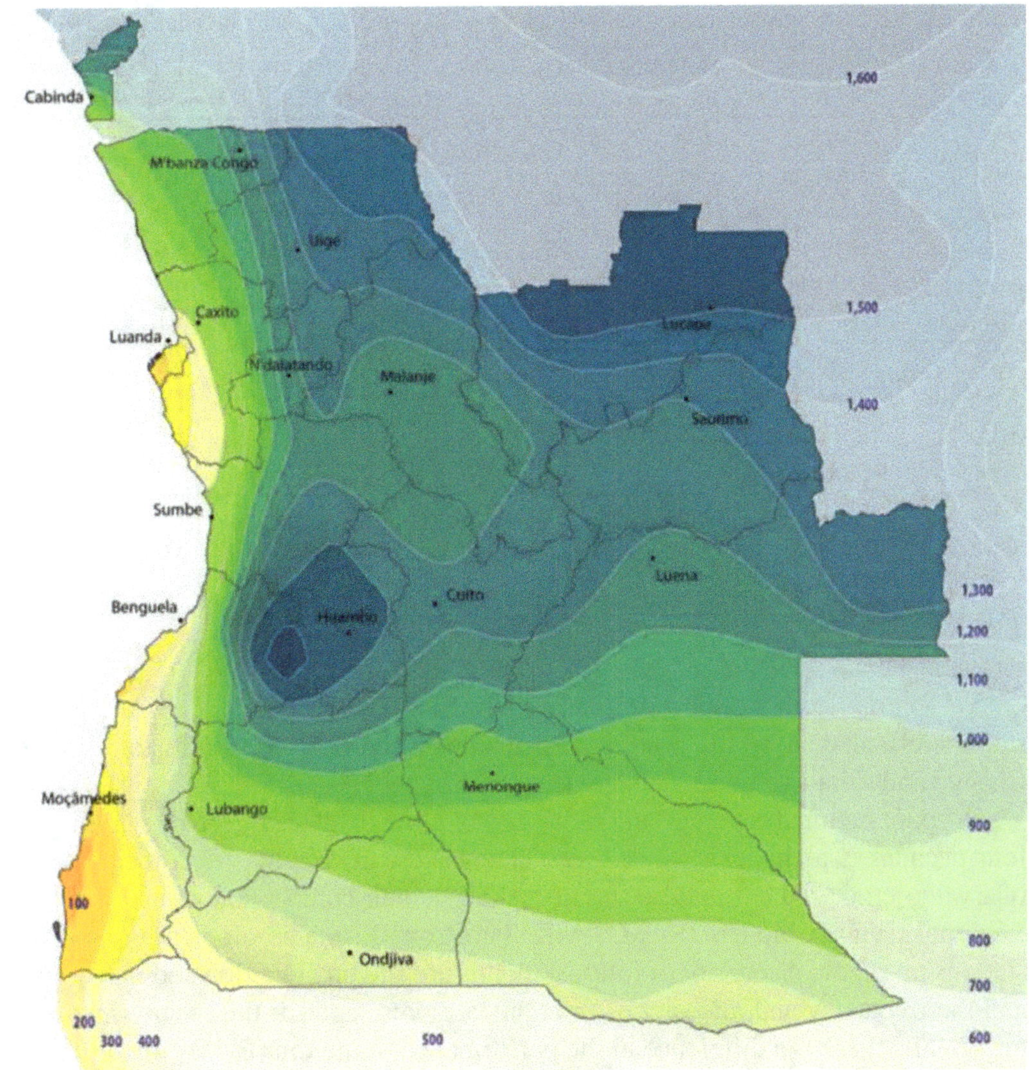

Fig. 2.7 Mean annual rainfall in Angola

Climate Change

Studies on the climate of Angola have been frustrated in recent years by the collapse of the extensive network of weather stations maintained during the colonial era by the then Meteorological Services of Angola. The publication by Silveira (1967) of recordings from 184 stations across all 18 provinces provides an invaluable record of the country's climate. According to the Ministry of Environment's Initial National Report to the UNFCCC (GoA 2013) the weather recording network collapsed from 225 'climatological posts' in 1974 to zero posts in 2010, while synoptic stations decreased from 29 in 1974 to 23 operational stations in 2010, 12 being automatic and 11 conventional. The network has since been strengthened by 22 automatic

Table 2.1 Representative climatic data following the Köppen climate classification system

Köppen symbol	Station	Altitude (m)	Precipitation (mm)	Mean of hottest month °C	Mean of coldest month °C
Aw	Belize	245	1612	26.7	22.2
Aw	Saurimo	1081	1355	23.8	20.3
Bsh	Ondgiva	1150	577	26.4	16.7
Bsh	Cuangar	1050	596	24.6	15.0
Bsh'	Chitado	1000	405	27.4	19.2
Bsh'	Luanda	44	405	27.0	20.1
Bwh	Moçâmedes	44	37	24.2	15.5
Bwh	Tômbwa	4	12	24.2	14.5
Bwh'	Benguela	7	184	26.3	18.0
Bwh'	Caraculo	440	123	26.4	17.2
Cwa	Menongue	1348	965	23.4	14.5
Cwa	Luena	1328	1182	22.7	17.0
Cwb	Huambo	1700	1210	20.6	15.7
Cwb	Lubango	1760	802	20.7	15.3

Data from Silveira (1967)

stations established by the Southern African Science Service Centre for Climate Change and Adaptive Land Management (SASSCAL).

The poor national coverage and reliability of climatic data collected over the past four decades is a challenge for climate change research. However, a recent study (Carvalho et al. 2017) provides the first analysis and comparison of a set of four Regional Climate Models (RCMs) with data from 12 meteorological monitoring stations in Angola. Scenarios of future temperature and precipitation anomaly trends and the frequency and intensity of droughts are presented for the twenty-first century. While there is a difference in the performance of the four RCMs, in particular for precipitation, consistent results were found for temperature projections, with an increase of up to 4.9 °C by 2100. The temperature increases are lowest for the northern coastal areas and highest for the southeast. In contrast to temperature rises, precipitation was projected to fall over the century, with an average of −2% across the country. Again, the strongest change was projected for the southeast, with decreases of up to −4%. Due principally to the projected increase in Sea Surface Temperatures by approximately 3 °C over the Atlantic during the twenty-first century, the central coastal region is expected to have a slight increase in precipitation.

Carvalho et al. (2017) highlight the extreme climate vulnerability of Angola, as previously noted by other studies (Brooks et al. 2005; Cain and Cain 2015). They conclude that climate change in Angola will bring stronger and more frequent droughts through the century, with impacts on water resources, agricultural productivity and wildfire potential. These factors will no doubt play out in negative ways on the current trends of land transformation and degradation as described by Mendelsohn (2019).

Biogeography, Biomes and Ecoregions

Overview

Angola's geographic location, geological history, climate and physiography account for its rich biological diversity. The comparative paucity of research focused on or within Angola explains the dependence of descriptions of the country's biogeography on broader regional reviews. While a full synthesis and interpretation of the evolution of the country's fauna and flora awaits development (Cotterill 2010, 2015), recent workers have advanced towards consensus on the main patterns, as discussed in general terms for terrestrial biota in this chapter, for marine systems by Kirkman and Nsingi (2019) and Weir (2019), and for freshwater fishes by Skelton (2019) in other chapters of this volume.

In brief, three marine ecoregions (Spalding et al. 2007) are within or overlap with Angola's marine environment, namely the Guinea South, Angolan, and Namib Ecoregions, the first two of which belong to the tropical Gulf of Guinea biogeographical province whereas the latter is part of the Benguela biogeographical province (Kirkman and Nsingi 2019). Most of Angola's EEZ falls within the Benguela Current Large Marine Ecosystem, with only Cabinda in the far north being included in the Guinea Current Large Marine Ecosystem.

The freshwater ecoregions of Africa have been classified and mapped by Thieme et al. (2005) and the eight ecoregions found within Angola are described in this volume (Skelton 2019). Skelton (2019) provides an elegant biogeographic model to explain the patterns and dynamics of freshwater fish faunas of Angola. Neither the floral nor vegetation patterns reflect the complexities and subtleties embedded in the ichthyological zoogeography of Angola, given the mobility of terrestrial plant dissemination. Here I will confine discussion to the terrestrial biota and ecosystems.

Angola lies between and within two major terrestrial biogeographic regions: the moist forests and savannas of the Congolian region; and the woodlands, savannas and floodplains of the Zambezian region. These two major divisions occupy over 97% of Angola. Gallery and escarpment forests of Congolian affinity penetrate southwards into the Zambezian savannas and woodlands of the Angolan *planalto* along deeply incised tributaries of the Congo Basin, and form a broken chain of forests southwards along the western escarpment. In the south, the extensive *Brachystegia/Julbernardia* miombo woodlands that occupy most of central Angola transition to *Baikiaea/Guibourtia/Burkea* savannas and woodlands. In the southwest, the arid *Acacia/Commiphora/Colophospermum* savannas, dwarf shrublands and desert of the Karoo-Namib region are found, penetrating northwards as a narrowing wedge along the coastal lowlands to Cabinda. The smallest of Africa's centres of botanical endemism – the *Podocarpus* Afromontane forests and montane grasslands – are represented by extremely restricted, relict patches in the mountains of the Benguela, Huambo and Huíla highlands.

Early Studies

Beyond general agreement on the above brief outline, botanists and zoologists have described and debated as many systems of biogeographic classification and of terminologies for Angola and for Africa as there are authors of the papers on the topic (Werger 1978). The pioneering works of Welwitsch (1859); Gossweiler and Mendonça (1939) and Barbosa (1970) provided the basis for several subsequent attempts to integrate the vegetation of Angola within a regional framework (Monteiro 1970; White 1971, 1983; Werger 1978). Zoogeographic classifications (Chapin 1932; Frade 1963; Monard 1937; Hellmich 1957; Crawford-Cabral 1983) are, with some minor exceptions, compatible with the overall systems of botanists (Werger 1978; Linder et al. 2012), (but see Branch et al. 2019, for comments on lizards). The Africa-wide synthesis of White (1983) is particularly useful in considering Angola's floristic (and in general terms, zoological) patterns and affinities. In broad terms, and following White's terminology, Angola includes representation of four 'regional centres of endemism'. They comprise the following centres with estimates of the percentage of their total area in Angola from Huntley (1974a, 2010):

- *Guineo-congolian regional centre of endemism* - mosaics of forests, thickets, tall grass savannas – 25.7% (This is Linder et al.'s Congolian Region and includes their Shaba sub-region);
- *Zambezian regional centre of endemism* – moist woodlands, savannas, grasslands and thickets – 71.6% (The Zambezian Region *sensu* Linder et al.);
- *Karoo-Namib regional centre of endemism* – desert, shrublands, arid savannas, woodlands and thickets – 2.6% (Most of this is placed in Linder et al.'s Southern African Region as their Southwest Angola sub-region); and
- *Afromontane archipelago-like regional centre of endemism* – forests, savannas and grasslands – 0.1%. (This is related to Linder et al.'s Ethiopian Region).

Statistical Regionalisation

Attempts have recently been made to use the massive databases of species distribution records held by museums and herbaria to bring objectivity and consistency to the classification of Africa's floral and faunal regions. A major step towards such regionalisation is provided by the statistical definition of biogeographical regions of sub-Saharan Africa by Linder et al. (2012). Using data for 1877 grid cells of one-degree resolution, the study included data for over a million records of 1103 species of mammals, 1790 species of birds, 769 species of amphibians, 480 species of reptiles and 5881 species of vascular plants. The databases were analysed using cluster analysis techniques to define biogeographical units that "comprise grid cells that are more similar in species composition to each other than to any other grid cells" (Linder et al. 2012). They proposed seven biogeographical regions for sub-Saharan

Africa: Congolian, Zambezian, Southern African, Ethiopian, Somalian, Sudanian and Saharan. Their analyses demonstrated that patterns of richness and endemism are positively and significantly correlated among plants, mammals, amphibians, birds and reptiles and with the overall biogeographical regions revealed by the sum of the data sets.

The use of modern cluster analysis techniques was taken further, at an Angolan level, by Rodrigues et al. (2015). Based on a cluster analysis of data for 9880 records of 140 species of ungulates, rodents and carnivores at a quarter degree resolution, the study found general congruence with that of Linder et al. (2012) and the earlier divisions of Angola's biogeography (Beja et al. 2019). Rodrigues et al. (2015) identify 18 indicator species for their four main divisions, which agree with the groupings based on field surveys undertaken in the 1970s (Huntley 1973) that also included the enclave of Cabinda, which was not included in the Rodrigues et al. (2015) analyses.

Both of the above very detailed and objective cluster analyses confirmed the general patterns of biogeographical regionalisation used for many decades across Africa, as described at the head of this section, even though terminology and detail of boundaries and transitions between regions differ from one author to the next. While objective, it is possible that the cluster analysis approach lacks the subtlety and flexibility of scale that classical expert systems permit. A particular challenge is the paucity of geo-referenced data for Angolan taxa, as experienced in a recent botanical analysis at inter-tropical scale (Droissart et al. 2018). Both cluster analyses and expert systems remain works in progress.

Biomes and Ecoregions

The chorological studies of White (1983) and statistical analyses of Linder et al. (2012) capture some of the evolutionary history and relationships of Africa's flora and fauna, but they do not fully reflect the continent's diversity of biomes, habitats and ecosystems – which are based on structural and functional rather than evolutionary relationships. The most comprehensive recent synthesis on African habitats (Burgess et al. 2004) has been widely adopted as a basis for conservation planning and is of use for any study of African biomes, ecoregions and habitats (MacKinnon et al. 2016). At the first level, a global classification and map of the world's ecoregions (Olson et al. 2001) was used to identify the nine biomes of Africa's three main biogeographic divisions (Palearctic, Afrotropical and Cape). The biome concept used was defined as "vegetation types with similar characteristics grouped together as habitats, and the broadest global habitat categories are called biomes" (Olson et al. 2001). Of the nine biomes recognised, seven are represented in Angola – the largest range of biomes represented in any African country. These are:

- Tropical and subtropical moist forests;
- Montane grasslands and shrublands;

- Tropical and subtropical grasslands, savannas, shrublands, and woodlands;
- Tropical and subtropical dry and broadleaf forests;
- Deserts and xeric shrublands;
- Mangroves; and
- Flooded grasslands and savannas.

Within the biomes, Burgess et al. (2004) defined a total of 119 terrestrial ecoregions for Africa and its islands. Ecoregions are defined as "large units of land or water that contain a distinct assemblage of species, habitats and processes, and whose boundaries attempt to depict the original extent of natural communities before major land-use change" (Dinerstein et al. 1995). It is impressive to note that based on the Burgess et al. (2004) assessment, Angola has not only the largest diversity of biomes, but also the second largest representation of ecoregional diversity in Africa (Table 2.2, Fig. 2.8).

Figure 2.8 (Burgess et al. 2004) provides a useful framework for the understanding of Angola's biodiversity patterns. Despite its coarse grain, it allows a general synthesis to be refined as new information becomes available. The relationship between biomes and ecoregions (*sensu* Burgess et al. 2004) and the vegetation types of Barbosa (1970) is summarised in Table 2.3. The very brief notes on key genera found within the Barbosa vegetation units provide an idea of the floristic composition that characterises the ecoregion. The photos presented in Fig. 2.9 provide examples of the main vegetation types and habitats

The Biological Importance of the Angolan Escarpment

While the classification of White (1983) and Linder et al. (2012) are useful at a continental scale, a more detailed and subtle analysis of the major biomes and habitat groupings is needed at a national scale for both research and conservation

Table 2.2 Representation of biomes and ecoregions in southern African countries

Country	Biomes	Ecoregions number and total (T)	T
Angola	7	8, 32, 42, 43, 49, 50, 51,55, 56, 63, 81, 82, 106, 109, 116	15
Botswana	3	54, 57, 58, 63, 68, 105	6
Congo Republic	3	8, 12, 13, 43,116	5
D.R. Congo	5	8, 13, 14, 15, 16, 17, 42, 43, 49, 50, 73, 116	12
Mozambique	3	21, 22, 52, 53, 54, 64, 76, 117	8
Namibia	3	51, 55, 58, 67, 105, 106, 107, 108, 109, 110	10
South Africa	5	22, 23, 24, 54, 57, 58, 77, 78, 79, 80, 89, 90, 91, 105, 108, 110, 117	17
Zambia	4	32, 50, 53, 54, 56, 63, 74	7
Zimbabwe	2	51, 53, 54, 57, 58, 76	6

From Burgess et al. (2004)

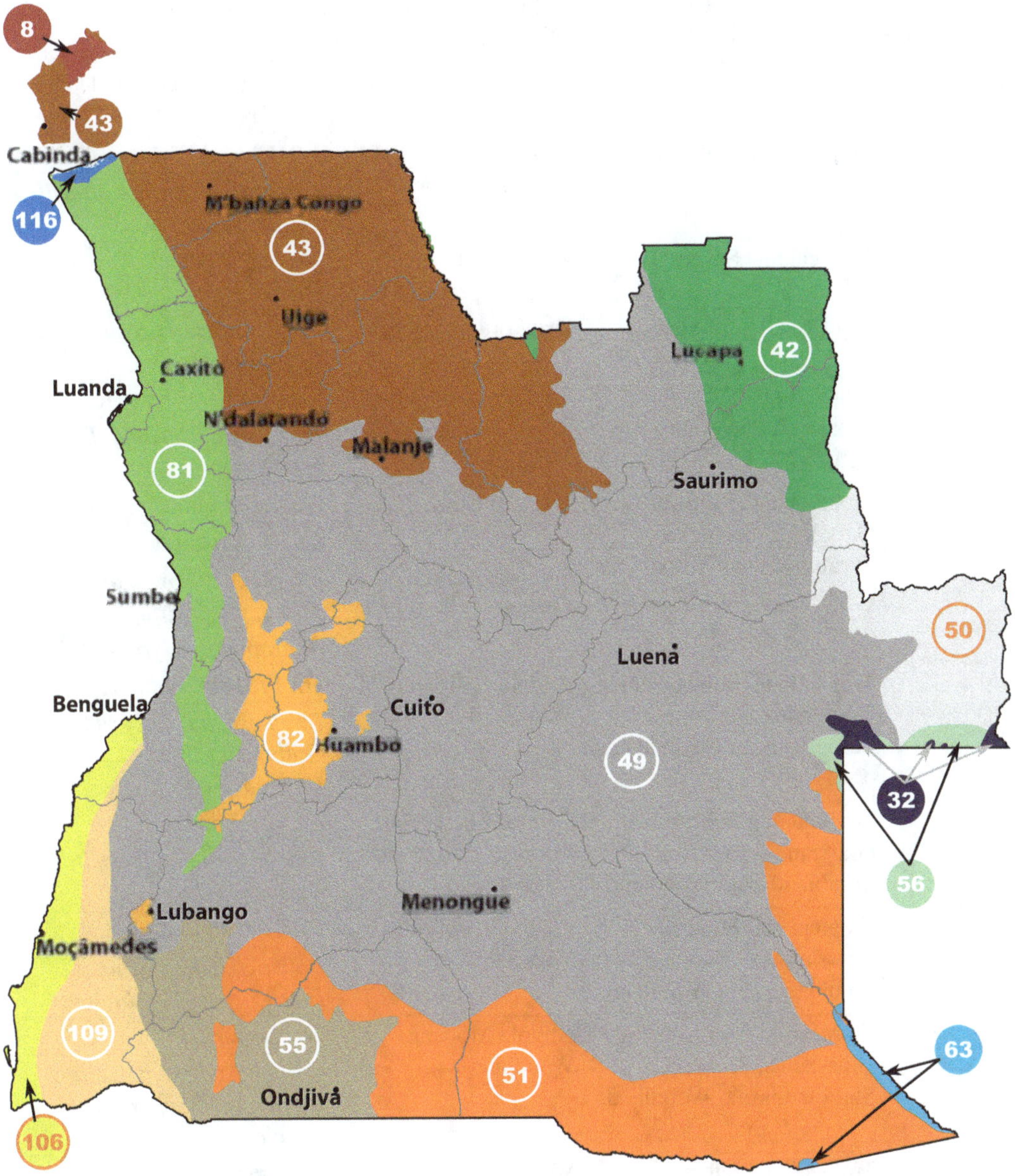

Fig. 2.8 Ecoregions of Angola
8 Atlantic Equatorial Coastal Forest • 32 Zambezian *Cryptosepalum* Dry Forest • 42 Southern Congolian Forest-Savanna Mosaic • 43 Western Congolian Forest-Savanna Mosaic • 49 Angolan Miombo Woodland • 50 Central Zambezian Miombo Woodland • 51 Zambezian *Baikiaea* Woodland • 55 Angola Mopane Woodland • 56 Western Zambezian Grassland • 63 Zambezian Flooded Grasslands • 81 Angolan Scarp Savanna and Woodland • 82 Angolan Montane Forest-Grassland Mosaic • 106 Kaokoveld Desert • 109 Namib Escarpment Woodlands • 116 Central African Mangroves. (After Burgess et al. 2004, map used with permission)

Table 2.3 African biomes and ecoregions (as defined by Burgess et al. 2004) and Angolan vegetation types (Barbosa 1970) with indicative genera

Ecoregion n°	Biome	Ecoregion	Barbosa n°, name and key genera
8	Tropical and Subtropical Forest	Atlantic Equatorial Coastal Forest	1,2. Closed Forest
			Gilbertiodendron, Librevillea, Tetraberlinia
32	Tropical and Subtropical	Zambezian *Cryptosepalum*	4. Closed Forest
	Dry Broadleaf Forest	Dry Forest	*Cryptosepalum, Brachystegia, Erythrophleum*
42	Tropical and Subtropical Grasslands, Savannas, Shrublands and Woodlands	Southern Congolian Forest-Savanna Mosaic	8. Forest-Savanna Mosaic
			Marquesia, Berlinia, Daniella, Hymenocardia
43	Tropical and Subtropical Grasslands, Savannas, Shrublands and Woodlands	Western Congolian	3. Closed Forest
			Celtis, Albizia, Celtis
		Forest-Savanna Mosaic	13. Thicket-Forest Mosaic
			Annona, Piliostigma, Andropogon, Hyparrhenia
49	Tropical and Subtropical Grasslands, Savannas, Shrublands and Woodlands	Angolan Miombo Woodland	16, 17, 18. Woodland
			Brachystegia, Julbernardia, Guibourtia, Burkea, Pterocarpus
50	Tropical and Subtropical Grasslands, Savannas, Shrublands and Woodlands	Central Zambezian Miombo Woodland	17, 19. Woodland
			Brachystegia, Julbernardia, Cryptosepalum
51	Tropical and Subtropical Grasslands, Savannas, Shrublands and Woodlands	Zambezian Baikiaea Woodland	25. Tree and Shrub Savanna
			Baikiaea, Guibourtia, Pterocarpus, Combretum
55	Tropical and Subtropical Grasslands, Savannas, Shrublands and Woodlands	Angola Mopane Woodland	20. Woodland
			Colophospermum, Croton, Combretum, Sclerocarya, Acacia
56	Tropical and Subtropical Grasslands, Savannas, Shrublands and Woodlands	Western Zambezian Grassland	31. Grasslands
			Loudetia, Monocymbium, Tristachya, Parinari, Syzygium
63	Flooded Grasslands and Savannas	Zambezian Flooded Grasslands	31. Grasslands
			Loudetia, Echinochloa, Oryza
81	Montane Grasslands and Shrublands	Angolan Scarp Savanna and Woodland	10, 11, 22, 23. Forest-Savanna-Woodland-Thicket Mosaic
			Adansonia, Acacia, Albizia, Celtis, Piliostigma
82	Montane Grasslands and Shrublands	Angolan Montane Forest-Grassland Mosaic	6, 32. Relict Forest, Grasslands
			Podocarpus, Apodytes, Pittosporum, Protea, Erica

(continued)

Table 2.3 (continued)

Ecoregion n°	Biome	Ecoregion	Barbosa n°, name and key genera
106	Deserts and Xeric Shrublands	Kaokoveld Desert	28, 29. Desert, Steppes
			Welwitschia, Zygophyllum, Stipagrostis, Odyssea
109	Deserts and Xeric Shrublands	Namib Escarpment Woodlands	27. Steppes
			Acacia Commiphora, Colophospermum, Sesamothamnus, Rhigozum
116	Mangroves	Central African Mangroves	14 A. Mangroves
			Rhizophora, Avicennia, Raphia, Elaeis

planning in Angola (Revermann and Finckh 2019). What is equally important in biogeographic analysis is the detection of patterns of endemism and diversity at dispersed scales – such as the Angolan Escarpment Zone – described by Hall (1960a) and subsequently recognised by many workers as of great biodiversity and evolutionary importance (Huntley 1973, 1974a, 2017; Hawkins 1993; Mills 2010; Clark et al. 2011). Indeed, each taxon-based account in this volume, on plants (Goyder and Gonçalves 2019), odonata (Kipping et al. 2019), lepidoptera (Mendes et al. 2019), fishes (Skelton 2019), birds (Dean et al. 2019), amphibians (Baptista et al. 2019), reptiles (Branch et al. 2019) and mammals (Beja et al. 2019) draws attention to the importance of the Angolan Escarpment as a centre of endemism and speciation. Hall (1960a) explained her recognition of the importance of the Angolan Escarpment as the major speciation hotspot for birds in Angola by it: (i) creating a barrier between arid-adapted species of the coastal plains and of the miombo woodlands of the plateau, (ii) creating a steep ecological gradient, and (iii) functioning as a refuge for moist forest specialists that were isolated here during the dry periods of the glacial cycles. Dean et al. (2019) note that 75% of Angola's endemic birds are found in this zone.

The Angolan Escarpment and the remote, isolated and fragmentary remnants of Afro-montane forests of the Angolan Highlands offer ideal testing grounds for biogeographic models, as recently explored by Vaz da Silva (2015). The Angolan Escarpment biogeographic unit awaits clear definition, description and demarcation, but its scientific importance is matched only by the vulnerability of its threatened forest habitats (Cáceres et al. 2015). Linder et al. (2012) similarly recognise the importance of the Angolan Escarpment, and that of the transition from Congolian to Zambezian regions along the northern border of Angola (which they place in their Shaba sub-region). High species replacement values are found across these biologically rich areas, emphasising the urgency for their protection.

Fig. 2.9 Examples of some of the Ecoregions of Angola with numbering as per map in Fig. 2.8 and summary in Table 2.3. 8. Maiombo Forest, Cabinda; 42. Congolian gallery forest and moist miombo woodlands and savanna grasslands, Lunda-Norte; 49. *Brachystegia/Julbernardia* woodland Luando Strict Nature Reserve, Malange; 51. *Baikiaea/Guibourtia* Woodland Mucusso, Cuando Cubango; 56. Wetlands of the Bulozi Floodplain, Moxico; 81. Angolan Escarpment at Serra da Chela, Tundavala, Huíla; 82. Remnant patches of Afromontane forest in ravines on Mount Moco, Huambo; 106 Grasslands of the intermontane plains of central Iona National Park, Namibe. (Photos: Bulozi – JM Mendelsohn, others by BJ Huntley)

Conclusions

This brief outline of the biogeography of Angola demonstrates the country's unusual diversity of landscapes, climates and ecoregions, with Angola embracing the highest number of biomes represented within any African state.

The many classifications and terminologies applied to Angola's biogeographic units over the past century have not yet resulted in a nationally adopted nomenclature for its biomes and habitats. This situation prevails despite the existence of strong traditions in Angola's ethnic groups of indigenous taxonomies for habitats, such as those of the Chokwe of the Lundas, that are as perfect and detailed as modern systems (Redinha 1961; Huntley 2015). Furthermore, while many vernacular terms (*mato de panda, anharas do alto, floresta cafeeira, muxitos, mulolas, chanas da borracha*, etc.) enjoy wide use, they are imprecise and inadequate for Angola's great diversity of biomes and habitats.

The absence of a uniform system of nomenclature limits the use of information attached to biological collections, which in most cases provide only site locality data, and more recently, geo-referencing. Several southern African countries have nationally accepted biome and vegetation maps (e.g. South Africa, Lesotho and Swaziland – Mucina and Rutherford 2006) with clear descriptors for each biome and vegetation unit, facilitating communication between researchers and conservation planners. As Angola re-assesses its biodiversity wealth, and the need to protect and sustainably utilise these resources, the development of a new map of its vegetation, ecosystems and biomes becomes a high priority. Equally urgent but similarly daunting is the study of the evolutionary processes and relationships of the biota of the Angolan Escarpment and Afromontane forests, and the effective protection of these fingerprints of the past.

References

Albuquerque S, Figueirôa S (2018) Depicting the invisible: Welwitsch's map of travellers in Africa. Earth Sci Hist 37:109–129

Amaral I (1969) 'Inselberge' (ou montes-ilhas) e superfícies de aplanação na bacia do Cubal da Hanha em Angola. Garcia da Orta 17:474–526

Azevedo AL (1972) Caracterizacão Sumária das Condições Ambientais de Angola. Universidade de Luanda, Luanda, 106 pp

Baptista N, Conradie W, Vaz Pinto P et al (2019) The amphibians of Angola: early studies and the current state of knowledge. In: Huntley BJ, Russo V, Lages F, Ferrand N (eds) Biodiversity of Angola. Science & conservation: a modern synthesis. Springer Nature, Cham

Barbosa LAG (1970) Carta Fitogeográfica da Angola. Instituto de Investigação Científica de Angola, Luanda, 343 pp

Beja P, Vaz Pinto P, Veríssimo L et al (2019) The mammals of Angola. In: Huntley BJ, Russo V, Lages F, Ferrand N (eds) Biodiversity of Angola. Science & conservation: a modern synthesis. Springer Nature, Cham

Branch WR, Vaz Pinto P, Baptista N et al (2019) The reptiles of Angola: history, diversity, endemism and hotspots. In: Huntley BJ, Russo V, Lages F, Ferrand N (eds) Biodiversity of Angola. Science & conservation: a modern synthesis. Springer Nature, Cham

Brooks N, Adger WN, Kelly PM (2005) The determinants of vulnerability and adaptive capacity at the national level and the implications for adaptation. Glob Environ Chang 15:151–163

Burgess N, Hales JD, Underwood E et al (2004) Terrestrial ecoregions of Africa and Madagascar – a conservation assessment. Island Press, Washington DC, 499 pp

Cáceres A, Melo M, Barlow J et al (2015) Threatened birds of the Angolan Central Escarpment: distribution and response to habitat change at Kumbira Forest. Oryx 49:727–734

Cáceres A, Melo M, Barlow J et al (2017) Drivers of bird diversity in an understudied African centre of endemism: the Angolan Escarpment Forest. Bird Conserv Int 27:256–268

Cain A, Cain A (2015) Climate change and land markets in coastal cities of Angola. In 2015 World Bank conference on land and poverty. The World Bank, Washington, DC

Carvalho SCP, Santos FD, Pulquério M (2017) Climate change scenarios for Angola: an analysis of precipitation and temperature projections using four RCMs. Int J Climatol 37:3398–3412

Chapin JP (1932) The birds of the Belgian Congo. Bull Am Mus Nat Hist 65:1–756

Choffat P (1888) Dr. Welwitsch: Quelques notes sur la géologie d'Angola coordonnées et annotées par Paul Choffat. Separata das Comunicações dos Serviços Geológicos de Portugal 19:1–24

Clark VR, Barker NP, Mucina L (2011) The Great Escarpment of southern Africa: a new frontier for biodiversity exploration. Biodivers Conserv 20:2543–2561

Costa FL (2006) O conhecimento geomorfológico de Angola. In: Moreira I (ed) Angola: Agricultura, Recursos Naturais e Desenvolvimento. ISA Press, Lisboa, pp 477–495

Cotterill FPD (2010) The evolutionary history and taxonomy of the Kobus leche species complex of south-central Africa in the context of Palaeo-drainage dynamics. Unpublished PhD thesis, University of Stellenbosch

Cotterill FPD (2015) Biogeographical overview of the Lunda region, northeast Angola. In: Huntley BJ, Francisco P (eds) Avaliação Rápida da Biodiversidade de Região da Lagoa Carumbo, Lunda-Norte – Angola/Rapid biodiversity assessment of the Carumbo Lagoon Area, Lunda-Norte – Angola. Ministério do Ambiente, Luanda, pp 77–99

Cotterill F, De Wit M (2011) Geoecodynamics and the Kalahari Epeirogeny: linking its genomic record, tree of life and palimpsest into a unified narrative of landscape evolution. S Afr J Geol 114:489–514

Crawford-Cabral J (1983) Esboço zoogeográfico de Angola. Unpublished manuscript, Lisbon, 50 pp + 13 maps

Dean WRJ (2000) The birds of Angola: an annotated checklist. BOU Checklist No. 18. British Ornithologists' Union. Tring, UK

Dean WRJ, Melo M, Mills MSL (2019) The Avifauna of Angola: richness, endemism and rarity. In: Huntley BJ, Russo V, Lages F, Ferrand N (eds) Biodiversity of Angola. Science & conservation: a modern synthesis. Springer Nature, Cham

Dinerstein E, Olson D, Graham A et al (1995) A conservation assessment of the ecoregions of Latin America and the Caribbean. World Bank, Washington, DC

Diniz AC, Aguiar FB (1966) Geomorfologia, solos e ruralismo de região central angolana. Agronomia Angolana 23:11–17

Diniz AC (1973) Características mesológicas de Angola. Missão de Inquéritos Agrícolas de Angola, Nova Lisboa

Diniz AC (1991) Angola, o meio físico e potencialidades agrárias. Instituto para a Cooperação Económica, Lisboa, 189 pp

Diniz AC (2006) Características mesológicas de Angola. Instituto Português de Apoio ao Desenvolvimento, Lisbon, 546 pp

Droissart V, Dauby G, Hardy OJ et al (2018) Beyond trees: biogeographical regionalization of tropical Africa. J Biogeogr 45:1153–1167

Frade F (1963) Linhas gerais da distribuição geográfica dos vertebrados em Angola. Memórias da Junta de Investigações do Ultramar 43:241–257

Feio M (1964) A evolução da escadaria de aplanações do sudoeste de Angola. Garcia da Orta 12:323–354

GoA (Government of Angola) (2013) Angola initial national communication to the United Nations framework convention on climate change. Ministry of the Environment, Luanda, 194 pp

Gossweiler J, Mendonça FA (1939) Carta Fitogeográfica de Angola. Ministério das Colónias, Lisboa, 242 pp

Goyder DJ, Gonçalves FMP (2019) The flora of Angola: collectors, richness and endemism. In: Huntley BJ, Russo V, Lages F, Ferrand N (eds) Biodiversity of Angola. Science & conservation: a modern synthesis. Springer Nature, Cham

Hall BP (1960a) The faunistic importance of the scarp of Angola. Ibis 102:420–442

Hall BP (1960b) The ecology and taxonomy of some Angolan birds. Bull Br Mus (Nat Hist) Zool 6:367–463

Hall BP, Moreau RE (1962) The rare birds of Africa. Bull Br Mus (Nat Hist) Zool 8:315–381

Hawkins F (1993) An integrated biodiversity conservation project under development: the ICBP Angola Scarp Project. In: Proceedings of the VIII Pan-African Ornithological Congress, pp 279–284. Kigali, Rwanda, 1992. Koninklijk Museum voor Midden-Afrika, Tervuren

Hellmich W (1957) Herpetologische Ergebnisse einer Forschungsreise in Angola. Veröffentlichungen der Zoologischen Staatssammlung München 5:1–92

Humbert H (1940) Zones et Étages de Végétation dans le Sud-Ouest de l'Angola. Compte-rendu Sommaire des Séances de la Societé de Biogéographie 17:47–57

Huntley BJ (1973) Distribution and Status of the Larger Mammals of Angola, with particular reference to Rare and Endangered species: First Progress Report. December 1973. Repartição Técnica da Fauna, Serviços de Veterinária, Luanda, Mimeograph report, 14 pp

Huntley BJ (1974a) Vegetation and Flora Conservation in Angola. Ecosystem Conservation Priorities in Angola. Ecologist's Report 22. Repartição Técnica da Fauna, Serviços de Veterinária, Luanda, Mimeograph report, 13pp

Huntley BJ (1974b) Ecosystem Conservation Priorities in Angola. Ecologist's Report 26. Repartição Técnica da Fauna, Serviços de Veterinária, Luanda, Mimeograph report

Huntley BJ (2010) Estratégia de Expansão da Rede de Áreas Protegidas da Angola/Proposals for an Angolan Protected Area Expansion Strategy (APAES). Unpublished report to the Ministry of Environment, Luanda, 28 pp + map

Huntley BJ (2015) Biophysical profile of Lunda-Norte. In: Huntley BJ, Francisco P (eds) Avaliação Rápida da Biodiversidade de Região da Lagoa Carumbo, Lunda-Norte – Angola / Rapid Biodiversity Assessment of the Carumbo Lagoon Area, Lunda-Norte – Angola. Ministério do Ambiente, Luanda, pp 31–75

Huntley BJ (2017) Wildlife at war in Angola: the rise and fall of an African Eden. Protea Book House, Pretoria, 432 pp

Huntley BJ, Matos EM (1994) Botanical diversity and its conservation in Angola. Strelitzia 1:53–74

Jessen O (1936) Reisen und Forschungen in Angola. Dietrich Reimer Verlag, Berlin

Jones A, Breuning-Madsen H, Brossard M et al (2013) Soil atlas of Africa. Publications Office of the European Union, Brussels

King LC (1962) Morphology of the Earth. Oliver & Boyd, London, 699 pp

Kipping J, Clausnitzer V, Fernandes Elizalde SRF et al (2019) The dragonflies and damselflies of Angola. In: Huntley BJ, Russo V, Lages F, Ferrand N (eds) Biodiversity of Angola. Science & conservation: a modern synthesis. Springer Nature, Cham

Kirkman SP, Nsingi KK (2019) Marine biodiversity of Angola: biogeography and conservation. In: Huntley BJ, Russo V, Lages F, Ferrand N (eds) Biodiversity of Angola. Science & conservation: a modern synthesis. Springer Nature, Cham

Linder HP, de Klerk HM, Born J et al (2012) The partitioning of Africa: statistically defined biogeographical regions in sub-Saharan Africa. J Biogeogr 39:1189–1925

MacKinnon J, Aveling C, Olivier R et al (2016) Inputs for an EU strategic approach to wildlife conservation in Africa – regional analysis. European Commission, Directorate-General for International Cooperation and Development, Brussels

Marques MM (1963) Notas sobre a geomorfologia da Angola 1. Significado morfológico de algumas 'anharas do alto'. Garcia da Orta 11:541–560

Maurin O, Davies TJ, Burrows JE et al (2014) Savanna fire and the origins of the 'underground forests' of Africa. New Phytol 204(1):201–214

Mateus O, Callapez P, Polcyn M et al (2019) Biodiversity in Angola through time: a paleontological perspective. In: Huntley BJ, Russo V, Lages F, Ferrand N (eds) Biodiversity of Angola. Science & conservation: a modern synthesis. Springer Nature, Cham

Mateus O, Jacobs LL, Schulp AS et al (2011) *Angolatitan adamastor*, a new sauropod dinosaur and the first record from Angola. Ann Braz Acad Sci 83(1):221–233

Mendelsohn JM (2019) Landscape changes in Angola. In: Huntley BJ, Russo V, Lages F, Ferrand N (eds) Biodiversity of Angola. Science & conservation: a modern synthesis. Springer Nature, Cham

Mendelsohn JM, Mendelsohn S (2018) Sudoeste de Angola: um Retrato da Terra e da Vida. South West Angola: a portrait of land and life. Raison, Windhoek

Mendelsohn J, Weber B (2013) An atlas and profile of Huambo: its environment and people. Development Workshop, Luanda, 80 pp

Mendelsohn J, Weber B (2015) Moxico: an atlas and profile of Moxico, Angola. Raison, Windhoek, 44 pp

Mendelsohn J, Jarvis A, Robertson T (2013) A profile and atlas of the Cuvelai-Etosha Basin. Raison & Gondwana Collection, Windhoek, 170 pp

Mendes L, Bivar-de-Sousa A, Williams M (2019) The butterflies and skippers of Angola. In: Huntley BJ, Russo V, Lages F, Ferrand N (eds) Biodiversity of Angola. Science & conservation: A modern synthesis. Springer Nature, Cham

Mills MSL (2010) Angola's central scarp forests: patterns of bird diversity and conservation threats. Biodivers Conserv 19:1883–1903

Mills MSL, Melo M, Vaz A (2013) The Namba mountains: new hope for Afromontane forest birds in Angola. Bird Conserv Int 23:159–167

Mills MSL, Olmos F, Melo M et al (2011) Mount Moco: its importance to the conservation of Swierstra's Francolin *Pternistis swierstrai* and the Afromontane avifauna of Angola. Bird Conserv Int 21:119–133

Monard A (1937) Contribution à l'herpétologie d'Angola. Arquivos do Museu Bocage 8:19–154

Monteiro RFR (1970) Estudo da Flora e da Vegetacão das Florestas Abertas do Planalto do Bié. Instituto de Investigacão Científica de Angola, Luanda, 352 pp

Mucina L, Rutherford MC (2006) The vegetation of South Africa, Lesotho and Swaziland. Strelitzia 19:1–807

Neto AG, Ricardo RP, Madeira M (2006) O alumínio nos solos de Angola. In: Moreira I (ed) Angola: Agricultura, Recursos Naturais e Desenvolvimento. ISA Press, Lisboa, pp 121–143

Olson DM, Dinerstein E, Wikramanayake ED et al (2001) Terrestrial ecoregions of the World: a new map of life on Earth. Bioscience 51:933–938

Peel MC, Finlayson BL, McMahon TA (2007) Updated world map of the Köppen-Geiger climate classification. Hydrol Earth Syst Sci 11:1633–1644

Redinha J (1961) Nomenclaturas nativas para as formações botânicas do nordeste de Angola. Agronomia Angolana 13:55–78

Revermann R, Finckh M (2019) Vegetation survey, classification and mapping in Angola. In: Huntley BJ, Russo V, Lages F, Ferrand N (eds) Biodiversity of Angola. Science & conservation: a modern synthesis. Springer Nature, Cham

Rodrigues P, Figueira R, Vaz Pinto P et al (2015) A biogeographical regionalization of Angolan mammals. Mammal Rev 45:103–116

Silveira MM (1967) Climas de Angola. Serviço Meteorólogico de Angola, Luanda, 44 pp

Skelton PH (2019) The Freshwater Fishes of Angola. In: Huntley BJ, Russo V, Lages F, Ferrand N (eds) Biodiversity of Angola. Science & conservation: a modern synthesis. Springer Nature, Cham

Spalding MD, Fox HE, Allen GR et al (2007) Marine ecoregions of the world: a bioregionalization of coastal and shelf areas. Bioscience 57:573–583

Thieme ML, Abell R, Stiassny ML et al (eds) (2005) Freshwater ecoregions of Africa and Madagascar: a conservation assessment. Island Press, Washington DC

Ucuassapi AP, Dias JCS (2006) Acerca da fertilidade dos solos de Angola. In: Moreira I (ed) Angola: Agricultura, Recursos Naturais e Desenvolvimento. ISA Press, Lisboa, pp 477–495

Vaz da Silva B (2015) Evolutionary history of the birds of the Angolan highlands – the missing piece to understand the biogeography of the Afromontane forests. MSc Thesis, University of Porto, Porto

Weir CR (2019) The whales and dolphins of Angola. In: Huntley BJ, Russo V, Lages F, Ferrand N (eds) Biodiversity of Angola. Science & conservation: a modern synthesis. Springer Nature, Cham

Welwitsch F (1859) Apontamentos phyto-geographicos sobre a Flora da Provincia de Angola na Africa Equinocial servindo de relatório preliminar acerca da exploração botanica da mesma provincia. Annaes do Conselho Ultramarino (Ser. 1):527–593

Werger MJA (1978) Biogeographical division of southern Africa. In: Werger MJA, van Bruggen AC (eds) Biogeography and ecology of Southern Africa. Junk, The Hague, pp 145–170

White F (1971) The taxonomic and ecological basis of chorology. Mitteilungen Botanischen Staatsammlung München 10:91–112

White F (1976) The underground forests of Africa: a preliminary review. The Gardens' Bull Singapore 24:57–71

White F (1983) The vegetation of Africa – a descriptive memoir to accompany the UNESCO/AETFAT/UNSO vegetation map of Africa. UNESCO, Paris, 356 pp

Zigelski P, Gomes A, Finckh M (2019) Suffrutex dominated ecosystems in Angola. In: Huntley BJ, Russo V, Lages F, Ferrand N (eds) Biodiversity of Angola. Science & conservation: a modern synthesis. Springer Nature, Cham

Chapter 3
Marine Biodiversity of Angola: Biogeography and Conservation

Stephen P. Kirkman and Kumbi Kilongo Nsingi

Abstract Some major physical and oceanographic features of the Angolan marine system include a narrow continental shelf, the warm, southward flowing Angola Current, the plume of the Congo River in the north and the Angola-Benguela Front in the south. Depth, substrate types and latitude have been shown to account for species differences in demersal faunal assemblages including fish, crustaceans, and cephalopods. The extremely narrow shelf between Tômbwa (15°48′S) and Benguela (12°33′S) may serve as a barrier for the spreading of shelf-occurring species between the far south, which is influenced by the Angola-Benguela Front, and the equatorial waters of the central and northern areas. A similar pattern is evident for coastal and shallow-water species, including fishes, intertidal invertebrates and seaweeds, with species that have temperate affinities found in the far south and tropical species further to the north. In general the fauna and flora of the littoral zone appears to be consistent with a pattern of relatively low diversity of the shore and near-shore areas, that is characteristic of West Africa, but paucity of data for Angola may make such comparisons of diversity with other areas inappropriate at this stage. The Congo River delta and many features that are interspersed along the coast such as estuaries and associated floodplains, wetlands, lagoons, salt marshes and mangroves, support a rich suite of species, many of which are rare, endemic, migratory, and/or threatened, and provide important ecosystem services. While the ecological value of many areas or features is recognised, lack of any legal protection in the form of marine protected areas (MPAs) has been identified as one of the main challenges facing conservation and sustainable use of Angola's marine and coastal biodiversity and habitats, in the face of multiple threats. A current process to identify and describe ecologically or biologically significant marine areas (EBSAs) could provide a foundation for designating some MPAs in future.

S. P. Kirkman (✉)
Department of Environmental Affairs, Oceans and Coastal Research, Cape Town, South Africa
e-mail: skirkman@environment.gov.za

K. K. Nsingi
Benguela Current Convention, Swakopmund, Namibia
e-mail: kkilongo@gmail.com

B. J. Huntley et al. (eds.), *Biodiversity of Angola*,
https://doi.org/10.1007/978-3-030-03083-4_3

Keywords Benguela current · Ecologically or biologically significant marine areas · Important bird areas · Fish · Marine protected areas · Marine spatial planning · Seaweed · Systematic conservation planning · West Africa

Physical and Oceanographic Context

The coastline of Angola, which is approximately 1650 km long, consists of sandy and rocky stretches of coastline, punctuated by numerous coastal features such as estuaries, mangroves, coastal lakes, wetlands and tidal flats (Harris et al. 2013). Between Rio Bero (north of Moçâmedes) in Namibe Province and to the north of Rio Coporolo in Benguela Province are rocky shores; the rest of the coastline is predominantly sandy although there are some scattered rocky shores further north of Lobito (Harris et al. 2013). The continental shelf, which extends to about 200 m depth, is relatively narrow especially near the south where it is as little as 6 km wide and very steep in parts of Namibe and Benguela, but it widens further north to 33 km width near the mouth of the Congo River, and in the south it widens a little between Tômbwa and Cunene (Figure 1, Bianchi 1992). The neritic zone (i.e. waters overlying the continental shelf) covers about one third of Angola's Exclusive Economic Zone (EEZ), which also includes extensive bathyal and abyssal zones, with depths up to 4000 m in the latter (Nsiangango et al. 2007).

Hutchings et al. (2009) describe the marine system of Angola's continental shelf area as a subtropical transition zone between the Equatorial Atlantic to the north and the Benguela's wind-driven upwelling system to the south. The conspicuous, dynamic but relatively shallow Angola-Benguela Front at 17°S in the south of Angola forms the boundary with the upwelling system, and the boundary to the north is near the plume of the Congo River. Seasons are well-defined and there is intermediate productivity; moderate to weak upwelling occurs year-round in the south and all along the coast in winter with strengthening of southeast trade winds. The major oceanographic feature of the system is the warm (>24 °C) Angola Current, which flows southward along the shelf and slope as an extension of the South Equatorial Counter-current, extending down to 200 m depth and with a mean flow at 50 m depth of 5–8 cm s^{-1} (Kopte et al. 2017). During winter and spring the Angola Current tends to retreat to the northwest and is replaced by slightly cooler waters; this is linked to the intensity of wind-driven upwelling off the Namibian coast (Meeuwis and Lutjeharms 1990; O'Toole 1980).

Other important drivers of the system are Kelvin waves propagating from the Equatorial Atlantic and the South Equatorial Counter Current (Florenchie et al. 2004; Shillington et al. 2006), as well as southward flow of brackish water with high nutrient loads from the Congo River outflow and solar heating (Veitch et al. 2010), Both result in stratification of the water column (Kirkman et al. 2016), with thermocline depth ranging from 10 m in the north down to about 50 m off central Angola (Bianchi 1992). Another feature of the Angolan system is the cold water Angola Dome, found offshore of the Angola Current. This is a cyclonic eddy that causes doming of the thermocline, centred near 10°S and 9°E (Lass et al. 2000). The Angola Dome has lower salinity and concentrations of oxygen than surrounding waters, but it does not exist in winter and its width and extension depend on the intensity and horizontal shear of southeasterly trade winds (Signorini et al. 1999). Phytoplankton production associated with the Angola Dome strongly influences the shelf ecosystem throughout northern Angola (Monteiro et al. 2008).

Biodiversity and Biogeography

There is a high diversity of demersal species in Angola relative to the temperate Benguela Ecosystem to the south, with species richness greatest at about 100 m depth according to research surveys (Kirkman et al. 2013). Demersal fish stocks are exploited by a multispecies bottom trawl fishery extending from southern to northern Angola, that exploits over 30 species of fish belonging to the families Sparidae (seabreams), Scianidae (croakers), Serranidae (groupers), Haemulidae (grunts) and Merlucidae (hakes). Some of the most commercially important species include Benguela Hake *Merluccius polli* and demersal sparid fish such as *Dentex* spp. (Kirkman et al. 2016); there is also bottom fishing for crustaceans, most importantly deep-sea crab *Chaceon maritae* and shrimps *Aristeus varidens* and *Parapenaeus longirostris* (Japp et al. 2011; Kirkman et al. 2016). The most important fish targeted by small pelagic fisheries include Kunene Horse Mackerel *Trachurus trecae* and *Sardinella* species, with most large pelagic fishing (tuna spp.) taking place in the south (Japp et al. 2011, Kirkman et al. 2016). Several of the above stocks are targeted both by industrial and artisanal fisheries (Duarte et al. 2005; Japp et al. 2011). The Angolan fisheries are described by Hutchings et al. (2009) as being of moderate intensity with stocks generally declining. There is also a rapidly growing local and foreign recreational shore-fishery sector in southern Angola targeting mainly Leerfish *Lichia amia*, West Coast Dusky Kob *Argyrosomus coronus* and Shad *Pomatomus saltatrix* (Potts et al. 2009).

Bianchi (1992) and later Nsiangango et al. (2007) studied the structure of demersal assemblages of the continental shelf and upper Angolan slope, including fish, crustaceans and cephalopods, based on trawl surveys. It was shown that thermal, depth-dependent stratification explained the main faunal groupings, with certain species generally restricted to waters shallower than where the lower limit of the thermocline meets the shelf, and others usually occurring in deeper waters than this. Species such as Shallow-water Croaker *Pteroscion peli*, Red Pandora *Pagellus bellotti*, Lesser African Threadfin *Galeiodes decadactylus* and Grunt *Pomadasys incisus* dominated in the shallower demersal water (coast to 100 m), with some sea breams in low densities, whereas deeper waters of the shelf and upper slope were dominated by such species as Splitfin *Synagrops microlepis*, Atlantic Green-eye *Chlorophthalmus atlanticus,* Angolan Dentex *D. angolensis* and *M. polli.* Within the different depth strata, substrate type and latitudinal gradients were the main factors affecting the composition of species assemblages, and a major latitudinal shift both in shallow- and deep-water assemblages was shown to occur in southern Angola between Tômbwa and Cunene where the shelf widens and where Large-eyed Dentex *D. macrophthalmus* dominated the catches. Bianchi (1992) related the shift to the southern limit of warmer equatorial waters, the presence of the Angola-Benguela Front where there is year-round upwelling and cooler waters, and the extremely narrow shelf to the north of Tômbwa (up to Benguela), which may serve as a barrier to the spreading of northern species to the south and vice versa.

While deep-water coral reefs have been documented for Angola's continental slope (Le Guilloux et al. 2009), shallow-water coral reefs are absent and in general the fauna and flora of littoral zone seems to be consistent with the pattern of relatively low diversity of the shore and near-shore areas of West Africa (John and Lawson 1991). Factors that could account for this include the lack of hard substrata (most of the shoreline being sandy), upwelling of cooler water in areas, high turbidity and sediment input from a massive river such as the Congo, or loss of species associated with reductions in sea-temperatures that considerably reduced the tropical zone during Pleistocene glaciations (van den Hoek 1975; John and Lawson 1991). However, while recent studies (e.g. Hutchings et al. 2007; Anderson et al. 2012) have added to the existing species lists (e.g. Lawson et al. 1975; Penrith 1978 and others) for coastal fishes, sandy beach macrofauna, rocky shore invertebrates and seaweeds, at this stage the paucity of information in Angola may not make comparison with other areas appropriate. In general the data that exist both for coastal and estuarine fishes (Whitfield 2005; Hutchings et al. 2007) but also offshore fish species (Kirkman et al. 2013; Yemane et al. 2015) of Angola show decreasing species richness from north to south, seemingly supporting the established trend of decreasing diversity with latitude as one moves polewards from tropical regions (e.g. Rex et al. 2000; Willig et al. 2003).

Based on the latitudinal distributions of intertidal fauna of rocky shores (Kensley and Penrith 1980), the southern limit of tropical biota has previously been reported to be around the border of Angola and Namibia. Lawson (1978) on the other hand, based on analyses of seaweed flora, considered Angola to be intermediate in nature between tropical and temperate. Results of surveys of intertidal invertebrates and seaweeds by Hutchings et al. (2007) however, showed that although there was a marked discontinuity between the biota of Angola and that of northern Namibia, which supports a cool-temperate intertidal flora up to nearby the Cunene River (Rull Lluch 2002), a number of taxa found in the south of Angola had temperate affinities. This led the authors to suggest that the inshore biota of the south of Angola may be intermediate in nature, and that of the north truly tropical. This is confirmed by Anderson et al. (2012) who conclude that the overall affinities of the Angolan seaweed flora is Tropical West African, but with a well-developed temperate element in southern Angola (from about 13°S) comprising mainly cooler-water species. Broadly, this supports the division of the Angolan coast into at least two sub-areas, with the more temperate south influenced by the cooler waters of the Angola-Benguela Front. This is similar to the division between demersal assemblages of north and south (Bianchi 1992) and also congruent with a break in the pelagic ecosystem of the inshore as determined from classification of key oceanographic variables and depth (Lagabrielle 2011). It also ties in with global mapping classifi-

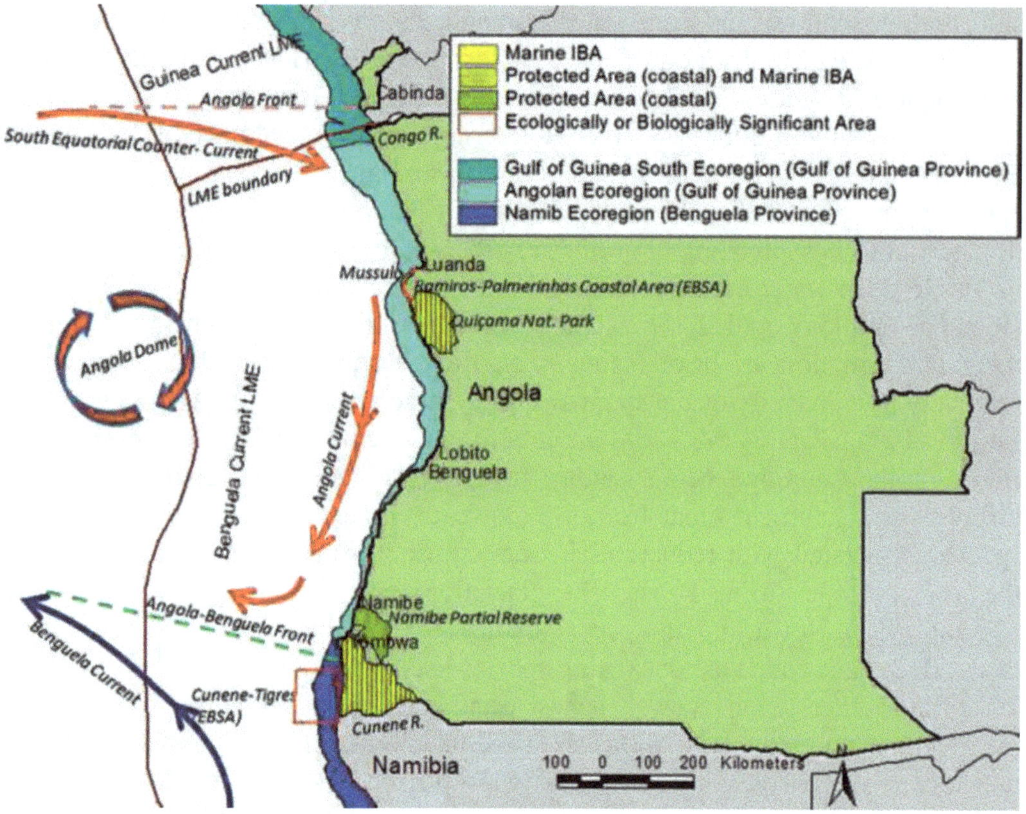

Fig. 3.1 Delineations of marine ecoregions (Spalding et al. 2007) and large marine ecosystems (Sherman 2014) that coincide with Angola. The ecoregions extend from the coast to the shelf edge. Also shown are recognised marine and coastal biodiversity areas and the approximate situations of important oceanographic processes

cation of coastal and shelf areas based on species distributions and levels of endemism of benthic and pelagic biota (Spalding et al. 2007; Briggs and Bowen 2012), that puts the divide between the temperate Benguela Province and the tropical Gulf of Guinea Province, near Moçâmedes (Fig. 3.1). Spalding et al. (2007) situate the majority of the Angolan EEZ in the Angolan ecoregion of the Gulf of Guinea Province, but include the area north of 6°30′S in the more tropical Gulf of Guinea south ecoregion. This is slightly incongruent with the mapping of large marine ecosystems (LMEs) of the world (which is expert- rather than data-derived), whereby most of Angola is included in the Benguela Current LME bounded to the north by the Angola Front (ca. 5°S), and only Cabinda in the far north included in the Guinea Current LME (Sherman 2014).

Marine and Coastal Biodiversity Hotspots, Threats and the Need for Protection

Whereas the Angolan coastal and shallow habitats are considered to be relatively low in biodiversity, coastal features such as the Congo River Delta, estuaries such as the Cuanza, Catumbela, Longa and Cunene, and associated floodplains, wetlands, lagoons, salt marshes and (north of Lobito) mangroves support a rich suite of species, often in high abundance (Hughes and Hughes 1992; van Niekerk et al. 2008; Harris et al. 2013). This includes several rare, endemic, migratory, and/or threatened fauna such as the African Manatee *Trichechus senegalensis*, turtle species and diverse waterbird species. Recognised ecosystem services of such features include (amongst others) providing habitat for important food-fish and crustacean species and their critical life stages (e.g. performing important nursery functions for many marine fishes) or providing plant species that are useful for medicinal, subsistence or construction purposes (Hughes and Hughes 1992; van Niekerk et al. 2008). While Angola is not currently a contracting party of the Ramsar Convention, some coastal wetland sites have been identified as being potential Ramsar sites, including Quiçama National Park between the Cuanza and Longa Rivers (Fig. 3.1), which is also a confirmed Important Bird and Biodiversity Area (IBA). Other confirmed coastal IBAs in Angola include Mussulo just south of Luanda and the Iona National Park in the south between the Cunene and Curoca Rivers. These IBAs are important for numerous waterbirds and are frequented by wintering seabird species that breed further south on the sub-continent such as Cape gannet *Morus capensis* (IUCN Red List – Endangered) and Damara tern *Sternula balaenarum* (Vulnerable) (Birdlife International 2002), the latter of which is known to also breed in the Iona National Park (Simmons 2010).

While the ecological value of these and other areas is recognised, the lack of formal protection of key biodiversity areas or features in Angola's marine and coastal environments has been noted as a concern (e.g. Tarr et al. 2007). As part of a regional systematic conservation planning (SCP) project involving all three member states of the Benguela Current Convention (BCC; a legally constituted collaborative mechanism representing Angola, Namibia and South Africa), Holness et al. (2014) showed that Angola is particularly poorly off in terms of spatial protection of its marine systems, with 102 out of 133 identified ecosystem types having no protection at all. Whereas some legislative protection of coastal ecosystem types in the Cuanza, Cunene and Tômbwa areas may be afforded by terrestrial national parks (Quiçama and Iona) or reserves (Namibe Partial Reserve), this may have value for conservation of the adjacent marine areas if effective management of these areas is achieved through the provision of increased human and financial resources. Holness et al. (2014) therefore opined that a programme of rapid expansion of protected areas for the Angolan marine systems is urgently required, and the ultimate product of their study was the prioritisation of sites for protection (ideally within a MPA network).

The current lack of marine protected areas (MPAs) was described by Tarr et al. (2007) as amongst the main challenges facing conservation and sustainable use of Angola's marine and coastal biodiversity and habitats, in light of multiple threats to the ecosystem that are likely to worsen over time. These threats include (but are not limited to) rapid, unplanned coastal urbanisation causing habitat destruction and a severe problem with waste management along the coast, particularly in the area of Luanda; escalation in over-exploitation of living marine resources related to rapid urbanisation and human migration to the coastal nodes, especially since the end of the civil war; industrial pollution caused e.g. by deposition of industrial wastes in catchment areas or cleaning of ships; offshore oil exploitation in the north, with potential for oil spills; loss of mangroves, which includes threats from pollution and wood collection for firewood and construction; rapid growth of the tourism industry; and impacts of climate change (Tarr et al. 2007; Heileman and O'Toole 2009).

With such threats in mind, Angola, like the other two member states of the BCC, has committed to implementing ecosystem-based management (EBM) of the marine environment to address responsible use of its ocean and its resources and put in practice the principles of sustainable development (BCC 2014). EBM is an integrative approach to management that takes into account all interactions in the ecosystem (including those involving human activities) and their cumulative impacts in space and time (Long et al. 2015). To be able to assist EBM with regard to the allocation and siting of ocean uses or protection measures, there is an initiative to implement marine spatial planning (MSP) in Angola and the other countries of the region (Kirkman et al. 2016). A pilot area for an experimental MSP project, covering an area of approximately 107,000 km^2 between south of Palmerinhas and the Tapado River mouth (GNC-OEM 2018), has recently been identified. A key element of the process is to identify and describe a network of Ecologically or Biologically Significant Marine Areas (EBSAs) - geographically or oceanographically discrete areas that have been identified as important for the services that they provide and for the healthy functioning of oceans (Dunstan et al. 2016) and to include these in marine spatial plans.

Currently, only two Angolan EBSAs have been described and subsequently endorsed by CBD (CBD 2014), namely the Ramiros-Palmerinhas Coastal Area which partly adjoins the Mussulo Peninsula south of Luanda, and the Cunene-Tigres EBSA which overlaps with northern Namibia and is adjacent to the Iona National Park on the Angola side (Fig. 3.1). The former includes estuaries with mangroves and salt marshes and has special importance for bird aggregations and breeding turtles. The latter includes the Cunene estuary and associated wetland as well as the Baía dos Tigres Island-Bay complex to the north of it, and has special importance for migratory birds and in terms of its nursery function for many marine species. Both of these areas have undergone thorough assessment processes with a view to expanding their areas in order to include other relevant features such as estuaries, sensitive coastline, canyons and seamounts.

Currently Angola is in the process of describing new potential EBSAs, in coastal and offshore areas, as part of a collaborative regional project with Namibia and South Africa, coordinated by the BCC (http://www.benguelacc.org). Currently, five new areas have been proposed as EBSAs which include coastal and offshore areas in the provinces of Cabinda, Zaire, Luanda, Cuanza-Sul and Namibe. Although EBSA status itself does not carry any conservation or protection interventions, legal protection is among the management measures that can be applied to secure the persistence of these special features and their ecosystem services. Therefore the process of expanding the EBSA network could provide a foundation for initiating a network of MPAs in Angola. In this regard, there is a recent project proposal for the establishment of the first MPA in Angola in the offshore area adjacent to the Iona National Park.

References

Anderson RJ, Bolton JJ, Smit AJ et al (2012) The seaweeds of Angola: the transition between tropical and temperate marine floras on the west coast of southern Africa. Afr J Mar Sci 34:1–13

BCC (Benguela Current Commission) (2014) Strategic action programme 2015–19. Swakopmund, Namibia, 36 pp. http://benguelacc.org/index.php/en/publications

Bianchi G (1992) Demersal assemblages of the continental shelf and upper slope of Angola. Mar Ecol Prog Ser 81:101–120

BirdLife International (2002) Important bird areas and potential Ramsar sites in Africa. BirdLife International, Cambridge, MA, 136 pp + appendices

Briggs JC, Bowen BW (2012) A realignment of marine biogeographic provinces with particular reference to fish distributions. J Biogeogr 39:12–30

CBD (Convention on Biological Diversity) (2014) Decision adopted by the Conference of the Parties to the Convention on Biological Diversity. XII/22. Marine and Coastal Biodiversity: Ecologically or Biologically Significant Marine Areas (EBSAs). Twelfth meeting of the Conference for the Parties, 6–17 October 2014, Pyeongchang, Republic of Korea. UNEP/CBD/COP/DEC/XII/22

Duarte A, Fielding P, Sowman M, et al (2005) Overview and analysis of socio-economic and fisheries information to promote the management of artisanal fisheries in the Benguela Current Large Marine Ecosystem (BCLME) region (Angola). Unpublished Final Report. Rep. No. LMRAFSE0301B. Cape Town Environmental Evaluation Unit, University of Cape Town, Cape Town

Dunstan PK, Bax NJ, Dambacher JM et al (2016) Using ecologically or biologically significant marine areas (EBSAs) to implement marine spatial planning. Ocean Coast Manag 121:116–127

Florenchie P, Reason CJC, Lutjeharms JRE et al (2004) Evolution of interannual warm and cold events in the Southeast Atlantic Ocean. J Clim 17:2318–2334

Grupo Nacional de Coordenação para o Ordenamento do Espaço Marinho (GNC-OEM) (2018). Relatório Preliminar sobre o Ordenamento do Espaço Marinho em Angola: Área Experimental Palmeirinhas – Tapado. Unpublished report

Harris L, Holness S, Nel R et al (2013) Intertidal habitat composition and regional-scale shoreline morphology along the Benguela coast. J Coast Conserv 17:143–154

Heileman S, O'Toole MJ (2009) I West and Central Africa: I-1 Benguela current LME. In: Sherman K, Hempel G (eds) The UNEP large marine ecosystems report: a perspective on changing conditions in LMEs of the World's regional seas. UNEP Regional Seas Report and Studies No. 182. United Nations Environment Programme, Nairobi, pp 103–115

Holness S, Kirkman S, Samaai T, et al (2014) Spatial biodiversity assessment and spatial management, including marine protected areas. Final report for the Benguela Current Commission project BEH 09-01, 105 pp + annexes

Hughes RH, Hughes JS (1992) A directory of African wetlands. IUCN/UNEP/WCMC, Gland/Cambridge/Nairobi/Cambridge, xxiv + 820 pp, 48 maps

Hutchings K, Clark B, Steffani, Anderson R (2007) Identification of communities, biotopes and species in the offshore areas and along the shoreline and in the shallow subtidal areas in the BCLME region. Section B. Angola field trip report. Final report for Benguela Current Large Marine Ecosystem Programme project BEHP/BAC/03/03

Hutchings L, van der Lingen CD, Shannon LJ et al (2009) The Benguela current: an ecosystem of four components. Prog Oceanogr 83:15–32

Japp DW, Purves MG, Wilkinson S (eds) (2011) State of stocks review. Report No. 2 (Updated by C Kirchner). Benguela Current Large Marine Ecosystem State of Stocks Report 2011, 105 pp

John DM, Lawson GW (1991) Littoral ecosystems of tropical western Africa. In: Mathieson AC, Nienhuis PH (eds) Ecosystems of the world, vol 24. London. New York, Tokyo, pp 297–321

Kensley B, Penrith ML (1980) The constitution of the fauna of rocky intertidal shores of south West Africa. Part III. The north coast from False Cape Frio to the Kunene River. Cimbebasia (Series A) 5:201–214

Kirkman SP, Blamey L, Lamont T et al (2016) Spatial characterisation of the Benguela ecosystem for ecosystem based management. Afr J Mar Sci 38:7–22

Kirkman SP, Yemane D, Kathena J et al (2013) Identifying and characterizing of demersal biodiversity hotspots in the BCLME: relevance in the light of global changes. ICES J Mar Sci 70:943–954

Kopte R, Brandt P, Dengler M et al (2017) The Angola current: flow and hydrographic characteristics as observed at 11°S. J Geophys Res Oceans 122:1177–1189

Lagabrielle E (2011) A pelagic bioregionalisation of the Benguela current system. Appendix 4 In: Holness S, Kirkman S, Samaai T, et al (2014) Spatial Biodiversity Assessment and Spatial Management, including Marine Protected Areas. Final report for the Benguela Current Commission project BEH 09-01, 105 pp + annexes

Lass HU, Schmidt M, Mohrholz V et al (2000) Hydrographic and current measurements in the area of the Angola-Benguela front. J Phys Oceanogr 30:2589–2609

Lawson GW (1978) The distribution of seaweed floras in the tropical and subtropical Atlantic Ocean: a quantitative approach. Bot J Linn Soc 76(3):177–193

Lawson GW, John DM, Price JH (1975) The marine algal flora of Angola: its distribution and affinities. Bot J Linn Soc 70(4):307–324

Le Guilloux E, Olu K, Bourillet JF et al (2009) First observations of deep-sea coral reefs along the Angola margin. Deep Sea Research Part II: Tropical Studies in Oceanography 56:2394–2403

Long RD, Charles A, Stephenson RL (2015) Key principles of marine ecosystem-based management. Mar Policy 57:53–60

Meeuwis JM, Lutjeharms JRE (1990) Surface thermal characteristics of the Angola-Benguela front. S Afr J Mar Sci 9:261–279

Monteiro PMS, van der Plas AK, Mélice J-L et al (2008) Interannual hypoxia variability in a coastal upwelling system: ocean-shelf exchange, climate and ecosystem-state implications. Deep-Sea Res I 55:435–450

Nsiangango S, Shine K, Clark B (2007) Identification of communities, biotopes and species in the offshore areas and along the shoreline and in the shallow subtidal areas in the BCLME region. Section C3. Biogeographic patterns and assemblages of demersal fishes on the coast of Angola. Final report for Benguela Current Large Marine Ecosystem Programme project BEHP/BAC/03/03

O'Toole MJ (1980) Seasonal distribution of temperature and salinity in the surface waters off south West Africa, 1972–1974. Investig Rep S Afr Sea Fish Inst 121:1–25

Penrith MJ (1978) An annotated check-list of the inshore fishes of southern Angola. Cimbebasia (Series A) 4:179–190

Potts WM, Childs AR, Sauer WHH et al (2009) Characteristics and economic contribution of a developing recreational fishery in southern Angola. Fish Manag Ecol 16:14–20

Rex MA, Stuart CT, Coyne G (2000) Latitudinal gradients of species richness in the deep-sea benthos of the North Atlantic. Proc Natl Acad Sci USA 97:4082–4085

Rull Lluch JR (2002) Marine benthic algae of Namibia. Sci Mar 66. (suppl. 3:5–256

Sherman K (2014) Toward ecosystem-based management (EBM) of the world's large marine ecosystems during climate change. Environ Dev 11:43–66

Shillington FA, Reason CJC, Duncombe Rae CM et al (2006) Large scale physical variability of the Benguela Current Large Marine Ecosystem (BCLME). In: Shannon V, Hempel G, Malanotte-Rizzoli P et al (eds) Benguela: predicting a large marine ecosystem, vol 14. Elsevier, Amsterdam, pp 49–70

Signorini SR, Murtuguddo RG, McClain CR et al (1999) Biological and physical signatures in the tropical and subtropical Atlantic. J Geophys Res 104:18367–18382

Simmons RE (2010) First breeding records for Damara Terns and density of other shorebirds along Angola's Namib Desert coast. Ostrich 81:19–23

Spalding MD, Fox HE, Allen GR et al (2007) Marine ecoregions of the world: a bioregionalization of coastal and shelf areas. Bioscience 57:573–583

Tarr P, Krugmann H, Russo V, Tarr J, et al (2007) Analysis of threats and challenges to marine biodiversity and marine habitats in Namibia and Angola. Final report for Benguela Current Large Marine Ecosystem Programme project BEHP/BTA/04/01. 132 pp + annexes

Van den Hoek C (1975) Phytogeographic provinces along the coast of the northern Atlantic Ocean. Phycologia 14:317–330

Van Niekerk L, Neto DS, Boyd AJ, et al (2008) Baseline surveying of species and biodiversity in estuarine habitats. BCLME project BEHP/BAC/03/04. 118 pp + appendices

Veitch JA, Penven P, Shillington F (2010) Modeling equilibrium dynamics of the Benguela Current System. J Phys Oceanogr 40:1942–1964

Whitfield AK (2005) Preliminary documentation and assessment of fish diversity in sub-Saharan African estuaries. Afr J Mar Sci 27(1):307–324

Willig MR, Kaufman DM, Stevens RD (2003) Latitudinal gradients of biodiversity: pattern, process, scale, and synthesis. Annu Rev Ecol Evol Syst 34:273–309

Yemane D, Mafwila SK, Kathena J et al (2015) Spatio-temporal trends in diversity of demersal fish species in the Benguela Current Large Marine Ecosystem (BCLME) region. Fish Oceanogr 24. (Suppl. 1:102–121

Chapter 4
The Fossil Record of Biodiversity in Angola Through Time: A Paleontological Perspective

Octávio Mateus, Pedro M. Callapez, Michael J. Polcyn, Anne S. Schulp, António Olímpio Gonçalves, and Louis L. Jacobs

Abstract This chapter provides an overview of the alpha paleobiodiversity of Angola based on the available fossil record that is limited to the sedimentary rocks, ranging in age from Precambrian to the present. The geological period with the highest paleobiodiversity in the Angolan fossil record is the Cretaceous, with more than 80% of the total known fossil taxa, especially marine molluscs, including ammonites as a majority among them. The vertebrates represent about 15% of the known fauna and about one tenth of them are species firstly described based on specimens from Angola.

O. Mateus (✉)
GeoBioTec, Faculdade de Ciências e Tecnologia, Universidade Nova de Lisboa, Lisbon, Portugal

Museu da Lourinhã, Rua João Luis de Moura, Lourinhã, Portugal
e-mail: omateus@fct.unl.pt

P. M. Callapez
CITEUC; Departamento de Ciências da Terra, Faculdade de Ciências e Tecnologia, Universidade de Coimbra, Coimbra, Portugal
e-mail: callapez@dct.uc.pt; jacobs@smu.edu

M. J. Polcyn · L. L. Jacobs
Roy M. Huffington Department of Earth Sciences, Southern Methodist University, Dallas, TX, USA
e-mail: mpolcyn@smu.edu

A. S. Schulp
Naturalis Biodiversity Center, Leiden, The Netherlands

Faculty of Earth and Life Sciences, VU University Amsterdam, Amsterdam, The Netherlands
e-mail: anne.schulp@naturalis.nl

A. O. Gonçalves
Departamento de Geologia, Faculdade de Ciências, Universidade Agostinho Neto, Luanda, Angola
e-mail: antonio.goncalves@geologia-uan.com

B. J. Huntley et al. (eds.), *Biodiversity of Angola*,
https://doi.org/10.1007/978-3-030-03083-4_4

Keywords Ammonites · Benguela Basin · Cenozoic · Cretaceous · Cuanza Basin · Dinosaur · Invertebrates · Mammals · Mollusca · Mosasaur · Namibe Basin · Paleobiodiversity · Pleistocene · Vertebrate · Plesiosaur · Turtle

Studies of Paleobiodiversity

The study of paleobiodiversity, i.e., the development of biodiversity through geological time, is challenging at multiple levels. In addition to the issues and biases affecting the study of the diversity of modern life, understanding paleobiodiversity faces extra challenges, mostly because of the dependency on the fossil record. Glimpses of entire ecosystems and clades may never reach the paleontologist's eyes if appropriate rocks of that exact time and space were not formed, or if formed, did not preserve fossils, or are eroded away or otherwise inaccessible (see Jackson and Johnson 2001; Crampton et al. 2003).

The study of life's diversity in the past is filtered by the remains that can leave traces and fossilize, remains that actually fossilized, fossils existing today, fossils accessible today, fossils collected (number of fossils accessible to scientists), and species recognised (Fig. 4.1). Moreover, the definition and discrimination of species in the fossil record can be problematic.

Angola has no known fossiliferous rocks from the Paleozoic (541–251 Ma – millions of years ago) nor the Jurassic (199–145 Ma) leaving only windows to life in the territory during the Triassic (251–199 Ma), Cretaceous (145–66 Ma), and Cenozoic (66 Ma -Present) (Fig. 4.2). The chances of finding Paleozoic or Jurassic fossils from Angola are essentially zero. Thus, within the last 550 Ma, the known

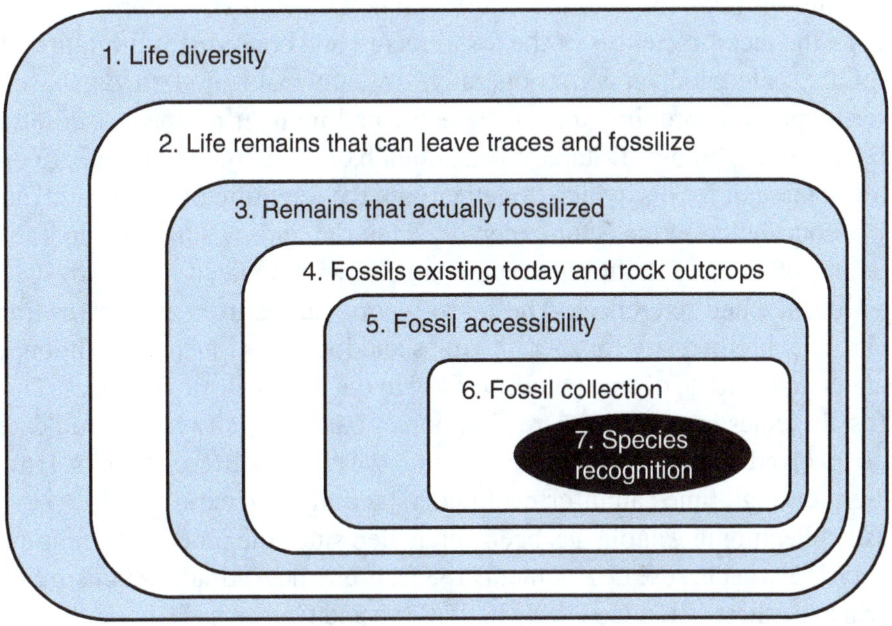

Fig. 4.1 The seven layers of filters of uneven preservation in the fossil record that obscure accurate reconstruction of paleobiodiversity

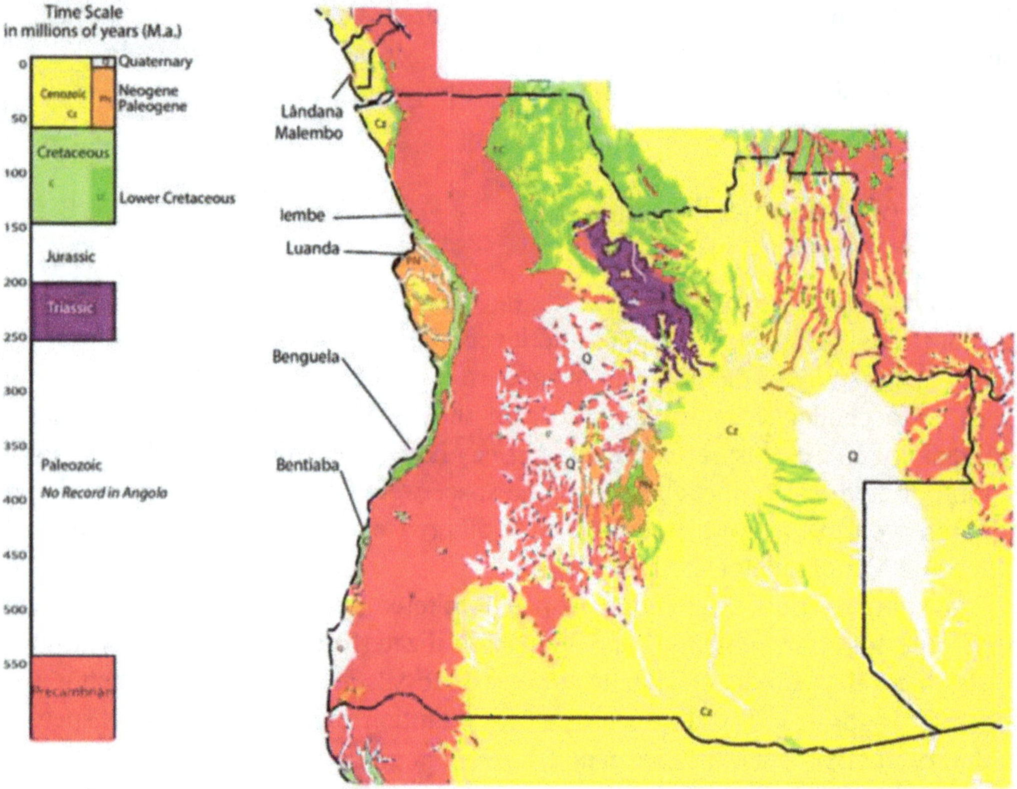

Fig. 4.2 The geological record of Angola, in a stratigraphic log (left) and geological map (right), leaves more than 350 Ma of blank geological record. Map extracted after Africa Geological Map 1:30.000.000 by U.S. Geological Survey, 2002 available at www.uni-koeln.de/sfb389/e/e1

rocks of Angola represent less than 196 million years of geologic time, leaving more than 354 million years (64% of the time) with no known fossil record.

Despite the incompleteness of the fossil record and consequent limitations to the study of the paleobiodiversity, cooperative research and modern databases can, however, improve approximation of the estimated number of species in the fossil record. The PaleoBiology Database (paleobiodb.org) is, by far, the most comprehensive database of fossils, which, together with the scientific literature and our own research, contributed to the Supplementary Material and its summary in Table 4.1 that compiles the list of the fossil taxa in Angola, with updated taxonomy, geological age, locality, and references. The fossil record can be used as a lower limit for the alpha paleobiodiversity for specific times and locations in Angola, although it is likely an underestimate of true paleodiversity in the vast majority of cases. The total of all fossil species is a gross underestimation of paleodiversity for the full extent of the time involved, exacerbated by missing intervals of fossiliferous rocks. However, the pattern through time can inform an understanding of general trends.

Fossil collecting in Angola has been conducted since the nineteenth century with Friederich Welwitsch, José de Anchieta (1885), Freire de Andrade, Augusto Eduardo Neuparth and others (Brandão 2008, 2010; Silva and Geirinhas 2010; Callapez et al. 2011; Masse and Laurent 2016). Numerous paleontologists have contributed to the

Table 4.1 Summary of the known fossil record of Angola

		Period							
COUNT (Genus/sp, Family Indet)		Triassic	Cretaceous	Paleogene	Neogene	Quaternary	Triassic	Grand Total	
Bacteria	Cyanobacteria		1					1	0%
Bacterial total			1					1	0 %
Plantae	Chlorophyta		1					1	0%
	Cycadophyta		1					1	0%
	Ginkgophyta		1					1	0%
	Pinophyta	1	1				1	2	0%
	Pteridosperma-tophyta	4					4	4	0%
	Rhodophyceae		2					2	0%
Plantae total		5	6				5	9	1%
Protista and Invertebrates	Foraminifera		207					207	16%
	Anthozoa		1	1	2	2		6	0%
	Brachiopoda		1					1	0%
	Mollusca		677	27	22	72		798	61%
	Echinodermata		61	1	4	1		67	5%
	Arthropoda	13	3	1	1	2	13	20	2%
Protista and invertebrates total		13	950	30	29	77	13	1102	84%
Vertebrata	Chondrichthyes	2	63	1	18		2	84	6%
	Actinopterygii	6	10		5		6	21	2%
	Sarcopterygii	3					3	3	0%
	Reptilia		21	5				26	2%
	Mammalia			12	1	54		67	5%
Vertebrata total		11	94	18	24	54	11	201	15%
Grand Total		**29**	**1052**	**48**	**53**	**131**	**29**	**1313**	**100%**
		2%	80%	4%	4%	10%	2%	100%	

understanding the paleobiodiversity of Angola, since the first explorations and studies: Fernando Mouta, Paul Choffat, Carlos Teixeira, Carlos Freire de Andrade, Edgard Casier, A Jamotte, M Leriche, Heitor de Carvalho, E Dartevelle, Miguel Telles Antunes, Alexandre Borges, Philippe Brebion, Gaspar Soares de Carvalho, Louis Dollo, Henri Douvillé, Otto Haas, Manuel Mascarenhas Neto, Arménio Tavares da Rocha, Gumerzindo Henriques da Silva, LF Spath, António Ferreira Soares, Maurice Collignon, among many others (see bibliography compiled by Nunes 1991). In vertebrate paleontology the work of Miguel Telles Antunes and co-authors is noteworthy. Today, various researchers work on the paleontology of Angola, among them is the team of the Projecto PaleoAngola (paleolabs.org/paleoangola), with regular yearly scientific expeditions since 2005 (Jacobs et al. 2006, 2016).

A Brief Geological History and Context of Angola

The most significant geological event governing the paleogeography of Angola is the opening of the South Atlantic Ocean, in which Africa and South America rifted apart beginning in the Early Cretaceous Epoch about 134 million years ago and the subsequent drifting apart of these continents as the South Atlantic grew (Guiraud and Maurin 1991; Buta-Neto et al. 2006; Quirk et al. 2013; Pérez-Díaz and Eagles 2017). After about 120 million years ago, marine deposition along the coast began to preserve fossils. Africa's place in Gondwana prior to this time resulted in the lack of a marine record for the entire Paleozoic Era and the consequent lack of a fossil record for that time.

As the South Atlantic opened it was colonised by species moving in from the southern ocean and, as a connection between the North Atlantic and South Atlantic oceans developed, from the north. Sea turtles (*Angolachelys mbaxi*) and mosasaurs (*Angolasaurus bocagei* and *Tylosaurus iembeensis*), with relatives to the north, first occur at about 88 million years ago. Along with them were plesiosaurs, probably with southern affinities. Washed into the sea and found with marine creatures, is the sauropod dinosaur *Angolatitan adamastor*, which is probably a remnant of a more broadly distributed Gondwanan dinosaur assemblage. At that time, coastal Angola lay approximately 10–12 degrees further south than today.

Geologic uplift along the coast resulted in the erosion of rock and loss of the fossil record from effected strata. Permian and Jurassic uplift eliminated those intervals from the terrestrial fossil record. Early and Late Cretaceous uplift also occurred, prior to burial of the remaining Cretaceous with up to 1.5–2 km of sediments, since removed by erosion accompanying mid-Cenozoic uplift beginning around 30 Ma and 20 Ma (Green and Machado 2015). Uplift from as young as 45,000 years ago also has been recorded along the Angolan coast (Walker et al. 2016), resulting in a large diversity of fossiliferous raised-beach deposits, frequently associated with pre-historic shell-middens with Paleolithic industries.

The Precambrian – The First Fossils

Precambrian Era (geologic time since the formation of Earth and prior to 541 Ma) rocks worldwide are mostly devoid of fossils as life was unicellular for most time and only macroscopic in the last stages. The shale-limestone series of the Bembe System in Angola includes dolomitic limestones, mostly devoid of fossils, but containing levels with concentrations of coalescent stromatolites (structures due to cyanophilic activity) attributable to the genera *Collenia* and *Conophyton* in Mavoio, Alto Zambeze, and Humpata (Vasconcelos 1951; Antunes 1970; Duarte et al. 2014).

In Angola, no fossils are known from the Paleozoic (Cambrian through Permian periods), which represents a time gap of more than 290 million years (Fig. 4.2).

The Triassic – Inner Basins

The Triassic is a geological period stretching from about 250 to 200 Ma. The beginning and end of this interval are both marked by mass extinction events, the older Permian-Triassic extinction event, concomitant with the Siberian Magmatic event, marking the initiation of the Triassic Period. The vast supercontinent Pangea existed until the Triassic, after which it gradually began to break-up apart, separating the two masses of land, Laurasia to the north and Gondwana to the south. The global climate during the Triassic Period was warm and dry, with deserts covering much of the interior of Pangea. The end of the Triassic was marked by another major mass extinction related with the Central Atlantic Magmatic Province and the early opening of the North Atlantic. Therapsids (a large group containing mammals and their extinct relatives) and archosaurs (dinosaurs, birds, crocodiles and their relatives) were the major terrestrial vertebrates during this time. The dinosaurs first appeared in the Triassic. The first true mammals, which are derived therapsids, evolved during this period, as well as the first flying vertebrates, the pterosaurs.

Triassic outcrops of Angola are restricted to the Baixa de Cassange (Cassange Depression) in Malanje and Lunda-Norte in rocks referred to the Karoo Supergroup geological unit.

Plants and Invertebrates

The Triassic paleoflora of Angola includes the extinct genera *Glossopteris*, *Sphenopteris*, and *Noeggerathiopsis* (Teixeira 1948a, 1961). A few fossils indicate a freshwater environment and a Triassic age. These include the coleopteran Coptoclavidae insect *Coptoclavia africana* and 12 conchostracan crustaceans: *Estheriella moutai* Leriche 1932, *E. cassambensis* Teixeira 1958, *Estheria anchietai* Teixeira 1947, *E. (Echinestheria) marimbensis* Marliére 1950, *E. (Euestheria) mangaliensis* Jones 1862, *Palaeolimnadiopsis reali* Teixeira 1958, *Palaeolimnadia*

(Palaeolimnadia) wianamattensis (Mitchell), *P. (Grandilimnadia) oesterleni* Tasch 1987, *P. (G.) africania* Tasch 1987, *Gabonestheria gabonensis* (Marliére 1950), *Cornia angolata* Tasch 1987, and *Estheriina (Nudusia)* cf. *rewanensis* Tasch 1979.

Vertebrates

The only Triassic vertebrates known from Angola are fishes, including the Elasmobranchii *Lissodus cassangensis,* the paleoniscoids *Perleidus lutoensis* Teixeira 1947, Palaeonisciformes canobiid *Marquesia moutai*, Halecostomi *Angolaichthys lerichei* Teixeira, the ray-finned *Teffichthys lehmani* and *T. lutoensis,* and the Sarcopterygii lungfish *Microceratodus angolensis*.

This faunal assemblage indicates a freshwater environment with insects and a nearly exclusively endemic fauna. Based on the fishes, a Lower Triassic age (252–247 Ma) is indicated for Lunda and Baixa de Cassange rocks (Murray 2000; Antunes et al. 1990). No tetrapods have been collected so far.

The Early Cretaceous – The Opening of South Atlantic

Most of the fossils and outcrops of Cretaceous age in Angola are in Mesozoic-Cenozoic basins of coastal Angola: the Cabinda, Zaire, Cuanza, Benguela, and Namibe sedimentary basins (Antunes 1964; Séranne and Anka 2005; Guiraud et al. 2010), bordered by basement rocks. Almost all formations are mostly marine except for the ichnofauna from the Catoca mine (Marzola et al. 2014; Mateus et al. 2017), in Lunda-Sul. The oldest fossiliferous Cretaceous formations seen in outcrop seem to be Barremian to Aptian lacustrine deposits that contain unidentified gastropods (Ceraldi and Green 2016).

Plants, Protists and Invertebrates

Cretaceous plant remains in Angola seem to be rare (see Supplementary Material), but several examples of unstudied field contexts with transitional facies, namely above the Cuvo units, might come to reveal new fossil sites. The same situation is likely for palynomorphs and dinoflagellates of Early Cretaceous and more recent ages. Plants are known from pollen taxa such as *Classopolis* sp. and *Eucommiidites* sp. *Pachypteris montenegroi* Teixeira 1948 is a Ginkgophyta Umkomasiaceae from Early Albian lagoonal sediments of the Cuanza Basin (Teixeira 1948b; Antunes 1964; Neto 1970; Nunes 1991) and first found in Angola. One species of Chlorophyta, one Cycadophyta, one Rhodophyceae, one Ginkgophyta, and two Pinophyta have been reported from Cabinda, Benguela and Cuanza Basin (Antunes 1964; Neto 1970; Nunes 1991; Araújo and Guimarães 1992; Tavares 2006).

More than 200 foraminifera taxa were identified in the Cretaceous of Angola, and many indicate an Albian age, such as *Globotruncana ventricosa* (Rocha 1984; Antunes and Cappetta 2002; Antunes 1964; Jacobs et al. 2006).

Crustaceans are known from the Decapoda *Parapirimela angolensis* Van Straelen 1937 from the Albian of Iela beach, Benguela Basin (Ferreira 1957; Van Straelen 1937; Antunes 1964) and the ostracods *Chloridella angolia* and *Petrobrasia tenuistriata longinsuela* from Quiçama, Cuanza Basin, (Berry 1939; Antunes 1964) and Cabinda (Araújo and Guimarães 1992), respectively.

The Early Cretaceous of Angola yielded fossils of seven species of Scleractinean corals, one of brachiopod and eight crustaceans.

Of the more than known 600 species of molluscs from the Cretaceous of Angola, the vast majority are ammonites, many are unique to Angola and received related specific epithets such as the *Anisoceras teixeirai, Durnovarites autunesi* Collignon 1978, *Durnovarites netoi, Elobiceras lobitoense* Spath 1922, *Hamitoides angolanus* (Tavares 2006), followed by bivalves such as *Neithea angoliensis* Newton 1917, and gastropods such as "*Cerithium*" *monteroi* Choffat. By far, the ammonites are the most relevant portion of the Cretaceous biodiversity of Angola (Tavares et al. 2007; Haas 1942, 1943) and also age-indicators for geologists and paleontologists. Haas (1942) described and reported many ammonites from the Albian, some as new species including *Hysteroceras falcicostatum* Haas 1942 and *H. intermedium* Haas 1942. See the Supplementary Material for the complete list.

Echinoderms, mostly echinoids, are remarkably common in the Early Cretaceous of Angola, with about 50 taxa known, but that number depends on the validity and synonymisation of the taxa addressed (see the Supplementary Material for the complete list). A few echinoderms received names after Angolan toponyms and researchers such as the *Douvillaster benguellensis* Loriol 1888 and *D. carvalhoi* Loriol 1888 and *Epiaster catumbelensis* Loriol 1888, and *Holaster domboensis* Loriol 1888 from the Lower Cretaceous of Dombe Grande, Catumbela and Praia da Hanha (Loriol 1888; Ferré and Granier 2001; Tavares 2006; Tavares et al. 2007).

Vertebrates

In the Catoca Diamond Mine, in Lunda-Sul Province, mammaliamorph, crocodylomorph, and sauropod tracks were discovered in Early Cretaceous crater lake sediments. One sauropod track has skin impressions preserved. These are the only fossil vertebrate tracks known in Angola. The most surprising feature is the unexpectedly large size of the mammaliamorph footprints considering the age (Mateus et al. 2017). The Catoca Diamond Mine has an eruption age of around 118 Ma (Aptian; Robles-Cruz et al. 2012). A fragment of a caudal vertebra of a sauropod dinosaur was recovered from Tzimbio, in northern Namibe, from strata that are likely Albian in age.

The Late Cretaceous – Marine Reptiles Flourish

The Late Cretaceous is the time of the Cretaceous Period between 100.5 Ma and 66 million years ago. It is subdivided into the Cenomanian, Turonian, Coniacian, Santonian, Campanian and Maastrichtian ages, from the oldest to the most recent. The climate was warmer than present, although with a cooling trend throughout the period. In the oceans, where the sea-level was much higher than today, mosasaurs (a group of marine lizards) suddenly appeared and underwent a spectacular evolutionary radiation (Polcyn et al. 2014). Modern grade sharks also appeared and plesiosaurs diversified. These predators fed on the numerous teleost fishes, which in turn evolved into new advanced and modern forms (Neoteleostei). Ichthyosaurs and pliosaurs (a group of short-necked plesiosaurs), on the other hand, became extinct during the Cenomanian-Turonian anoxic event (Schlanger et al. 1987) and are not known from Angola. The end of Cretaceous is marked by the mass extinction of some three-quarters of plant and animal species on Earth, known as the Cretaceous-Paleogene (K/Pg) event (Archibald et al. 2010).

Protists and Invertebrates

The Late Cretaceous Foraminifera of Angola were studied by various researchers including Ferreira and Rocha (1957), Lapão and Simões (1972), Rocha (1984) and Blake et al. (1996) who list more than 180 taxa. Foraminifera are known from 15 taxa of Granuloreticulosea, such as the Gavelinellidae *Anomalia berthelini* from the Aptian to Cenomanian of Cabinda, Cuanza and Benguela Basins (Araújo and Guimarães 1992; Rocha 1984).

Molluscs of the Late Cretaceous count more than 240 known taxa, mostly ammonites (see Supplementary Material for the full list). Some were named after Angolan toponyms or after paleontologists that worked in Angola (Borges 1946; Carvalho 1961; Haas 1943; Howarth 1965; Cooper 1972, 1982; Cooper and Kennedy 1979): *Acera choffati* Rennie 1945, *Axonoceras angolanum* Haas 1943, *Didymoceras* cf. *angolaense* Haughton 1924 (Howarth 1965), *Eutrephoceras egitoense* Miller and Carpenter 1956, *Kitchinites angolaensis* Howarth 1965, *Libycoceras dandense* Howarth 1965, *Lucina egitoensis* Rennie 1945, *L. angolensis* Rennie 1929, *Mammites mocamedensis* Howarth 1966, *Nostoceras mariatheresianum* Haas 1943, *Oiophyllites angolaensis* Spath 1953, *Prionocyclus carvalhoi* Howarth 1966, *Protacanthoceras angolaense* Spath 1931, *Prohysteroceras hanhaense* (Haas 1942), *P. angolaense* (Boule et al. 1907), *Protocardia moutai* Rennie 1945, *Pseudocalycoceras angolaense* (Spath 1931), *Pseudomelania salenasensis* Rennie, *Pterotrigonia borgesi* Rennie 1945, *Mortoniceras (Angolaites) stolikzcai* (Spath 1922), *Mortoniceras (?) rochai* Collignon 1978, *M. (Deiradoceras) reali* Collignon 1978, *Collignoniceras (Selwynoceras) reali* Collignon 1978, and *Solenoceras bembense* Haas 1943. The ammonites are found in almost all Upper Cretaceous sections, including those of the Quissonde Formation from the beaches of Quimbala, Chamure, Cabeça da Baleia, Egito in Benguela Basin, Teba, Bembe in Cuanza Basin, Bentiaba and Salinas in Namibe, and Iembe in Bengo Province (Segundo et al. 2014).

Rennie (1945) describes ten species of gastropods and bivalves from Cabeça da Baleia, Egito-Praia: *Trigonia (Scabrotrigonia) borgesi, Lucina egitoensis, Protocardia moutai, Pseudomelania egitoensis, Confusiscala angolensis, Acirsa (Plesioacirsa?) egitoensis, Dicroloma (Perissoptera) o'donnelli, Paleopsephaea o'donnelli, Acera choffati,* and *Ringicula moutai* (Lapão and Pereira 1971).

More than ten echinoderm taxa are known from the Late Cretaceous of Angola, mostly echinoids, some received Angola-related names, such as the Toxasteridae *Epiaster angolensis* Haughton 1924, collected 150 m below the Itombe Formation at Zenza do Itombe, and *E. carvalhoi* Dartevelle 1953 (Haughton 1924; Kier and Lawson 1978; Néraudeau and Mathey 2000), *Leiostomaster angolanus* Greyling and Cooper 1995, Palaeostomatidae and *Tholaster carvalhoi* Greyling and Cooper 1995, and Holasteridae from the Middle Campanian of Egito Praia, near Quimbala.

Vertebrates

In Angola, a peak of vertebrate paleobiodiversity is observed in the Late Cretaceous, with more than 100 taxa recognised (Antunes 1964; Antunes and Cappetta 2002; Jacobs et al. 2006, 2009, 2016). This peak is in comparison with other time intervals and is likely an artifact of the inadequacies of the fossil record rather than a statement of biological reality. Most known Late Cretaceous sedimentary rocks from Angola are marine, thus the fossil assemblage also reflects this ecosystem. Non-marine organisms are occasionally drifted into a marine setting, including the sauropod dinosaur *Angolatitan adamastor* Mateus et al. (2011) from Iembe, in Bengo Province, originally considered Turonian in age but following the discovery of the ammonite *Protexanites* sp. at Iembe, it is now considered Coniacian. Pterosaurs are known from the Maastrichtian marine sediments of Bentiaba, in Namibe Province. *Angolatitan adamastor* is the first dinosaur found in Angola. Its generic name means 'Titan of Angola' and the specific name refers to Adamastor made famous by Luís de Camões in the *Lusíadas*. It was 13 m long and lived in an arid environment (Mateus et al. 2011). A number of bones of large flying reptiles, pterosaurs, collected in the marine sediments of the Maastrichtian of Bentiaba are the only know pterosaur remains in sub-saharian Africa for this stage.

The fossil record of chondrichthyans shows a peak in Late Cretaceous deposits with 64 species known, including taxa only known in Angola or with Angolan toponyms such as *Angolabatis angolensis, A. benguelaensis, Chlamydoselachus gracilis, Cretascymnus quimbalaensis*, and *Echinorhinus lapaoi* named by Antunes and Cappetta (2002). The diversity of chondrichthyans includes representatives of most of the Cretaceous lineages: Hexanchiformes, Squaliformes, Rajiformes, Orectolobiformes, Odontaspidida, Lamniformes, Carcharhiniformes, and Squaliformes (Antunes and Cappetta 2002).

In contrast to the diversity of chondrichthyans, the Late Cretaceous bony ray-finned fish from Angola are represented by ten taxa only, most likely due to the difficulty of identification of isolated remains, in comparison with the abundance of shark and ray teeth. The teleost genus *Enchodus* is known from various species (*Enchodus bursauxi, E. crenulatus, E. elegans, E. faujasi* and *E. libycus*). Other teleosts include *Eodiaphyodus lerichei, Pseudoegertonia bebianoi*, and *Stephanodus libycus* from the Late Campanian and Maastrichtian in Benguela and Namibe provinces (Antunes and Cappetta 2002) (Fig. 4.3).

Two Upper Cretaceous fossil localities from Angola are remarkably rich in large reptiles: Iembe in Bengo Province (Fig. 4.4) of Coniacian age (~88 Ma), and Bentiaba in Namibe (Maastrichtian). The Coniacian provided the bizarre durophagous cryptodiran turtle *Angolachelys mbaxi* Mateus et al. (2009) that justified the erection of its own clade Angolachelonia, and the sauropod dinosaur *Angolatitan adamastor.* Indeterminate plesiosaur remains have also been recovered from this region. Mosasaurs are represented by *Angolasaurus bocagei* (Fig. 4.3), *Tylosaurus iembeensis* Antunes (1964) and an indeterminate halisaurine.

The richest vertebrate locality in Angola is near Bentiaba in Namibe province (Strganac et al. 2014, 2015a, b). The reptile list includes the cheloniid turtles *Euclastes* sp., *Protostega* sp. and *Toxochelys* sp. Terrestrial reptiles include isolated remains that allow undetermined large pterosaurs, an undetermined sauropod based on a metapodial, and a possible hadrosaur based on an isolated phalanx (Mateus et al. 2012). The squamate mosasaurs are, by far, the most abundant and speciose group of tetrapods, and include *Globidens phosphaticus*, *Prognathodon kianda*, *Halisaurus* sp., *Mosasaurus* sp., *Phosphorosaurus* sp., and '*Platecarpus*' *ptychodon* (Polcyn et al. 2007, 2010, 2014; Schulp et al. 2006, 2008, 2013). The plesiosaur

Fig. 4.3 The mosasaur *Angolasaurus bocagei* skull and anterior postcrania (Museum of Geology, Universidade Agostinho Neto). (Photo: Hillsman Jackson, Southern Methodist University)

Fig. 4.4 Reconstruction of the fauna during the Late Cretaceous, based on the Coniacian fauna of Iembe. Illustration using acrylic painting with brush by Fabio Pastori

Elasmosauridae are also abundant at Bentiaba where two taxa are known: Aristonectinae indet. and *Cardiocorax mukulu* (Araújo et al. 2015a, b).

The Paleogene – Mammals Take Over

The Paleogene Period begins at the end of the Cretaceous (66 Ma) and lasted until the Neogene Period 23.03 Ma. The Paleogene is the interval of Earth's history in which mammals diversified and flourished after the K/Pg mass extinction, when most large reptiles, and belemnites and ammonites went extinct. There is a dearth of terrestrial vertebrates, especially mammals, from the Paleogene of Angola. During the global Paleocene-Eocene Thermal Maximum (PETM), 55.8 million years ago, there occurred a sudden change of the climate that marked the end of the Paleocene and the beginning of the Eocene, one of the most significant periods of climate change in the Cenozoic era (Zachos et al. 2005). The best known Paleogene geologic section in Angola is that of Lândana in Cabinda Province. A recent study of the Paleocene and Eocene biota and strata of Lândana indicated that the PETM is missing from that section, either because the event was too short to be recorded at the sampling interval used or that it falls within one of the several stratigraphic gaps documented in the Lândana section (Solé et al. 2018).

Protista and Invertebrates

Among the protists, foraminifera stand out as the most important and well-known taxonomic group in the Angolan marine series of Paleogene age, due to their importance for biostratigraphic correlations in offshore oil drilling and their equivalence with landward outcrops. Important works include those of Rocha (1973) and Kender et al. (2008), which include many Eocene and Oligocene characteristic taxa such as: *Cyclammina* cf. *compressa*, *Nonion centrosulcatum*, *Cassidulina subglobosa*, *Globigerina ampliapertura*, *Bolivina* cf. *pygmaea*, *Bulimina alsatica*, *B. kacksonensis* and *B. nkomi*.

In the Paleogene of Angola, molluscs remain the most speciose clade of invertebrates, comprising at least three nautiloids, 14 bivalves, and 21 gastropods (see Supplementary Material). The bivalves are mainly known from the Eocene (Lutetian) Quimbriz Formation along the Luculo River (Tavares et al. 2007) and include *Leda africae, Noetia veatchi, 'Cardium' luculensis, 'C.' sandigii, Crassatella schoonoverae, Lucina* cf. *landanensis, Macrocallista palmerae, Metis olssoni, Pitar quimbrizensis, P. quipayensis, Protonoetia nigeriensis, Raetomya schweinfurthi, Venericardia angolae,* and *V. heroyi* (Tavares et al. 2007). The three nautiloids are *Cimomia landanensis, Deltoidonautilus caheni, Hercoglossa diderrichi* from the Danian of Cabinda Basin, Landana (Soares 1965), and the gastropods are also mainly known from the Eocene, Lutetian of Luculo River (Tavares et al. 2007): *Ficula roscheni, Fulguroficus harrisi, Pleurotoma angolae, P. rebeccae, Polinices (Neverita) angolae, Ringicula hughesae, Sinum dusenberryi, Surcula* cf. *ingens,* and *Turricula (Knefastia) angolensis.*

Other groups such as Anthozoa, Arthropoda and Echinodermata, exist but are low in numbers (Dartevelle 1953).

Vertebrates

Paleogene marine sediments have been studied mostly in the Benguela Basin and in Cabinda. Adnet et al. (2009) recognized a new species of Eocene lamniform shark, *Xiphodolamia serrata*, from Benguela. Sharks and bony fishes from Cabinda have been listed by Solé et al. (2018), and Taverne (2016) has provided new information about osteoglossid fishes from Lândana. The tetrapods from the Paleogene of Angola are mostly from the fossil sites of Lândana (Solé et al. 2018) and Malembo in Cabinda. The section begins with Lower Paleocene strata at Lândana. Reptiles comprise the dyrosaurid crocodylomorph *Congosaurus bequaerti,* and indeterminate crocodilians (Jouve and Schwarz 2004; Schwarz 2003; Schwarz et al. 2006), the turtles *Taphrosphys congolensis*, a toxochelyid, and *Cabindachelys landanensis* (Myers et al. 2017). Along the Chiloango River, Cabinda, a vertebra of the snake *Palaeophis* was reported by Antunes (1964).

The youngest portion of the Cabinda section is at Malembo Point, south of Lândana, which was originally considered Miocene. The Malembo mammal fauna is comparable to the Early Oligocene fauna of the Fayum, Egypt because of the presence of the embrithopod *Arsinoitherium*, hyracoids such as *Geniohyus* aff. *Mirus* and *Bunohyrax* aff. *Fajumensis*, the proboscidean cf. *Phiomia* or *Hemimastodon,* the sirenian *Halitherium,* and a reported anthropoid canine (Hooijer 1963; Pickford 1986; Jacobs et al. 2016). Recent findings by *Projecto PaleoAngola* in Malembo include a ptolemaiidan molar more similar to *Kelba* from the Miocene of Songhor, Kenya (19.5 Ma) than to Fayum *Ptolemaia*, and an isolated premolar tooth of a large primate comparable in size to that of a female gorilla, likely an undescribed taxon (Jacobs et al. 2016), and not representing any of the numerous Fayum primate taxa. In addition, *Arsinoitherium* is now known from the latest Oligocene of Kenya. The presence of a Kelba-like ptolemaiadan, a unique primate, and Arsinoitherium may indicate a latest Oligocene or even earliest Miocene age for Malembo Point. The assemblage certainly has differences from the Fatum fauna and may indicate the presence of a lowland West Africa faunal province near the Paleogene-Neogene boundary in age and distinct from other regions such as the East African Rift Valley or the Fayum.

The Neogene – The Founding of Modern Biodiversity

The Neogene began about 23 Ma and extends until the Pleistocene (1.8 Ma). It is divided into Miocene (23 Ma to 5.3 Ma) and Pliocene (5.3 Ma to 2.6 Ma), from the oldest to the more recent. This period saw the expansion of the large mammals, and the appearance of hominids. In the Miocene the climate warms again and grasslands and savannas spread. In the Pliocene, the Earth had become similar to the one we know today.

Protists and Invertebrates

The Neogene foraminifera of the Angolan coastal basins, including those from the Miocene Quifandongo series of the Cuanza Basin, are known from a diversity of planktonic and benthic taxa widely used in oil industry offshore correlations or as palaeoenvironmental indicators. Rocha (1957), Graham et al. (1965), Mcmillan and Fourie (1999) and Kender et al. (2009), among others describe the essentials of these West African foraminiferal assemblages, which include planktonic taxa such as *Globigerina praebulloides*, *Globigerinella obesa*, *Globigerinoides bisphericus*, *G. immaturus*, *G. trilobus*, *Globorotalia peripheroronda* and *Orbulina bilobata*.

Invertebrates are surprisingly poorly known and comprise, at least, the nautiloid mollusc *Aturia luculoensis,* the decapod crustacean *Callianassa floridana* from the Miocene Burdigalian of Cabinda Basin (Newton 1917), several taxa of Miocene echinoid echinoderms such as *Clypeaster borgesi, Echinolampas antunesi, Rotula deciesdigitata, Rotuloidea vieirai, Amphiope neuparthi*, and *Plagiobrissus* sp. (Loriol 1905; Dartevelle 1953; Gonçalves 1971; Kroh 2010; Silva and Pereira 2014; Pereira and Stara 2018), and two species of anthozoan corals *Flabellum extensum* and *Stylophora raristella* (Chevalier 1970). Nevertheless, bivalve and gastropod molluscs are undoubtedly the most diverse and abundant invertebrate taxa of the marine Neogene of Luanda, Benguela and Namibe, with several rich fossil sites, some of them presently in course of study. The molluscan faunas of these coastal basin areas, including new species such as *Pereiraea africana*, *Clavatula loandensis* or *Chlamys silvai* Antunes (1964), were the focus of Douvillé (1933), Keller (1934), Dartevelle (1952, 1953), Dartevelle and Roger (1954), Soares (1961, 1962), Silva (1962), Silva and Soares (1962), Antunes (1964), and more recently Antunes (1984), Lozouet and Gourgues (1995), among others.

Vertebrates

Mammals are known from Benguela and the Cuanza provinces where Projecto PaleoAngola collected skulls of fossil baleen whales. An odontocete has also been found from Barra da Cuanza.

In Angola the most abundant group of Neogene vertebrates are the Elasmobranchii chondrichthyans (18 taxa; see Supplementary Material). The following taxa are from the Pliocene of Farol das Lagostas (Cuanza Basin): *Aetobatus, Carcharhinus egertoni, Carcharhinus priscus, Carcharias taurus, Carcharocles megalodon, Carcharodon carcharias, Galeocerdo cuvier, Hemipristis serra, Isurus benedeni, Isurus oxyrinchus, Mitsukurina, Myliobatis, Negaprion brevirostris, Paragaleus, Pristis, Pteromylaeus bovina, Rhinoptera brasiliensis,* and *Sphyrna zygaena* (Antunes 1964). Five bony fishes are known, the actinopterygians *Cybium, Sparus, Sphyraena barracuda, Tachysurus,* and *Tetrodon*.

The Quaternary – The Dominance of Humans

The Quaternary (2.6 Ma to present, including the Pleistocene and Holocene) is the third geological period of the Cenozoic era and the most recent in the geological time scale. This period is characterised by the return of glaciations at higher elevations and latitudes, the dominant role of the genus *Homo* in all terrestrial habitats and the extinction of much of the megafauna.

Invertebrates

In Angola, the Quaternary biodiversity is again marked by the high number of mollusc taxa (73 or more), of which 29 are bivalves such as *Arcopsis afra, Barbatia complanata, Cardium indicum, Chama crenulata, Glycymeris concentrica, Lutraria senegalensis, Noetiella congoensis, Ungulina cuneata* from the Middle Pleistocene of Pipas (Namibe Basin) and 44 gastropods mainly known from Namibe Basin, such as *Cantharus viverratus*, *Columbella adansoni*, *Conus babaensis*, *Siphonaria capensis*, and *Terebra senegalensis* (Miller and Carpenter 1956; Sessa et al. 2013). Other invertebrates such as corals, arthropods and echinoderms are known but reduced to a handful of known taxa, such as *Cladangia carvalhoi* from the Pleistocene of Salinas de Bero, Saco, Namibe (Wood 1973). In most situations they occur in a variety of raised-beach and lagoonal deposits related to coastal uplift and major sea-level changes (Carvalho 1961). The post-glacial Holocene is marked by the accretion of sand-spits and deltaic facies with rich coquinas, including the bivalve *Senilia senilis* as a typical species (Dinis et al. 2016). See Supplementary Material for the updated list.

Vertebrates

A remarkable fossilized jaw (dentary bone) of Blue Whale *Balaenoptera musculus* present in the *Museu Nacional de História Natural* in Luanda measuring nearly seven meters in length is not only the largest known fossilized bone but also one of the largest whales, thus animals, ever recorded. Large land mammals, including *Bubalus*, *Syncerus* cf. *nanus* Boddaert; *Phacochoerus* sp., *Equus*, *Hippotigris* cf. *zebra* have been reported from the site called *Cemitério dos Ossos*, north of Luanda (Antunes 1961).

The caves of Humpata, in Huíla Province, in southern Angola, are formed in Chela Dolomite that hosts fossiliferous caves and fissures (Amaral 1973; Antunes 1965; Arambourg and Mouta 1952; França 1964; Mouta 1950). Pickford et al. (1990, 1992, 1994) listed taxa of mammals from the Humpata Caves. These include the insectivore *Crocidura*, a Macroscelididae, the chiropteran *Rhinolophus, Miniopterus, Nycteris*, 19 genera of rodents (*Uranomys, Acomys, Dasymys, Aethomys, Thallomys, Zelotomys, Mus, Pelomys, Malacomys, Praomys, Grammomys, Dendromus, Steatomys, Petromyscus, Tatera, Otomys, Cryptomys, Graphiurus*, and *Hystrix*), the lagomorph *Serengetilagus*, Mustelidae, Viverridae, Canidae, and the Hyaenidae cf. *Chasmoporthetes*, the Hyracoidea *Gigantohyrax* and *Procavia,* Rhinocerotidae, Equidae, Suidae *Metridiochoerus andrewsi* and the Bovidae Hippotragini and *Connochaetes.*

The most thoroughly studied of the Humpata fossils are those of the extinct baboon. Cercopithecid primates from Humpata caves include *Soromandrillus quadratirostris*, cf. *Theropithecus* sp., and *Cercopithecoides* sp. (dated as ca. 2.0–3.0 Ma) (Minkoff 1972; Jablonski 1994; Jablonski and Frost 2010; Gilbert 2013).

Pleistocene deposits in Namibe provided remains of fossilized ostrich *Struthio* eggshells, artiodactyl bones and numerous human artifacts - including Acheulean hand axes (amygdaloid bifaces) suggesting the presence of early humans, such as *Homo ergaster* or *H. erectus*, whereas bones of early humans are not known in Angola.

Final Remarks on the Fossil Record and Paleobiodiversity

Measuring paleobiodiversity is challenging due to the sparse and limited availability of data, compared with modern extant faunas. The paleobiodiversity of Angola is mostly known from Cretaceous and Cenozoic fossils that comprise 90% or more of all known fossil records of Angolan taxa (see Table 4.1 and Supplementary Material). The vast preponderance of the fossil taxa is marine which is consistent with the geological settings and paleogeography, related to the opening of the South Atlantic and repeated inundation of the Angolan continental margin.

For this study, we compiled a list of taxa using species, genus or the lowest known taxonomical clade reported in the scientific literature for Angola (see Supplementary Material). Of the resulting list of more than 1300 fossil taxa, many may require systematic revision and the final number will depend on the validity of the taxonomy.

By far the most speciose group are the molluscs (about 61% of taxa, more than half being Cretaceous ammonites) and Cretaceous foraminiferans (16%), followed by vertebrates with about 15% of taxa. Chondrichthyes and mammals represent 6% and 5% of taxa, respectively.

About 10% of vertebrate taxa listed are unique or were first recognised in Angola, most of them receiving species names after localities in Angola or of geologists that worked in the country. According to current knowledge, at least 67 taxa (6.1%) of invertebrates are endemic or were first mentioned from Angola.

References

Adnet S, Hosseinzadeh R, Antunes MT et al (2009) Review of the enigmatic Eocene shark genus *Xiphodolamia* (Chondrichthyes, Lamniformes) and description of a new species recovered from Angola, Iran and Jordan. J Afr Earth Sci 55(3–4):197–204

Amaral L (1973) Nota sobre o "karst" ou carso do Planalto do Humpata (Huila), no Sudoeste de Angola. Garcia de Orta 1:29–36

Anchieta J (1885) Traços geológicos da África Occidental Portuguesa. Tipografia Progresso, Benguela, 15 pp

Antunes MT (1961) A jazida de vertebrados fósseis do Farol das Lagostas: II Paleontologia. Boletim dos Serviços de Geologia e Minas de Angola 3:1–18

Antunes MT (1964) O Neocretácico e o Cenozóico do litoral de Angola. Junta de Investigações Ultramar, Lisboa, 255 pp

Antunes MT (1965) Sur la faune de vertébrés du Pléistocène de Leba, Humpata (Angola). Actes du Ve Congrès Panafricain de Préhistoire et de l'Étude du Quaternaire. Tenerife, 127–128

Antunes MT (1970) Paleontologia de Angola. In: Curso de Geologia do Ultramar. Junta de Investigações do Ultramar, Lisboa, pp 126–143

Antunes MT (1984) Étude d'une faune gastéropodes miocène récoltés par M. M Feio dans le Sud de l'Angola Comunicações dos Serviços Geológicos de Portugal 70(1):126–128

Antunes MT, Cappetta H (2002) Sélaciens du Crétacé (Albien-Maastrichtien) d'Angola. *Palaeontographica*, Abteilung A 264 (5–6):85–146

Antunes MT, Maisey JG, Marques MM, et al (1990) Triassic fishes from the Cassange depression (R.P. de Angola). Ciências da Terra (UNL), special number 1:1–64

Arambourg C, Mouta F (1952) Les grottes et fentes à ossements du sud de l'Angola. Actes du IIème Congrès Panafricain de Préhistoire d'Alger 12:301–301

Araújo AG, Guimarães F (1992) Geologia de Angola, Notícia explicativa da Carta Geológica à escala 1: 1 000 000. Serviço Geológico de Angola, Luanda, 140 pp

Araújo R, Polcyn MJ, Lindgren J et al (2015a) New aristonectine elasmosaurid plesiosaur specimens from the Early Maastrichtian of Angola and comments on paedomorphism in plesiosaurs. Neth J Geosci 94(1):93–108

Araújo R, Polcyn MJ, Schulp AS et al (2015b) A new elasmosaurid from the early Maastrichtian of Angola and the implications of girdle morphology on swimming style in plesiosaurs. Neth J Geosci 94(1):109–120

Archibald JD, Clemens WA, Padian K et al (2010) Cretaceous extinctions: multiple causes. Science 328(5981):973–973

Berry CT (1939) A summary of the fossil Crustacea of the Order Stomatopoda, and a description of a new species from Angola. Am Midl Nat 21(2):461–471

Blake DB, Breton G, Gofas S (1996) A new genus and species of Asteriidae (Asteroidea; Echinodermata) from the Upper Cretaceous (Coniacian) of Angola, Africa. Paläontol Z 70(1–2):181–187

Borges A (1946) A costa de Angola da Baía da Lucira à Foz do Bentiaba (entre Benguela e Mossâmedes). Boletim da Sociedade Geológica de Portugal 5(3):141–150

Boule M, Lemoine P, Thevenin A (1907) Paléontologie de Madagascar. III Céphalopodes crétacés des environs de Diègo-Suarez. Annales de Paléontologie 2:1–56

Brandão JM (2008) "Missão Geológica de Angola": contextos e emergência. Memórias e Notícias, new series 3:285–292

Brandão JM (2010) O "Museu de Geologia Colonial" das Comissões Geológicas de Portugal: contexto e memória. Revista Brasileira de História da Ciência 3(2):184–199

Buta-Neto A, Tavares TS, Quesne D et al (2006) Synthèse préliminaire des travaux menés sur le bassin de Benguela (Sud Angola): implications sédimentologiques et structurales. Áfr Geosci Rev 13(3):239–250

Callapez PM, Gomes CR, Serrano Pinto M et al (2011) O contributo do Museu e Laboratório Mineralógico e Geológico da Universidade de Coimbra para os estudos de Paleontologia Africana. In: Neves LF, Pereira AC, Gomes CR, Pereira LCG, Tavares AO (eds) Modelação de Sistemas Geológicos. Homenagem ao Professor Doutor Manuel Maria Godinho. Laboratório de Radioactividade Natural da Universidade de Coimbra, Coimbra, pp 159–174

Carvalho GS (1961) Geologia do deserto de Moçâmedes, (Angola): Uma contribuição para o conhecimento dos problemas da orla sedimentar de Moçâmedes. Memórias da Junta de Investigações do Ultramar 26:1–227

Ceraldi TS, Green D (2016) Evolution of the South Atlantic lacustrine deposits in response to Early Cretaceous rifting, subsidence and lake hydrology. In: Ceraldi S, Hodgkinson RA, Backe G (eds) Petroleum geoscience of the West Africa Margin. Geological Society, London, Special Publications, 438, doi: https://doi.org/10.1144/SP438.10

Chevalier JP (1970) Les Madreporaires du Neogene et du Quaternaire de l'Angola [Neogene and Quaternary corals from Angola]. Annalen Koninklijk Museum voor Midden-Afrika 8: Geologische Wetenschappen 68:13–33

Collignon M (1978) Ammonites du Crétacé Moyen-Supérieur de l'Angola. 2° Centenário Academia das Ciências. Estudos de Geologia e Paleontologia e de Micologia, Academia das Ciências, Lisboa, pp 1–75

Cooper MR (1972) The Cretaceous stratigraphy of San Nicolau and Salinas, Angola. Ann S Af Mus 6(8):245–251

Cooper MR (1982) Lower Cretaceous (Middle Albian) ammonites from Dombe Grande, Angola. Ann S Af Mus 89:265–314

Cooper MR, Kennedy WJ (1979) Upper most Albian (Stoliczkaia dispar zone) ammonites from the Angolan littoral. Ann S Af Mus 77:175–308

Crampton JS, Beu AG, Cooper RA et al (2003) Estimating the rock volume bias in paleobiodiversity studies. Science 301(5631):358–360

Dartevelle E (1952) Echinides fossiles du Congo et de l'Angola. Partie 1: Introduction historique et stratigraphique. Annales du Musée Royal du Congo Belge, série 8, Sciences Géologiques 12:1–70

Dartevelle E (1953) Echinides fossiles du Congo et de l'Angola. Partie 2: description systématique des échinides fossiles du Congo et de l'Angola. Annales du Muséum Royal du Congo Belge, série 8, Sciences Géologiques 13:1–240

Dartevelle E, Roger J (1954) Contribution à la connaissance de la faune du Miocène de l'Angola. Comunicações dos Serviços Gelógicos de Portugal 35:227–312

Dinis P, Huvi J, Cascalho J, Garzanti E, Vermeesch P, Callapez P (2016) Sand-spits systems from Benguela region (SW Angola). An analysis of sediment sources and dispersal from textural and compositional data. J Afr Stud 117:181–192

Douvillé H (1933) Le Tertiaire de Loanda. Boletim do Museo e Laboratorio Mineralógico e Geológico da Universidade de Lisboa, 1:63–118

Duarte LV, Callapez PM, Kalukembe A, et al (2014) Do Proterozoico da Serra da Leba (Planalto da Humpata) ao Cretácico da Bacia de Benguela (Angola). A geologia de lugares com elevado valor paisagístico. Comunicações Geológicas 101(Especial III):1255–1259

Ferré B, Granier B (2001) Albian roveacrinids from the southern Congo Basin off Angola. J S Am Earth Sci 14:219–235

Ferreira JM, Rocha AT (1957) Foraminíferos do Senoniano de Catumbela (Angola). Garcia de Orta 5(3):517–545

Ferreira OV (1957) Acerca de *Parapirimela angolensis* Van Straelen nas Camadas de Iela, Angola. Comunicações dos Serviços Geológicos de Portugal 38:465–468

França JC (1964) Nota preliminar sobre uma gruta pré-histórica do Planalto da Humpata (Angola). Junta de Investigações do Ultramar 2(50):59–67

Gilbert CC (2013) Cladistic analysis of extant and fossil African papionins using craniodental data. J Hum Evol 64(5):399–433

Gonçalves F (1971) *Echinolampas antunesi*, nov. sp. Cassidulidae, échinide nouveau du Miocène de la région de Luanda, Angola. Revista da Faculdade de Ciências, C – Ciências Naturais 16(2):307–310

Graham JJ, Klasz I, Rerat D (1965) Quelques importants foraminifères du Tertiaire du Gabon (Afrique Equatoriale). Revue de Micropalèontologie 8:71–84

Green PF, Machado V (2015) Pre-rift and synrift exhumation, post-rift subsidence and exhumation of the onshore Namib1e margin of Angola revealed from apatite fission track analysis. In: Sabato Ceraldi T, Hodgkinson RA, Backe G (eds) Petroleum geoscience of the West Africa Margin. Geological Society, London, Special Publications 438, pp 99–118

Greyling MR, Cooper MR (1995) Two new irregular echinoids from the Upper Cretaceous (mid-Campanian) of Angola. Durban Museum Novitates 20(1):63–71

Guiraud R, Maurin JC (1991) Le rifting en Afrique au Crétacé Inférieur: Synthèse structural, mise en évidence de deux phases dans la genèse des bassins, relations avec les ouvertures océaniques péri-africaines. Bulletin de la Société Géologique de France 165(5):811–823

Guiraud M, Buta-Neto A, Quesne D (2010) Segmentation and differential post-rift uplift at the Angola margin as recorded by the transform – rifted Benguela and oblique-to-orthogonal-rifted Kwanza basins. Mar Pet Geol 27:1040–1068

Haas O (1942) The vernay collection of cretaceous (Albian) ammonites from Angola. Bull Am Mus Nat Hist 81(1):1–224

Haas O (1943) Some abnormally coiled ammonites from the Upper Cretaceous of Angola. Am Mus Novit 1222:1–17

Haughton SH (1924) Notes sur quelques fossiles crétacés de l'Angola (Céphalopodes et Échinides). Comunicações dos Serviços Geológicos de Portugal 15:79–106

Hooijer DA (1963) Miocene mammals of Congo. Annales du Museum Royal d'Afrique Centrale, Series 8 Sciences Géologiques 46:1–77

Howarth MK (1965) Cretaceous ammonites and nautiloids from Angola. Bull Br Mus Nat Hist Geol 10:335–412

Howarth MK (1966) A mid-Turonian ammonite fauna from the Moçâmedes desert, Angola. Garcia de Orta 14(2):217–228

Jablonski NG (1994) New fossil cercopithecid remains from the Humpata Plateau, southern Angola. Am J Phys Anthropol 94(4):435–464

Jablonski NG, Frost S (2010) Cercopithecoidea. In: Werdelin L, Sanders WJ (eds) Cenozoic Mammals of Africa. University of California Press, Berkeley, pp 393–428

Jackson JB, Johnson KG (2001) Measuring past biodiversity. Science 293(5539):2401–2404

Jacobs LL, Mateus O, Polcyn MJ et al (2006) The occurrence and geological setting of Cretaceous dinosaurs, mosasaurs, plesiosaurs, and turtles from Angola. J Paleontol Soc Korea 22:91–110

Jacobs LL, Mateus O, Polcyn MJ et al (2009) Cretaceous paleogeography, paleoclimatology, and amniote biogeography of the low and mid-latitude South Atlantic Ocean. Bull Geol Soc Fr 180(4):333–341

Jacobs LL, Polcyn MJ, Mateus O et al (2016) Post-Gondwana Africa and the vertebrate history of the Angolan Atlantic Coast. Mem Mus Vic 74:343–362

Jones TR (1862) A Monograph of the fossil Estheriae. Palaeontol Soc 14:1–134

Jouve S, Schwarz D (2004) *Congosaurus bequaerti*, a Paleocene dyrosaurid (Crocodyliformes; Mesoeucrocodylia) from Landana (Angola). Bulletin de l'Institut Royal des Sciences Naturelles de Belgique, Sciences de la Terre 74:129–146

Keller A (1934) Contribution a la géologie de l'Angola. Le Tertiaire dc Luanda. Description des especes, Mollusque. Lamellibranches. Boletim do Museo e Lab. Mineral. e Geol. do Universidade de Lisboa, la Ser, 3, pp 219–250

Kender S., Kaminski MA, Jones, BW (2008) Oligocene deep-water agglutinated foraminifera from the Congo Fan, offshore Angola: Palaeoenvironments and assemblage distributions. In: Kaminski MA, Coccioni R (eds) Proceedings of the seventh international workshop on agglutinated foraminifera. Grzybowski Foundation Special Publication 13, London, pp 107–156

Kender S, Kaminski MA, Jones BW (2009) Early to middle Miocene foraminifera from the deep-sea Congo Fan, offshore Angola. Micropaleontology 54(6):477–568

Kier PM, Lawson M (1978) Index of living and fossil echinoids, 1924-1970. Smithson Contrib Paleobiol 34:1–182

Kroh A (2010) Index of Living and Fossil Echinoids 1971-2008. Annalen des Naturhistorischen Museums in Wien 112:195–470

Lapão LGP, Pereira ES (1971) Notícia explicativa Carta Geológica de Angola, escala 1:100 000, folha n° 206, Egito Praia. Direcção Provincial dos Serviços de Geologia e Minas, Luanda, 42 pp

Lapão LGP, Simões MC (1972) Notícia Explicativa da Carta Geológica de Angola, escala 1:100 000, Folha n°184, Novo Redondo. Direcção Provincial dos Serviços de Geologia e Minas, Luanda, 54 pp

Leriche M (1932) Sur les premiers fossiles découverts au nord de l'Angola, dans le prolongement des couches du Lubilash et des couches du Lualaba. Association Française pour l'Avancement des Sciences, Compte Rendu 56éme session: 1–6

Loriol P (1888) Matériaux pour l'étude stratigraphique et paléontologique de la province d'Angola. Description des Echinides. Mémoires de la Société de Physique et d'histoire naturelle de Genève 30(2):97–114

Loriol P (1905) Notes pour servir à l'étude des échinodermes. Fasc. 2 (3). Georg Editeur, Bâle/Genève, pp 119–146

Lozouet P, Gourgues D (1995) Senilia (Bivalvia: Arcidae) et Anazola (Gastropoda: Olividae) dans le Miocène d'Angola et de France, témoins d'une paléo-province Ouest-Africaine. Haliotis, 24, pp 101–108

Marliére R (1950) Ostracodes and Phyllopodes au Système du Karroo au Congo Belge et les régions avoisinantes. Annales du Muséum Royal du Congo Belge, Sciences Géologiques, Série 8(6):1–43

Marzola M, Mateus O, Schulp A, et al (2014) Early Cretaceous tracks of a large mammaliamorph, a crocodylomorph, and dinosaurs from an Angolan diamond mine. J Vertebr Paleontol, Program and Abstracts, 181

Masse P, Laurent O (2016) Geological exploration of Angola from Sumbe to Namibe: a review at the frontier between Geology, natural resources and the history of Geology. C R Geosci 348(1):80–88

Mateus O, Jacobs LL, Polcyn MJ et al (2009) The oldest African eucryptodiran turtle from the Cretaceous of Angola. Acta Paleontol Pol 54:581–588

Mateus O, Jacobs LL, Schulp AS et al (2011) *Angolatitan adamastor*, a new sauropod dinosaur and the first record from Angola. Anais da Academia Brasileira de Ciências 83(1):221–233

Mateus O, Polcyn MJ, Jacobs LL, et al (2012) Cretaceous amniotes from Angola: dinosaurs, pterosaurs, mosasaurs, plesiosaurs, and turtles. V Jornadas Internacionales sobre Paleontología de Dinosaurios y su Entorno, Salas de los Infantes, Burgos, pp 75–105

Mateus O, Marzola M, Schulp AS et al (2017) Angolan ichnosite in a diamond mine shows the presence of a large terrestrial mammaliamorph, a crocodylomorph, and sauropod dinosaurs in the Early Cretaceous of Africa. Palaeogeogr Palaeoclimatol Palaeoecol 471:220–232

Mcmillan IK, Fourie A (1999) Kwanza Basin coastal stratigraphy with atlas of Albian to Holocene Foraminifera species. De Beers Marine (Pty) Limited, Luanda, 167 pp

Miller AK, Carpenter LB (1956) Cretaceous and Tertiary Nautiloids from Angola. Estudos, Ensaios e Documentos da Junta de Investigações do Ultramar 21:1–48

Minkoff EC (1972) A fossil baboon from Angola, with a note on *Australopithecus*. J Paleontol 46(6):836–844

Mouta F (1950) Sur la présence du Quaternaire ancien dans les hauts plateaux du Sud de l'Angola (Humpata, Leba). Compte Rendu sommaire des Scéances de la Société géologique de France 14:261–262

Murray AM (2000) The palaeozoic, mesozoic and early cenozoic fishes of Africa. Fish Fish 1(2):111–145

Myers TS, Polcyn MJ, Mateus O et al (2017) A new durophagous stem cheloniid turtle from the lower Paleocene of Cabinda, Angola. Pap Palaeontol 2017:1–16

Néraudeau D, Mathey B (2000) Biogeography and diversity of South Atlantic Cretaceous echinoids: implications for circulation patterns. Palaeogeogr Palaeoclimatol Palaeoecol 156(1–2):71–88

Neto MGM (1970) O sedimentar costeiro de Angola. Algumas notas sobre o estado actual do seu conhecimento. In: Curso de Geologia do Ultramar, vol 2. Publicações da Junta de Investigações do Ultramar, Lisboa, pp 193–232

Newton RB (1917) On some Cretaceous Brachiopoda and Mollusca from Angola, Portuguese West Africa. Earth and environmental science. Trans R Soc Edinb 51(3):561–580

Nunes AF (1991) A Investigação Geológico-Mineira em Angola. Ministérios dos Negócios Estrangeiros, Ministério das Finanças, Instituto para a Cooperação Económica, Lisboa, 387 pp

Pereira P, Stara P (2018) Redefinition of *Amphiope neuparthi* de Loriol, 1905 (Echinoidea, Astriclypeidae) from the early-middle Miocene of Angola. Comunicações Geológicas, 104(1): in press

Pérez-Díaz L, Eagles G (2017) South Atlantic paleobathymetry since early Cretaceous. Sci Rep 7:11819. https://doi.org/10.1038/s41598-017-11959-7

Pickford M (1986) Première découverte d'une faune mammalienne terrestre paléogène d'Afrique sub-saharienne. Comptes rendus de l'Académie des Sciences Paris Série II 19:1205–1210

Pickford M, Fernandez T, Aço S (1990) Nouvelles découvertes de remplissages de fissures à primates dans le 'Planalto da Humpata', Huilà, Sud de l'Angola. Comptes Rendus de l'Académie des Sciences Paris, Série II 310:843–848

Pickford M, Mein P, Senut B (1992) Primate-bearing Plio-Pleistocene cave deposits of Humpata, southern Angola. Hum Evol 7:17–33

Pickford M, Mein P, Senut B (1994) Fossiliferous Neogene karst fillings in Angola, Botswana and Namibia. S Afr J Sci 90:227–230

Polcyn M, Jacobs L, Schulp A, et al (2007) *Halisaurus* (Squamata: Mosasauridae) from the Maastrichtian of Angola. J Vertebr Paleontol 27(Suppl. to 3):130A

Polcyn MJ, Jacobs LL, Schulp AS et al (2010) The North African Mosasaur *Globidens phosphaticus* from the Maastrichtian of Angola. Hist Biol 22(1–3):175–185

Polcyn MJ, Jacobs LL, Araújo R et al (2014) Physical drivers of mosasaur evolution. Palaeogeogr Palaeoclimatol Palaeoecol 400:17–27

Quirk DG, Hertle M, Jeppesen JW, et al (2013) Rifting, subsidence and continental break-up above a mantle plume in the central South Atlantic. In: Mohriak WU, Danforth A, Post PJ, et al (eds) Conjugate Divergent Margins. Geological Society, London, Special Publication, 369: 185–214

Rennie JVL (1929) Cretaceous fossils from Angola (Lamellibranchia and Gastropoda). Ann S Afr Mus 28:1–54

Rennie JVL (1945) Lamelibrânquios e gastrópodes do Cretácico Superior de Angola (vol. 1). Junta das Missões Geográficas e de Investigações Coloniais 1:1–141

Robles-Cruz SE, Escayola M, Jackson S et al (2012) U-Pb SHRIMP geochronology of zircon from the Catoca kimberlite, Angola: Implications for diamond exploration. Chem Geol 310-311:137–147. https://doi.org/10.1016/j.chemgeo.2012.04.001

Rocha AT (1957) Contribuição para o estudo dos foraminíferos do Terciário de Luanda. Garcia de Orta 5(2):297–312

Rocha AT (1973) Contribution à l'étude des foraminifères paléogènes du bassin du Cuanza (Angola). Memórias e Trabalhos do Instituto de Investigação Científica de Angola 12:1–309

Rocha AT (1984) Notas micropaleontológicas sobre as formações sedimentares da orla meso-cenozóica de Angola - V. O Maestrichtiano inferior da mancha de Cabeça da Baleia (a norte de Egito-Praia). Garcia de Orta, Série Geológica 7(1–2):97–108

Schlanger SO, Arthur MA, Jenkyns HC et al (1987) The Cenomanian-Turonian Oceanic Anoxic Event, I. Stratigraphy and distribution of organic carbon-rich beds and the marine δ13C excursion. Geol Soc Lond, Spec Publ 26(1):371–399

Schulp AS, Polcyn MJ, Mateus O et al (2006) New mosasaur material from the Maastrichtian of Angola, with notes on the phylogeny, distribution and palaeoecology of the genus *Prognathodon*. On Maastricht Mosasaurs. Publicaties van het Natuurhistorisch Genootschap in Limburg 45(1):57–67

Schulp AS, Polcyn MJ, Mateus O et al (2008) A new species of *Prognathodon* (Squamata, Mosasauridae) from the Maastrichtian of Angola, and the affinities of the mosasaur genus Liodon. In: Proceedings of the Second Mosasaur Meeting, vol 3. Fort Hays State University Hays, Kansas, pp 1–12

Schulp AS, Polcyn MJ, Mateus O et al (2013) Two rare mosasaurs from the Maastrichtian of Angola and the Netherlands. Neth J Geosci 92(1):3–10

Schwarz D (2003) A procoelous crocodilian vertebra from the lower Tertiary of Central Africa (Cabinda enclave, Angola). Neues Jahrbuch für Geologie und Paläontologie, Monatshefte 2003:376–384

Schwarz D, Frey E, Martin T (2006) The postcranial skeleton of the Hyposaurinae (Dyrosauridae; Crocodyliformes). Palaeontology 49(4):695–718

Segundo J, Duarte LV, Callapez PM (2014) Lithostratigraphy of the Quissonde Formation marl-limestone succession (Albian) of the Ponta do Jomba-Praia do Binge sector (Benguela Basin, Angola). Comunicações Geológicas 101(Especial III):567–571

Séranne M, Anka Z (2005) South Atlantic continental margins of Africa: a comparison of the tectonic vs climate interplay on the evolution of equatorial west Africa and SW Africa margins. J Afr Earth Sci 43(1–3):283–300

Sessa JA, Callapez PM, Dinis PA et al (2013) Paleoenvironmental and paleobiogeographical implications of a Middle Pleistocene mollusc assemblage from the marine terraces of Baía das Pipas, southwest Angola. J Paleontol 87(6):1016–1079

Silva GH (1962) Fósseis do Miocénico de Luanda (Angola). Associação Portuguesa para o Progresso das Ciências. Actas do XXVI Congresso Luso-Espanhol (Porto, 22–26 June, 1962), sections II and IV, 3 pp

Silva R, Geirinhas F (2010) Colecções geológicas das antigas Províncias Ultramarinas Portuguesas arquivadas na Litoteca do LNEG. e-Terra 15(4):1–4

Silva R, Pereira P (2014) Redescoberta dos equinodermes fósseis das coleções históricas ultramarinas do LNEG. Comunicações Geológicas 101(Especial III):1379–1382

Silva GH, Soares AF (1962) Contribuição para o conhecimento da fauna miocénica de S. Pedro da Barra e do Farol das Lagostas (Luanda, Angola). Garcia de Orta 9:721–736

Soares AF (1961) Nouvelle espèce de *Chlamys* du Miocène de la région de Luanda (Angola). Memórias e Notícias 51:1–6

Soares AF (1962) Nota sobre alguns lamelibrânquios e gastrópodes do Miocénico de Luanda (Angola). Memórias e Notícias 53:31.35

Soares AF (1965) Contribuição para o estudo dos lamelibrânquios Cretácicos da região de Moçâmedes. Serviços de Geologia e Mina de Angola Boletim 11:137–168

Solé F, Noiret C, Desmares D et al (2018) Reassessment of historical sections from the Paleogene marine margin of the Congo Basin reveals an almost complete absence of Danian deposits. Geosci Front. https://doi.org/10.1016/j.gsf.2018.06.002

Spath LF (1922) On Cretaceous ammonites from Angola, collected by Prof. J.W. Gregory, D.Sc., F.R.S. Trans R Soc Edinb Earth Sci 53:91–160

Spath LF (1931) On cretaceous Ammonoidea from Angola, collected by Pr. J.W. Gregory. Trans Geol Soc Edinb. 53 (1):91–160

Spath LF (1953) The Upper Cretaceous cephalopod fauna of Grahamland. Sci Rep Falkland Islands Depend Surv 3:1–60

Strganac C, Salminen J, Jacobs LL et al (2014) Carbon isotope stratigraphy, magnetostratigraphy, and 40Ar/39Ar age of the Cretaceous South Atlantic coast, Namibe Basin, Angola. J Afr Earth Sci 99:452–462

Strganac C, Jacobs LL, Polcyn M et al (2015a) Stable oxygen isotope chemostratigraphy and paleotemperature regime of mosasaurs at Bentiaba, Angola. Neth J Geosci 94(1):137–143

Strganac C, Jacobs LL, Polcyn MJ et al (2015b) Geological setting and paleoecology of the Upper Cretaceous Bench 19 marine vertebrate bonebed at Bentiaba, Angola. Neth J Geosci 94(1):121–136

Tasch P (1979) Permian and Triassic Conchostraca from the Bowen Basin (with a note on a Carboniferous leaiid from the Drummond Basin), Queensland. Aust Bureau Mineral Resour Geolo Geophys Bull 185:33–44

Tasch P (1987) Fossil Conchostraca of the Southern Hemisphere and continental drift: Paleontology, biostratigraphy, and dispersal. Memoir of the Geological Society of America 165:1–290

Tavares T (2006) Ammonites et échinides de l'Albien du bassin de Benguela (Angola). Systématique, biostratigraphie, paléogéographie et paléoenvironnement. Unpublished PhD Thesis, Université de Bourgogne, Dijon, 389 pp

Tavares T, Meister C, Duarte-Morais ML et al (2007) Albian ammonites of the Benguela Basin (Angola): a biostratigraphic framework. S Afr J Geol 110(1):137–156

Taverne L (2016) New data on the osteoglossid fishes (Teleostei, Osteoglossiformes) from the marine Danian (Paleocene) of Landana (Cabinda Enclave, Angola). Geo-Eco-Trop 40(4):297–304

Teixeira C (1947) Contribuição para o conhecimento geológico do Karroo da África Portuguesa. I-Sobre a flora fóssil do Karroo da região de Téte. Anais da Junta de Investigações do Ultramar, Estudos de Geologia e Paleontologia 2(2):9–28

Teixeira C (1948a) Fósseis vegetais do Karroo de Angola. Boletim da Sociedade Geológica de Portugal 7(1–2):73–77

Teixeira C (1948b) Vegetais fósseis do grés do Quilungo. Anais da Junta de Investigações Coloniais 2:85–92

Teixeira C (1958) Note paléontologique sur le Karroo de la Lunda, Angola. Boletim da Sociedade Geológica de Portugal 12(3):83–92

Teixeira C (1961) Paleontological notes on the Karroo of the Lunda (Angola). Garcia de Orta 9(2):307–311

Van Straelen V (1937) Parapirimela angolensis. Brachyure nouveau du Miocène de l'Angola. Bulletin du Musée Royal d'Histoire Naturelle de Belgique 8(5):1–4

Vasconcelos P (1951) Sur la découverte d'algues fossiles dans les terrains anciens de l'Angola. Int Geol Congr XVIIIth Sess 14:288–293

Walker RT, Telfer M, Kahle RL et al (2016) Rapid mantle-driven uplift along the Angolan margin in the late Quaternary. Nat Geosci 9(12):909–914. https://doi.org/10.1038/NGEO2835

Wood RC (1973) Fossil marine turtle remains from the Paleocene of the Congo. Annales du Musée Royal d'Afrique Centrale, Sciences Geologiques 75:1–28

Zachos JC, Röhl U, Schellenberg SA et al (2005) Rapid acidification of the ocean during the Paleocene-Eocene thermal maximum. Science 308(5728):1611–1615

Part II
Flora, Vegetation and Landscape Change

Chapter 5
The Flora of Angola: Collectors, Richness and Endemism

David J. Goyder and Francisco Maiato P. Gonçalves

Abstract Angola is botanically rich and floristically diverse, but is still very unevenly explored with very few collections from the eastern half of the country. We present an overview of historical and current botanical activity in Angola, and point to some areas of future research. Approximately 6850 species are native to Angola and the level of endemism is around 14.8%. An additional 230 naturalised species have been recorded, four of which are regarded as highly invasive. We draw attention to the paucity of IUCN Red List assessments of extinction risk for Angolan vascular plants and note that the endemic aquatic genus *Angolaea* (Podostemaceae), not currently assessed, is at high risk of extinction as a result of dams built on the Cuanza river for hydro-electric power generation. Recent initiatives to document areas of high conservation concern have added many new country and provincial records and are starting to fill geographic gaps in collections coverage.

Keywords Botanical collectors · Botanical diversity · Botanical history · Gossweiler · Invasive species · Welwitsch

D. J. Goyder (✉)
Herbarium, Royal Botanic Gardens, Kew, Richmond, Surrey, UK

National Geographic Okavango Wilderness Project, Wild Bird Trust, Parktown, Gauteng, South Africa
e-mail: D.Goyder@kew.org

F. M. P. Gonçalves
National Geographic Okavango Wilderness Project, Wild Bird Trust, Parktown, Gauteng, South Africa

Instituto Superior de Ciências da Educação da Huíla, Lubango, Angola

University of Hamburg, Institute for Plant Science and Microbiology, Hamburg, Germany
e-mail: francisco.maiato@gmail.com

B. J. Huntley et al. (eds.), *Biodiversity of Angola*,
https://doi.org/10.1007/978-3-030-03083-4_5

History of Botanical Exploration in Angola

It appears that the earliest extant botanical collections from Angola date from either 1669 (Exell 1939; Martins 1994) or more probably 1696 (Dandy 1958; Exell 1962; Mendonça 1962; Figueiredo et al. 2008), and were made by Mason in the Luanda region, and by John Kirckwood in Cabinda. These reached Hans Sloane whose plant and insect collections formed the core of the British Museum (now the Natural History Museum), London, via James Petiver who encouraged surgeons on English ships to send him natural history collections from their overseas travels. Other Pre-Linnean collections from Angola in the Sloane Herbarium were made by Gladman and William Browne (Fig. 5.1). The earliest known Portuguese collector was the naturalist Joaquim José da Silva, who collected along the Angolan coast and the western escarpment between 1783 and 1804. This material was taken from Lisbon to Paris, where it now resides, in 1808 during the Napoleonic Peninsula War (Mendonça 1962; Figueiredo et al. 2008).

Mendonça (1962) presents a historical account of plant collectors in Angola, giving helpful insights to the itineraries of a number of early expeditions. A more complete list of collectors is given by Figueiredo et al. (2008), which volume also includes a useful listing of references relevant to the study of the flora of Angola.

Fig. 5.1 One of the earliest herbarium specimens collected in Angola, in 1706 or 1707, by W Browne and now housed in the Sloane Herbarium at the Natural History Museum in London. The New World starch crop – cassava *Manihot esculenta* (Euphorbiaceae)

Most eighteenth and early nineteenth century explorers visited only coastal regions of Angola, but by the 1850s, botanists and explorers were starting to document plants from more elevated parts of the interior. Friedrich Welwitsch, who spent 6 years in Angola, amassed over 8000 collections of plants representing around 5000 species, of which around 1000 were new to science (Albuquerque 2008; Albuquerque et al. 2009; Albuquerque and Figueirôa 2018). He spent his first year in Angola in the coastal zone between the mouth of the Rio Sembo ('Quizembo') just north of Ambriz, and the mouth of the Cuanza. In September 1854 he embarked on a three-year excursion, initially following the Bengo River and reaching Golungo Alto (Cuanza-Norte). He based himself eventually at Sange from where he made excursions to Ndalatando ('Cazengo') and the banks of the Luinha. In October 1856 he arrived at Pungo Andongo (Malange) where he was based for the next eight months, making collections from Pedras Negras, Pedras de Guinga and localities along the Cuanza River – the furthest point he reached upstream was Quissonde, south of Malange. After an extended period back in Luanda, he headed south via Benguela to Namibe ('Little Fish Bay') in June 1859, gradually extending his journeys along the coast to Cabo Negro, the port of Pinda (probably Tômbua) and Baía dos Tigres. In October 1859 he headed inland from Namibe, following the Rio Giraul ('Maiombo river') to Bumbo on the slopes of the Serra da Chela. He was based at Lopollo on the Huíla plateau until 1860. In 1866, José Anchieta moved to Angola and was based at Caconda on the Huíla plateau. And by the 1880s, missionaries such as José Maria Antunes and Eugène Dekindt, and collectors such as Francisco Newton and Henry Johnston were also making significant collections from this region.

Three nineteenth century German expeditions to the Congo travelled through Angola – Pechuël-Lösche's 1873 Loango Expedition with Paul Güssfeldt and Hermann Soyaux started from Cabinda; Pogge, Buchner and Wissmann's Cassai Expedition made collections from Malange and the Lundas (Mona Quimbundo, Saurimo, Cuango River) in 1876; while Teucsz and Mechow's Cuango Expedition made collections from Dondo (Cuanza-Norte), Pungo Andongo and Malange (Malange), and the Cuango river (Uíge) in 1879–1881. A fourth German expedition, the Kunene-Sambesi Expedition, left Namibe on 11 August 1899 and travelled east, through present-day Cunene and Cuando Cubango provinces, reaching the Cuando River in March 1900 before returning to Namibe in June of that year. Over 1000 collections were made on this expedition by the botanist Hugo Baum (Warburg 1903; Figueiredo et al. 2009b).

The first half of the twentieth century was dominated by the efforts of Kew-trained Swiss botanist John Gossweiler who in the course of 50 years' work collected in all of Angola's provinces, and amassed over 14,000 collections. His final 2 years' collections, in 1946 and 1948, were from the remote northeast of the country, and formed the basis of Cavaco's Flora of Lunda (Cavaco 1959). Other significant colonial era collectors included Portuguese and British participants of the *Missões Botânicas* such as Luiz Carrisso, Francisco Mendonça, Arthur Exell and Francisco de Sousa (as well as John Gossweiler), whose work formed the basis of early parts of the *Conspectus Florae Angolensis*, and the first vegetation map of Angola

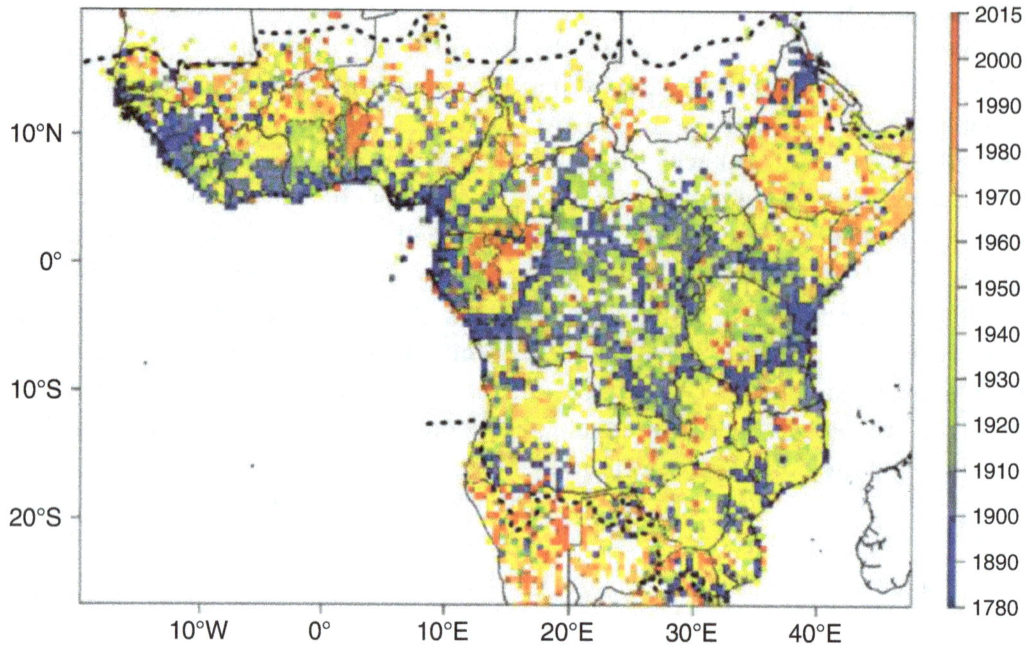

Fig. 5.2 Time lapse of botanical collecting history across tropical Africa. The map represents the date of the first botanical collection made within each 0.5° sampling unit. Dashed lines represent the limits for tropical Africa as defined by Sosef et al. (Used with permission from Sosef et al. 2017: http://rainbio.cesab.org)

(Gossweiler and Mendonça 1939). There are too many other collectors from 1950–1975 to list (see Figueiredo et al. 2008), but two specialist collections are here noted – Hans Hess's aquatic and wetland plants from many of the rivers of western Angola in 1950–1952 are now housed principally in Zurich, and Larry Leach and IC Cannell who travelled up the arid and semi-arid coastal plain between 1967 and 1973, focussed mostly on the succulent flora. After Angolan independence in 1975 and the commencement of the long-running civil war, collection of plants essentially ground to a halt until the end of the twentieth century. Several recent collecting programmes will be described in a later section of this paper. Despite Gossweiler and his successor Brito Teixeira's efforts to survey little known regions of Angola, plant collection coverage and intensity is skewed heavily to the western half of the country, and large parts of Moxico, Cuando Cubango, the Lundas and Uíge are still devoid of collections (Sosef et al. 2017: Fig. 5.2; http://rainbio.cesab.org).

Floristic Diversity and Endemism

Under the leadership of Estrela Figueiredo and Gideon Smith, thirty-two authors from around the world compiled the first checklist of vascular plants for Angola (Figueiredo and Smith 2008; Smith and Figueiredo 2017). A total of 6735 native species were recorded with an additional 226 non-native species. The exotic flora of

Angola was documented by Gossweiler (1948, 1949, 1950). Four of these alien species pose particular threats as they are highly invasive in Angola (Rejmánek et al. 2017). Forty-four additional species have been described or entered onto the International Plant Names Index since publication of Figueiredo and Smith (2008), and inventories in Lunda-Norte (see below) and elsewhere added a further 70 or so species to the Angolan list. So the current estimate of the vascular plants native to Angola is around 6850 species. Current accepted nomenclature for plants can be checked on the African Plants Database (2018), and local plant names in Gossweiler (1953) and Figueiredo and Smith (2012).

Figueiredo et al. (2009a) reported that 997 species (14.8%) are endemic to the country. This percentage is considerably lower than the estimate of 27.3% by Exell and Gonçalves (1973) based on a limited sample of the flora, or studies of individual families of plants where 19% of both Rubiaceae (Figueiredo 2008) and legume species (Soares et al. 2009) were recorded as endemics. Several genera are endemic to Angola, including *Calanda* K.Schum. and *Ganguelia* Robbr. (Rubiaceae); *Carrissoa* Baker f. (Leguminosae); and *Angolaea* Wedd. (Podostemaceae) – the latter now possibly extinct as it was described from the Cambambe rapids on the now heavily dammed Cuanza River.

Legumes (934 spp.), grasses (526 spp.), Compositae (463 spp.) and Rubiaceae (444 spp.) are the most diverse families in the flora, and *Crotalaria* L. and *Euphorbia* L. each have more than 40 Angolan endemic species.

Two of the six tropical African centres of endemism identified by Linder (2001) fall partially or entirely within Angola. A recent analysis of RAINBIO data (Droissart et al. 2018) identifies the western Angolan highlands as a distinct floristic bioregion, although the limited data preclude statements on the remainder of the country. The Huíla plateau consistently stands out as being rich in endemic species (Exell and Gonçalves 1973, Brenan 1978: 472, Linder 2001) and Soares et al. (2009) record 83 endemic legumes from the province. For Rubiaceae, Cabinda has the highest level of diversity with 175 species, but Huíla possesses the most endemics (Figueiredo 2008). Figueiredo (2008) also demonstrates that for Rubiaceae, Huíla is the most intensively collected province. However, our experience is that many of these collections have not necessarily been well studied. Clark et al. (2011) state that the western highlands of Angola comprise the least well-documented stretch of the Great Escarpment of southern Africa.

The western margin of the Huíla Plateau reaches its highest elevation along the Lubango Escarpment of the Serra da Chela and runs in a southwesterly direction from near Tundavala c. 15 km NW of Lubango to Bimbe c. 20 km NW of Humpata. It reaches a height of just over 2200 m and Goyder et al. (in prep.) estimate around 200 species are endemic to this area. However, as other mountains further to the north are surveyed botanically, some of these supposed local endemics may prove to be more widely distributed than currently thought.

Linder's (2001) second area of high species diversity and endemism, the Zambezi-Congo watershed, encompasses eastern Angola, northern Zambia and the Katanga region of the DR Congo. This area has not been well documented in Angola.

Biogeography, Regional Centres of Endemism and Vegetation

With its extremes of landform, climate and rainfall, Angola is host to six of White's (1983) phytochoria, or regional centres of endemism.

Outliers of the Guineo-Congolian forests in Cabinda, Uíge and Cuanza-Norte are progressively smaller in area to the south, ending in the isolated coffee forests of Gabela and Cumbira in Cuanza-Sul. The northward-draining tributaries of the Cuango and Cassai rivers in Uíge and Lunda-Norte have fingers of pure Congolian forest along them. However, much of northern Angola forms a transition zone between Guineo-Congolian vegetation and Zambezian – the latter covers the rest of the country with the exception of the fragmented Afromontane centre of endemism at higher elevations, and the more arid Karoo-Namib and Kalahari-Highveld zones in the southwest.

Geologically, the eastern half of Angola is notable for its deep deposits of Kalahari sand, while to the west crystalline rocks predominate. Marine sediments and recent sands cover the coastal plain (Huntley and Matos 1994; Huntley 2019). The coastal plain is arid in the south due to the cold, upwelling Benguela current, and semi-arid further to the north. Most of the rainfall occurs on the escarpment and the plateau, again with a steady increase to the north. Central Angolan headwaters of major river systems drain into the Okavango (Cuito and Cubango), the Indian Ocean (Cuando, Lungué Bungo and Zambezi) and the Atlantic (Cassai, Cuango, Cuanza and Cunene).

The standard work for vegetation is Barbosa's (1970) *Carta Fitogeográfica de Angola* which recognises 32 vegetation types ranging from desert to moist evergreen and swamp forests. Huntley and Matos (1994) present a concise summary. Barbosa's vegetation map built on the painstaking pioneering work of Gossweiler and Mendonça (1939) – a major contribution that reached a wider audience through the extended English summary by Airy-Shaw (1947).

Angola has a diverse seaweed flora and 169 species have been recorded (Lawson et al. 1975; Anderson et al. 2012). Biogeographically, Angola's marine algae group with those of tropical West Africa, but with a well-developed southern element from around 13°S comprising mainly cooler-water species from the Benguela Marine Province of Namibia and western South Africa.

Recent Botanical Survey Initiatives

In 1968, Angola had only three National Parks (Quiçama, Cameia and Iona) and two Nature reserves (Mupa and Luando), plus a number of forest and game reserves (Teixeira 1968a). Between 1971 and 1975 a programme of field surveys was undertaken to identify areas of high importance for biodiversity conservation (Huntley 1973, 1974; Huntley and Matos 1994). These were supplemented by fieldwork in Huíla, Namibe, Cuanza-Sul and Huambo (Huntley 2009; Mills et al.

2011), and synthesised into an 'Angolan Protected Area Expansion Strategy – APAES' (Huntley 2010). The APAES report was submitted to the Angolan Ministry of Environment in 2010, and formed the basis for the proposals approved by the Angolan *Conselho do Ministros* on 28th April 2011 (GoA 2011).

Much of the recent botanical activity in Angola has focused on the eleven areas highlighted in this conservation planning document. The areas proposed for protection were: Maiombe (Cabinda), Serra do Pingano (Uíge), Lagoa Carumbo (Lunda-Norte), Serra Mbango (Malange), Gabela and Cumbira Forests (Cuanza-Sul), Morro Namba (Cuanza-Sul), Morro Moco (Huambo) Serra da Neve (Namibe), Serra da Chela (Huíla) and Luiana (Cuando Cubango). A listing of post-Independence botanical collectors in Angola is given in Appendix, following the format used for earlier collectors used by Figueiredo et al. (2008).

A collaborative Rapid Biodiversity Assessment and training expedition to the Huíla Plateau and to Iona National Park, with 30 scientific participants from 10 countries and with 15 Angolan students, was convened in 2009. Over 2700 botanical collections were made and deposited in the National Herbarium, Pretoria with duplicates deposited in the ISCED-Huíla Herbarium in Lubango (Huntley 2009).

In northern Angola botanical surveys have been initiated in the moist coffee forests of Serra do Pingano, and more widely in Uíge Province, by a team from Dresden in cooperation with the Universidade Kimpa Vita (Lautenschläger and Neinhuis 2014; Neinhuis and Lautenschläger 2014). These have resulted in a revised list of bryophytes for Angola (Müller 2014; Müller et al. 2018), the description of new species of vascular plant (Abrahamczyk et al. 2016), and ethnobotanical assessments (Göhre et al. 2016; Mawunu et al. 2016; Heinze et al. 2017; Lautenschläger et al. 2018). In total, about 820 species were identified; several of these are new records for Angola.

Lagoa Carumbo and the Luxico, Luele and Lovua valleys were surveyed by a team from Kew, the Ministry of the Environment and Agostinho Neto University, Luanda in 2011, and again in 2013, trebling the known flora of Lunda-Norte as compared to Cavaco (1959) – the combined report documents 752 taxa including 72 additions to the flora of Angola, and 22 potential new species (Darbyshire et al. 2014; Cheek et al. 2015). This part of Lunda-Norte has Congolian swamp forest in the river valleys, moist miombo woodland on the slopes, and Zambezian savanna grasslands on the plateau.

The isolated patch of Guineo-Congolian forest at Cumbira was the subject of a rapid botanical assessment with more than a hundred botanical specimens collected, including new Guineo-Congolian records for Angola and species potentially new to science (Gonçalves and Goyder 2016).

Plants collected from Mount Namba are currently being studied by the Kew/ Lubango team – this work may inform studies on the Lubango Escarpment further to the south. Both share a mosaic of Afromontane forest, grassland and miombo woodland habitats, although most of Lubango's woody vegetation is now heavily degraded. Comparisons with the much better preserved vegetation on Mount Namba might inform habitat restoration initiatives in the area.

Serra da Neve and Serra da Chela were visited briefly in 2013 as part of a wider floristic survey of the Angolan Escarpment led by a team from Rhodes University in South Africa, ISCED-Huíla in Lubango, and Kew. One or two new species have been published from these collections (Hind and Goyder 2014), but wider analysis of the flora is still on-going. Through the German-funded Southern African Science Service Centre for Climate Change and Adaptive Land Management (SASSCAL) project, researchers at the Lubango Herbarium are working on vegetation classification of the woodlands of Huíla Province, towards a new vegetation map for the region (Chisingui et al. 2018). A checklist of the Huíla flora is one of the expected early outputs.

In addition to the Protected Areas Expansion Strategy sites mentioned above, three cross-border initiatives have focused on the catchment of the Okavango system in Angola, Namibia and Botswana in recent years. Botswana's flagship wetland ecosystem – the Okavango Delta – is dependent entirely on the two main Angolan tributaries (Cuito and Cubango) for its hydrology. The Southern Africa Regional Environmental Program (SAREP) and OKACOM organised fieldwork in Cuando Cubango in 2013 with botanists from Kew and the University of Botswana. About 350 collections were made from the southeast corner of Angola, as far east as the Cuando river, thus contributing to the documentation of the Luiana proposed protected area. The Future Okavango (TFO) project led by a research team from Hamburg focused on two research sites in Angola (Cusseque, Bié Province; Caiundo, Cuando Cubango Province) both in the more westerly Cubango catchment, one in Namibia (Mashare), and Seronga in Botswana. This project contributed significantly to a better understanding of Angolan miombo and *Baikiaea-Burkea* woodlands in terms of recovery following disturbance caused by shifting cultivation (Wallenfang et al. 2015, Gonçalves et al. 2018, Gonçalves et al. 2017). A checklist of woody species and geoxylic suffrutices in the grasslands of south-central Angola was provided, documenting potential new species and new records for the country (Gonçalves et al. 2016; Revermann et al. 2017, 2018). Further vegetation and ecological studies are published in Oldeland et al. (2013).

The easterly Cuito and Cuanavale catchment has been the focus of the National Geographic Okavango Wilderness Project from 2015 onwards. Surveys in the upper Cubango were initiated in 2017. To date, over 1300 plant collections have been made by a Kew, South African and Angolan team, who have recorded 417 species of vascular plant from the high-rainfall upper Cuito and Cuanavale drainage system, and 176 from the lower rainfall zones further south (e.g., Fig. 5.3). Over 100 new provincial records were reported for Moxico, with a further 24 for Cuando Cubango, underlining how poorly documented and understood this vast and sparsely inhabited part of Angola is, even now (Goyder et al. 2018). Baseline botanical collection data such as these feed into wider biodiversity assessments of the area and provide vital evidence in building a case to protect the headwaters of not only the Okavango system, but other major river systems originating in central Angola (NGOWP 2018).

Fig. 5.3 Some plants collected during recent fieldwork in central and eastern Angola as part of the National Geographic Okavango Wilderness Project. Top to bottom, left to right: *Protea poggei* subsp. *haemantha* (Proteaceae); *Clerodendrum baumii* (Lamiaceae); *Erythrina baumii* (Leguminosae); *Monotes gossweileri* (Dipterocarpaceae); *Gloriosa sessiliflora* (Colchicaceae); *Raphionacme michelii* (Apocynaceae). All photos: David Goyder

Future Botanical Work

Almost every botanical survey made in recent years in Angola has revealed undescribed species and new country or provincial records. Eastern and northern provinces are in most need of collecting programmes and botanical documentation. Most national parks lack basic botanical inventories. To give one example, Teixeira's (1968b) work on plant diversity in Bicuar National Park (Huíla Province) resulted in the recognition of six vegetation types in the park. But recent SASSCAL-funded surveys revealed species unaccounted for by Figueiredo and Smith (2008), underlining the need for more botanical surveys in both existing and newly proposed areas of conservation concern.

Analysis of the collections from recent surveys is starting to reveal little-documented areas of endemism. The Lubango Escarpment is one obvious focus, but so too is the highly leached high-rainfall Kalahari sand system of Moxico Province and adjacent area that has its own peculiar and little-understood flora.

Only 399 species of vascular plant in Angola have been formally assessed for extinction risk through the IUCN Red List system (IUCN 2018), and a mere 36 of these appear in threatened categories. None of the genera listed in an earlier section of this paper as Angolan endemics have been assessed. Much work is needed in this area.

Four Angolan institutions are listed in Index Herbariorum (Thiers, continually updated), LUAI (ex-*Centro Nacional de Investigação Cientifica* (CNIC), Luanda), LUA (*Instituto de Investigação Agronómica*, Huambo), LUBA (*Instituto Superior de Ciencias da Educação*, Lubango), and DIA (*Museu do Dundo*). While the Dundo Museum has been refurbished and reopened to the public in 2012, it appears that the herbarium collections formerly housed there no longer exist. The LUA herbarium contains 40,000 collections. It was evacuated to Luanda in 1995, and has now returned to Huambo, but is in poor condition and funds are needed to employ well-trained young staff to conserve, rehabilitate and work on this important collection. LUAI contains 35,000 collections and LUBA around 50,000. There are ongoing digitisation programmes at both institutions that will make these collections more widely accessible.

Outside of Angola, Portuguese institutes in Coimbra (COI) and Lisbon (LISC, LISU) hold the largest collections of Angolan plants, an estimated 90,000 collections (Figueiredo and César 2008). 8700 of Gossweiler's Angolan collections are housed at COI and these are available online. The collections at LISC are also available digitally, and are now being incorporated into the Lisbon University herbarium LISU. Most other herbaria with significant Angolan holdings have only digitised their type collections, although mass digitisation of entire national collections has made material in the Paris Natural History Museum (P) and Leiden's Naturalis (L, WAG, U) accessible. In the UK, the Natural History Museum (BM) and Royal Botanic Gardens, Kew (K) in London – both of which contain significant Angolan holdings, and Royal Botanic Gardens, Edinburgh (E) have plans to follow suit. In Germany, the collection of Technische Universität Dresden (DR) comprises 2400 specimens, kept separately from the main herbarium. The Future Okavango project has augmented Hamburg's (HBG) Angolan collections by around 2000 numbers.

Once these combined resources are available online, georeferencing the Angolan material should be a priority. Such collections data could then be used in a variety of projects or programmes. Georeferenced specimen data underpins IUCN conservation assessments, for example, and these in turn inform Important Plant Area designations (Darbyshire et al. 2017) and other forms of conservation planning.

Acknowledgements We are delighted to acknowledge the support of the former Minister of Environment, Dr. Fátima Jardim, and the present Minister, Dr. Paula Francisco, at the Ministério do Ambiente, Luanda, in our attempts to provide the botanical evidence for the conservation of Angola's unique flora. We are grateful to Thea Lautenschläger for providing biographical and other information relating to projects in Uíge Province.

Appendix

Post-Independence collectors in Angola. Entries follow a format developed from Figueiredo et al. (2008).

Surname, first names (birth–death); **C:** period when collecting in Angola; **H:** herbaria [abbreviations after Thiers, continuously updated; FC-UAN = Faculdade de Ciências, Universidade Agostinho Neto, Luanda; INBAC = Instituto Nacional da Biodiversidade e Áreas de Conservação of the Ministério do Ambiente, Luanda]; **L:** provinces abbreviated after Figueiredo and Smith 2008: principal localities; **B:** biographical information.

Alcochete, António (1963–)
C: 1991; **H:** K; **L:** CU HI NA; **B:** Angolan botanist, collected with Gerrard, Matos and Newman.

Baragwanath, S.
C: 1994. **H:** PRE.

Barker, Nigel P.(1962-)
C: 2013, 2015, 2017; **H:** GRA, INBAC, K, LUBA, PRE; **L:** CC HI NA: Lubango Escarpment, Mt. Tchivira, Serra da Neve, Mundondo Plateau, Okavango, Cuito and Longa Rivers; **B:** South African Professor of Plant Science at University of Pretoria, formerly at Rhodes University.

Bester, Stoffel Petrus (Pieter) (1969–)
C: 2009, 2015; **H:** GRA, INBAC, K, LUBA, PRE; **L:** CC CU HI NA: Iona, Lubango Escarpment, Bicuar, Okavango, Cuito and Longa Rivers; **B:** South African botanist based at PRE.

Bruyns, Peter Vincent (1957–)
C: 2006, 2007; **H:** BOL, E, K, NBG, PRE; **L:** BE HI NA: Lubango Escarpment and coastal plain; **B:** South African mathematician and botanist with particular interest in succulent plants.

Cardoso, João Francisco (1974–)
C: 2005, 2006; **H:** LISC, LUAI; **L:** HI NA: Serra da Leba, Virei, Caraculo, Cainde; **B:** Agronomist with Agostinho Neto University.

Cheek, Martin Roy (1960–)
C: 2012; **H:** K; **L:** CA; **B:** British botanist at Royal Botanic Gardens Kew, specialist on West African flora.

Clark, Vincent Ralph (1977–)
C: 2013; **H:** GRA, K, LUBA, PRE; **L:** HI NA: Lubango Escarpment, Mt. Tchivira, Serra da Neve; **B:** South African botanist.

Cooper, C.E.
C: 1997; **H:** PRE.

Crawford, Frances Mary (1981–)
C: 2009, 2011; **H:** INBAC, K, PRE; **L:** HI LN NA: Lucapa, Lagoa Carumbo, Iona, Lubango Escarpment; **B:** British botanist, Curator of WIND herbarium, formerly at Royal Botanic Gardens, Kew; collected with Darbyshire and Goyder in LN.

Daniel, José Maria (1943–2015)
C: 1964–2008; **H:** LUBA, LUA, LUAI, L: Collected in all Angolan Provinces; **B:** Angolan botanist at Lubango Herbarium until his retirement; collected with Huntley, Matos and Gonçalves.

Darbyshire, Iain Andrew (1976–)
C: 2011, 2013; **H:** INBAC, K, LISC; **L:** LN: Lucapa, Lagoa Carumbo; **B:** British botanist at Royal Botanic Gardens, Kew; collected with Crawford, Gomes, Goyder & Kodo.

Dexter, Kyle Graham (1980–)
C: 2017–; **H:** E, COLO, LUBA, WIND; **L:** CU HI NA; **B:** Senior Lecturer at University of Edinburgh and Associate Researcher at Royal Botanic Garden Edinburgh.

Ditsch, Barbara (1961–)
C: 2013, 2015 **H:** DR, LUA; **L:** UI: Serra do Pingano, Municipality of Uíge, Kimbele, Damba, Mucaba; **B:** German botanist at Dresden Botanic Garden.

Finckh, Manfred (1963–)
C: 2011–; **H:** HGB, LUBA, WIND; BI CC HA HI MO: Chitembo (Cusseque), Caiundo, Cachingues, Savate, Cuangar, Bicuar National Park, Cameia National Park, Tundavala Observatory under TFO and SASSCAL Projects; **B:** Ecologist at University of Hamburg, Germany.

Francisco, Domingos Mumbundu (1974–)
C: 2008–; **H:** LISC, LUAI, LUBA; **L:** CA CC LA MA NA ZA: Barra do Cuanza, Iona, Cangandala, Quiçama National Parks; **B:** Angolan botanist at Universidade Agostinho Neto, Centro de Botânica, LUAI Herbarium.

Frisby, Arnold.
C: 2016, 2017; **H:** INBAC, K, LUBA, PRE; **L:** BI CC: Cubango and Cuito Rivers; **B:** South African botanist at University of Pretoria.

Gerrard, Jacqueline
C: 1991; **H:** K; **L:** CU HI NA.

Godinho, Elizeth
C: 2013; **H:** INBAC, K, LISC; **L:** LN: Lagoa Carumbo; **B:** Angolan botanist at INBAC; collected with Darbyshire, Goyder and Kodo.

Göhre, Anne (1990–).
C: 2014–2016; **H:** B, BR, BONN, P; **L:** UI: Municipality of Uíge, Kimbele, Damba, Mucaba; **B:** German botanist at Dresden Botanic Garden.

Gomes, Amândio Luís (1971–).
C: 2010–; **H:** FC-UAN, INBAC, K, LISC, LUAI, LUBA; **L:** BE BI BO CC CN CS HA LN ZA: Lucapa, Lagoa Carumbo, Chitembo (Cusseque), Tundavala Observatory under TFO and SASSCAL Projects; **B:** Angolan botanist at Universidade Agostinho Neto, Luanda; collected with Crawford, Darbyshire and Goyder in LN.

Gonçalves, Francisco Maiato Pedro (1982–).
C: 2008–; **H:** HBG, INBAC, K, LUBA; **L:** BI CC CU CS HA HI LA NA MO: Chitembo (Cusseque), Cumbira forest, Mt. Namba, Lubango Escarpment, Okavango headwaters, Huíla Province SASSCAL Project; **B:** Angolan botanist at Lubango Herbarium, ISCED Huíla, Lubango.

Goyder, David John (1959–)
C: 2011–; **H:** GRA, INBAC, K, LUBA, PRE; **L:** BI CC CS HI LN MO NA: Cumbira, Mt. Namba, Serra da Neve, Lubango Escarpment, Mt. Tchivira, Okavango headwaters, Lucapa, Lagoa Carumbo; **B:** British botanist at Royal Botanic Gardens, Kew; collected with Crawford, Darbyshire, Godinho, Gomes and Kodo in LN, with Barker and Clark on the western escarpment, with Gonçalves in CS and Okavango headwaters, with Barker, Bester, Frisby and Janks in CC.

Harris, Timothy (1982–)
C: 2013; **H:** K, LUAI, PSUB, WIND; **L:** CC: Okavango, Cuito and Cuando Rivers; **B:** British botanist; collected with Murray-Hudson.

Heinze, Christin (1993–).
C: 2014–2017; **H:** DR, LUA; **L:** CN: all municipalities; **B:** German botanist at Technische Universität Dresden.

Janks, Matthew.
C: 2015; **H:** GRA, INBAC, LUBA, PRE; **L:** CC: Okavango, Cuito and Longa Rivers; **B:** South African botanist; collected with Barker, Bester & Goyder.

Jürgens, Norbert (1953–)
C: 2008–; **H:** HGB, WIND, LUBA; **L:** CU HI NA; **B:** Professor at Institute for Plant Science and Microbiology, University of Hamburg, Germany.

Kodo, Felipe
C: 2013; **H:** INBAC, K, LISC; **L:** LN: Lagoa Carumbo; **B:** Angolan botanist at INBAC; collected with Darbyshire, Godinho and Goyder.

Lautenschläger, Thea (1980–)
C: 2012–2018; **H:** DR, LUA; **L:** UI: Municipality of Uíge, Mucaba, Maquela do Zombo, Quitexe, Milunga, Sanza Pombo, Kimbele, Ambuila, Songo, Bungo, Bembe, Puri, Negage, Altocauale, Damba; **B:** German botanist at Technische Universität Dresden.

Luís, José Camôngua (1984–)
C: 2015–. **H:** K, LUBA; **L:** CS HI: Lubango Escarpment, Mt. Namba; **B:** Angolan botanist.

Maiato, Francisco
See **Gonçalves, Francisco Maiato Pedro**.

Manning, Stephen D.
C: 1986–1998.

Matos, Elizabeth (Liz), Merle (1938–)
C: 1975–; **B:** British botanist, founder and director of Angola's National Plant Genetic Resources Centre, Agostinho Neto University, Luanda. Retired in 2008.

Mawunu, Monizi (1973–)
C: 2013–2018; **H:** DR, LUA; **L:** UI, whole province; **B:** Angolan botanist at Universidade Kimpa Vita.

Müller, Frank (1966–)
C: 2015; **H:** DR, LUA; **L:** UI: Municipality of Uíge, Songo, Mucaba; **B:** German botanist at Technische Universität Dresden.

Murray-Hudson, Frances
C: 2013; **H:** K, LUAI, PSUB, WIND; **L:** CC: Okavango, Cuito and Cuando Rivers; **B:** Volunteer at Peter Smith University of Botswana Herbarium (PSUB); collected with Harris.

Neinhuis, Christoph (1962–)
C: 2012–2018; **H:** DR, LUA; **L:** UI: Municipality of Uíge, Mucaba, Maquela do Zombo, Quitexe, Milunga, Sanza Pombo; **B:** German botanist at Technische Universität Dresden, director of the Botanical Garden TU Dresden.

Newman, Mark Fleming (1959–)
C: 1991; **H:** K; **L:** CU HI NA; **B:** British botanist at Royal Botanic Garden Edinburgh. In 1991, at the Seed Bank, Royal Botanic Gardens, Kew; collected with Alcochete, Gerrard and Matos, mainly for seeds, with herbarium voucher specimens for identification.

Rejmánek, Marcel (Marek/Marc) (1946–)
C: 2014; **H:** LUBA, STE; **L:** BE BO CN CS HA HI MA NA UI; **B:** Czech botanist based at University of California, Davis, working on biological invasions. Conducted a rapid inventory of invasive plants in Angola in 2014 with Huntley, Roux and Richardson.

Revermann, Rasmus (1979–)
C: 2011–; **H:** HGB, WIND, LUBA; **L:** BI CC HA HI: Chitembo (Cusseque), Caiundo, Cachingues, Savate, Cuangar under TFO and SASSCAL Projects; **B:** Ecologist at University of Hamburg, Germany.

Roux, Jacobus Petrus (Koos) (1954–2013)
C: 2001; **H:** PRE; **B:** South African Pteridophyte specialist.

Tripp, Erin Anne (1979–)
C: 2017–; **H:** COLO, E, LUBA, WIND; **L:** CU HI NA; **B:** Researcher at Colorado Herbarium, University of Colorado.

References

Abrahamczyk S, Janssens S, Xixima L et al (2016) *Impatiens pinganoensis* (Balsaminaceae), a new species from Angola. Phytotaxa 261:240–250

African Plant Database version 3.4.0 (2018) Conservatoire et Jardin Botaniques de la Ville de Genève and South African National Biodiversity Institute, Pretoria. http://www.ville-ge.ch/musinfo/bd/cjb/africa

Airy-Shaw H (1947) The vegetation of Angola. J Ecol 35:23–48

Albuquerque S (2008) Friedrich Welwitsch. Figueiredo E, Smith GF Plants of Angola/Plantas de Angola. *Strelitzia* 22: 2–3

Albuquerque S, Figueirôa S (2018) Depicting the invisible: Welwitsch's map of travellers in Africa. Earth Sci Hist 37(1):109–129

Albuquerque S, Brummitt RK, Figueiredo E (2009) Typification of names based on the Angolan collections of Friedrich Welwitsch. Taxon 58:641–646

Anderson RJ, Bolton JJ, Smit AJ et al (2012) The seaweeds of Angola: the transition between tropical and temperate marine floras on the west coast of southern Africa. Afr J Mar Sci 34:1–13

Barbosa LAG (1970) Carta Fitogeográfica de Angola. Instituto de Investigação Científica de Angola, Luanda

Brenan JPM (1978) Some aspects of the phytogeography of tropical Africa. Ann Mo Bot Gard 65(2):437–478

Cavaco A (1959) Contribution à l'Étude de la Flore de la Lunda d'Après les Récoltes de Gossweiler (1946–1948*)*. Publicações Culturais da Companhia de Diamantes de Angola 42, 230 pp

Cheek M, Lopez Poveda L, Darbyshire I (2015) *Ledermanniella lunda* sp. nov. (Podostemaceae) of Lunda-Norte, Angola. Kew Bull 70:10

Chisingui AV, Gonçalves FMP, Tchamba JJ et al (2018) Vegetation survey of the woodlands of Huíla Province. Biodivers Ecol 6:426–437

Clark VR, Barker NP, Mucina L (2011) The Great Escarpment of southern Africa: a new frontier for biodiversity exploration. Biodivers Conserv 20:2543–2561

Dandy JE (1958) The Sloane Herbarium. An annotated list of the horti sicci composing it; with biographical accounts of the principal contributors. British Museum (Natural History), London

Darbyshire I, Goyder D, Crawford F, et al (2014) Update to the Report on the Rapid Botanical Survey of the Lagoa Carumbo Region, Lunda-Norte Prov., Angola for the Angolan Ministry of the Environment, following further field studies in 2013, incl. Appendix 2: checklist to the flowering plants, gymnosperms and pteridophytes of Lunda-Norte Prov, Angola. Ministério do Ambiente, Luanda

Darbyshire I, Anderson S, Asatryan A et al (2017) Important Plant Areas: revised selection criteria for a global approach to plant conservation. Biodivers Conserv 26:1767–1800

Droissart V, Dauby G, Hardy OJ et al (2018) Beyond trees: biogeographical regionalization of tropical Africa. J Biogeogr 2018:1–15

Exell AW (1939) Notes on the flora of Angola. IV. 1. Collections from Angola in the Sloane Herbarium. J Bot 77:146–147

Exell AW (1962) Pre-Linnean collections in the Sloane Herbarium from Africa south of the Sahara. In: Fernandes A (ed) Comptes Rendus de la IVe Réunion Plénière de l'Association pour l'Étude Taxonomique de la Flore d'Afrique Tropicale (Lisbonne et Coïmbre, 16–23 Septembre, 1960). Junta de Investigações do Ultramar, Lisbon, pp 47–49

Exell AW, Gonçalves ML (1973) A statistical analysis of a sample of the flora of Angola. Garcia de Orta, Série de Botânica 1:105–128

Figueiredo E (2008) The Rubiaceae of Angola. Bot J Linn Soc 156:537–638

Figueiredo E, César J (2008) Herbaria with collections from Angola/Herbários com colecções de Angola. Strelitzia 22:11–12

Figueiredo E, Smith GF (eds) (2008) Plants of Angola/Plantas de Angola. Strelitzia 22:1–279

Figueiredo E, Smith GF (2012) Common names of Angolan plants. Inhlaba Books, Pretoria

Figueiredo E, Matos S, Cardoso JF et al (2008) List of collectors/Lista de colectores. Strelitzia 22:4–11

Figueiredo E, Smith GF, César J (2009a) The flora of Angola: first record of diversity and endemism. Taxon 58:233–236

Figueiredo E, Soares M, Siebert G et al (2009b) The botany of the Cunene-Zambezi Expedition with notes on Hugo Baum (1867-1950). Bothalia 39:185–211

GoA (Government of Angola) (2011) Plano Estratégico da Rede Nacional de Áreas de Conservação de Angola. Direcção Nacional da Biodiversidade, Ministério do Ambiente, Luanda, 35 pp

Göhre A, Toto-Nienguesse AB, Futuro M et al (2016) Plants from disturbed savannah vegetation and their usage by Bakongo tribes in Uíge, Northern Angola. J Ethnobiol Ethnomed 12:42

Gonçalves FMP, Goyder DJ (2016) A brief botanical survey into Kumbira forest, an isolated patch of Guineo-Congolian biome. PhytoKeys 65:1–14

Gonçalves FMP, Tchamba JJ, Goyder DJ (2016) *Schistostephium crataegifolium* (Compositae: Anthemideae), a new generic record for Angola. Bothalia 46:a2029

Gonçalves FMP, Revermann R, Gomes AL, et al (2017) Tree species diversity and composition of Miombo woodlands in south-central Angola, a chronosequence of forest recovery after shifting cultivation. Int J For Res 2017(Article ID 6202093), 13 pp

Gonçalves FMP, Revermann R, Cachissapa MJ, et al (2018) Species diversity, population structure and regeneration of woody species in fallows and mature stands of tropical woodlands of SE Angola. J Forest Res. Published online 13 January 2018

Gossweiler J (1948) Flora exótica de Angola. Nomes vulgares e origem das plantas cultivadas ou sub-espontâneas. Agronomia Angolana 1:121–198

Gossweiler J (1949) Flora exótica de Angola. Nomes vulgares e origem das plantas cultivadas ou sub-espontâneas. Agronomia Angolana 2:173–255

Gossweiler J (1950) Flora exótica de Angola. Nomes vulgares e origem das plantas cultivadas ou sub-espontâneas. Agronomia Angolana 3:143–167

Gossweiler J (1953) Nomes indígenas das plantas de Angola. Agronomia Angolana 7:1–587

Gossweiler J, Mendonça FA (1939) Carta Fitogeográfica de Angola. Ministério das Colónias, Lisboa, 242 pp

Goyder DJ, Barker N, Bester SP et al (2018) The Cuito catchment of the Okavango system: a vascular plant checklist for the Angolan headwaters. PhytoKeys 113:1–31. https://doi.org/10.3897/phytokeys.113.30439

Heinze C, Ditsch B, Congo MF et al (2017) First Ethnobotanical Analysis of Useful Plants in Cuanza Norte North Angola. Res Rev J Bot Sci 6:44

Hind DJN, Goyder DJ (2014) *Stomatanthes tundavalaensis* (Compositae: Eupatorieae: Eupatoriinae), a new species from Huíla Province, Angola, and a synopsis of the African species of *Stomatanthes*. Kew Bull 69(9545):1–9

Huntley BJ (1973) Proposals for the creation of a strict nature reserve in the Maiombe forest of Cabinda. Report 16. Repartição Técnica da Fauna, Serviços de Veterinária, Luanda, Mimeograph report, 10 pp

Huntley BJ (1974) Ecosystem conservation priorities in Angola. Report 28. Repartição Técnica da Fauna, Serviços de Veterinária, Luanda, Mimeograph report, 22 pp

Huntley BJ (2009) SANBI/ISCED/UAN Angolan biodiversity assessment capacity building project. Report on Pilot Project. Unpublished Report to Ministry of Environment, Luanda, 97 pp, 27 figures

Huntley BJ (2010) Estratégia de Expansão de Rede da Áreas Protegidas da Angola/Proposals for an Angolan Protected Area Expansion Strategy (APAES). Unpublished Report to the Ministry of Environment, Luanda, 28 pp, map

Huntley BJ (2019) Angola in Outline: Physiography, Climate and Patterns of Biodiversity. In: Huntley BJ, Russo V, Lages F, Ferrand N (eds) Biodiversity of Angola. Science & conservation: a modern synthesis. Springer Nature, Cham

Huntley BJ, Matos EM (1994) Botanical diversity and its conservation in Angola. Strelitzia 1:53–74

IUCN (2018) The IUCN Red List of Threatened Species. Ver. 2017-3. http://www.iucnredlist.org. Downloaded on 27 March 2018

Lautenschläger T, Neinhuis C (eds) (2014) Riquezas Naturais de Uíge – uma Breve Introdução Sobre o Estado Atual, a Utilização, a Ameaça e a Preservação da Biodiversidade. Technische Universität Dresden, Dresden

Lautenschläger T, Monizi M, Pedro M et al (2018) First large-scale ethnobotanical survey in the province Uíge, northern Angola. J Ethnobiol Ethnomed 14:51

Lawson GW, John DM, Price JH (1975) The marine algal flora of Angola: its distribution and affinities. Bot J Linn Soc 70:307–324

Linder HP (2001) Plant diversity and endemism in sub-Saharan tropical Africa. J Biogeogr 28:169–182

Martins ES (1994) John Gossweiler. Contribuição da sua obra para o conhecimento da flora angolana. Garcia de Orta, Série de Botânica 12:39–68

Mawunu M, Bongo K, Eduardo A et al (2016) Contribution à la connaissance des produits forestiers non ligneux de la Municipalité d'Ambuila (Uíge, Angola): Les plantes sauvages comestibles [Contribution to the knowledge of no-timber forest products of Ambuila Municipality (Uíge, Angola): The wild edible plants]. Int J Innov Sci Res 26:190–204

Mendonça FA (1962) Botanical collectors in Angola. In: Fernandes A (ed) Comptes Rendus de la IVe Réunion Plénière de l'Association pour l'Étude Taxonomique de la Flore d'Afrique Tropicale (Lisbonne et Coïmbre, 16–23 septembre, 1960). Junta de Investigações do Ultramar, Lisbon, pp 111–121

Mills MSL, Olmos F, Melo M et al (2011) Mount Moco: its importance to the conservation of Swierstra's Francolin *Pternistis swierstrai* and the Afromontane avifauna of Angola. Bird Conserv Int 21:119–133

Müller F (2014) About 150 years after Welwitsch – a first more extensive list of new bryophyte records for Angola. Nova Hedwigia 100:487–505

Müller F, Sollman P, Lautenschläger T (2018) A new synonym of *Weissia jamaicensis* (Pottiaceae, Bryophyta) and an extension of the range of the species from the Neotropics to the Palaeotropics. Plant Fungal Syst 63(1):1–5

Neinhuis C, Lautenschläger T (2014) The potentially natural vegetation in Uíge province and its current status – arguments for a protected area in the Serra do Pingano and adjacent areas. Unpublished Report to Ministry of Environment, Luanda, 64 pp

NGOWP (National Geographic Okavango Wilderness Project) (2018) *Initial findings from exploration of the upper catchments of the Cuito, Cuanavale, and Cuando Rivers, May 2015 to* December 2016. Report prepared for and submitted to the Ministério do Ambiente of the Republic of Angola, the Ministry of Environment, Wildlife and Tourism Botswana, and the Ministry of Environment and Tourism of the Republic of Namibia. Available (as Report 1) from: http://www.wildbirdtrust.com/owp-publications/

Oldeland J, Erb C, Finckh M, Jürgens N (eds) (2013) Environmental assessments in the Okavango region. Biodivers Ecol 5:1–418

Rejmánek M, Huntley BJ, le Roux JJ, Richardson DM (2017) A rapid survey of the invasive plant species in western Angola. Afr J Ecol 55:56–69

Revermann R, Gonçalves FM, Gomes AL et al (2017) Woody species of the miombo woodlands and geoxylic grasslands of the Cusseque area, south-central Angola. Check List 13:2030

Revermann R, Oldenland J, Gonçalves FM et al (2018) Dry tropical forests and woodlands of the Cubango basin in southern Africa – First classification and assessment of their woody species diversity. Phytocoenologia 48:23–50

Smith GF, Figueiredo E (2017) Determining the residence status of widespread plant species: studies in the flora of Angola. Afr J Ecol 55:710–713

Soares M, Abreu J, Nunes H et al (2009) The Leguminosae of Angola: diversity and endemism. Syst Geogr Plants 77:141–212

Sosef MSM, Dauby G, Blach-Overgaard A et al (2017) Exploring the floristic diversity of tropical Africa. BMC Biol 15:15

Teixeira JB (1968a) Angola. In: Hedberg I, Hedberg O (eds.) Conservation of vegetation in Africa south of the Sahara. Proceedings of a symposium held at the 6th plenary meeting of the "Association pour l'Etude Taxonomique de la Flore d'Afrique Tropicale" (A.E.T.F.A.T.) in Uppsala, Sept. 12th–16th, 1966. *Acta Phytogeographica Suecica* 54:193–197

Teixeira JB (1968b) Parque Nacional do Bicuar. Carta da Vegetação (1ª aproximação) e Memória Descritiva. Instituto de Investigação Agronómica de Angola, Nova Lisboa

Thiers B (continuously updated). Index Herbariorum: A global directory of public herbaria and associated staff. New York Botanical Garden's Virtual Herbarium. http://sweetgum.nybg.org/science/ih/

Wallenfang J, Finckh M, Oldeland J et al (2015) Impact of shifting cultivation on dense tropical woodlands in southeast Angola. Trop Conserv Sci 8:863–892

Warburg O (1903) Kunene-Sambesi-Expedition. Kolonial-Wirtschaftliches Komitee, Berlin

White F (1983) The Vegetation of Africa – A Descriptive Memoir to Accompany the UNESCO/AETFAT/UNSO Vegetation Map of Africa. UNESCO, Paris 356 pp

Chapter 6
Vegetation Survey, Classification and Mapping in Angola

Rasmus Revermann and Manfred Finckh

Abstract Spatial information about plant species composition and the distribution of vegetation types is an essential baseline for natural resource management planning. In Angola, the first countrywide vegetation map was elaborated by Gossweiler in 1939. Subsequently, Barbosa published a revised map with much higher detail in 1970 and his work has remained the main reference for the vegetation of Angola until today. However, these early maps were expert drawn and were not based on systematic surveys. Instead, the delimitation of vegetation units was based on many years of field observations and also incorporated results of local studies carried out by other authors. In spite the rich history of the scientific exploration of Angola's vegetation in colonial times, quantitative and plot based studies were rare. After the end of the armed conflict, new vegetation surveys making use of new methodological developments in numerical approaches to vegetation classification in combination with modern remote sensing imagery have provided spatial information of unprecedented detail. However, vast areas of the country still remain seriously understudied. At the same time, sustainable land management strategies are urgently needed due to the increasing pressure on natural resources driven by socio-economic development and global change, thus calling for a new era of vegetation surveys that will enable data-based landuse and conservation planning in Angola.

Keywords Conservation · Landuse planning · Natural resources · Plant communities · Remote sensing

---

R. Revermann (✉) · M. Finckh
Institute for Plant Science and Microbiology, University of Hamburg, Hamburg, Germany
e-mail: rasmus.revermann@gmail.com; manfred.finckh@uni-hamburg.de

B. J. Huntley et al. (eds.), *Biodiversity of Angola*,
https://doi.org/10.1007/978-3-030-03083-4_6

Introduction

Knowledge on the spatial distribution of vegetation and its species composition is paramount for any kind of natural resource management and conservation planning. Vegetation serves as habitat for other organismic groups and is the source of energy in the ecosystem. As such, vegetation integrates many ecological processes and reflects patterns of topography, geology, soil, hydrology and climate. Thus, vegetation classification is ideal to provide an aggregated image of the landscape and its ecological communities.

Historical Exploration of Vegetation Patterns in Angola

First reports on the vegetation of Angola were directly linked to the floristic exploration of the country, as outlined by Goyder and Gonçalves (2019). Scientific missions during colonial times in Angola served several purposes: on the one hand they should chart the potential for economic exploitation and development while on the other hand they may also have been used to demonstrate the supremacy of the colonial power (Gago et al. 2016). The expedition by the geographer Jessen (1936) provided a first sketch of the vegetation along the routes of his transects through western Angola. Jessen's work remains a classic as he was among the first to document the landscape and ecosystem properties of the region. However, it is hardly read today as it is only available in German.

The systematic descriptions of the vegetation of Angola started with Gossweiler and Mendonça's (1939) phytogeographical map of Angola. The often-cited English summary by Shaw (1947) contributed much to the recognition of Gossweiler's work internationally. The map is based on the combined structural and ecological approach to vegetation classification developed by Brockmann-Jerosch and Rübel (1912) in Zurich. Thus, in a first level of classification they categorised the vegetation according to woodiness and persistence into the three categories Lignosa (woody), Herbosa (herbaceous) and Deserta (land surfaces without permanent vegetation cover). The next step of the classification included climatic and edaphic factors, as well as leaf traits, leading for instance to five sub-categories of woody vegetation called Pluviilignosa, Laurilignosa, Durilignosa, Ericilignosa, Aestililignosa and Hiemilignosa. Stand structure was the main criterion for the next categories, dividing the afore-mentioned categories between tall forests (-silva) and dense but low forests (-fruticeta) (e.g. Pluviisilva vs. Pluviifruticeta or Durisilva vs. Durifruticeta). Below this third level we finally find floristically defined vegetation units, albeit mostly named after one or two dominant species. Similar structural criteria were used for the sub-classification of the Herbosa and Deserta.

The vegetation map used this rather rigid classification scheme for the 19 main mapping units. However, the resulting map apparently did not fully satisfy the authors, who then applied 29 additional symbols to indicate occurrence of small-scale vegetation units, of transition zones and of species that appeared to be of special interest to the authors – a very nice real world example of dealing with rigid mapping manuals. However, neglecting these small methodological inconsistencies, the map by Gossweiler & Mendonça presented the first overall picture of the vegetation of Angola, a first approach towards a systematic compilation of observations of phytogeographical patterns and a first attempt at ecological interpretation. While many of the mapped polygons seem outdated in times of modern earth observation, the number of observed details in remote parts of Angola is still surprising for today's botanists. The authors were probably the first to report for Angola on invasive species, on seed dispersal by bats, on the morphological plasticity of the genus Syzygium and on many other current scientific topics. Also quite astonishing was the classification of the suffrutex grassland within the woody vegetation types (Ericifruticeta), more than 30 years before White (1976) published the groundbreaking paper on the 'Underground forests of Africa'.

The next important integrating step towards a synthesis on the vegetation units of Angola and their spatial distribution was Grandvaux Barbosa's phytogeographical map of Angola (Barbosa 1970). His work can be seen as a continuation and extension of the Gossweiler approach. The map clearly benefited from several regional studies that had been carried out in the meantime (see below) and of course also from Barbosa's own knowledge gained during several field missions throughout the country and his extensive experience of similar vegetation types found in Mozambique. As ancillary information Barbosa included descriptions of the main soil types and climatic zones of Angola.

The mapping approach adopted by Barbosa was to some degree harmonised with the parallel efforts of the Flora Zambesiaca map and the UNESCO initiative mapping the vegetation of Africa. The first level of classification differentiates the vegetation based on the formation, i.e. deals with the physiognomy of the vegetation such as closed forests, forest savanna mosaics, woodlands etc. and beyond that includes azonal edaphic vegetation units such as mangrove stands and coastal dune vegetation. In the second level of classification vegetation types are distinguished according to dominant species. In total, the map by Barbosa displays 32 main vegetation types and the descriptive text accompanying the map provides details on over 100 subordinate types (for a brief summary in English see Barbosa 1971).

The result was a good overview of the main vegetation types of Angola, in terms of spatial patterns much superior to the first attempt by Gossweiler and Mendonça (1939). Until today the vegetation units of the map by Barbosa (1970) constitute the foundation for the Angolan part of most continental or global scale vegetation maps (see below). However, due to Barbosa's floristic rather than ecological emphasis, the report on the vegetation units did neither contribute much to a better understanding of the ecology of the main vegetation patterns, nor did it make use of a modern classification concept based on plant communities.

Shortly after Barbosa's vegetation map, Diniz (1973) published a monograph on the physical properties of the agricultural zones of Angola. Included in this monograph, are soil and vegetation maps for 36 agricultural zones, albeit in rather fragmented components and without an overall map of the country. The vegetation classification scheme he used is not clearly defined, somewhere in between those of Gossweiler and Barbosa, but sometimes with more detail than Barbosa (1970). The main achievement of Diniz (1973) is that he assembled sound environmental information (with a focus on geology and soils) for all delimited agricultural zones. However, due to the lack of a seamless map and his unclear classification approach his contribution to the knowledge on the vegetation of Angola did not receive much attention in the subsequent scientific literature and due to the violent conflicts following Angola's independence in 1975, Barbosa's work has remained the main reference on the vegetation of Angola.

Integration of the Vegetation Map of Angola Within Continental Scale Maps

The next important step for a better understanding of Angolan vegetation was the UNESCO/AETFAT/UNSO initiative for a Vegetation Map of Africa (UNESCO/AETFAT/UNSO 1981), compiled and described by White (1983). For Angola the continental map is largely based on the units supplied by Barbosa (1970) but they were subject to further generalisation resulting in only 14 mapping units compared to Barbosa's 32 vegetation types. However, the important achievement of White's map lies in the fact that it inserted Angolan vegetation in a common conceptual and methodological framework with the vegetation of neighbouring countries and the African continent as a whole. As such, the UNESCO (UNESCO/AETFAT/UNSO 1981)/(UNESCO 1981) map and White's (1983) description established the now widely used term 'miombo woodlands' in our scientific and geographic frameworks, allowing thus for the comparison of Angola's ecosystems with similar vegetation types throughout Africa. Although Barbosa and White provided seamless maps covering the entire country, the level of information supporting the mapping units varies strongly and for some units, especially in the more remote eastern parts of the country, barely any details are given. All these early maps are based on expert knowledge and no quantitative data was involved in the process of map making.

The Vegetation Map of Africa then again was the main baseline for the WWF's approximation of the world's terrestrial ecoregions (Olson et al. 2001) in so far as the African continent was concerned. Although without a presentation of a systematic biogeographical database, the map of the terrestrial ecoregions currently constitutes the most used baseline map for strategic conservation planning on a continental and subcontinental scale (e.g. MacKinnon et al. 2016). The availability of modern remote sensing techniques has allowed the generation of continental or global land cover products, i.e. GlobCover, MODIS/Terra Land Cover, GlobLand30 or the map of African ecosystems by Sayre et al. (2013). However, these maps display structural vegetation types only and provide no floristic information.

Regional and Local Studies on Vegetation Composition

Early plot based studies were conducted by Ilse von Nolde on the Planalto de Quela (von Nolde 1938a, b, c). Since the mid-1950s, several local studies based on missions assessing natural resources were carried out at the regional level in Angola. Monteiro studied the forest resources in Moxico (Monteiro 1957), in the northern Maiombe and Dembos forests (Monteiro 1962, 1965a, b, 1967), and in Bié (Monteiro 1970a) contributing to our knowledge on species composition in the respective forest types. Monteiro's (1970a, b) work in Bié needs to be highlighted as for Angola he implemented new methods in mapping the vegetation. His map of the woody vegetation of the province of Bié is not drawn based on pure observations but is based on quantitative vegetation plot data. He collected data on species composition in 144 vegetation relevés sized 30 m × 30 m that were subject to a vegetation classification based on vegetation tables. The mapping process was guided by aerial photography, quite an advanced approach for its time. Menezes (1965, 1971) undertook phytosociological studies and produced local vegetation maps in pastoral ecosystems of the Cunene Province. Teixeira elaborated vegetation maps for two of the main protected areas of Angola, the Quiçama and Bicuar national parks (Teixeira et al. 1967; Teixeira 1968). A few years later, Huntley produced a much more detailed map of the Quiçama National Park in 1972 at a scale of 1:100000 depicting 28 plant communities (Huntley 1972). Aguiar and Diniz (1972) mapped the vegetation of the western plateau of Cela. Coelho explored the potential of forestry in Cuando Cubango and elaborated a classification of the lower Cubango Basin into 32 forestry zones (Coelho 1964, 1967). Santos (1982) used a transect approach, so called 'itinerários florísticos', in order to generate an expert-drawn vegetation map for Cuando Cubango Province (Fig. 6.1).

Modern Approaches to Vegetation Mapping and Classification

This early period of vegetation mapping and classification was followed by the absence of any such activities for the coming decades due to the long-lasting armed conflict in the country. During this period significant methodological advances were made in vegetation ecology and phytosociology as well as in remote sensing techniques. The advent of computers allowed the development of new methodological tools to semi-automatically classify large amounts of multivariate vegetation plot data based on objective criteria. As such, vegetation classification moved away from the subjective assignments of vegetation types to more formalised data analysis. Similarly, remote sensing imagery became readily available often at no cost and in unprecedented temporal and spatial resolution. Thus, new numerical methods together with modern remote sensing products have the potential to provide a much more detailed and objective picture of vegetation and plant diversity patterns than expert drawn maps and arbitrarily assigned vegetation types of earlier times (Fig. 6.2).

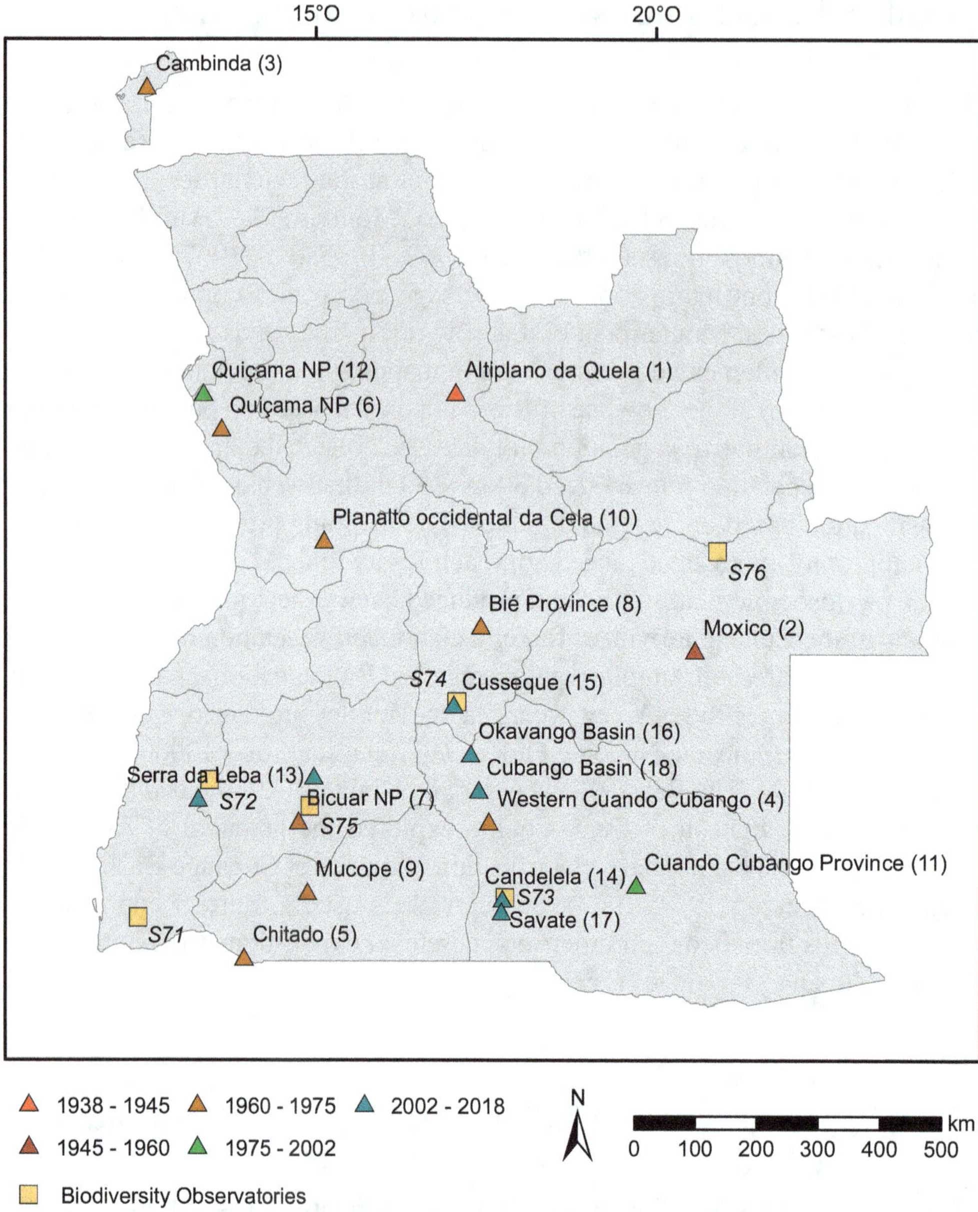

Fig. 6.1 Location of regional and local studies on vegetation composition, vegetation classification or vegetation mapping approaches according to the year the study was published. The country-wide maps by Gossweiler and Mendonça (1939), Barbosa (1970) and Diniz (1973) are not depicted. (1) von Nolde 1938a, b, c (2) Monteiro 1957 (3) Monteiro 1962 (4) Coelho 1964 (5) Menezes 1965 (6) Teixeira et al. 1967, Huntley 1972 unpublished (7) Teixeira 1968 (8) Monteiro 1970a (9) Menezes 1971 (10) Diniz and Aguiar (1968) (11) dos Santos 1982 (12) De Bruyn and Eberle 2001 (13) Cardoso et al. 2006 (14) Revermann and Finckh 2013a (15) Revermann et al. 2013, Schneibel et al. 2013, Gonçalves et al. 2017 (16) Revermann and Finckh 2013b, Stellmes et al. 2013 (17) Wallenfang et al. 2015 (18) Revermann 2016, Revermann et al. 2018a (19) Chisingui et al. 2018. Furthermore, the six biodiversity observatories installed by the SASSCAL project are shown: Espinheira (S71), Tundavala (S72), Candelela (S73), Cusseque (S74), Bicuar National Park (S75), Cameia National Park (S76)

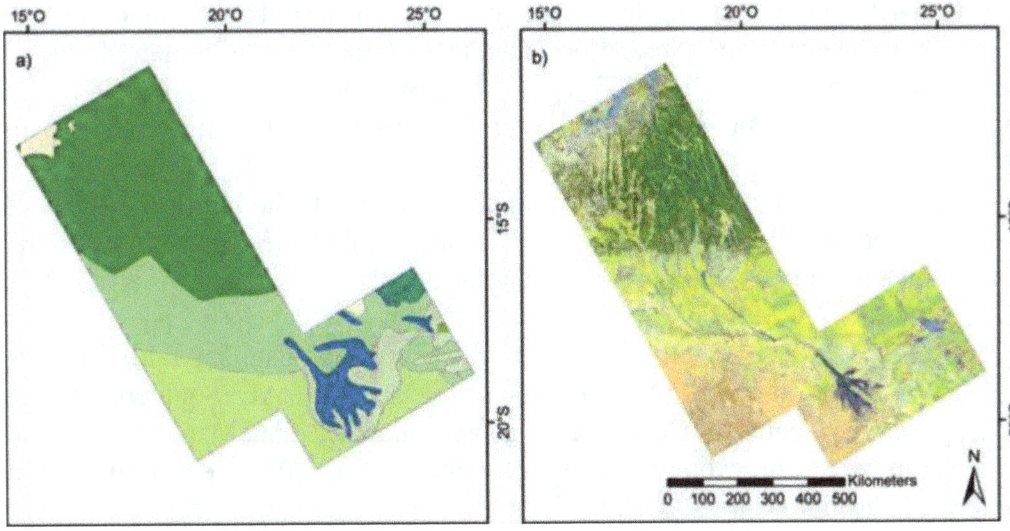

Fig. 6.2 Maps for the Okavango Basin located in southeast Angola and extending into Namibia and northern Botswana. (**a**) shows the ecoregions as defined by Olson et al (2001) which are largely based on the vegetation maps of Barbosa (1970) and White (1983), (**b**) Vegetation map produced for the same area by The Future Okavango project based on unsupervised classification of land surface phenology metrics derived from 16–day MODIS EVI time series from the years 2000–2011 (Stellmes et al. 2013) and interpreted using the information of vegetation plots stored in the vegetation database of the Okavango Basin (Revermann and Finckh 2013b; Revermann et al. 2016a). For an explanation of the vegetation units depicted in the maps please refer to the original publications

Recent years have seen increasing activity in the investigation of vegetation patterns at the local and regional scale. During the years 1995–2002 some vegetation surveys were carried out in the Quiçama National Park south of Luanda, for which the map elaborated by Huntley in the year 1972 served as a baseline. The activities aimed at gathering data for the re-establishment of the national park and to develop management strategies (Jeffery et al. 1996). De Bruyn and Eberle (2001) studied a small fenced of area in the north of the park where they collected 74 relevés and identified four plant communities including eight subcommunities. Additional quantitative data was collected to investigate grazing and browsing capacities. Cardoso et al. (2006) studied the vegetation communities along the steep altitudinal gradient of the Serra da Leba near Lubango.

Within The Future Okavango (TFO, www.future-okavango.org) project detailed investigations have been carried out in the Okavango (Cubango) River Basin. The project team assembled a vegetation database containing vegetation relevé data on all terrestrial vegetation types within the Okavango Basin (Revermann et al. 2016a). The plot design followed the standards implemented for woodland vegetation in the southern neighbour countries, i.e. a nested plot design of one small 10 m × 10 m plot in a large 20 m × 50 m plot (Strohbach 2001; Jürgens et al. 2012). Based on this data, classifications for local study sites based on numerical classification approaches have been published (Revermann and Finckh 2013a; Wallenfang et al. 2015) and a first classification of the terrestrial vegetation of the entire Cubango Basin was elaborated (Revermann et al. 2018a).

The vegetation database of the Okavango Basin was also the foundation to produce a first vegetation map based on quantitative ground data for the Okavango Basin (Fig. 6.2b Revermann and Finckh 2013b; Stellmes et al. 2013) and allowed modelling the α-diversity of vascular plants for the same region (Revermann et al. 2016b).

Based on vegetation relevés various studies have investigated the impact of land use on vegetation (Revermann et al. 2017) and studied the regeneration of the vegetation after land use had ceased (Wallenfang et al. 2015; Gonçalves et al. 2017, 2018).

Presently a number of vegetation classification and mapping initiatives are underway in the framework of the research project SASSCAL (Southern African Science Service Centre for Climate Change and Adaptive Land Management). For a compilation of project outcomes see Revermann et al. 2018b, e.g. in the Huíla Province (Chisingui et al. 2018) and along the coastal plain from the Cunene River to Benguela including Iona National Park (Jürgens et al. in prep.). The same project includes six newly implemented biodiversity observatories (http://www.sasscalobservationnet.org/), depicted on Fig. 6.1. The standardised monitoring of the 1 km^2 sites (Jürgens et al. 2012) will allow the long term monitoring of changes in plant species composition and plant diversity. Zigelski et al. (2018) present first analyses of the data gathered on such a biodiversity observatory in the Cameia National Park.

Outlook: A Call for a New Vegetation Survey of Angola

Vegetation and natural resources in general are under strong pressure from the increasing demands of a growing population and the transition from traditional lifestyles to modern consumerism (cf. Pröpper et al. 2015). The main drivers of deforestation and degradation of woodlands and the general loss of pristine vegetation cover in Angola are the clearing of new fields for shifting cultivation, industrialised agricultural schemes and the production of charcoal (Cabral et al. 2010; Hansen et al. 2013; Schneibel et al. 2013, 2016, 2018; Röder et al. 2015; Wallenfang et al. 2015; Mendelsohn 2019). Without adequate knowledge of the spatial distribution and extent of vegetation types, their species composition and the environmental drivers of vegetation patterns (climate, geology, soils, landuse) sound landuse management is not feasible. Thus, a nationwide vegetation survey based on quantitative, plot level data is urgently needed. Combined with remote sensing data and ecological modelling tools an accurate vegetation map can be produced serving the needs of conservationists, planners, entrepreneurs and scientists alike. A successful vegetation survey however relies on good taxonomic knowledge, current flora compendia and plant identification guides. Functioning and strengthened herbaria are also of great importance building the capacity of the future generation of field ecologists and environmental scientists.

References

Aguiar FQB, Diniz AC (1972) *Carta de Vegetação do Planalto Ocidental da Cela: Estudo Interpretativo*. Série Científica N° 26. Instituto de Investigação Agronómica de Angola, Nova Lisboa

Barbosa LAG (1970) Carta Fitogeográfica de Angola. Instituto de Investigação Científica de Angola, Luanda, 323 pp

Barbosa LAG (1971) Phytogeographical map of Angola. Mitteilungen der Botanischen Staatssammlung München 10:114–115

Brockmann-Jerosch H, Rübel E (1912) Die Einteilung der Pflanzengesellschaften nach okologisch-physiognomischen Gesichtspunkten. Engelmann, Leipzig, 72 pp

Cabral AIR, Vasconcelos MJ, Oom D et al (2010) Spatial dynamics and quantification of deforestation in the central-plateau woodlands of Angola (1990–2009). Appl Geogr 31(3):1185–1193

Cardoso J, Duarte M, Costa E et al (2006) Communidades vegetais da Serra da Leba. In: Moreira I (ed) Angola: Agricultura, Recursos Naturais e Desenvolvimento. ISA Press, Lisbon, pp 205–223

Chisingui AV, Gonçalves FMP, Tchamba JJ et al (2018) Vegetation survey of the woodlands of Huíla province. In: Revermann R, Krewenka KM, Schmiedel U et al (eds) Climate change and adaptive land management in southern Africa – assessments, changes, challenges, and solutions, Biodivers & Ecol, vol 6. Klaus Hess Publishers, Göttingen & Windhoek, pp 426–437

Coelho H (1964) Contribuição para o Conhecimento da Composição Florística e Possibilidades de uma Zona Compreendida entre os rios Cubango, Cueio e Quatir. Agronomia Angolana 20:49–82

Coelho H (1967) Zonagem Florestal do Distrito do Cuando-Cubango. Primeiros elementos. Agronomia Angolana 26:3–28

De Bruyn PJN, Eberle D (2001) An ecological study of the plant communities of the fenced sector of the Quiçama National Park, Angola, with Management Recommendations. B.Sc. (Hons) Thesis. University of Pretoria, Pretoria

Diniz AC (1973) Características mesológicas de Angola. Missão de Inquéritos Agrícolas de Angola, Nova Lisboa, 482 pp

Diniz AC, Aguiar FB (1968) Regiões Naturais de Angola. Série Científica N° 2. Instituto de Investigação Agronómica de Angola, Nova Lisboa, 6 pp + 1 map

Gago MM, Macedo M, Castelo C (2016) Surveying Angola, São Tomé and Timor: experts and transnational practices. In: Serrão JV, Freire D, Fernández L (eds) Old and new worlds: the global challenges of rural history, Conference eBook. ISCTE – Instituto Universitário de Lisboa. Centro de Investigação e Estudos de Sociologia, Lisboa

Gonçalves FMP, Revermann R, Gomes AL, et al. (2017) Tree species diversity and composition of Miombo woodlands in south-central Angola, a chronosequence of forest recovery after shifting cultivation. Int J For Res 2017(6202093), 13 pp.

Gonçalves FMP, Revermann R, Cachissapa MJ, et al (2018) Species diversity, population structure and regeneration of woody species in fallows and mature stands of tropical woodlands of SE Angola. J For Res. Published online 13 January 2018

Gossweiler J, Mendonça FA (1939) Carta Fitogeográfica de Angola. Ministério das Colónias, Lisbon, 242 pp

Goyder DJ, Gonçalves FMP (2019) The flora of Angola: collectors, richness and endemism. In: Huntley BJ, Russo V, Lages F, Ferrand N (eds) Biodiversity of Angola. Science & conservation: a modern synthesis. Springer Nature, Cham

Hansen MC, Potapov PV, Moore R et al (2013) High-resolution global maps of 21st-century forest cover change. Science 342(6160):850–853

Huntley BJ (1972) Parque Nacional da Quiçama. Carta da Vegetação, 1° Aproximação Julho 1972. Ecologist's report 22. Repartição Técnica da Fauna, Serviços de Veterinária, Luanda, Mimeograph report

Jeffery RF, van der Waal C, Radloff F (1996) An ecological evaluation with management guidelines for the re-establishment of the Quiçama National Park, Angola. B.Sc. (Hons) thesis. University of Pretoria, Pretoria

Jessen O (1936) Reisen und Forschungen in Angola. Dietrich Reimer Verlag, Berlin, 397 pp

Jürgens N, Schmiedel U, Haarmeyer DH et al (2012) The BIOTA biodiversity observatories in Africa-a standardized framework for large-scale environmental monitoring. Environ Monit Assess 184(2):655–678

MacKinnon J, Aveling C, Olivier R et al (2016) Inputs for an EU strategic approach to wildlife conservation in Africa – regional analysis. European Commission, Directorate-General for International Cooperation And Development, Brussels, 494 pp

Mendelsohn JM (2019) Landscape changes in Angola. In: Huntley BJ, Russo V, Lages F, Ferrand N (eds) Biodiversity of Angola. Science & conservation: a modern synthesis. Springer Nature, Cham

Menezes JA (1965) Estudo fitosociológico e características das pastagens da região do Chitado. Boletim do Instituto de Investigação Científica de Angola 2(2):137–181

Menezes JA (1971) Estudo fitoecológico da região de Mucope e carta da vegetação. Boletim Instituto de Investigação Científica de Angola 8(2):7–54

Monteiro RFR (1957) Aspectos da exploração florestal no distrito do Moxico. Garcia de Orta 5(1):129–146

Monteiro RFR (1962) Le massif forestier du Mayombe angolais. Revue Bois et Forêts des Tropiques 82:3–17

Monteiro RFR (1965a) A formação florestal dos Dembos. Boletim do Instituto de Investigação Científica de Angola 2(1):71–82

Monteiro RFR (1965b) Correlação entre as florestas do Maiombe e dos Dembos. Indicação de factores predominantes. Boletim do Instituto de Investigação Científica de Angola 1(2):257–265

Monteiro RFR (1967) Essências Florestais de Angola. Estudo das Suas Madeiras. Espécies do Maiombe. Instituto de Investigação Científica de Angola, Luanda

Monteiro RFR (1970a) Estudo da Flora e da Vegetação das Florestas Abertas do Plantalto do Bié. Instituto de Investigação Científica de Angola, Luanda, 352 pp

Monteiro RFR (1970b) Alguns Elementos de Interese Ecológico da Flora Lenhosa do Planalto do Bié (Angola). Instituto de Investigação Científica de Angola, Luanda, 166 pp

Olson DM, Dinerstein E, Wikramanayake ED et al (2001) Terrestrial ecoregions of the world: a new map of life on earth. Bioscience 51(11):933–938

Pröpper M, Gröngröft A, Finckh M et al (2015) The future Okavango – findings, scenarios and recommendations for action. Research project final synthesis report 2010-2015. Hamburg, 190 pp

Revermann R (2016) Analysis of vegetation and plant diversity patterns in the Okavango basin at different spatial scales – integration of field based methods, remote sensing information and ecological modelling. PhD thesis, University of Hamburg Hamburg, 295 pp

Revermann R, Finckh M (2013a) Caiundo – vegetation. Biodivers Ecol 5:91–96

Revermann R, Finckh M (2013b) Okavango basin – vegetation. Biodivers Ecol 5:29–35

Revermann R, Gomes A, Gonçalves FM et al (2013) Cusseque – vegetation. Biodivers Ecol 5:59–63

Revermann R, Gomes AL, Gonçalves FM et al (2016a) Vegetation database of the Okavango basin. Phytocoenologia 46(1):103–104

Revermann R, Finckh M, Stellmes M et al (2016b) Linking land surface phenology and vegetation-plot databases to model terrestrial plant alpha diversity of the Okavango Basin. Remote Sens 8:370

Revermann R, Wallenfang J, Oldeland J et al (2017) Species richness and evenness respond to diverging land-use patterns – a cross-border study of dry tropical woodlands in southern Africa. Afr J Ecol 55:152–161

Revermann R, Oldeland J, Gonçalvess FM et al (2018a) Dry tropical forests and woodlands of the Cubango Basin in southern Africa – first classification and assessment of their woody species diversity. Phytocoenologia 48(1):23–50

Revermann R, Krewenka KM, Schmiedel U et al (eds) (2018b) Climate change and adaptive land management in southern Africa – assessments, changes, challenges, and solutions. Biodivers Ecol 6:1–497

Röder A, Pröpper M, Stellmes M et al (2015) Assessing urban growth and rural land use transformations in a cross-border situation in northern Namibia and southern Angola. Land Use Policy 42:340–354

Santos RM (1982) Itinerários Florísticos e Carta da Vegetacão do Cuando Cubango. Instituto de Investigação Científica Tropical, Lisbon, 265 pp
Sayre R, Comer P, Hak J et al (2013) A new map of standardized terrestrial ecosystems of Africa. Association of American Geographers, Washington, DC, 24 pp
Schneibel A, Stellmes M, Revermann R et al (2013) Agricultural expansion during the post-civil war period in southern Angola based on bi-temporal Landsat data. Biodivers Ecol 5:311–320
Schneibel A, Stellmes M, Röder A et al (2016) Evaluating the trade-off between food and timber resulting from the conversion of Miombo forests to agricultural land in Angola using multi-temporal Landsat data. Sci Total Environ 548-549:390–401
Schneibel A, Röder A, Stellmes M et al (2018) Long-term land use change analysis in south-central Angola. Assessing the trade-off between major ecosystem services with remote sensing data. Biodivers Ecol 6:360–367
Shaw HKA (1947) The vegetation of Angola. J Ecol 35(1):23–48
Stellmes M, Frantz D, Finckh M et al (2013) Okavango basin – earth observation. Biodivers Ecol 5:23–27
Strohbach BJ (2001) Vegetation survey of Namibia. J Namibia Sci Soc 49:93–124
Teixeira JB (1968) Parque Nacional do Bicuar. Carta da vegetação (1ª aproximação) e Memória Descritiva. Instituto de Investigação Agronómica de Angola, Nova Lisboa
Teixeira JB, Matos GC, Sousa JNB (1967) Parque Nacional da Quiçama. Carta da Vegetação e Memória Descritiva. Instituto de Investigação Agronómica de Angola, Nova Lisboa
UNESCO/AETFAT/UNSO (1981) Vegetation map of Africa – scale 1:5 000 000. In: White F (ed) UNESCO, Paris
von Nolde I (1938a) Probeflächen verschiedener Savannenformationen im Hochland von Quela in Angola. Notizblatt des Botanischen Gartens und Museums zu Berlin-Dahlem 14:298–311
von Nolde I (1938b) Probeflächen verschiedener Waldformationen aus dem Hochland von Quela in Angola. Notizblatt des Botanischen Gartens und Museums zu Berlin-Dahlem 14:483–486
von Nolde I (1938c) Botanische Studie über das Hochland von Quela in Angola. Feddes Repertorium Beihefte 101:35
Wallenfang J, Finckh M, Oldeland J et al (2015) Impact of shifting cultivation on dense tropical woodlands in southeast Angola. Trop Conserva Sci 8(4):863–892
White F (1976) The underground forests of Africa: a preliminary review. Gard Bull Singapore 11:57–71
White F (1983) The vegetation of Africa – a descriptive memoir to accompany the Unesco/AETFAT/UNSO vegetation map of Africa. UNESCO, Paris, 356 pp
Zigelski P, Lages F, Finckh M (2018) Seasonal changes of biodiversity patterns and habitat conditions in a flooded savanna – the Cameia national park biodiversity observatory in the upper Zambezi catchment, Angola. In: Revermann R, Krewenka KM, Schmiedel U et al (eds) Climate change and adaptive land management in southern Africa – assessments, changes, challenges, and solutions, Biodivers Ecol, vol 6. Klaus Hess Publishers, Göttingen & Windhoek, pp 438–447

Chapter 7
Suffrutex Dominated Ecosystems in Angola

Paulina Zigelski, Amândio Gomes, and Manfred Finckh

Abstract A small-scale mosaic of miombo woodlands and open, seasonally inundated grasslands is a typical aspect of the Zambezian phytochorion that extends into the eastern and central parts of Angola. The grasslands are home to so-called 'underground trees' or *geoxylic suffrutices*, a life form with massive underground wooden structures. Some (but not all) of the *geoxylic suffrutices* occur also in open woodland types. These iconic dwarf shrubs evolved in many plant families under similar environmental pressures, converting the Zambezian phytochorion into a unique evolutionary laboratory. In this chapter we assemble the current knowledge on distribution, diversity, ecology and evolutionary history of geoxylic suffrutices and suffrutex-grasslands in Angola and highlight their conservation values and challenges.

Keywords Endemism · Geoxyles · Miombo · Phytochorion · Underground forests · Vegetation

Introduction

Open grassy vegetation is a common aspect of Angolan landscapes and is a characteristic part of the Zambezian phytochorion. Grasses are the most conspicuous element of these landscapes towards the end of the rainy season, whereas at the onset of the rainy season many woody species of so called *geoxylic suffrutices* or 'underground trees' (Davy 1922; White 1976) dominate the aspect of the vegetation. Thus, in vast areas of central and eastern Angola, the open 'grasslands' are de-facto co-dominated by grasses and geoxylic suffrutices. Closely intertwined with miombo

P. Zigelski (✉) · M. Finckh
Institute for Plant Science and Microbiology, University of Hamburg, Hamburg, Germany
e-mail: paulina.meller@gmx.de; paulina.meller@studium.uni-hamburg.de; manfred.finckh@uni-hamburg.de

A. Gomes
Institute for Plant Science and Microbiology, University of Hamburg, Hamburg, Germany

Faculty of Sciences, Agostinho Neto University, Luanda, Angola
e-mail: amandiogomes2@hotmail.com

B. J. Huntley et al. (eds.), *Biodiversity of Angola*,
https://doi.org/10.1007/978-3-030-03083-4_7

woodlands and with wetlands, suffrutex-grasslands constitute one of the main and most particular ecosystem types of Angola. According to Mayaux et al. (2004), they cover at least 70,080 km^2 or 5.6% of the Angolan territory (not including the small scale woodland suffrutex-grassland mosaics of the central Angolan plateau).

The geoxylic suffrutex life form is marked by proportionally massive underground woody organs, in literature often termed as *lignotuber, xylopodia* or *woody rhizomes*. Annual shoots sprout readily from the buds on these perennial woody organs, bearing leaves, inflorescences and fruits before they die back after the end of the rainy season. Coexistence of grasses and suffrutices is made possible by occupation of different ecological niches together with phase-delayed activity periods (i.e. main assimilation/flowering/fruiting time) that reduces competition.

Exploration of Geoxylic Grasslands

The first authors who indicated the distribution and ecological particularity of suffrutex-grasslands in Angola were Gossweiler and Mendonça (1939), who classified them as heathland-like woodlands ('Ericilignosa'). They already noted the main differentiation between the *Cryptosepalum* spp. dominated suffrutex communities ('Anharas de Ongote') on ferralitic and psammoferralitic soils and the vegetation types characterised by *Parinari capensis* and the Apocynaceae *Landolphia thollonii* and *L. camptoloba* on leached sandy soils ('Chanas da Borracha'). They had also already observed the strong thermic oscillations of which at least the 'Anharas de Ongote' are subject (see below) and commented on the generative cycle of *Cryptosepalum maraviense* from flowering to fruiting in the dry season (and thus, being inverse to the generative cycle of the C4-grasses).

Using a different mapping and classification approach, typical suffrutex-grasslands mostly on sandy soils were again mapped and described by Barbosa (1970) as 'Chanas da Borracha' (alluding to the presences of species of the genus *Landolphia*), 'Chanas da Cameia', and 'Anharas do Alto'. The *Cryptosepalum* spp. dominated 'Anharas de Ongote' on ferralitic soils are described (but not depicted on the map) as being inserted in the main miombo types of the Angolan plateau. However, he describes the typical spatial pattern, i.e. how they appear close to the headwaters of the small tributaries and then follow the watercourses in narrow or broad fringes downstream. Gossweiler and Mendonça (1939) as well as Barbosa (1970), treated these ecosystems as particular site specific plant communities closely linked to woodland ecosystems, and not as grass-dominated savannas.

White (1983), however, mapped and described only the sandy 'Chanas' as 'Kalahari and dambo-edge suffrutex grassland' in the context of the 'Zambezian edaphic grassland', but did not refer to the 'Anharas de Ongote' which constitute a key (but small scale) element of the miombo ecosystems of the Angolan Plateau. Even in his prominent suffrutex review, White (1976) focuses solely on the 'Chanas' in the range of the Zambezi Graben and neither mentions (psammo-) ferralitic

'Anharas', nor lists their dominant key species *Cryptosepalum maraviense* and *C. exfoliatum* ssp. *suffruticans* in his suffrutex list. He certainly recognises a transition zone between Zambezian and Guineo-Congolian floras that spans over central and northern Angola (where the 'Anharas' are included) (White 1983). However, he did not recognise the importance and floristic singularity of the ferralitic suffrutex-grasslands dominated by *Cryptosepalum* spp.

Suffrutex Flora and Endemism

The suffrutex life form appears in many different floristic groups and obviously evolved convergently. A similar center of geoxyle diversity has been reported from the Brazilian Cerrado. Today, 198 species from 40 families are listed for the western Zambezian phytochorion (White 1976; Maurin et al. 2014, own data), but an even higher number is expected as floristic exploration of the region is still poor and new species might be found (see Goyder and Gonçalves 2019). In some cases suffrutices are considered a dwarf variety or subspecies of a closely related tree species (e.g. *Gymnosporia senegalensis* var. *stuhlmanniana*, *Syzygium guineense* ssp. *huillense*) and hence classified as such and not as one species, although the genetic relatedness between tree and dwarf form is rarely investigated. On the other hand, not all dwarf forms are obligate suffrutices; some can facultatively outgrow the dwarf state if protected from environmental stressors (White 1976), for instance *Oldfieldia dactylophylla* or *Syzygium guineense* ssp. *macrocarpum* (Zigelski et al. 2018).

Within the suffrutex communities of the Zambezian phytochorion, the Rubiaceae have the highest number of described taxa (46), followed by Anacardiaceae (22) and Lamiaceae (14). Table 7.1 lists all families with known geoxylic suffrutex taxa occurring in Angola and gives examples of common geoxyles for each family. Furthermore, Fig. 7.1 shows some examples and aspects of suffrutex species given in Table 7.1. The unique Zambezian geoxylic flora with a high number of endemic species (Brenan 1978; White 1983; Frost 1996) is a consequence of challenging environmental conditions, as illustrated further below. According to Figueiredo and Smith's catalogue of Angolan plants (2008) and our list of suffrutices (Table 7.1), 121 of the 198 suffrutex species occurring in the Zambezian phytochorion are known from Angola (61%). Of these 121 species 12 are endemic to Angola (10%).

Environmental Conditions of Suffrutex-Grasslands Through the Year

The substrate strongly influences the species composition of the suffrutex-grasslands. In Angola geoxylic suffrutices occur on (a) well-drained arenosols which are found as seasonally flooded savannas in the Zambezi Graben of the

Table 7.1 List of plant families with geoxylic suffrutices in the Zambezian phytochorion

Plant family	N°	Species common in Angola	Angolan endemics
Rubiaceae	46	*Pygmaeothamnus zeyheri* (Sond.) Robyns, *Pachystigma pygmaeum* (Schltr.) Robyns	2, e.g. *Leptactina prostrata*
Anacardiaceae	22	*Lannea edulis* (Sond.) Engl., *Rhus arenaria* Engl.	3, e.g. *Lannea gossweileri*
Lamiaceae	14	*Clerodendrum ternatum* Schinz, *Vitex madiensis* ssp. *milanjensis* (Britten) F.White	
Fabaceae-Papilionioideae	13	*Erythrina baumii* Harms, *Abrus melanospermum* ssp. *suffruticosus* Hassk.	3, e.g. *Adenodolichos mendesii*
Proteaceae	11	*Protea micans* ssp. *trichophylla* (Engl. & Gilg) Chisumpa & Brummitt	1, *Protea paludosa* (Hiern) Engl.
Ochnaceae	9	*Ochna arenaria* De Wild. & T. Durand, *Ochna manikensis* De Wild.	
Passifloraceae	7	*Paropsia brazzaeana* Baill.	
Fabaceae-Detarioideae	6	*Cryptosepalum maraviense* Oliv., *C. exfoliatum* ssp. *suffruticans* (P.A.Duvign.)	
Apocynaceae	5	*Chamaeclitandra henriquesiana* (Hallier f.) Pichon	1, *Landolphia gossweileri*
Ebenaceae	5	*Diospyros chamaethamnus* Mildbr, *Euclea crispa* (Thunb.) Gürke	
Celastraceae	4	*Gymnosporia senegalensis* var. *stuhlmanniana* Loes.	
Dichapetalaceae	4	*Dichapetalum cymosum* (Hook.) Engl.	
Fabaceae-Caesalpinioideae	4	*Entada arenaria* Schinz	
Myrtaceae	4	*Syzygium guineense* ssp. *huillense,* (Hiern) F. White *Eugenia malangensis* (O.Hoffm.) Nied.	
Tiliaceae	4	*Grewia herbaceae* Hiern	
Combretaceae	3	*Combretum platypetalum* Welw. ex M. A. Lawson	2, e.g. *Combretum argyrotrichum*
Euphorbiaceae	3	*Sclerocroton oblongifolius* (Müll.Arg.) Kruijt & Roebers	
Loganiaceae	3	*Strychnos gossweileri* Exell	
Annonaceae	2	*Annona stenophylla* ssp. *nana* (Exell) N. Robson	
Apiaceae	2	*Steganotaenia hockii* (C. Norman) C. Norman	
Chrysobalanaceae	2	*Parinari capensis* Harv., *Magnistipula sapinii* De Wild.	
Meliaceae	2	*Trichilia quadrivalvis* C.DC.	
Moraceae	2	*Ficus pygmaea* Welw. ex Hiern	
Myricaceae	2	*Morella serrata* (Lam.) Killick	
Phyllanthaceae	2	*Phyllanthus welwitschianus* Müll.Arg.	
Ranunculaceae	2	*Clematis villosa* DC.	

(continued)

Table 7.1 (continued)

Plant family	N°	Species common in Angola	Angolan endemics
Achariaceae	1	*Caloncoba suffruticosa* (Milne-Redh.) Exell & Sleumer	
Anisophyllaceae	1	*Anisophyllea quangensis* Engl. ex Henriq.	
Clusiaceae	1	*Garcinia buchneri* Engl.	
Dilleniaceae	1	*Tetracera masuiana* De Wild. & T. Durand	
Fabaceae-Caesalpinioideae	1	*Bauhinia mendoncae* Torre & Hillc.	
Hypericaceae	1	*Psorospermum mechowii* Engl.	
Ixonanthaceae	1	*Phyllocosmus lemaireanus* (De Wild. & T. Durand) T. Durand & H. Durand	
Lecythidaceae	1	*Napoleonaea gossweileri* Baker f.	
Linaceae	1	*Hugonia gossweileri* Baker f. & Exell	
Malpighiaceae	1	*Sphedamnocarpus angolensis* (A. Juss.) Planch. ex Oliv.	
Malvaceae	1	*Hibiscus rhodanthus* Gürke	
Melastomaceae	1	*Heterotis canescens* (E. Mey. ex Graham) Jacq.-Fél.	
Picrodendraceae	1	*Oldfieldia dactylophylla* (Welw. ex Oliv.) J.Léonard	
Rhamnaceae	1	*Ziziphus zeyheriana* Sond.	
Urticaceae	1	*Pouzolzia parasitica* (Forssk.) Schweinf.	

N°: overall number of Suffrutex species in the Zambezian phytochorion; examples of species occurring in Angola are given for each family. Compilation of families and species according to White (1976), Maurin et al. (2014) and own data

Moxíco province or as sandy alluvial deposits on fossil river terraces along the valleys of the southern slopes of the Angolan plateau (Fig. 7.2a); (b) on psammo- ferralitic plinthisols as they frequently occur on the Bíe Plateau in central Angola. The suffrutex-grasslands on ferralitic soils mostly occur on mid- and foot-slopes and are embedded within a matrix of miombo woodland (Fig. 7.2b).

Environmental conditions in suffrutex-grasslands change dramatically throughout the year. The most perceived stresses are man-made fires in the dry season (May–October) which are mostly deployed to induce resprouting for livestock fodder or to facilitate hunting (Hall 1984). Depending on fire intensity, which in turn depends mostly on fuel load, ambient temperature and wind (Govender et al. 2006), such fires can completely burn unprotected aboveground biomass.

Another abiotic stress occurring mostly in the early dry season (June–August) is nocturnal frost, peaking immediately before sunrise. At this time of year masses of cold dry air from southern latitudes intrude into south-central Africa (Tyson and Preston-Whyte 2000). As depressions accumulate confluent cold air, the undulating topography of the Angolan highlands facilitates frequent radiation frost especially in valleys (Revermann and Finckh 2013; Finckh et al. 2016). Up to 44 frost events per dry season (with a minimum temperature of −7.5 °C) were recorded by Finckh et al. (2016), with a temperature span of up to 40 degrees within 12 h. Most woody

Fig. 7.1 Common Angolan suffrutex species. (**a**) *Ochna arenaria* (Ochnaceae), fruiting and growing on sandy sediments of the Bíe Plateau. (**b**) *Syzygium guineense* ssp. *huillense* (Myrtaceae) flowering in the dry season and growing on sandy soils of the Bíe Plateau. (**c**) *Lannea edulis* (Anacardiaceae), bearing edible fruits, growing on Kalahari sands in southeast Angola. (**d**) *Hibiscus rodanthus* (Malvaceae), growing on Kalahari sands in southeast Angola and flowering in the rainy season. (**e**) *Landolphia gossweileri* (Apocynaceae), typical element of the 'Chanas da Borracha', growing on sandy soils of the Bíe Plateau and bearing edible fruits. (**f**) *Phyllanthus welwitschianus* (Phyllanthaceae), growing on sandy soils of the Bíe Plateau and flowering in the rainy season. (**g**) *Cryptosepalum exfoliatum* ssp. *suffruticans* (Fabaceae – Detarioideae) with excavated rootstocks, typical element of the 'Anharas de Ongote', growing on psammoferralitic soils of the Bíe Plateau. (**h**) *Parinari capensis* (Chrysobalanaceae), typical element of the 'Chanas da Borracha', growing on slightly elevated termite mounds in flooded savannas of the Cameia National Park, Moxico Province

Fig. 7.2 Typical geoxylic suffrutex grasslands of Angola. (**a**) 'Chanas da Cameia' in the Cameia National Park, Moxíco Province, during dry season in June. The slightly elevated termite mounds provide habitat for several geoxyle species that avoid the low-lying areas that are waterlogged from January to May. (**b**) 'Anharas de Ongote' in the Sovi Valley on the southern slopes of the Bíe Plateau, in August. The mid- and footslopes are dominated by suffrutex-grassland with the characteristic reddish and green patches of the fresh leaves of *Cryptosepalum maraviense*, whereas the wetlands in the drainage lines are covered mostly by Cyperaceae (background, in dark green)

species from tropical background (including geoxylic suffrutices) are sensitive to frost, their leaves wilt or their shoots die-off entirely.

The geoxylic suffrutex species seem to be triggered by the destruction of their shoots by frost and/or fire, as they readily resprout after these disturbances and in most cases already start flowering in the dry season. The suffrutices therefore have often already finished their generative cycle when the grasses start to cover them.

The suffrutex-grasslands of the sandy plains in eastern Angola are furthermore subject to seasonal flooding in the late rainy and early dry season (January–May), leading, for example, in the Cameia National Park to standing water up to 0.5 m deep. Whereas grass species dominate the sites which are inundated for several months, suffrutex species seem to avoid fully waterlogged sites and grow patchily on slightly elevated termite mounds (Fig. 7.2a) or other well drained sites.

The dominant grass species seem to profit from inundation. Their tufts develop massively in the middle of the rainy season and they flower and bear fruits throughout the flooding season (own observations).

Knowledge Gaps on the Evolution of the Geoxylic Suffrutices and the Formation of Suffrutex-Grasslands

A common observation within suffrutex ecosystems is the resemblance (Meerts 2017) and assumed close relatedness of suffrutex species to tree species that occur in forests and woodlands. The indigenous people (e.g. the Chokwe in eastern Angola) in many cases recognise the similarity and relatedness and use similar local names for such pairs, for instance Muhaua and Mupaua for the tree and suffrutex forms of *Syzygium guineense* Willd. DC. The striking fact that the suffrutex life form was developed by several plant families independently and at roughly the same time (Maurin et al. 2014) indicates a common driver that triggered its convergent evolution.

Grassy biomes emerged in Africa in the late Miocene approximately 10 mya (Cerling et al. 1997; Keeley and Rundel 2005; Herbert et al. 2016). This period is characterised by global climatic fluctuations which led to cooler, drier conditions, to a drop of atmospheric CO_2 concentrations and particularly to pronounced precipitation seasonality (i.e. wet and dry seasons) in southern Africa (Pagani et al. 1999). As a consequence, humid tropical forests retreated to more favorable sites further north and were replaced by more open dry and seasonal tropical forest ecosystems like the miombo (Bonnefille 2011). In parts where miombo landscapes prevail today, canopies were disrupted and allowed the establishment of open ecosystems embedded in woodland matrices. These open ecosystems were then rapidly occupied by light-demanding C4-grasses and the evolving geoxylic suffrutices.

It is still an open discussion why open suffrutex-grasslands are able to persist within the woodlands (or vice versa). It is however likely that rainfall seasonality and the above described abiotic stresses that characterise the suffrutex-grasslands play a major role in their establishment and maintenance (Sankaran et al. 2005; Staver et al. 2011).

Savanna ecologists tend to see fire as the main driver for grassland formation. On the one hand frequent fires prevent tree establishment if saplings cannot outgrow the reach of the flames and are destroyed therein. For woodlands in eastern South Africa, a fire free time period of at least 5 years is necessary for many tree species to escape the 'fire trap' (Sankaran et al. 2004; Gignoux et al. 2009). This time window, allowing for successful reestablishment of trees, is rarely achieved in Angolan grasslands, at least nowadays (Schneibel et al. 2013; Stellmes et al. 2013). C4-savanna grasses, however, respond positively to periodic burning and resprout within weeks (Bond and Keeley 2005), thus being able to colonise seasonally burnt sites.

Forest ecologists, on the other hand, attribute the frequent short duration frost events in the dry season for preventing tree recruitment in the open areas (Finckh et al. 2016). As the list of suffrutices (Table 7.1) shows, mainly (but not exclusively) tropical families or genera evolved suffrutex life forms. Frost is deleterious to most tropical tree taxa, as they have not developed physiological adaptations to this 'un-tropical' stress factor, thus showing little or no frost tolerance (Sakai and Larcher 2012). As the suffrutex-grasslands are typically situated in particularly frost prone sites (depressions), tree taxa that are not adapted to frost are being filtered out of such environments.

In any case, a promising strategy to cope with seasonally returning thermic stress (by frost or fire) is to protect sensitive organs (buds) by hiding them underground. Tree species relocated their woody biomass and regenerative buds belowground at the expense of growth height and were thus able to cope with frost and fire prone sites (White 1976; Maurin et al. 2014; Finckh et al. 2016). Even shallow soil depths of less than 10 cm are sufficient to alleviate thermic stresses (Revermann and Finckh 2013). The high number of tropical genera and families that contribute to the suffrutex flora show how successful this strategy is for frost sensitive and fire susceptible taxa, in order to survive the adverse conditions of the open grasslands.

Concomitantly other evolutionary advantages of the geoxylic life form have been discussed, for instance poor edaphic conditions, as favoured by White (1976). He considered the low nutrient status of the leached and locally seasonal waterlogged soils on Kalahari sands as a likely cause for the lack of regular trees and the suffrutication of them as means of compensation. However, trees as well as suffrutices often grow on the same or similarly poor soils, with comparable physical and chemical properties (Gröngröft et al. 2013); forests and grasslands are not separated by edaphic boundaries but follow topographic rather than edaphic logics.

The waterlogging argument on the other hand would imply that the woody underground organs show adaptations to inundation, for instance aerenchymatic tissue or adventitious roots (Parolin 2008). Anatomical analyses of the rootstocks of four common suffrutex species however did not provide any support for aerenchymatic tissue nor other adaptations to inundation (Sanguino 2015). Moreover, in seasonally flooded savannas suffrutices avoid inundated sites. This is even the case for

Syzygium guineense ssp. *huillense*, a suffrutex closely related to a tree species that grows along and in rivers and floodplains (Coates Palgrave 2002; Meerts and Hasson 2016).

To summarise, so far the main environmental driver for the astonishing radiation of geoxylic suffrutices has not been conclusively identified. The emergence of the suffrutex grassland at the end of the Pliocene and the peak of radiation at the beginning of the Pleistocene is clearly related to climatic seasonality and pronounced dry seasons. Dry seasons, however, did not only provide the necessary dry fuel for fire but also provided the atmospheric conditions for nocturnal frost events – the seasonality argument, thus, does not tip the balance toward fire or frost.

Conservation Value and Conservation Challenges

Various studies recognise the high floristic singularity of the Zambezian phytochorion and suffrutex-grasslands with its unique life forms contribute prominently to its high number of endemic species (Clayton and Cope 1980; White 1983). The high degree of suffrutex-grassland endemics within the Zambezian phytochorion as well as within Angola is a consequence of a unique setting of environmental drivers like nutrient poor soils, frequent frosts and fires or precipitation seasonality in a small-scale heterogeneous landscape (Linder 2001). Thus, the Zambezian phytochorion can be seen as an evolutionary laboratory that promoted the evolution of many specialised plant species, e.g. suffrutices, orchids and grasses.

Suffrutex-grasslands are sometimes misunderstood as 'degraded forests', overlooking their naturalness. Through this misconception they are listed as sites for reforestation in order to recover presumably lost forests and to sequestrate atmospheric CO_2 (Parr et al. 2014). However, the well-intentioned act of reforestation would in fact destroy biodiverse natural ecosystems (Bond 2016). A lack of understanding, however, frustrates the development of appropriate conservation measures for the suffrutex grasslands today and in the future. The rebuilding process in Angola also has risks, happening at a rapid pace and shaping the landscape to human demands with limited consideration for sustainable management (Pröpper et al. 2015). Flooded savannas in the Moxíco Province for instance are targeted for large-scale agro-industrial development (ANGOP 2017). Not even National Parks offer adequate protection to ecosystems in this area, as the first rice schemes emerged during 2016 within the limits of Cameia National Park (own observation). Deficiencies in communication and cooperation between different ministries and governance levels aggravate such problems.

Outlook

Many questions still remain to be answered around the enigmatic life form of the geoxylic suffrutices. In order to efficiently safeguard suffrutex-grasslands, we need to understand the evolutionary drivers and evolutionary processes shaping these ecosystems. For instance, a thorough understanding of the evolutionary drivers and the response of suffrutices to them would help to assess how current environmental conditions affect the Zambezian ecosystems and how landscape shaping processes work. Moreover, investigations about genetic patterns of suffrutices and close tree-relatives would give insight to speciation processes, means of propagation (clonal or sexual) and evolutionary history. Also, ecophysiological or morphological measurements would contribute another perspective from which to assess how suffrutices react to environmental stresses and change processes. All these facets are currently the subjects of incipient research.

References

ANGOP (2017) Cameia prepara mais de mil hectares para cultivar arroz. Agência Angola Press, Moxico, 16.07.2017. http://www.angop.ao/angola/pt_pt/noticias/economia/2017/6/28/Moxico-Cameia-prepara-mais-mil-hectares-para-cultivar-arroz,be825477-10c7-472d-b6e3-cb1ed5d25974.html. Accessed 22 Aug 2017

Barbosa LAG (1970) Carta fitogeográfica de Angola. Instituto de Investigação Científica de Angola, Luanda

Bond WJ (2016) Ancient grasslands at risk. Science 351(6269):120–122

Bond W, Keeley J (2005) Fire as a global 'herbivore': the ecology and evolution of flammable ecosystems. Trends Ecol Evol 20(7):387–394

Bonnefille R (2011) Rainforest responses to past climatic changes in tropical Africa. In: Tropical rainforest responses to climatic change. Springer, Berlin/Heidelberg, pp 125–184

Brenan JP (1978) Some aspects of the phytogeography of tropical Africa. Ann Mo Bot Gard 65(2):437–478

Cerling TE, Harris JM, MacFadden BJ et al (1997) Global vegetation change through the Miocene/Pliocene boundary. Nature 389(6647):153–158

Clayton WD, Cope TA (1980) The chorology of Old World species of Gramineae. Kew Bull 35(1):135–171

Coates Palgrave K (2002) Trees of southern Africa, 3rd edn. New edition revised and updated by Meg Coates Palgrave, Struik, Cape Town, pp 833–836

Davy JB (1922) The suffrutescent habit as an adaptation to environment. J Ecol 10(2):211–219

Figueiredo E, Smith G (eds) (2008) Plants of Angola: Plantas de Angola. Strelitzia 22:1–279

Finckh M, Revermann R, Aidar MP (2016) Climate refugees going underground–a response to Maurin et al. (2014). New Phytol 209(3):904–909

Frost P (1996) The ecology of miombo woodlands. In: Campbell B (ed) The Miombo in transition: woodlands and welfare in Africa. CIFOR, Jakarta, pp 11–57

Gignoux J, Lahoreau G, Julliard R et al (2009) Establishment and early persistence of tree seedlings in an annually burned savanna. J Ecol 97(3):484–495

Gossweiler J, Mendonça FA (1939) Carta fitogeografica de Angola. Ministério das Colónias, Lisboa, p 242

Govender N, Trollope WS, Van Wilgen BW (2006) The effect of fire season, fire frequency, rainfall and management on fire intensity in savanna vegetation in South Africa. J Appl Ecol 43(4):748–758

Goyder DJ, Gonçalves FPM (2019) The Flora of Angola: collectors, richness and endemism. In: Huntley BJ, Russo V, Lages F, Ferrand N (eds) Biodiversity of Angola. Science & conservation: a modern synthesis. Springer

Gröngröft A, Luther-Mosebach J, Landschreiber L et al (2013) Cusseque – soils. Biodivers Ecol 5:51–54

Hall M (1984) Man's historical and traditional use of fire in southern Africa. In: de Booysen PV, Tainton NM (eds) Ecological effects of fire in South African ecosystems. Springer, Berlin/Heidelberg, pp 40–52

Herbert TD, Lawrence KT, Tzanova A et al (2016) Late Miocene global cooling and the rise of modern ecosystems. Nat Geosci 9(11):843–847

Keeley JE, Rundel PW (2005) Fire and the Miocene expansion of C4 grasslands: miocene C4 grassland expansion. Ecol Lett 8(7):683–690

Linder HP (2001) Plant diversity and endemism in sub-Saharan tropical Africa. J Biogeogr 28(2):169–182

Maurin O, Davies TJ, Burrows JE et al (2014) Savanna fire and the origins of the 'underground forests' of Africa. New Phytol 204(1):201–214

Mayaux P, Bartholomé E, Fritz S et al (2004) A new land-cover map of Africa for the year 2000. J Biogeogr 31(6):861–877

Meerts P (2017) Geoxylic suffrutices of African savannas: short but remarkably similar to trees. J Trop Ecol 33(4):1–4

Meerts PJ, Hasson M (2016) Arbres et arbustes du Haut-Katanga. National Botanic Garden of Belgium

Pagani M, Freeman KH, Arthur MA (1999) Late Miocene atmospheric CO2 concentrations and the expansion of C4 grasses. Science 285(5429):876–879

Parolin P (2008) Submerged in darkness: adaptations to prolonged submergence by woody species of the Amazonian floodplains. Ann Bot 103(2):359–376

Parr CL, Lehmann CER, Bond WJ et al (2014) Tropical grassy biomes: misunderstood, neglected, and under threat. Trends Ecol Evol 29(4):205–213

Pröpper M, Gröngröft A, Finckh M et al (2015) The future Okavango: findings, scenarios and recommendations for action: research project final synthesis report 2010–2015. University of Hamburg-Biocentre Klein Flottbek, pp 53–129

Revermann R, Finckh M (2013) Cusseque – micro-climatic conditions. Biodivers Ecol 5:47–50

Sakai A, Larcher W (2012) Frost survival of plants: responses and adaptation to freezing stress, vol 62. Springer, New York, pp 138–173

Sanguino G (2015) Wood anatomy and adaptation strategies of suffrutescent shrubs in South-Central Angola. MSc. thesis, Universität Hamburg, Hamburg, 68 pp

Sankaran M, Ratnam J, Hanan NP (2004) Tree-grass coexistence in savannas revisited – insights from an examination of assumptions and mechanisms invoked in existing models. Ecol Lett 7(6):480–490

Sankaran M, Hanan NP, Scholes RJ et al (2005) Determinants of woody cover in African savannas. Nature 438(7069):846–849

Schneibel A, Stellmes M, Frantz D et al (2013) Cusseque – earth observation. Biodivers Ecol 5:55–57

Staver AC, Archibald S, Levin SA (2011) The global extent and determinants of savanna and forest as alternative biome states. Science 334(6053):230–232

Stellmes M, Frantz D, Finckh M et al (2013) Okavango basin – earth observation. Biodivers Ecol 5:23–27

Tyson PD, Preston-Whyte RA (2000) Weather and climate of Southern Africa, 2nd edn. Oxford University Press Southern Africa, Cape Town Chapter 10–12

White F (1976) The underground forests of Africa: a preliminary review. Gard Bull Singap 29:57–71

White F (1983) The Zambezian regional centre of endemism. In: White F (ed) The vegetation of Africa – a descriptive memoir to accompany the UNESCO/AETFAT/UNSO vegetation map of Africa. UNESCO, Paris, 356 pp

Zigelski P, Lages F, Finckh M (2018) Seasonal changes of biodiversity patterns and habitat conditions in a flooded savanna – the Cameia National Park Biodiversity Observatory in the Upper Zambezi catchment, Angola. Biodivers Ecol 6:438–447

Chapter 8
Landscape Changes in Angola

John M. Mendelsohn

Abstract Landscape changes in Angola are dominated by woodland and forest losses due to clearing for crops, bush fires (which convert woodland into shrubland) and the harvesting of fuel (as wood and charcoal) and timber. Rates of clearing for small-scale dryland crops are high over much of Angola as a result of poor soil fertility. Erosion is also a severe problem, which has caused widespread losses of topsoil, soils nutrients and ground water. Rates of erosion are greatest in areas with steep slopes, sparse plant cover and high numbers of people, as well as around diamond mines in Lunda-Norte. Patterns of river flow and water quality have been changed, largely as a result of soil erosion and plant cover loss, as well as large irrigation schemes and dams. High rates of urban growth and the production of untreated urban waste have led to large concentrations of contamination around towns. Further research is needed, for example to assess the environmental impacts of the fishing and petroleum industries offshore, the effects of large volumes of urban waste being washed into and down major rivers to the sea, and landscape changes in an around areas of highland forests and grasslands that support populations of rare and endemic species.

Keywords Bushmeat · Charcoal · Deforestation · Fire · Land transformation · Mining impacts · River flows · Shifting cultivation · Soil erosion · Urbanisation

Introduction

Angola is a developing country, its development occurring in multiple ways in different areas of the country and affecting a variety of natural resources. Some changes and developments are likely to accelerate as the country seeks to diversify its economy and reduce dependence on revenues from oil and diamonds. It is also likely that the changes will contribute to such global trends as loss of biodiversity and land degradation.

J. M. Mendelsohn (✉)
RAISON (Research & Information Services of Namibia), Windhoek, Namibia
e-mail: john@raison.com.na

B. J. Huntley et al. (eds.), *Biodiversity of Angola*,
https://doi.org/10.1007/978-3-030-03083-4_8

This brief review provides perspectives and information on changes to Angola's terrestrial landscapes, particularly in the southern half of the country. There are three sections to the chapter, the first of which describes the major kinds of landscape change. The second is an account of conditions that drive changes, both ultimately and proximately. Finally, areas most affected by major changes are identified in the third part of the paper.

Major Changes

Woodland and Forest Loss

Losses of woodland are by far the most obvious and conspicuous of changes in Angola. Much of this has been due to clearing for small-scale crop farming, particularly of dry-land crops, and large-scale commercial agriculture (including relatively small areas of exotic tree plantations). Other losses have come from the harvesting of charcoal, wood fuel, timber production (both for commercial and domestic uses), and runaway bush fires. On a smaller scale, swathes of riverine forest have been removed to give miners access to alluvial diamonds in rivers in Lunda-Norte.

As a result of all these losses, large areas of forest and savanna are now grasslands or shrublands. For example, the greater part of Huambo and Angola's central *planalto* was originally wooded, and 78.4% of the province of Huambo was covered in miombo woodland in 2002. In 13 years that figure had dropped in 2015 to 48.3%, amounting to the loss of some 1.265 million ha, 63.2% of which was converted from forest to crop land (Palacios et al. 2015). Similar losses in western Cuando Cubango, eastern Huíla and eastern Huambo have been documented by Schneibel et al. (2013), and elsewhere in Huíla and the Cuvelai drainage in Cunene (Mendelsohn and Mendelsohn 2018).

A countrywide perspective on the loss of forest or tree canopy cover is presented in Fig. 8.1. Several relevant features are visible in this image. First is the open, deforested expanse stretching southwest to northeast across western Huíla, southwestern Huambo and western Bié. Much of this area of highlands was cleared for crops between the 1950s and 1970s, although grasslands (*anharas do alto*) probably always dominated high altitude areas of the central *planalto* above about 1900 m above sea level. Substantial areas were cleared at the same time in parts of Cuanza-Norte, Cuanza-Sul and Malange, but their boundaries are not easily defined.

Second is the clearing of woodlands around urban areas. Many had already been cleared of tree cover by 2000, after which clearings expanded as trees were removed progressively further from the town centres, a trend illustrated by Schneibel et al. (2018). Examples of recent clearings between 2000 and 2015 are conspicuous as 'red bands' around Dundo, Menongue, Luena, Malange, Cafunfo, Cubal and Caimbambo in Fig. 8.1. Much of the deforestation is clear-felling for dryland fields by residents, while other trees are removed for charcoal production, wood fuel and timber.

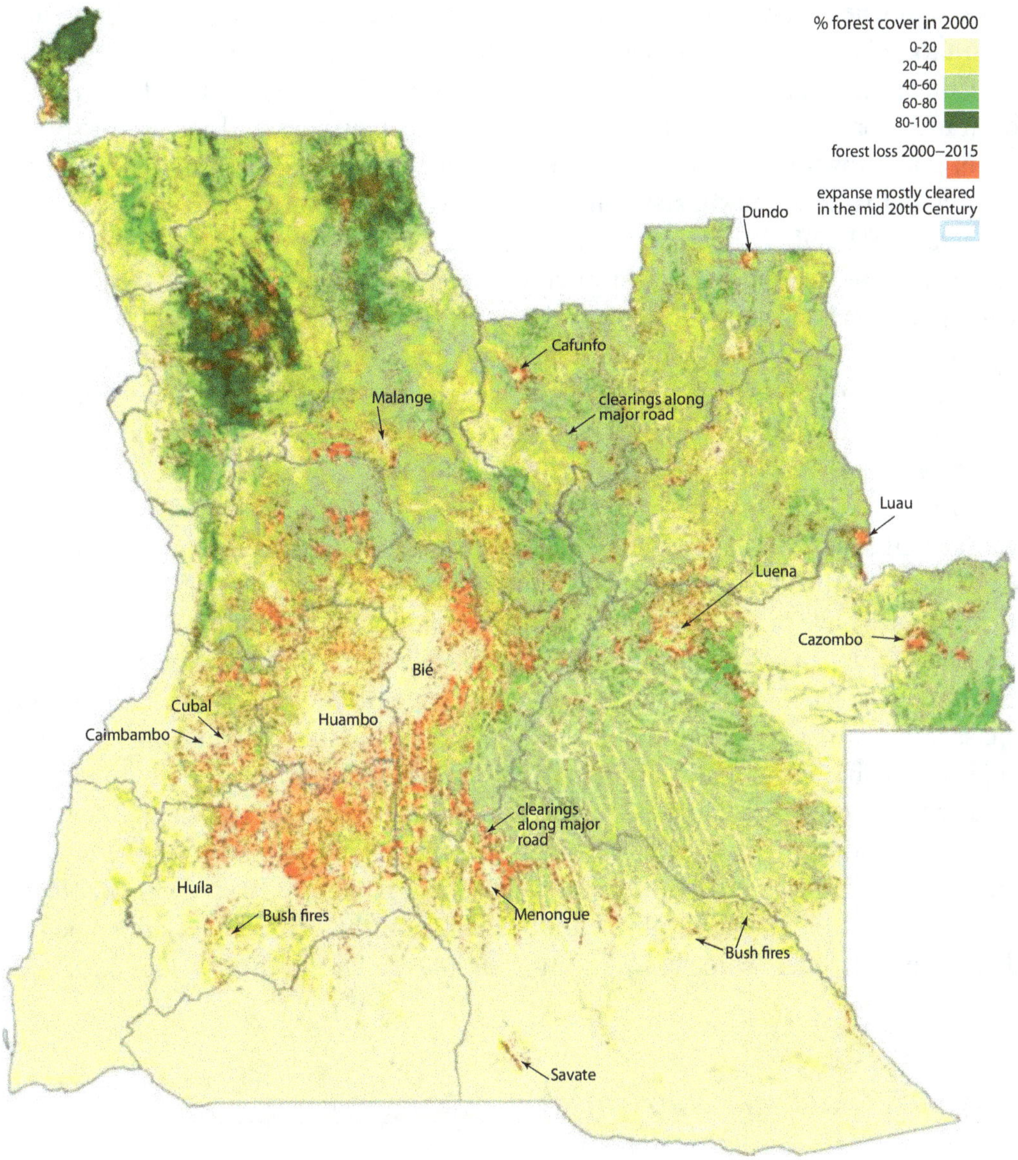

Fig. 8.1 Forest or tree canopy loss between 2000 and 2015 derived from data described by Hansen et al. (2013), updated and available from http://earthenginepartners.appspot.com/science-2013-global-forest. Percentage forest cover in the year 2000 is shown in shades of green. Red areas are those which, by 2015, had lost all the forest or canopy cover that still remained in 2000. (Source: Hansen/UMD/Google/NASA)

Third, the concentration of clearings along major roads where many rural families choose to settle is visible, but requires closer inspection of Fig. 8.1. Most losses of tree cover here are also due to clear-felling for dryland crops. Local residents also produce charcoal on a large scale, particularly along roads frequently travelled by trucks that can transport large volumes of charcoal to urban markets. However, the effects of charcoal – and timber – harvesting are seldom visible in satellite images of tree or canopy cover because harvesters typically remove only larger, taller trees, leaving smaller trees and shrubs which present a seemingly intact canopy of woodland when viewed from high above. After some years of regrowth, harvesters return to fell those bigger individual trees that produce good charcoal.

Timber has been harvested on a substantial scale for many years. Most of it has been used for the construction of domestic homes, palisades and fences, or sold as exported hardwood. The harvesting of selected species and large individual trees has evidently increased substantially in recent years, and further increases are to be expected (ANGOP 2017). Conversely, the use of poles for houses, palisades and fences may be declining, at least in certain areas where people increasingly build with home-made or bought bricks, and fence with wire (Calunga et al. 2015).

Trees were evidently harvested in large numbers to fuel railway engines running between Benguela and Huambo, and perhaps elsewhere, in the early twentieth century (Silva 2008). There are also reports of Zambezi teak *Baikiaea plurijuga* and *Marquesia macroura* timber being used for sleepers on the *Caminho do Ferro Moçâmedes* (CFM) and *Caminho de Ferro Benguela* (CFB) lines, respectively, while indigenous woodlands were cleared to make way for the many eucalyptus plantations established along the CFB line.

Bush fires have major effects on woodlands, particularly in limiting the growth of trees and shrubs in savannas. Indeed, fires maintain the 'balance' between grass and trees that characterise savannas. However, hot, intense runaway fires set by people are seemingly more frequent than before. The fiercest of fires kill all plants, old sizeable trees being burnt and scarred year after year until they eventually die. Large areas have thus been converted from woodland and forest into shrubland, particularly in southern Angola (Fig. 8.2). Much of Cuando Cubango and parts of Moxico are mosaics of open woodland separated along sharp margins from dense woodland and forest. As a probable result of fire, the edges of the dense cover are smoothed and often rounded, in some cases creating circular patches of forest (Fig. 8.3).

Soil Loss (Bulk and Nutrients)

At least three areas appear to have lost large volumes of soil and soil nutrients. The first is the central *planalto* and surrounding higher areas of ferralsol soils. In the catchment of the Cunene River, erosion has been greatest in areas that are densely populated, extensively cultivated with dryland crops, largely cleared of plant cover and that have at least moderate slopes (Fig. 8.4). The catchments of other major rivers (Cuando, Queve, Quicombo, Catumbela, Guvrire and Coporolo, for example) that drain the central catchment are likewise eroded, particularly where slopes are steep and plant cover is sparse. Similar, more concentrated effects are seen in cities where inadequate management of storm water has led to the formation of erosion gullies, many of them damaging urban roads, houses and other infrastructure.

Second is in Lunda-Norte where open-cast mining leads to considerable volumes of soil (probably also ferralsols) being washed into rivers that flow north into the Congo Basin (Fig. 8.5; see Ferreira-Baptista et al. 2018).

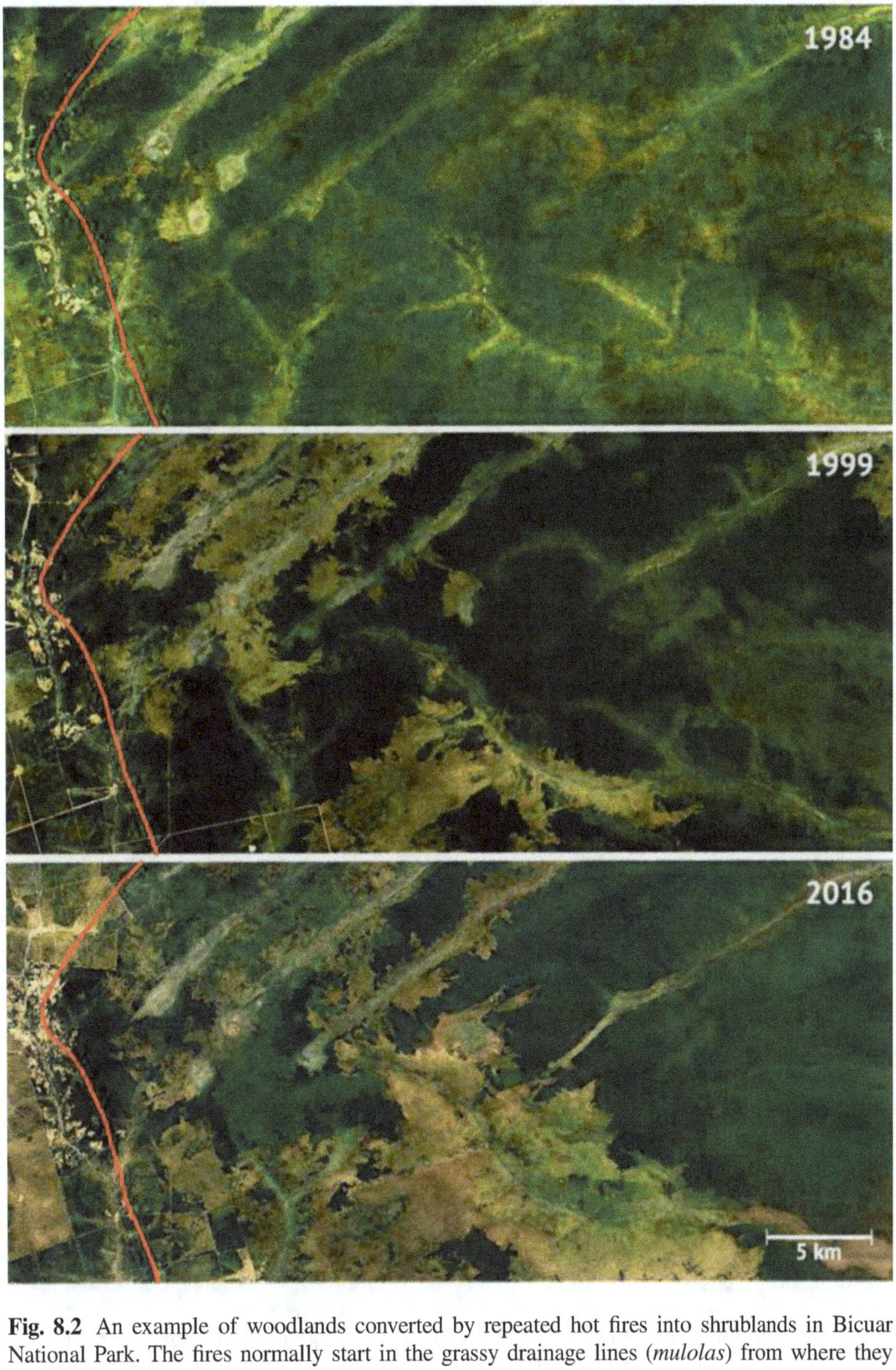

Fig. 8.2 An example of woodlands converted by repeated hot fires into shrublands in Bicuar National Park. The fires normally start in the grassy drainage lines (*mulolas*) from where they spread into the surrounding woodlands. With the same areas being burnt by fierce fires every few years, large areas of woodland (dark greenish zones) have progressively been turned into shrublands (pale areas). These satellite images from Google Earth (LandSat/Copernicus) were taken between 1984 and 2016, and viewed from about 15.3 South, 14.4 East. The red line marks the western border of Bicuar National Park. (From Mendelsohn and Mendelsohn 2018)

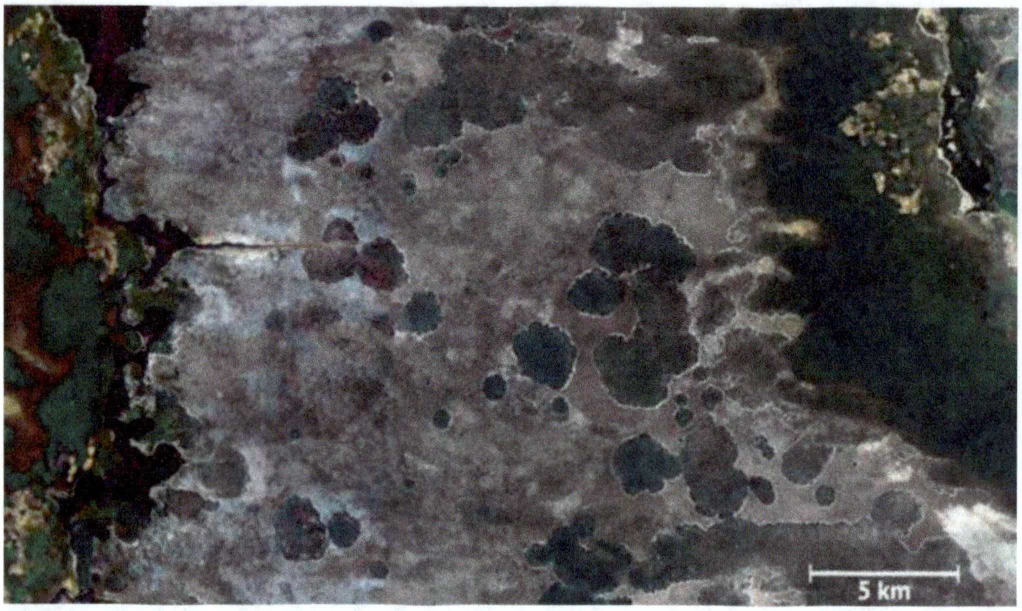

Fig. 8.3 Patches of open woodland (pale grey areas) and dense miombo forest (dark green) between the Longa and Sovi rivers in Cuando Cubango. The forest margins have probably been sharpened and smoothed by bush fires. Isolated blocks are so rounded and reminiscent of the Namib Desert's fairy circles that they may be called 'fairy forests'. (The image was taken from Google Earth (LandSat/Copernicus) as viewed from about 15.4 South, 18.9 East)

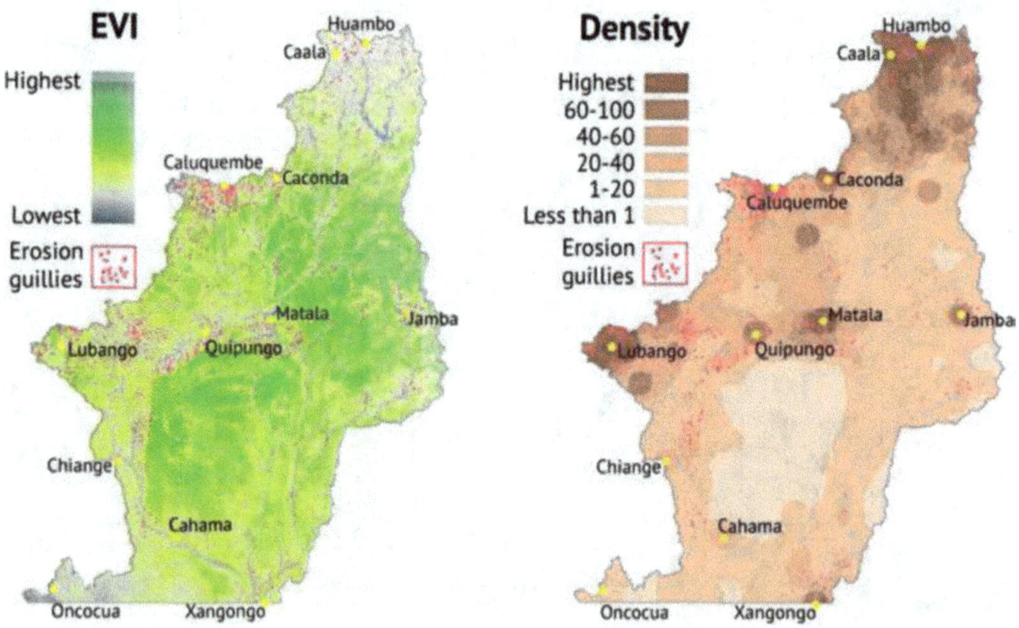

Fig. 8.4 The distribution of erosion gullies in relation to vegetation cover (Enhanced Vegetation Index – EVI) and population density in the catchment of the Cunene River between Huambo in the north and Xangongo in the south. (Adapted from Mendelsohn and Mendelsohn 2018)

Fig. 8.5 Mining impacts on Angolan rivers. Top left and right: The confluence of the clear Cassai River and the turbid Lubembe River carrying suspended sediments from open-cast diamond mining in Lunda-Norte. The confluence is in the DRC about 80 km north of Angola's border. The left photograph was taken on 30 May 2007, while the right image from Google Earth was taken 10 years later on 21 May 2017, viewed from 6.62 South, 21.07 East. Bottom left: The confluence of the Calonga and Cunene rivers at Quiteve (16.02 South, 15.20 East), showing the volumes of eroded sediments from upstream in the Cunene catchment. By contrast, the clear waters of the Calonga mainly come from areas where arenosols predominate, where few people live and where large areas of woodland have not been cleared for dryland agriculture. Bottom right: Erosion from open-cast mining along the Luachimo River 22 km north of Lucapa. (The image from Google Earth was taken in May 2017 as viewed from 8.23 South, 20.77 East)

There is a likely net loss of certain soil nutrients in the third area, which is where bush fires are frequent and/or intense, predominantly so in Cuando Cubango, Moxico and the Lunda provinces (Figs. 8.6 and 8.7). Fires often result in the loss of nitrogen, phosphorus and organic carbon, although cooler fires also facilitate the release of nutrients from plant matter into the soil (Jain et al. 2008). A study comparing open and dense woodland near Savate (see Fig. 8.1), found much lower nutrient levels in open than dense woodland soils (Wallenfang et al. 2015). This stark difference was probably a consequence of the open areas being burnt often and intensely, while the dense woodlands were seldom burnt (Stellmes et al. 2013).

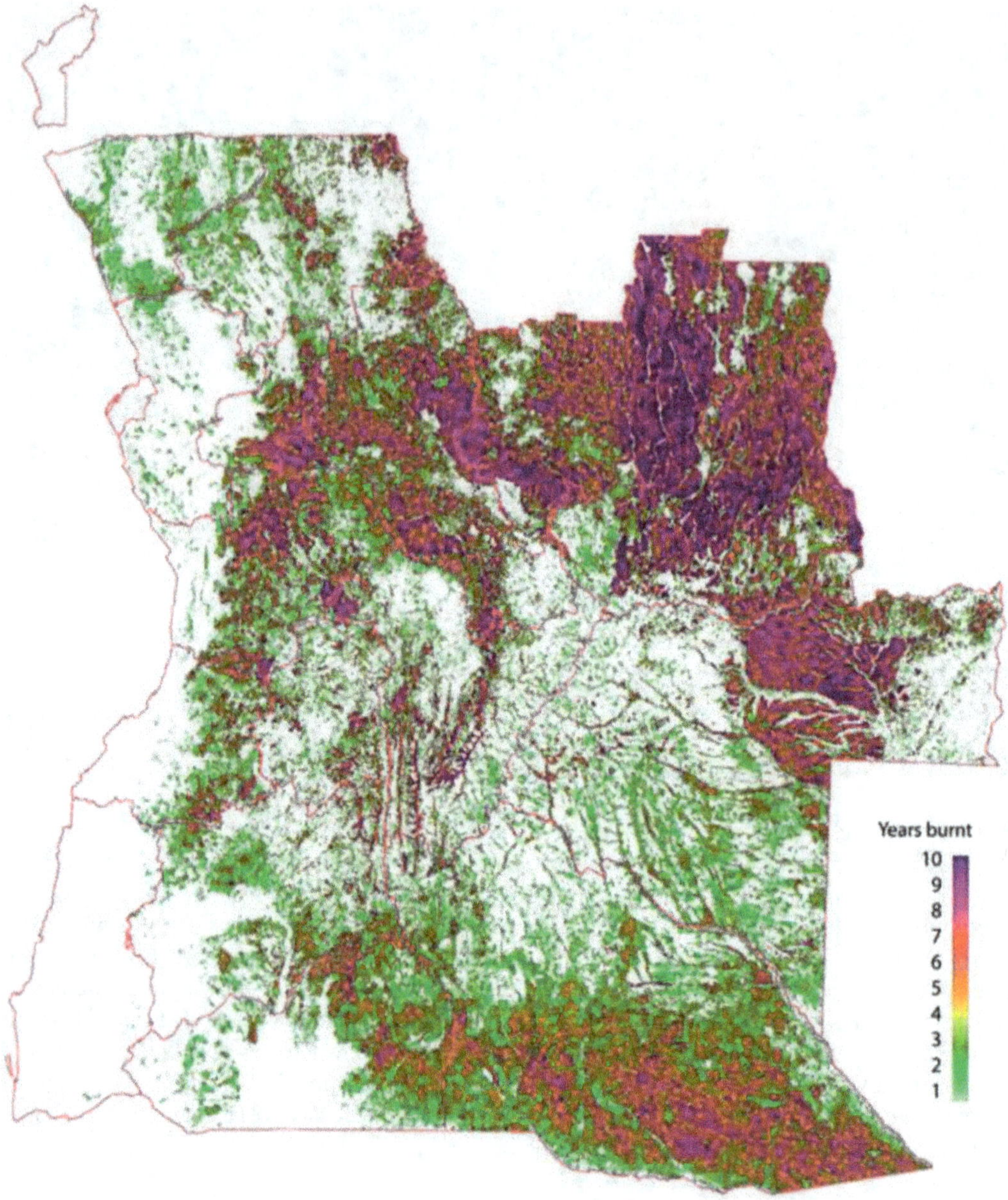

Fig. 8.6 The frequency of fires expressed as the number of years each area of 500 by 500 m burnt between 2000 and 2010. (From Archibald et al. (2010) and data available at http://wamis.meraka.org.za/products/firefrequency-map)

Water Flows and Quality

Discharges and the quality of water have changed significantly in certain rivers, and in a number of ways. The most obvious changes are in the heavy sediment loads which impair the functioning of aquatic animals and plants that require well-lit waters, and reduce the capacity of dams. For example, eroded sediments washed down the Cunene River have evidently accumulated in Gove and Matala dams to such an extent that their production of hydro-electricity has declined (António 2017).

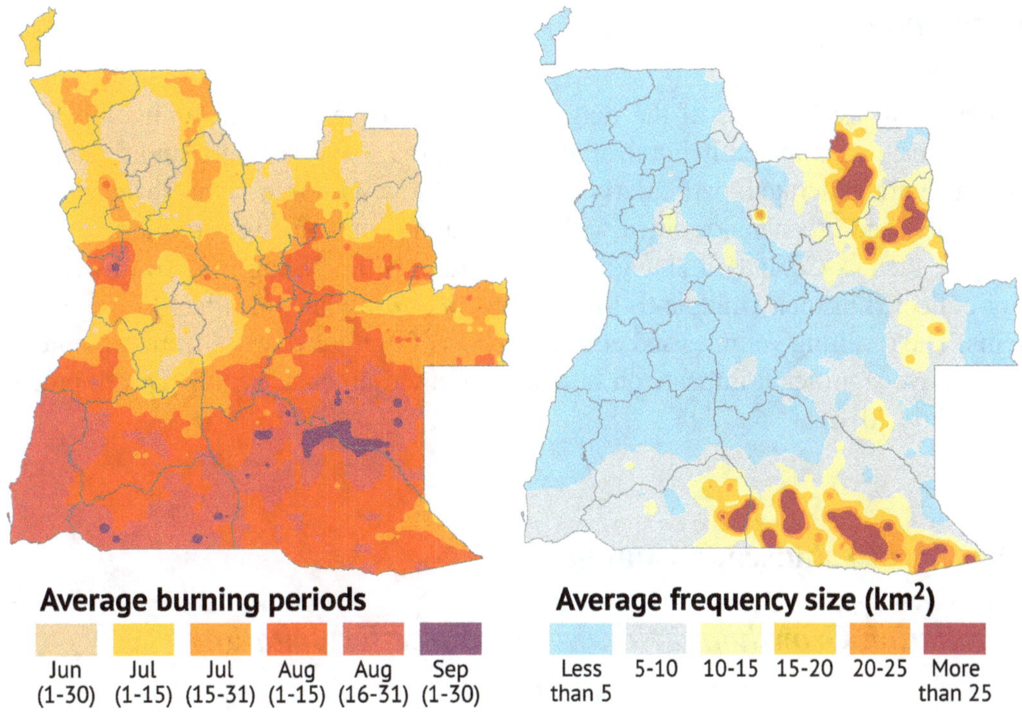

Fig. 8.7 Left: the seasonality of fires, reflected by the average period of the year when fires were recorded. Right: the average size of fires. (From Archibald et al. (2010) and data available at http://wamis.meraka.org.za/products/firefrequency-map)

River flows, soil moisture levels and groundwater recharge have been affected by losses of plant cover. Sheets of surface flows after heavy rain have increased in bare areas, causing higher river flows and probabilities of flooding, especially in seasons with above average rainfall. For example, the clearing of plant cover in the catchment of the Guvrire River around Caimbambo and Cubal (Fig. 8.1) is considered to have increased the risk and frequency of flooding at the river mouth in the city of Benguela (Development Workshop 2016).

A different impact of plant cover loss and erosion may affect the Cuvelai. Many residents there believe that surface flows down the floodplains (*chanas*) are now slower and wider than before because eroded sediments deposited in the shallow channels have further reduced their depths and slopes (Calunga et al. 2015).

Reductions in plant cover result in lower volumes of rain water being trapped or impeded, thus reducing seepage into the top soil to replenish soil moisture and recharge local aquifers. With lower soil moisture, seepage to sustain river flows during the dry season also declines. This is a likely – and at least partial – explanation for flows of the Cunene River at Ruacana dropping to less than 10 cubic metres/second in September 2017. Such low levels were only recorded previously during extreme drought years in 1993–1994 and 1994–1995 (Mendelsohn and Mendelsohn 2018).

Contamination

Quantitative assessments of the magnitude of environmental contamination from urban waste are apparently not available for Angola. However, substantial volumes of waste are generated, particularly in Luanda (now with more than seven million residents) and other major cities with populations approaching a million or more people, such as Cabinda, Lubango, Lobito, Huambo and Benguela. Solid waste is not collected in many middle to low-income *bairros*, which also lack sewage systems. The resulting volumes and concentrations of untreated waste from these large cities have significant impacts on human and environmental health (Development Workshop 2016).

Drivers of Landscape Change

Population Growth and Natural Resource Exploitation

As elsewhere in the world, but particularly in developing countries, most changes have been driven by rising demands for natural resources to meet the needs of Angola's growing population and increasing consumption per capita. The country's population rose from about 6 million people in 1970 to almost 26 million in 2014, which amounts to an annual growth rate of 3.4%. Over the same period a high proportion of the population shifted from rural to urban areas and their accompanying economies. Most urban residents live in low income areas where they generally lack piped water supplies, electricity, secure tenure, sewage systems and solid waste collection services. For example, 85% of Luanda residents live in such areas, with similar percentages in Lobito (90%), Cabinda (86%) and Benguela (92%) (Development Workshop 2016). Similar conditions and proportions hold in Huambo, Malange, Cuito and Ndalatando.

Urban consumption patterns differ from those in rural areas, but one difference of particular interest concerns the use of fuels for cooking. Rural homes generally use wood collected from around their homes, whereas the majority of urban people use purchased charcoal because alternative fuels are more expensive or not available in towns. Their supplies of charcoal all come from rural areas, in particular from poor families that harvest and then sell bags of charcoal along roads leading to major urban areas. The supply of charcoal is therefore an informal one that generates incomes for many rural homes. For both consumers and suppliers this seems to be an ideal market, providing affordable fuel for urban consumers and incomes for rural families, that often being their only monetary income. Similar market arrangements hold for supplies of bush meat from rural suppliers to urban consumers.

Another new, sometimes surprising economic link between urban and rural areas involves investments in cattle by wealthier town folk. Their cattle (and sometimes goats and sheep) are placed in rural areas where they are normally tended by relatives and held as savings or capital, with the best returns coming from large stock numbers (Gomes 2012). Owners are thus encouraged to have as many animals as possible, which places added pressures on forage, water and the limited resources available to poor rural residents (who seldom have other incomes).

Food Production

An abundance of wealth, much of it derived from the boom in oil revenues, has provided resources to develop large-scale agricultural projects, often with limited or no environmental impact assessments. For example, several new irrigation schemes have been developed along the Cunene River. If and when the farms are fully developed, downstream stretches of the Cunene could be dry for much of the year. Elsewhere, tens of thousands of hectares of woodland and forest have been cleared in recent years, one example being the Angola Biocom project which has 70,106 ha allocated to produce sugar, ethanol fuel and electricity south of Malange (Angola Biocom 2017).

Clearing for small-scale dryland crop production has caused much of the loss of woodland and forest in Angola. The *rate* at which trees are cleared is however driven by four related, but arguably separate factors. First is the need to feed a growing number of rural residents. Second is the need for farmers to abandon their fields after several years of use and to clear new fields (which will produce better yields than those that have had their supplies of nutrients exhausted). Third is the general low-input/low-output crop production strategy adopted and adapted for dryland agriculture, which means that fertilisers are seldom used to replenish soil nutrients. Fourth is the poor quality of soils available for dryland farming (Ucuassapi and Dias 2006; Asanzi et al. 2006; Wallenfang et al. 2015). Indeed, the relative lack of nutrients and moisture in soils is arguably the most important factor driving the rapid rate at which Angolan woodlands and forests are cleared, as well as the very slow rate of recovery.

Grassland and Woodland Fires

Much of Angola's vegetation has been moulded by frequent fire. This is particularly true for savanna woodlands, the grasslands of the *anharas de ongote* in the central highlands and grassy *chanas da borrachas* in the Lunda provinces. Most woody plants in the latter habitats and many in open woodlands are geoxylic suffrutices, their growth forms adapted to survive frequent hot fires (see Zigelski et al. 2019).

Fire therefore has major impacts on Angola's vegetation, and any changes in fire regimes are likely to result in landscape changes. Against that background, and the widely held assumption that burning has increased in frequency, the following information is provided on fires in Angola.

Fires are recorded most frequently in grasslands of the Lunda provinces, Malange, and the Bulozi Floodplains in Moxico, and in open, savanna woodlands in Cuando Cubango (Fig. 8.6). Additionally, fires are frequent in highland grasslands (*anharas do alto*) distributed between Serra da Chela (near Lubango) in the south, the Benguela, Huambo and Huíla highlands and higher elevations in northern Cuanza-Sul and southwestern Malange.

Almost all fires are in the dry season between late April and early November. However, those in northern Angola and the central highlands burn considerably earlier than in the south (Fig. 8.7). Grass fuel is likely to contain more moisture in June than later in August, with the likely result that the earlier fires are cooler, less intense and probably less damaging to vegetation than later, hotter burns in southern Angola. That seems true for the large fires in the tall grasses that comprise the widespread *chanas da borrachas* in the Lunda provinces (Huntley 2017). Similar trends were found by Stellmes et al. (2013) within the river basins of the Cubango and Cuito Rivers where fires in the northern catchment areas were both earlier and less intense than those in the south. That trend also roughly corresponded to land cover, with dense miombo woodland in the northern areas and open savanna woodland, often called *Baikiaea-Burkea* woodland in the southern zones.

Areas of Major and Widespread Landscape Change

Figure 8.8 provides perspectives on the distribution of the major landscape changes described in this chapter. The majority of changes are in and around the central plateau (*planalto*) and to the north in parts of Cuanza-Norte, Bengo and Uíge. The effects of fire are probably most severe in Cuando Cubango, although the large (but probably cooler) fires that are so frequent in Lunda provinces may too have major effects on those extensive grasslands.

The landscape changes around towns are limited to the 18 provincial capitals shown as dark red circles in Fig. 8.8. But landscape changes around many other large towns need to be recorded.

Future Needs for Research and Documentation

Large volumes of waste are washed into the Atlantic, both close to major coastal cities – such as Luanda, Benguela and Cabinda – and down large rivers that drain large areas of the country, such as the Cunene, Cuanza and Queve Rivers. As far as is known, the volumes, nature and impacts of the waste have not been assessed. The same is true for impacts on populations of fish and other marine animals which are harvested from Angolan and foreign vessels that operate offshore, where their activities and impacts are not monitored.

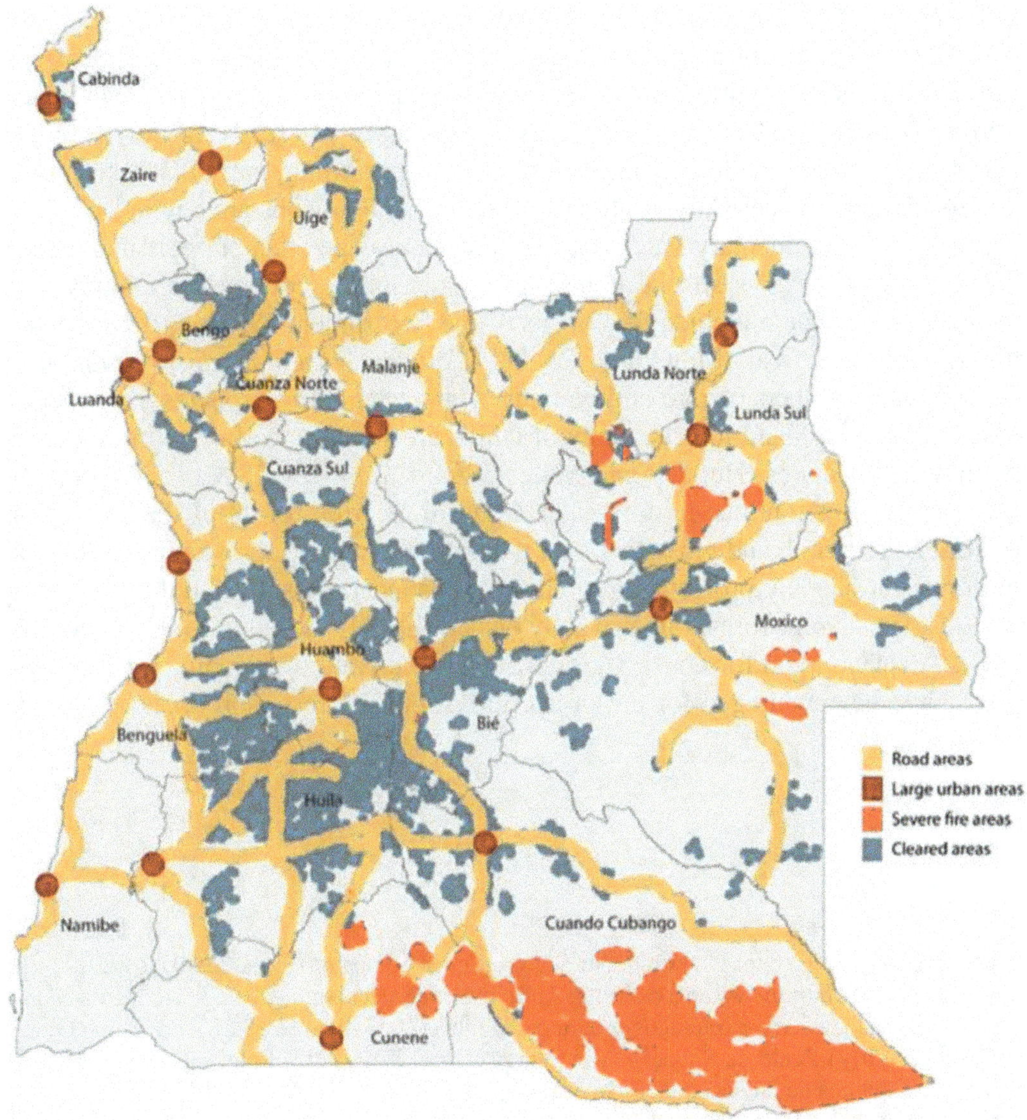

Fig. 8.8 Areas in Angola where substantial landscape changes have occurred in recent years as a result of woodland and forest clearing, and of bush fires, and in areas around main roads and towns where major changes have occurred, or are likely to occur. Woodland and forest clearings are large contiguous areas shown in Fig. 8.1. Severe fire areas are those where fires burnt in five or more years between 2000 and 2010, where fires normally burn in August and September when grass is driest and where fires are normally large (> 20 km^2), as derived from data shown in Figs. 8.6 and 8.7. Zones where people settle, farm and harvest wood are usually within 10 km either side of roads or within 15 km of major towns

The construction of very large dams on the Cuanza River is likely to have affected the functioning of that river. However, I am not aware of assessments of those effects, either by individual dams or the cumulative impounding of large volumes of water.

There are other activities and areas of concern, for example the offshore impacts of exploration and exploitation by the petroleum industry; large-scale logging in Cabinda and more recently Moxico and Cuando Cubango; pollution of river water used for washing and other domestic uses, especially where rivers flow through large towns; and pesticide contamination from crop farming, particularly from big commercial farms where large volumes of agricultural poisons are applied.

Finally, the fragmentary patchwork of mini-landscapes that support many species and which deserve special conservation measures, requires more study and documentation. These include the forests of the Escarpment Zone *(Faixa subplanaltica)* and Marginal Mountain Chain (*Cadeia Marginal de Montanhas*). Considerable numbers of rare and endemic plants and animal species are concentrated in these highland forests, many of which are small, covering no more than a few hundred hectares (Huntley and Matos 1994; Cáceres et al. 2014). The forests have shrunk, and continue to do so as a result of clearing for crops, harvesting of timber and charcoal, and grassland fires that kill trees on the forest edges. None of the forests are legally protected, and all are surrounded by substantial numbers of rural residents. Some forests are privately owned and their owners should be encouraged to manage them for conservation. Likewise, private ownership and management could be encouraged for the protection of other forests and areas of special value.

References

Angola Biocom (2017) http://www.biocom-angola.com/en/company. Accessed 30 Nov 2017

ANGOP (2017) http://www.angop.ao/angola/pt_pt/noticias/economia/2017/5/24/Angola-Mais-228-mil-madeira-serao-explorados-este-ano,40579b4d-10d3-4bed-b2e5-2751edb213eb.html

António PS (2017) Ponto de Situação Albufeira do Gove 2012–2017. Relatório de PRODEL – Empresa Pública de Produção de Electricidade, Luanda

Archibald S, Scholes R, Roy D et al (2010) Southern African fire regimes as revealed by remote sensing. Int J Wildland Fire 19:861–878

Asanzi C, Kiala D, Cesar J et al (2006) Food production in the Planalto of southern Angola. Soil Sci 171:81–820

Cáceres A, Melo M, Barlow J et al (2014) Threatened birds of the Angolan Central Escarpment: distribution and response to habitat change at Kumbira Forest. Oryx 49:727–734

Calunga P, Haludilu T, Mendelsohn J et al (2015) Vulnerabilidade na Bacia do Cuvelai/Vulnerability in the Cuvelai Basin, Angola. Development Workshop, Luanda

Development Workshop (2016) Water resource management under changing climate in Angola's coastal settlements. Project Number: 107025–001. Final technical report to the International Development Research Centre (IDRC), Canada

Ferreira-Baptista L, Manuel J, Aguiar PF et al (2018) Impact of mining on the environment and water resources in northeastern Angola. Biodivers Ecol 6:155–159

Gomes AF (2012) O Gado na Agricultura Familiar Praticada no Sudoeste de Angola – Meios de Vida e Vulnerabilidade dos Grupos Domésticos Pastoralistas e Agro-pastoralistas. PhD thesis, Technical University of Lisbon, Lisbon

Hansen MC, Potapov PV, Moore R et al (2013) High-resolution global maps of 21st-century forest cover change. Science 342:850–853

Huntley BJ (2017) Wildlife at war in Angola: the rise and fall of an African Eden. Protea Book House, Pretoria, 432 pp

Huntley BJ, Matos EM (1994) Botanical diversity and its conservation in Angola. Strelitzia 1:53–74

Jain TB, Gould W, Graham RT et al (2008) A soil burn severity index for understanding soil-fire relations in tropical forests. Ambio 37:563–568

Mendelsohn JM, Mendelsohn S (2018) Sudoeste de Angola: um Retrato da Terra e da Vida. South West Angola: a Portrait of Land and Life. Raison, Windhoek

Palacios G, Lara-Gomez M, Márquez A et al (2015) Spatial dynamic and quantification of deforestation and degradation in Miombo Forest of Huambo Province (Angola) during the period 2002–2015. SASSCAL Proceedings, Huambo, 182 pp

Schneibel A, Stellmes M, Revermann R et al (2013) Agricultural expansion during the post-civil war period in southern Angola based on bi-temporal Landsat data. Biodivers Ecol 5:311–319

Schneibel A, Röder A, Stellmes M et al (2018) Long-term land use change analysis in south-central Angola. Assessing the trade-off between major ecosystem services with remote sensing data. Biodivers Ecol 6:360–367

Silva ERS (2008) Companhia do Caminho de Ferro de Benguela: uma História Sucinta da sua Formação e Desenvolvimento. Lisbon. https://sites.google.com/site/cfbumahistoriasucinta/

Stellmes M, Frantz D, Finckh M et al (2013) Fire frequency, fire seasonality and fire intensity within the Okavango region derived from MODIS fire products. Biodivers Ecol 5:351–362

Ucuassapi AP, Dias JCS (2006) Acerca da fertilidade dos solos de Angola. In: Moreira I (ed) Angola: Agricultura, Recursos Naturais e Desenvolvimento. ISA Press, Lisboa, pp 477–495

Wallenfang J, Finckh M, Oldeland J et al (2015) Impact of shifting cultivation on dense tropical woodlands in southeast Angola. Trop Conserv Sci 8:863–892

Zigelski P, Gomes A, Finckh M (2019) Suffrutex dominated ecosystems in Angola. In: Huntley BJ, Russo V, Lages F, Ferrand N (eds) Biodiversity of Angola. Science & conservation: A modern synthesis. Springer, Cham

Part III
Invertebrate Diversity: Environmental Indicators

Chapter 9
The Dragonflies and Damselflies of Angola: An Updated Synthesis

Jens Kipping, Viola Clausnitzer, Sara R. F. Fernandes Elizalde, and Klaas-Douwe B. Dijkstra

Abstract Prior to 2012, only 158 species of Odonata were known from Angola. Surveys in 2012 and 2013 added 76 species and further additions in 2016 brought the national total to 236 species. This was published earlier in 2017 as the checklist of the dragonflies and damselflies (Odonata) of Angola by the same authors (Kipping et al. Afr Invertebr 58 (I):65–91. https://africaninvertebrates.pensoft.net/article/11382/, 2017) on which this chapter is based. Records obtained in 2017 and 2018 and a survey by two of the authors in December 2017 led to the discovery of 25 additional species, of which several are undescribed. We provide a revised checklist here comprising 260 species and discuss the history of research, the biogeography of the fauna with endemism and the potential for further discoveries. The national total is likely to be above 300 species. This would make Angola one of the richest countries for Odonata in Africa.

Keywords Africa · Biogeography · Checklist · Conservation · Endemism

---

J. Kipping (✉)
BioCart Ökologische Gutachten, Taucha/Leipzig, Germany
e-mail: biocartkipping@email.de

V. Clausnitzer
Senckenberg Museum for Natural History, Görlitz, Görlitz, Germany
e-mail: Viola.Clausnitzer@senckenberg.de

S. R. F. Fernandes Elizalde
SASSCAL – BID GBIF, Instituto de Investigação Agronómica, Huambo, Angola
e-mail: kikas.sara@gmail.com

K.-D. B. Dijkstra
Naturalis Biodiversity Center, Leiden, The Netherlands
e-mail: kd.dijkstra@naturalis.nl

B. J. Huntley et al. (eds.), *Biodiversity of Angola*,
https://doi.org/10.1007/978-3-030-03083-4_9

Introduction

Given the country's size, diverse landscapes, climatic regimes and habitats Angola is likely to be one of the richest in Odonata species in Africa. However, Angola's biodiversity is very poorly known, with comparatively limited research before independence in 1975 halting altogether in the three decades of civil war and unrest that followed. Research coverage is also limited for Odonata, with much of the north and east never surveyed at all (Clausnitzer et al. 2012). The potentially very species-rich highland catchments of the Congo, Cuanza, Cubango (Okavango) and Zambezi rivers are almost unknown and may hold many undescribed species. The whole Angolan part of the extensively marshy Cuando River and almost the entire Cuito River system are also largely unsurveyed.

History of Odonata Research in Angola

Research on Odonata began in July 1928, when the Swiss zoologist Albert Monard embarked on the first of his two expeditions to Angola, which lasted until February 1929. Monard was a curator at the Natural Museum of La-Chaux-de-Fonds in Switzerland with a broad interest in nature who mainly collected vertebrates and plants. Ris (1931) identified 27 and described four species from Monard's first expedition.

With the death of Ris, Monard submitted the Odonata from his second expedition (April 1932 to October 1933) to Cynthia Longfield at the British Museum (now the Natural History Museum) in London, who had published several records obtained by Karl Jordan from Mount Moco in 1934 (Longfield 1936). Longfield (1947) identified 77 species from Monard's new material and described 13 new species and two new genera. She also dealt with the Odonata held at the Dundo Museum in northern Angola, first revising the genus *Orthetrum* based on the long series available (Longfield 1955) and later listing 61 species from the collection, including three new species (Longfield 1959).

Elliott CG Pinhey (1961a, b) described five new species of Gomphidae from northern Angola received from António de Barros Machado of the Dundo Museum. While Longfield (1959) stated that the Dundo collection "shows the usual scarcity of the genera Gomphidae", Pinhey (1961a) noted it "was particularly notable for the number of Gomphids." Possibly Machado split the material between the two authors. It is uncertain whether the material was collected in Dundo or only held there, as most records lack details on collector, date and precise locality. However, Pinhey (1961b) did detail collecting in localities around Dundo, suggesting that all material came from this part of Lunda-Norte Province. The collector was probably Machado himself. No-one has worked on this collection since and its state is thus unknown.

Elliot Pinhey was curator at the National Museum of Zimbabwe from 1955 until 1975 and while he collected intensively in adjacent countries, he only visited Angola twice (Vick et al. 2001). In April and May 1963 Pinhey participated in an expedition to northwestern Zambia, also visiting an area east of Caianda and the Lutchigena River in Angola directly adjacent to the Ikelenge Pedicle of Zambia, where he recorded 26 species (Pinhey 1964, 1974, 1984). His second excursion into Angola went to an area between Luanda and the Duque de Bragança Falls on the Lucala River (now known as Calandula Falls) in October 1964 with records of 32 species (Pinhey 1965).

Pinhey further treated the material of three collectors, describing a species in honour of each of them. Edward S Ross of the California Academy of Sciences collected between Cuchi and Dondo in 1957 and 1958 (Pinhey 1966), the American expert of mammal behaviour Richard Estes in central Angola in 1970 (Pinhey 1971a), and Ivan Bampton around the Serra da Chela and Tundavala in 1973 (Pinhey 1975). In the 1975 paper he also repeated records from Pinhey (1964, 1965) and Longfield (1947), and provided a gazetteer, causing confusion about the precise locality of some sites. The correct historic collecting sites could be verified with the gazetteer of Mendes et al. (2013).

Various collectors gathered about 1000 specimens in the collection of the *Instituto de Investigação Agronómica* in Huambo between 1950 and 1974. These records were never published but this will be done shortly by Sara F Elizalde and David Elizalde as a GBIF dataset.

After Angola's independence in 1975 there was a long break in field research, with only a few records by various collectors. Namely in the two decades between 1980 and 2000 not a single record of Odonata is available. Some years after the end of the civil war a renaissance of research began, resulting in a growing number of records (Fig. 9.1). All localities with available Odonata records distinguished in the three periods (a) pre-independence 1928–1974, (b) after independence 1975–2001 and (c) after the end of civil war 2002-today are shown in Fig. 9.2.

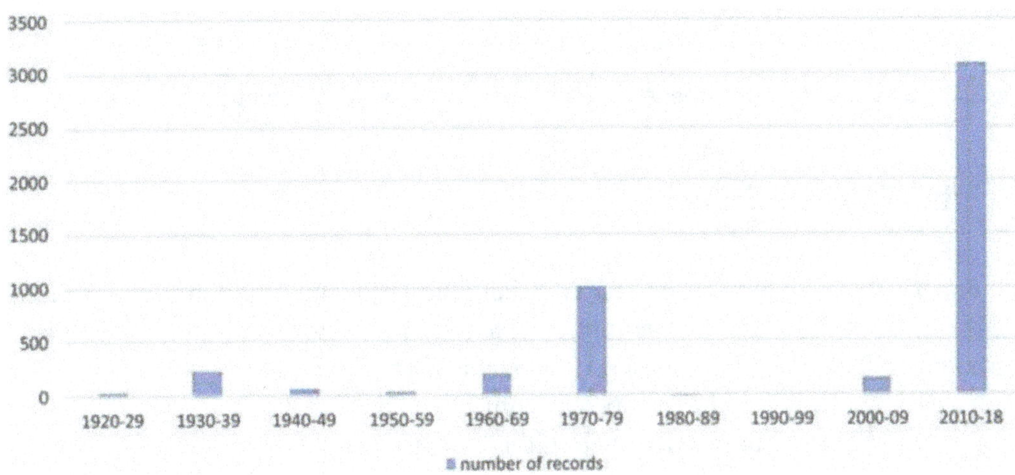

Fig. 9.1 Number of Odonata records from Angola over past decades

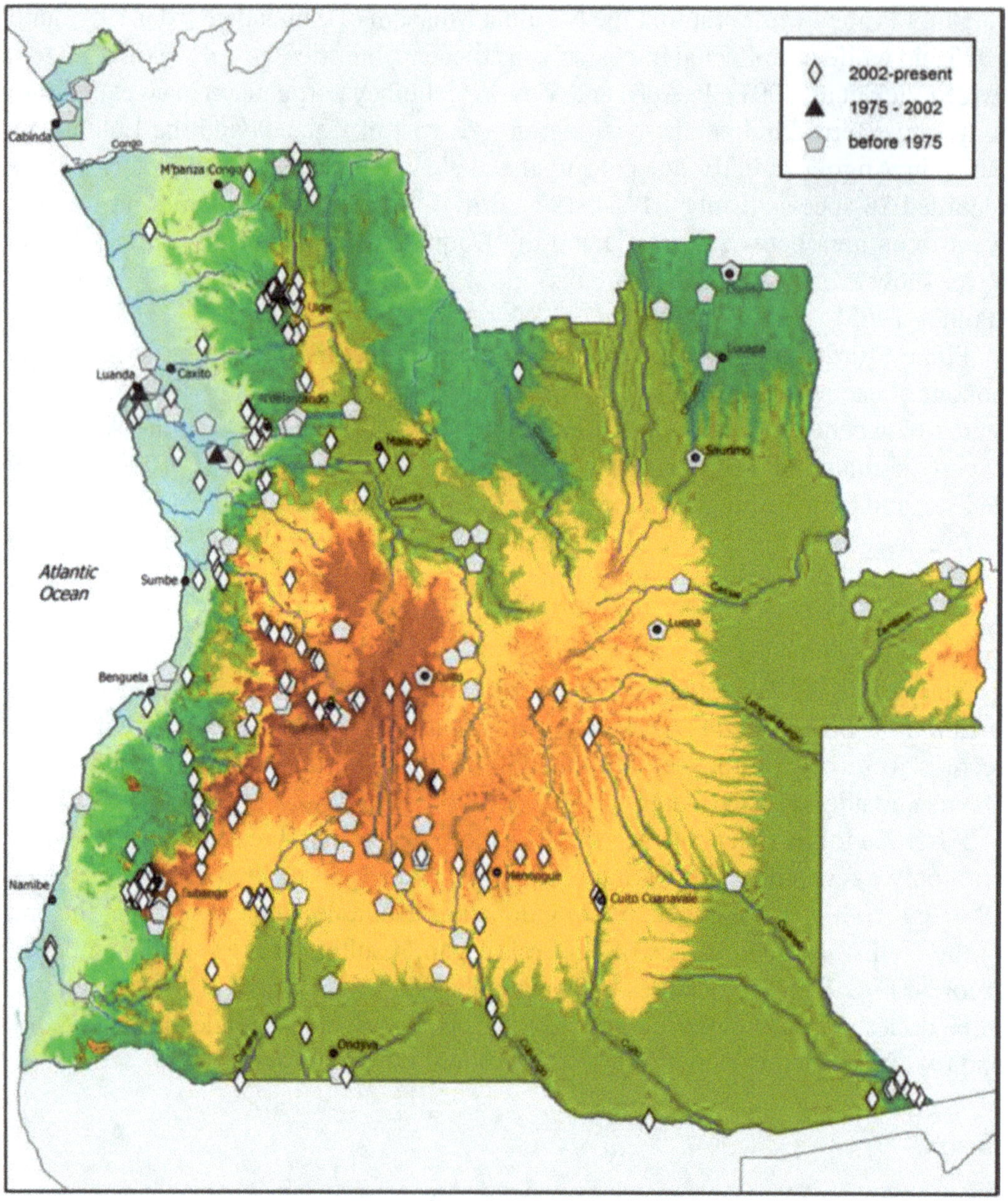

Fig. 9.2 Records of Odonata from Angola before 1975, before 2002 and up to 2018

Origin of Recent Data

In January 2009, an expedition led by Brian Huntley visited the Serra da Chela in southwestern Angola and the Namib Desert to the south. During that survey Warwick Tarboton collected and photographed Odonata around Humpata (7 field days).

Jens Kipping surveyed the upper catchment of the Okavango (Cubango) River on the SAREP (Southern African Regional Environmental Program) Expedition from 5 to 22 May 2012 (18 field days). A second SAREP survey visited southeastern Angola with the Cubango and Cuando River floodplains in April 2013.

Viola Clausnitzer and K-D B Dijkstra in collaboration with the Universidade Kimpa Vita (Uíge) and the Technical University of Dresden (Germany) surveyed around Uíge, Negage and Ndalatando in northern Angola in the wet season from 13 November to 1 December 2012 (19 days). Dijkstra revisited this area in the dry season, from 26 September to 5 October 2013 (10 days).

From 27 November to 10 December 2016 (14 field days), Manfred Haacks and colleagues of SASSCAL (Southern African Science Service Centre for Climate Change and Adaptive Land Management) visited Bicuar NP and a few other places in southern Angola.

Sara F Elizalde and David Elizalde, Chris Hines, André Günther, Raik Moritz and Jens Kipping surveyed the Serra da Chela around Lubango and the mountain range stretching from Huambo northwards to Gabela from 30 November to 19 December 2017 (20 field days).

Sara F Elizalde, Chris Hines, Rogério Ferreira and other experts provided many photographic records from 2016 to 2018.

The National Geographic Okavango Wilderness Project (NGOWP 2018) gathered scattered data on Odonata which has not yet been fully considered, except for some field photographs and exceptional records provided by John Mendelsohn.

Apart from the field surveys the authors also examined the Angolan collections and type material in the Natural History Museum in London, the National History Museum of Zimbabwe in Bulawayo (Dijkstra 2007a, b), the Royal Museum for Central Africa in Tervuren, Belgium and the *Instituto de Investigação Agronómica* in Huambo, Angola. All records are kept in the Odonata Database of Africa – ODA (Kipping et al. 2009) and mapped per species on African Dragonflies and Damselflies Online – ADDO (visit http://addo.adu.org.za/ also for further information about all mentioned species).

Odonata Species Recorded in Angola

From all the historic sources mentioned above, 152 species of Odonata were known to occur in Angola until 2009. Some of the formerly published species had to be deleted from the country list in the light of new taxonomic knowledge and after careful validation of all records (see Kipping et al. 2017).

In 2009 Warwick Tarboton recorded 47 species of Odonata at the Serra da Chela of which five were recorded in Angola for the first time and one was new to science (Tarboton 2009, Dijkstra et al. 2015). The first SAREP Expedition in 2012 yielded 87 species, 17 of them new to the country list and two new to science (Kipping 2012, Dijkstra et al. 2015). One additional species new for the country came from a second SAREP Expedition in April 2013 of which all collected specimens were examined. The first expedition to Uíge, Negage and Ndalatando resulted in 138 species, of which 43 were recorded for the first time in Angola and five were new to science. The second visit produced 86 species, adding another 15 to the national list. With the surveys from 2009 to 2013 and a careful review of the historic data, the known odonate fauna of Angola had increased from 152 species in the year 2009 to 234 species in 2013: an increase of about one-third with only 54 days in the field. Two species were added in 2016 by photographs made by Chris Hines and specimens from the collection of the *Instituto de Investigação Agronómica* in Huambo provided by Sara F Elizalde. The state of knowledge at the end of 2016 was published as the checklist of the dragonflies and damselflies of Angola by Kipping et al. in early 2017 (free download: https://africaninvertebrates.pensoft.net/article/11382/).

The SASSCAL expedition in November–December 2016 recorded 44 species, amongst them one new species for Angola. The latest survey in December 2017 yielded 88 species of which 10 were new for the country list, amongst them probably three species new to science. A further 14 species new for Angola were recorded only in 2017 and early 2018 by Chris Hines and colleagues mostly in northern Angola.

The updated checklist of the Odonata of Angola, now of 260 species, is provided in Appendix 1. The ODA database now holds about 4900 Angolan records from more than 400 localities. All species of Appendix 1 are reliably recorded from Angolan territory. Footnotes will give further information about the 25 additional species to the updated country list and one species that was deleted from it.

There are 15 more species listed in Appendix 2 that are known to occur at rivers bordering the country with Namibia and Zambia. The Namibian bank of the Okavango River is very well surveyed (Suhling and Martens 2007, 2014) and most of the mentioned species derive from this river. These species were technically not found on the Angolan riverbank and therefore not included in the checklist. But naturally they belong to the country's fauna.

Composition

Angola's rich dragonfly fauna expresses its geographic position, size and diversity. Its territory, especially in the north, falls within a region with an estimated highly diverse fauna (Fig. 9.3). Dijkstra et al. (2011) observed that roughly half of tropical African species occur predominantly within the extensive lowland forests of western and central Africa, a quarter is associated with the eastern and southern part dominated by highlands, while the remaining quarter occurs in open habitats throughout much of the Afrotropics. Indeed, about half of Angola's species are widespread across the continent and its exceptional diversity can be attributed to two major sources. Almost 30% are confined to forest habitats in the north, mostly below 1000 m altitude. Nine species confined to the Lower Guinea, the forest area that stretches between the Congo Basin and Atlantic Ocean from Cameroon to Gabon and western Congo, reaching their southern limit in northwestern Angola. Nearly 20% favour the swamps, grasslands, miombo woodlands and gallery forests that stretch eastwards, mostly above 1000 m asl. This fauna is concentrated in Katanga and northern Zambia but has now been proven to extend across to the

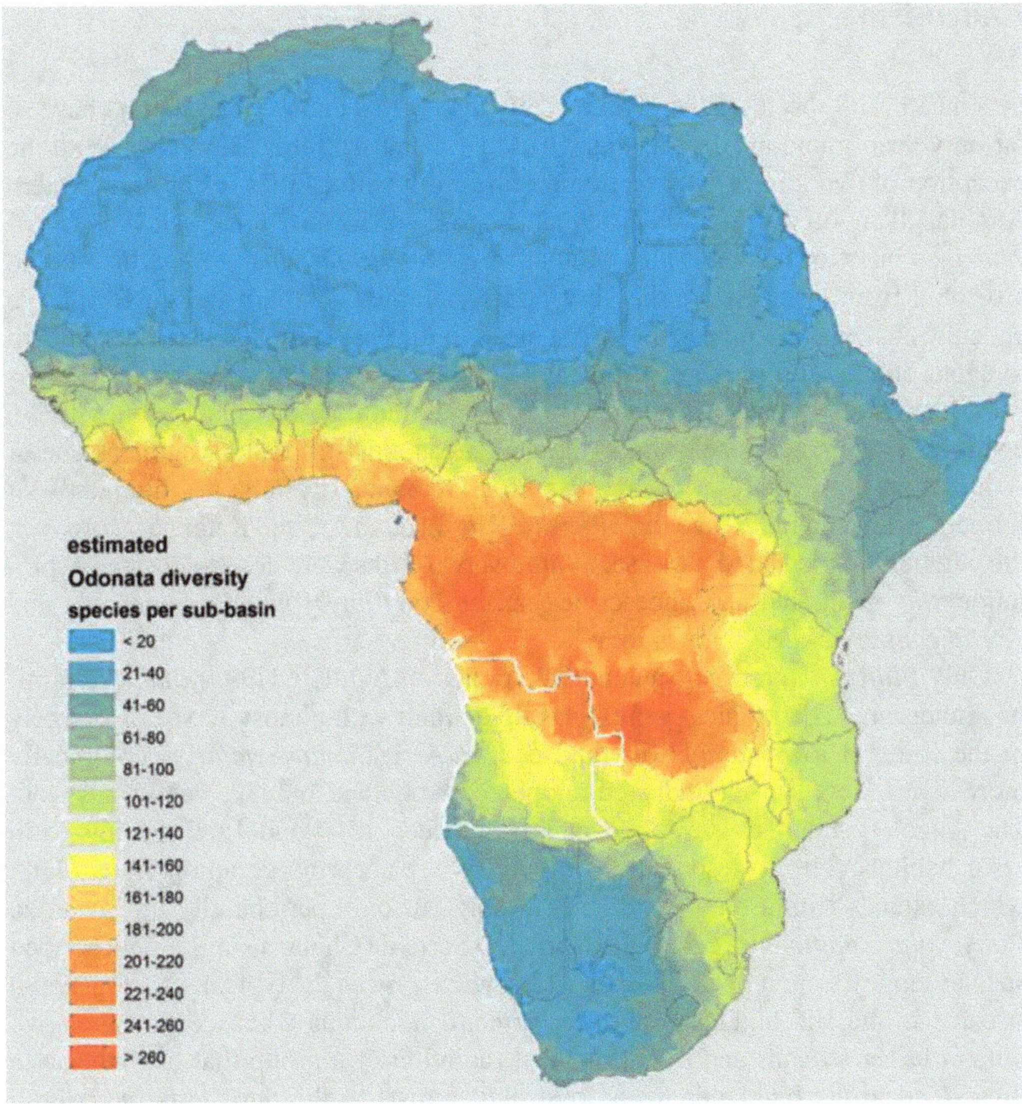

Fig. 9.3 Spatial estimation of Odonata diversity in continental Africa, based on the summation of the inferred ranges of all 770 species known; mapped as the number of species per Hydro1K basin. (Adapted from Clausnitzer et al. 2012). Angola is outlined in white

Angolan upland. This was confirmed by the discovery of *Orthetrum kafwi* at two localities in Cuanza-Sul in December 2017. The species was until then only known from Upemba NP in southern DRC. A number of palustrine species, e.g. *Anax bangweuluensis, Pinheyagrion angolicum*, *Pseudagrion deningi* and *P. rufostigma*, prefer larger marsh areas and swamps as in the Okavango Delta of Botswana (see Kipping 2010) and spreading north into Bié highlands with the headwaters of that river system. The discovery of *Trithemis integra* near Uíge is also of special interest, as it had seemed to be endemic to the Albertine Rift, being known previously only from western Tanzania and Uganda and eastern DRC.

Endemism

Seventeen valid species and several recently discovered undescribed species have so far only been found in Angola (Figs. 10.4, 10. 5 and 10.6 for examples). With the exception of two known only from their alleged type localities in far northeastern Angola, all are limited to the central plateau: the type locality for *Platycypha rubriventris* is questionable as it may be that of *Pseudagrion dundoense*, which could also be a river species from the very poorly sampled southern Congo Basin. No endemics have been found below 1200 m asl in the east, although some drop down to about 500 m west of the escarpment. While the proportion of endemics (7%) is lower than for Ethiopia (12 endemics; 11%) and South Africa (30 endemics; 18%), countries that also enclose distinct highland areas, this still ranks Angola as one of Africa's greatest centres of endemism for Odonata, rivalling the highlands of Cameroon (13 endemics) and the Albertine Rift, Eastern Arc and Katanga. Moreover, the number is expected to increase, as almost two-fifths were described since exploration was reinitiated and undescribed species of *Platycypha*, *Paragomphus* and *Tetrathemis* are already known to us.

Only *Platycypha* presents an endemic radiation. While *Chlorocypha* (the family's other large Afrotropical genus) has diversified with almost 30 species largely in the forested lowlands of west and central Africa, *Platycypha* is ecologically more diverse, with species adapted to open, submontane and lake habitats as well. The Angolan endemics are found mainly between 1300 and 1800 m altitude in open habitats. The widespread *P. angolensis* replaces the common *P. caligata*, which extends from South Africa to Ethiopia but only peripherally into Angola. *Platycypha bamptoni* is probably confined to Serra da Chela; a similar undescribed species appears more widespread. *Platycypha crocea* is typical of very small streams in the Bié highlands and escarpment mountains whereas the other two inhabit larger streams and rivers. A local radiation of a group that has otherwise diversified in the highlands to the east, and forests to the north, fits the overall affinities of Angola's endemic Odonata both geographically and ecologically (Kipping et al. 2017) (Fig. 9.4).

The four endemic *Pseudagrion* species have separate origins but similar links: the nearest relatives of *P. angolense* and *P. estesi* appear to be the rainforest species *P. grilloti* Legrand, 1987 and *P. kibalense* respectively. The former is limited to Congo and Gabon, but the latter extends to Cameroon and Uganda. *P. sarepi* is closely related to *P. fisheri* and *P. greeni*, both of which extend from Angola into Zambia. While these species belong to the genus's A-group, the B-group species *P. dundoense* is known only from Dundo and may not be endemic at all (see above).

Notogomphus kimpavita is the sister-species of *N. praetorius* found in highlands across southern Africa (including Angola), while *Eleuthemis eogaster* is nearest to an unnamed species from Gabon (Dijkstra et al. 2015).

Molecular data for *Umma femina* and *Onychogomphus rossii* are not available yet. By its unusual habitus and colouration *Umma femina* (see Fig. 9.5) is a very distinct member of the genus. It is definitely the odonate flagship species of the

Fig. 9.4 Photographs of some of Angola's (near) endemic dragonflies and damselflies. (**a**) Sarep Sprite (*Pseudagrion sarepi*), (**b**) Blue Wisp (*Agriocnemis angolensis*) that just extends into Namibia and Zambia, (**c**) Angola Claspertail (*Onychogomphus rossii*), (**d**) Angola Longleg (*Notogomphus kimpavita*), (**e**) Sunrise Firebelly (*Eleuthemis eogaster*), (**f**) Angola Micmac (*Micromacromia flava*). (All males, photographs a–d by J Kipping, e–f by K-DB Dijkstra)

Angolan highlands and probably also most threatened. The morphology of *O. rossii* is close to other pale *Onychogomphus* species from the open plateaus stretching from Angola to Zambia and Katanga.

Thus, like the majority of Angola's Odonata, most endemics probably originated quite recently and proximally from the forests to the north and open habitats to the east. However, some affinities are unresolved and potentially more distant: *Agriocnemis toto* and especially *A. canuango* have no obvious close relatives (Dijkstra et al. 2015), while the near-endemic *A. angolensis* and *A. bumhilli* are probably related to each other but even more distinct overall (Kipping et al. 2017).

Fig. 9.5 Photographs of some of Angola's endemic damselflies. (**a**) Angola Blue Jewel (*Platycypha crocea*), (**b**) Highland Blue Jewel (*Platycypha bamptoni*), (**c**) undescribed Blue Jewel (*Platycypha* sp. nov.), (**d**) Angola Dancing Jewel (*Platycypha angolensis*), (**e**) Angola Sparklewing (*Umma femina*), (**f**) Stout Threadtail (*Elattoneura tarbotonorum*), (**g**) Angola Sprite (*Pseudagrion angolense*), (**h**) Estes's Sprite (*Pseudagrion estesi*). (All males, photographs by J Kipping)

These data suggest that Angola may be the centre of diversification of this genus, which includes Africa's smallest damselflies. The morphologically very distinct *Aciagrion rarum* is only known from very few specimens from Lunda Sul Province in the northeast and molecular data is not available yet.

Micromacromia flava is morphologically nearest *M. miraculosa* (Förster 1906), known only from the East Usambara Mountains of north-eastern Tanzania and the only one of four *Micromacromia* species adapted to non-forest habitats, being strongly pruinose with maturity. *Elattoneura tarbotonorum* may be closest to *E. frenulata* of southwestern South Africa (Dijkstra et al. 2015): after its discovery at the Serra da Chela in 2009 it was found more widespread in December 2017 along the mountain range stretching north into Cuanza-Sul.

Potential for Discovery

If we compare the tallies for the well-studied neighbouring countries of Zambia and Namibia, the total number of species in Angola should lie somewhat above 300, meaning that less than 80% of the fauna is currently known. All Odonata expeditions in modern times surveyed areas that are easily accessible.

Additions can be expected throughout the country, but especially in the remote regions on the eastern and particularly northern border, as species diversity is expected to be extraordinarily high in the transition to the Congolian rainforest (Fig. 9.3). The province of Lunda-Norte with only 92 recorded species and 162 records should be the richest area for discovery, around Dundo where exploration began in the 1950s. Generally, all the northern and eastern provinces are largely unsurveyed and the discrepancy between the amount of available data, the number of known species and the expected diversity is extremely high. This applies also to the provinces of Lunda-Sul (10 species, 11 records), Zaire (17 species, 21 records), Malanje (35 species, 152 records) and Moxico (46 species, 51 records). An exception is Uíge where recent surveys increased the number of known species to 145 from 820 records gathered.

The central highlands can also yield more surprises, like the discovery of additional endemic species, with three areas being especially notable. Firstly, despite having most records, the north-south directed mountain range that lies entirely above 1600 m asl and includes the Serra do Chilengue, Serra da Chela and Angola's highest peak at Mount Moco (2620 m asl) is poorly sampled as the large gaps in Fig. 9.2 illustrates.

Secondly, except for its extreme northern and southern ends, the western escarpment has only been surveyed recently, which already led to the discovery of an undescribed *Paragomphus* species from Cumbira Forest. Even more easily accessible provinces such as Bengo and Cuanza-Sul will prove to be much richer in species than currently known. The potential of these mountains is illustrated by the discovery of a spectacular and unique but unknown species by Chris Hines and Rogério Ferreira in May 2018 (Fig. 9.6). Two males at a stream that flows off the

Fig. 9.6 An undescribed species that probably belongs to *Trithemis*, although the extensive markings and dense veins in the wings are unusual even for that highly diverse genus. Two males were observed at a stream running off Namba Mountains in Cuanza-Sul. (Photograph by R Ferreira)

Namba Mountains in Cuanza-Sul were photographed but not collected. These mountains reach over 2000 m in altitude and harbour larger pockets of Afromontane forest than Mt. Moco (Mills et al. 2013). They are known for their plant endemism and fieldwork there will definitely lead to the discovery of more endemic Odonata.

Thirdly, an extensive plateau at 1200–1600 m altitude stretches east from the Bié Highlands. Except for its southern edge, this area shared between Bié and Moxico Provinces, which is almost as large as Uganda (or the United Kingdom), has almost no records. A few collections from the NGS Okavango Wilderness Project suggest that more new species for the country and for science can be expected here. These deep Kalahari sands are the 'watertower' of Angola and its neighbours, incorporating the headwaters of the Cuito, Cuando, Chicapa, Cuango, Cuanza and large tributaries of the Congo and Zambezi such as the Cassai and Lungué-Bungo. The sources of the vast catchments of the Congo, Cuanza, Okavango and Zambezi meet in a small area between Munhango and Cangonga. Watersheds are prone to endemism (Dijkstra et al. 2011) and this region is the top priority for further research.

Studying insect collections of Angola's museums will be also a valuable source of more records and possibly even to get insight into past conditions in the light of the recent landscape change. Of special interest is the Dundo Museum that holds many interesting specimens of Odonata, some of which have been published. This remarkable collection has not been studied since independence but survived the civil war. There is also material dispersed over several museums in Europe and probably also in private collections, mainly in Portugal.

New species are most likely to be found among genera prone to narrow (highland) ranges, i.e. with known Angolan endemics like *Platycypha* and *Pseudagrion*, but also *Agriocnemis*, *Elattoneura*, *Notogomphus* and *Paragomphus*. Also possible is the discovery of endemics in genera that are well represented across the country and continent, and that have highland endemics elsewhere but not in Angola, such as *Africallagma*, *Neodythemis* and *Orthetrum*. However, given the biogeographic diversity of Angola's fauna and endemics, we could expect greater surprises. Among forest genera with no known Angolan endemic, *Allocnemis* seems most likely to reveal one, e.g. on the escarpment. The presence (or local endemism) of distinctly Lower Guinean genera like *Neurolestes*, *Africocypha*, *Pentaphlebia* and *Stenocnemis* seems less likely, but the Lower Guinean *Stenocypha gracilis* (Karsch 1899) has four endemic relatives in the Albertine Rift and the sister-taxon of the Upper and Lower Guinean *Tragogomphus* is *Nepogomphoides stuhlmanni* (Karsch 1899) in the Eastern Arc, suggesting an Angolan taxon is possible.

Some typical African highland genera are notably absent from Angola. *Atoconeura* is most likely to be present above 1400 m asl, being found in Zambia, Katanga, the Lower Guinea and Albertine Rift. However, its absence also from South Africa suggests historical factors may have been limiting, e.g. that the highlands were too harsh in cooler periods and too isolated when habitats were suitable (Dijkstra 2006).

This might not apply to *Proischnura*, present in South Africa as well as Cameroon and the Albertine Rift. However, that genus is absent from Katanga and northern Zambia, which lies lower and thus possibly provided no stepping-stone to the mountains of Angola. Kipping et al. (2017) also noted the absence of *Zosteraeschna* and *Pinheyschna*, which have a similar range (although the latter does occur in Katanga and northern Zambia), but isolated populations of *Z. minuscula* (McLachlan 1895) and *P. subpupillata* (McLachlan 1896) were discovered in the Serra da Chela in southern Angola in December 2017.

Conservation

Our findings show that Angola's wealth of aquatic habitats harbours a rich freshwater fauna. Although large areas are relatively untouched, Angola's rapid economic and population growth will have a tremendous impact on the environment and thus human well-being in the future. In the light of this, Angola's development should consider (1) the establishment of sewage works in cities and larger villages; (2) a stop to deforestation, especially along stream courses; (3) restoration of deforested water catchments; (4) village-level awareness campaigns for sustainable use of freshwater sources, e.g. no detergents and waste dumping in rivers; (5) biodiversity surveys and monitoring to feed into a national conservation plan.

With the exception of four species, all endemics are currently considered Data Deficient for the IUCN Red List of Threatened Species. *Platycypha angolensis*, *Pseudagrion angolense* and *Micromacromia flava* are Near Threatened because, while they seem fairly widespread, their dependence on relatively natural habitats may put them at risk as human development progresses. Only *Umma femina* is now listed as threatened. It is currently known from only a few sites in the fairly densely populated highlands around Lubango and seems to inhabit exclusively the smaller and cooler highland streams. There is much development in this densely settled region and increasing pressure on those habitats by grazing, deforestation and urbanisation. As it seems to prefer cool mountain streams we can assume additional risks from climate change and it is therefore thought to be Vulnerable to extinction. More research on all endemic species' statuses and ecology is urgently required.

Angola has an exceptional fauna of dragonflies and damselflies, as well as many valuable rivers and wetlands. Odonata are excellent indicators of the health and biodiversity of both the freshwater and terrestrial realm. As the biological survey of Angola advances, they should be a priority taxon.

Acknowledgements We are grateful to Her Excellency Madame Minister of Environment Dr. Paula C Francisco Coelho (MINAMB) for making the SAREP survey in southern Angola possible, to Dr. Chris Brooks of SAREP for the preparation and organization of the 2012 survey, to Marta Alexandre Zumbo (MINAMB), Maria Helena Loa (MINAMB), Julius Bravo (MINAMB), Francisco de Almeida (INIP), Manuel Domingos (INIP) and Gabriel Cabinda (Agriculture and Rural Development and Fisheries) for their help in organization and management on the 2012 tour, and to Vince Shacks and Werner Conradie for collecting specimens on the second SAREP survey in 2013. We thank Alvaro Bruno Toto Nienguesso, the driving force behind biodiversity research in Uíge Province, Angola, Prof Dr. Neinhuis and Dr. Thea Lautenschlaeger from TU Dresden for inviting us to the field survey in Uíge province. Part of the fieldwork in Angola was supported by a travel fund from the German Academic Exchange Service (DAAD). We thank Dr. Aristófanes Pontes, director of the *Instituto Nacional da Biodiversidade e Áreas de Conservação* (INBAC) for the support during the December 2017 expedition, by providing the necessary permits. Chris Hines provided many valuable photographic records. Further records were provided by Warwick Tarboton, Dr. Manfred Haacks (SASSCAL), John Mendelsohn (RAISON) and Rogério Ferreira. The latter also gave permission to use his wonderful photograph.

Appendices

Appendix 1

Checklist of Odonata recorded from Angola.

– see Taxonomic comments in Kipping et al. (2017); ## – see footnotes following this table;

(V) Validation of species: "1!" new national record made by the authors; "1!!" new national record made by the authors and addendum to Kipping et al. (2017); "1" records obtained by authors and confirming existing records; "2" specimens kept in collections (identification confirmed or primary types); "3" literature records, regarded as reliable because specimens were described well or location agrees with known biogeographic pattern; "4!!" new national record made by other persons and addendum to Kipping et al. (2017); ** – range restricted to Angola; * – range restricted to Angola with very few exceptions (see Endemism in the discussion).

(RL) Global conservation status according to the IUCN Red List of Threatened Species (2016): CR (Critically Endangered), DD (Data-Deficient), EN (Endangered), NT (Near-Threatened), VU (Vulnerable), LC (Least Concern), NE (Not Evaluated)

Scientific name	English name	V	RL
Lestidae			
Lestes amicus (Martin, 1910)	Yellow-winged Spreadwing	1	LC
Lestes dissimulans (Fraser, 1955)	Cryptic Spreadwing	1	LC
Lestes pallidus (Rambur, 1842)	Pallid Spreadwing	1	LC
Lestes pinheyi (Fraser, 1955)	Pinhey's Spreadwing	1	LC
Lestes plagiatus (Burmeister, 1839)	Highland Spreadwing	1	LC
Lestes tridens (McLachlan, 1895)	Spotted Spreadwing	1	LC
Lestes virgatus (Burmeister, 1839)	Smoky Spreadwing	3	LC
Calopterygidae			
Phaon camerunensis (Sjöstedt, 1900)	Emerald Demoiselle	1!	LC
Phaon iridipennis (Burmeister, 1839)	Glistening Demoiselle	1	LC
Sapho orichalcea (McLachlan, 1869)#	Mountain Bluewing	1!	LC
Umma electa (Longfield, 1933)	Metallic Sparklewing	1	LC
Umma femina (Longfield, 1947)	Angola Sparklewing	1**	VU
Umma longistigma (Selys, 1869)	Bare-bellied Sparklewing	1	LC
Umma mesostigma (Selys, 1879)	Hairy-bellied Sparklewing	1!	LC
Chlorocyphidae			
Chlorocypha aphrodite (Le Roi, 1915)##	Blue Jewel	4!!	LC
Chlorocypha cancellata (Selys, 1879)	Exquisite Jewel	1!	LC
Chlorocypha curta (Hagen in Selys, 1853)	Blue-tipped Jewel	1!	LC
Chlorocypha cyanifrons (Selys, 1873)	Blue-fronted Jewel	1!	LC
Chlorocypha fabamacula (Pinhey, 1961)	Spotted Jewel	1	LC
Chlorocypha victoriae (Förster, 1914)	Victoria's Jewel	1	LC
Platycypha angolensis (Longfield, 1959)	Angola Dancing Jewel	1**	NT
Platycypha bamptoni (Pinhey, 1975)#	Highland Blue Jewel	1**	NE
Platycypha cf. *bamptoni* (Pinhey, 1975)#	(near Highland Blue Jewel)	1!**	NE
Platycypha caligata (Selys, 1853)#	Common Dancing Jewel	2	LC
Platycypha crocea (Longfield, 1947)#	Angola Blue Jewel	1**	LC
Platycypha rubriventris (Pinhey, 1975)#	Red-bellied Blue Jewel	2**	DD
Platycypha rufitibia (Pinhey, 1961)	Beautiful Jewel	1	LC
Platycnemididae			
Allocnemis nigripes (Selys, 1886)	Rainbow Yellowwing	1	LC
Allocnemis pauli (Longfield, 1936)	Orange-tipped Yellowwing	1!	LC
Copera congolensis (Martin, 1908)	Congo Featherleg	1!	LC
Elattoneura acuta (Kimmins, 1938)	Red Threadtail	1!	LC
Elattoneura cellularis (Grünberg, 1902)#	Zambezi Threadtail	3	LC
Elattoneura cf. *glauca* (Selys, 1860)#	(near Common Threadtail)	1	LC
Elattoneura lliba (Legrand, 1985)	Eastern Stream Threadtail	1!	LC
Elattoneura tarbotonorum (Dijkstra, 2015)#	Stout Threadtail	1**	DD
Mesocnemis singularis (Karsch, 1891)##	Common Riverjack	1!!	LC
Mesocnemis cf. *singularis* (Karsch, 1891)#	(near Common Riverjack)	1!	NE
Coenagrionidae			
Aciagrion africanum (Martin, 1908)	Blue Slim	1	LC
Aciagrion macrootithenae (Pinhey, 1972)	Awl-tipped Slim	3	DD

(continued)

Scientific name	English name	V	RL
Aciagrion nodosum (Pinhey, 1964)	Cryptic Slim	1!	LC
Aciagrion rarum (Longfield, 1947)	Tiny Slim	2**	DD
Aciagrion steeleae (Kimmins, 1955)	Swamp Slim	3	LC
Aciagrion zambiense (Pinhey, 1972)	Zambia Slim	3	DD
Africallagma fractum (Ris, 1921)	Slender Bluet	1	LC
Africallagma glaucum (Burmeister, 1839)	Swamp Bluet	1	LC
Africallagma sinuatum (Ris, 1921)##	Peak Bluet	4!!	LC
Africallagma subtile (Ris, 1921)##	Fragile Bluet	1!!	LC
Africallagma vaginale (Sjöstedt, 1917)	Forest Bluet	1!	LC
Agriocnemis angolensis (Longfield, 1947)	Blue Wisp	1*	LC
Agriocnemis bumhilli (Kipping, Suhling & Martens, 2012)	Bumhill Wisp	1!*	LC
Agriocnemis canuango (Dijkstra, 2015)	Bog Wisp	1!**	DD
Agriocnemis exilis (Selys, 1872)	Little Wisp	1	LC
Agriocnemis forcipata (Le Roi, 1915)	Greater Pincer-tailed Wisp	1	LC
Agriocnemis gratiosa (Gerstäcker, 1891)##	Gracious Wisp	4!!	LC
Agriocnemis cf. *maclachlani* (Selys, 1877)#	(near Forest Wisp)	1!	LC
Agriocnemis pinheyi (Balinsky, 1963)##	Pinhey's Wisp	1!!	LC
Agriocnemis ruberrima (Balinsky, 1961)	Orange Wisp	1!	LC
Agriocnemis toto (Dijkstra, 2015)	Bruno's Wisp	1!**	DD
Agriocnemis victoria (Fraser, 1928)	Lesser Pincer-tailed Wisp	1	LC
Azuragrion nigridorsum (Selys, 1876)	Sailing Bluet	1	LC
Ceriagrion annulatum (Fraser, 1955)	Green-eyed Citril	1!	LC
Ceriagrion bakeri (Fraser, 1941)	Blue-fronted Citril	3	LC
Ceriagrion corallinum (Campion, 1914)	Green-fronted Citril	1	LC
Ceriagrion glabrum (Burmeister, 1839)	Common Citril	1	LC
Ceriagrion junceum (Dijkstra & Kipping, 2015)	Spikerush Citril	1!	LC
Ceriagrion platystigma (Fraser, 1941)	Variable Citril	1	LC
Ceriagrion sakejii (Pinhey, 1963)	Cream-sided Citril	1!	LC
Ceriagrion suave (Ris, 1921)	Plain Citril	1	LC
Ceriagrion whellani (Longfield, 1952)	Yellow-faced Citril	1!	LC
Ischnura senegalensis (Rambur, 1842)	Tropical Bluetail	1	LC
Pinheyagrion angolicum (Pinhey, 1966)	Pinhey's Bluet	1	LC
Pseudagrion (A) *angolense* (Selys, 1876)	Angola Sprite	1**	NT
Pseudagrion (A) *coeruleipunctum* (Pinhey, 1964)	Pretty Sprite	3	LC
Pseudagrion (A) *estesi* (Pinhey, 1971)	Estes's Sprite	1**	LC
Pseudagrion (A) *fisheri* (Pinhey, 1961)	Dark-tailed Sprite	3	LC
Pseudagrion (A) *greeni* (Pinhey, 1961)	Clasper-tailed Sprite	1	LC
Pseudagrion (A) *inconspicuum* (Ris, 1931)	Little Sprite	1	LC
Pseudagrion (A) *kersteni* (Gerstäcker, 1869)	Powder-faced Sprite	1	LC
Pseudagrion (A) *kibalense* (Longfield, 1959)	Forest Sprite	1	LC
Pseudagrion (A) *makabusiense* (Pinhey, 1950)	Green-striped Sprite	3	LC
Pseudagrion (A) *melanicterum* (Selys, 1876)	Farmbush Sprite	1	LC
Pseudagrion (A) *salisburyense* (Ris, 1921)	Slate Sprite	1	LC

(continued)

Scientific name	English name	V	RL
Pseudagrion (A) *sarepi* (Kipping & Dijkstra, 2015)	Sarep Sprite	1!**	DD
Pseudagrion (A) *serrulatum* (Karsch, 1894)	Superb Sprite	1!	LC
Pseudagrion (A) *simonae* (Legrand, 1987)	Wide-striped Sprite	1!	LC
Pseudagrion (A) *simplicilaminatum* (Carletti & Terzani, 1997)##	Blue Slim Sprite	4!!	LC
Pseudagrion (B) *acaciae* (Förster, 1906)	Acacia Sprite	1	LC
Pseudagrion (B) *camerunense* (Karsch, 1899)##	Yellow-fronted Sprite	4!!	LC
Pseudagrion (B) *coeleste* (Longfield, 1947)	Catshead Sprite	1	LC
Pseudagrion (B) *deningi* (Pinhey, 1961)	Dark Sprite	1!	LC
Pseudagrion (B) *dundoense* (Longfield, 1959)	Dundo Sprite	2**	DD
Pseudagrion (B) *glaucescens* (Selys, 1876)	Blue-green Sprite	1	LC
Pseudagrion (B) *hamoni* (Fraser, 1955)	Swarthy Sprite	1!	LC
Pseudagrion (B) *helenae* (Balinsky, 1964)	Little Blue Sprite	1!	LC
Pseudagrion (B) *isidromorai* (Compte Sart, 1967)	Large Blue Sprite	1!	LC
Pseudagrion (B) *massaicum* (Sjöstedt, 1909)	Masai Sprite	1	LC
Pseudagrion (B) *rufostigma* (Longfield, 1947)	Ruby Sprite	1	LC
Pseudagrion (B) *sjoestedti* (Förster, 1906)	Variable Sprite	1	LC
Pseudagrion (B) *sublacteum* (Karsch, 1893)	Cherry-eye Sprite	1	LC
Aeshnidae			
Afroaeschna scotias (Pinhey, 1952)	Shadow Hawker	1!	LC
Anaciaeschna triangulifera (McLachlan, 1896)##	Evening Hawker	4!!	LC
Anax bangweuluensis (Kimmins, 1955)##	Swamp Emperor	4!!	NT
Anax congoliath (Fraser, 1953)	Dark Emperor	1!	LC
Anax ephippiger (Burmeister, 1839)	Vagrant Emperor	1	LC
Anax imperator (Leach, 1815)	Blue Emperor	1	LC
Anax speratus (Hagen, 1867)	Eastern Orange Emperor	1	LC
Anax tristis (Hagen, 1867)	Black Emperor	1	LC
Gynacantha (A) *sextans* (McLachlan, 1896)	Dark-rayed Duskhawker	3	LC
Gynacantha (A) *vesiculata* (Karsch, 1891)	Lesser Girdled Duskhawker	3	LC
Gynacantha (B) *bullata* (Karsch, 1891)	Black-kneed Duskhawker	1	LC
Gynacantha (B) *manderica* (Grünberg, 1902)	Little Duskhawker	3	LC
Heliaeschna cynthiae (Fraser, 1939)##	Blade-tipped Duskhawker	4!!	LC
Heliaeschna fuliginosa (Karsch, 1893)	Black-banded Duskhawker	1	LC
Heliaeschna ugandica (McLachlan, 1896)	Uganda Duskhawker	3	LC
Pinheyschna subpupillata (McLachlan, 1896)##	Stream Hawker	1!!	LC
Zosteraeschna minuscula (McLachlan, 1895)##	Friendly Hawker	1!!	LC
Gomphidae			
Crenigomphus cf. *cornutus* (Pinhey, 1956)#	(near Horned Talontail)	1!	LC
Diastatomma selysi (Schouteden, 1934)	Common Hoetail	3	LC
Diastatomma soror (Schouteden, 1934)	Painted Hoetail	3	LC
Gomphidia quarrei (Schouteden, 1934)	Southern Fingertail	3	LC
Ictinogomphus dundoensis (Pinhey, 1961)	Swamp Tigertail	1	LC
Ictinogomphus ferox (Rambur, 1842)	Common Tigertail	1	LC
Ictinogomphus regisalberti (Schouteden, 1934)	Congo Tigertail	3	LC

(continued)

Scientific name	English name	V	RL
Lestinogomphus calcaratus (Dijkstra, 2015)	Spurred Fairytail	1!	LC
Libyogomphus tenaculatus (Fraser, 1926)	Large Horntail	1!	LC
Mastigogomphus chapini (Klots, 1944)#	Western Snorkeltail	2	LC
Mastigogomphus dissimilis (Cammaerts, 2004)##	Southern Snorkeltail	2	LC
Microgomphus cf. *nyassicus* (Grünberg, 1902)#	(near Eastern Scissortail)	1!	LC
Neurogomphus alius (Cammaerts, 2004)	Large Siphontail	1!	LC
Notogomphus kimpavita (Dijkstra & Clausnitzer, 2015)	Angola Longleg	1!**	DD
Notogomphus praetorius (Selys, 1878)	Yellowjack Longleg	2	LC
Notogomphus spinosus (Karsch, 1890)	Jungle Longleg	1!	LC
Onychogomphus rossii (Pinhey, 1966)	Angola Claspertail	1**	DD
Onychogomphus cf. *styx* (Pinhey, 1961)#	(near Northern Dark Claspertail)	1!	LC
Paragomphus abnormis (Karsch, 1890)	Humdrum Hooktail	1!	LC
Paragomphus cognatus (Rambur, 1842)	Rock Hooktail	1!!	LC
Paragomphus cf. *darwalli* (Dijkstra, Mézière & Papazian, 2015)#	(near Darwall's Hooktail)	1!	DD
Paragomphus genei (Selys, 1841)	Common Hooktail	1	LC
Paragomphus machadoi (Pinhey, 1961)	Forest Hooktail	2	LC
Paragomphus cf. *nigroviridis* (Cammaerts, 1969)#	(near Black-and-green Hooktail)	1!	LC
Paragomphus sabicus (Pinhey, 1950)##	Flapper Hooktail	1!!	LC
Paragomphus sp. nov. ##	(Hooktail, undescribed species)	1!!**	NE
Phyllogomphus annulus (Klots, 1944)	Crested Leaftail	1	LC
Phyllogomphus selysi (Schouteden, 1933)	Bold Leaftail	3	LC
Macromiidae			
Phyllomacromia aureozona (Pinhey, 1966)	Golden-banded Cruiser	1!	LC
Phyllomacromia contumax (Selys, 1879)	Two-banded Cruiser	1!	LC
Phyllomacromia hervei (Legrand, 1980)	River Cruiser	1!	LC
Phyllomacromia melania (Selys, 1871)	Sombre Cruiser	1	LC
Phyllomacromia overlaeti (Schouteden, 1934)	Clubbed Cruiser	3	LC
Phyllomacromia paula (Karsch, 1892)	Greater Double-spined Cruiser	3	LC
Phyllomacromia picta (Hagen in Selys, 1871)	Darting Cruiser	3	LC
Phyllomacromia unifasciata (Fraser, 1954)	Golden-eyed Cruiser	3	LC
Libellulidae			
Acisoma inflatum (Selys, 1882)	Stout Pintail	1	LC
Acisoma trifidum (Kirby, 1889)	Pied Pintail	1	LC
Aethiothemis bequaerti (Ris, 1919)	Skimmer-like Flasher	1	LC
Aethiothemis ellioti (Lieftinck, 1969)	Plump Flasher	1!	LC
Aethiothemis mediofasciata (Ris, 1931)#	Orange Flasher	2	LC
Aethiothemis solitaria (Martin, 1908)	Pearly Flasher	1	LC
Aethriamanta rezia (Kirby, 1889)	Pygmy Basker	1	LC
Brachythemis lacustris (Kirby, 1889)	Red Groundling	1	LC

(continued)

Scientific name	English name	V	RL
Brachythemis leucosticta (Burmeister, 1839)	Southern Banded Groundling	1	LC
Bradinopyga strachani (Kirby, 1900)[##]	Red Rockdweller	1!!	LC
Chalcostephia flavifrons (Kirby, 1889)	Inspector	1!	LC
Crocothemis brevistigma (Pinhey, 1961)	Spotted Scarlet	1!	LC
Crocothemis divisa (Baumann, 1898)	Rock Scarlet	1	LC
Crocothemis erythraea (Brullé, 1832)	Broad Scarlet	1	LC
Crocothemis sanguinolenta (Burmeister, 1839)	Little Scarlet	1	LC
Cyanothemis simpsoni (Ris, 1915)	Bluebolt	1!	LC
Diplacodes deminuta (Lieftinck, 1969)	Little Percher	1	LC
Diplacodes lefebvrii (Rambur, 1842)	Black Percher	1	LC
Diplacodes luminans (Karsch, 1893)	Barbet Percher	1	LC
Diplacodes pumila (Dijkstra, 2006)	Dwarf Percher	1!	LC
Eleuthemis eogaster (Dijkstra, 2015)	Sunrise Firebelly	1!**	DD
Eleuthemis libera (Dijkstra & Kipping, 2015)	Free Firebelly	1!	DD
Hadrothemis camarensis (Kirby, 1889)	Saddled Jungleskimmer	3	LC
Hadrothemis coacta (Karsch, 1891)	Robust Jungleskimmer	1!	LC
Hadrothemis defecta (Karsch, 1891)	Scarlet Jungleskimmer	3	LC
Hemistigma albipunctum (Rambur, 1842)	African Piedspot	1	LC
Malgassophlebia bispina (Fraser, 1958)	Ringed Leaftipper	1!	LC
Micromacromia camerunica (Karsch, 1890)	Stream Micmac	1!	LC
Micromacromia flava (Longfield, 1947)	Angola Micmac	1**	NT
Neodythemis afra (Ris, 1909)	Seepage Junglewatcher	1!	LC
Neodythemis klingi (Karsch, 1890)	Stream Junglewatcher	1!	LC
Nesciothemis cf. *farinosa* (Förster, 1898)[#]	(near Eastern Blacktail)	1	LC
Nesciothemis fitzgeraldi (Longfield, 1955)	Lesser Peppertail	1!	LC
Notiothemis jonesi (Ris, 1919)[##]	Eastern Forestwatcher	1!!	LC
Notiothemis robertsi (Fraser, 1944)	Western Forestwatcher	1!	LC
Olpogastra lugubris (Karsch, 1895)	Bottletail	1	LC
Orthetrum abbotti (Calvert, 1892)	Little Skimmer	1	LC
Orthetrum austeni (Kirby, 1900)	Giant Skimmer	1	LC
Orthetrum brachiale (Palisot de Beauvois, 1817)	Banded Skimmer	1	LC
Orthetrum caffrum (Burmeister, 1839)	Two-striped Skimmer	1	LC
Orthetrum chrysostigma (Burmeister, 1839)	Epaulet Skimmer	1	LC
Orthetrum guineense (Ris, 1910)	Guinea Skimmer	1	LC
Orthetrum hintzi (Schmidt, 1951)	Dark-shouldered Skimmer	1	LC
Orthetrum icteromelas (Ris, 1910)	Spectacled Skimmer	1	LC
Orthetrum julia (Kirby, 1900)	Julia Skimmer	1	LC
Orthetrum kafwi (Dijkstra, 2015)[##]	Bog Skimmer	1!!	DD
Orthetrum machadoi (Longfield, 1955)	Highland Skimmer	1	LC
Orthetrum macrostigma (Longfield, 1947)	Sharkfin Skimmer	1	LC
Orthetrum microstigma (Ris, 1911)	Farmbush Skimmer	1	LC
Orthetrum monardi (Schmidt, 1951)	Woodland Skimmer	1	LC
Orthetrum robustum (Balinsky, 1965)	Robust Skimmer	1!	LC

(continued)

Scientific name	English name	V	RL
Orthetrum saegeri (Pinhey, 1966)	Eastern Mushroom Skimmer	1!	LC
Orthetrum stemmale (Burmeister, 1839)	Bold Skimmer	1	LC
Orthetrum trinacria (Selys, 1841)	Long Skimmer	1	LC
Oxythemis phoenicosceles (Ris, 1910)	Pepperpants	1!	LC
Palpopleura albifrons (Legrand, 1979)	Pale-faced Widow	1!	LC
Palpopleura deceptor (Calvert, 1899)	Deceptive Widow	3	LC
Palpopleura jucunda (Rambur, 1842)	Yellow-veined Widow	1	LC
Palpopleura lucia (Drury, 1773)	Lucia Widow	1	LC
Palpopleura portia (Drury, 1773)	Portia Widow	1	LC
Pantala flavescens (Fabricius, 1798)	Wandering Glider	1	LC
Porpax asperipes (Karsch, 1896)	Powdered Pricklyleg	1	LC
Porpax risi (Pinhey, 1958)	Highland Pricklyleg	1	LC
Rhyothemis fenestrina (Rambur, 1842)	Skylight Flutterer	1	LC
Rhyothemis mariposa (Ris, 1913)	Butterfly Flutterer	2	LC
Rhyothemis cf. *notata* (Fabricius, 1781)[##]	(near Veiled Flutterer)	4!!	LC
Rhyothemis semihyalina (Desjardins, 1832)	Phantom Flutterer	1!	LC
Sympetrum fonscolombii (Selys, 1840)	Nomad	1	LC
Tetrathemis camerunensis (Sjöstedt, 1900)	Forest Elf	2	LC
Tetrathemis fraseri (Legrand, 1977)	Treefall Elf	1!	LC
Tetrathemis polleni (Selys, 1869)	Black-splashed Elf	2	LC
Tetrathemis sp. nov. [##]	(Elf, undescribed species)	4!!**	NE
Thermochoria equivocata (Kirby, 1889)	Dash-winged Piedface	1!	LC
Tholymis tillarga (Fabricius, 1798)	Twister	1	LC
Tramea basilaris (Palisot de Beauvois, 1817)	Keyhole Glider	1	LC
Trithemis aconita (Lieftinck, 1969)	Halfshade Dropwing	1!	LC
Trithemis aenea (Pinhey, 1961)[##]	Bronze Dropwing	4!!	LC
Trithemis annulata (Palisot de Beauvois, 1807)	Violet Dropwing	1	LC
Trithemis anomala (Pinhey, 1956)	Striped Dropwing	1!	LC
Trithemis apicalis (Fraser, 1954)	Furtive Dropwing	1!	LC
Trithemis arteriosa (Burmeister, 1839)	Red-veined Dropwing	1	LC
Trithemis basitincta (Ris, 1912)	Jungle Dropwing	1!	LC
Trithemis dichroa (Karsch, 1893)	Black Dropwing	1	LC
Trithemis dorsalis (Rambur, 1842)	Highland Dropwing	1	LC
Trithemis cf. *dubia* (Fraser, 1954)[#]	(near Sleek Dropwing)	1!	DD
Trithemis furva (Karsch, 1899)	Navy Dropwing	1	LC
Trithemis imitata (Pinhey, 1961)[#]	Northern Fluttering Dropwing	1!	LC
Trithemis integra (Dijkstra, 2007)	Albertine Dropwing	1!	LC
Trithemis kirbyi (Selys, 1891)	Orange-winged Dropwing	1	LC
Trithemis leakeyi (Pinhey, 1956)	Mealy Dropwing	1!	LC
Trithemis monardi (Ris, 1931)[#]	Southern Fluttering Dropwing	1	LC
Trithemis nuptialis (Karsch, 1894)	Hairy-legged Dropwing	1	LC
Trithemis palustris (Damm & Hadrys, 2009)[#]	Marsh Dropwing	1!	LC

(continued)

Scientific name	English name	V	RL
Trithemis pluvialis (Förster, 1906)	Russet Dropwing	1	LC
Trithemis pruinata (Karsch, 1899)	Cobalt Dropwing	1!	LC
Trithemis stictica (Burmeister, 1839)	Jaunty Dropwing	1	LC
Trithemis werneri (Ris, 1912)	Elegant Dropwing	3	LC
Trithemis sp. nov. (Fig 9.6)[##]	(Dropwing, undescribed species)	4!!**	NE
Urothemis assignata (Selys, 1872)	Red Basker	1	LC
Urothemis edwardsii (Selys, 1849)	Blue Basker	1	LC
Urothemis venata (Dijkstra & Mézière, 2015)[##]	Red-veined Basker	4 !!	LC
Zygonoides fuelleborni (Grünberg, 1902)	Southern Riverking	3	LC
Zygonyx denticulatus (Dijkstra & Kipping, 2015)	Pale Cascader	1!	LC
Zygonyx eusebia (Ris, 1912)	Imperial Cascader	3	LC
Zygonyx flavicosta (Sjöstedt, 1900)[##]	Ensign Cascader	1	LC
Zygonyx natalensis (Martin, 1900)	Blue Cascader	1	LC
Zygonyx regisalberti (Schouteden, 1934)	Regal Cascader	1	LC
Zygonyx torridus (Kirby, 1889)	Ringed Cascader	1	LC

Notes on new country records (by J Kipping and S F Elizalde unless stated otherwise)

Chlorocypha aphrodite – male photographed by C Hines near Lucala north of Uíge in June 2017.

Mesocnemis singularis – first record of true *M. singularis* (see Kipping et al. 2017) from the Angolan bank of the Cunene River in December 2017.

Africallagma sinuatum – single male photographed by C Hines near Cambondo, Cuanza-Norte Province in February 2017.

Africallagma subtile – several collected at marshy floodplains of the Yevedula River, 20 km northwest of Caconda, Benguela Province in December 2017.

Agriocnemis gratiosa – several collected by M Haacks from Bicuar NP, Huíla Province in December 2016.

Agriocnemis pinheyi – several collected at a marsh northwest of Caconda, Benguela Province in December 2017.

Pseudagrion (A) *simplicilaminatum* – male photographed by C Hines near Lucala north of Uíge in June 2017.

Pseudagrion (B) *camerunense* – male photographed by C Hines in Cuanza River floodplains south of Luanda in January 2018.

Anaciaeschna triangulifera – female photographed by C Hines in Cuanza River floodplains south of Luanda in June 2017. Westernmost record; nearest locality is Ikelenge in northwestern Zambia, about 1200 km to the east.

Anax bangweuluensis – teneral male photographed by J Mendelsohn at Lake Saliakembo, Moxico Province in October 2017. The Cuito River links to the nearest known population in the Okavango Delta, Botswana, about 750 km away.

Heliaeschna cynthiae – female and two males recorded by C Hines at the Rio Nzadi and near Quicunga in Uíge Province in June 2017.

Pinheyschna subpupillata – many observed and collected at Tchiamena River near Lubango and Neve River near Humpata on Serra da Chela, Huíla Province in

December 2017. Presumably isolated population; widespread in South Africa, with another isolated population on border of Mozambique and Zimbabwe. With the new finding of this species a former record of a female *P. rileyi* (Calvert, 1892) from Tundavala (Pinhey 1975) became more doubtful and the species is therefore deleted from the country list.

Zosteraeschna minuscula – male collected at the Tchiamena River near Lubango on Serra da Chela, Huíla Province in December 2017. Northernmost record; widespread in South Africa but with scattered records in Namibia and eastern Botswana.

Mastigogomphus dissimilis – the *Instituto de Investigação Agronómica* in Huambo has one male from Nova Sintra (Catabola), Bié Province from October 1973, coll. L Amorim.

Paragomphus cognatus – presence in Angola was uncertain due to lack of reliable material (Kipping et al. 2017), but several males collected at Tchiamena, Leba and Neve Rivers in the Serra da Chela in December 2017.

Paragomphus sabicus – common at the Rio Coporolo, north of Chongoroi, Benguela Province in December 2017.

Paragomphus sp. nov. – two males collected at the Uiri River near Conda, Cuanza-Sul Province in December 2017 belong to an undescribed species similar to *P. cognatus* but darker and with stouter paraprocts and more curved cerci.

Bradinopyga strachani – the *Instituto de Investigação Agronómica* in Huambo has three males from Ndalatando, Cuanza-Norte Province from March 1973, coll. U Passos. Several also collected at Rio Mussenju, south of Quilengues, Benguela Province in December 2017 and photographed by R Ferreira at Calandula Falls, Lunda-Norte Province in June 2018.

Notiothemis jonesi – male was collected in Lubango, Huíla Province in December 2017.

Orthetrum kafwi – several males and females collected at boggy streams and bogs in the highlands around Cassongue, Cuanza-Sul Province in December 2017. Previously know only from the type locality in the Upemba National Park in Katanga, which lies 1400 km to the east.

Rhyothemis cf. *notata* – male photographed by J Mendelsohn at Sacangombe near the Cuito River source in Moxico Province in November 2011. The black markings in the forewings reach only to the nodus and in the hindwings halfway the nodus and pterostigma, which is much less than even the palest variation of *R. notata* illustrated by Dijkstra & Clausnitzer (2014). The habitat is open, while true *R. notata* favour rainforest conditions. This species therefore needs to be verified with specimens.

Tetrathemis sp. nov. – several males photographed by C Hines in dry forest near Cambondo, Cuanza-Norte Province in March 2017. Differs from *T. fraseri* by the smoky wings and shape of the very hairy cerci.

Trithemis aenea – photographed by C Hines near Lucala north of Uíge in June 2017.

Trithemis sp. nov. – see Fig. 9.6 and main text.

Urothemis venata – photographed by Carel van der Merwe in the Cuango area, Cuanza-Norte Province in May 2017.

Appendix 2

Odonata recorded from rivers bordering Angola that most likely also occur in Angola.

Scientific name	English name	Nearest occurrence
Coenagrionidae		
Pseudagrion (A) *spernatum* (Selys, 1881)	Upland Sprite	At Jimbe and other rivers in Ikelenge Pedicle of north-western Zambia.
Pseudagrion (B) *assegaii* (Pinhey, 1950)	Assegai Sprite	Cuando River in Namibian Caprivi Strip.
Pseudagrion (B) *sudanicum* (Le Roi, 1915)	Blue-sided Sprite	Okavango and Cuando Rivers in Namibian Caprivi Strip.
Gomphidae		
Crenigomphus kavangoensis (Suhling & Marais, 2010)	Kavango Talontail	Okavango River in Namibia.
Lestinogomphus angustus (Martin, 1911)	Common Fairytail	Cunene, Okavango and Cuando Rivers in northern Namibia.
Lestinogomphus silkeae (Kipping, 2010)	Silke's Fairytail	One locality on the southern bank of the Okavango River near Rundu, Namibia.
Paragomphus cataractae (Pinhey, 1963)	Cataract Hooktail	Waterfalls and rapids of the Cunene and Okavango Rivers in northern Namibia.
Paragomphus elpidius (Ris, 1921)	Corkscrew Hooktail	Cunene, Okavango and Cuando River in northern Namibia and the Ikelenge Pedicle of Zambia.
Neurogomphus cocytius Cammaerts, 2004	Kokytos Siphontail	Okavango River in northern Namibia.
Libellulidae		
Parazyxomma flavicans (Martin, 1908)	Banded Duskdarter	Okavango and Cuando Rivers in northern Namibia.
Trithemis aequalis (Lieftinck, 1969)	Swamp Dropwing	Okavango and Cuando Rivers in the Namibian Caprivi.
Trithemis donaldsoni (Calvert, 1899)	Denim Dropwing	Okavango and Cunene Rivers in northern Namibia.
Trithemis hecate (Ris, 1912)	Silhouette Dropwing	Common along the Cunene, Okavango and Cuando Rivers in northern Namibia.
Trithemis morrisoni (Damm & Hadrys, 2009)	Rapids Dropwing	Okavango and Cuando Rivers in the Namibian Caprivi.
Trithetrum navasi (Lacroix, 1921)	Fiery Darter	Cunene, Okavango and Cuando Rivers in northern Namibia.

References

Clausnitzer V, Koch R, Dijkstra K-DB et al (2012) Focus on African freshwaters: hotspots of dragonfly diversity and conservation concern. Front Ecol Environ 10:129–134

Damm S, Hadrys H (2009) *Trithemis morrisoni* sp. nov. and *Trithemis palustris* sp. nov. from the Okavango and Upper Zambezi Floodplains previously hidden under *T. stictica* (Odonata: Libellulidae). Int J Odonatol 12(1):131–145

Dijkstra K-DB (2006) The *Atoconeura* problem revisited: taxonomy, phylogeny and biogeography of a dragonfly genus in the highlands of Africa (Odonata, Libellulidae). Tijdschrift voor Entomologie 149:121–144

Dijkstra K-DB (2007a) The name-bearing types of Odonata held in the Natural History Museum of Zimbabwe, with systematic notes on Afrotropical taxa. Part 1: introduction and Anisoptera. Int J Odonatol 10(1):1–29

Dijkstra K-DB (2007b) The name-bearing types of Odonata held in the Natural History Museum of Zimbabwe, with systematic notes on Afrotropical taxa. Part 2: Zygoptera and description of new species. Int J Odonatol 10(2):137–170

Dijkstra K-DB, Clausnitzer V (2014) The dragonflies and damselflies of Eastern Africa: handbook for all Odonata from Sudan to Zimbabwe. . Studies in afrotropical zoology 298. Royal Museums for Central Africa, Tervuren, 263 pp

Dijkstra K-DB, Kipping J, Mézière N (2015) Sixty new dragonfly and damselfly species from Africa (Odonata). Odonatologica 44(4):447–678

Dijkstra K-DB, Boudot J-P, Clausnitzer V et al (2011) Chapter 5. Dragonflies and damselflies of Africa (Odonata): history, diversity, distribution, and conservation. In: Darwall WRT, Smith KG, Allen DJ et al (eds) The diversity of life in African freshwaters: under water, under threat. An analysis of the status and distribution of freshwater species throughout mainland Africa. IUCN, Cambridge and Gland, 347 pp

Kipping J (2010) The dragonflies and damselflies of Botswana – an annotated checklist with notes on distribution, phenology, habitats and Red List status of the species (Insecta: Odonata). Mauritiana (Altenburg) 21:126–204

Kipping J (2012) Southern African Regional Environmental Program (SAREP) – first biodiversity field survey upper Cubango (Okavango) catchment, Angola, May 2012 – Dragonflies & Damselflies (Insecta: Odonata). Expert Report:1–108

Kipping J, Clausnitzer V, Fernandes Elizalde SRF et al (2017) The dragonflies and damselflies (Odonata) of Angola. Afr Invertebr 58(I):65–91 https://africaninvertebrates.pensoft.net/article/11382/

Kipping J, Dijkstra K-DB, Clausnitzer V et al (2009) Odonata Database of Africa (ODA). Agrion 13:20–23

Longfield C (1936) Studies on African Odonata, with synonymy and descriptions of new species and subspecies. Trans R Entomol Soc Lond 85:467–499

Longfield C (1947) The Odonata of South Angola: results of the Mission Scientifiques Suisses 1928–29, 1932–33. Arquivos do Museu Bocage 16:1–31

Longfield C (1955) The Odonata of North Angola, part 1. Publicações Culturais, Companhia de Diamantes de Angola 27:11–64

Longfield C (1959) The Odonata of North Angola, part 2. Publicações Culturais, Companhia de Diamantes de Angola 45:16–42

Mendes LF, Bivar-de-Sousa A, Figueira R et al (2013) Gazetteer of the Angolan localities known for beetles (Coleoptera) and butterflies (Lepidoptera: Papilionoidea). Boletim da Sociedade Portuguesa de Entomologia 228(VIII–14):257–292

Mills MSL, Melo M, Vaz A (2013) The Namba mountains: new hope for Afromontane forest birds in Angola. Bird Conserv Int 23:159–167

NGOWP – National Geographic Okavango Wilderness Project (2018) Initial Findings from Exploration of the Upper Catchments of the Cuito, Cuanavale and Cuando Rivers in Central and South-Eastern Angola (May 2015 to December 2016). National Geographic Okavango Wilderness Project, 352 pp

Pinhey ECG (1961a) A collection of Odonata from Dundo, Angola with the descriptions of two new species of Gomphids. Publicações Culturais, Companhia de Diamantes de Angola 56:71–78

Pinhey ECG (1961b) Some dragonflies (Odonata) from Angola and descriptions of three new species of the family Gomphidae. Publicações Culturais, Companhia de Diamantes de Angola 56:79–86

Pinhey ECG (1964) Dragonflies (Odonata) of the Angola-Congo borders of Rhodesia. Publicações Culturais, Companhia de Diamantes de Angola 63:97–130

Pinhey ECG (1965) Odonata from Luanda and the Lucala River, Angola. Revista de Biologia 5:159–164

Pinhey ECG (1966) New distributional records for African Odonata and notes on a few larvae. Arnoldia Rhodesia 2(26):1–5

Pinhey ECG (1971a) Notes on the genus *Pseudagrion* Selys (Odonata: Coenagrionidae). Arnoldia Rhodesia 5(6):1–4

Pinhey ECG (1971b) Odonata collected in Republique Centre-Africaine by R. Pujol. Arnoldia Rhodesia 5(18):1–16

Pinhey ECG (1974) A revision of the African *Agriocnemis* Selys and *Mortonagrion* Fraser (Odonata: Coenagrionidae). Occasional Papers of the National Monuments of Rhodesia B 5/4:171–278

Pinhey ECG (1975) A collection of Odonata from Angola. Arnoldia Rhodesia 7(23):1–16

Pinhey ECG (1984) A check-list of the Odonata of Zimbabwe and Zambia. Smithersia 3:1–64

Ris F (1931) Odonata aus Süd-Angola. Revue Suisse Zoologie 38(7):97–112

Suhling F, Martens A (2007) Dragonflies and damselflies of Namibia. Gamsberg Macmillan, Windhoek, 280 pp

Suhling F, Martens A (2014) Distribution maps and checklist of Namibian Odonata. Libellula Suppl 13:107–175

Tarboton W (2009) A dragonfly survey of the Humpata District. In: Huntley BJ (ed) Projecto de estudo da biodiversidade de Angola. (Biodiversity Rapid Assessment – Huíla /Namibe) Report on Pilot project. SANBI, Cape Town, 3 pp

Vick GS, Chelmick DG, Martens A (2001) In memory of Elliot Charles Gordon Pinhey (10 July 1910 – 7 May 1999). Odonatologica 30:1–11

Chapter 10
The Butterflies and Skippers (Lepidoptera: Papilionoidea) of Angola: An Updated Checklist

Luís F. Mendes, A. Bivar-de-Sousa, and Mark C. Williams

Abstract Presently, 792 species/subspecies of butterflies and skippers (Lepidoptera: Papilionoidea) are known from Angola, a country with a rich diversity of habitats, but where extensive areas remain unsurveyed and where systematic collecting programmes have not been undertaken. Only three species were known from Angola in 1820. From the beginning of the twenty-first century, many new species have been described and more than 220 faunistic novelties have been assigned. As a whole, of the 792 taxa now listed for Angola, 57 species/subspecies are endemic and almost the same number are known to be near-endemics, shared by Angola and by one or another neighbouring country. The Nymphalidae are the most diverse family. The Lycaenidae and Papilionidae have the highest levels of endemism. A revised checklist with taxonomic and ecological notes is presented and the development of knowledge of the superfamily over time in Angola is analysed.

Keywords Africa · Conservation · Ecology · Endemism · Taxonomy

L. F. Mendes (✉)
Museu Nacional de História Natural e da Ciência, Universidade de Lisboa, Lisboa, Portugal

CIBIO, Centro de Investigação em Biodiversidade e Recursos Genéticos, Vairão, Portugal
e-mail: luisfmendes22@gmail.com

A. Bivar-de-Sousa
Museu Nacional de História Natural e da Ciência, Universidade de Lisboa, Lisboa, Portugal

Sociedade Portuguesa de Entomologia, Lisboa, Portugal
e-mail: abivarsousa@gmail.com

M. C. Williams
Pretoria University, Pretoria, South Africa
e-mail: lepidochrysops@gmail.com

B. J. Huntley et al. (eds.), *Biodiversity of Angola*,
https://doi.org/10.1007/978-3-030-03083-4_10

Introduction

Angola is a large country of 1,246,700 km^2, notable for its great diversity of physiography, climates, habitats and resultant biodiversity). The country includes seven biomes and 15 ecoregions, ranging from equatorial rainforests of the northwest (Cabinda) and along the northern border with the Democratic Republic of Congo, through the moist miombo woodlands and savannas of the central plateaus, to the dry forests and woodlands of the southeast, and to the arid shrublands and Namib Desert of the southwest. Isolated forests with Congolian affinities are found along the Angolan Escarpment, and similar remnant patches of Afromontane forests are found on some of the highest mountains such as Mount Moco and Mount Namba.

Despite the fact that at the beginning of the nineteenth century only a few species of butterflies and skippers (Insecta: Lepidoptera: Papilionoidea) were recorded from Angola, today a large number of taxa (at least 792 species and subspecies: Fig. 10.1, Table 10.1 and Appendix) are known to occur in the country. However, extensive areas are still poorly surveyed for butterflies, or have not been surveyed at all (Fig. 10.2). This applies in particular to the southern provinces of Namibe, Cunene and Cuando Cubango and the northwestern province of Zaire as well as most of southern Moxico. Furthermore, the Baixa de Cassanje (Malanje), separated from surrounding areas by steep escarpments, appears to have distinctive vegetation and may produce some interesting butterflies. Although most of the localities where but-

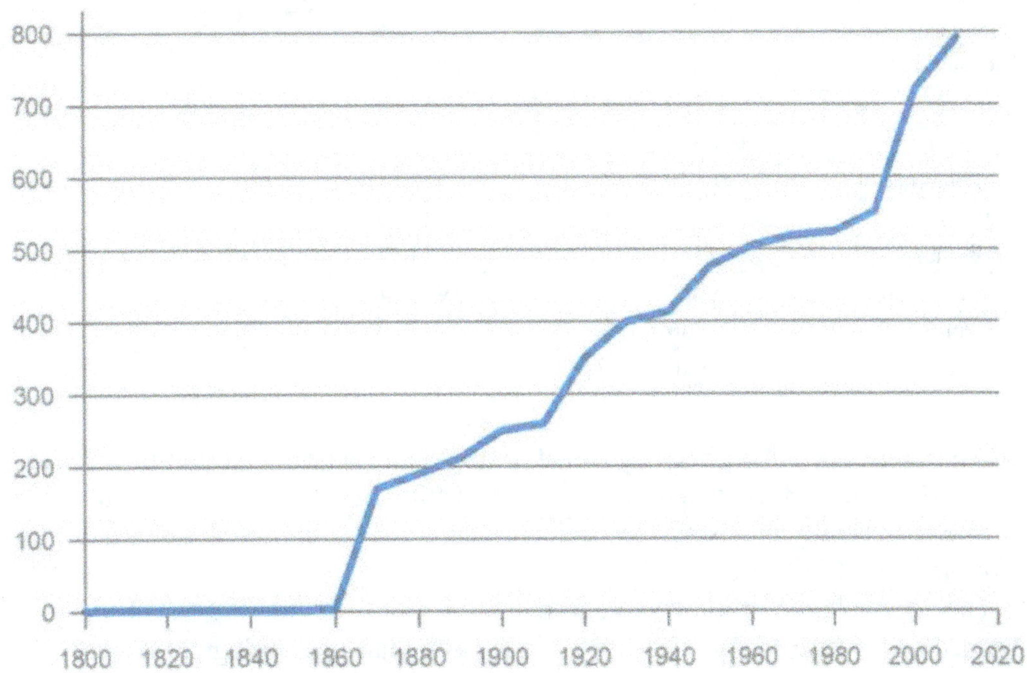

Fig. 10.1 Cumulative number of species/subspecies of Papilionoidea reported from Angola from 1801–1819 (first records) to the recent decade – 2011–2017 – according to Appendix. For practical reasons, species which first reference to the country was untraceable (marked in Appendix with a ▲) were included in the decade 2001–2010; species that are now assigned as faunistic novelties to Angola (marked in Appendix with a ◘) are included in the last decade (2011–2017)

Table 10.1 Number of species of Papilionoidea families and subfamilies known to occur in the Afrotropical Region and Angola (with % of Afrotropical species present in the country), and number of species endemic to Angola (with % of endemism shown)

Family	Afrotropical	Angola	Endemism
Subfamily	N°	N° I %	N I %
HESPERIIDAE	618	134 I 22	5 I 3.7
Coeliadinae	21	7 I 33	
Pyrginae	216	48 I 22	
Heteropterinae	27	5 I 19	
Hesperiinae	334	74 I 22	
PAPILIONIDAE	101	33 I 33	3 I 9.1
PIERIDAE	200	67 I 34	5 I 7.5
Pseudopontiinae	5	2 I 40	
Coliadinae	14	8 I 57	
Pierinae	181	57 I 32	
LYCAENIDAE	1837	210 I 11	18 I 8.6
Miletinae	119	11 I 9	
Poritiinae	658	53 I 8	
Theclinae	301	42 I 14	
Aphnaeinae	260	21 I 8	
Polyommatinae	496	83 I 17	
RIODINIDAE	15	4 I 27	0 I 0
NYMPHALIDAE	1634	344 I 21	26 I 7.6
Libytheinae	5	2 I 40	
Danainae	26	9 I 35	
Satyrinae	347	50 I 14	
Charaxinae	190	56 I 30	
Apaturinae	3	1 I 33	
Nymphalinae	73	35 I 48	
Cyrestinae	1	1 I 100	
Biblidinae	31	16 I 52	
Limenitidinae	702	97 I 14	
Heliconiinae	256	77 I 30	
TOTALS	4405	792 I 18	

terflies and skippers have been collected in Angola have been determined (Mendes et al. 2013b), some localities previously reported for a few species remain untraced despite searches by us, using the detailed maps of the *Junta de Investigações do Ultramar* (JIU 1948–1963).

The accumulation of knowledge in regard to Angolan butterflies has been constrained by several factors. The two largest Angolan entomological collections, deposited in the *Museu do Dundo* (Lunda-Norte) and in the *Instituto de Investigação Agronómica* (Huambo), have never been studied in detail. In addition little fieldwork was carried out in Angola during the post-independence period because of the protracted civil war. Finally, the vastness of the country and the difficulty in accessing many remote regions has impeded progress.

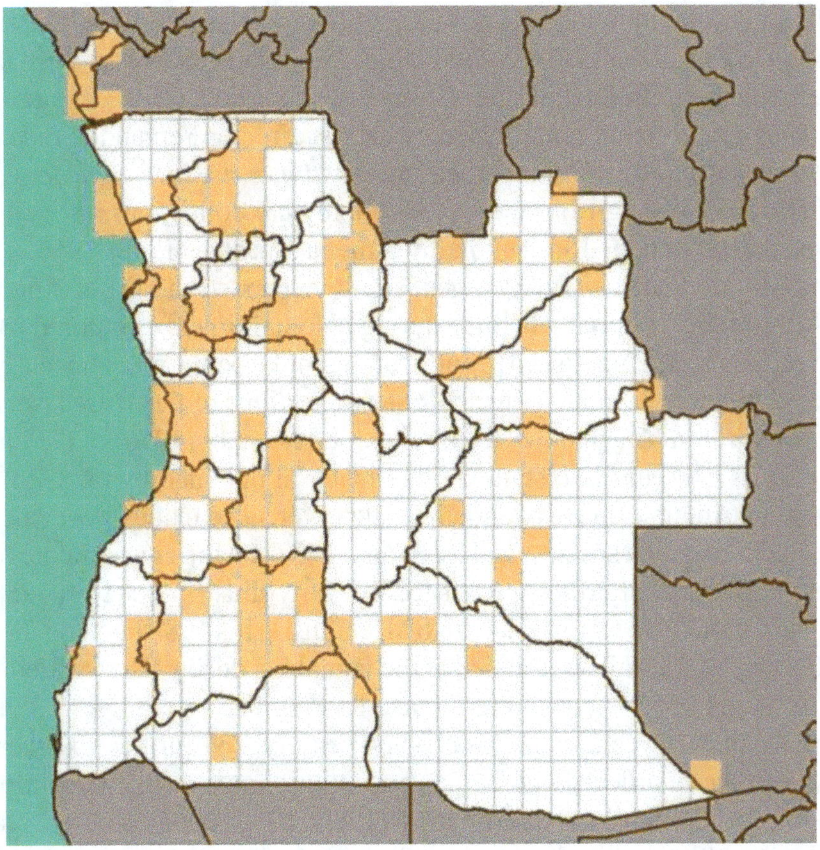

Fig. 10.2 Map of Angola showing, marked in orange, the known areas surveyed for the Papilionoidea from the beginning of their study, in the nineteenth century to the present day – each square ca. 33 × 33 km. The collecting pressure varies across the country, from 'squares' where samples were obtained only once, in passing, to others where the collectors were based for months

Until recently the Hesperiidae (skippers) were placed in the superfamily Hesperioidea, separate from the rest of the butterflies, which were placed in the superfamily Papilionoidea. However, today the skippers and butterflies are all placed in the Papilionoidea (e.g. Heikkilä et al. 2012). The classification used for butterflies in this chapter is based on Williams (2018), Espeland et al. (2018) and Dhungel and Wahlberg (2018). Six families of butterflies are represented in Angola, namely Papilionidae, Hesperiidae, Pieridae, Riodinidae, Lycaenidae and Nymphalidae.

History of Research on the Papilionoidea of Angola

The first known reference to butterflies obtained in Angola is by Latreille and Godart (1819), who reported the presence of *Colotis euippe* (Linnaeus, 1758) and described *Acraea parrhasia servona*. In the decade between 1871 and 1880, Druce (1875) reported about 90 species from Angola for the first time, a number of these being descriptions of species new to science. By the end of the nineteenth century a total of 214 butterfly taxa were known from Angola.

The first contributions to our knowledge of Angolan butterflies by Portuguese researchers were only made in the middle nineteen hundreds. These were the result of the activities in Angola of the *Centro de Zoologia* (CZ) of the *Junta de Investigações do Ultramar*, coordinated by its first director Fernando Frade. In 2014 this research institution was renamed the *Instituto de Investigação Científica Tropical* (IICT). Working from these large zoological collections, as well as from further specimens obtained in Angola by Amélia Bacelar (1948, 1956, 1958a, b, 1961) and Miguel Ladeiro (1956), considerably expanded the list of Angolan butterflies. Most of this material, obtained during colonial times, was stored in Lisbon, with corrections to the published identifications only being made recently. All of this material has now been integrated into the collections of the *Museu Nacional de História Natural e da Ciência* (MUHNAC). Albert Monard (1956) of the La Chaux-de-Fonds Swiss Museum also studied other material obtained by the CZ missions. Significant contributions in the twentieth century were also made by Weymer (1901) on the southern Angolan species, and by Evans (1937) on the Hesperiidae. All of the then known Angolan butterflies were listed by Aurivillius (in Seitz) in 1928. All of the Angolan Charaxinae were dealt with by Henning in his 1988 book on the African taxa of this family. The 339 taxa added to the faunal list during the twentieth century brought the total to 553 known butterfly taxa for Angola.

During the first 18 years of the twenty-first century, 239 further taxa were added to the total. In the first decade of the present century, most of the new information was due to several contributions by Libert (1999, 2000, 2004) on the Lycaenidae, and by Gardiner (2004). The latter author listed taxa from the southeastern Cuando Cubango province, which borders the Caprivi Strip of Namibia. Cuando Cubango and the easternmost province of Moxico are the only provinces in Angola with Zambezian fauna. To these taxa we add our own contributions (Bivar-de-Sousa and Mendes 2006, 2007, 2009a, b; Mendes and Bivar-de-Sousa 2006a, b, 2007a, b, 2009a, b, c, d). Over the last 8 years 33 species were described as new or recorded for the first time from Angola by Mendes and Bivar-de-Sousa (2012, 2017) Mendes et al. (2013a, 2017, 2018), Bivar-de-Sousa and Mendes (2014) and Bivar-de-Sousa et al. (2017), Turlin and Vingerhoedt (2013) and Pierre and Bernaud (2013). Finally, 66 further taxa are now recorded as faunistic novelties for the country (Appendix). The current total number of butterfly taxa for Angola now stands at 792.

Sources Consulted for the Checklist

In preparing this revised checklist of the Papilionoidea, the following collections of Angolan butterflies held by institutions in Portugal were examined: *Museu Nacional de História Natural e da Ciência* (MUHNAC) in Lisbon, *Museu de História Natural da Universidade do Porto* (MHNC-UP), *Liceu Nun'Álvares* in the Caldas da Saúde and the Singeverga Order of St Benedict Abbey in Areias. Major contributions to these collections were made by A Bivar-de-Sousa (Luanda district and Cuanza-Norte, Cuanza-Sul and Moxico Provinces), António Figueira, (northwestern Angola), Mário Macedo (northern Angola), Passos de Carvalho (Huambo and

Cuanza-Norte Provinces), Carneiro Mendes and Pessoa Guerreiro. The Angolan insect collection of Nozolino de Azevedo (mainly Huambo Province), maintained and made available by his widow, was also studied.

The collections in the MUHNAC in Lisbon were destroyed by a fire in March 1978. However, prior to the fire BS had studied some of the material and published his findings. In 1995 LM studied the collections, mainly of Barros Machado and Luna de Carvalho, in the Dundo Museum in Angola but there was insufficient time to do a detailed analysis. We did not inspect the entomological collections in the former *Instituto de Investigação Agronómica* de Angola, collected mainly by Passos de Carvalho, but they are apparently in good condition. No entomological collections were found by LM, in 1995 and 2013, at the *Museu de História Natural de Luanda*. Material collected from 2010 to 2014 by Ruben Capela and Carmen Van-Dúnen Santos of Agostinho Neto University, Luanda and Artur Serrano of the Faculty of Sciences, Lisbon University, was examined by us.

In addition, images of live specimens published by Lautenschläger and Neinhhuis (2014) were examined, as were several images presented by Jorge Palmeirim of Lisbon University and Pedro Vaz Pinto of the Kissama Foundation.

Taxa Excluded from the Checklist

A number of taxa have erroneously been reported to occur in Angola. This was due mainly to probable misidentifications or mislabelled specimens. Some older records are omitted because the known range of the taxon is unlikely to include Angola. A list of the omitted taxa is given below.

- Hesperiidae: *Eretis djaelaelae* (Wallengren, 1857), *Metisella metis* (Linnaeus, 1764), *Kedestes chaca* (Trimen, 1873), *Platylesches chamaeleon* (Mabille, 1891).
- Papilionidae: *Papilio menestheus* Drury, 1773, *Graphium taboranus* (Oberthür, 1886), *Graphium (Arisbe) junodi* (Trimen, 1893).
- Pieridae: *Eurema brigitta* (Stoll, 1780), *Colotis chrysonome* (Klug, 1829), *Colotis ephyia* (Klug, 1829), *Belenois theora* (Doubleday, 1846), *Mylothris rubricosta* (Mabille, 1890), *Mylothris similis* Lathy, 1906.
- Lycaenidae: *Telipna acraea* (Westwood, 1851), *Cooksonia abri* Collins & Larsen, 2008, *Mimacraea darwinia* Butler, 1872, *Liptena bassae* Bethune-Baker, 1926, *Aethiopana honorius honorius* (Fabricius, 1793), *Stempfferia uniformis* (Kirby, 1887), *Stempfferia dorothea* (Bethune-Baker, 1904), *Oxylides faunus* (Drury, 1773), *Dapidodigma hymen* (Fabricius, 1775), *Aloeides molomo* (Trimen, 1870), *Leptomyrina lara* (Linnaeus, 1764), *Deudorix livia* (Klug, 1834), *Neurellipes onias* (Hulstaert, 1924), *Zintha hintza* (Trimen, 1864).
- Riodinidae: *Afriodinia caeca semicaeca* (Riley, 1932), *Afriodinia gerontes* (Fabricius, 1781).
- Nymphalidae: *Bicyclus milyas* (Hewitson, 1864), *Ypthima congoana* Overlaet, 1955, *Charaxes jahlusa argynnides* Westwood, 1864, *Junonia touhilimasa*

Vuillot, 1892, *Neptis continuata* Holland, 1892, *Neptis strigata* Aurivillius, 1894, *Evena oberthueri* (Karsch, 1894), *Euriphene atrovirens* (Mabille, 1878), *Bebearia mardania* (Fabricius, 1793), *Euphaedra morini* Hecq, 1983, *Euphaedra xypete* (Hewitson, 1865), *Euphaedra campaspe* (Felder & Felder, 1867), *Euphaedra inanum* (Butler, 1873), *Euphaedra eupalus* (Fabricius, 1781).

A Revised Checklist of the Papilionoidea of Angola

A revised and annotated checklist of the Papilionoidea of Angola (Appendix) confirms the presence of at least 792 taxa in the country. Their presence is based mainly on verification by the authors of this chapter. Some taxa, recorded by other authors, are accepted because they were, with rare exceptions, reported by more than one author, are based on reliable literature records, or because Angola falls within their putative geographical range. In the Checklist, the first reference to their occurrence in Angola is given, followed by the sources of validation of the record and their preferred habitat(s). Occasionally more than one subspecies of a particular species occurs in the country. This is due to both the size and ecological diversity of Angola. A number of forests, especially gallery-forests, are independent of each other as are the fragmented forests of the Angolan Escarpment. In addition, the southeastern parts of Moxico and the Cuando Cubango provinces are part of the Zambesi Basin; consequently their fauna has affinities with that of eastern Africa.

As far as habitats are concerned the great majority of the Angolan Papilionoidea, as might be expected, occur in forest, both wet and dry (Appendix). However, the Hesperiidae and Pieridae appear to be almost as diverse in moist woodland (miombo) and dry woodland as they are in forest. The number of Pieridae in dry woodland and miombo is similar, while the number of species in dry woodland, arid shrubland and grassland surpasses that of wet forest. The subfamily Nymphalinae is more diverse in miombo than forest and equally diverse in savanna. The Heliconiinae (Nymphalidae) in savanna are almost as diverse as they are in wet forest.

Composition, Diversity and Endemism

All six families and all of the subfamilies (except the Lycaeninae, Leach, 1815) of Afrotropical butterflies are represented in Angola (Table 10.1).

One genus and 56 species/subspecies of Papilionoidea are endemic to Angola, many of which were described over the last few decades. The endemic genus *Mashunoides*, Mendes and Bivar-de-Sousa 2009a, b, c, d (Nymphalidae: Satyrinae) is confined to Cuando Cubango Province, in the ecotone between miombo and savanna/dry woodland mosaic. Endemism rates for Angolan butterfly families are highest for the Papilionidae and Lycaenidae and lowest for the Hesperiidae and Riodinidae (Tables 10.1 and 10.2). Examples of endemic species are illustrated in Fig. 10.3.

Table 10.2 Endemic butterfly species and subspecies in Angola

Family	Endemic species	Endemic subspecies
Hesperiidae	*Eagris multiplagata* *Abantis bergeri*	*Calleagris jamesoni ansorgei* *Eretis herewardi rotundimacula* *Spialia colotes colotes*
Papilionidae	*Papilio bacelarae* *Papilio chitondensis*	*Papilio macinnoni benguellae*
Pieridae	*Mylothris carvalhoi*	*Appias epaphia angolensis* *Appias phaola uigensis* *Appias sylvia ribeiroi* *Mylothris spica gabela*
Lycaenidae	*Alaena rosei* *Cooksonia nozolinoi* *Falcuna lacteata* *Deloneura barca* *Aloeides angolensis* *Zeritis krystyna* *Cupidesthes vidua* *Uranothauma nozolinoi Lepidochrysops ansorgei* *Lepidochrysops flavisquamosa* *Lepidochrysops fulvescens* *Lepidochrysops hawker* *Lepidochrysops nacrescens* *Lepidochrysops reichenowi*	*Liptena homeyeri straminea* *Falcuna libyssa angolensis* *Cigarits modestus modestus* *Leptomyrina henningi angolensis*
Nymphalidae	*Brakefieldia angolensis* *Brakefieldia ochracea* *Neita bikuarica* *Mashunoides carneiromendesi* *Charaxes figuerai* *Charaxes ehmckei* *Precis larseni* *Bebearia hassoni* *Euphaedra divoides* *Euphaedra uigensis* *Acraea bellona* *Acraea lapidorum,* *Acraea onerata*	*Amauris crawshayi angola,* *Amauris dannfelti dannfelti* *Charaxes fulvescens rubenarturi,* *Charaxes macclouni carvalhoi,* *Charaxes lucretius saldanhai* *Charaxes jahlusa angolensis* *Charaxes minor karinae* *Charaxes trajanus bambi* *Palla ussheri hassoni* *Sevenia occidentalium penricei* *Euphaedra harpalyce commineura* *Acraea violarum anchietai*

Conservation

Because butterflies are sensitive to changing environmental conditions and are taxonomically well known, they are valuable as indicators of ecological dynamics. They are also key drivers of ecological processes. In particular, adult butterflies are active pollinators of many plants and the imagos and larvae are an important source of nutrition for a diverse range of vertebrate and invertebrate predators and insect parasitoids. Their conservation importance is also due to their positive and occasionally negative economic impacts. Although humans utilise mainly moth caterpillars as a

Fig. 10.3 The holotype specimens of endemic Angolan papilionoidea: Left to right, top to bottom (*V* Ventral, *D* Dorsal): 1. *Abantis bergeri* male D (Mendes and Bivar-de-Sousa 2009a, b, c, d), 2. *Eagris multiplagata* male V (Bivar-de-Sousa and Mendes 2007), 3. *Cooksonia nozolinoi* female D (Mendes and Bivar-de-Sousa 2007), 4. *Papilio bacelarae* male D (Bivar-de-Sousa and Mendes 2009a, b), 5. *Mashunoides carneiromendesi* male V (Mendes and Bivar-de-Sousa 2009a, b, c, d), 6. *Charaxes jahlusa angolensis* male D (Mendes et al. 2017), 7. *Euxanthe trajanus bambi* male D (Bivar-de-Sousa and Mendes 2006), 8. *Euphaedra (Euphaedrana) divoides* male V (Bivar- de-Sousa and Mendes 2018)

food source, the larvae of the skipper *Coeliades libeon* is much appreciated. A limited number of butterfly species are agricultural pests, including *Papilio demodocus* (young citrus orchards), *Lampides boeticus* (cultivated Leguminosae) and *Acraea acerata* (sweet-potatoes). A few species, such as *Pyrrhochalcia iphis* and *Zophopetes dysmephila*, may cause damage in coconut and oil-palm plantations.

In terms of species of conservation concern, information on the status of Angolan butterflies is very limited. Many species of Angolan butterflies are obviously abundant and widespread, both within and outside the country. Those taxa that appear to be rare and/or more localised may be genuinely rare or local but this may simply reflect a paucity of information. This makes it difficult or impossible to propose rational conservation measures at present. The urgent need for more fieldwork, particularly in regard to the endemic taxa, is thus highlighted. In the meantime habitat conservation, especially with respect to isolated forest patches, can be considered as part of a wider effort to conserve both the fauna and flora of the country.

Potential Future Discoveries and Research

Considering the number of taxa new to science described in the last few decades there are almost certainly further undiscovered butterfly taxa in Angola. Vast areas of the country remain unexplored, mainly because of inaccessibility and post-independence political instability. Not only will new taxa be found but also known taxa from bordering countries will be added to the list of Angolan butterflies during future fieldwork. This work will also improve our knowledge in regard to the distribution of the taxa in the country. Finally, almost nothing is known about the habitats, behaviours, early stages and larval host plants of Angolan butterflies, making these fertile areas for future research on the fauna. More information concerning all the endemic taxa is urgently needed in order to determine conservation priorities.

Appendix

Checklist of Papilionoidea recorded from Angola (by Family I Sub-family).

For the species' authors and references, Bivar-de-Sousa is abbreviated as BS and Mendes as M; (*) = Taxa with Angola as type locality; ◘ = Taxa now reported as new to Angola; ▲ = Previous references existing but not traced (several species when reported from Angola were assigned under names today considered synonyms, others at species level – the corresponding Angolan endemics were described later).

V = Validation of the taxon – 1: Endemic (restricted to Angola); 2: Restricted to Angola and to the neighbouring northern countries – Gabon, Congo and/or DRC; 3: Restricted to Angola and to the neighbouring eastern countries – southern DRC (former Shaba) and/or Zambia; 4: Restricted to Angola and to the southern and south-eastern neighbouring countries – Namibia and/or Botswana; 5: Species/subspecies with material studied by the authors; 6: Taxa exclusively known in the country from previously collected unstudied material – known from bibliographic references only. H – Preferred habitat: A: Humid forest – primary and secondary wet forest, gallery and riverine forest, forest edge; B: Dry forest, including dry forest and savanna mosaics; C: Brachystegia woodland (miombo) and other woodland; D: Mixed savanna with or without trees; E: Swampy areas, including the northeastern moist thicket and savanna mosaic; F: Arid shrubland and grassland; G: Rocky hillsides; H: Ubiquitous or almost ubiquitous; X: Caterpillars (almost) monophagous, the imagos range dependent on the presence of host-plants

Taxon	First reference for Angola	V	H
Hesperiidae | Coeliadinae			
Coeliades bixana (Evans, 1940)	Evans, 1937, as *C. bixae*	6	A
Coeliades c. chalybe (Westwood, 1852)	Evans, 1937	5	A
Coeliades libeon (Druce, 1875)	Druce, 1875 (*)	5	B,C
Coeliades f. forestan (Stoll, 1782)	Ladeiro, 1959	5	H
Coeliades hanno (Plötz, 1879)	Evans, 1937	6	A
Coeliades pisistratus (Fabricius, 1793)	Bacelar, 1948	6	D
Pyrrhochalcia iphis dejongi (Collins & Larsen, 2008)	Bacelar, 1956, as *P. iphis*	2,5	A
Hesperiidae | Pyrginae			
Apallaga rutilans (Mabille, 1877)	M et al., 2013a	5	A
Apallaga h. homeyeri (Plötz, 1880)	Plötz, 1880 (*)	6	A
Celaenorrhinus p. proxima (Mabille, 1877)	M & BS, 2009a, b, c, d	5	A
Tagiades flesus (Fabricius, 1781)	Evans, 1937	5	A,B
Eagris lucetia (Hewitson, 1875)	Aurivillius, 1928	5	A
Eagris decastigma fuscosa (Holland, 1893	M et al., 2013a	5	A
Eagris tigris liberti (Collins & Larsen, 2005)	Evans, 1937, as *E. tigris*	6	A
Eagris h. hereus (Druce, 1875)	Druce, 1875 (*)	6	A
Eagris t. tetrastigma (Mabille, 1891)	M et al., 2013a	5	A
Eagris multiplagata (BS & M, 2007)	BS & M, 2007 (*)	1,5	A
Ortholexis hollandi (Druce, 1909) f. *karschi* (Evans, 1937)	Evans, 1937	6	A
Calleagris hollandi (Butler, 1897)	Evans, 1937	5	C
Calleagris jamesoni ansorgei (Evans, 1951)	Weymer, 1901 (*), as *C. jamesoni*	1,5	C
Calleagris l. lacteus (Mabille, 1877)	Bacelar, 1961	6	A
Eretis lugens (Rogenhofer, 1891)	Larsen, 2005	5	D
Eretis herewardi rotundimacula (Evans, 1937)	Evans, 1937 (*)	1,5	C
Eretis melania (Mabille, 1891)	Evans, 1937	5	C,D

(continued)

Taxon	First reference for Angola	V	H
Sarangesa loelius (Mabille, 1877)	Bacelar, 1948	5	C
Sarangesa l. lucidella (Mabille, 1891)	M et al., 2013a	5	D
Sarangesa motozi (Wallengren, 1857)	Aurivillius, 1928	5	A,C
Sarangesa phidyle (Walker, 1870)	Evans, 1937	5	B,D
Sarangesa s. seineri (Strand, 1909)	Evans, 1937	5	D
Sarangesa p. pandaensis (Joicey & Talbot, 1921)	Evans, 1937	3,5	C
Sarangesa bouvieri (Mabille, 1877)	Evans, 1937	6	B
Sarangesa brigida sanaga (Miller, 1964)	M et al., 2013a	5	A
Sarangesa maculata (Mabille, 1891)	M et al., 2013a	5	A,B,C
Triskelionia tricerata (Mabille, 1891)	M et al., 2013a	6	A
Caprona cassualala (Bethune-Baker, 1911)	Bethune-Baker, 1911 (*)	5	D,F
Caprona pillaana (Wallengren, 1857)	M & BS, 2013a	5	D,F
Netrobalane canopus (Trimen, 1864)	M & BS, 2013a	5	D,A
Leucochitonea levubu (Wallengren, 1857)	Weymer, 1901	5	C,D
Abantis tettensis (Hopffer, 1855)	Aurivillius, 1928	5	C,D
Abantis bergeri (M & BS, 2009a, b, c, d)	M & BS, 2009a, b, c, d (*)	1,5	C
Abantis paradisea (Butler, 1870)	Weymer, 1901	6	C
Abantis zambesiaca (Westwood, 1874)	Weymer, 1901	5	D,C
Abantis contigua (Evans, 1937)	Evans, 1937	5	C
Abantis venosa (Trimen, 1889)	M et al., 2013a	5	C
Abantis vidua (Weymer, 1901)	Weymer, 1901	3	C
Spialia m. mafa (Trimen, 1870)	Weymer, 1901	6	D
Spialia spio (Linnaeus, 1764)	Weymer, 1901	5	D
Spialia delagoae (Trimen, 1898)	Larsen, 1996	5	D
Spialia c. colotes (Druce, 1875)	Druce, 1875 (*)	1,5	D,B
Spialia colotes transvaaliae (Trimen, 1889)	▲	5	D
Spialia ferax (Wallengren, 1863)	M et al., 2013a	5	C,D
Spialia dromus (Plötz, 1884)	Weymer, 1901	5	C,D
Spialia p. ploetzi (Aurivillius, 1891)	Evans, 1937, as *S. rebeli*	6	A
Spialia secessus (Trimen, 1891)	Trimen, 1891 (*)	5	F
Gomalia e. elma (Trimen, 1862)	Aurivilius, 1928	5	D
Hesperiidae | Heteropterinae			
Metisella m. midas (Butler, 1894)	Monard, 1956	5	E
Metisella a. angolana (Karsch, 1896)	Karsch, 1896 (*)	5	B
Metisella willemi (Wallengren, 1857)	M et al., 2013a, b	5	C
Metisella meninx (Trimen, 1873)	Evans, 1937	6	E
Lepella lepeletier (Latreille, 1824)	Druce, 1875	5	F
Hesperiidae | Hesperiinae			
Astictopterus abjecta (Snellen, 1872)	Snellen, 1872 (*)	6	A
Astictopterus punctulata (Butler, 1895)	M et al., 2013a	5	C
Kedestes mohozutza (Wallengren, 1857)	Monard, 1956	5	D

(continued)

Taxon	First reference for Angola	V	H
Kedestes nerva paola (Plötz, 1884)	Plötz, 1884 (*)	5	A
Kedestes brunneostriga (Plötz, 1884)	Plötz, 1884 (*)	5	C
Kedestes straeleni (Evans, 1956)	M et al., 2013a	5	C
Kedestes l. lema (Neave, 1910)	Evans, 1937	3	C
Kedestes callicles (Hewitson, 1868)	Aurivillius, 1928	5	C
Gorgyra mocquerysii (Holland, 1896)	Evans, 1937	5	A
Gorgyra diversata (Evans, 1937)	Evans, 1937	6	A
Ceratrichia nothus makomensis (Strand, 1913)	M et al., 2013a	5	A
Ceratrichia punctata (Holland, 1896)	Evans, 1937	6	A
Teniorhinus harona (Westwood, 1881)	Weymer, 1901, as *Oxypalpus ruso*	5	C
Teniorhinus ignita (Mabille, 1877)	Monard, 1956	5	B
Pardaleodes edipus (Stoll, 1781)	Bacelar, 1948	6	A
Pardaleodes i. incerta (Snellen, 1872)	Evans, 1937	5	A,D
Pardaleodes sator pusiella (Mabille, 1877)	Mabille, 1877 (*)	5	A
Pardaleodes t. tibullus (Fabricius, 1793)	M & BS, 2009	5	A
Acada biseriata (Mabille, 1893)	Evans, 1937	5	C
Parosmodes lentiginosa (Holland, 1896)	Evans, 1937	5	A
Parosmodes m. morantii (Trimen, 1873)	Weymer, 1901	5	C/D
Osmodes laronia (Hewitson, 1868)	Druce, 1875	6	A
Osmodes thora (Plötz, 1884)	Evans,1937	6	A
Acleros mackenii olaus (Plötz, 1884)	Druce, 1875, the species	5	A
Acleros nigrapex (Strand, 1913)	M et al., 2013a	5	A
Acleros ploetzi (Mabille, 1889)	M & BS, 2009	5	A
Semalea arela (Mabille, 1891)	M et al., 2013a	5	A
Semalea pulvina (Plötz, 1879)	M & BS, 2009	5	A
Semalea sextilis (Plötz, 1886)	M et al., 2013a	5	A
Hypoleucis o. ophiusa (Hewitson, 1866)	M & BS, 2009	5	A
Meza indusiata (Mabille, 1891)	Larsen, 2005	5	A,B
Meza meza (Hewitson, 1877)	Hewitson, 1877 (*)	5	A
Meza c. cybeutes (Holland, 1894)	Evans, 1937	6	A
Meza mabillei (Holland, 1893)	M et al., 2013a	5	A
Paronymus ligora (Hewitson, 1876)	Hewitson, 1876 (*)	6	A
Andronymus n. neander (Plötz, 1884)	Evans, 1937	5	A
Andronymus c. caesar (Fabricius, 1793)	Aurivillius, 1928 as *A. caesar*	5	A
Andronymus caesar philander (Hopffer, 1855)	Aurivillius, 1928 as *A. caesar*	6	A
Andronymus hero (Evans, 1937)	Evans, 1937	5	A
Andronymus helles (Evans, 1937)	Evans, 1937	5	A
Chondrolepis niveicornis (Plötz, 1882)	Plötz, 1882 (*)	5	E
Zophopetes dysmephila (Trimen, 1868)	Aurivillius, 1928, as *Z. schultzi*	6	A,D
Zophopetes cerymica (Hewitson, 1867)	M & BS, 2009	5	X
Gamia shelleyi (Sharpe, 1890)	M et al., 2013a	5	A

(continued)

Taxon	First reference for Angola	V	H
Gretna cylinda (Hewitson, 1876)	Aurivillius, 1928 (*)	5	A,C
Gretna waga (Plötz, 1886)	M et al., 2013a	5	A,C
Pteroteinon laufella (Hewitson, 1868)	Druce, 1875	5	B,C
Pteroteinon caenira (Hewitson, 1867)	M & BS, 2009	5	B
Pteroteinon concaenira (Belcastro & Larsen, 1996)	M et al., 2013a	6	B
Leona maracanda (Hewitson, 1876)	Hewitson, 1876 (*)	5	A
Caenides dacela (Hewitson, 1876)	Williams, 2007	5	A
Monza cretacea (Snellen, 1872)	Evans, 1937	5	B
Fresna nyassae (Hewitson, 1878)	Aurivillius, 1928	5	C,D
Platylesches langa (Evans, 1937)	M et al., 2013a	6	C
Platylesches moritili (Wallengren, 1857)	Trimen, 1891	5	C,D
Platylesches robustus (Neave, 1910)	M et al., 2013a	6	C
Platylesches cf. *batangae* (Holland, 1894)	▲	5	B
Brusa allardi (Berger, 1967)	M et al., 2013a	6	C
Zenonia zeno (Trimen, 1864)	Plötz, 1883, as *Hesperia coanza*	5	A,C
Pelopidas m. mathias (Fabricius, 1798)	Evans, 1937	5	C
Pelopidas thrax (Hübner, 1821)	Gardiner, 2004	5	D
Borbo fallax (Gaede, 1916)	M & BS, 2009a, b, c, d	5	D
Borbo fanta (Evans, 1937)	Evans, 1937	5	D
Borbo sirena (Evans, 1937)	M et al., 2013a	5	A,B,C
Borbo b. borbonica (Boisduval, 1833)	▲	5	D
Borbo detecta (Trimen, 1893)	Weymer, 1901	5	B
Larsenia gemella (Mabille, 1884)	Evans, 1937	5	D
Borbo micans (Holland, 1896)	M & BS, 2009	5	E
Larsenia perobscura (Druce, 1812)	M et al., 2013a	5	D
Borbo f. fatuellus (Hopffer, 1855)	M & BS, 2009	5	B,C
Larsenia holtzi (Plötz, 1883)	Plötz, 1883 (*)	5	D
Parnara monasi (Trimen, 1889)	Evans, 1937	5	E
Afrogegenes hottentota (Latreille, 1824)	Weymer, 1901	5	C,D
Afrogegenes letterstedti (Wallengren, 1857)	Evans, 1937	5	D
Gegenes pumilio gambica (Mabille, 1878)	M & BS, 2009	5	D
Papilionidae			
Papilio a. antimachus (Drury, 1782)	Carvalho, 1962	5	A
Papilio zalmoxis (Hewitson, 1864)	BS, 1983	5	A
Papilio bacelarae (BS & M., 2009)	BS & M, 2009 (*)	1,5	A
Papilio f. filaprae (Suffert, 1904)	Druce, 1875, as *P. cypraeophila*	5	A
Papilio m. mechowi (Dewitz, 1881)	Dewitz, 1881 (*)	5	A
Papilio mechowianus (Dewitz, 1885)	Aurivillius, 1928	6	A
Papilio zenobia (Fabricius, 1775)	BS & Fernandes, 1966	5	A
Papilio cynorta (Fabricius, 1793)	Druce, 1875	5	A
Papilio echerioides homeyeri (Plötz, 1880)	Plötz, 1880 (*)	5	A,C

(continued)

Taxon	First reference for Angola	V	H
Papilio chitondensis (BS & Fernandes, 1966)	BS & Fernandes, 1966 (*)	1,5	B,C
Papilio chrapkowskoides nurettini (Koçak, 1983)	Bacelar, 1956, as *P. bromius*	5	A
Papilio n. nireus (Linnaeus, 1758)	Druce, 1875	5	H
Papilio nireus lyaeus (Doubleday, 1845)	Ladeiro, 1956	5	A,D
Papilio sosia pulchra (Berger, 1950)	BS & Fernandes, 1966	5	A
Papilio mackinnoni benguellae (Jordan, 1908)	Jordan, 1908 (*)	1,5	A
Papilio d. dardanus (Brown, 1776)	Druce, 1875	5	C
Papilio phorcas congoanus (Rothschild, 1896)	BS & Fernandes, 1964	5	A
Papilio h. hesperus (Westwood, 1843)	Aurivillius, 1928	5	A
Papilio l. lormieri (Distant, 1874)	Bacelar, 1956	5	A
Papilio d. demodocus (Esper, 1798)	Weymer, 1901	5	H
Graphium a. angolanus (Goeze, 1779)	Goeze, 1779 (*)	5	H
Graphium schaffgotschi (Niepelt, 1927)	Villiers, 1979	4,5	D
Graphium ridleyanus (White, 1843)	Druce, 1875	5	A
Graphium latreillianus theorini (Aurivillius,1881)	Aurivillius, 1928	5	A
Graphium tynderaeus (Fabricius, 1793)	BS & Fernandes, 1964	5	A
Graphium a. almansor (Honrath, 1884)	Aurivillius, 1928	5	A,C
Graphium ucalegonides (Staudinger, 1884)	Smith & Vane-Wright, 2001	2	A
Graphium h. hachei (Dewitz, 1881)	Dewitz, 1881 (*)	5	A
Graphium poggianus (Honrath, 1884)	Aurivillius, 1928	3	A
Graphium u. ucalegon (Hewitson, 1865)	Bacelar, 1956	5	A
Graphium l. leonidas (Fabricius, 1793)	Druce, 1875	5	H
Graphium antheus (Cramer, 1779)	Druce, 1875	5	A,D
Graphium p. policenes (Cramer, 1775)	Druce, 1875	5	A,C
Graphium p. porthaon (Hewitson, 1865)	Gardiner, 2004	6	B,D
PIERIDAE | Pseudopontinae			
Pseudopontia paradoxa (Felder & Felder, 1869)	◘	5	A
Pseudopontia australis (Dixey, 1923)	Snellen, 1882, as *P. paradoxa*	3,5	A
PIERIDAE | Coliadinae			
Catopsilia florella (Fabricius, 1775)	Weymer, 1901	5	H
Colias e. electo (Linnaeus, 1763)	Gardiner, 2004	6	D
Colias electo hecate (Strecker, 1905)	Bacelar, 1948	5	F
Eurema b. brigitta (Stoll, 1780)	Butler, 1871	5	D,H
Eurema desjardinsi regularis (Butler, 1876)	Mabille, 1877	5	D,C
Eurema floricola leonis (Butler, 1886)	Trimen, 1891, as *E. floricola*	6	B
Eurema hapale (Mabille, 1882)	Ladeiro, 1956	5	A,B
Eurema hecabe solifera (Butler, 1875)	Butler, 1875 (*)	5	D

(continued)

Taxon	First reference for Angola	V	H
Eurema senegalensis (Boisduval, 1836)	Butler, 1871	5	A
Pieridae | Pierinae			
Pinacopteryx e. eriphia (Godart, 1819)	Butler, 1871	5	B,D
Nepheronia a. argia (Fabricius, 1775)	Druce, 1875	5	A
Nepheronia b. buquetii (Boisduval, 1836)	Druce, 1875	5	B,D
Nepheronia p. pharis (Boisduval, 1836)	Aurivillius, 1928	5	B
Nepheronia thalassina verulanus (Ward, 1871)	Bacelar, 1958a, b	5	B
Eronia cleodora Hübner, 1823	Aurivillius, 1928	6	D
Afrodryas leda (Boisduval, 1847)	Bacelar, 1961	5	B
Teracolus a. agoye (Wallengren, 1857)	Weymer, 1901	5	D
Colotis calais williami (Henning & Henning, 1994)	Willis, 2009	6	D
Colotis antevippe gavisa (Wallengren, 1857)	Trimen, 1891	5	D
Colotis celimene pholoe (Wallengren, 1860)	Talbot, 1939	4,5	F
Colotis annae walkeri (Butler, 1884)	Butler, 1884 (*)	4,5	F
Colotis doubledayi (Hopffer, 1862)	Hopffer, 1862 (*)	5	D
Colotis e. euippe (Linnaeus, 1758)	Latreille & Godart, 1819	5	B
Colotis euippe mediata (Talbot, 1939)	Talbot, 1939	5	C,D
Colotis evagore antigone (Boisduval, 1836)	Druce, 1875	5	B
Colotis e. evenina (Wallengren, 1857)	Trimen, 1891	5	F
Colotis ione (Godart, 1819)	Bacelar, 1961	6	D
Colotis regina (Trimen, 1863)	Trimen, 1891	5	D
Colotis vesta rhodesinus (Butler, 1894)	Bacelar, 1958a, b	5	D
Teracolus e. eris (Klug, 1829)	Druce, 1875	5	D
Teracolus subfasciatus (Swainson, 1833)	Aurivillius, 1928	5	C
Belenois aurota (Fabricius, 1793)	Trimen, 1891	5	H
Belenois calypso dentigera (Butler, 1888)	Druce, 1875, as *B. calypso*	5	A
Belenois welwitschii welwitschii (Rogenhofer, 1890)	Rogenhofer, 1890 (*)	5	C
Belenois crawshayi (Butler, 1894)	Aurivillius, 1928	5	C,D
Belenois creona severina (Stoll, 1781)	Butler, 1871	5	H
Belenois g. gidica (Godart, 1819)	▲	5	D
Belenois r. rubrosignata (Weymer, 1901)	Weymer, 1901 (*)	3,5	C
Belenois s. solilucis (Butler, 1874)	Butler, 1874 (*)	5	A
Belenois sudanensis mayumbana (Berger, 1981)	◘	2,5	A
Belenois sudanensis pseudodentigera (Berger, 1981)	◘	5	C
Belenois theuszi (Dewitz, 1889)	Dewitz, 1889 (*)	5	A
Pieridae | Pierinae			
Belenois t. thysa (Hopffer, 1855)	◘	5	C

(continued)

Taxon	First reference for Angola	V	H
Belenois thysa meldolae (Butler, 1872)	Butler, 1872 (*)	5	A,B
Dixeia capricornus falkensteinii (Dewitz, 1879)	Dewitz, 1879 (*)	2,5	B,C
Dixeia sp.	◘	5	A
Dixeia pigea (Boisduval, 1836)	Aurivillius, 1928	5	C
Pontia h. helice (Linnaeus, 1764)	Willis, 2009	5	D
Appias epaphia angolensis (M & BS, 2006)	M & BS, 2006 (*)	1,5	A,B
Appias perlucens (Butler, 1898)	Butler, 1898 (*)	5	A
Appias phaola uigensis (M & BS, 2006)	M & BS, 2006 (*)	1,5	A
Appias s. sabina (Felder & Felder, 1865)	Druce, 1875	5	A
Appias sylvia nyassana (Butler, 1897)	Druce, 1875, as *Belenois*	6	A?
Appias sylvia ribeiroi (M & BS, 2006)	M & BS, 2006 (*)	1,5	A
Leptosia a. alcesta (Stoll, 1782)	Druce, 1875	5	A
Leptosia h. hybrida (Bernardi, 1952)	◘	5	A,C
Leptosia n. nupta (Butler, 1873)	Butler, 1873 (*)	5	A
Leptosia wigginsi pseudalcesta (Bernardi, 1965)	◘	5	A
Mylothris carvalhoi (M & BS, 2009)	M & BS, 2009 (*)	1,5	A
Mylothris mavunda (Hancock & Heath, 1985)	Koçak & Kemal, 2009	3	A?
Mylothris a. agathina (Cramer, 1779)	Trimen, 1891	5	C
Mylothris asphodelus (Butler, 1888)	Aurivillius, 1928	5	A
Mylothris elodina diva (Berger, 1954)	Berger, 1981	2,5	C
Mylothris poppea (Cramer, 1777)	Druce, 1875	5	A
Mylothris rembina (Plötz, 1880)	Talbot, 1944	6	A
Mylothris rhodope (Fabricius, 1775)	Talbot, 1944	5	D
Mylothris rueppellii rhodesiana (Riley, 1921)	Talbot, 1944	5	C
Mylothris spica gabela (Berger, 1979)	Berger, 1979 (*)	1,6	A
Mylothris sulphurea (Aurivillius, 1895)	◘	5	A
Mylothris y. cf. *yulei* (Butler, 1897)	◘	5	A
Lycaenidae | Miletinae			
Euliphyra mirifica (Holland, 1890)	Larsen, 2005	6	A
Aslauga m. marshalli (Butler, 1899)	Larsen, 2005	6	A
Megalopaplpus zymna (Westwood, 1851)	Ackery et al., 1995	5	A
Spalgis l. lemolea (Druce, 1890)	Ladeiro, 1956	5	A,C
Lachnocnema angolanus (Libert, 1996)	Libert, 1996 b (*)	5	A,D
Lachnocnema bamptoni (Libert, 1996)	Libert, 1996 b (*)	6	C
Lachnocnema bibulus (Fabricius, 1793)	Libert, 1996 b	5	A,C,D
Lachnocnema emperamus (Snellen, 1872)	Snellen, 1872	5	A
Lachnocnema intermedia (Libert, 1996)	Ladeiro, 1956, as *L. durbani* (*)	5	C
Lachnocnema laches (Fabricius, 1793)	Libert, 1996 a	5	A,C
Lachnocnema r. regularis (Libert, 1996)	Libert, 1996 c	6	C?

(continued)

Taxon	First reference for Angola	V	H
LYCAENIDAE | Poritiinae			
Alaena amazoula congoana (Aurivillius, 1914)	Aurivillius, 1914 (*)	6	G
Alaena rosei (Vane-Wright, 1980)	Vane-Wright, 1980 (*)	1,5	G
Pentila maculata pardalena (Druce, 1910)	Stempffer & Bennett, 1961	6	A
Pentila amenaida (Hewitson, 1873)	Hewitson, 1873 (*)	5	A
Pentila pauli benguellana (Stempffer & Bennett, 1961)	Stempffer & Bennett, 1961 (*)	5	A,B
Pentila t. tachyroides (Dewitz, 1879)	Dewitz, 1879 (*)	5	A
Telipna acraeoides (Grose-Smith & Kirby, 1890)	Grose-Smith & Kirby, 1890 (*)	5	A
Telipna a. albofasciata (Aurivillius, 1910)	Libert, 2005	6	A
Telipna atrinervis (Hulstaert, 1924)	◘	5	A
Telipna cuypersi (Libert, 2005)	Libert, 2005	6	A
Telipna nyanza katangae (Stempffer, 1961)	Libert, 2005	6	A
Telipna s. sanguinea (Plötz, 1880)	Aurivillius, 1928	5	A
Ornipholidotus gabonensis (Stempffer, 1947)	▲	6	A
Ornipholidotus perfragilis (Holland, 1890)	Libert, 2005	6	A
Ornipholidotus ugandae goodi (Libert, 2000)	Libert, 2005	6	A
Cooksonia nozolinoi (M & BS, 2007)	M & BS, 2007 (*)	1,5	C
Mimacraea charmian (Grose-Smith & Kirby, 1890)	Grose-Smith & Kirby, 1890 (*)	6	A
Mimacraea landbecki (Druce, 1910)	Libert, 2000 b	5	A
Mimacraea marshalli (Trimen, 1898)	Libert, 2000 b	5	A
Mimeresia debora deborula (Aurivillius, 1899)	◘	5	A
Eresiomera osheba (Holland, 1890)	▲	5	A
Citrinophila e. erastus (Hewitson, 1866)	Aurivillius, 1928	6	A
Cnodontes vansomereni (Stempffer & Bennett, 1953)	▲	5	D
Liptena evanescens (Kirby, 1887)	◘	5	A
Liptena fatima (Kirby, 1890)	◘	5	A
Liptena h. homeyeri (Dewitz, 1884)	▲	6	A
Liptena homeyeri straminea (Stempffer, Bennett & May, 1974)	Stempffer, Bennett & May, 1974 (*)	1,6	A
Liptena parva (Kirby, 1887)	◘	5	A
Liptena undularis (Hewitson, 1866)	Druce, 1875	5	A
Liptena xanthostola xantha (Grose-Smith, 1901)	Larsen, 2005	6	A
Falcuna h. hollandii (Aurivillius, 1895)	Ackery et al., 1995	6	A
Falcuna lacteata (Stempffer & Bennett, 1963)	Stempffer & Bennett, 1963 (*)	1,6	A

(continued)

Taxon	First reference for Angola	V	H
Falcuna libyssa angolensis (Stempffer & Bennett, 1963)	Stempffer & Bennett, 1963 (*)	1,5	A
Falcuna s. synesia (Hulstaert, 1924)	Stempffer & Bennett, 1963 (*)	2	A
Tetrarhanis ilala etoumbi (Stempffer, 1964)	■	5	A
Tetrarhanis i. ilma (Hewitson, 1873)	Hewitson, 1873 (*)	6	A
Larinopoda lircaea (Hewitson, 1866)	Stempffer, 1957	6	A
Larinopoda tera (Hewitson, 1873)	Aurivillius, 1928	5	A
Hewitsonia bitjeana (Bethune-Baker, 1915)	■	5	A
Hewitsonia k. kirbyi (Dewitz, 1879)	Dewitz, 1879 (*)	6	A
Cerautola ceraunia (Hewitson, 1873)	Larsen, 2005	6	A
Cerautola crowleyi leucographa (Libert, 1999)	Libert, 1999	6	A
Hewitola hewitsonii (Mabille, 1877)	Mabille, 1877 (*)	6	A
Cerautola miranda vidua (Talbot, 1935)	Bacelar, 1958a, b	5	A
Epitola posthumus (Fabricius, 1793)	Bacelar, 1956	5	A
Epitola urania (Kirby, 1887)	Libert, 1999	6	A
Hypophytala h. hyetta (Hewitson, 1873)	Hewitson, 1873 (*)	2	A
Stempfferia cercene (Hewitson, 1873)	Hewitson, 1873 (*)	6	A
Stempfferia cinerea (Berger, 1981)	Libert, 1999	2	A
Stempfferia michelae centralis (Libert, 1999)	Libert, 1999	5	A
Deloneura barca (Grose-Smith, 1901)	Grose-Smith, 1901 (*)	1,6	C?,D?
Deloneura cf. *subfusca* (Hawker-Smith, 1933)	■	5	F
Epitolina dispar (Kirby, 1887)	Larsen, 2005	6	A
Epitolina melissa (Druce, 1888)	Larsen, 2005	6	A
LYCAENIDAE | Theclinae			
Myrina s. silenus (Fabricius, 1775)	Druce, 1875	5	B,D
Myrina silenus ficedula (Trimen, 1879)	Gardiner, 2004	6	D
Oxylides binza (Berger, 1981)	Druce, 1875 as *O. faunus*	2,5	A
LYCAENIDAE | Theclinae			
Oxylides feminina stempfferi (Berger, 1981)	Libert, 2004	6	A
Syrmoptera amasa (Hewitson, 1869)	Libert, 2004	6	A
Syrmoptera homeyerii (Dewitz, 1879)	Dewitz, 1879 (*)	6	A
Dapidodigma demeter nuptus (Clench, 1961)	Larsen, 2005	3,5	A
LYCAENIDAE | Aphnaeinae			
Lipaphneus a. cf. *aderna* (Plötz, 1880)	■	5	A
Crudaria leroma (Wallengren, 1857)	Gardiner, 2004	6	D
Aloeides angolensis (Tite & Dickson, 1973)	Tite & Dickson, 1973 (*)	1,6	F
Aphnaeus erikssoni (Trimen, 1891)	Trimen, 1891 (*)	5	C

(continued)

Taxon	First reference for Angola	V	H
Aphnaeus orcas (Drury, 1782)	Larsen, 2005	6	A
Aphnaeus affinis (Riley, 1921)	Libert, 2013		
Erikssonia acraeina (Trimen, 1891)	Trimen, 1891 (*)	6	D
Pseudaletis a. agrippina (Druce, 1888)	▲	5	A
Cigaritis ella (Hewitson, 1865)	Gardiner, 2004	6	D
Cigaritis phanes (Trimen, 1873)	Weymer, 1901	6	F
Cigaritis homeyeri (Dewitz, 1887)	Aurivillius, 1928	5	C
Cigaritis m. modestus (Trimen, 1891)	Trimen, 1891 (*)	1,4,5	A,C
Cigaritis mozambica (Bertoloni, 1850)	▲	6	D
Cigaritis natalensis (Westwood, 1851)	Ladeiro, 1956	6	C,D
Cigaritis trimeni congolanus (Dufrane, 1954)	▲	2,5	A
Zeritis fontainei (Stempffer, 1956)	Willis, 2009	6	C,G?
Zeritis krystyna (D'Abrera, 1980)	D'Abrera, 1980 (*)	1,6	C?
Zeritis sorhagenii (Dewitz, 1879)	Dewitz, 1879 (*)	6	C?
Axiocerces a. amanga (Westwood, 1881)	Trimen, 1891	5	C
Axiocerces bambana orichalcea (Henning & Henning, 1996)	◘	5	B
Axiocerces amanga baumi (Weymer, 1901)	Weymer, 1901 (*)	,5	C
Axiocerces t. tjoanae (Wallengren, 1857)	Henning & Henning, 1996	6	B
Iolaus hemicyanus barnsi (Joicey & Talbot, 1921)	◘	2,5	A
Iolaus i. iasis (Hewitson, 1865)	Larsen, 2005	6	A
Iolaus mimosae rhodosense (Stempffer & Bennett, 1959)	Gardiner, 2004	6	D
Iolaus obscura (Aurivillius, 1923)	◘	4,5	D
Iolaus violacea (Riley, 1928)	Riley, 1928 (*)	5	C
Iolaus pallene (Wallengren, 1857)	Gardiner, 2004	6	D
Iolaus trimeni (Wallengren, 1875)	Ackery et al., 1995	5	C,D
Iolaus iturensis (Joicey & Talbot, 1921)	▲	6	C
Iolaus parasilanus mabillei (Riley, 1928)	Riley, 1928 (*)	2	A
Iolaus s. silarus (Druce, 1885)	Gardiner, 2004	5	C,D
Iolaus t. timon (Fabricius, 1787)	Ackery et al., 1995	6	A
Hemiolaus vividus (Pinhey, 1962)	Aurivillius, 1928, as *caeculus*	5	C,D
Stugeta bowkeri maria (Suffert, 1904)	Druce, 1875, as *bowkeri*	3,5	B
Stugeta bowkeri tearei (Dikson, 1980)	Gardiner, 2004	6	A,C,E
Hypolycaena a. antifaunus (Westwood, 1851)	Druce, 1875	5	A
Hypolycaena h. hatita (Hewitson, 1865)	Druce, 1875	5	A
Hypolycaena l. lebona (Hewitson, 1865)	Druce, 1875	6	A
Hypolycaena naara (Hewitson, 1873)	Hewitson, 1873 (*)	6	A
Hypolycaena nigra (Bethune-Baker, 1914)	M & BS, 2012	5	A

(continued)

Taxon	First reference for Angola	V	H
Hypolycaena p. philippus (Fabricius, 1793)	Druce, 1875	5	D
Hypolycaena buxtoni spurcus (Talbot, 1929 M & BS., 2012)	M & BS, 2012	5	F?
Pilodeudorix badhami (Carcasson, 1961)	Libert, 2004	6	?
Pilodeudorix caerulea (Druce, 1890)	Libert, 2004	5	C,D
Pilodeudorix pseudoderitas (Stempffer, 1964)	Larsen, 2005	6	A
Pilodeudorix zeloides (Butler, 1901)	Libert, 2004	6	C
Paradeudorix cobaltina (Stempffer, 1964)	Larsen, 2005	6	A
Leptomyrina henningi angolensis (M & BS, 2009)	M & BS, 2009 (*)	1,5	B
Pilodeudorix deritas (Hewitson, 1874)	Hewitson, 1874 (*)	5	A
Pilodeudorix m. mera (Hewitson, 1873)	Hewitson, 1873 (*)	5	A
Pilodeudorix otraeda genuba (Hewitson, 1875)	▲	6	A
Hypomyrina nomenia (Hewitson, 1874)	Larsen, 2005	5	A
Deudorix antalus (Hopffer, 1855)	Bacelar, 1948	5	D,G
Deudorix caliginosa (Lathy, 1903)	▲	6	C
Deudorix dinochares (Grose-Smith, 1887)	Gardiner, 2004	5	C,D
Deudorix cf. *diocles* (Hewitson, 1869)	Libert, 2004	5	C
Deudorix lorisona coffea (Jackson, 1966)	Libert, 2004	5	A,C
Capys c. connexiva (Butler, 1896)	Henning & Henning, 1988	6	X
Lycaenidae | Polyommatine			
Anthene akoae (Libert, 2010)	Libert. 2010	6	?
Anthene alberta (Bethune-Baker, 1910)	Aurivillius, 1928	5	A,C
Anthene a. amarah (Guérin-Méneville, 1849)	Stempffer, 1957	5	D
Anthene lvida livida (Trimen, 1881)	Gardiner, 2004	6	D
Anthene c. crawshayi (Butler, 1899)	◘	5	D
Anthene d. definita (Butler, 1899)	Gardiner, 2004	5	A
Anthene larydas (Cramer, 1780)	Weymer, 1901	5	A
Anthene l. ligures (Hewitson, 1874)	Hewitson, 1874 (*)	6	A
Anthene liodes (Hewitson, 1874)	Aurivillius, 1909	6	A,D
Anthene l. lunulata (Trimen, 1894)	Trimen, 1894 (*)	5	D
Anthene nigropunctata (Bethune-Baker, 1910)	Gardiner, 2004	6	?
Anthene princeps (Butler, 1876)	Gardiner, 2004	5	A,D
Anthene r. rubricinctus (Holland, 1891)	Aurivillius, 1928	6	A
Anthene sylvanus (Drury, 1773)	Aurivillius, 1928	6	A
Anthene talboti (Stempffer, 1936)	Libert, 2010	6	D
Neurellipes flavomaculatus (Grose-Smith & Kirby, 1893)	Aurivillius, 1928	6	A
Neurellipes lachares (Hewitson, 1878)	Larsen, 2005	6	A
Neurellipes onias (Hulstaert, 1924)	Willis, 2009	6	?

(continued)

Taxon	First reference for Angola	V	H
Neurellipes pyroptera (Aurivillius, 1895)	Libert, 2010	6	A
Neurypexina lyzanius (Hewitson, 1874)	Druce, 1875	6	A
Triclema lacides (Hewitson, 1874)	Hewitson, 1874 (*)	6	A
Triclema lucretilis (Hewitson, 1874)	Stempffer, 1957	6	A
Triclema cf. *nigeriae* (Aurivillius, 1905)	Libert, 2010	5	D
Monile g. gemmifera (Neave, 1910)	Libert, 2010	6	A
Cupidesthes vidua (Talbot, 1929)	Talbot, 1929 (*)	1,6	C?
Pseudonacaduba aethiops (Mabille, 1877)	Mabille, 1877 (*)	5	A
Pseudonacaduba s. sichela (Wallengren, 1857)	Weymer, 1901	5	D
Lampides boeticus (Linnaeus, 1767)	Butler, 1871	5	H
Uranothauma falkensteinii (Dewitz, 1879)	Dewitz, 1879 (*)	5	A?
Uranothauma antinorii cf. *felthami* (Stevenson, 1934)	BS & M, 2007	5	A
Uranothauma h. heritsia (Hewitson, 1876)	Stempffer, 1957	5	A,D
Uranothauma nozolinoi (BS & M., 2007)	BS & M, 2007 (*)	1,5	C
Uranothauma poggei (Dewitz, 1879)	Dewitz, 1879 (*)	5	A,C,D
Uranothauma c. cyara (Hewitson, 1876)	Hewitson, 1876 (*)	5	A?
Cacyreus lingeus (Stoll, 1782)	Gardiner, 2004	5	B,D
Cacyreus marshalli (Butler, 1898)	Gardiner, 2004	6	D
Cacyreus virilis (Aurivillius, 1924)	Aurivillius, 1924 (*)	6	D
Leptotes babaulti (Stempffer, 1935)	Stempffer, 1957	5	A,D
Leptotes brevidentatus (Tite, 1958)	Tite, 1958 (* – in part)	5	A,C,D
Leptotes jeanneli (Stempffer, 1935)	Stempffer, 1957	5	A,D
Leptotes p. pirithous (Linnaeus, 1767)	Snellen, 1882	5	D
Leptotes p. pulchra (Murray, 1874)	Gardiner, 2004	5	E
Tuxentius calice (Hopffer, 1855)	Gardiner, 2004	5	E
Tuxentius c. carana (Hewitson, 1876)	Hewitson, 1876 (*)	5	A
Tuxentius margaritaceus (Sharpe, 1892)	Larsen, 2005	6	A
Tuxentius m. melaena (Trimen, 1887)	Aurivillius, 1928	6	C,D
Tarucus sybaris linearis (Aurivillius, 1924)	Aurivillius, 1928	6	D
Actizera lucida (Trimen, 1883)	Stempffer, 1957	5	F
Eicochrysops eicotrochilus (Bethune-Baker, 1924)	▲	6	C
Eicochrysops hippocrates (Fabricius, 1793)	Gardiner, 2004	5	A,B,D
Eicochrysops messapus mahallakoaena (Wallengren, 1857)	Weymer, 1901, as *messapus*	6	C
Cupidopsis c. cissus (Godart, 1824)	Ladeiro, 1956	5	E
Cupidopsis j. jobates (Hopffer, 1855)	Stempffer, 1957	5	D
Euchrysops barkeri (Trimen, 1893)	Aurivillius, 1928	6	C

(continued)

Taxon	First reference for Angola	V	H
Euchrysops malathana (Boisduval, 1833)	Stempffer, 1957	5	C,D
Euchrysops osiris (Hopffer, 1855)	Druce, 1875	5	C,D
Euchrysops subpallida (Bethune-Baker, 1923)	Gardiner, 2004	6	F,G
Lepidochrysops abyssiniensis loveni (Aurivillius, 1921)			
Lepidochrysops ansorgei (Tite, 1959)	Tite, 1959 (*)	1,5	C,D
Lepidochrysops chlouages (Bethune-Baker, 1923)	▲	5	A,D
Lepidochrysops flavisquamosa (Tite, 1959)	Tite, 1959 (*)	1,6	C
Lepidochrysops fulvescens (Tite, 1961)	Tite, 1961 (*)	1,6	C
Lepidochrysops g. glauca (Trimen, 1887)	Bacelar, 1948	5	D
Lepidochrysops hawkeri (Talbot, 1929)	Talbot, 1929 (*)	1,5	C?
Lepidochrysops nacrescens (Tite, 1961)	Tite, 1961 (*)	1,6	C
Lepidochrysops reichenowi (Dewitz, 1879)	Dewitz, 1879 (*)	1,6	?
Thermoniphas distincta (Talbot, 1935)	◘	5	C
Thermoniphas p. plurilimbata (Karsch, 1895)	◘	5	A
Thermoniphas t. togara (Plötz, 1880)	▲	6	A
Oboronia guessfeldtii (Dewitz, 1879)	Larsen, 1991	5	B
Oboronia pseudopunctatus (Strand, 1912)	▲	6	A
Oboronia punctatus (Dewitz, 1879)		5	A
Actizera lucida (Trimen, 1883)	Willis, 2009	6	C
Brephidium metophis (Wallengren, 1860)	Willis, 2009	6	?
Azanus isis (Drury, 1773)	Bacelar, 1948	5	A
Azanus jesous (Guérin-Méneville, 1849)	Trimen, 1891	6	D
Azanus mirza (Plötz, 1880)	Stempffer, 1957	5	D,A
Azanus moriqua (Wallengren, 1857)	Weymer, 1901	5	D
Azanus natalensis (Trimen, 1887)	Bacelar, 1948	5	D
Azanus ubaldus (Stoll, 1782)	Bacelar, 1948	5	F
Chilades trochylus (Freyer, 1844)	Ladeiro, 1956	5	D
Zizeeria k. knysna (Trimen, 1862)	Ladeiro, 1956	5	D
Zizina otis antanossa (Mabille, 1877)	▲	5	D
Zizula hylax (Fabricius, 1775)	Gardiner, 2004	5	H
Riodinidae			
Afriodinia dewitzi (Aurivillius, 1899)	▲	6	A
Afriodinia intermedia (Aurivillius, 1895)	Larsen, 2005	6	A
Afriodinia r. rogersi (Druce, 1878)	Druce, 1878 (*)	5	A
Afriodinia tantalus caerulea (Riley, 1932)	Druce, 1875 as *tantalus*	6	A
Nymphalidae | Libytheinae			
Libythea labdaca (Westwood, 1851)	Snellen, 1882	5	A
Libythea laius (Trimen, 1879)	Gardiner, 2004	5	A

(continued)

Taxon	First reference for Angola	V	H
NYMPHALIDAE | Danainae			
Danaus c. orientis (Aurivillius, 1909)	Butler, 1871	5	H
Tirumala petiverana (Doubleday, 1847)	Butler, 1866, as *Danais leonora*	5	D
Amauris n. niavius (Linnaeus, 1758)	Aurivillius, 1928	5	A,B,D
Amauris t. tartarea (Mabille, 1876)	Mabille, 1876 (*)	5	A
Amauris crawshayi angola (Bethune-Baker, 1914)	Bethune-Baker, 1914 (*)	1,6	A
Amauris h. hecate (Butler, 1866)	◘	5	A
Amauris d. dannfelti (Aurivillius, 1891)	Aurivillius, 1891 (*)	1,5	A,C
Amauris h. hyalites (Butler, 1874)	Butler, 1874 (*)	5	A
Amauris vashti (Butler, 1869)	◘	5	A
NYMPHALIDAE | Satyrinae			
Gnophodes betsimena parmeno (Doubleday, 1849)	Aurivillius, 1928	5	A
Gnophodes chelys (Fabicius, 1793)	Aurivillius, 1928	5	A
Melanitis leda (Linnaeus, 1758)	Bacelar, 1948	5	B,H
Elymnias b. bammakoo (Westwood, 1851)	Druce, 1875	5	A
Bicyclus iccius (Hewitson, 1865)	Larsen, 2005	6	A
Bicyclus sebetus (Hewitson, 1877)	Aurivillius, 1928	5	A
Bicyclus s. saussurei (Dewitz, 1879)	Dewitz, 1879 (*)	3,5	A
Bicyclus s. suffusa (Riley, 1921)	▲	3,5	C
Bicyclus taenias (Hewitson, 1877)	◘	5	A,B
Bicyclus nachtetis (Condamin, 1965)	▲	3	A
Bicyclus technatis (Hewitson, 1877)	Larsen, 2005	6	A
Bicyclus vulgaris (Butler, 1868)	Druce, 1875	5	A,D
Bicyclus moyses (Condamin & Fox, 1964)	Condamin & Fox, 1964	5	A
Bicyclus sandace (Hewitson, 1877)	Bacelar, 1958a, b	5	A,D
Bicyclus auricruda fulgida (Fox, 1963)	Aurivillius, 1928, as *auricruda*	5	A
Bicyclus collinsi (Aduse-Poku et al., 2009)	Hewitson, 1873	5	B,D
Bicyclus angulosa selousi (Trimen, 1895)	Condamin, 1963	5	C,D
Bicyclus campus (Karsch, 1893)	▲	6	A
Bicyclus a. anynana (Butler, 1879)	Gardiner, 2004	6	D
Bicyclus anynana centralis (Condamin, 1968)	Condamin, 1968	5	A
Bicyclus cottrelli (van Son, 1952)	▲	5	C
Bicyclus s. safitza (Westwood, 1850)	Butler, 1871, as *Mycalesis caffra*	5	D
Bicyclus funebris (Guérin-Méneville, 1844)	Ladeiro, 1956	5	A,B,D
Bicyclus istaris (Plötz, 1880)	▲	5	A
Bicyclus lamani (Aurivillius, 1900)	Bacelar, 1958a, b	5	A
Bicyclus golo (Aurivillius, 1893)	Monard, 1956	5	A
Bicyclus s. smithi (Aurivillius, 1899)	▲	5	A
Bicyclus vansoni (Condamin, 1965)	Condamin, 1965	5	C

(continued)

Taxon	First reference for Angola	V	H
Bicyclus buea (Strand, 1912)	Larsen, 2005	5	A
Bicyclus sanaos (Hewitson, 1866)	Druce, 1875	5	A
Hallelesis asochis congoensis (Joicey & Talbot, 1921)	Druce, 1875, as *asochis*	6	E
Brakefieldia angolensis (Kielland, 1994)	Kielland, 1994 (*)	1,6	C
Brakefieldia p. phaea (Karsch, 1894)	Kielland, 1994	5	C
Brakefieldia simonsii (Butler, 1877)	Gardiner, 2004	5	D,F
Brakefieldia centralis (Aurivillius, 1903)	Ackery et al., 1995	3	C
Brakefieldia ochracea (Lathy, 1906)	Lathy, 1906 (*)	1,5	C
Brakefieldia eliasis (Hewitson, 1866)	Druce, 1875	3,5	A,B
Mashuna upemba (Overlaet, 1955)	◘	5	E
Ypthima a. asterope (Klug, 1832)	Druce, 1875	5	D
Ypthima asterope hereroica (van Son, 1955)	Gardiner, 2004	6	D
Ypthima c. condamini (Kielland, 1982)	Larsen, 2005	6	C,D
Ypthima granulosa (Butler, 1883)	Ladeiro, 1956	6	C?,D
Ypthima recta (Overlaet, 1955)	Kielland, 1982	6	A
Ypthima doleta (Kyrby, 1880)	Aurivillius, 1928	5	A,B
Ypthima i. impura (Elwes & Edwards, 1893)	Aurivillius, 1928	5	A,D
Ypthima impura paupera (Ungemach, 1932)	Gardiner, 2004	6	A,D
Ypthima praestans (Overlaet, 1954)	▲	5	A
Ypthima pulchra (Overlaet, 1954)	▲	5	A
Ypthima diplommata (Overlaet, 1954)	Kielland, 1982	3,6	New data
Ypthimomorpha itonia (Hewitson, 1865)	Aurivillius, 1928	5	D,E
Neita bikuarica (M & BS, 2006)	M & BS, 2006 (*)	1,5	C
Neocoenyra cooksoni (Druce, 1907)	▲	6	C
Mashunoides carneiromendesi (M & BS, 2009)	M & BS, 2009 (*)	1,5	D
NYMPHALIDAE ǀ Charaxinae			
Charaxes fulvescens rubenarturi (BS & M, 2017)	BS *et al.*, 2017 (*)	1,5	A
Charaxes varanes vologeses (Mabille, 1876)	Mabille, 1876, sub *Palla* (*)	5	B,D
Charaxes candiope (Godart, 1824)	Druce, 1875	5	B,D
Charaxes cynthia kinduana (Le Cerf, 1923)	Aurivillius, 1928	5	A,D
Charaxes macclounii (Butler, 1895)	van Someren, 1970	6	A
Charaxes macclouni carvalhoi (BS, 1983)	BS, 1983 (*)	1,5	A
Charaxes protoclea protonothodes (van Someren, 1971)	Aurivillius, 1928, as *protoclea*	5	A,B
Charaxes lucretius saldanhai (BS, 1983)	BS, 1983 (*)	1,5	A

(continued)

Taxon	First reference for Angola	V	H
Charaxes brutus angustus (Rothschild, 1900)	Druce, 1875, as *brutus*	5	A,B,D
Charaxes brutus natalensis (Staudinger, 1885)	van Someren, 1970	6	A,B,D
Charaxes c. castor (Cramer, 1775)	Druce, 1875	5	A,B
Charaxes druceanus proximans (Joicey & Talbot, 1922)	Aurivillius, 1928	5	A,D
Charaxes eudoxus mechowi (Rothschild, 1900)	Rothschild, 1900 (*)	5	A,D
Charaxes eudoxus mitchelli (Plantrou & Howarth, 1977)	◘	5	A,D
Charaxes s. saturnus (Butler, 1866)	Druce, 1875, as *C. pelias brunnescens*	5	A,D
Charaxes p. pollux (Cramer, 1775)	Druce, 1875	5	A,C
Charaxes numenes aequatorialis (van Someren, 1972)	Aurivillius, 1928, as *numenes*	5	A
Charaxes tiridates tiridatinus (Röber, 1936)	Druce, 1875, the species	5	A
Charaxes ameliae amelina (Joicey & Talbot, 1925)	◘	5	C
Charaxes b. bohemani (Felder & Felder, 1859)	Druce, 1875	5	D
Charaxes p. pythodoris (Hewitson, 1873)	Hewitson, 1873 (*)	5	B
Charaxes smaragdalis leopoldi (Ghesquiére, 1933)	van Someren, 1964	2	A
Charaxes zingha (Stoll, 1780)	Bacelar, 1958a, b	5	A
Charaxes a. achaemenes (Felder & Felder, 1867)	van Someren, 1970	5	C,D
Charaxes e. etesipe (Godart, 1824)		5	A,E
Charaxes p. penricei (Rothschild, 1900)	Rothschild, 1900 (*)	5	C,D
Charaxes penricei dealbata (van Someren, 1966)	van Someren, 1966 (*)	2,5	A
Charaxes jahlusa angolensis (M & BS, 2017)	BS et al., 2017 (*)	1,5	A
Charaxes eupale latimargo (Joicey & Talbot, 1921)	Druce, 1875 as *eupale*	5	A
Charaxes minor karinae (Bouyer, 1999)	Bouyer, 1999 (*)	1,5	A
Charaxes anticlea proadusta (van Someren, 1971)	Aurivillius, 1928, as *anticlea*	5	A
Charaxes h. hildebrandti (Dewitz, 1879)	Dewitz, 1879 (*)	5	A
Charaxes hildebrandti katangensis (Talbot, 1928)	BS & M, 2014	5	A
Charaxes g. guderiana (Dewitz, 1879)	Dewitz, 1879 (*)	5	A,D
Charaxes brainei (van Son, 1966)	Henning, 1988	4	D
Charaxes catachrous (van Someren & Jackson, 1952)	◘	5	A

(continued)

Taxon	First reference for Angola	V	H
Charaxes cedreatis (Hewitson, 1874)	Aurivillius, 1928	5	A,C
Charaxes diversiforma (van Someren & Jackson, 1957)	van Someren, 1969	5	A
Charaxes etheocles silvestris (Turlin, 2011)	Druce, 1875, as *C. ephyra*	5	A,B
Charaxes figueirai (BS & M, 2014)	BS & M, 2014 (*)	1,5	C,D
Charaxes fulgurata (Aurivillius, 1899)	Aurivillius, 1899 (*)	5	C
Charaxes howarthi (Minig, 1976)	Henning, 1988	5	C
Charaxes phaeus (Hewitson, 1877)	Gardiner, 2004	6	C,D
Charaxes variata (van Someren, 1969)	■	3,5	C
Charaxes p. paphianus (Ward, 1871)	Aurivillius, 1928	5	A
Charaxes pleione congoensis (Plantrou, 1989)	van Someren 1974, as *pleione*	5	A
Charaxes ehmckei (Homeyer & Dewitz, 1882)	Homeyer & Dewitz, 1882 (*)	1,5	A,B
Charaxes zoolina (Westwood, 1850)	Gardiner, 2004	6	A
Charaxes kahldeni (Homeyer & Dewitz, 1882)	Homeyer & Dewitz, 1882 (*)	5	A
Charaxes n. nichetes (Grose-Smith, 1883)	Aurivillius, 1928	5	A
Charaxes nichetes pantherinus Rousseau-(Decelle, 1934)	■	5	A
Charaxes lycurgus (Fabricius, 1793)	Plantrou, 1978	5	A
Charaxes zelica rougeoti (Plantrou, 1978)	Plantrou, 1978	5	A
Charaxes doubledayi (Aurivillius, 1899)	Bacelar, 1956	5	A
Palla decius (Cramer, 1777)	Druce, 1875	5	A
Palla publius centralis (van Someren, 1975)	■	5	A
Palla ussheri hassoni (Turlin & Vingerhoedt, 2013)	Turlin & Vingerhoedt, 2013 (*)	1,5	A
Palla violinitens coniger (Butler, 1896)	Aurivillius, 1928	5	A
Charaxes c. crossleyi (Ward, 1871)	Aurivillius, 1928	6	A
Charaxes eurinome ansellica (Butler, 1870)	Butler, 1870 (*)	5	B
Charaxes trajanus bambi BS & M, 2007	BS & M, 2007 (*)	1,5	A
NYMPHALIDAE | Apaturinae			
Apaturopsis c. cleocharis (Hewitson, 1873)	Hewitson, 1873 (*)	6	A
NYMPHALIDAE | Nymphalinae			
Kallimoides rumia jadyae (Fox, 1968)	Druce, 1875, as *rumia*	5	A
Vanessula milca buechneri (Dewitz, 1887)	▲	6	A
Antanartia d. delius (Drury, 1782)	▲	5	A
Vanessa cardui (Linnaeus, 1758)	Snellen, 1882	5	C,D
Precis antilope (Feisthamel, 1850)	Monard, 1956	5	C,D
Precis a. archesia (Cramer, 1779)	Aurivillius, 1928	5	C,D
Precis c. ceryne (Boisduval, 1847)	Druce, 1875	5	A,B,E

(continued)

Taxon	First reference for Angola	V	H
Precis coelestina (Dewitz, 1879)	Dewitz, 1879 (*)	5	A
Precis octavia sesamus (Trimen, 1883)	Druce, 1875, as	5	C
Precis pelarga (Fabricius, 1775)	Ladeiro, 1956	5	A,C
Precis actia (Distant, 1880)	Aurivillius, 1928	5	A,C
Precis s. sinuata (Plötz, 1880)	Bacelar, 1956	5	A,C
Precis rauana silvicola (Schultze, 1916)	◘	5	A
Precis larseni manuscript name – (M et al., 2018)	M et al., 2018 (*)	1,5	A,C
Hypolimnas a. anthedon (Doubleday, 1845)	Druce, 1875, sub *Diadema*	5	A,C
Hypolimnas misippus (Linnaeus, 1764)	Druce, 1875	5	H
Hypolimnas d. dinarcha (Hewitson, 1865)	Bacelar, 1958a, b	5	A
Hypolimnas m. monteironis (Druce, 1874)	Druce, 1874 (*)	5	A
Hypolimnas s. salmacis (Drury, 1773)	Druce, 1875	5	A
Salamis c. cacta (Fabricius, 1793)	Ackery et al., 1995	5	A
Protogoniomorpha anacardii ansorgei (Rothschild, 1904)	Rothschild, 1904 (*)	5	A,C
Protogoniomorpha parhassus (Drury, 1782)	Druce, 1875, as *Diadema salamis*	5	A,C
Protogoniomorpha t. temora (Felder & Felder, 1867)	Aurivillius, 1928	6	A
Junonia artaxia (Hewitson, 1864)	Aurivillius, 1928	5	C,D
Junonia hierta crebrene (Trimen, 1870)	Butler, 1871	5	A,C,D
Junonia n. natalica (Felder & Felder, 1860)	Gardiner, 2004	6	D
Junonia natalica angolensis (Rothschild, 1918)	Rothschild, 1918 (*)	3,5	A,C
Junonia o. oenone (Linnaeus, 1758)	Butler, 1871	5	H
Junonia orythia madagascariensis (Guenée, 1865)	Bacelar, 1948	5	A,C,D
Junonia sophia infracta (Butler, 1888)	Bacelar, 1948	5	A
Junonia stygia (Aurivillius, 1894)	Aurivillius, 1928	5	A
Junonia w. westermanni (Westwood, 1870)	Hewitson, 1873	5	A
Junonia terea elgiva (Hewitson, 1864)	Aurivillius, 1928	5	A
Junonia ansorgei (Rothschild, 1899)	◘	5	A
Junonia cymodoce lugens (Schultze, 1912)	Aurivillius, 1909, as *cymodoce*	5	A
Catacroptera c. cloanthe (Stoll, 1781)	Butler, 1871	5	A,C,D
NYMPHALIDAE | Cyrestinae			
Cyrestis c. camillus (Fabricius, 1781)	Ladeiro, 1956	5	A
NYMPHALIDAE | Biblidinae			
Biblia anvatara crameri (Aurivillius, 1894)	Bacelar, 1948	5	C,D
Byblia ilithyia (Drury, 1773)	Snellen, 1882	5	A,D

(continued)

Taxon	First reference for Angola	V	H
Mesoxantha ethosea ethoseoides (Rebel, 1914)	Aurivillius, 1928	5	A
Ariadne albifascia (Joicey & Talbot, 1921)	▲	5	A
Ariadne enotrea archeri (Carcasson, 1958)	Carcasson, 1958 (*)	5	A
Neptidiopsis ophione nucleata (Grünberg, 1911)	Aurivillius, 1928	5	A
Eurytela dryope angulata (Aurivillius, 1899)	Butler, 1871, as *dryope*	5	A,C,D
Eurytela h. hiarbas (Drury, 1782)	Druce, 1875	5	A
Sevenia boisduvali omissa (Rothschild, 1918)	▲	5	A
Sevenia occidentalium penricei (Rothschild & Jordan, 1903)	Rothschild & Jordan, 1903 (*)	1,5	A
Sevenia consors (Rothschild & Jordan, 1903)	Rothschild & Jordan, 1903 (*)	5	C
Sevenia t. trimeni (Aurivillius, 1899)	Aurivillius, 1899, as *Crenis natalensis* var. *trimeni* (*)	5	A,C
Sevenia umbrina (Karsch, 1892)	▲	5	
Sevenia amulia intermedia (Carcasson, 1961)	Aurivillius, 1928, as *amulia*	5	A,C,E
Sevenia benguelae (Chapman, 1872)	Aurivillius, 1928	5	A,C,D
Sevenia p. pechueli (Dewitz, 1879)	Dewitz, 1879 (*)	5	C
NYMPHALIDAE | Limenitidinae			
Harma theobene superna (Fox, 1968)	Fox, 1968	5	A
Cymothoe o. oemilius (Doumet, 1859)	Bacelar, 1958a, b	5	A
Cymothoe b. beckeri (Herrich-Schaeffer, 1858)	Druce, 1875	5	A
Cymothoe haynae fumosa (Staudinger, 1896)	Bacelar, 1956	2,5	A
Cymothoe confusa (Aurivillius, 1887)	Bacelar, 1956	5	A
Cymothoe lucasii cloetensi (Seeldrayers, 1896)	◘	5	A
Cymothoe h. harmilla (Hewitson, 1874)	Larsen, 2005	6	A
Cymothoe h. hesiodotus (Hewitson, 1869)	Aurivillius, 1928	6	A
Cymothoe h. hypatha (Hewitson, 1866)	Druce, 1875	5	A
Cymothoe lurida hesione (Weymer, 1907)	Druce, 1875, as *lurida*	5	A
Cymothoe altisidora (Hewitson, 1869)	Aurivillius, 1898	5	A
Cymothoe capella (Ward, 1871)	Bacelar, 1956	5	A
Cymothoe caenis (Drury, 1773)	Druce, 1874	5	A
Cymothoe jodutta ciceronis (Ward, 1871)	Bacelar, 1956	5	A
Cymothoe jodutta ehmckei (Dewitz, 1887)	◘	5	A
Cymothoe cf. *c. coccinata* (Hewitson, 1874)	◘	5	A

(continued)

Taxon	First reference for Angola	V	H
Cymothoe excelsa deltoides (Overlaet, 1944)	D'Abrera, 1980	3,5	A
Cymothoe s. sangaris (Godart, 1824)	Druce, 1875	5	A
Pseudoneptis bugandensis ianthe (Hemming, 1964)	Snellen, 1882, as *bugandensis*	5	A
Pseudacraea eurytus eurytus (Linnaeus, 1758)	Druce, 1875	5	A
Pseudacraea d. dolomena (Hewitson, 1865)	Aurivillius,1928	6	A
Pseudacraea b. boisduvalii (Doubleday, 1845)	Druce, 1875	5	A
Pseudacraea kuenowii gottbergi (Dewitz, 1884)	Williams, 2007, as *kuenowii*	6	A
Pseudacraea lucretia protracta (Butler, 1874)	Butler, 1874 (*)	5	A
Pseudacraea poggei (Dewitz, 1879)	Gardiner, 2004	5	A,D
Pseudacraea semire (Cramer, 1779)	Druce, 1875	5	A
Neptis saclava marpessa (Hopffer, 1855)	Druce, 1875	5	A,B,C
Neptis nemetes margueriteae (Fox, 1968)	Butler, 1871, the species	6	C,D
Neptis gratiosa (Overlaet, 1955)	▲	5	C
Neptis jordani (Neave, 1910)	Gardiner, 2004	6	D
Neptis kiriakoffi (Overlaet, 1955)	◘	5	A,C
Neptis laeta (Overlaet, 1955)	Gardiner, 2004	5	C
Neptis morosa (Overlaet, 1955)	Larsen, 2005	5	A,C
Neptis s. serena (Overlaet, 1955)	Gardiner, 2004	5	A,C,E
Neptis alta (Overlaet, 1955)	Gardiner, 2004	5	C
Neptis constantiae kaumba (Condamin, 1966)	◘	5	A
Neptis nysiades (Hewitson, 1868)	Aurivillius, 1928	5	A
Neptis nicomedes (Hewitson, 1874)	Hewitson, 1874 (*)	5	A
Neptis quintilla (Mabille, 1890)	Larsen, 2005	5	A
Neptis a. agouale (Pierre-Baltus, 1978)	◘	5	A
Neptis melicerta (Drury, 1773)	Aurivillius, 1928	5	A,D
Neptis nebrodes (Hewitson, 1874)	Hewitson, 1874 (*)	6	A
Neptis nicoteles (Hewitson, 1874)	Hewitson, 1874 (*)	6	A
Neptis e. exaleuca (Karsch, 1894)	◘	5	A
Evena crithea (Drury, 1773)	Fox, 1968	5	A
Evena angustatum (Felder & Felder, 1867)	◘	5	A
Euryphura c. chalcis (Felder & Felder, 1860)	Bacelar, 1956, as *E. fulminea*	5	A
Euryphura plautilla Hewitson, 1865	Druce, 1875	5	A
Euryphura concordia (Hopffer, 1855)	Aurivillius, 1928	5	C,E
Hamanumida daedalus (Fabricius, 1775)	Druce, 1875, as *Aterica meleagris*	5	C,D,F
Pseudargynnis hegemone (Godart, 1819)	Druce, 1874, as *Aterica clorana*	5	E

(continued)

Taxon	First reference for Angola	V	H
Aterica g. extensa (Heron, 1909)	Druce, 1875, as *Aterica cupavia*	5	A,B,C
Cynandra opis bernardii (Lagnel, 1967)	Druce, 1875, as *Aterica afer*	5	A
Euriphene barombina (Aurivillius, 1894)	Larsen, 2005	5	A
Euriphene iris (Aurivillius, 1903)	▲	5	C
Euriphene plagiata (Aurivillius, 1897)	Larsen, 2005	6	A
Euriphene saphirina trioculata (Talbot, 1927)	▲	6	A
Euriphene t. tadema (Hewitson, 1866)	◘	5	A
Euriphene gambiae gabonica (Bernardi, 1966)	Bacelar, 1958a, b	6	A
Bebearia phantasia concolor (Hecq, 1988)	Druce, 1875	6	A
Bebearia languida (Schultze, 1920)	▲	5	A
Bebearia a. absolon (Fabricius, 1793)	Bacelar, 1958a, b	5	A
Bebearia micans (Aurivillius, 1899)	◘	5	A
Bebearia zonara (Butler, 1871)	◘	5	A
Bebearia oxione squalida (Talbot, 1928)	Aurivillius, 1909, as *oxione*		A
Bebearia cocalia katera (van Someren, 1939)	◘	5	A
Bebearia guineensis (Felder & Felder, 1867)	Holmes, 2001	6	A
Bebearia sophus aruunda (Overlaet, 1955)	Druce, 1875, as *sophus*	5	A
Bebearia plistonax (Hewitson, 1874)	Hewitson, 1874 (*)	5	F
Bebearia hassoni (Hecq, 1998)	Hecq, 1998 (*)	1	A
Euphaedra medon celestis (Hecq, 1986)	Butler, 1871, as *medon*	5	A
Euphaedra z. zaddachii (Dewitz, 1879)	Dewitz, 1879 (*)	5	A
Euphaedra cf. *sinuosa* (Hecq, 1974)	◘	6	A
Euphaedra diffusa diffusa (Gaede, 1916)	◘	6	A
Euphaedra ansorgei (Rothschild, 1918)	◘	5	A
Euphaedra p. permixtum (Butler, 1873)	Bacelar, 1956	5	A
Euphaedra divoides (BS & M, 2018) (Manuscript name)	Staudinger, 1886, as *E. themis* var. *innocentia* (?)	1,5	A
Euphaedra adonina spectacularis (Hecq, 1997)	Bacelar, 1956	5	A
Euphaedra ceres electra (Hecq, 1983)	Butler, 1871, as *ceres*	5	A
Euphaedra fontainei (Hecq, 1977)	▲	5	A
Euphaedra v. viridicaerulea (Bartel, 1905)	◘	5	A
Euphaedra preussiana robusta (Hecq, 1983)	◘	5	A
Euphaedra rezia (Hewitson, 1866)	Bacelar, 1956	5	A
Euphaedra albofasciata (Berger, 1981)	◘	5	A
Euphaedra disjuncta virens (Hecq, 1984)	◘	5	A
Euphaedra mayumbensis (Hecq, 1984)	◘	5	A

(continued)

Taxon	First reference for Angola	V	H
Euphaedra p. preussi (Staudinger, 1891)	Aurivillius, 1928	6	A
Euphaedra uigensis (BS & M, 2017)	BS & M, 2017 (*)	1,5	A
Euphaedra c. castanoides (Hecq, 1985)	◘	5	A,F
Euphaedra coprates (Druce, 1875)	Druce, 1875 (*)	5	A
Euphaedra e. eleus (Drury, 1782)	Aurivillius, 1928	5	A
Euphaedra simplex (Hecq, 1978)	◘	5	A
Euphaedra ruspina (Hewitson, 1865)	Druce, 1875	5	A,C
Euphaedra harpalyce comminuera (Hecq, 1999)	Hecq, 1999 (*)	1.5	A
Euphaedra harpalyce spatiosa (Mabille, 1876)	Mabille, 1876 (*)	6	A
Euphaedra losinga wardi (Druce, 1874)	Druce, 1874	5	A
Euphaedra losinga limita (Hecq, 1978)	◘	3,5	A
Euptera mocquerysi (Staudinger, 1893)	Bacelar, 1956, as *E. falsathyma*	5	A
Nymphalidae | Heliconiinae			
Acraea r. rogersi (Hewitson, 1873)	Hewitson, 1873 (*)	5	B,D
Acraea e. egina (Cramer, 1775)	Druce, 1875	5	A,C
Acraea acrita ambigua (Trimen, 1891)	Weymer, 1901, as *A. bella*	6	C,D
Acraea bellona (Weymer, 1908)	Weymer, 1908 (*)	1,5	C
Acraea periphanes (Oberthür, 1893)	Le Doux, 1923	5	C,E
Acraea asema (Hewitson, 1877)	Trimen, 1891	6	C
Acraea o. omrora (Trimen, 1894)	Trimen, 1894 (*)	1,6	C
Acraea violarum anchietai (M & BS, 2017)	M & BS, 2017 (*)	1,5	C,D
Acraea buettneri (Rogenhofer, 1890)	Trimen, 1891, as *A. felina*	5	A,C
Acraea cepheus (Linnaeus, 1758)	Eltringham, 1912	5	A,C
Acraea atolmis (Westwood, 1881)	Aurivillius, 1928, morph *acontias*	5	C
Acraea bailundensis (Wichgraf, 1918)	Wichgraf, 1918 (*)	3,5	C
Acraea diogenes (Suffert, 1904)	▲	3,5	D
Acraea guillemei (Oberthür, 1893)	Aurivillius, 1928	6	C
Acraea lapidorum (Pierre, 1988)	Pierre, 1988 (*)	1,5	C
Acraea onerata (Trimen, 1891)	Trimen, 1891 (*)	1,4	C,D
Acraea aglaonice (Westwood, 1881)	Gardiner, 2004	6	D
Acraea atergatis (Westwood, 1881)	Aurivillius, 1928	5	D
Acraea axina (Westwood, 1881)	Aurivillius, 1928	5	C
Acraea c. caldarena (Hewitson, 1877)	Gardiner, 2004	5	C
Acraea ella (Eltringham, 1911)	Eltringham, 1911 (*)	4,5	C
Acraea lygus (Druce, 1875)	Druce, 1875 (*)	5	C, D
Acraea natalica (Boisduval, 1847)	Aurivillius, 1928	5	A,C
Acraea oncaea (Hopffer, 1855)	◘	6	C
Acraea pseudegina (Westwood, 1852)	Druce, 1875	5	A
Acraea pudorella (Aurivillius, 1899)	▲	5	C
Acraea stenobea (Wallengren, 1860)	Trimen, 1891	6	C,D
Acraea anemosa (Hewitson, 1865)	Aurivillius, 1928	5	C,D

(continued)

Taxon	First reference for Angola	V	H
Acraea p. pseudolycia (Butler, 1874)	Butler, 1874 (*)	3,5	A,C
Acraea acara melanophanes (Le Cerf, 1927)	Pierre & Bernaud, 2013	5	D
Acraea z. zetes (Linnaeus, 1758)	Druce, 1875	5	C,D
Acraea admatha (Hewitson, 1865)	Pierre, 1979	6	A
Acraea endoscota (Le Doux, 1928)	Larsen, 2005	6	A
Acrae l. leucographa (Ribbe, 1889)	Larsen, 2005	6	A
Acraea q. quirina (Fabricius, 1781)	▲	5	A
Acraea camaena (Drury, 1773)	Larsen, 2005	6	B
Acraea n. neobule (Doubleday, 1843)	Druce, 1875	5	A,C,D
Acraea eugenia ochreata (Grünberg, 1910)	Larsen, 2005	5	A
Acraea brainei (Henning, 1986)	Ackery et al., 1995	6	?
Acraea e. epaea (Cramer, 1779)	Aurivillius, 1928	5	A
Acraea formosa (Butler, 1874)	Butler, 1874 (*)	5	A,B
Acraea l. leopoldina (Aurivillius, 1895)	▲	3	A
Acraea p. poggei (Dewitz, 1879)	Dewitz, 1879 (*)	6	C
Acraea alcinoe camerunica (Aurivillius, 1893)	▲	6	A
Acraea umbra macarioides (Aurivillius, 1893)	Aurivillius, 1928	5	A,B
Acraea consanguinea intermedia (Aurivillius, 1899)	Le Doux, 1937	2,5	A
Acraea excisa (Butler, 1874)	Bacelar, 1948	5	A
Acraea pseuderyta (Godman & Salvin, 1890)	Aurivillius, 1928	5	A
Acraea vestalis congoensis (Le Doux, 1937)	◘	5	A
Telchinia p. perenna (Doubleday, 1847)	Aurivillius, 1928	5	A
Telchinia p. penelope (Staudinger, 1896)	◘	5	A
Telchinia o. oreas (Sharpe, 1891)	Lathy, 1906	6	C
Telchinia circeis (Drury, 1782)	Larsen, 2005	6	A
Telchinia parrhasia servona (Godart, 1819)	Godart, 1819 (*)	6	A
Telchinia peneleos pelasgia (Grose-Smith, 1900)	Larsen, 2005	5	A,C
Telchinia p. pharsalus (Ward, 1871)	Aurivillius, 1928	5	A,B
Telchinia encedana (Pierre, 1976)	Pierre, 1976	6	E
Telchinia e. encedon (Linnaeus, 1758)	Druce, 1875	5	A,C,D
Telchinia alciope (Hewitson, 1852)	Bacelar, 1956	5	A
Telchinia a. aurivillii (Staudinger, 1896)	▲	6	A
Telchinia esebria (Hewitson, 1861)	Butler, 1874	5	B,C
Telchinia j. jodutta (Fabricius, 1793)	Druce, 1875	5	A
Telchinia lycoa (Godart, 1819)	Druce, 1875	5	A,C
Telchinia serena (Fabricius, 1775)	Butler, 1871	5	A,C,D

(continued)

Taxon	First reference for Angola	V	H
Telchinia v. ventura (Hewitson, 1877)	Monard, 1956	6	E
Telchinia acerata (Hewitson, 1874)	Snellen, 1882	5	A,C
Telchinia oberthueri (Butler, 1895)	Bacelar, 1948	6	A
Telchinia sotikensis karschi (Aurivillius, 1899)	Aurivillius, 1928	5	A
Telchinia b. bonasia (Fabricius, 1775)	Druce, 1875	5	A,C,D
Telchinia uvui balina (Karsch, 1892)	Larsen, 2005	6	A
Telchinia o. orestia (Hewitson, 1874)	Snellen, 1882	5	A
Telchinia p. pentapolis (Ward, 1871)	▲	6	A
Telchinia induna imduna (Trimen, 1895)		5	C
Telchinia r. rahira (Boisduval, 1833)	Druce, 1875	5	C,D,E
Telchnia mirifica (Lathy, 1906)	Lathy, 1906 (*)	3	C
Lachnoptera anticlia (Hübner, 1819)	Bacelar, 1958a, b, as *L. iole*	6	A
Phalanta e. eurytis (Doubleday, 1847)	Bacelar, 1956	5	A
Phalanta phalantha aethiopica (Rothschild & Jordan, 1903)	Druce, 1875	5	C,D

References

Ackery PR, Smith CR, Vane Wright RI (1995) Carcasson's African Butterflies. An annotated catalogue of the Papilionoidea and Hesperioidea of the Afrotropical Region. CSIRO, Victoria, i–xii + 803 pp

Aduse-Poku K, Vingerhoedt E, Wahlberg N (2009) Out-of-Africa again: A phylogenetic hypothesis of the genus Charaxes (Lepidoptera: Nymphalidae) based on five genes region. Mol Phylogenet Evol 53:463–478

Aurivillius C In Seitz A (1928) Les Macrolepidopteres du Globe. Les Macrolepidoptères de la Faune Ethiopienne 13(4):615 pp + 80 pl. E. Le Moult, Paris

Bacelar A (1948) Lepidópteros de África principalmente das colónias portuguesas (colecção do Museu Bocage). Arquivos do Museu Bocage 19:165–207

Bacelar A (1956) Lepidópteros (Rhopalocera) de Buco Zau, enclave de Cabinda, Angola. Anais da Junta de Investigações do Ultramar 11(3):175–197

Bacelar A (1958a) Alguns Lepidópteros (Rhopalocera) do enclave de Cabinda. Revista portuguesa de Zoologia e Biologia Geral 1(2/3):197–217

Bacelar A (1958b) Alguns Lepidópteros (Rhopalocera) da África Ocidental portuguesa. Revista portuguesa de Zoologia e Biologia Geral 1(4):311–330

Bacelar A (1961) Lepidópteros do Bié (Rhopalocera) da colecção do Colégio de São Bento, em Luso (Angola). Memórias da Junta de Investigações do Ultramar (2)23:61–81

Berger LA (1979) Espèces peu connues et descriptions de nouvelles sous-espèces de Mylothris (Lepidoptera Pieridae). Revue de Zoologie Africaine 93(1):1–9

Bethune-Baker GT (1914) Notes on the taxonomic value of genital armature in Lepidoptera. Trans Entomol Soc London 1914:314–337

Bivar-de-Sousa A (1983) Contribuição para o conhecimento dos lepidópteros de Angola (3ª nota). Dados sobre a ocorrência do género Charaxes (Lep. Nymphalidae) em Angola (1ª parte). Actas do I Congresso Ibérico de Entomologia 1:107–119

Bivar-de-Sousa A, Fernandes JA (1964) Contribuição para o conhecimento dos Lepidópteros de Angola. Boletim da Sociedade Portuguesa de Ciências Naturais (2) 10(25):104–115

Bivar-de-Sousa A, Mendes LF (2006) On the genus Euxanthe Hübner, 1819 in Angola, with description of a new subspecies (Lepidoptera, Nymphalidae, Charaxinae). Nouvelle Revue d'Entomologie 23(4):369–376

Bivar-de-Sousa A, Mendes LF (2007) New data on the Uranothauma from Angola, with description of a new species (Lepidoptera: Lycaenidae: Polyommatinae). Boletin Sociedad Entomológica Aragonesa 41:73–76

Bivar-de-Sousa A, Mendes LF (2009a) On a new species of the genus Princeps Hübner, (1807) from Cabinda (Angola) (Lepidoptera: Papilionidae). SHILAP, Revista de Lepidopterologia 37(147):313–318

Bivar-de-Sousa A, Mendes LF (2009b) New data on the Angolan Charaxes of the "etheocles group" with description of a new species (Lepidoptera, Nymphalidae, Charaxinae). Boletim da Sociedade portuguesa de Entomologia 8(15)(229):293–309

Bivar-de-Sousa A, Mendes LF (2014) New data on the Angolan Charaxesof the "etheocles group" with description of a new species (Lepidoptera, Nymphalidae, Charaxinae). Boletim da Sociedade portuguesa de Entomologia 229(8–15):293–309

Bivar-de-Sousa A, Mendes LF, Vasconcelos S (2017) Description of one new species and one new subspecies of Nymphalidae from Angola (Lepidoptera: Papilionoidea). SHILAP, Revista de Lepidopterologia 45(178):227–236

Bivar-de-Sousa A, Mendes LF (2018, in press) The "themis group" of Euphaedra (Euphaedrana) in Angola. Revision and description of one new species (Lepidoptera: Nymphalidae: Limenetidinae). Boletim da Sociedade portuguesa de Entomologia

Bouyer T (1999) Note sur les Charaxes du «groupe eupale» avec description d'une nouvelle sous-espèce (Lepidoptera Nymphalidae). Entomologia Africana 4(1):37–40

Butler AG (1872) On a small collection of butterflies from Angola. Proc Zool Soc London 1871:721–725

Carvalho EL (1962) Alguns Papilionídeos da Lunda (Lepidoptera). Publicações Culturais da Companhia dos Diamantes de Angola 60:163–170

Condamin M (1966) Mise au point sur les Neptis au facies d' «14avann» (Lepidoptera: Nymphalidae). Bulletin de l'Institut Fondamental de l'Afrique Noire 28(A)(3):1008–10029

D'Abrera B (1980) Butterflies of the afrotropical region. Lansdowe Ed., Melbourne, pp i–xx + 1–593

Dhungel B, Wahlberg N (2018) Molecular systematics of the subfamily Limenitidinae (Lepidoptera: Nymphalidae). PeerJ 6:e4311

Druce H (1875) A list of the collections of diurnal Lepidoptera made by J. J. Monteiro in Angola with description of some new species. Proc Zool Soc London 27:406–417

Eltringham H (1912) A monograph of the African species of the genus Acraea Fab., with a supplement of those of the Oriental Region. Trans Entomol Soc London 1912(1):1–374

Espeland M, Breinholt J, Willmott KR et al (2018) A comprehensive and dated phylogenomic analysis of butterflies. Curr Biol 28:1–9

Evans WH (1937) Catalog of African Hesperiidae (Indicating the classification and nomenclature adopted in the British Museum). British Museum, London, 212pp + 30 pl

Gardiner A (2004) Chapter 10. Butterflies of the four corners area. In: Timberlake JR, Childers SL (eds) Biodiversity of the Four Corners Area: Technical Review, vol 15. Occasional Publications on Biodiversity, Bulawayo, pp 381–397

Hecq J (1997) Euphaedra. Lambillionea & Hecq, Tervuren & Mont-Sur-Marchienne, 121 pp + 48 pl

Heikkilä M, Kaila L, Mutanen M et al (2012) Cretaceous origin and repeated Tertiary diversification of the redefined butterflies. Proc R Soc Lond B 279:1093–1099

Henning SF (1988) The Charaxes Butterflies of Africa. Aloe Books, Johannesburg, 457 pp

Holmes CWN (2001) A reaprisal of the Bebearia mardania complex (Lepidoptera, Nymphalidae). Trop Zool 14:31–62

Homeyer A, Dewitz H (1882) Drei neue westafrikanische Charaxes. Berliner Entomologisches Zeitschrift 26(II):381–383

JIU (1948–1963) Cartas do levantamento aerofotogramétrico de Angola (escala 1: 250,000). Junta de Investigações do Ultramar: folhas:8–471

Kielland J (1982) Revision of the genus *Ypthima* in the Ethiopian Region excluding Madagascar (Lepidoptera, Satyridae). Tijdschrift voor Entomologie 125(5):99–154

Kielland J (1994) A revision of the genus *Henotesia* (excluding Madagascar and other Indian Ocean islands) (Lepidoptera Satyridae). Lambillionea 94(2):235–273

Ladeiro JM (1956) Lepidópteros de Angola (estudo de uma colecção oferecida ao Museu Zoológico de Coimbra). Anais da Junta de Investigações do Ultramar 11(3):151–172

Larsen TB (2005) Butterflies of West Africa. Text volume: 1–595. Apollo Books, Stenstrup

Latreille PA, Godart JB (1819) Encyclopédie Méthodologique. Histoire Naturelle (Zoologie), 9, Entomologie, Paris, i–iv + 1–328

Lautenschläger T, Neinhuis C (eds) (2014) Riquezas Naturais do Uíge – Uma breve Introdução sobre o estado atual. A utilização, a Ameaça e a Preservação da Biodiversidade. Technische Universität Dresden, Dresden, 125 pp

Libert M (1999) Revision des Epitola (ls). *Révision des genres* Epitola *Westwood,* Hypophytala *Clench et* Stempfferia Jackson et description de trois nouveaux genres (Lepidoptera Lycaenidae). ABRI & Lambillionea, Nairobi/Tervuren, pp 1–219, pl. I–XVI

Libert M (2000) Révision du genre Mimacraea Butler, avec description de quatre nouvelles espèces et deux nouvelles sous-espèces (Lepidoptera, Lycaenidae). Lambillionea & ABRI, Tervuren/ Nairobi, pp 1–73

Libert M (2004) Revision du genre *Oxylides* Hübner (Lepidoptera, Lycaenidae). Lambillionea 104(2):143–158

Mendes LF, Bivar-de-Sousa A (2006a) A new species of *Neita* van Son (Nymphalidae, Satyrinae) from southern Angola. Boletin Sociedad Entomológica Aragonesa 39:95–96

Mendes LF, Bivar-de-Sousa A (2006b) Notes and descriptions of Afrotropical *Appias* butterflies (Lepidoptera: Pieridae). Boletin Sociedad Entomológica Aragonesa 39:151–160

Mendes LF, Bivar-de-Sousa A (2007a) New species of *Cooksonia* Druce, 1905 from Angola (Lepidoptera: Lycaenidae, Lipteninae). SHILAP, Revista de Lepidopterologia 35(138):265–268

Mendes LF, Bivar-de-Sousa A (2007b) On the genus Eagris Guenée, 1863 in Angola (Lepidoptera: Hesperiidae). SHILAP, Revista de Lepidopterologia 35(139):311–316

Mendes LF, Bivar-de-Sousa A (2009a) Description of a new species of *Mylothris* from northern Angola (Lepidoptera Pieridae). Bolletino della Societá Entomológica Italiana 141(1):55–58

Mendes LF, Bivar-de-Sousa A (2009b) New account on the butterflies of Angola. The genus Leptomyrina (Lepidoptera Lycaenidae). Bolletino della Societá Entomológica Italiana 141(2):109–112

Mendes LF, Bivar-de-Sousa A (2009c) On a new south-eastern Angolan Satyrine butterfly belonging to a new genus (Lepidoptera, Nymphalidae). Entomologia Africana 14(2):5–8

Mendes LF, Bivar-de-Sousa A (2009d) The genus *Abantis* Hopffer, 1855 in Angola and description of a new species (Lepidoptera: Hesperiidae, Pyrginae). SHILAP, Revista de Lepidopterologia 37(147):313–318

Mendes LF, Bivar-de-Sousa A (2012) Notes on the species of *Hypolycaena* (Lepidoptera, Lycaenidae, Theclinae) known to occur in Angola. Boletin Sociedad Entomológica Aragonesa 50:193–197

Mendes LF, Bivar-de-Sousa A, Figueira R (2013a) Butterflies of Angola / Borboletas diurnas de Angola. Lepidoptera. Papilionoidea, I. Hesperiidae, Papilionidae. IICT and CIBIO, Lisboa and Porto, 288 pp

Mendes LF, Bivar-de-Sousa A, Figueira R et al (2013b) Gazetteer of the Angolan localities known for beetles (Coleoptera) and butterflies (Lepidoptera: Papilionoidea). Boletim da Sociedade Portuguesa de Entomologia 8(14/228):257–290

Mendes LF, Bivar-de-Sousa A, Vasconcelos S et al (2017) Description of two new subspecies and notes on *Charaxes* Ochenheimer, 1816 of Angola (Lepidoptera: Nymphalidae). SHILAP, Revista de Lepidopterologia 45(178):299–315

Mendes LF, Bivar-de-Sousa A, Vasconcelos S (2018, in press) On the butterflies of genus *Precis* Hübner, 1819 from Angola and description of a new species (Lepidoptera: Nymphalidae: Nymphalinae). SHILAP Revista de Lepidopterologia

Monard A (1956) Compendium Entomologicum Angolae – 1. Insecta. VI – Ord. Lepidoptera. Anais da Junta de Investigações do Ultramar 11(3):119–128

NHM – Natural History Museum (2004). Wallowtails. Available at: http://internet.nhm.ac.uk/cgi-bin/perth/wallowtails/list.dsml

Pierre J (1988) Les Acraea du super-group «egina»: Révision et phylogénie (Lepidoptera : Nymphalidae). Annales de la Société Entomologique de France 24(3):263–287

Pierre J, Bernaud D (2013) Butterflies of the world. NymphalidaeXXIII. Acraea subgenus Acraea. Goecke & Evers, Keltern, 39:1–8, pl. 1–28

Rothschild W, Jordan K (1903) Lepidoptera collected by Oscar Neumann in north-east Africa. Novitates Zoologicae 10(3):491–542

Smith CR, Vane-Wright RI (2001) A review of the Afrotropical species of the genus Graphium (Lepidoptera: Rhopalocera: Papilionidae). Bull Nat His Mus London (Entomol) 70(2):503–719

Stempffer H (1957) Les Lépidoptères de l'Afrique noire française. Lycaenidés. Initiations Africaines 14(3):1–228

Stempffer H, Bennett NH (1963) A new genus of Lipteninae (Lepidoptera: Lycaenidae). Bull Brit Mus (Nat His) (Entomol) 13:171–194

Talbot G (1944) A preliminary revision of the genus *Mylothris* Hübn. (1819) (Lep. Rhop. Pieridae). T Roy Ent Soc London 94(2):155–185

Tite GE (1959) New species and notes on the genus *Lepidochrysops* (Lepidoptera, Lycaenidae). The Entomologist 92:158–163

Tite GE (1961) New species of the genus *Lepidochrysops* (Lepidoptera, Lycaenidae). The Entomologist 94:21–25

Tite GE, Dickson CGC (1973) The genus *Aloeides* and allied genera. Bull Brit Mus (Nat His) (Entomol) 29:227–280

Trimen R (1891) On butterflies collected in tropical south-western Africa by Mr. A. W. Eriksson. Proc Zool Soc 1891:59–107

Turlin B, Vingerhoedt E (2013) Butterflies of the World, supl. 23. Les Charaxinae de la faune Afrotropicale. Les genres Palla et Euxanthe. Nymphalidae: Charaxinae: Pallini et Euxanthini. Goecke & Evers, Keltern

Weymer G (1901) Beitrag zur Lepidopterofauna von Angola. Entomologischen Zeitschriften, Stuttgart 15(17):61–64, 65–67, 69–70

Weymer G (1908) Einige neuer Lepidopteren des Deutschen Entom. National-Museums, gesammelt von Dr. F. Cr. Wellman in Benguella. Deutsche Entomologische Zeitschrift 1908:507–413

Willis CK (2009) Amist the butterflies of southwestern Angola. Metamorphosis 20(3):74–88

Williams MC (2018) Afrotropical butterflies. Available at: http://www.lepsocafrica.org/?p=publications&s=atb

Part IV
Vertebrates: Distribution and Diversity

Chapter 11
The Freshwater Fishes of Angola

Paul H. Skelton

Abstract The discovery and exploration of Angolan freshwater fishes was largely effected by foreign scientists on expeditions organised by European and North American parties. Current knowledge of Angolan freshwater fishes is briefly described according to the main drainage systems that include Cabinda, Lower Congo, Angolan Coastal region including the Cuanza, the southern Congo tributaries, the Zambezi, Okavango, Cunene and Cuvelai drainages. A biogeographic model to explain the freshwater fish fauna of Angola is presented. The need for the conservation of Angolan freshwater fishes will rise with rapidly increasing pressures on aquatic ecosystems from urbanisation, dams for power, agriculture and human needs, habitat destruction from mining and deforestation, pollution, the introduction of alien species and overfishing.

Keywords Africa · Cuanza · Cunene · Cuvelai · Okavango · Southern Congo · Zambezi

Historical Review

> Despite Poll's work (1967) over a very limited area, Angola remains a poorly known region in which there remains much to be discovered (Lévêque and Paugy 2017a: 93)

The quotation above sums up the current state of knowledge for the freshwater fishes of Angola. Poll (1967) is a landmark publication that reviews the historical literature and records the known species and their distribution within the major river basins of the country at that time. No other account of Angolan fishes as a whole has been published. The current situation of a poorly known region is due to a number of factors including the historical neglect of scientific exploration by the colonial

P. H. Skelton (✉)
South African Institute for Aquatic Biodiversity (SAIAB), Grahamstown, South Africa

Wild Bird Trust, National Geographic Okavango Wilderness Project, Hogsback, South Africa
e-mail: P.skelton@saiab.ac.za

B. J. Huntley et al. (eds.), *Biodiversity of Angola*,
https://doi.org/10.1007/978-3-030-03083-4_11

authorities, widely dispersed collections in international institutions from various expeditions, the relative inaccessibility to scientists and collectors of the inland rivers and biologically rich areas, and the difficulties of aquatic exploration relative to terrestrial fauna. The fact that there is no national Angolan depository for wet collections such as fishes fostered by local expertise is a further hindrance to discovery. This aspect is fundamental to effective and sustained scientific productivity in any endeavor such as ichthyology (Skelton and Swartz 2011). This accentuates the situation for Angolan freshwater fishes when it is recognised that Poll's (1967) account rested largely on the collection in the Diamond Company of Angola (DIAMANG) museum in Dundo, which to a large extent is a product of the industrial diamond mining activity in the mainly local drainages.

There are four distinct phases of scientific discovery of Angolan freshwater fishes, Phase 1 – early explorations in the second half of the nineteenth century; Phase 2 – scientific expeditions in the twentieth century until World War II; Phase 3 – post WWII to Angolan independence in 1975; and, Phase 4 – post independence investigations.

Although several of Castelnau's (1861) Lake Ngami fishes occur in the Angolan reaches of the Okavango River system, the discovery and scientific description of Angolan freshwater fishes was initiated by Steindachner (1866) describing a collection derived from the Atlantic coastal rivers. Steindachner's species include some iconic species such as his *Kneria angolensis, Clarias angolensis* and *Enteromius kessleri* that help define the Angolan Atlantic coastal fauna. Guimarães (1884) working with specimens in the Lisbon Museum (subsequently lost in the fire of 1978) (Saldanha 1978) submitted by the Portuguese explorer José Alberto de Oliveira Anchieta provided detailed descriptions and illustrations of three species taken from the Cunene and the Curoca Rivers from 1873–1884, viz. *Schilbe steindachneri, Mormyrus anchietae* and *Enteromius mattozi.*

The second phase of discovery (early twentieth century) is marked by a series of expedition reports that include freshwater fishes. Boulenger's (1909–1916) catalogue of fishes in the British Museum (Natural History) provided the basis of Angolan freshwater fish fauna. Again the fauna included collections such as Woosnam's Okavango collection described by Boulenger (1911) that includes species which occur in Angolan reaches. Boulenger (1910) described a collection by Ansorge from the Cuanza and Bengo Rivers that set the scene for considering the uniqueness of the fauna of these rivers of the Atlantic coast. Other notable expeditions that included descriptions of freshwater fishes are the Vernay Angola Expedition of 1925 (Nichols and Boulton 1927), the Gray African Expedition of 1929 (Fowler 1930), the Vernay-Lang Kalahari Expedition of 1930 (Fowler 1935), the Swiss Scientific Mission to Angola 1928–1929 and 1932–1933 (Pellegrin 1936), and Karl Jordan's Expedition to South West Africa and Angola of 1933–1934 (Trewavas 1936). All these expeditions realised new species but were somewhat limited in geographical scope to the southern Atlantic coastal rivers and to the upland western reaches of the Cubango (Okavango) tributaries, the Oshana-Etosha system and the plateau reaches of the Cuanza. This restriction emanates from the access realised by the Benguella Railway constructed from 1903 to 1928 from Lobito port to Huambo and beyond (Ball 2015).

The third phase of scientific exploration of the freshwater fishes of Angola after the WWII up until independence in 1975 is significant in that studies into ecological aspects as well as the beginnings of a synthesis of the Angolan fauna occurred. Ladiges and Voelker (1961) studied the fish fauna of the Longa River in the Angolan watershed highlands. In addition to providing an ecological description and zonation of the river they described a few new species – *Kneria maydelli* from the Cunene, *Enteromius* (as *Barbus*) *roussellei* and *Chiloglanis sardinhai*. Ladiges (1964) followed up this article with an account of the zoogeography and ecology of Angolan freshwater fishes based on a present/absent list of fishes in the Angolan Coastal region, the Cunene, the Okavango Basin, and the Zambezi. Trewavas (1973) recorded the cichlids of the Cuanza and Bengo Rivers that exposed the independent derivation of the cichlid fauna of the Cuanza River in terms of the inland and coastal reaches. An unpublished collection by Graham Bell-Cross from the Okavango and the Cunene basins was deposited in the NHM in 1965, and this together with collections made by Mike Penrith from the State Museum in Windhoek, Namibia, provided specimens essential for Greenwood's (1984) revision of serranochromine species. Mike Penrith's collections from the Cunene and Okavango in the early 1970s led to a few descriptions of new species by Penrith (1970) and Penrith (1973).

A major milestone account of Angolan freshwater fishes was Max Poll's (1967) *Contribution à la Faune Ichthyologique de l'Angola* – based largely on the extensive collections made by Barros Machado and others and lavishly illustrated with excellent fish drawings as well as a gallery of photographs drawn from the Dundo Museum in Lunda-Norte. Poll (1967) summarised the history of freshwater ichthyology and provided a full checklist of 264 species in 18 families and 54 genera as then recorded from the inland waters of the country (excluding the Cabinda enclave). A faunistic and zoogeographical account considered five ichthyological regions (see below). Acknowledging a clearly incomplete inventory Poll listed the diversity of his regions as follows: The Congo tributaries with 121 species are richest and most diverse with characteristic families and genera known from the Congo Basin. Next in diversity was the western Atlantic coastal region with 109 species, followed by the Zambezi (62 species) but Poll pointed out that Bell-Cross had then recently recorded 77 species from the upper Zambezi, also of tropical diversity but distinct in character from the Congolean rivers. The Okavango-Cubango (57 species) reflected its close connections to the Zambezi as well as to the Cunene (55 species) in the west. The Cunene presented a mixed fauna of both the Zambezian elements as well as Atlantic coastal nature.

Poll (1967) pointed out and summarised a few notable ichthyological characteristics of the Angolan fauna – there was no pronounced endemic character to the fauna as a whole. The occurrence of lungfish (*Protopterus*) in Angola is known only from records in Congo tributaries and from Cabinda, but Poll mentions that he was shown a photograph by Ladiges of *Protopterus annectens brieni* from the Cubango region (see this recorded in Ladiges 1964: 265). Such occurrence of lungfish in the Cubango or Okavango system has not yet been confirmed in spite of extensive collecting in that drainage. Polypterids are restricted to Congo tributaries as are freshwater clupeids (however marine or estuarine species also occur in coastal Atlantic rivers). The presence of kneriids is a distinct feature of the fauna especially of the escarpment reaches of rivers of the coastal region. Mormyrid diversity (36 species) is relatively high, most especially in the southern Congo rivers. Characins (17 species) are less diverse but there is an equivalent representation of Citharinids (16 species). The largest family represented in the country is the Cyprinidae (79 species) and this is especially notable for the Atlantic coastal drainages (43 species) that is even richer than the Congo tributaries (27 species). However the chedrins (*Raiamas* and *Opsaridium* and *Engraulicypris*) are poorly represented – two species in Congo tributaries, one in the Zambezian region and one in the Atlantic coastal drainages. Of the catfishes, the claroteids (10 species) are a presence as are the clariids (17 species) of which the majority (11 species) are represented across different provinces. Other catfish families present include schilbeids (eight species) mochokids (15 species), amphiliids (six species) and one malapterurid. Cyprinodonts are relatively few (eight species) but show a particular relationship across the Cassai-Zambezi watershed. Cichlids (31 species) are well represented but not as well as the cyprinids. They are however more endemic in nature, in particular the Atlantic coastal fauna (19 species with eight endemic). The anabantids (three species) and mastacembelids (three species) are poorly represented.

The last phase of ichthyological exploration informing on the fishes of Angola, since independence in 1975, includes several taxonomic or systematic articles (e.g. Greenwood 1984, Musilová et al. 2013); published river faunal accounts (Skelton et al. 1985; Hay et al. 1997) and several unrestricted informal fish survey reports emanating from specific projects (Bills et al. 2012, 2013; Skelton et al. 2016). These surveys have exposed several new species to the fauna and together with phylogenetic studies on a wide range of lineages that include Angolan representatives, have led to a vastly improved understanding of the distributional nuances that give explanation to improved biogeographical insights.

Freshwater Drainages and Ecoregions of Angola

The drainages of Angola include southern source tributaries of the Congo, western source tributaries of the Zambezi, coastal rivers to the Atlantic from the Chiloango in Cabinda to the Cunene in the south, and the endorheic Etosha and Okavango Basin drainages in the south (Fig. 11.1). The watershed between the Congo system and the coastal Atlantic and Zambezian rivers is a major ichthyofaunal divide of considerable biogeographic significance (Poll 1967, Jubb 1967, Skelton 1994, Snoeks et al. 2011, Paugy et al. 2017).

The freshwater fishes of Angola fall within four major African ichthyological provinces (Fig. 11.1) – Lower Guinea, Congolese, Angolan (coastal) or Cuanza and Zambezian (Roberts 1975, Snoeks et al. 2011, Lévêque and Paugy 2017b).

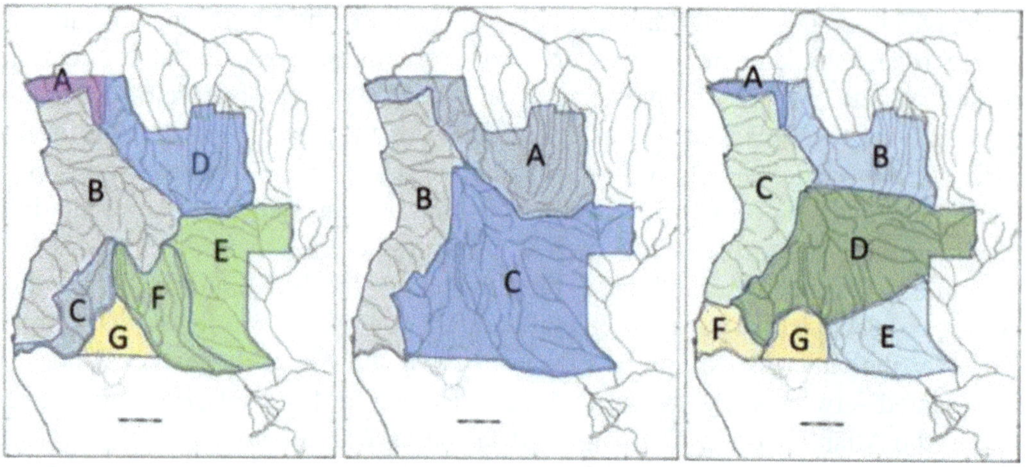

Fig. 11.1 *Left*: **Main drainage basins of Angola**. A: Lower Congo, B: Cuanzan or Atlantic Coastal, C: Cunene, D: Southwest Congo E: (west) Upper Zambezi, F: Okavango, G: Cuvelai. Chiloango in Cabinda not illustrated. *Center*: **Ichthyological provinces in Angola**, modified from Lévêque and Paugy (2017b) to include upper Cuanza and upper Cunene in the Zambezian Province. A: Congolian, B: Angolan or Cuanzan, C: Zambezian. Cabinda and the Chiloango River in the Lower Guinea Province not illustrated. *Right*: **Aquatic ecoregions in Angola**, modified, after Thieme et al. (2005). A – Lower Congo, B – Cassai, C – Cuanza, D – Zambezian headwaters, E – Okavango and Upper Zambezian floodplains, F – Namib coastal, G – Etosha. Southern West Coastal Equatorial (Cabinda) not illustrated

Previously Poll (1967) considered the freshwater fishes of Angola to be from five ichthyological regions drawn along watershed lines – Congo tributaries, Zambezi region, Angolan (western) coastal drainages excluding the Cunene, the Cubango-Okavango, and the Cunene. Thieme et al. (2005) defined ecoregions as "a large area containing a distinct assemblage of natural communities and species, whose boundaries approximate the original extent of natural communities before major land use change", and often reach across watershed lines. The Thieme et al. (2005) aquatic ecoregion map for Africa included Angolan inland waters within eight aquatic ecoregions as follows (Fig. 11.1): Floodplains, Swamps and Lakes – Region 12 Okavango Floodplains, Region 16 Upper Zambezi Floodplains; Moist Forest Rivers – Region 22 Lower Congo, Region 29 Southern West Coastal Equatorial; Savannah Dry Forest – Region 63 Cuanza, Region 76 Upper Zambezian headwaters; Xeric Systems – Region 82 Etosha, Region 88 Namib Coastal. Both the Ichthyological Provinces and the Ecoregions are convenient categories to consider the freshwater fishes of Angola.

Cabinda

Pellegrin (1928) recorded 28 species from the Chiloango River from Republic of Congo (formerly 'French' Congo). The freshwater fishes of Lower Guinea, Central West Africa that includes Cabinda were considered in detail through the two volumes edited by Stiassny et al. (2007). This rich Central West African fauna includes 555 species in 147 genera and 38 families of which 78 species, 52 genera and 25 families have been recorded from the Chiloango (also Shiloango) River in Cabinda (Appendix 1). This Atlantic coastal river is clearly boosted by the large number of marine and estuarine species that enter freshwaters sporadically or regularly (Whitfield 2007). A number of species from here such as *Enteromius holotaenia, Enteromius musumbi, Aplocheilichthys spilauchen* and *Oreochromis angolensis*, and estuarine species of marine origin also occur in the lower reaches of Angolan Atlantic coastal rivers, some to as far south as the Cunene River (Penrith 1982, Hay et al. 1997). Fowler (1930) recorded a number of species in the Academy of Natural Sciences of Philadelphia collection taken from the Chiloango region as it was known at the time. The fauna from the Chiloango in Angolan territory is likely underrepresented in most groups due to lack of sampling.

Lower Congo

There are no records available of fishes collected in Angolan waters of the Lower Congo mainstream or of the southern bank tributaries. The largest of these tributaries is the Inkisi River of which the fish fauna is known from the studies of Wamuini Lunkayilakio et al. (2010) supplemented by the description of new species described in association with that work (Wamuini Lunkayilakio and Vreven 2008, 2010). Based on these studies it is likely that most of the species in the DRC from the reaches above the Sanga Falls at least are likely to occur in Angola as well. The nature of the likely fauna of this neglected area of Angola as far as fish exploration is concerned (Appendix 2) indicates that the species are essentially of Congolian or Lower Guinean affinity with a few endemic species indicative of the isolation of fauna in the river reaches above the Sanga Falls. The widespread presence of *Oreochromis niloticus* is attributed to introduction for aquaculture (Wamuini Lunkayilakio et al. 2010).

Cuanza and Atlantic Coastal Rivers

Poll (1967) listed 110 species in 32 genera and 15 families from the Atlantic coastal region that includes the Cuanza River, which is revised (Appendix 3) to 105 species in 45 genera and 17 families in the light of more recent surveys in the Cuanza. There are very few species recorded from the Angolan Coastal rivers other than the Cuanza, and in areas north of the Bengo to the mouth of the Congo records from Angola are practically non-existent. Devaere et al. (2007) record *Channallabes apus* as being described from this region. Fowler (1930) noted species of the Cuanza and the Bengo rivers received from the British Museum on exchange, in many instances as described by Boulenger (1910) or as recorded in Boulenger

(1909–1916). Trewavas (1936) recorded and described seven species from a headwater stream of the Cuvo River arising on Mount Moco including the only *Amphilius* species (*Amphilius lentiginosus*) described from the region. A second undescribed *Amphilius* species has been recorded in the Cuanza (South African Institute for Aquatic Biodiversity – SAIAB – collection). Both these species differ in key morphological characteristics from the *Amphilius* of the Zambezian region that indicate their faunal connections lie primarily with the Lower Guinean or Congolean regions. Trewavas (1936) also described species from the Longa (*Enteromius breviceps*), the Catumbela (*Enteromius dorsolineatus, E evansi*), and the Balombo (*Enteromius dorsolineatus*). Pellegrin (1936) described the fishes collected by two Swiss expeditions (1928–1929 and 1932) made under the direction of Monard from the *Musée d'Histoire Naturelle de la Chaux-de-Fonds* included two species, *Enteromius kessleri, Clarias dumerilii*, that were drawn from the Cueve, with the bulk of the collections coming from the Cunene, the Cuvelai and the Cubango. Ladiges and Voelker (1961) described *Kneria maydelli* from the Cunene, and *Enteromius rousellei* and *Chiloglanis sardinhai* from the Longa. Poll (1967) described *Kneria sjolandersi* and *Chiloglanis angolensis* from the Bero, to the north of the Cunene. Trewavas (1973) recorded *Oreochromis angolensis* and *Tilapia cabrae* from the Bengo. Bills et al. (2012) made a small collection from the upper reaches of the Cueve River that included species of the following genera – *Petrocephalus, Enteromius, Labeobarbus, Micralestes, Amphilius, Chiloglanis, Clarias, Pharyngochromis, Thoracochromis, Tilapia, Coptodon,* and *Mastacembelus*. The list is typically 'Zambezian' and the species positively identified are closely linked to the upper Cuanzan and Cubango fauna. The indication from these references is therefore that the Angolan Coastal fauna is a mix of Lower Guinean (along the coastal plain) and Zambezian (above the escarpment) with some Congolean elements in the upper Cuanza/Lucala (see below).

The 'Cuanzan or Angolan Coastal ' ichthyofaunal region is drawn primarily on what is known of the fishes of the Cuanza River as described by Boulenger (1910) and in Boulenger's (1909–1916) catalogue of fishes in the British Museum (Natural History) now the Natural History Museum (NHM). Fowler's (1930) account of the fishes of the Gray African Expedition in 1929 included records from the Bengo and the lower Cuanza, but also a collection of species from Chouzo on the upper reaches of the Cutato-Cuanza tributary, that provided a first strong indication that the fauna of these reaches is 'Zambezian' in character and different to those from the coastal reaches as reported by Boulenger (1910) and others. This association was later reiterated by Trewavas (1973) when considering the cichlid species of the Cuanza and Bengo rivers and has been firmly supported by the extensive surveys conducted by SAIAB and INIP (*Instituto Nacional de Investigação Pesqueira*) between 2005 and 2010. The current assessment records at least 102 species, some of which are undescribed (Appendix 3). The collections indicate that the river basin is even more heterogenous in fish faunal characteristics than simply 'lower' and 'upper' and the different zones distinguishable include (1) the lower reaches from the escarpment base to the sea, (2) the escarpment reaches, (3) the upper Cuanza and (4) the Lucala tributary, itself probably sub-zoned into the middle and upper reaches separated by the Calandula Falls (formerly 'Duque de Bragança' Falls).

Two ecophysiological components derive the fishes of the lower Cuanza: a diverse Tropical West African or Lower Guinean brackish water or marine component, and secondly the primary and secondary freshwater fishes. The known Tropical West African brackish water fishes from the system are generally widespread species and do not include endemics. Some species such as the Bull Shark (*Carcharhinus leucas*) and the Atlantic Tarpon (*Megalops atlanticus*) are well known as gamefish from this river. Two clupeid species include the recognised freshwater species (*Pellonula vorax* and *Odaxothrissa ansorgii*) and probably other brackish water forms. One haemulid (*Pomadasys* sp.) and one polynemid threadfin, possibly *Polydactylus quadrifilis* as known from Central West Africa (Snoeks and Vreven 2007), have been recorded (SAIAB records). Mullets (Mugilidae), as yet unidentified at species level, are present as are the sleepers (Eleotridae) and gobies (*Awaous* and *Periopthalmus*). Two pipefish have been positively identified: *Enneacampus ansorgii* and *Microphis brachyurus aculeatus*. The tonguefish *Cynoglossus senegalensis* was collected in the downstream reaches.

The freshwater species of this lower zone are mostly widespread species that also occur in coastal reaches of rivers to the north, well into the adjacent Lower Guinean Province and beyond, and many probably also to the south. An example of this is *Parailia occidentalis* that has a range through to the Senegal River in West Africa (de Vos 1995). The species that occur are found generally throughout the region to the escarpment, with a few ascending into middle Cuanza sections. Other characteristic species in this zone include mormyrids of the genera *Hippopotamyrus, Marcusenius* and *Petrocephalus*, the alestid *Alestes ansorgii*, cyprinids of the genus *Labeo,* two *Enteromius* species (*E. holotaenia* and *E. musumbi*), and several distinctive claroteid catfishes (two *Chrysichthys* species *C. acutirostris* and *C. ansorgii)*, as well as *Schilbe bocagii*, and the widespread *Clarias gariepinus*. The *Chrysichthys* species confirm the West Africa coastal affinities of the assemblage as the genus is not known from the upper reaches nor from the upper Zambezian floodplain fauna. The cichlid fauna, as detailed systematically by Trewavas (1973) is in part also restricted to the zone – *Oreochomis angolensis, Hemichromis angolensis* and *Tilapia* cf. *cabrae.* The range of the procatopodid lampeye *Aplocheilichthys spilauchen* previously known from the Senegal River to the Bengo River has been extended to the Cuanza. The absence of the anabantid genus *Ctenopoma* from this zone is remarkable.

The Escarpment Zone of the Cuanza is characterised by a stepwise series of rapids, cascades and falls interspersed by rocky pools and runs. The fish fauna of this important zone for hydropower development is rich but relatively poorly known or described. The SAIAB-INIP collections are extensive and indicate that few species from the coastal zone penetrate high up into the zone. This is most probably partly an artefact of the Cambambe Dam near the base, in existence for several decades, which has likely affected the natural penetration of many species. The major freshwater families are represented; the smaller cyprinids, various catfish families, and cichlids are particularly well represented. The generic composition includes: *Hippopotamyrus, Petrocephalus, Marcusenius, Parakneria, Enteromius, Labeobarbus, Labeo, Raiamas, Brycinus, Rhabdalestes, Hepsetus, Schilbe, Chrysichthys, Clarias, Clariallabes, Parauchenoglanis, Chiloglanis, Synodontis, Micropanchax, Hemichromis, Pharyngochromis, Pseudocrenilabrus, Serranochromis, Tilapia, Oreochromis*, and *Mastacembelus*. Only a single *Labeobarbus* species was recorded in this zone during the survey and it also occurs in the Lucala tributary. Boulenger (1910) recorded two *Labeobarbus* species from the Cuanza at Dondo – *L. rocadasi* and *L gulielmi*. A unique morphotype of *Labeobarbus* with an extremely pointed tiny mouth, collected during the Capanda pre-impoundment surveys is present in the Luanda Museum (pers. obs., Fig. 11.2) and is likely to be an undescribed species.

The Upper Cuanza extends from a waterfall on the mainstream above the Capanda dam to the watershed and consists largely of relatively low-gradient floodplain rivers on Kalahari sand formations, similar to that of the upper reaches of the Zambezi and Okavango systems in Angola. Characteristic genera from this zone are: *Hippopotamyrus, Petrocephalus, Marcusenius, Parakneria, Enteromius, Labeobarbus, Labeo, Brycinus, Rhabdalestes, Hepsetus, Schilbe, Chrysichthys, Doumea, Clarias, Clariallabes, Parauchenoglanis, Chiloglanis, Synodontis, Micropanchax, Hemichromis, Pharyngochromis, Pseudocrenilabrus, Serranochromis, Tilapia, Oreochromis,* and *Mastacembelus*. Fishes from Chouzo in the upper Cuanza described by Fowler (1930), include species such as *Marcusenius angolensis, Hepsetus cuvieri, Labeo rocadasi, Enteromius evansi* (type locality), *Enteromius lujae* (identity of this species is still debated but the same species occurs in the Okavango headwaters), *Clarias gariepinus, Clarias theodorae* (as *C. fouloni*), *Clarias ngamensis* (as *Dinotopterus prentissgrayi*), *Ctenopoma machadoi* (type locality), *Serranochromis macrocephalus* (as *Tilapia acuticeps*, see Trewavas 1973). Norman (1923) described *Synodontis laessoei*, synonymised with *Synodontis nigromaculatus* by Poll (1971), as the only species of this genus in the Cuanza, a contrast to the specious lineage in the Okavango-Zambezi region (Day et al. 2009, Pinton et al. 2013). Few species characteristic of the upper Cuanza are found beyond the zone within the basin. This agrees with the notion that the fauna in this zone is historically and biogeographically an integral part of the 'Zambezian' fauna

Fig. 11.2 An extraordinary undescribed *Labeobarbus* species from the Cuanza River in the Luanda Museum, 2005. (Photo PH Skelton)

(Trewavas 1973). Ladiges (1964) and Poll (1967) showed this to be general for the fauna as a whole, and specific studies on species like *Hepsetus cuvieri* (Zengeya et al. 2011) and cichlids like *Serranochromis* and *Tilapia sparrmanii* (Musilová et al. 2013) confirm this relationship. Recent surveys across the watershed between the Cuanza and the Okavango indicate that a number of other species like *Parakneria fortuita*, and several *Enteromius* species like *E mocoensis, E evansi, E breviceps, E brevidorsalis* occur in streams on either side and have helped to define the Upper Zambezi headwaters ecoregion that embraces this trans-system conformance.

An early indication that the Lucala River, a major tributary that joins the system in the lower reaches, is exceptional for its fishes was the fauna collected by Ansorge using a wide range of methods including explosives (Boulenger 1910). It is however only in the escarpment and upper reaches that such exception occurs. An assemblage of large fishes of the genus *Labeobarbus* in particular is outstanding, and Boulenger (1910) described 12 species now in *Labeobarbus* (Vreven et al. 2016), all of which remain valid at this time. In addition to these species, unpublished barcode studies conducted by SAIAB on the fauna indicates that several lineages in the system are restricted to the Lucala, including an *Alestes, Pharyngochromis, Serranochromis, Tilapia,* two *Enteromius species*, a *Parakneria, Hippopotamyrus*, and a undescribed *Congoglanis*.

The significance of the use of explosives in assembling Ansorge's collection described by Boulenger (1910) is that it included a number of large mainstream species that otherwise are extremely difficult to collect. The assemblage of large *Labeobarbus* described in the paper has defined the Cuanza Basin since that time. The overall faunal characteristics of the Lucala include species of the following genera: *Hippopotamyrus, Petrocephalus, Kneria, Alestes, Enteromius, Labeobarbus, Labeo, Raiamas, Amphilius, Congoglanis, Schilbe, Clarias, Chiloglanis, Synodontis, Micropanchax, Pharyngochromis, Serranochromis, Tilapia,* and *Mastacembelus*. The fauna of the upper reaches is poorly known. Only a single collection made by SAIAB was drawn from the Lucala above the Calandula Falls. This limited sample is not sufficient to gauge the full character of the zone, but does indicate a degree of continuity with the Middle Lucala zone, and differing through the absence of major elements like the *Labeobarbus* species so characteristic of the latter. The physical character of the upper reaches suggests there is a zonal distinction in the ecological character and thus the faunal elements. The known fauna includes species of the following genera: *Hippopotamyrus, Petrocephalus, Parakneria, 'Barbus', Enteromius, Amphilius, Congoglanis, Clarias, Micropanchax, Pharyngochromis, Serranochromis.* Little else can be stated at this point except that an investigation into the fauna is highly desirable given the unique nature of the Middle Lucala.

The Lucala catchment shares its watershed with tributaries of the Congo-Cuango River and is likely one of the underlying reasons for its unique character. A high degree of endemicity to this catchment is therefore evident and with further taxonomic investigation likely to be upheld and enhanced.

Cunene

Poll (1967), from the ichthyological perspective, treated the Cunene River system as a separate entity to the Atlantic coastal region, whereas it has been regarded as part of the Zambezian Province by Roberts (1975), part of the 'Angola' ichthyofaunal province by Lévêque and Paugy (2017a, b), and divided as part of the Namib aquatic ecoregion and part upper Zambezian headwaters ecoregions by Thieme et al. (2005). The reason for these varied treatments is that the river system is geo-eco-historically complex. Thus it has a dual geomorphological origin (the upper reaches being a natural part of the Kalahari Basin that has been captured by an Atlantic coastal river) and environmentally the lower reaches sit within the 'xeric' Namib region and the inland upper reaches within the savanna dry forest environs.

The fishes of the Cunene River are relatively well documented, starting with *Schilbe steindachneri* (a synonym of *S. intermedius)* and *Mormyrus anchietae* (a synonym of *M. lacerda*) described by Guimarães (1884), and summarised in the most recent checklist by Hay et al. (1997). Excluding the more strictly marine families there are 82 species recorded from the Cunene (Appendix 3). Hay et al. (1997) also record the broad distribution of species within the system according to three sections, the upper reaches down to Ruacana Falls, a middle section down to Epupa Falls and the lower river from below Epupa Falls to the mouth. Of 65 species recorded above Ruacana Falls 13 are restricted to that section. At least one species, *Marcusenius deserti*, is restricted to the lower reaches close to the coast (Kramer et al. 2016). Apart from the several marine species recorded in the extreme lower reaches by Penrith (1970) and Hay et al. (1997) that reflect a southernmost extension of the tropical (Lower Guinean) fauna, the general composition is clearly Zambezian in character. There are few representatives indicative of the Angolan (Cuanzan) Province, e.g. *Enteromius mattozi* (described by Guimarães (1884) from the Curoca River to the north of the Cunene). Pellegrin's (1921) *Enteromius* (formerly *Barbus) rohani*, probably a synonym of *E. mattozi,* was likely taken from the Caculovar River, a tributary of the Cunene, and not from the Lomba (neither the Zambezi as Pellegrin claimed, nor the Longa coastal Atlantic as suggested by Poll 1967). *Enteromius argenteus* is another minnow that has been reported from the Cunene but whose identity is unconfirmed – and is likely to be juveniles of *E. mattozi* (Skelton Unpublished Data).

There are also several isolated endemics from the system such as *Marcusenius deserti, Marcusenius magnoculis, Marcusenius multisquamatus, Hippopotamyrus longilateralis, Engraulicypris howesi, Zaireichthys cuneneensis, Orthochromis machadoi, Thoracochromis albolabrus, Thoracochromis buysi*, that suggests a degree of isolation probably reflecting older biogeographic connections. The absence of certain conspicuous families or genera such as *Parakneria, Labeobarbus, Opsaridium, Hydrocynus, Parauchenoglanis, Amphilius, Hemichromis,* and *Mastacembelus* is also noteworthy and perhaps indicative of a lack of more recent connections with the Zambezian and Cuanza systems.

Cassai and Southern Congo Rivers

Collections from the Lulua River, a tributary of the Cassai in Congo by Fowler (1930) whilst not strictly in Angola, probably pertain to Angola as well. Thus although not the only source, Poll (1967) is the current practical published source for the fishes of the southern Congo river tributaries in Angola. There are three main tributaries draining the region, from the east the Cassai including the Luangwe, the Cuilu (or Kwilu) and the Cuango. Poll (1967: 18–23) plotted the records of the fishes of each of these in his distribution table, recording 108, 28 and 37 species respectively and in the addendum supplemented the Cassai with three species and the Cuango with 24 species. The figure for the southern catchments of the Congo in Angola is now estimated at around 162 species (Appendix 3). The Cuilo and Cuango faunas are most evidently far from well explored. The Cassai fish fauna is better represented but still poorly explored, and includes species both typical of the Congo (e.g. *Polypterus ornatipinnis, Channallabes apus,* several mormyrid species, *Bryconaethiops microstoma, Alestes grandisquamis, Distichodus fasciolatus, Distichodus lusosso, Mastacembelus congicus*), and many species found also in the Upper Zambezi or the Okavango (e.g. *Hydrocynus vittatus, Hepsetus cuvieri, Pollimyrus castelnaui, Enteromius brevidorsalis, Parauchenoglanis ngamensis, Clarias stappersii, Clarias theodorae, Schilbe yangambianus, Micropanchax katangae, Oreochromis andersonii, Coptodon rendalli, Tilapia sparrmanii, Tilapia ruweti, Hemichromis elongatus, Serranochromis microcephalus, Serranochromis robustus jallae, Pseudocrenilabrus philander, Ctenopoma multispine, Microctenopoma intermedium*). The presence of *Dundocharax bidentatus* in the Cassai and the rare Zambezian endemic not yet found in Angola, *Neolebias lozii* are further good indicators of geographical connection. The strong Cassai-Zambezian faunal association is attributed to the clear evidence of hydrological pattern that the upper Cassai was formerly part of the Upper Zambezi system (Bell-Cross 1965).

Zambezian-Cuando-Cubango Headwaters and Floodplains

There is sufficient direct connection between the Zambezi, the Cuando and the Okavango river basins and similarity of the fish faunas in each to consider these under a single heading.

The Zambezi headwaters in Angola drain the Kalahari sand formation over an extensive divide with the Cassai to form major floodplains known as the Bulozi Floodplains. There are a number of lakes associated with the drainage including the largest freshwater lake in Angola, Lake Dilolo. The Okavango drainage is divided into two branches, the Cuito-Cuanavale in the east and the Cubango in the west. The Cuito-Cuanavale drains Kalahari sand formations giving rise to extensive low-gradient seepage bog and floodplain rivers in slump valleys extending into miombo savanna woodlands in the upper reaches. There are several lakes in these headwaters.

The Cubango branch arises as several relatively steep gradient rocky rivers in the Angolan highlands on the Bié plateaux before descending to the low-gradient reaches along the Namibian border to join with the Cuito before crossing to Botswana and forming the mostly endorheic Okavango Delta. The watershed of the system is shared with the Cuando, the Zambezi (mainly the Lungwe-Bungo), the Cueve-Cuanza, and the Cuanza as well as the Cunene and Cuvelai oshanas in the west.

The fishes of the Upper Zambezi are well studied and documented (e.g. Jackson 1961, Jubb 1961, 1967, Balon 1974, Bell-Cross and Minshull 1988, Tweddle 2010) with numbers now estimated at around 100–120 species (Appendix 3; Tweddle et al. 2004), possibly with as many as 20–25 undescribed. However published records from the Angolan territory are sparse, and limited in the published literature to Poll's (1967) 41 species (against his checklist of 62) taken mostly from two localities close to the watershed (Lagoa Calundo and the Longa-Luena tributary). Recent collections from the source reaches of Zambezian tributaries in Angola made by the National Geographic Okavango Wilderness Project (NGOWP 2018) are still being assessed but include 39 species from 12 families that have been included in the checklist of fishes from this region (Appendix 3). One notable new record is *Enteromius chiumbeensis* described by Poll (1967) from the Chiumbe River a tributary of the Cassai, reinforcing the close connections between these adjacent trans-watershed systems.

The upper Zambezian fish fauna is distinctive in several respects, most notably for the relatively speciose endemic *Synodontis* catfishes and the serranochromine cichlids (Trewavas 1964, Bell-Cross 1975, Greenwood 1993, Day et al. 2009; Pinton et al. 2013). To a large extent, in Angola, the fauna is ecologically tuned to the extensive seepage and floodplain drainages within a band of miombo savanna woodland on Kalahari sand deposits. Overall the known Angolan Upper Zambezi fish fauna is similar to that of the better-studied (in Angola) Okavango Basin fishes (often with the same or closely related species e.g., mormyrids of the genera *Hippopotamyrus, Marcusenius, Petrocephalus, Pollimyrus* – Kramer et al. 2003, 2004, 2012, 2014, and *Zaireichthys* species – Eccles et al. 2011). Whilst there are a few endemics, only one, *Paramormyrops jacksoni* Poll 1967 is restricted to Angola. The isolated *Neolebias lozii* is known only from the Barotse floodplains in Zambia.

Fishes of the Cuando-Linyanti-Chobe system have not been reported on from the Angolan section of that Zambezi tributary but van der Waal and Skelton (1984) provided a checklist of fishes in the Cuando River in Namibian waters. The 56 species recorded were all also found in the Zambezi system in Namibia. The Pallid Sand Catlet, *Zaireichthys pallidus* Eccles et al. (2011) is described from the Cuando but is not restricted to that system. Kramer et al. (2014) described a new species of *Pollimyrus* from the Cuando, a species possibly endemic to that tributary. Recent collections by the National Geographic Okavango Wilderness Project (NGOWP/SAIAB) from the upper reaches of the Cuando in Angola further inform the list of species (Appendix 3).

The fishes of the Okavango Basin have been studied and reported on in the literature for over 150 years since Castelnau (1861) described 14 species from Lake

Ngami, including the iconic Tigerfish (*Hydrocynus vittatus*) the Southern African Pike (*Hepsetus cuvieri*), the large Blunttooth Catfish (*Clarius ngamensis*) and the Three Spot Bream (*Oreochromis andersonii*). Fifty years later Boulenger (1911) reported on a collection from the Okavango-Lake Ngami made by RB Woosnam and described six new species including one named for Castelnau – *Pollimyrus castelnaui*. These fishes were all included in Gilchrist and Thompson (1913, 1917) and Boulenger (1909–1916). Fowler (1935) described a collection made from the Delta by the Vernay-Lang Expedition of 1930. Pellegrin (1936) described fishes collected by two Swiss expeditions of 1929 and 1933 from the Cunene, the Cuvelai and the Cubango. Barnard (1948) described in detail a collection from Rundu, Namibia. The results of all these efforts were summarised in checklists published by Poll (1967), Jubb (1967), Jubb and Gaigher (1971) and Skelton et al. (1985). More recently surveys of Angolan Okavango Basin rivers have been made (Bills et al. 2012, 2013, Skelton et al. 2016) that have reached little-explored areas, encountered additional species and provide for a more complete assessment of the fishes and their intra-basin distributions.

The additional species recently discovered include new species of *Clariallabes*, several serranochromine cichlids, and a dwarf climbing perch (*Microctenopoma* sp). Recent distribution records extend the range of several species from the Congo tributaries or in the case of *Clypeobarbus bellcrossi* from Zambezi headwaters in Zambia to the Okavango. Congolean species such as *Marcusenius moorii* (Günther) and *Enteromius chicapaensis* (Poll), and *Nannocharax lineostriatus* (Poll), and several *Micropanchax* as *M. luluae, M. nigrolateralis, M. lineolateralis*. The known range of a number of species from the Atlantic Coastal and Cuanza systems has been extended to the Okavango, e.g. *Enteromius breviceps, E. brevidorsalis, E. evansi, E. mocoensis, E. greenwoodi*. A new understanding of the complex distribution of the twin species *Enteromius trimaculatus* and *E. poechii* has also been reached – the former being found in the Cunene and the extreme upper reaches of the Cubango in place of the latter which is widespread in the downstream floodplain reaches of the Okavango and Upper Zambezi system.

Cuvelai

The Cuvelai drainage lies in a triangle between the Okavango in the east and the Cunene in the west and the streams known as 'iishanas' are intermittent, only flowing during periods of sustained rainfall into the endorheic Etosha Pan in Namibia (van der Waal 1991, Hipondoka et al. 2018). The 1929 and 1932–1933 Swiss expeditions to Angola collected the following species from Mupa (Pellegrin 1936): *Marcusenius altidorsalis* (?), *Mormyrus lacerda, Enteromius paludinosus, Tilapia sparrmanii*, and *Pseudocrenilabrus philander*. Seventeen species, all conforming to Cunene fauna, have been confirmed from the western iishanas of the system by Hipondoka et al. (2018), and connections with the Cunene substantiated through remote sensing techniques. Four widespread pioneering species are consistently present in collections, viz., *Clarias gariepinus, Enteromius paludinosus, Oreochromis andersonii* and *Pseudocrenilabrus philander* and several others are common – *Clarias ngamensis, Schilbe intermedius* and *Enteromius trimaculatus*.

Biogeography

The biogeography of Angolan freshwater fishes is closely tied to the geomorphology and the geomorphological history of the territory. In brief, Angola consists of a narrow coastal plain, a distinct escarpment and an interior plateau that is being eroded most rapidly from the Congo Basin. The coastal plain consists of a series of rivers flowing from the escarpment or – in the case of the Congo in the north, the Cuanza in the middle and the Cunene in the south – where the escarpment has been penetrated, from the interior plateau or the Congo Basin. The fish fauna of the coastal plain is primarily a southern extension of the tropical coastal fauna of West Africa and Central West Africa. River connections along this narrow strip are either via sea-level fluctuations or via river captures between watersheds, either as adjacent systems or via extended reaches through captured inland drainages that are not determined by the coastal gradients and processes. According to Lévêque and Paugy (2017a,b) the primary direction of dispersal of the coastal west African fauna was northwards from the Congo. Present day ocean currents off Angola are counter clockwise (http://oceancurrents.rsmas.miami.edu) and it is possibly only inshore counter currents that might have facilitated faunal dispersal southwards from the Congo, especially after the capture and penetration of the Congo Basin by the Lower Congo in the late Cretaceous (Flügel et al. 2015). Such would certainly explain much of the marine derived elements of the region. Given favourable currents it is likely also that the considerable volumes of freshwater entering the sea from the Congo at various times would facilitate even freshwater fishes down the coast and might explain the presence of such species as *Enteromius musumbi, Physailia occidentalis, Chysichthys spp, Oreochromis angolensis* and *Aplocheilichthys spilauchen* in the Angolan region. An alternative and complementary explanation for some freshwater faunal elements such as *Marcusenius deserti* and *Raiamas ansorgii* of the Angolan Coastal reaches is that it is primarily derived via the Cuanzan and Cunene gateways through capture of portions of the Kalahari Basin drainage. It is not only the Cuanza and the Cunene that have breached the escarpment but also the Cuvo and the Longa and possibly others, as is evident in the list of freshwater fishes reported from these lesser rivers (see above).

The evolution of the extensive Kalahari Basin is certainly key to understanding the majority of the freshwater fish fauna of Angola. Haddon and McCarthy (2005), Key et al. (2015), Moore and Larkin (2001), and Moore et al. (2012) sketch the evolution of the Kalahari Basin and its drainage since the breakup of Gondwanaland and the isolation of Africa in the late Cretaceous. Following rifting, the continental margins were probably elevated and this formed an escarpment that separated the narrow coastal plain from the elevated Kalahari sedimentary basin that was drained primarily by the palaeo-upper Zambezi, the predominant system in the Angolan region (Fig. 11.3). The western portion of the system flowed from the escarpment highlands of the extreme northwest of the basin, now part of the Cuanza, generally southeast through the Limpopo valley to the Indian Ocean. The eastern parts of the upper Zambezi reached northeastwards to as far as pre-rift East African plateaus and

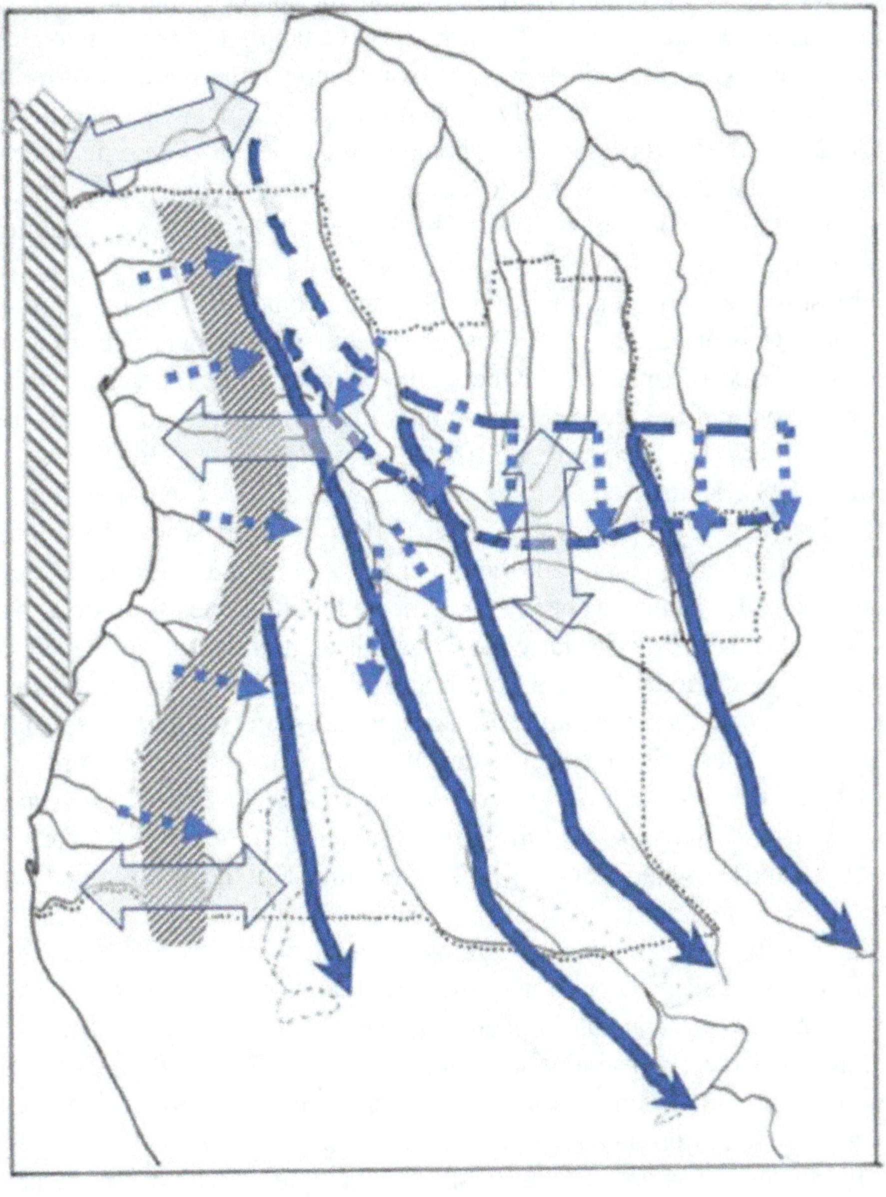

Fig. 11.3 A diagrammatic model for the post-Cretaceous biogeography of Angolan freshwater fishes. Angolan border – fine dotted line; present day drainage – thin lines; present day inter basin watersheds – open dotted lines; paleo drainage lines – thick extended arrows; paleo and present escarpment retreat – dashed arrows; paleo and present south and southwestern Congo Basin watershed – thick dashed lines; Angolan escarpment – right slanted hash; gateway drainage captures – large open bi-directional arrows. Coastal dispersal of fishes – large left-slanted bi-directions arrow. The model is based on geomorphological interpretations by Flügel et al. (2015), Haddon and McCarthy (2005), Moore and Larkin (2001), Moore et al. (2012), and others

included the proto-Luangwa and the proto-Chambeshi-Kafue-upper Zambezi as well as the Okavango. These drained into an interior basin to form, at times, a mega palaeo lake – Palaeolake Magadigadi (Burroughs et al. 2009, Moore and Larkin 2001, Moore et al. 2012, Podgorski et al. 2013). The proto-Cunene consisted of an upper portion draining endorheically to the Etosha basin. The most significant events in the history of the Kalahari Basin were firstly the downwarping and back-tilting of drainage coupled with upwarping along the southern margins that severed the initial Indian Ocean outlet via the Limpopo; the tapping of the Congo Basin by the lower Congo River that advanced the erosion and southern retreat of the northern watershed of the basin, especially in the northeast (Luapula-Chambeshi) and, in the Angola area, the Cassai-Zambezi. The dismemberment and tapping of drainage portions from the Kalahari Basin to coastal outlets including the Cuanza, the Cunene, and the Zambezi also affects the biogeographical history significantly (Moore and Larkin 2001, Moore et al. 2012, Key et al. 2015).

The most profound biogeographic significance to emerge from this geomorphological narrative is that the Kalahari Basin has been an evolutionary basin for fishes over a long period of time. The evidence is exemplified in the serranochromine cichlid radiation and the clade of *Synodontis* catfish and the radiation of several mormyrid genera that characterise the Zambezian fauna (Bell-Cross 1975, Greenwood 1984, Kramer et al. 2003, 2004, Day et al. 2009, Kramer and Swartz 2010, Kramer et al. 2012, Schwarzer et al. 2012, Pinton et al. 2013, Kramer and Wink 2013). The strong identity of the upper Zambezian fauna further exemplifies this notion. That the fauna has been supplemented with species from neighbouring ichthyological provinces, especially the Congo, is also evident in species or genera with internally restricted distributions such as *Hepsetus cuvieri, Hydrocynus vittatus, Parauchenoglanis ngamensis, Mastacembelus, Hemichromis elongatus, Amphilius* and others. The broader distributions of some species into basins like the east coast rivers (e.g. *Enteromius bifrenatus, Microctenopoma intermedium, Clarias theodorae, Brycinus lateralis*) gives biological credence to the former east coast linkage and subsequent drainage dismemberment on the proto-upper Zambezi (Skelton 1994, 2001).

There are other emerging details of biogeographical interest to Angola that will in time lead to a detailed accounting of the origins and development of the freshwater fishes. Thus the presence of doumeine catfishes in the Cuanza, southwest of the Congo, indicates clearly insemination from the Congo. The flock of *Labeobarbus* species in the Lucala-Cuanza probably also indicates a Congolian insemination. However the assumption that all traffic was from the Congo is not necessarily correct and *Neolebias bidentatus* in the Cassai, for example, as with other 'Zambezian' elements in that system, more likely reflects a Zambezian (i.e. Kalahari) insemination to the Congo. This, in essence, is the basis of the 'Upper Zambezi headwater' freshwater ecoregion (Fig. 11.1: basin C).

Conservation

Angola is an emerging African economy with a rapidly growing human population and increasing demand on freshwater resources. The rapid population growth and expansion of urban areas in places such as Luanda but also in the more rural districts (Mendelsohn and Weber 2015) is placing an ever increasing stress on the environment, especially that of the rivers for which such urban growth centres are dependent on for water and power. Although many Angolan rivers are relatively unregulated there are dams on several systems such as the three major hydroelectric dams on the Cuanza. A further four hydroelectric dams are planned for the escarpment section of this system alone. In the case of certain transboundary rivers like the Okavango, the threat of increased river regulation is of serious concern to the integrity of the Okavango Delta in Botswana, a World Heritage and Ramsar site (King and Chonguic 2016).

Diamond mining activities along the southern Congo tributaries have had environmental impacts of unknown severity as practically no public investigations or information is available.

With human populations, urbanisation and development comes pollution and other direct threats to aquatic life such as fishing and the introduction of invasive alien species. Few alien fishes have been recorded from Angola, but two species that have been introduced are *Oreochromis mossambicus* (SAIAB, in the Cuanza) and *Oreochromis niloticus* in Cabinda and, as recently confirmed, in the upper Cubango. The threats these particular species pose as aliens is well documented (e.g. Wise et al. 2007, Zengeya et al. 2013, Bbole et al. 2014). This is the first record of an alien species with high impact potential in the Okavango system and the threat posed is transboundary in nature. Potential transboundary threats from outside Angola include that of alien crayfish from the Zambezi (Nunes et al. 2016).

Indigenous fishery practices in Angola include a range of gear ranging from simple traps to elaborate fishing fences and walls (Poll 1967, Mendelsohn and Weber 2015). In places such traditional practices are still in evidence (Fig. 11.4 *top*), but elsewhere traditional practices are being replaced by modern gear such as monofilament gillnets and mosquito-net seines (Fig. 11.4 *bottom*) that are excessively destructive and unsustainable (Tweddle et al. 2015).

The current IUCN redlist assessments for Angolan freshwater fishes (Appendix 3) reflects the relatively weak knowledge of the species – a third of the known species are either not assessed or are Data Deficient (DD). One species (*Oreochromis lepidurus*) is listed as Endangered (see Moelants 2010), three are Vulnerable (1%) and 185 (65%) are Least Concern. The endangered species is a Lower Congo endemic found mainly in the DRC and is primarily threatened by oil pollution derived from boats. The Vulnerable species are also cichlids of the genus *Oreochromis* – *O. andersonii* (see Marshall and Tweddle 2007) and *O. macrochir* (see Marshall and Tweddle 2007), both are threatened through hybridisation from the alien invasive species *Oreochromis niloticus*. The latter species has recently been confirmed as present in Angola, within the Okavango catchment and its impact

Fig. 11.4 *Top* – traditional fishing fence on the Cacuchi River, 2012 (Photo PH Skelton). *Bottom* – Drying fish caught with monofilament gillnets on the Cuito River, 2015. (Photo G Neef)

on the native *Oreochromis* is now an imminent threat. Given the situation of rapidly escalating changes to the natural aquatic environment in Angola it is likely that the IUCN redlist score for the country will rise rapidly.

Acknowledgements I am supported in my research by the Director and staff of SAIAB, in particular Roger Bills and members of the collections division, administration staff, and by Maditaba Meltaf in the library for the provision of literature. Steve Boyes and John Hilton of the Wild Bird Trust have provided me with excellent opportunity to study fishes in Angola since 2015. I have been supported in the field and laboratory by Adjani Costa, Roger Bills, Ben van der Waal, Götz Neef and others of the National Geographic Okavango Wilderness Project. SAIAB engagement with Angolan fishes was initiated in 2005 in partnership with INIP (Instituto Nacional de Investigação Pesqueira). Ernst R Swartz (SAIAB) and D Neto (INIP) were instrumental in opening the channels of new knowledge on Angolan freshwater fishes.

Appendices

Appendix 1

Freshwater and brackish water fishes of Shiloango River, Cabinda, as recorded by Stiassny et al. (2007)

Species	Author & Date
Clupeidae	
Pellonula vorax	Günther, 1868
Mormyridae	
Isichthys henryi	Gill, 1863
Marcusenius moorii	Günther, 1863
Paramormyrops kingsleyae	(Günther, 1863)
Brienomyrus brachyistius	(Gill, 1862)
Hepsetidae	
Hepsetus lineatus	(Pellegrin, 1926)
Alestidae	
Brycinus longipinnis	(Günther, 1864)
Brycinus macrolepidotus	Valenciennes, 1850
Brycinus kingsleyae	(Günther, 1896)
Nannopetersius ansorgii	(Boulenger, 1910)
Distichodontidae	
Distichodus notospilus	Günther, 1867
Eugnathichthys macroterolepis	Boulenger, 1899
Nannaethiops unitaeniatus	Günther, 1872
Nannocharax parvus	Pellegrin, 1906
Neolebias ansorgii	Boulenger, 1912
Neolebias spilotaenia	Boulenger, 1912

(continued)

Species	Author & Date
Cyprinidae	
Enteromius carens	(Boulenger, 1912)
Enteromius jae	(Boulenger, 1903)
Enteromius guirali	(Thominot, 1886)
Enteromius callipterus	(Boulenger, 1907)
Enteromius camptacanthus	(Bleeker, 1863)
Enteromius rubrostigma	(Poll & Lambert, 1964)
Enteromius holotaenia	(Boulenger, 1904)
Labeobarbus aspius	(Boulenger, 1912)
Labeobarbus cardozoi	(Boulenger, 1912)
Labeobarbus roylii	(Boulenger, 1912)
Labeobarbus batesii	(Boulenger, 1903)
Labeobarbus sandersi	(Boulenger, 1912)
Labeo batesii	Boulenger, 1911
Labeo lukulae	Boulenger, 1902
Opsaridium ubangiense	(Pellegrin, 1901)
Ariidae	
Arius latiscutatus	Günther, 1864
Claroteidae	
Anaspidoglanis macrostoma	(Pellegrin, 1909)
Parauchenoglanis altipinnis	(Boulenger, 1911)
Chrysichthys auratus	(Geoffroy Saint-Hilaire, 1809)
Chrysichthys nigrodigittatus	(Lacépède, 1803)
Schilbeidae	
Parailia occidentalis	(Pellegrin, 1901)
Pareutropius debauwi	(Boulenger, 1900)
Clariidae	
Clarias angolensis	Steindachner, 1866
Clarias gabonensis	Günther, 1867
Malapteruridae	
Malapterurus beninensis	Murray, 1855
Procatopodidae	
Aplocheilichthys spilauchen	(Duméril, 1861)
Plataplochilus loemensis	(Pellegrin, 1924)
Nothobranchiidae	
Epiplatys singa	(Boulenger, 1899)
Aphyosemion escherischi	(Ahl, 1924)
Anabantidae	
Ctenopoma nigropannosum	Reichenow, 1875
Microctenopoma ansorgii	(Boulenger, 1912)
Microctenopoma nanum	(Günther, 1896)
Microctenopoma congicum	(Boulenger, 1887)

(continued)

Species	Author & Date
Cichlidae	
Pelvicachromis subocellatus	(Günther, 1872)
Chilochromis duponti	Boulenger, 1902
Coptodon tholloni	(Sauvage, 1884)
Pelmatolapia cabrae	(Boulenger, 1899)
Coptodon guineensis	(Günther, 1862)
Oreochromis schwebischi	(Sauvage, 1884)
Sarotherodon nigripinnis	(Guichenot, 1861)
Lutjanidae	
Lutjanus dentatus	(Duméril, 1861)
Monodactylidae	
Monodactylus sebae	(Cuvier, 1829)
Polynemidae	
Polydactylus quadrifilis	(Cuvier, 1829)
Mugilidae	
Mugil bananensis	(Pellegrin, 1927)
Neochelon falcipinnis	(Valenciennes, 1836)
Chelon dumerili	(Steindachner, 1870)
Eleotridae	
Eleotris daganensis	Steindachner, 1870
Eleotris senegalensis	Steindachner, 1870
Eleotris vittata	Duméril, 1861
Bostrychus africanus	(Steindachner, 1879)
Dormitator lebretonis	(Steindachner, 1870)
Gobiidae	
Periopthalmus barbarus	(Linnaeus, 1766)
Gobionellus occidentalis	(Boulenger, 1909)
Bathygobius soporator	(Valenciennes, 1837)
Bathygobius casamancus	(Rochebrune, 1880)
Nematogobius maindroni	(Sauvage, 1880)
Microdesmidae	
Microdesmus aethiopicus	(Chabanaud, 1927)
Mastacembelidae	
Mastacembelus shiloangoensis	(Vreven, 2004)
Mastacembelus niger	Sauvage, 1879
Syngnathidae	
Enneacampus ansorgii	(Boulenger, 1910)
Microphis aculeatus	(Kaup, 1856)
Cynoglossidae	
Cynoglossus senegalensis	(Kaup, 1858)
Citharichthys stampflii	(Steindachner, 1894)

Appendix 2

Freshwater fishes of the Inkisi River DRC, from above the Sangha waterfalls, after Wamuini Lunkayilakio et al. (2010)

Species	Author & Date
Mormyridae	
Hippopotamyrus cf. ansorgii	(Boulenger, 1905)
Paramormyrops cf. kingsleyae	(Günther, 1896)
Paramormyrops cf. sphekodes	(Sauvage, 1879)
Cyprinidae	
Enteromius miolepis	(Boulenger, 1902)
Enteromius unitaeniatus	(Günther, 1867)
Enteromius vandersti	(Poll, 1945)
Garra congoensis	Poll, 1959
Labeo macrostomus	Boulenger, 1898
Labeobarbus sp. nov.	
Labeobarbus boulengeri	Vreven, Musschoot, Snoeks & Schliewen, 2016
Labeobarbus robertsi	(Banister, 1984)
Raiamas kheeli	Stiassny, Schelly & Schliewen, 2006
Alestidae	
Nannopetersius mutambuei	Wamuini Lunkayilakio & Vreven, 2008
Claroteidae	
Parauchenoglanis balayi	(Sauvage, 1879)
Clariidae	
Clarias angolensis	Steindachner, 1866
Clarias buthupogon	Sauvage, 1879
Clarias camerunensis	Lönnberg, 1895
Clarias gariepinus	(Burchell, 1822)
Clarias gabonensis	Günther, 1867
Schilbeidae	
Schilbe zairensis	de Vos, 1995
Cichlidae	
Haplochromis snoeksi	Wamuini Lunkayilakio & Vreven, 2010
Hemichromis elongatus	(Guichenot, 1861)
Oreochromis niloticus	(Linnaeus, 1758)
Sarotherodon galilaeus	(Linnaeus, 1758)
Coptodon tholloni	(Sauvage, 1884)
Anabantidae	
Ctenopoma nigropannosum	Reichenow, 1875
Chanidae	
Parachanna obscura	(Günther, 1861)

Appendix 3

Freshwater fishes of the (A) Cuanza (Atlantic coastal), (C) southern Congo, (Z) Upper Zambezian, (O) Okavango, and (K) Cunene basins in Angola, after Poll (1967) with updated adjustments for taxonomy and known records by the author

Species	Author & Date	A	C	Z	O	K	I[a]
Protopteridae							
Protopterus aethiopicus	Heckel, 1851		x				DD
Protopterus dolloi	Boulenger, 1900		x				LC
Polypteridae							
Polypterus ornatipinnis	Boulenger, 1902		x				LC
Clupeidae							
Pellonula vorax	Günther, 1868	x	x				LC
Odaxothrissa ansorgii	Boulenger, 1910	x	x				LC
Kneriidae							
Kneria angolensis	Steindachner, 1866	x	x	?			LC
Kneria maydelli	Ladiges & Voelker, 1961					x	LC
Kneria polli	Trewavas, 1936	x	x				LC
Kneria sjolandersi	Poll, 1967	x					DD
Kneria ansorgii	(Boulenger, 1910)	x	x				DD
Parakneria marmorata	(Norman, 1923)	x					DD
Parakneria vilhenae	Poll, 1965		x				DD
Parakneria fortuita	Penrith, 1973	x		x	x		DD
Mormyridae							
Mormyrops attenuatus	Boulenger, 1898		x				LC
Mormyrops anguilloides	(Linnaeus, 1758)		x				LC
Petrocephalus okavagoensis	Kramer et al., 2012			x	x		NE
Petrocephalus magnitrunci	Kramer et al., 2012				x		NE
Petrocephalus magnoculis	Kramer et al., 2012					x	NE
Petrocephalus longicapitis	Kramer et al., 2012			x	x		NE
Petrocephalus christyi	Boulenger, 1920		x				NE
Petrocephalus cunganus	Boulenger, 1910	x					DD
Petrocephalus micropthalmus	Pellegrin, 1909		x				LC
Petrocephalus simus	Sauvage, 1879	x	x	?			LC
Hippopotamyrus ansorgii	(Boulenger, 1905)	x		x	x		LC
Hippopotamyrus longilateralis	Kramer & Swartz, 2010					x	NE
Pollimyrus brevis	(Boulenger, 1913)		x				LC
Pollimyrus castelnaui	(Boulenger, 1911)			x	x	x	LC
Pollimyrus cuandoensis	Kramer, van der Bank & Wink, 2013			x			NE
Pollimyrus marianne	Kramer et al., 2003			x			NE
Cyphomyrus cubangoensis	(Pellegrin, 1936)			x	x		NE

(continued)

Species	Author & Date	A	C	Z	O	K	I[a]
Cyphomyrus psittacus	(Boulenger, 1897)		x				LC
Paramormyrops jacksoni	(Poll, 1967)			x			DD
Marcusenius altisambesi	Kramer et al., 2007		x	x	x		LC
Hippopotamyrus pappenheimi	(Boulenger, 1910)	x					LC
Heteromormyrus pauciradiatus	(Steindachner, 1866)	x					DD
Pollimyrus tumifrons	(Boulenger, 1902)		x				NE
Marcusenius desertus	Kramer, vanderBank & Wink, 2016					x	NE
Marcusenius multisquamatus	Kramer & Wink, 2013					x	NE
Marcusenius angolensis	(Boulenger, 1905)	x	x	x	x	x	LC
Marcusenius cuangoanus	(Poll, 1967)		x				VU
Marcusenius dundoensis	(Poll, 1967)		x				DD
Marcusenius moorii	(Günther, 1867)		x				LC
Marcusenius stanleyanus	(Boulenger, 1897)		x				LC
Campylomormyrus alces	(Boulenger, 1920)		x				LC
Campylomormyrus cassaicus	(Poll, 1967)		x				DD
Campylomormyrus elephas	(Boulenger, 1898)		x				LC
Campylomormyrus numenius	(Boulenger, 1898)		x				LC
Campylomormyrus luapulaensis	(David & Poll, 1937)		x				DD
Campylomormyrus rhynchophorus	(Boulenger, 1898)		x				LC
Campylomormyrus tshokwe	(Poll, 1967)		x				LC
Gnathonemus barbatus	Poll, 1967		x				DD
Gnathonemus petersii	(Günther, 1862)		x				LC
Mormyrus caballus	Boulenger, 1898		x				NE
Mormyrus lacerda	Castelnau, 1861	x		x	x	x	LC
Mormyrus rume	Valenciennes, 1847		x				NE
Cyprinidae							
Garra dembeensis	(Rüppell, 1835)		x				LC
Clypeobarbus bellcrossi	(Jubb, 1965)			x	x		DD
Coptostomabarbus wittei	David & Poll, 1937			x	x		LC
Enteromius afrovernayi	(Nichols & Boulton, 1927)		x	x	x	x	LC
Enteromius amphigramma	(Boulenger, 1903)	x					
Enteromius ansorgii	(Boulenger, 1904)		x				LC
Enteromius argenteus	(Günther, 1868)	x				x	LC
Enteromius barotseensis	(Pellegrin, 1920)			x	x	x	LC
Enteromius barnardi	(Jubb, 1965)			x	x	x	LC
Enteromius bifrenatus	(Fowler, 1935)			x	x	x	LC
Enteromius breviceps	(Trewavas, 1936)	x			x	x	LC
Enteromius brevidorsalis	(Boulenger, 1915)	x	x	x	x	x	LC
Enteromius brevilateralis	(Poll, 1967)	x	x				DD
Enteromius caudosignatus	(Poll, 1967)		x				DD

(continued)

Species	Author & Date	A	C	Z	O	K	I[a]
Enteromius chicapaensis	(Poll, 1967)		x		x		LC
Enteromius chiumbeensis	(Pellegrin, 1936)		x		x		LC
Enteromius dorsolineatus	(Trewavas, 1936)	x				x	LC
Enteromius eutaenia	(Boulenger, 1904)	x	x	x	x	x	DD
Enteromius evansi	(Fowler, 1930)	x			x		LC
Enteromius fasciolatus	(Günther, 1868)	x	x	x	x	x	LC
Enteromius greenwoodi	(Poll, 1967)	x			x		DD
Enteromius haasianus	(David, 1936)	x	x	x	x		LC
Enteromius holotaenia	(Boulenger, 1904)	x	x				LC
Enteromius kerstenii	(Peters, 1868)			x	x	x	LC
Enteromius kessleri	(Steindachneri, 1866)	x	x	x	x		LC
Enteromius lineomaculatus	(Boulenger, 1903)		x	x	x	x	LC
Enteromius lujae	(Boulenger, 1913)	x	x	x	x	x	DD
Enteromius machadoi	(Poll, 1967)		x				DD
Enteromius mattozi	(Guimarães, 1884)	x	x			x	LC
Enteromius mediosquamatus	(Poll, 1967)		x				DD
Enteromius miolepis	(Boulenger, 1902)		x	x	x		LC
Enteromius mocoensis	(Trewavas, 1936)	x			x		DD
Enteromius multilineatus	(Worthington, 1933)		x	x	x	x	LC
Enteromius musumbi	(Boulenger, 1910)	x					LC
Enteromius paludinosus	(Peters, 1852)	x	x	x	x	x	LC
Enteromius petchkovski	(Poll, 1967)		x				DD
Enteromius poechii	(Steindachneri, 1911)		?	x	x	x	LC
Enteromius radiatus	(Peters, 1853)	x	x	x	x	x	LC
Enteromius rousellei	(Ladiges & Voelker, 1961)		x				DD
Enteromius thamalakanensis	(Fowler, 1935)			x	x	x	LC
Enteromius trimaculatus	(Peters, 1852)	x	x		x	x	LC
Enteromius unitaeniatus	(Günther, 1867)	x	x	x	x	x	LC
Enteromius cf viviparus	(Weber, 1897)			x	x	x	NE
Enteromius wellmani	(Boulenger, 1911)	x					DD
Labeobarbus caudovittatus	(Boulenger, 1902)		x				LC
Labeobarbus codringtonii	(Boulenger, 1908)		x	x	x		LC
Labeobarbus ensis	(Boulenger, 1910)	x					LC
Labeobarbus gulielmi	(Boulenger, 1910)	x					DD
Labeobarbus girardi	(Boulenger, 1910)	x					DD
Labeobarbus jubbi	(Poll, 1967)		x				DD
Labeobarbus lucius	(Boulenger, 1910)	x					DD
Labeobarbus marequensis (Cassai)	(Smith, 1841)		x				LC
Labeobarbus nanningsi	de Beaufort, 1933	x	x				DD
Labeobarbus rhinophorus	(Boulenger, 1910)	x					DD
Labeobarbus rocadasi	(Boulenger, 1910)	x					DD
Labeobarbus rosae	(Boulenger, 1910)	x					DD

(continued)

Species	Author & Date	A	C	Z	O	K	I[a]
Labeobarbus ansorgii	(Boulenger, 1906)	x					LC
Labeobarbus ensifer	(Boulenger, 1910)	x					LC
Labeobarbus boulengeri	Vreven et al., 2016	x					NE
Labeobarbus macrolepidotus	(Pellegrin, 1928)		x				LC
Labeobarbus steindachneri	(Boulenger, 1910)	x					LC
Labeobarbus stenostomata	(Boulenger, 1910)	x					DD
Labeobarbus varicostoma	(Boulenger, 1910)	x					DD
Labeo annectens	Boulenger, 1903	x	x				LC
Labeo ansorgii	Boulenger, 1907	x	x			x	LC
Labeo chariensis	Pellegrin, 1904		x				LC
Labeo cylindricus	Peters, 1852		x	x	x		LC
Labeo greeni	Boulenger, 1902		x		?		LC
Labeo lineatus	Boulenger, 1898		x				LC
Labeo longipinnis	Boulenger, 1898		x				LC
Labeo macrostoma	Boulenger, 1898		x				LC
Labeo parvus	Boulenger, 1902	x	x				LC
Labeo rocadasi	Boulenger, 1907	x					LC
Labeo ruddi	Boulenger, 1907					x	LC
Labeo velifer	Boulenger, 1898		x				NE
Labeo weeksii	Boulenger, 1909		x				LC
Engraulicypris howesi	Ridden, Bills & Villet, 2016					x	NE
Opsaridium zambezense	(Peters, 1852)		x	x	x		LC
Raiamas ansorgii	(Boulenger, 1910)	x					DD
Raiamas christyi	(Boulenger, 1920)		x				LC
Hepsetidae							
Hepsetus cuvieri	(Castelnau, 1861)	x	x	x	x	x	NE
Alestidae							
Bryconaethiops microstoma	Günther, 1873		x				LC
Alestes macropthalmus	Günther, 1867		x				LC
Brycinus kingsleyae	(Günther, 1896)		x				LC
Brycinus grandisquamis	(Boulenger, 1899)		x				LC
Brycinus humilis	(Boulenger, 1905)	x	x				DD
Brycinus imberi	(Peters, 1852)	?	x				LC
Brycinus lateralis	(Boulenger, 1900)		x	x	x	x	LC
Micralestes acutidens	(Peters, 1852)		x	x	x		LC
Micralestes argyrotaenia	Trewavas, 1936					x	LC
Micralestes humilis	Boulenger, 1899		x				LC
Nannopetersius ansorgii	(Boulenger, 1910)	x					LC
Rhabdalestes maunensis	(Fowler, 1935)			x	x	x	LC
Hydrocynus vittatus	Castelnau, 1861		x	x	x		LC
Distichodontidae							
Distichodus fasciolatus	Boulenger, 1898		x				LC
Distichodus lusosso	Schilthuis, 1891		x				LC
Distichodus maculatus	Boulenger, 1898		x				LC

(continued)

Species	Author & Date	A	C	Z	O	K	I[a]
Distichodus notospilus	Günther, 1867		x				LC
Distichodus sexfasciatus	Boulenger, 1897		x				LC
Nannocharax macropterus	Pellegrin, 1926		x	x	x		LC
Nannocharax procatopus	Boulenger, 1920		x				LC
Nannocharax angolensis	(Poll, 1967)		x				LC
Nannocharax lineostriatus	(Poll, 1967)		x	x	x		DD
Nannocharax machadoi	(Poll, 1967)			x	x	x	LC
Nannocharax multifasciatus	Boulenger, 1923			x	x	x	DD
Dundocharax bidentatus	Poll, 1967		x				DD
Claroteidae							
Chrysichthys ansorgii	Boulenger, 1910	x					LC
Chrysichthys bocagii	Boulenger, 1910	x					LC
Chrysichthys cranchii	(Leach, 1818)		x				LC
Chrysichthys delhezi	Boulenger, 1899		x				LC
Chrysichthys macropterus	Boulenger, 1920		x				DD
Chrysichthys nigrodigitatus	(Lacepède, 1803)	x					LC
Parauchenoglanis ngamensis	(Boulenger, 1911)		x	x	x		LC
Amphiliidae							
Zaireichthys dorae	(Poll, 1967)		x				DD
Zaireichthys flavomaculatus	(Pellegrin, 1926)		x				DD
Zaireichthys pallidus	Eccles, Tweddle & Skelton, 2011			x	x		NE
Zaireichthys conspicuus	Eccles, Tweddle & Skelton, 2011			x	x		NE
Zaireichthys kavangoensis	Eccles, Tweddle & Skelton, 2011				x		NE
Zaireichthys kunenensis	Eccles, Tweddle & Skelton, 2011					x	NE
Congoglanis alula	(Nichols & Griscom, 1917)		x				LC
Doumea angolensis	Boulenger, 1906	x					LC
Congoglanis howesi	Vari, Ferraris & Skelton, 2012		x				NE
Congoglanis sp.		x					NE
Amphilius lentiginosus	Trewavas, 1936	x	?				DD
Amphilius cubangoensis	Pellegrin, 1936			x	x		NE
Phractura macrura	Poll, 1967		x				DD
Phractura scaphyrhynchura	(Vaillant, 1886)		x				LC
Malapteruridae							
Malapterurus monsembeensis	Roberts, 2000		x				LC
Clariidae							
Heterobranchus longifilis	Valenciennes, 1840		x				LC
Channallabes apus	(Günther, 1873)	x	x				LC
Clarias angolensis	Steindachner, 1866	x	x				LC
Clarias buthupogon	Sauvage, 1879		x				LC

(continued)

Species	Author & Date	A	C	Z	O	K	I[a]
Clarias dumerilii	Steindachner, 1866	x	x			x	LC
Clarias platycephalus	Boulenger, 1902		x				NE
Clarias gariepinus	(Burchell, 1822)	x	x	x	x	x	LC
Clarias ngamensis	Castelnau, 1861	x	x	x	x	x	LC
Clarias nigromarmoratus	Poll, 1967		x				LC
Clarias stappersii	Boulenger, 1915	x	x	x	x	x	LC
Clarias liocephalus	Boulenger, 1898		x	x	x	x	LC
Clarias theodorae	Weber, 1897		x	x	x	x	LC
Clariallabes heterocephalus	Poll, 1967		x				LC
Clariallabes variabilis	Pellegrin, 1926		x				LC
Clariallabes platyprosopos	Jubb, 1965			x	x		LC
Clariallabes sp					x		NE
Platyclarias machadoi	Poll, 1977		x				DD
Schilbeidae							
Parailia occidentalis	(Pellegrin, 1901)	x					LC
Schilbe intermedium	Rüppell, 1832		x	x	x	x	LC
Schilbe angolensis	(De Vos, 1984)	x					DD
Schilbe ansorgii	(Boulenger, 1910)	x					LC
Schilbe bocagii	(Guimarães, 1884)	x					LC
Schilbe grenfelli	(Boulenger, 1900)		x				LC
Schilbe yangambianus	(Poll, 1954)		x	x			LC
Mochokidae							
Synodontis laessoei	Norman, 1923	x					DD
Synodontis leopardinus	Pellegrin, 1914			x	x	x	LC
Synodontis longirostris	Boulenger, 1902		x				LC
Synodontis macrostigma	Boulenger, 1911			x	x	x	LC
Synodontis macrostoma	Skelton & White, 1990			x	x	x	LC
Synodontis nigromaculatus	Boulenger, 1905		x	x	x	x	LC
Synodontis ornatipinnis	Boulenger, 1899	x	x				LC
Synodontis thamalakanensis	Fowler, 1935			x	x	x	LC
Synodontis woosnami	Boulenger, 1911			x	x	x	LC
Synodontis vanderwaali	Skelton & White, 1990			x	x	x	LC
Chiloglanis angolensis	Poll, 1967	x				x	DD
Chiloglanis fasciatus	Pellegrin, 1936			x	x		LC
Chiloglanis lukugae	Poll, 1944		x				LC
Chiloglanis micropogon	Poll, 1952		x				NE
Chiloglanis sardinhai	Ladiges & Voelker, 1961	x					LC
Euchilichthys astatodon	(Pellegrin, 1928)		x				LC
Euchilichthys royauxi	Boulenger, 1902		x				LC
Atopochilus macrocephalus	Boulenger, 1906		x				DD
Chiloglanis sp. (dark)				x	x		NE
Chiloglanis sp. (gold)				x	x		NE

(continued)

Species	Author & Date	A	C	Z	O	K	I[a]
Procatopodidae							
Aplocheilichthys spilauchen	(Duméril, 1861)	x					LC
Micropanchax hutereaui	(Boulenger, 1913)		x	x	x		LC
Micropanchax johnstonii	(Günther, 1894)		x	x	x	x	LC
Micropanchax katangae	(Boulenger, 1912)		x	x	x	x	LC
Micropanchax luluae	(Fowler, 1930)		x		x		NE
Micropanchax macrurus	(Boulenger, 1904)	x	x			x	LC
Micropanchax mediolateralis	(Poll, 1967)		x		x		LC
Micropanchax myaposae	(Boulenger, 1908)	x					LC
Micropanchax nigrolateralis	(Poll, 1967)		x		x		DD
Micropanchax 'pigmy'				x	x		NE
Cichlidae							
Hemichromis elongatus	(Guichenot, 1861)	x	x	x	x		LC
Hemichromis angolensis	Steindachner, 1865	x					NE
Pharyngochromis acuticeps	(Steindachner, 1866)	x		x	x	x	LC
Pseudocrenilabrus philander	(Weber, 1897)	x	x	x	x	x	LC
Oreochromis andersonii	(Castelnau, 1861)			x	x	x	VU
Oreochromis macrochir	(Boulenger, 1912)		x	x	x	x	VU
Oreochromis angolensis	(Trewavas, 1973)	x					LC
Coptodon rendalli	(Boulenger, 1897)	x	x	x	x	x	LC
Pelmatolapia cabrae	(Boulenger, 1899)	x	x				LC
Oreochromis lepidurus	(Boulenger, 1899)	x	x				EN
Oreochromis schwebischi	(Sauvage, 1884)	x	x				LC
Tilapia sparrmanii	Smith, 1840	x	x	x	x	x	LC
Tilapia ruweti	(Poll & Thys van den Audenaerde, 1965)	x	x	x			LC
Serranochromis altus	Winemiller & Kelso-Winemiller, 1991			x	x		LC
Serranochromis angusticeps	(Boulenger, 1907)		x	x	x	x	LC
Serranochromis longimanus	(Boulenger, 1911)			x	x		LC
Serranochromis macrocephalus	(Boulenger, 1899)	x	x	x	x	x	LC
Serranochromis robustus jallae	(Boulenger, 1864)		x	x	x	x	LC
Serranochromis thumbergi	(Castelnau, 1861)		?	x	x	x	LC
Sargochromis greenwoodi	(Bell-Cross, 1975)				x	x	LC
Sargochromis carlottae	(Boulenger, 1905)			x	x		LC
Sargochromis giardi	(Pellegrin, 1903)			x	x	x	LC
Sargochromis coulteri	(Bell-Cross, 1975)					x	LC
Sargochromis codringtonii	(Boulenger, 1908)			x	x	x	LC
Thoracochromis lucullae	(Boulenger, 1913)	x					LC
Orthochromis machadoi	(Poll, 1967)					x	LC
Sargochromis thysi	(Poll, 1967)		x				DD
Chetia welwitschi	(Boulenger, 1898)	x				x	DD

(continued)

Species	Author & Date	A	C	Z	O	K	I[a]
Chetia gracilis	(Greenwood, 1984)				x		LC
Thoracochromis albolabrus	(Trewavas & Thys vd Audenaerde, 1969)					x	LC
Thoracochromis buysi	(Penrith, 1970)					x	LC
Anabantidae							
Ctenopoma machadoi	(Fowler, 1930)	x					LC
Ctenopoma multispine	Peters, 1844		x	x	x	x	LC
Microctenopoma intermedium	(Pellegrin, 1920)		x	x	x		LC
Microctenopoma sp.		x			x		NE
Mastacembelidae							
Mastacembelus ansorgii	Boulenger, 1905	x					DD
Mastacembelus niger	Sauvage, 1879		x				LC
Mastacembelus congicus	Boulenger, 1896		x				LC
Mastacembelus frenatus	Boulenger, 1901			x	x		LC
Mastacembelus sp.		x					NE
Eleotridae							
Eleotris vittata	Duméril, 1861					x	LC
Dormitator lebretonis	(Steindachner, 1870)					x	NE
Gobiidae							
Awaous lateristriga	(Duméril, 1861)					x	NE
Nematogobius maindroni	(Sauvage, 1880)					x	NE
Ctenogobius lepturus	(Pfaff, 1933)					x	NE
Periophthalmus barbarus	(Linnaeus, 1766)	x					LC
Syngnathidae							
Enneacampus ansorgii	(Boulenger, 1910)	x					LC
Enneacampus kaupi	(Bleeker, 1863)	x					LC
	TOTALS	**104**	**161**	**93**	**103**	**82**	

IUCN status (I) as recorded by Darwell et al. (2011) and IUCN (2018). The table is for tentative indications of distribution and IUCN status

DD data deficient, *EN* endangered, *LC* least concern, *NE* not evaluated, *VU* vulnerable

[a]IUCN Red List Categories Codes

References

Ball P (2015) Benguela – more than just a current. The Heritage Portal, p 13. http://www.theheritageportal.co.za/article/Benguela-more-just-current

Balon EK (1974) Fishes from the edge of Victoria Falls, Africa: demise of a physical barrier for downstream invasions. Copeia 1974(3):643–660

Barnard KH (1948) Report on a collection of fishes from the Okavango River, with notes on Zambesi fishes. Ann S Afr Mus 36:407–458

Bbole I, Katongo C, Deines AM et al (2014) Hybridization between non-indigenous *Oreochromis niloticus* and native *Oreochromis* species in the lower Kafue River and its potential impacts on fishery. J Ecol Nat Environ 6(6):215–225

Bell-Cross G (1965) Movement of fish across the Congo-Zambezi watershed in the Mwinilunga district of Northern Rhodesia. Proceedings of the Central African Scientific and Medical Congress, Lusaka, 1963, pp 415–424

Bell-Cross G (1975) A revision of certain *Haplochromis* species (Pisces: Cichlidae) of Central Africa. Occas Pap Natl Mus Monuments Rhod Ser B 5(7):405–464

Bell-Cross G, Minshull JL (1988) The fishes of Zimbabwe. National Museums and Monuments of Zimbabwe, Harare

Bills IR, Skelton PH, Almeida F (2012) A survey of the fishes of the upper Okavango system in Angola. SAIAB Investigational Report 73, 61 pp

Bills IR, Mazungula N, Almeida F (2013) A survey of the fishes of upper Okavango River system in Angola. SAIAB Investigational Report 74, 21 pp

Boulenger GA (1909–1916) Catalogue of the fresh-water of Africa in the British Museum (Natural History), Vol 1 (1909) Vol 2 (1910), Vol 3 (1915), Vol 4 (1916). Trustees of the British Museum, London

Boulenger GA (1910) LXI.–on a large collection of fishes made by Dr. W. J. Ansorge in the Quanza and Bengo Rivers, Angola. Ann Mag Nat Hist 6(36):537–561

Boulenger GA (1911) V. on a collection of fishes from the Lake Ngami Basin, Bechuanaland. Trans Zool Soc London 18(5):399–418, pls XXXVIII-XLIII

Burrough SL, Thomas DSG, Bailey RM (2009) Mega-lake in the Kalahari: a late Pleistocene record of the Palaeolake Magadigadi system. Quat Sci Rev 28:1392–1411

Castelnau F (1861) Mémoire sur les Poissons de l'Afrique Australe. J-B Baillière et Fils, Paris, p 78

Day JJ, Bills R, Friel JP (2009) Lacustrine radiation in African *Synodontis* catfish. J Evol Biol 22:805–817

De Vos LDG (1995) A systematic revision of the African Schilbeidae (Teleostei, Siluriformes). With an annotated bibliography. Annalen Zoologische Wetenschappen 271:1–450

Devaere S, Adriaens D, Verraes W (2007) *Channallabes sanghaensis* sp.n. a new anguilliform catfish from the Congo River basin, with some comments on other anguilliform clariids (Teleostei, Siluriformes). Belg J Zool 137:17–26

Eccles DH, Tweddle D, Skelton PH (2011) Eight new species in the dwarf catfish genus *Zaireichthys* (Siluriformes: Amphiliidae). Smithiana Bull 13:3–28

Flügel TJ, Eckardt FD, Cotterill FPD (2015) Chapter 15: the present day drainage patterns of the Congo river system and their Neogene evolution. In: de Wit MJ et al (eds) Geology and resource potential of the Congo basin, Regional geology reviews. Springer, Berlin/Heidelberg, pp 315–337

Fowler HW (1930) The fresh-water fishes obtained by the gray African expedition – 1929. With notes on other species in the academy collection. Proc Acad Natl Sci Phila 82:27–83

Fowler HW (1935) Scientific results of the Vernay-Lang Kalahari Expedition, March to September, 1930. The freshwater fishes. Ann Transv Mus 16(2):251–293

Gilchrist JDF, Thompson WW (1913) The freshwater fishes of South Africa. Ann S Afr Mus 11(5):321–463

Gilchrist JDF, Thompson WW (1917) The freshwater fishes of South Africa (continued). Ann S Afr Mus 11(6):465–575

Greenwood PH (1984) The haplochromine species (Teleostei, Cichlidae) of the Cunene and certain other Angolan rivers. Bull Brit Mus (Nat Hist) 47(4):187–239

Greenwood PH (1993) A review of the serranochromine cichlid fish genera *Pharyngochromis, Sargochromis, Serranochromis and Chetia* (Teleostei, Labroidei). Bull Brit Mus (Nat Hist) 59:33–44

Guimarães ARP (1884) 1. Diagnoses de trois nouveaux poisons d'Angola. J Sci Math Phys Lisboa 37:1–10

Haddon IG, McCarthy TS (2005) The Mesozoic–Cenozoic interior sag basins of Central Africa: the late-cretaceous–Cenozoic Kalahari and Okavango basins. J Afr Earth Sci 43:316–333

Hay CJ, van Zyl BJ, van der Bank FH et al (1997) A survey of the fishes of the Kunene River, Namibia. Modoqua 19:129–141

Hipondoka MHT, van der Waal BCW, Ndeutapo MH, Hango L (2018) Sources of fish in the ephemeral western iishana region of the Cuvelai–Etosha Basin in Angola and Namibia. Afr J Aquat Sci 43(3):199–214.https://doi.org/10.2989/16085914.2018.1506310

Jackson PBN (1961) The fishes of northern Rhodesia: a checklist of indigenous species. Department of Game and Fisheries, Lusaka

Jubb RA (1961) An illustrated guide to the freshwater fishes of the Zambezi River, Lake Kariba, Pungwe, Sabi, Lundi and Limpopo Rivers. Stuart Manning, Bulawayo

Jubb RA (1967) The freshwater fishes of southern Africa. AA Balkema, Cape Town

Jubb RA, Gaigher IG (1971) Checklist of the fishes of Botswana. Arnoldia, Rhodesia 5(97):1–22

Key RM, Cotterill FPD, Moore AE (2015) The Zambezi river: an archive of tectonic events linked to the amalgamation and disruption of Gondwana and subsequent evolution of the African plate. S Afr J Geol 118:425–438

King J, Chonguic E (2016) Integrated management of the Cubango-Okavango River basin. Ecohydrol Hydrobiol 16:263–271

Kramer B, Swartz ER (2010) A new species of slender Stonebasher within the *Hippopotamyrus ansorgii* complex from the Cunene River in southern Africa (Teleostei: Mormyriformes). J Nat Hist 44(35–36):2213–2242

Kramer B, Wink M (2013) East–west differentiation in the *Marcusenius macrolepidotus* species complex in southern Africa: the description of a new species for the lower Cunene River, Namibia (Teleostei: Mormyridae). J Nat Hist 47(35–36):2327–2362

Kramer B, van der Bank FH, Flint N et al (2003) Evidence for parapatric speciation in the Mormyrid fish, *Pollimyrus castelnaui* (Boulenger, 1911), from the Okavango–upper Zambezi River systems: *P. marianne* sp. nov., defined by electric organ discharges, morphology and genetics. Environ Biol Fish 77:47–70

Kramer B, van der Bank FH, Wink M (2004) The *Hippopotamyrus ansorgii* species complex in the upper Zambezi River system with a description of a new species, *H. szaboi* (Mormyridae). Zool Scr 33:1–18

Kramer B, Bills IR, Skelton PH et al (2012) A critical revision of the churchill snoutfish, genus *Petrocephalus* Marcusen, 1854 (Actinopterygii: Teleostei: Mormyridae), from southern and eastern Africa, with the recognition of *Petrocephalus tanensis*, and the description of five new species. J Nat Hist 46:2179–2258

Kramer B, van der Bank H, Wink M (2014) Marked differentiation in a new species of dwarf stonebasher, *Pollimyrus cuandoensis* sp. nov. (Mormyridae: Teleostei), from a contact zone with two sibling species of the Okavango and Zambezi rivers. J Nat Hist 48(7–8):429–463

Kramer B, van der Bank FH, Wink M (2016) *Marcusenius desertus* sp. nov. (Teleostei: Mormyridae), a mormyrid fish from the Namib desert. Afr J Aquat Sci 41(1):1–18

Ladiges W (1964) Beiträge zur zoogeographie und Oekologie der süßwasserfische Angolas. Die Mitteilungen aus dem Hamburgischen Zoologischen Museum und Institut 61:221–272

Ladiges W, Voelker J (1961) Untersuchungen über die Fishfauna in Gebirgsgewässern des Wasserscheidenhochlands in Angola. Die Mitteilungen aus dem Hamburgischen Zoologischen Museum und Institut 59:117–140

Lévêque C, Paugy D (2017a) General characteristics of ichthyological fauna. In: Paugy D, Lévêque C, Otero O (eds.) The inland water fishes of Africa, diversity, ecology and human use. IRD Éditions, Paris, & Royal Museum for Central Africa, Tervuren, pp 83–96

Lévêque C, Paugy D (2017b) Geographical distribution and Affinities of African freshwater fishes. In: Paugy D, Lévêque C, Otero O (eds) The inland water fishes of Africa, diversity, ecology and human use. IRD Éditions. France, & Royal Museum for Central Africa, Belgium, pp 97–114

Marshall BE, Tweddle D (2007) *Oreochromis macrochir*. The IUCN Red List of Threatened Species 2007: e.T63336A12659168

Mendelsohn J, Weber B (2015) An atlas and profile of Moxico, Angola. RAISON, Windhoek

Moelants T 2010. *Oreochromis lepidurus*. The IUCN Red List of Threatened Species2010: e.T182875A7991695

Moore AE, Larkin PA (2001) Drainage evolution in south-Central Africa since the break-up of Gondwana. S Afr J Geol 104:47–68

Moore AE, Cotterill FPD, Eckardt FD (2012) The evolution and ages of Makgadikgadi palaeo-lakes: Consilient evidence from Kalahari drainage evolution. S Afr J Geol 115:385–413

Musilová Z, Kalous L, Petrtýl M et al (2013) Cichlid fishes in the Angolan headwaters region: molecular evidence of the ichthyofaunal contact between the Cuanza and Okavango-Zambezi systems. PLoS One 8(5):e65047

NGOWP – National Geographic Okavango Wilderness Project (2018) Initial findings from exploration of the upper catchments of the Cuito, Cuanavale and Cuando Rivers in Central and South-Eastern Angola (May 2015 to December 2016). National Geographic Okavango Wilderness Project, 352 pp

Nichols JT, Boulton R (1927) Three new minnows of the genus *Barbus*, and a new characin from the Vernay Angola expedition. Am Mus Novit 264:1–8

Norman JR (1923) A new cyprinoid fish from Tanganyika territory, and two new fishes from Angola. Ann Mag Nat Hist 12(72):694–696

Nunes AL, Douthwaite RJ, Tyser B et al (2016) Invasive crayfish threaten Okavango Delta. Front Ecol Environ 14(5):237–238

Paugy D, Lévêque C, Otero O (eds) (2017) The inland water fishes of Africa, IRD Éditions. Institut de Recherche pour de Developpement/RMCA Royal Museum for Central Africa, Paris/Tervuren

Pellegrin J (1921) Description d'un Barbeau nouveau de l'Angola. Bull Soc Zool Fr 46:118–120

Pellegrin J (1928) Poissons du Chiloango et du Congo receuillis par l'expédition du Dr Schouteden (1920–1922). Annales du Musée Royal du Congo Belge, Zoologie Série 1 3(1):1–50

Pellegrin J (1936) Contribution à l'ichthyologie de l'Angola. Arquivos do Museu Bocage 7:45–62

Penrith M-L (1970) Report on a small collection of fishes from the Kunene River mouth. Cimbebasia Series A 1:165–176

Penrith MJ (1973) A new species of *Parakneria* from Angola (Pisces: Kneriidae). Cimbebasia Series A 11:131–135

Penrith MJ (1982) Additions to the checklist of southern African freshwater fishes and a gazetteer of south-western Angolan collecting localities. J Limnol Soc South Afr 8(2):71–75

Pinton A, Agnèse J-F, Paugy D, Otero O (2013) A large-scale phylogeny of *Synodontis* (Mochokidae, Siluriformes) reveals the influence of geological events on continental diversity during the Cenozoic. Mol Phylogenet Evol 66:1027–1040

Podgorski JE, Green AG, Kgotlhang L et al (2013) Paleo-megalake and paleo-megafan in southern Africa. Geology 11:1155–1158

Poll M (1967) Contribution à la Faune Ichthyologique de l'Angola. Publicações Culturais 75 75. Companhia dos Diamentes de Angola (DIAMANG), Lisbon, 381 pp

Poll M (1971) Révision des *Synodontis* Africains (Famille Mochocidae). Annales Musée Royal de l'Afrique Centrale Serie IN-8 Sciences Zoologiques No. 191. Musée Royal de l'Afrique Centrale, Tervuren, 497 pp

Roberts TC (1975) Geographical distribution of African freshwater fishes. Zool J Linnean Soc 57(4):249–319

Saldanha L (1978) Museu Bocage. Copeia 1978(4):739–740

Schwarzer J, Swartz ER, Vreven E et al (2012) Repeated trans-watershed hybridization among haplochromine cichlids (Cichlidae) was triggered by Neogene landscape evolution. Proc R Soc London, Ser B 279:4389–4398

Skelton PH (1994) Diversity and distribution of freshwater fishes in East and Southern Africa. Annales Musée Royal de l'Afrique Centrale, Sciences Zoologiques 275:95–131

Skelton PH (2001) A complete guide to the freshwater fishes of Southern Africa. Struik, Cape Town

Skelton PH, Swartz ER (2011) Walking the tightrope: trends in African freshwater systematic ichthyology. J Fish Biol 79:1413–1435

Skelton PH, Bruton MN, Merron GS et al (1985) The fishes of the Okavango drainage system in Angola, South West Africa and Botswana: taxonomy and distribution. Ichthyol. Bull. JLB Smith Inst Ichthyol 50:1–21

Skelton PH, Neef G, Costa A (2016) Into the wilderness expedition 2015: the fishes. SAIAB Investigational Report No 75, 49 pp

Snoeks J, Vreven EJ (2007) Chapter 38: Polynemidae, 445-449 in: Stiassny, MLJ, Teugels GG, Hopkins CD (eds) The fresh and brackish water fishes of lower Guinea, west-Central Africa. Collection Faune et Flore tropicales 42, vol 2. Institut de recherché pour le développement, Paris, France/Muséum national d'histoire naturelle, Paris, France/Musée royal de l'Afrique Centrale, Tervuren

Snoeks J, Harrison IJ, Stiassny MLJ (2011) Chapter 3: The status and distribution of freshwater fishes. In: Darwall WRT, Smith KG, Allen DJ, Holland RA, Harrison IJ, Brooks EGE (eds) The diversity of life in African freshwaters: under water, under threat. An analysis of the status and distribution of freshwater species throughout mainland Africa. IUCN, Cambridge/Gland, pp 42–73

Steindachner F (1866) Ichthyologische Mittheilungen. (IX.) [With subtitles I-VI.]. Verh Zool Bot Ges Wien 16:761–796

Stiassny MLJ, Teugels GG, Hopkins CD (eds) (2007) The fresh and brackish water fishes of Lower Guinea, West-Central Africa. Collection Faune et Flore Tropicales 42, Volume 1 and 2. IRD & Muséum National d'Histoire Naturelle, Paris & Musée Royal de l'Afrique Centrale, Tervuren

Thieme ML, Abell R, Stiassny ML et al (eds) (2005) Freshwater ecoregions of Africa and Madagascar, a conservation assessment. Island Press, Washington

Trewavas E (1936) Dr. Karl Jordan's expedition to south-West Africa and Angola: the fresh-water fishes. Novitates Zoologicae 40:63–74

Trewavas E (1964) A revision of the genus *Serranochromis* Regan (Pisces, Cichlidae). Annales Musée Royal de l'Afrique Centrale Serie IN-8 Sciences Zoologiques No. 125, Musée Royal de l'Afrique Centrale, Tervuren, 58 pp

Trewavas E (1973) A new species of cichlid fishes of rivers Quanza and Bengo, Angola, with a list of the known Cichlidae of these rivers and a note on *Pseudocrenilabrus natalensis* fowler. Bull Brit Mus (Nat Hist) 25(1):28–37

Tweddle D (2010) Overview of the Zambezi River system: its history, fish fauna, fisheries, and conservation. Aquat Ecosyst Health Manage 13(3):224–240

Tweddle D, Skelton, PH, van der Waal et al (2004) Aquatic biodiversity survey "four corners" transboundary natural resources management area. SAIAB Investigational Report No 71 202 pp

Tweddle D, Cowx IG, Peel RA et al (2015) Challenges in fisheries management in the Zambezi, one of the great rivers of Africa. Fish Manag Ecol 22:99–111

Van der Waal BCW (1991) A survey of the fisheries in Kavango, Namibia. Modoqua 17(2):113–122

Van der Waal BCW, Skelton PH (1984) Checklist of fishes of Caprivi. Modoqua 13(4):303–321

Vreven EJ, Musschoot T, Snoeks J et al (2016) The African hexaploid Torini (Cypriniformes: Cyprinidae): review of a tumultuous history. Zool J Linnean Soc 177(2):231–305

Wamuini Lunkayilakio S, Vreven E (2010) *'Haplochromis' snoeksi*, a new species from the Inkisi River basin, lower Congo (Perciformes: Cichlidae). Ichthyol Explor Freshwaters 21(3):279–287

Wamuini Lunkayilakio SW, Vreven E (2008) *Nannopetersius mutambuei* (Characiformes: Alestidae), a new species from the Inkisi River basin, Democratic Republic of Congo. Ichthyol Explor Freshwaters 19:367–376

Wamuini Lunkayilakio S, Vreven E, Vandewalle P et al (2010) Contribution à la connaissance de l'ichtyofaune de l'Inkisi au Bas-Congo (RD du Congo). Cybium 34(1):83–91

Whitfield AK (2007) Estuary associated fish species. In: Stiassny MLJ, Teugels GG, Hopkins CD (eds) The fresh and brackish water fishes of Lower Guinea, West-Central Africa. Collection

Faune et Flore Tropicales 42, vol 1. IRD & Muséum National d'Histoire Naturelle, Paris & Musée Royal de l'Afrique Centrale, Tervuren, pp 46-56

Wise RM, van Wilgen BW, Hill MP et al (2007) The economic impact and appropriate management of selected invasive alien species on the African continent. Final report for GISP. CSIR report number CSIR/RBSD/ER/2007/0044/C

Zengeya TA, Decru E, Vreven EJ (2011) Revalidation of *Hepsetus cuvieri* (Castelnau, 1861) (Characiformes: Hepsetidae) from the Quanza, Zambezi and southern part of the Congo ichthyofaunal provinces. J Nat Hist 45:1723–1744

Zengeya TA, Robertson MP, Booth AJ et al (2013) Qualitative ecological risk assessment of the invasive Nile tilapia, *Oreochromis niloticus* in a sub-tropical African river system (Limpopo river, South Africa). Aquat Conserv Mar Freshwat Ecosyst 23:51–64

Chapter 12
The Amphibians of Angola: Early Studies and the Current State of Knowledge

Ninda Baptista, Werner Conradie, Pedro Vaz Pinto, and William R. Branch

Abstract Angolan amphibians have been studied since the mid-nineteenth century by explorers and scientists from all over the western world, and collections have been deposited in around 20 museums and institutions in Europe, Northern America, and Africa. A significant interruption of this study occurred during Angola's liberation struggle and civil war for nearly four decades and, as a consequence, knowledge about the country's biodiversity became outdated with critical gaps. Since 2009, a new era in Angolan biodiversity studies started as expeditions scattered in southwest-

N. Baptista (✉)
Instituto Superior de Ciências da Educação da Huíla, Rua Sarmento Rodrigues, Lubango, Angola

National Geographic Okavango Wilderness Project, Wild Bird Trust, Parktown, Gauteng, South Africa

CIBIO-InBIO, Centro de Investigação em Biodiversidade e Recursos Genéticos, Laboratório Associado, Universidade do Porto, Vairão, Portugal
e-mail: nindabaptista@gmail.com

W. Conradie
National Geographic Okavango Wilderness Project, Wild Bird Trust, Hogsback, South Africa

School of Natural Resource Management, Nelson Mandela University, George, South Africa

Port Elizabeth Museum (Bayworld), Humewood, South Africa
e-mail: werner@bayworld.co.za

P. Vaz Pinto
Fundação Kissama, Luanda, Angola

CIBIO-InBIO, Centro de Investigação em Biodiversidade e Recursos Genéticos, Universidade do Porto, Campus de Vairão, Vairão, Portugal
e-mail: pedrovazpinto@gmail.com

W. R. Branch (deceased)
National Geographic Okavango Wilderness Project, Wild Bird Trust, Parktown, Gauteng, South Africa

Department of Zoology, Nelson Mandela University, Port Elizabeth, South Africa

B. J. Huntley et al. (eds.), *Biodiversity of Angola*,
https://doi.org/10.1007/978-3-030-03083-4_12

ern, northeastern, southeastern, and northwestern Angola lead to exciting discoveries, including new records for the country, descriptions of new species, range extensions and taxonomical updates. Currently 111 amphibian species are listed for the country (of which 21 are endemic), but this number is an underestimate and the various unresolved taxonomical issues challenge the study of every other aspect of this group. The Angolan amphibian fauna remains one of the most poorly known in Africa and much still has to be done in order to understand its diversity, evolution and conservation needs. An overview of existing knowledge of Angolan amphibians is presented, including an updated checklist for the country, comments on problematic groups, endemic species, biogeography, recent findings, and priority research topics.

Keywords Angolan escarpment · Checklist · Endemism · Herpetology · Research priorities · Taxonomy

Introduction

Amphibians are a fascinatingly diverse group that plays crucial ecological roles (Beard et al. 2002; Davic and Welsh 2004; Regester et al. 2006) and are useful as indicators of ecosystem health (Waddle 2006), thus the relevance of their study surpasses herpetological curiosity. Despite the fact that the rate of description of amphibian species in the world is continuously increasing, current taxonomic research is still insufficient to properly inform conservation planning (Köhler et al. 2005; Brito 2010).

Like other groups presented in this book, Angolan amphibians are among the most poorly known in Africa (Conradie et al. 2016). To study this group it is necessary to deal with historical as well as scientific issues including: many species are known from holotypes collected more than a century ago and which may have been subsequently lost; collection localities had old colonial names, some no longer used and others confused with homonyms; a considerable amount of early literature is written in diverse languages (Portuguese, French, German, English and even Latin) and is not easily accessible; and many names used for Angolan taxa have been lost in synonymies and their current status remains problematic. Overviews of the history and evolution of the southern African amphibian taxonomy exist, mentioning Angolan taxa briefly (Poynton 1964; Channing 1999; Du Preez and Carruthers 2009, 2017). This chapter focuses on Angola, and the compiled information is intended to serve as a baseline that facilitates the study of this group. It consists of an essentially chronological summary of the studies of Angolan amphibians since the very first to the most recent findings, presents a checklist of species, and identifies some of the most evident challenges and exciting research priorities. Given the complicated status of many names available for Angolan taxa, species considered as valid in this review follow Frost (2018). An Atlas of historical and bibliographic records of Angolan herpetofauna has been released subsequent to the compilation of information for this chapter (Marques et al. 2018).

Early Beginnings

The European exploration and settlement in Africa resulted in the discovery of strange and wondrous animals. As these were sent in increasing numbers to European centres of learning and study, they stimulated the departure of expeditions to explore the Angolan flora and fauna by Portugal and by other nations. The exotic collections obtained by these explorers were then shipped to their home countries, and so, in the nineteenth century, the study of amphibians from Angola started in Europe. This was the case for the rest of southern Africa, the only exception being South Africa, which in the early 1800s already had Andrew Smith, a British explorer and researcher, based in the Cape (Channing 1999; Branch and Bauer 2005).

In 1866, José Vicente Barbosa du Bocage made the first list of amphibians and reptiles from Angola based on assorted specimens deposited in the Natural History Museum of Lisbon (Bocage 1866a, b). It documented only 19 amphibian species, eight of which were new to science and which Bocage (1866b) described *(Hyperolius cinnamomeoventris, H. tristis, H. fuscigula, H. quinquevittatus, H. steindachneri, Rana (=Ptychadena) subpunctata, Rana (=Amietia) angolensis, Bufo funereus (=Sclerophrys funerea)*. The material came from two expeditions, one by José de Anchieta in 1864 to Cabinda, and the other from Duque de Bragança (now Calandula) by Pinheiro Bayão.

During this period, Europeans were exploring Angola, either on their own initiative, or on behalf of various institutions that promoted scientific expeditions to Angola. Publications from this era consist essentially of descriptions of new species and new distribution records for known species. The renowned Austrian explorer and botanist Friedrich Martin Josef Welwitsch (1806–1872) explored Angola for the Portuguese government, arriving in 1853 and undertaking almost a decade of strenuous exploring and collecting. After his return to Europe his collections were donated to the British Museum, later shared with Portugal, and the Angolan amphibians were reported on by Günther (1865), who described new species of reed frog *(Hyperolius nasutus, H. parallelus)*.

Collections from the Austrian frigate Novara were deposited in the Natural History Museum of Vienna and studied by Steindachner (1867), who described *Ptychadena porosissima* and *Hyperolius bocagei* from no precise locality. Anchieta persisted in his extensive exploration of Angola, and Bocage (1867, 1873, 1879a, b, 1882, 1893, 1897b) examined his specimens, as well as the herpetological collections of Capello & Ivens (Bocage 1879a, b), describing *Hylambates (=Leptopelis) anchietae, Hylambates (=Leptopelis) cynnamomeus,* and *Rappia (=Hyperolius) benguellensis* among other species currently not valid. The German explorers von Homeyer, who collected in Pungo-Andongo, and von Mechow, who collected in Malanje and Cuango, had their specimens deposited in the Zoological Museum of Berlin and studied by Peters (1877, 1882), who described *Bufo buchneri* from Cabinda. Boulenger (1882) studied the material from the British Museum and described *Tomopterna tuberculosa*, and Rochebrune (1885) described four new *Hyperolius* species from Cabinda *(H. lucani, H. maestus, H. protchei, H. rhizophilus)*.

Bocage (1895a) compiled the extant information about the herpetology of Angola and Congo, using all the above-mentioned references, except for Rochebrune's (1885). A total of 40 species of amphibians were listed for Angola. Even today, more than a century after its release, this work is still a valuable reference on Angolan herpetology. After this, Bocage published several other findings (Bocage 1895b, 1896a, b, 1897a, b), mostly from Anchieta's new collections, with new locality records for many frogs, and the description of a new pygmy toad, *Bufo (=Poyntonophrynus) dombensis*.

From 1898 to 1906, José Júlio Bethencourt Ferreira studied Angolan material collected by Anchieta, Francisco Newton and Pereira do Nascimento (Ferreira 1897, 1900, 1904, 1906), and described new species (*Rappia (=Afrixalus) osorioi, Arthroleptis carquejai, Rappia (=Hyperolius) nobrei*) and some species and varieties that have been subsequently synonymised.

From 1903 to 1905, William John Ansorge collected considerable material in northern, central and southwestern Angola. The collected amphibians are deposited in the British Museum, and were studied by Boulenger (1905, 1907a, b). *Arthroleptis (=Phrynobatrachus) parvulus, Arthroleptis xenochirus, Rana (=Ptychadena) ansorgii, Rana (=Tomopterna) cryptotis*, and *Rana (=Ptychadena) bunoderma* were all described from this material.

A number of expeditions in Angola included herpetological surveys, and had their reptiles studied, but the amphibians were not reported. Examples of this are the Rohan-Chabot Mission (1912–1914), which explored the south of Angola and had its specimens deposited in the Paris Natural History Museum, and the Vernay Angola Expedition (1925), from which the large collection is housed in the American Museum of Natural History.

Analysing material from the Berlin Zoological Museum, Ahl (1925) described *Hylarthroleptis (=Phrynobatrachus) brevipalmatus* from Angola, and several species of reed frogs, two of which are endemic to Angola (*Hyperolius bicolor, Hyperolius gularis)* and others which have later been synonymised into larger species complexes such as *Hyperolius parallelus* complex (*H. angolensis, H. huillensis, H. microstictus), Hyperolius marmoratus* complex (*H. decoratus, H. marungaensis),* and *Hyperolius platyceps* complex (*H. angolanus*).

In 1930–1931, the Pulitzer-Angola Expedition surveyed southwestern and central areas of the country. Over 400 specimens of amphibians were collected and deposited in the Carnegie Museum, in the United States of America. These were studied by Karl Patterson Schmidt (1936), who reported on 17 species. Although no new species were described, some were synonymised and others revived from synonymy leading the author to highlight the importance of understanding the Angolan fauna for clarifying African amphibian taxonomy.

During two trips to central and southern Angola (1928–1929 and 1932–1933) Albert Monard made important collections of amphibians and reptiles, as well as other groups. The herpetological material was deposited in the La Chaux-de-Fonds Museum, Switzerland. Monard (1937) provided an updated compilation of Angolan amphibians with a revision of the existing literature (including Ahl, Bocage, Boulenger and Schmidt's publications), as well as his own findings. Five new species of frog were described: *Hyperolius cinereus, Cassiniopsis (=Kassina) kuvan-*

gensis, Rana (=Ptychadena) keilingi, Hyperolius erythromelanus, Rana (=Ptychadena) buneli, the last two now considered synonyms of *H. paralleus* and *Ptychadena bunoderma,* respectively. In total, 80 species of amphibians were mentioned, meaning that in the four decades since Bocage's (1895a) first synthesis the known frog species for Angola had doubled.

In 1933–1934, Karl Jordan's expedition to South West Africa (now Namibia) and Angola surveyed localities on the Angolan escarpment (Congulo and Quirimbo) and afromontane forest (Mount Moco) (Jordan 1936). This material is deposited in the British Museum and the herpetofauna studied by Parker (1936). One new species of treefrog (*Leptopelis jordani*) and a new subspecies of white-lipped frog (*Rana (=Amnirana) albolabris acustirostris*) were described from this expedition. As the name *acustirostris* was preoccupied, Mertens (1938b) proposed the replacement name *Rana (=Amnirana) albolabris parkeriana*, which was later elevated to a full species by Perret (1977). Both these species remain known only from their type localities, and are escarpment-endemics.

In the 1930s W Schack visited Angola and made a collection of amphibians which were deposited in the *Natur-Museum Senckenberg*, Frankfurt, and studied by Mertens (1938a), who recorded only eight species, none of which was new.

In 1952–1954, within the scope of the Hamburg Museum expeditions, GA von Maydell made significant herpetological collections from north to south of Angola. The reptiles were studied by Walter Hellmich (1957a), but the amphibians have never been studied until recently (Ceríaco et al. 2014b). Hellmich made a trip to the Angolan region of Entre-Rios, and reported on new localities for frog species (Hellmich 1957b), also commenting on the Angolan biogeography.

From 1957 to 1959, the Portuguese Mission of Apiarian Studies of the Overseas collected amphibians especially in central and eastern Angola (Luando and Cameia), which were deposited in the Zoology Center of the Institute of Tropical Scientific Research, in Lisbon. These were studied only decades later, by Clara Ruas (1996, 2002).

Raymond F Laurent worked extensively on the herpetofauna of the Congo Basin. He studied material from Museu do Dundo, Lunda-Norte, including the extensive collection made in southwestern Angola by the Museum Director, António Barros Machado. During this period, he recorded several new frogs for Angola (Laurent 1950, 1954, 1964), and described four new species (*Ptychadena grandisonae, P. guibei, P. perplicata* and *Hyperolius vilhenai*).

In 1971 and 1974, Wulf Haacke, from the then Transvaal Museum, South Africa, made two trips to Angola to search mainly for geckos, but incidentally collecting amphibians that were later studied by John Poynton (Poynton and Haacke 1993).

Until the 1970s, zoological expeditions surveyed mostly southwestern and central parts of the country, which were more easily accessible than the inland plateau and the moist forests of the north. Herpetological knowledge about the northeastern region was greatly improved by Laurent's studies. The most poorly studied areas of Angola remained the northwest (the region of Zaire and Uíge provinces, and northern Malange, Bengo, and Cuanza-Norte provinces), followed by the southeastern 'lands at the end of the world', a commonly used expression that refers to the very remote and extensive regions of Moxico and Cuando Cubango provinces.

Recent History and Increase of Information

For almost three decades, in the period between Angola's independence and the end of the civil war (1975–2002), the country's instability precluded virtually all field surveys. Every amphibian publication dating from this period involved taxonomic revisions based on existing literature and museum collections, e.g. Perret's (1976) revision of the amphibians, particularly types, deposited in the Lisbon Museum of Natural History. This has become an extremely valuable work given the subsequent loss of these important specimens following the fire that destroyed the museum in 1978.

A key for the identification of Angolan amphibians mainly based on literature revision, including all the species listed for Angola at the time, was published (Cei 1977). With dichotomous keys, drawings, and insights on the Angolan amphibian biogeography, it was intended to make Angolan amphibian identification more accessible to the general public and particularly to students. Poynton (1964) published a faunal study of the southern African amphibians, which referred to Angolan material. This was later updated from 1985 to 1991, when Poynton & Broadley published *Amphibia Zambesiaca*, a series of papers (Poynton and Broadley 1985a, b, 1987, 1988, 1991) that addressed in detail all the amphibian families occurring in the Zambezi drainage region, including many that extend into Angola. The publication of a toponymic index of the zoological collections made in Angola (Crawford-Cabral and Mesquitela 1989) was a valuable contribution to the study of vertebrates of the country. It provided an overview of the zoological collections performed in Angola and studies related to these expeditions, including a section of type localities and the list of described vertebrates per locality, which lists amphibian species, subspecies and varieties.

In 1993, Poynton & Haacke described the first new Angolan amphibian species in decades: *Bufo (= Pontynophrynus) grandisonae*, based on Haacke's expeditions of the 1970s. In 1996, the re-examination of Monard's collection of amphibians from 1928, revealed an 'enigmatic' ranid originally identified as *Aubria subsigillata* that could not be assigned to any known genus (Perret 1996), but which was later assigned to *Aubria masako* (Channing 2001) following features described by Ohler (1996). A comprehensive revision of the Angolan amphibians and mapping of each species' distribution based on museum and literature records was made by Ruas (1996), providing taxonomic comments on some species, but not addressing the Hyperoliidae family (then including the current Leptopelinae subfamily). Ruas (2002) described in detail the contents of the amphibian collection deposited in the Zoology Center of the Institute of Tropical Scientific Research in Lisbon, again excluding the Hyperoliidae and Leptopelinae, which are still to be examined. Channing (1999) discussed aspects of Angolan amphibian taxonomy within a southern African historical perspective. Blanc and Frétey (2000) analysed the biogeography, species richness and endemism of the central African and Angolan amphibians, based on the number of species per country. They highlighted the discrepancy in species richness among genera in Angola, with *Bufo* (currently *Mertensophryne, Sclerophrys* and *Poyntonophrynus*), *Hyperolius* and *Ptychadena* being the most specious genera, which totalled 42 species, almost half of the species known for the country at the time (86).

Only in 2009 did Angolan-international collaboration lead to a new era of field surveys, initiated with an expedition to Huíla and Namibe provinces in southwestern Angola. This trip, organised by Brian Huntley, can be considered as a historical landmark for research on Angolan biodiversity. Numerous groups were surveyed (plants, invertebrates, mammals, birds, reptiles and amphibians). A new escarpment-endemic reed frog, *Hyperolius chelaensis,* was described from Serra da Chela (Conradie et al. 2012), and the colourful ashy reed frog, *Hyperolius cinereus* Monard 1937 was rediscovered (Conradie et al. 2013). Later in the same year, Alan Channing and Pedro Vaz Pinto surveyed Cangandala National Park and made a trip to Calandula, revisiting this important type locality of several amphibian species, and rediscovered *Hyperolius steindachneri* Bocage, 1866 in Angola (Channing and Vaz Pinto Unpublished Data). The material obtained from these trips was important for a number of taxonomic revisions. The Angolan river frog *Amietia angolensis,* previously thought to be widespread in Africa, was found to occur only in Angola (Channing and Baptista 2013; Channing et al. 2016), reed frogs of the *Hyperolius nasutus* complex (Channing et al. 2013) were shown to include numerous cryptic species, with possibly four occurring in Angola, and the *Hyperolius cinnamomeoventris* complex was split into different sister clades (Schick et al. 2010).

Another Angolan international expedition, again organised by Brian Huntley in 2011, visited the unexplored Lagoa Carumbo, Angola's second largest freshwater lake, in Lunda Norte province. Preliminary findings revealed a complex herpetofauna (Branch and Conradie 2015), with the description of the new *Hyperolius raymondi* (Conradie et al. 2013), and the addition of two new country records: *Amnirana* cf. *lepus* and *Hyperolius pardalis*.

Two books, *Treefroogs of Africa* (Schiøtz 1999) and *Amphibians of Central and Southern Africa* (Channing 2001) address the Angolan territory, providing species identification keys, colour photographs, and distribution maps. In 2011, a book on the central African and Angolan amphibians was released (Frétey et al. 2011). It addressed Angolan fauna only briefly, providing a species list (without discussion), and synthesis of species and habitat/biogeographical associations. In *Tadpoles of Africa* (Channing et al. 2012), the larvae of several species occurring in the country are described, and the description of *Leptopelis anchietae* and *Ptychadena porosissima* tadpoles are based on Angolan specimens. The popular book *Frogs of Southern Africa – A Complete Guide* (Du Preez and Carruthers 2009, 2017) provides descriptions of species, morphology, distribution, behaviour, and has advertisement calls available for many species. It has been recently updated to a cell phone app. "Frogs of Southern Africa" and has relevant information about species that also occur in Angola.

In 2012 and 2013, studies of the lower catchments of the Cubango, Cuito and Cuando rivers in southeastern Angola were organised by the Southern Africa Regional Environmental Program (SAREP), funded by the USAID, and included herpetological surveys. Preliminary results have been published (Brooks 2012, 2013), as well as an annotated checklist of the herpetofauna of the region (Conradie et al. 2016).

In 2013, a partnership between the Kimpa Vita University in Uíge, the Technical University Dresden and Senckenberg Natural History Collections, Dresden, promoted herpetological surveys in the extremely poorly known Serra do Pingano ecosystem and surrounding forest fragments in Uíge Province. Two forest species, *Trichobatrachus robustus* and *Xenopus andrei*, typical of the Congo Basin, were added to the country's list (Ernst et al. 2014, 2015). Both these observations represented southern range extensions of hundreds of kilometers. Additional important discoveries from this survey await formal publication, and will certainly increase current knowledge of the taxonomy and biogeography of Angolan amphibians, as well as highlight the exceptional biodiversity of northern Angola (Ernst pers. comm.).

Since 2013, a project of the California Academy of Sciences in collaboration with the National Institute of Biodiversity and Conservation Areas (INBAC), Angola, initiated a study of the Angolan herpetofauna, including the development of an atlas of the Angolan amphibians and reptiles, based on literature, analysis of museum collections from several countries, and new findings (Marques et al. 2014, 2018). The Angolan type material deposited in the Porto Museum was studied, and the nomenclature and taxonomy of hyperoliids, *Leptopelis* and *Arthroleptis* described by Ferreira were discussed (Ceríaco et al. 2014a). Analysis of amphibians collected in the Capanda Dam surroundings in Malanje (Ceríaco et al. 2014a) included a possible record of *Kassina maculosa,* which if confirmed would be the first for the country. In a study of the Namibe Province herpetofauna, *Tomopterna damarensis* was recorded for the first time for Angola (Ceríaco et al. 2016a; Heinicke et al. 2017), and a new species of pygmy toad has been described from Serra da Neve (Ceríaco et al. 2018a). A booklet on the herpetofauna of the Cangandala National Park in Malanje (Ceríaco et al. 2016c) was also released, followed by a scientific publication on the same subject (Ceríaco et al. 2018b). Research on the project's findings and surveys to additional regions in Angola are ongoing.

In 2015 the Wild Bird Trust, supported by the National Geographic Society, organised Angolan expeditions associated with the Okavango Wilderness Project. Herpetological surveys took place in the headwaters of the Cuito, Cuanavale, Cubango and Cuando rivers and other river sources in the region in both wet and dry seasons. Whilst some of these results have been published (Conradie et al. 2016), the project is ongoing but already two new country records (*Kassinula wittei* and *Leptopelis* cf. *parvus*), numerous range extensions for Angolan herpetofauna, and a number of candidate new species of amphibians have been identified.

Within the Southern African Science Service Centre for Climate Change and Adaptive Land Management (SASSCAL) project, research on herpetology is being undertaken by the *Instituto Superior de Ciências da Educação* (ISCED)-Huíla. Observatories have been implemented in Tundavala, Bicuar National Park, Cameia National Park, Iona National Park, Candelela and Cusseque (Jürgens et al. 2018). Opportunistic surveys of herpetofauna are made at all observatories (SASSCAL ObservationNet 2017), herpetofauna monitoring has been carried out at the Tundavala observatory since 2016 (Baptista et al. 2018), and a checklist of Bicuar National Park herpetofauna compiled (Baptista et al. in press). Additionally, in collaboration with Fundação Kissama, herpetological surveys have been made at several sites in Huíla Province, and throughout Angola, with emphasis along the Angolan escarpment: Cuanza-Norte, Cuanza-Sul (Cumbira) and Huíla Provinces. An Angolan herpetofauna archive is being developed at ISCED Huíla, and research undertaken in conjunction with these projects.

International and National Resources

Given the scarcity and the difficulties in obtaining information about Angolan amphibians, the compilation and listing of existing information sources is relevant. Table 12.1 lists generalist on-line platforms with relevant information about amphibians that include Angolan species, as well as a list of institutions known to have significant Angolan material in their assets.

The Current State of Knowledge on Angolan Amphibians

Despite some progress made during the last decade, the Angolan herpetofauna remains one of the most poorly known in Africa (Conradie et al. 2016). This lack of information becomes more evident when contrasted with the comprehensive information compilations regarding adjacent Namibia, which include updated lists of species (Herrmann and Branch 2013) and analysis of habitat availability, species richness and conservation (Curtis et al. 1998). For Angola, even basic information, such as accurate species checklists for the country, is absent. Existing information is scattered in recent and historical publications, many of which are not easily accessible. The recent Atlas of Angolan herpetofauna (Marques et al. 2018) contributes to filling this gap. Figure 12.1 shows the localities where amphibians have been collected before and after independence. Although recent surveys have filled some gaps, many areas remain unsurveyed. Figure 12.2 depicts some of the amphibian diversity present in Angola.

Checklist of Angolan Amphibians

Currently only 111 species are recorded from Angola (Appendix). Marques et al. (2018) list 117 species for the country. This discrepancy results from the use of different criteria for synonymies, and of a conservative approach of the present authors not incuding unconfirmed records, which are discussed elsewhere in this chapter. Both these totals are considered to be underestimates, given the country's size and habitat richness, including the southern desert, the tropical northern forests, the unique escarpment and the extensive plateau, many areas of which remain unsurveyed. This becomes more evident when compared with a country of similar size such as South Africa, whose herpetofauna is the best studied in Africa and which is considerably drier and cooler (and therefore less suitable for amphibians) than Angola, and yet it has 128 species (Frost 2018), and new species continue to be discovered (Turner and Channing 2017; Minter et al. 2017).

Table 12.1 List of relevant websites with information regarding Angolan amphibians, and collections where Angolan amphibian specimens are deposited, according to available literature

On-line platforms and mobile phone apps	
Amphibian Species of the World: http://research.amnh.org/vz/herpetology/amphibia/	
AmphibiaWeb http://amphibiaweb.org/	
IUCN Red List http://www.iucnredlist.org/initiatives/amphibians	
Frogs of Southern Africa https://play.google.com/store/apps/details?id=com.coolideas.eproducts.safrogs	
Collections where amphibians from Angola are deposited	
Angola	Instituto Nacional para a Biodiversidade e áreas de Conservação, Ministério do Ambiente (INBAC/MINAMB)[a]
	Museu do Dundo (MD)
	Museu Nacional de História Natural (Luanda)[a]
	Southern African Science Service Centre for Climate Change and Adaptive Land Management (SASSCAL) / Instituto Superior de Ciências da Educação da Huíla (ISCED-Huíla)[a]
Austria	Imperial Natural History Museum (K.K. Museum) / Natural History Museum of Vienna (NHMW)
France	National Museum of Natural History (Paris) (MNHNP)
Germany	Berlin Zoological Museum (ZMB – Zoologisches Museum)[a]
	Forschungsinstitut und Naturmuseum Senckenberg (SMF)
	Hamburg Museum (ZMH – Zoologisches Museum für Hamburg)
	Senckenberg Natural History Collections Dresden (MTD – Museum für Tierkunde Dresden)[a]
Portugal	Centro de Zoologia do Instituto de Investigação Científica Tropical, Lisbon (IICT)
	Museu de História Natural na Universidade do Porto (MUP)
	Museu Nacional de História Natural e da Ciência, formerly Museu Bocage, Lisbon (MBL) – collections destroyed on the 1978 fire
South Africa	Ditsong National Museum of Natural History (formerly Transvaal Museum) (TMP), Pretoria
	Port Elizabeth Museum at Bayworld (PEM)[a]
	South African Institute for Aquatic Biodiversity (SAIAB)[a], Grahamstown
Spain	Estación Biológica de Doñana (EBD-CSIC), Sevilla
Switzerland	Musée de la Chaux-de-Fonds (LCFM)
	Museum d'histoire naturelle de la Ville de Genève (MHNG – Geneva Natural History Museum)
United Kingdom	Natural History Museum, London (NHMUK, formerly British Museum)
	Natural History Museum at Tring
United States of America	Carnegie Museum of Natural History (CM), Pittsburgh
	California Academy of Sciences (CAS), San Francisco[a]
	American Museum of Natural History (AMNH), New York[a]
	Academy of Natural Sciences of Philadelphia (ANSP), Philadelphia
	Field Museum of Natural History (FMNH), Chicago
	Museum of Comparative Zoology (MCZ), Harvard University, Cambridge, Massachusets
	National Museum of Natural History, Smithsonian Institution (NMNH), Washington, D.C.

[a]indicates the institutions containing specimens from recent (post-1975) surveys

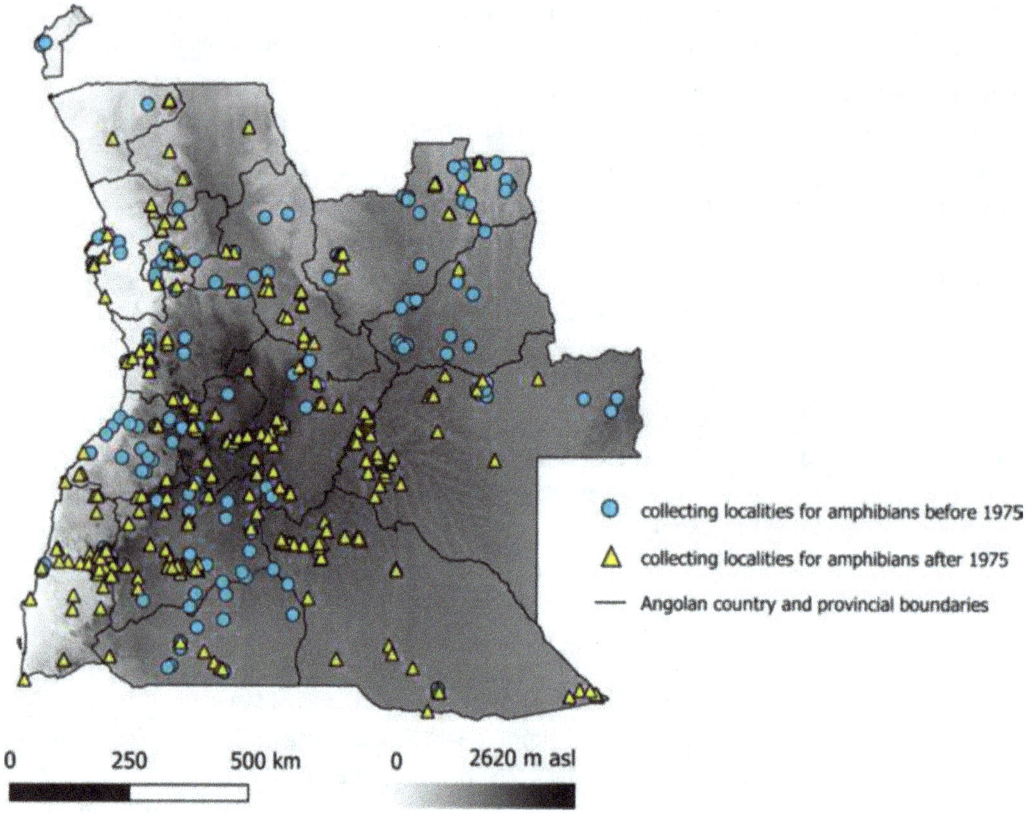

Fig. 12.1 Map with collecting localities for amphibians. Blue circles represent surveys before 1975 (based on literature records), and yellow triangles represent surveys after 1975 (literature records, localities from the 2009 and 2011 expeditions, SAREP and NGOWP trips to southeastern Angola, surveys in the scope of the SASSCAL Project and Fundação Kissama work, and Senckenberg Technical University, Dresden)

Records That Require Confirmation

A number of unconfirmed records for Angola require further investigation. These include *Leptopelis notatus* (Laurent 1964), *Ptychadena schillukorum* (Channing 2001), and *Kassina maculosa* (Ceríaco et al. 2014a). Monard (1937) noted one specimen of *Aubria subsigillata* from Caquindo that Perret (1996) could not confidently associate with any known genus, but that Channing (2001) considered to be *A. masako*. However, the latter is a closed-canopy forest species that is not expected to occur in southern Angola. The stated locality is either in error or the specimen deserves further investigation. *Phrynobatrachus dispar* was recorded from Cabinda by Peters (1877, as *Arthroleptis dispar*), but this species originates from São Tomé and Príncipe islands (Uyeda et al. 2007; Frost et al. 2018), and it is therefore likely that the Angolan record refers to another species. *Hyperolius nitidulus* was also recorded from Angola (Peters 1877), but was described from Nigeria and is currently considered to extend south only to Cameroon (Amiet 2012). *Hyperolius ocellatus* has been described both from Angola and Fernando Pó, but the type locality was later restricted to Fernando Pó (Perret 1975) which leaves Angolan specimens

Fig. 12.2 Representative of some of the families of frogs present in Angola. **1** Congulo Forest Tree Frog (*Leptopelis jordani*) from Congulo. **2** Dombe Pigmy Toad (*Poyntonophrynus dombensis*) from Meva. **3** Kuvango Kassina (*Kassina kuvangensis*) from Cuanavale River Source. **4** Spot-bellied Grass Frog (*Ptychadena subpunctata*) from Cameia National Park. **5** Marbled Rubber Frog (*Phrynomantis annectens*) from Meva. **6** Marbled Snout-Burrower (*Hemisus marmoratus*) from Bicuar National Park. **7** Angolan Reed Frog (*Hyperolius* cf. *parallelus*) from Quilengues. **8** Rain frog (*Breviceps* sp. nov.) from Cuando River Source. (Photo credits – N Baptista: **4**,**6**,**7**; P Vaz Pinto: **1**,**2**,**5**; W Conradie: **3**,**8**)

with no appliable name. *Phrynobatrachus auritus* was recorded from Cabinda by Peters (1877) as *Arthroleptis plicatus*, but the validity of this synonymy for Cabinda requires further study. A number of species recorded from Angola were presumably misidentified as the currently known species' range does not include Angola, including: *Phrynobatrachus minutus* recorded by Ruas (1996), but which is now restricted to Ethiopia; *Hyperolius microps* recorded by Bocage (1895) and Monard (1937), now restricted to Eastern Africa; *Hyperolius multifasciatus* Ahl 1931 which was included provisionally by Monard (1937), but placed in the synonymy of *H. kivuensis* Ahl 1931, by Pickersgill (2007); and *Xenopus calcaratus* recorded by Peters (1877), but now restricted to western Africa. Specimens of *Ptychadena* cf. *aequiplicata*, which occurs approximately 50 km from the Cabinda Enclave (Nagy et al. 2013), exist in the AMNH collection, but their identity requires confirmation (Ernst pers. comm.).

Species Likely to Occur in Angola But Not Yet Recorded

The ranges of many species occurring in adjacent countries (Namibia, Zambia and the Democratic Republic of the Congo, DRC) are likely to extend into Angola and are discussed below. A significant example is that of caecilians (Order Gymnophiona), which are known from the Congo Basin but have not been recorded in Angola, including Cabinda. Species that have been recorded close to the Angolan border and that are likely to occur in the country are listed below.

Caecilians (Gymnophiona)

The Gaboon Caecilian (*Geotrypetes seraphini* (Duméril, 1859)) and the Congo Caecilian (*Herpele squalostoma* (Stutchbury, 1836)) have both been recorded from the extreme western DRC, in Mayombe, River Minkala, Vemba-Minionzi, around Kidima, around 40 km from the Angolan border (Scheinberg and Fong 2017), and are likely to occur in this poorly known region.

Frogs and Toads (Anura)

Arthroleptidae

Cryptic Tree Frog (*Leptopelis parbocagii* Poynton and Broadley, 1987). This tree frog occurs in northern Mwinilunga district, northwest Zambia, less than 50 km from Cazombo, eastern Angola (Schiøtz and Van Daele 2003), and may occur on the Angolan side of the border.

Breviciptidae

Power's Rain Frog (*Breviceps poweri* Parker, 1934). This rainfrog was found in southwestern Zambia, less than 100 km from the Angolan border (Pietersen et al. 2017), and can be expected in Angola.

Bufonidae

Beira Pygmy Toad (*Poyntonophrynus beiranus* (Loveridge, 1932)). Recorded from southwestern Zambia near the Angolan border (Poynton and Broadley 1991) and may occur in Angola.

Northern Pygmy Toad (*Poyntonophrynus fenoulheti* (Hewitt and Methuen, 1913)). This pigmy toad is recorded from Caprivi Strip in northeastern Namibia (Channing and Griffin 1993) and southwestern Zambia (Pietersen et al. 2017), less than 100 km from the Angolan border, and its presence is expected in southeastern Angola.

Hemisotidae

Perret's Snout-burrower (*Hemisus perreti* Laurent, 1972). Recorded in Singa Mbamba, Mayumbe (Royal Museum for Central Africa 2017) and in the region of Kipanzu, Tshela (MHNG 2017) both in the Bas-Congo Province, DRC, in close proximity to the Cabinda enclave, and it is therefore expected to occur in Cabinda.

Barotse Snout-burrower (*Hemisus barotseensis* Channing and Broadley 2002). Described from the Barotse floodplain near Mongu, in southwestern Zambia, 120 km east of the Angolan border, but may occur in suitable floodplain habitat along the eastern Zambezi drainage.

Hyperoliidae

Foulassi Spiny Reed Frog (*Afrixalus paradorsalis* (Perret, 1960)). This hyperoliid was found in Luango-Nzambi, DRC, around 50 km from the Cabinda Enclave (Nagy et al. 2013) and is likely to occur in Angola.

Rainforest Reed Frog (*Hyperolius tuberculatus* (Mocquard, 1897)). Also found in Luango-Nzambi, DRC (Nagy et al. 2013) and likely to occur at least in Cabinda.

Kachalola Reed Frog (*Hyperolius kachalolae* Schiøtz, 1975). Known from Mwinilunga district, in northwestern Zambia (Schiøtz and Van Daele 2003), less than 50 km from the eastern Angolan border.

Hyperolius major Laurent, 1957. This reed frog occurs in Mwinilunga district, in northwestern Zambia, less than 50 km of Cazombo (Poynton and Broadley 1991; Schiøtz and Van Daele 2003), eastern Angola.

Phrynobatrachidae

Golden Puddle Frog (*Phrynobatrachus auritus* Boulenger, 1900). This species of puddle frog was found in Luki, DRC, only 20 km north of Angola (Nagy et al. 2013) and may occur in the country.

Horned Puddle Frog (*Phrynobatrachus* sp. aff. *cornutus* (Boulenger, 1906)), also found in Luki, DRC (Nagy et al. 2013) and likely to occur in Angola.

Pipidae

Gaboon Dwarf Clawed Frog (*Hymenochirus* sp. aff. *feae* Boulenger 1906), and *Xenopus (Silurana)* sp. This Dwarf Clawed Frog and an unidentified species of clawed frog were found in Luki, DRC, 20 km north of the Angolan border (Nagy et al. 2013) and are expected in Angolan territory.

Fraser's Clawed Frog (*Xenopus* cf. *fraseri* Boulenger, 1905). This clawed frog has been found in Luki, DRC, 20 km north of the Angolan border (Nagy et al. 2013) and is expected in Angola, although the records of these species are considered to need a critical revision (Ernst et al. 2015).

Common Platanna (*Xenopus laevis* (Daudin, 1802)). Recorded in Luki, DRC, 20 km north of the Angolan border, and in Tsumba-Kituti (Nagy et al. 2013) and might occur in Angola.

Ptychadenidae

Dark Grass Frog (*Ptychadena obscura* (Schmidt and Inger, 1959)). This species has been recorded in the Ikelenge pedicle, northern Mwinilunga district, northwestern Zambia, in close proximity to the Angolan eastern border (Poynton and Broadley 1991).

Mapacha Grass Frog (*Ptychadena* cf. *mapacha* Channing, 1993). This Grass Frog is described from the Caprivi Strip in Namibia, near southeastern Angola (Channing 1993). It has also been recorded about 80 km east of Rundu (Haacke 1999), near Vicota, around 30 km south of the Angolan border (Ceríaco et al. 2016a), and in southwestern Zambia (Pietersen et al. 2017). Conradie et al. (2016) collected a series of *Ptychadena* at Jamba provisionally assigned to *P.* cf. *mossambica*, but mentioned that the specimens might be referable to *P. mapacha*. All these records suggest that this species may occur in southeastern Angola.

Perret's Grass Frog (*Ptychadena* cf. *perreti* Guibé and Lamotte, 1958). This grass frog was found in Nkamuna, in the Bas-Congo province of DRC, near Angola (Nagy et al. 2013).

Pyxicephalidae

Boettger's Dainty Frog (*Cacosternum boettgeri* (Boulenger, 1882)). This species has been recorded near the Angolan border in northern Namibia in Caprivi Strip and in Omusati province (Channing and Griffin 1993), and Southern Province in Zambia (Broadley 1971) and may occur in Angolan territory.

Knocking Sand Frog (*Tomopterna krugerensis* Passmore and Carruthers, 1975). This frog has been recorded in northern Namibia close to the Angolan border (Channing and Griffin 1993).

Tandy's Sand Frog (*Tomopterna tandyi* Channing and Bogart, 1996). Recorded from northern Namibia near the Angolan border (Coetzer 2017), and may occur in southwestern Angola. A recent fing of *Tomopterna* has been made in Bicuar National Park and its identification as *T. tandyi* is under discussion (Baptista et al. in press).

Rhacophoridae

Southern Foam Nest Frog (*Chiromantis xerampelina* Peters, 1854). Recorded from Caprivi Strip in northern Namibia (Channing and Griffin 1993) and from south-eastern Zambia (Broadley 1971; Pietersen et al. 2017), and therefore expected in southeastern Angola. It is recorded from southwestern Angola (Schiøtz 1999), but the original source of this record is unknown. This odd distribution record requires confirmation.

Western Foam-nest Tree Frog (*Chiromantis rufescens* (Günther, 1869)). This species is known from near Boma, close to the northern bank of the Congo River (Royal Belgian Institute of Natural Sciences 2017) and may occur in Angola.

According to Frost (2018), based on distribution and species' habitat affinities, around 20 additional species are expected in the country, mostly from the northern forests and expected in the Cabinda enclave in northern Angola. These are generalist assumptions that do not necessarily take into account actual proximity to the Angolan border. These include arthroleptids Silver Long-fingered Frog (*Cardioglossa leucomystax* (Boulenger, 1903)), Kala Forest Treefrog (*Leptopelis aubryioides* (Andersson, 1907)), Victoria Forest Treefrog (*Leptopelis boulengeri* (Werner, 1898)), Red Treefrog (*Leptopelis rufus* Reichenow, 1874)); bufonids [High Tropical Forest Toad (*Sclerophrys latifrons* (Boulenger, 1900))], hyperoliids [African Wart Frog (*Acanthixalus spinosus* (Buchholz and Peters, 1875)); Greshoff's Wax Frog (*Cryptothylax greshoffii* (Schilthuis 1889) with an unconfirmed record from northwestern Angola (Schiøtz 1999)), Olive Striped Frog (*Phlyctimantis leonardi* (Boulenger, 1906), ptychadenids [Savanna Grass Frog (*Ptychadena superciliaris* (Günther, 1858)], and pipids [Western Dwarf Clawed Frog (*Hymenochirus curtipes* Noble, 1924), False Fraser's Clawed Frog (*Xenopus allofraseri* Evans, Carter, Greenbaum, et al., 2015)].

Hidden Among the Unknown – Angolan Tadpoles

An important and often neglected component of studying amphibians is knowledge of their larvae. Unlike adult frogs, whose activity is quite dependent on appropriate weather conditions, breeding season, and nocturnal activity for most species, tadpoles can be easily found in water bodies, during the day, and throughout the year in some cases. The study of tadpoles includes not only morphology, but also microhabitat requirements, ecology, behaviour, feeding habits, predator-prey interactions, etc. Although they look similar at first glance, tadpole morphology usually allows the identification to genus, and a more precise analysis can often lead to species identification.

Early literature regarding southern African tadpoles often covers species occurring in Angola (Van Dijk 1966, 1971). Channing et al. (2012) provide a comprehensive review of the knowledge about African tadpoles with keys to the identification of genera and detailed description of species. Given the poorly known status of Angolan amphibians, it is not surprising that very little is known about Angolan tadpoles. Out of the 99 Angolan species that have tadpoles (i. e., *Breviceps* and *Arthroleptis* genera not included), the tadpoles of only 44 species have been described, and only those of *Ptychadena porosissima* (Channing et al. 2012), *Amietia angolensis* (Channing et al. 2016) and of the recent discoveries of the endemic *Hyperolius chelaensis* (Conradie et al. 2012), *H. cinereus* and *H. raymondi* (Conradie et al. 2013) are based on Angolan material. A recent description of *Leptopelis anchietae* tadpoles is also based on Angolan material (Channing et al. 2012), but it was not encountered with adult specimens, and was based on the association with the first description of that tadpole (Lamotte and Perret 1961), which was based on a specimen from Cameroon that may involve another species. A list of the Angolan frogs with undescribed tadpoles (Table 12.2) includes some of the more common local species.

Comments on Selected Groups

As a consequence of the current poor knowledge of Angola's amphibians, the taxonomic status of many species in the checklist remains unresolved. Some of these are discussed in this section, as well as recent discoveries from ongoing studies.

Species Complexes and Species with Unclear Boundaries

Some morphologically similar species display variation in calls or habitat and are considered to form a complex of closely-related species, and the resolution of their taxonomic status and distribution requires comprehensive investigation. This is exemplified by the *Hyperolius marmoratus/viridiflavus* complex in Africa, in which 15 names from Angola have been synonymised *(Hyperolius cinctiventris, H. decoratus, H. huillensis, H. insignis, H. marungaensis, H. microstictus, H. pliciferus, H. vermiculatus, Rappia cinctiventris, R. marmorata marginata, R. m. paralella, R. m. variegata, R. plicifera, R. toulsonii, H. m. alborufus)*. *Hyperolius parallelus* is closely related to this complex, and has several Angolan taxa in its synonymy (*H. angolensis, H. marmoratus* var. *angolensis, H. erythromelanus, H. toulsonii, Rappia marmorata huillensis, R. m. insignis, R. m. taeniolata)*. Other difficult groups are the *Hyperolius platyceps* complex, with four names currently subsumed within it (*Hyperolius angolanus, Rappia platyceps* var. *angolensis,*

Table 12.2 Angolan frog species with undescribed tadpoles

Leptopelis bocagii (Günther, 1865)	*Hyperolius platyceps* (Boulenger, 1900)
Leptopelis cynnamomeus (Bocage, 1893)	*Hyperolius polli* (Laurent, 1943)
Leptopelis jordani (Parker, 1936)	*Hyperolius protchei* (Rochebrune, 1885)
Leptopelis marginatus (Bocage, 1895)	*Hyperolius quinquevittatus* (Bocage, 1866)
Leptopelis parvus (Schmidt and Inger, 1959)	*Hyperolius rhizophilus* (Rochebrune, 1885)
Mertensophryne melanopleura (Schmidt and Inger, 1959)	*Hyperolius steindachneri* (Bocage, 1866) *Hyperolius vilhenai* (Laurent, 1964)
Mertensophryne mocquardi (Angel, 1924)	*Kassinula wittei* (Laurent, 1940)
Poyntonophrynus grandisonae (Poynton and Haacke, 1993)	*Phrynomantis affinis* (Boulenger, 1901) *Phrynobatrachus brevipalmatus* (Ahl, 1925)
Poyntonophrynus kavangensis (Poynton and Broadley, 1988)	*Phrynobatrachus cryptotis* (Schmidt and Inger, 1959)
Poyntonophrynus pachnodes. (Ceríaco, Marques, Bandeira et al. 2018a)	*Phrynobatrachus parvulus* (Boulenger, 1905) *Xenopus andrei* (Loumont, 1983)
Sclerophrys buchneri (Peters, 1882)	*Xenopus petersii* (Bocage, 1895)
Afrixalus osorioi (Ferreira, 1906) *Afrixalus fulvovittatus* (Cope, 1861)	*Xenopus epitropicalis* (Fischberg, Colombelli, and Picard, 1982)
Afrixalus wittei (Laurent, 1941).	*Hildebrandtia ornatissima* (Bocage, 1879)
Hyperolius adspersus (Peters, 1877)	*Ptychadena ansorgii* (Boulenger, 1905)
Hyperolius benguellensis (Bocage, 1893)	*Ptychadena bunoderma* (Boulenger, 1907)
Hyperolius bicolor (Ahl, 1931)	*Ptychadena grandisonae* (Laurent, 1954)
Hyperolius bocagei (Steindachner, 1867)	*Ptychadena guibei* (Laurent, 1954)
Hyperolius cinnamomeoventris (Bocage, 1866)	*Ptychadena keilingi* (Monard, 1937)
Hyperolius fasciatus (Ferreira, 1906)	*Ptychadena perplicata* (Laurent, 1964)
Hyperolius ferreirai (Noble, 1924)	*Ptychadena taenioscelis* (Laurent, 1954)
Hyperolius fuscigula (Bocage, 1866)	*Ptychadena upembae* (Schmidt and Inger, 1959)
Hyperolius gularis Ahl, 1931	*Ptychadena uzungwensis* (Loveridge, 1932)
Hyperolius langi (Noble, 1924) *Hyperolius lucani* (Rochebrune, 1885)	*Tomopterna damarensis* (Dawood and Channing, 2002)
Hyperolius maestus (Rochebrune, 1885)	*Tomopterna tuberculosa* (Boulenger, 1882)
Hyperolius nobrei (Ferreira, 1906) *Hyperolius parallelus* (Günther, 1858)	*Amnirana parkeriana* (Mertens, 1938)

Hyperolius fasciatus, Hyperolius ferreirai (originally *Rappia bivittata*)), and the super-cryptic *Hyperolius nasutus* complex. Currently this is represented in Angola by at least four species *(H. adspersus, H. benguellensis, H. dartevellei, H. nasutus)* (Channing et al. 2013) and additional names that have been synonymised *(H. punctulatus, Rappia punctulata)* (Channing et al. 2013) or not assigned to any known species occurring in Angola *(H. microps)*.

Typical toads are another problematic group. Formerly known as *Bufo,* which was cosmopolitan in distribution and included the majority of bufonids, the genus was partitioned with African typical toads transferred to *Amietophrynus* (Frost et al. 2006), and more recently renamed in the reinstated genus *Sclerophrys* (Poynton et al. 2016). Seven species of typical toad occur in Angola (see Table 12.2). The mysterious *S. buchneri,* known only from the holotype from northeastern Angola, is considered as a valid species (Frost 2018), but synonymy with *S. funerea* has been suggested and requires further studies (Tandy and Keith 1972). Apart from *S. lemairii* which is easily distinguishable morphologically from the remaining species, distinction between the other *Sclerophrys* is difficult, even between the most common species. Hybridisation between *Sclerophrys* species has been documented and discussed (Guttman 1967; Passmore 1972; Cunningham and Cherry 2004) and may further complicate identification. The red coloration of the interior thigh and parotid gland development are features commonly used to distinguish the often sympatric *S. pusilla*, *S. gutturalis* and *S. regularis* (Du Preez and Carruthers 2017), but do not distinguish these species in Angola. It is likely that cryptic diversity exists, and understanding of the genus and delimitation of species boundaries requires an integrative approach with comprehensive surveys, analysis of advertisement calls and genetic studies.

Grass frogs, *Ptychadena* spp., are a challenging genus. At least 15 species of this specious genus are represented in Angola (Appendix). *P. mascareniensis*, a large species complex widespread in Africa and Madagascar, has been recently partitioned (Dehling and Sinsch 2013b) with *Ptychadena nilotica* in much of continental Africa, including Angola (Zimkus et al. 2017). Difficulties in distinguishing *Ptychadena* species have been discussed (Poynton and Broadley 1985b; Dehling and Sinsch 2013a, b), although coloration features such as triangular patch on the head, pattern of the interior thigh (Poynton 1970) and several morphometric and morphological features enable species identification (Dehling and Sinsch 2013a, b). Species distinction in Angola is not clear, and in a recent study as many as six different species of *Ptychadena* were found in the same region (Conradie et al. 2016).

Rainfrogs in Angola are known from a single species, *Breviceps adspersus*. However, analysis of material from Angola and adjacent regions has revealed that the Angolan form has features of *B. mossambicus* and may indicate an undescribed Angolan species (Poynton and Broadley 1985a, 1991).

Groups that remain not fully understood such as *Phrynobatrachus* (Zimkus et al. 2010), *Xenopus* (Furman et al. 2015) and *Amnirana* (Jongsma et al. 2018), all have species widespread in Africa with type localities from Angola, and the resolution of their taxonomy depends on detailed studies in Angola.

Species Synonymised with No Clear Justification

A number of putative Angolan species currently placed in synonymy require reassessment as they may represent hidden diversity currently placed under a different name. Cases are mentioned in the previous section, especially in the *Hyperolius* genus. Other examples of this include the placement of *Hylambates (=Leptopelis) angolensis* in the synonymy of *Leptopelis bocagii.* This resulted from comparison between adult and juvenile specimens (Perret 1976) that may not be comparable. *Hylambates bocagei* var. *leucopunctata* Ferreira 1904, has also been placed in the synonymy of *Leptopelis bocagii* (Ceríaco et al. 2014b) and this also requires further investigation as the well developed finger pads in the type specimen of *H. b. leucopunctata* suggests an arboreal habit, very different from the ground-dwelling habits of *L. bocagii,* which lacks pads on fingers or toes.

Species with Questionable Distributions

Some species described from Angola have widespread distributions throughout Africa and inhabit diverse habitats, suggesting that cryptic diversity may be involved (see examples in Endemism section, below). The classic example of this is the Common River Frog, *Amietia angolensis*, which was considered widespread in the continent, but which was discovered to be in fact a complex of cryptic species, with true *A. angolensis* being restricted to Angola (Channing and Baptista 2013). Another potential example is *Afrixalus osorioi,* which was described from western Angola and remains known in the country only from the type locality, whereas the closest other records are in DRC, nearly 1000 km away from the type locality. Other examples include *Ptychadena porosissima, Leptopelis cynnamomeus, L. bocagii, Hyperolius bocagei,* and highlight the earlier comments that study of Angolan amphibians is crucial for solving many problems in African amphibian taxonomy.

Recent Discoveries and Ongoing Studies

The endemic Anchieta's Treefrog, *Leptopelis anchietae* and Congulo Forest Treefrog, *Leptopelis jordani* have been rediscovered in the Angolan escarpment (Baptista et al. 2017), and together with other frogs belonging to the genus *Kassina, Arthroleptis* and *Amnirana* found in the region, their conservation and taxonomic status are being investigated (Baptista et al. in prep.). Further to this, ongoing studies (Baptista et al. in prep.) are assessing: a candidate new species of *Schismaderma*; the taxonomic status of *Hildebrandtia ornatissima* from the Angolan central plateau, previously discussed by Boulenger (1919); the status of *Hyperolius punctulatus* from the Cuanza River (currently in the synonymy of *Hyperolius nasutus*); and the status of various populations of morphologically distinct pygmy toads that cannot be assigned to known *Poyntonophrynus* species. During the 2011 expedition to Lagoa Carumbo, a large white-lipped frog was morphologically assigned to the *Amnirana lepus* group (Branch and Conradie 2015). This assigment has been confirmed in a phylogeny of the genus (Jongsma et al. 2018), and further studies are underway to address the taxonomical status of the Angolan population (Conradie pers. comm.). On the SAREP (2012/3) and the NGOWP (2016/7) expeditions to southeastern Angola, numerous candidate new species were discovered, in the genera *Phrynobatrachus*, *Breviceps* and *Amnirana,* and are currently under investigation (Conradie pers. comm.). The new country records of *Kassinula wittei* and *Leptopelis* cf. *parvus* are being studied to determine if they conform to the nominal forms from northern Zambia and southern DRC, respectively (Conradie pers. comm.). During recent independent surveys conducted in the northern Angolan provinces of Uíge (Ernst et al. 2014, 2015) and Zaire (Vaz Pinto and Baptista Unpublished Data), two different *Alexteroon* spp. were discovered. The taxonomic status of these, the first Angolan records for this poorly-known hyperoliid genus, are under investigation with the Uíge species tentatively assigned to the nominal species *A. hypsiphonus,* whilst the Zaire discovery has affinities to *A. obstetricans*. The material awaits formal taxonomic assignment pending analysis of type material.

Biogeography

Angola is one of the most biogeographically rich countries in Africa (Huntley 1974, 2019). Geomorphologically, the country can be divided into various regions, including the western lowlands of the Coast Belt, the Transition Zone which includes the escarpment, the Marginal Mountain Chain, the Old (Highland) Plateau, also known as central plateau, which progressively decreases in altitude to the east, where the Congo Basin in the north and the Zambezi-Cubango Basin in the south are located (Huntley 1974). Each of these regions have several biome associations, with habitats ranging from the tropical rainforests on the Maiombe region in the north, to the Namib Desert in the south, one of the oldest deserts in the world (Huntley 1974, 2019). This complexity is reflected in the country's diverse fauna.

The difficulties in establishing clear biogeographic regions for amphibians is demonstrated by Poynton and Broadley (1991), in their thorough analysis of the biogeography of the Zambezian amphibians. For Angolan amphibians, which are much more poorly known, this difficulty is immensely increased. The biogeography of Angolan amphibians can only be assessed after major taxonomic issues are resolved, which in some cases requires the revision of entire genera (Cei 1977; Blanc and Frétey 2000). In early studies of the Angolan herpetofauna, several attempts were made to group species according to the distributions known at the time, and these will be summarised below.

Bocage (1895a) made the first grouping, distinguishing a northern and southern region, each divided into coastal, intermediate and high-altitude zones, and listing species occurring in each block. Monard (1937) used humidity to explain the higher diversity of amphibian species in central Angola (a high-rainfall region), compared to the south. He divided Angolan amphibians into four groups: (i) pan-African species (4% of the country's species; such as *Rana mascareniensis (=Ptychadena nilotica)*, and *Bufo (=Sclerophrys) regularis;* (ii) southern species (10%) which reached their northern limit in Angola, such as *Pyxicephalus adspersus*; (iii) tropical species (40%), from western, central and eastern Africa, highlighting the central African tropics as the most significant influence, and including *Rana (=Amnirana) albolabris* and *Rana (=Hoplobatrachus) occipitalis*; and (iv) endemic species (46%), most of which are no longer considered endemic (see Endemism section).

Based on the species known from Angola at the time, Cei (1977) organised Angolan amphibians in three questionable groups, each with affinities to different habitats and regions: (i) the northern and northeastern forests and savannas, (ii) the plateau, and (iii) the arid and semi-arid regions of the coast and of the south, providing a map to delineate those areas. The first area is wide, with northern and northeastern limits in the Congo, Cuanza and Cassai rivers (in Zaire, Uíge, Malanje and Lunda-Norte), and extending to the southeast through Moxico and Cuando Cubango. Examples of species within this group are *Arthroleptis carquejai* and *Hyperolius steindachneri.* The second region corresponds to the south of Congo and Cuango rivers and comprises the southern tropics: Cuanza-Norte, Cuanza-Sul, Huambo, Bié, Malanje and Huíla provinces. Characteristic species in this group include *Hildebrandtia ornatissima, Hyperolius cinereus, Hyperolius quinquevittatus, Leptopelis anchietae*. The third and southernmost region comprises the arid sections of Benguela, Namibe, and Cunene provinces. The fauna on this group is related to that of the Namib, Kalahari, and Namaqualand regions, and can be exemplified by *Pyxicephalus adspersus* and *Poyntonophrynus dombensis*.

Surprisingly, the Great Escarpment of Angola has not been considered in any of these studies. This escarpment is part of a much larger geomorphological unit that dominates the African subcontinent and extends into western Angola, where it acts as a barrier between the dry coastline and the inland plateau. Due to its climatic and topographic peculiarities, it promotes isolation and thus speciation (Huntley 1974). It is a well-documented center of endemism for birds (Hall 1960), and although the escarpment herpetofauna is poorly understood, its endemism potential for herpetofauna has been highlighted (Laurent 1964, Clark et al. 2011, Baptista et al. 2018, Branch et al. 2019), and endemic amphibian species are known from the region *(Leptopelis jordani, L. marginatus, Amnirana parkeriana* and *Hyperolius chelaensis)*. Bordering the Angolan escarpment to the east, the highlands of the ancient massif include patches of Afromontane forest. These consist of islands of relic cool moist Afromontane forest with great biogeographic interest (Huntley 1974), and also potential for endemism.

Inland to the escarpment zone, the plateau is broadly dominated by miombo woodlands, and its fauna often has influences from adjacent regions. Boundaries between regions are not always clear or well understood. Some of these uncertainties have been mentioned in early studies and still require explanation. Hellmich (1957b) referred to the difficulty in establishing geographical limits between the moist forests of the north and the central plateau. An example of this is the penetration of forest species in association with riverine habitats along the northern rivers of Angola. He also noted that faunal boundaries between the slopes of the eastern plateau and the flatlands between Cassai and Cuando were not clear, with the presence of 'pockets' of herpetological elements typical of the south on the central plateau. Laurent (1964) referred to the known affinities between the species of Katanga, in southeastern DRC, with the species of the Lundas and Moxico in Angola.

All of these early biogeographic regions and the species assigned to them need to be re-evaluated with updated taxonomy, accurate species distributions, and in association with the study of phylogenetic relationships among the various amphibian families and genera occurring in Angola. The confirmation of ancestral relationships within these groups is a prerequisite for testing hypotheses about the timing and environmental correlates of amphibian movement and speciation across the Angolan landscape.

Endemism

The originality of the Angolan amphibians due to the richness of endemic species has been highlighted (Blanc and Frétey 2000). Angola's more unique amphibians are also the most poorly known. There are 21 species of amphibians endemic to Angola, of which about 75% are only known from the type locality or type specimens (Table 12.3). Many have not been found for decades, and in some cases for over 100 years. Most of these species are classified as Data Deficient in the IUCN Red List (IUCN 2017).

A number of endemic taxa have been mentioned in the literature but still await formal description: *Hyperolius* sp. I, *Hyperolius* sp. II, *Hyperolius* sp. III (Monard 1937), possibly unknown genus (Perret 1996), and as the taxonomic studies on Angolan amphibians progress, more endemic species will very likely be discovered. In contrast, many early species that were considered endemic have now been relegated to the synonymy of wide-ranging species. Monard (1937), for instance, considered nearly half (46%) of the 80 Angolan species he considered to occur in the country as endemic. However, of the 37 endemic species he identified, only eight are still recognised. Sixteen of these former 'endemics' have been synonymised with other species; e.g. *Leptopelis angolensis (= L. bocagii)*, *Rana buneli (= Ptychadena bunoderma)*, *Hyperolius seabrai (= H. bocagei)*, *Hyperolius angolanus, H. ferreirai, H. fasciatus* (all *=Hyperolius platyceps*), *H. pliciferus, H. vermiculatus, H. marungaensis* (all *=Hyperolus marmoratus*), *H. angolensis, H. erythromelanus, H. toulsonii* (all *=Hyperolius parallelus*), *H. punctulatus (=Hyperolius nasutus)*, *Rana myotympanum (=Hildebrandtia ornatissima)*, *Rana cacondana* and *R. signata (=Tomopterna tuberculosa)*. Many of these synonymies have poor justification, and whilst some names may reflect regional variation, others

Table 12.3 List of amphibian species endemic to Angola, with IUCN Red List Category (*LC* least concern, *DD* data deficient, *N/A* not assessed), and marked (X) when known only from the type locality

Common name	Scientific name	IUCN	TYPE
Angola River frog	*Amietia angolensis* (Bocage, 1866)	LC	
Parker's white-lipped frog	*Amnirana parkeriana* (Mertens, 1938)	DD	X
Cambondo squeaker	*Arthroleptis carquejai* (Ferreira, 1906)	DD	X
Angola ornate frog	*Hildebrandtia ornatissima* (Bocage, 1879)	DD	
Two-colored reed frog	*Hyperolius bicolor* (Ahl, 1931)	DD	X
Chela Mountain Reed Frog	*Hyperolius chelaensis* (Conradie et al., 2012)	N/A	X
Monard's Reed Frog	*Hyperolius cinereus* (Monard, 1937)	LC	
Brown-throated Reed Frog	*Hyperolius fuscigula* (Bocage, 1866)	DD	X
Loanda Reed Frog	*Hyperolius gularis* (Ahl, 1931)	DD	X
Landana Reed Frog	*Hyperolius lucani* (Rochebrune, 1885)	DD	X
Cabinda Reed Frog	*Hyperolius maestus* (Rochebrune, 1885)	DD	X
Nobre's Reed Frog	*Hyperolius nobrei* (Ferreira, 1906)	N/A	X
Rochebrune's Reed Frog	*Hyperolius protchei* (Rochebrune, 1885)	DD	X
Raymond's Reed Frog	*Hyperolius raymondi* (Conradie et al., 2013)	N/A	
African Reed Frog	*Hyperolius rhizophilus* (Rochebrune, 1885)	DD	X
Luita River Reed Frog	*Hyperolius vilhenai* (Laurent, 1964)	DD	X
Congulo Forest Treefrog	*Leptopelis jordani* (Parker, 1936)	DD	X
Quissange Forest Treefrog	*Leptopelis marginatus* (Bocage, 1895)	DD	X
Ahl's Puddle Frog	*Phrynobatrachus brevipalmatus* (Ahl, 1925)	DD	X
Grandison's Pygmy Toad	*Poyntonophrynus grandisonae* (Poynton and Haacke, 1993)	DD	X
Serra da Neve Pygmy Toad	*Poyntonophrynus pachnodes* (Ceríaco, Marques, Bandeira et al., 2018a)	N/A	X

Taxonomy follows Frost (2018)

referred to species found in other countries may not be conspecific (see Comments on selected groups). All deserve careful re-examination.

At least four species (*Leptopelis marginatus, L. jordani, Amnirana parkeriana,* and *Hyperolius chelaensis*) are escarpment-endemics, and others are plateau-endemics (*Hildebrandtia ornatissima, H. cinereus*). However, in order to effectively protect Angolan endemic amphibians and their habitats, further studies are needed to reveal the relations between endemic amphibians and particular habitat, and also the importance of other potential areas of endemism (e.g. relic Afromontane forest patches, isolated mountains such as Serra da Neve, the Angolan escarpment).

Directions for Future Research in Angola

Detailed species lists for a country are an essential baseline tool for understanding biodiversity, its distribution and conservation status. The confusing status of Angolan amphibian taxonomy has been discussed in previous sections and demonstrates how studying taxonomy forms the bedrock for resolving the many pressing questions regarding Angolan amphibian conservation and biology.

A critical first taxonomic step is to revisit the type localities of all the species described from the country to obtain new topotypical material. This is particularly important for the 15 species described by Bocage *(Amietia angolensis, Hyperolius benguellensis, H. cinnamomeoventris, H. fuscigula, H. quinquevittatus, H. steindachneri, Ptychadena anchietae, P. subpunctata, Sclerophrys funerea, Leptopelis anchietae, L. cynnamomeus, L. marginatus, Hildebrandtia ornatissima, Poyntonophrynus dombensis, Xenopus petersii)*, for which many of the type specimens were lost in the fire that destroyed the collections of the Natural History Museum of Lisbon, and for which the original descriptions are the only available source of information. Possibly also lost are the type specimens of several Angolan endemics described by Rochebrune *(Hyperolius lucani, H. maestus, H. protchei, H. rhizophilus)* (Frost 2018), which have very vague descriptions. For many species, neotypes may need to be designated in order to stabilise their taxonomy. Integrative taxonomic studies, including analysis of genetic material, advertisement calls, adult and larval morphology, habitat associations and natural history are crucial to bring Angolan studies into the new millennium.

Many regions of Angola have never been surveyed for amphibians (see Fig. 12.1). Surveying these areas would greatly improve understanding of amphibian distributions, habitat associations and relative abundance, but are also critical for making assessments on their conservation status in terms of IUCN criteria. Priority areas include the northwestern provinces (Uíge and Zaire), the extensive wetlands of Moxico, the escarpment and the adjacent Afromontane forest patches that are rich in endemic birds (Hall 1960), other vertebrates (Crawford-Cabral 1966; Clark et al. 2011) and also probably amphibians, and for which the urgent need of studies has been highlighted (Laurent 1964; Clark et al. 2011).

The controversial frog from Caquindo (Perret 1996) for which genus assignment lacks consensus (see *Records that require confirmation*), still has to be recollected and its true affinities resolved. This could enrich Angolan herpetology possibly with a new endemic genus. This begs the question – how much remains to be discovered about Angolan amphibians? It also shows how the analysis of extant collections can contribute significantly to the knowledge of the country's fauna. Collections that remain to be studied include those from the Rohan-Chabot Mission, the Vernay Angola Expedition, and the Leptopelinae and Hyperoliidae from the Portuguese Mission of Apiarian Studies of the Overseas.

Another important step to furthering amphibian knowledge is studying the biology of individual species. Some studies are available for iconic species such as the Dombe Pigmy toad *Poyntonophrynus dombensis* (Channing and Vences 1999), based on individuals from Namibia, and Lemaire's toad *Sclerophrys lemairii,* the first study of this kind made in Angolan territory (Conradie and Bills 2017). However, this is still missing for many species, and understanding their natural history, reproduction strategies, breeding sites, breeding seasons, behaviour, habitat and microhabitat requirements, both for adults and tadpole stages, are key for an effective planning of species conservation. All of this is even more relevant for the extremely poorly known Angolan endemics.

Conservation-driven studies about Angolan amphibians require awareness of potential threats to biodiversity, particularly those resulting from habitat loss and climate change. Habitat degradation as a result of exploitation of natural resources and associated with industrialisation have increased dramatically in Angola in recent decades and will affect amphibians. The implementation of monitoring programmes are crucial for documenting and understanding this relation. Research about the appearance and effect of global amphibian diseases such as the chytrid fungus *(Batrachochytrium dendrobatidis)*, viruses *(Ranavirus* spp.*)*, and other pathogens, are lacking in Angola, even though they are threatening amphibians around the world and are reported from neighbouring countries (Greenbaum et al. 2014).

The study of Angolan amphibians is a broad and important component of biodiversity studies, for which many baseline questions remain unanswered, and exciting discoveries are still to be made. This becomes more evident when confronted with the fact that Angolan fauna is among the least studied in Africa. Increasing public awareness about amphibians and their importance is necessary for their conservation, and requires developing local knowledge and expertise, as well as constructing functional amphibian collections in national archives. These are essential steps for understanding and protecting this rich, diverse and ecologically important group. This is even more urgent in an era where an "amphibian decline crisis" is happening around the world (Beebee and Griffiths 2005), and where this decline is known to have major consequences in ecosystem function (Whiles et al. 2006).

Acknowledgements The writing of this chapter was made possible through a convergence of efforts and projects. SASSCAL Project (sponsored by the German Federal Ministry of Education and Research (BMBF) under promotion number 01LG1201M); Conservation Leadership Programme (Project CLP ID: F01245015: Conserving Angolan Scarp Forests: a Holistic Approach for Kumbira Forest); National Geographic/Okavango Wilderness Project (NGOWP); South Africa's National Research Foundation (2009–2017, WRB), National Geographic Society (Explorer Grant 2011, WRB); Fundação para a Ciência e Tecnologia (contract SFRH/PD/BD/140810/2018, NB); and Wild Bird Trust 2015–2018. Particular thanks go to Fernanda Lages (ISCED Huíla), Brian Huntley (South Africa), John Hilton and Rainer Von Brandis (Wild Bird Trust) for logistical and administrative support.

Appendix 1

Checklist of the amphibians recorded in Angola, based on historical records and on confirmed records from recent surveys. Taxonomy follows Frost (2018). Unconfirmed records are not included. To avoid redundancy, records included in existing compilations (e.g. Monard 1937; Ruas 1996) are mentioned under the compilation reference, and the original reference(s) is not included in the list

Common name	Species	References
Family Arthroleptidae		
Carqueja's Squeaker	*Arthroleptis carquejai* (Ferreira, 1906)	Ferreira (1906)
Lameer's Squeaker	*Arthroleptis lameerei* (De Witte, 1921)	Laurent (1964) and Ruas (1996)
Tanganyika Screeching Frog	*Arthroleptis spinalis* (Boulenger, 1919)	Laurent (1950)
Common Squeaker	*Arthroleptis stenodactylus* (Pfeffer, 1893)	Laurent (1964), Ruas (1996) and Conradie et al. (unpub. data)
Variable Squeaker	*Arthroleptis variabilis* (Matschie, 1893)	Baptista and Vaz Pinto (unpub. data)
Plain Squeaker	*Arthroleptis xenochirus* (Boulenger, 1905)	Monard (1937), Laurent (1964), Ruas (1996), Ceríaco et al. (2018b), Conradie et al. (unpub. data), Baptista and Vaz Pinto (unpub. data), and Ernst (unpub. data)
Anchieta's Treefrog	*Leptopelis anchietae* (Bocage, 1873)	Bocage (1895), Boulenger (1905), Schmidt (1936), Monard (1937), Laurent (1964), Conradie et al. (2016), Baptista et al. (2018), (in prep.) and Ernst (unpub. data)
Gaboon Forest Treefrog	*Leptopelis aubryi* (Duméril, 1856)	Peters (1887) and Laurent (1954)
Bocage's Burrowing Treefrog	*Leptopelis bocagii* (Günther, 1865)	Bocage (1895), Monard (1937), Hellmich (1957b), Laurent (1954, 1964), Ceríaco et al. (2018b), Baptista et al (2018, in prep), Baptista and Vaz Pinto (unpub. data) and Conradie et al. (unpub. data)
Efulen Forest Treefrog	*Leptopelis calcaratus* (Boulenger, 1906)	Baptista and Vaz Pinto (unpub. data)
Cinnamon Treefrog	*Leptopelis cynnamomeus* (Bocage, 1893)	Bocage (1895), Monard (1937) and Laurent (1964)
Congulo Forest Treefrog	*Leptopelis jordani* (Parker, 1936)	Parker (1936) and Baptista et al. (2017)
Quissange Forest Treefrog	*Leptopelis marginatus* (Bocage, 1895)	Bocage (1895)
Kanole Forest Treefrog	*Leptopelis* cf. *parvus* (Schmidt and Inger, 1959)	Conradie et al. (unpub. data)
Rusty Forest Treefrog	*Leptopelis viridis* (Günther, 1869)	Boulenger (1882) and Bocage (1895)
Hairy Frog	*Trichobatrachus robustus* (Boulenger, 1900)	Ernst et al. (2014)

(continued)

Common name	Species	References
Family Brevicipitidae		
Common Rain Frog	*Breviceps* cf. *adspersus* (Peters, 1882)	Bocage (1895), Monard (1937), Hellmich (1957b), Laurent (1964), Ruas (1996) and Conradie et al. (unpub. data)
Family Bufonidae		
Dark-sided Forest Toad	*Mertensophryne melanopleura* (Schmidt and Inger, 1959)	Ruas (1996)
Mocquard's Forest Toad	*Mertensophryne mocquardi* (Angel, 1924)	Monard (1937)
Dombe Pygmy Toad	*Poyntonophrynus dombensis* (Bocage, 1895)	Bocage (1895), Poynton and Haake (1993), Ceríaco et al. (2018a) and Vaz Pinto and Branch (unpub. data)
Grandison's Pygmy Toad	*Poyntonophrynus grandisonae* (Poynton and Haacke, 1993)	Poynton and Haacke (1993) and Ceríaco et al. (2018a)
Kavango Pygmy Toad Toad	*Poyntonophrynus kavangensis* (Poynton and Broadley, 1988)	Poynton and Haacke (1993), Ruas (1996) and Vaz Pinto (unpub. data)
Serra da Neve Pygmy Toad	*Poyntonophrynus pachnodes* (Ceríaco et al. in press.)	Ceríaco et al. (2018a)
Red Toad	*Schismaderma carens* (Smith, 1848)	Monard (1937), Ruas (1996) and Baptista and Vaz Pinto (unpub. data)
Buchner's Toad	*Sclerophrys buchneri* (Peters, 1882)	Peters (1882)
Somber Toad	*Sclerophrys funerea* (Bocage, 1866)	Bocage (1895), Monard (1937), Laurent (1954, 1964), Ruas (1996) and Conradie et al. (2016)
Guttural Toad	*Sclerophrys gutturalis* (Power, 1927)	Ruas (1996, 2002), Conradie et al. (2016, unpub. data), Baptista et al. (2018), Baptista and Vaz Pinto (unpub. data)
Lemaire's Toad	*Sclerophrys lemairii* (Boulenger, 1901)	Laurent (1950, 1964), Ruas (1996) and Conradie et al. (2016)
Western Olive Toad	*Sclerophrys poweri* (Hewitt, 1935)	Conradie et al. (2016) and Baptista et al. (in press)
Merten's Striped Toad	*Sclerophrys pusilla* (Mertens, 1937)	Ruas (1996, 2002), Conradie et al. (2016), Poynton et al. (2016), Ceríaco et al. (2018b), Baptista et al. (2018) and Baptista and Vaz Pinto (unpub. data)
Common Toad	*Sclerophrys regularis* (Reuss, 1833)	Bocage (1895), Monard (1937), Laurent (1964), Ruas (1996), Ceríaco et al. (2014b) and Vaz Pinto and Baptista (unpub. data)
Family Dicroglossidae		
Crowned Bullfrog	*Hoplobatrachus occipitalis* (Günther, 1858)	Bocage (1895), Monard (1937), Hellmich (1957b) and Ruas (1996), Baptista (unpub. data)

(continued)

Common name	Species	References
Family Hemisotidae		
Guinea Snout-burrower	*Hemisus guineensis* (Cope, 1865)	Laurent (1964), Ceríaco et al. (2018b), Conradie et al. (unpub. data) and Baptista and Vaz Pinto (unpub. data)
Marbled Snout-burrower	*Hemisus marmoratus* (Peters, 1854)	Bocage (1895), Monard (1937), Hellmich (1957b), Ruas (1996) and Baptista et al. (in prep.)
Family Hyperoliidae		
Striped Spiny Reed Frog	*Afrixalus dorsalis* (Peters, 1875)	Laurent (1964) and Baptista and Vaz Pinto (unpub. data)
Four Lined Reed Frog	*Afrixalus fulvovittatus* (Cope, 1861)	Bocage (1866a) and Ferreira (1904)
Osorio's Spiny Reed Frog	*Afrixalus osorioi* (Ferreira, 1906)	Ferreira (1906), Baptista and Vaz Pinto (unpub. data) and Ernst (unpub. data)
Four-Lined Spiny Reed Frog	*Afrixalus quadrivittatus* (Werner, 1908)	Peters (1887) and Perret (1976)
De Witte's Spiny Reed Frog	*Afrixalus wittei* (Laurent, 1941)	Ceríaco et al. (2018b), Baptista and Vaz Pinto (unpub. data) and Ernst (unpub. data)
Sprinkled Long Reed Frog	*Hyperolius adspersus* (Peters, 1877)	Laurent (1964)
Benguela Long Reed Frog	*Hyperolius benguellensis* (Bocage, 1893)	Bocage (1895), Ferreira (1906), Monard (1937), Laurent (1950), Channing et al. (2013), Conradie et al. (2016) and Baptista et al. (2018)
Two-colored Reed Frog	*Hyperolius bicolor* (Ahl, 1931)	Ahl (1931)
Bocage's Reed Frog	*Hyperolius bocagei* (Steindachner, 1867)	Monard (1937), Laurent (1950, 1954, 1964), Ceríaco et al. (2014b), Conradie (unpub. data), Baptista and Vaz Pinto (unpub. data) and Ernst (unpub. data)
Chela Mountain Reed Frog	*Hyperolius chelaensis* (Conradie et al. 2012)	Conradie et al. (2012)
Monard's Reed Frog	*Hyperolius cinereus* (Monard, 1937)	Monard (1937), Conradie et al. (2016), Baptista et al. (2018), Baptista and Vaz Pinto (unpub. data) and Conradie et al. (unpub. data)
Dimorphic Reed Frog	*Hyperolius cinnamomeoventris* (Bocage, 1866)	Monard (1937), Laurent (1950, 1954, 1964), Ceríaco et al. (2016c), (2018b) and Baptista and Vaz Pinto (unpub. data)
Variable Reed Frog	*Hyperolius concolor* (Hallowell, 1844)	Monard (1937)
Dartevelle's Long Reed frog	*Hyperolius dartevellei* (Laurent, 1943)	Laurent (1964) and Channing et al. (2013)
Brown-throated Reed Frog	*Hyperolius fuscigula* (Bocage, 1866)	Bocage (1866)
Family Hemisotidae (cont.)		
Loanda Reed Frog	*Hyperolius gularis* (Ahl, 1931)	Ahl (1931)
Kivu Reed Frog	*Hyperolius kivuensis* (Ahl, 1931)	Laurent (1950, 1954)

(continued)

Common name	Species	References
Lang's Reed Frog	*Hyperolius langi* (Noble, 1924)	Monard (1937)
Landana Reed Frog	*Hyperolius lucani* (Rochebrune, 1885)	Rochebrune (1885)
Cabinda Reed Frog	*Hyperolius maestus* (Rochebrune, 1885)	Rochebrune (1885)
Marbled Reed Frog	*Hyperolius marmoratus* (Rapp, 1842)	Boulenger (1882), Bocage (1895) and Monard (1937)
Large-nosed Long Reed Frog	*Hyperolius nasutus* (Günther, 1865)	Bocage (1895), Monard (1937), Laurent (1950, 1954, 1964), Hellmich (1957b), Baptista and Vaz Pinto (unpub. data) and Ceríaco et al. (2018b)
Nobre's Reed Frog	*Hyperolius nobrei* (Ferreira, 1906)	Ferreira (1906)
Angolan Reed Frog	*Hyperolius parallelus* (Günther, 1858)	Monard (1937), Laurent (1950, 1954, 1964), Ceríaco et al. (2018b), Conradie et al. (unpub. data), Baptista et al. (2018) and Baptista and Vaz Pinto (unpub. data)
Leopard Reed Frog	*Hyperolius pardalis* (Laurent, 1948)	Conradie (unpub. data)
Rio Luinha Reed Frog	*Hyperolius platyceps* (Boulenger, 1900)	Monard (1937), Laurent (1950, 1954) and Baptista and Vaz Pinto (unpub. data)
Tshimbulu Reed Frog	*Hyperolius polli* (Laurent, 1943)	Laurent (1954)
Rochebrune's Reed Frog	*Hyperolius protchei* (Rochebrune, 1885)	Rochebrune (1885)
Five-striped Reed Frog	*Hyperolius quinquevittatus* (Bocage, 1866)	Bocage (1895), Laurent (1950, 1954) and Baptista and Vaz Pinto (unpub. data)
Raymond's Reed Frog	*Hyperolius raymondi* (Conradie et al. 2013)	Conradie et al. (2013)
African Reed Frog	*Hyperolius rhizophilus* (Rochebrune, 1885)	Rochebrune (1885)
Steindachner's Reed Frog	*Hyperolius steindachneri* (Bocage, 1866)	Bocage (1895), Monard (1937), Laurent (1950, 1954, 1964), Poynton and Haacke (1993) and Channing and Vaz Pinto (unpub. data)
Luita River Reed Frog	*Hyperolius vilhenai* (Laurent, 1964)	Laurent (1964)
Kuvangu Kassina	*Kassina kuvangensis* (Monard, 1937)	Monard (1937) and Conradie et al. (2016, unpub. data)
Family Hemisotidae (cont.)		
Bubbling Kassina	*Kassina senegalensis* (Duméril and Bibron, 1841)	Monard (1937), Laurent (1954, 1964), Poynton and Haacke (1993), Conradie et al. (2016), Baptista et al. (2018), Baptista and Vaz Pinto (unpub. data), Conradie et al. (unpub. data) and Ernst (unpub. data)
De Witte's Clicking Frog	*Kassinula wittei* (Laurent, 1940)	Conradie et. al (unpub. data)

(continued)

Common name	Species	References
Family Microhylidae		
Spotted Rubber Frog	*Phrynomantis affinis* (Boulenger, 1901)	Laurent (1964)
Marbled Rubber Frog	*Phrynomantis annectens* (Werner, 1910)	Ruas (1996) and Vaz Pinto and Branch (unpub. data)
Banded Rubber Frog	*Phrynomantis bifasciatus* (Smith, 1847)	Boulenger (1882), Monard (1937), Ruas (1996), Channing (unpub. data) and Baptista et al. (unpub. data)
Family Phrynobatrachidae		
Ahl's Puddle Frog	*Phrynobatrachus brevipalmatus* (Ahl, 1925)	Ahl (1925)
Cryptic Puddle Frog	*Phrynobatrachus cryptotis* (Schmidt and Inger, 1959)	Laurent (1964)
Mababe Puddle Frog	*Phrynobatrachus mababiensis* (FitzSimons, 1932)	Poynton and Haacke (1993), Conradie et al. (2016, unpub. data) and Baptista and Vaz Pinto (unpub. data)
Snoring Puddle Frog	*Phrynobatrachus natalensis* (Smith, 1849)	Bocage (1895), Monard (1937), Hellmich (1957b), Ruas (1996), Conradie et al. (2016, unpub. data), Ceríaco et al. (2018b), Baptista et al. (2018, in press) and Baptista and Vaz Pinto (unpub. data)
Loanda River Frog	*Phrynobatrachus parvulus* (Boulenger, 1905)	Ruas (1996), Baptista and Vaz Pinto (unpub. data) and Conradie et al. (unpub. data)
Family Pipidae		
Andre's Clawed Frog	*Xenopus andrei* (Loumont, 1983)	Ernst et al. (2015)
Southern Tropical Clawed Frog	*Xenopus epitropicalis* (Fischberg et al., 1982)	Laurent (1950, 1954) and Klein (unpub. data)
Müller's Clawed Frog	*Xenopus muelleri* (Peters, 1844)	Conradie et al. (2016)
Peters' Clawed Frog	*Xenopus petersii* (Bocage, 1895)	Bocage (1895), Monard (1937), Hellmich (1957b), Ruas (1996), Baptista et al. (2018), Baptista and Vaz Pinto (unpub. data), Ceríaco et al. (2018b) and Ernst (unpub. data)
Power's Clawed Frog	*Xenopus poweri* (Hewitt, 1927)	Conradie et al. (2016)
Family Pipidae (cont.)		
Clawed Frog	*Xenopus* sp.	Laurent (1950)
Family Ptychadenidae		
Common Ornate Frog	*Hildebrandtia ornata* (Peters, 1878)	Poynton and Haacke (1993)
Angola Ornate Frog	*Hildebrandtia ornatissima* (Bocage, 1879)	Bocage (1895), Monard (1937) and Baptista and Vaz Pinto (unpub. data)
Anchieta's Grass Frog	*Ptychadena anchietae* (Bocage, 1868)	Ruas (1996), Ceríaco et al. (In press.), Baptista et al (2018) and Baptista and Vaz Pinto (unpub. data)
Ansorge's Grass Frog	*Ptychadena ansorgii* (Boulenger, 1905)	Monard (1937) and Ruas (1996)

(continued)

Common name	Species	References
Rough Grass Frog	*Ptychadena bunoderma* (Boulenger, 1907)	Monard (1937), Ruas (1996) and Conradie et al. (unpub. data)
Grandison's Grass Frog	*Ptychadena grandisonae* (Laurent, 1954)	Ruas (1996)
Guibe's Grass Frog	*Ptychadena guibei* (Laurent, 1954)	Ruas (1996), Ceríaco et al. (in press), Conradie et al. (2016) and Baptista and Vaz Pinto (unpub. data)
Keiling's Grass Frog	*Ptychadena keilingi* (Monard, 1937)	Ruas (1996) and Conradie et al. (unpub. data)
Mozambique Grass Frog	*Ptychadena* cf. *mossambica* (Peters, 1854)	Conradie et al. (2016) and Conradie (unpub. data)
Nile Grass Frog	*Ptychadena nilotica* (Seetzen, 1855)	Monard (1937), Schmidt and Inger (1959), Ruas (1996), Conradie et al. (2016), Dehling and Sinsch (2013b) and Zimkus et al. (2017)
Sharp-nosed Grass Frog	*Ptychadena oxyrhynchus* (Smith, 1849)	Monard (1937), Hellmich (1957b), Ruas (1996), Ceríaco et al. (2018b), Conradie et al. (2016) and Baptista (unpub. data)
Many-Grass Frog	*Ptychadena perplicata* (Laurent, 1964)	Laurent (1964)
Striped Grass Frog	*Ptychadena porosissima* (Steindachner, 1867)	Ruas (1996), Conradie et al. (unpub. data) and Channing et al. (2012)
Spot-bellied Grass Frog	*Ptychadena subpunctata* (Bocage, 1866)	Ruas (1996), Conradie et al. (2016) and Baptista (unpub. data)
Small Grass Frog	*Ptychadena taenioscelis* (Laurent, 1954)	Ruas (1996) and Conradie et al. (2016)
Upemba Grass Frog	*Ptychadena upembae* (Schmidt and Inger, 1959)	Ruas (1996)
Udzungwa Grass Frog	*Ptychadena uzungwensis* (Loveridge, 1932)	Ruas (1996) and Conradie et al. (2016, unpub. data)
Family Pyxicephalidae		
Angola River Frog	*Amietia angolensis* (Bocage, 1866)	Bocage (1895), Monard (1937), Ruas (1996), Channing and Baptista (2013), Ceríaco et al. (2016b), Channing et al. (2016), Baptista et al. (2018) and Baptista and Vaz Pinto (unpub. data)
African Bullfrog	*Pyxicephalus adspersus* (Tschudi, 1838)	Monard (1937) and Ruas (1996)
Cryptic Sand Frog	*Tomopterna cryptotis* (Boulenger, 1907)	Monard (1937), Ruas (1996), Conradie et al. (2016) and Baptista et al. (in press)
Damaraland Sand Frog	*Tomopterna damarensis* (Dawood and Channing, 2002)	Ceríaco et al. (2016a) and Heinicke et al. (2017)
Rough Sand Frog	*Tomopterna tuberculosa* (Boulenger, 1882)	Bocage (1895), Monard (1937), Ruas (1996), Baptista et al (2018, unpub. data) and Conradie et al. (unpub. data)
Family Ranidae		
Forest White-lipped Frog	*Amnirana albolabris* (Hallowell, 1856)	Bocage (1895), Monard (1937), Ruas (1996) and Jongsma et al. (2018)

(continued)

Common name	Species	References
Darling's White-lipped Frog	*Amnirana darlingi* (Boulenger, 1902)	Monard (1937), Laurent (1964), Ruas (1996), Ceríaco et al. (2018b) Branch and Conradie (2015) and Conradie et al. (unpub. data)
Lemaire's White-lipped Frog	*Amnirana lemairei* (De Witte, 1921)	Laurent (1964), Ruas (1996) and Baptista and Vaz Pinto (unpub. data)
Andersson's White-lipped Frog	*Amnirana* cf. *lepus* (Andersson, 1903)	Branch and Conradie (2015)
Parker's White-lipped Frog	*Amnirana parkeriana* (Mertens, 1938)	Mertens (1938)

References

Ahl E (1925 "1923") Ueber neue afrikanische Frösche der Familie Ranidae. Sitzungsberichte der Gesellschaft Naturforschender Freunde zu Berlin 1923:96–106

Ahl E (1931) Amphibia, Anura III, Polypedatidae. Das Tierreich 55: xvi + 477

Amiet JL (2012) Les Rainettes du Cameroun (Amphibiens Anoures). La Nef des Livres, Saint-Nazaire, 591 pp

Baptista N, António T, Branch WR The herpetofauna of Bicuar National Park and surrounds, southwestern Angola: a first description and preliminary checklist. Amphibian & Reptile Conservation. (in press).

Baptista N, Vaz Pinto P, Ernst R et al (2017) Cryptic diversity in treefrogs (*Leptopelis*) of the Angolan escarpment – fitting the pieces together. 13th conference of the Herpetological Association of Africa, Bonamanzi, South Africa

Baptista N, António T, Branch WR (2018) Amphibians and reptiles of the Tundavala region of the Angolan Escarpment. In: Revermann R, Krewenka KM, Schmiedel U et al (eds) Climate change and adaptive land management in Southern Africa – assessments, changes, challenges, and solutions, Biodiversity & ecology, vol 6, pp 397–403

Beard KH, Vogt KA, Kulmatiski A (2002) Top-down effects of a terrestrial frog on forest nutrient dynamics. Oecologia 133(4):583–593

Beebee TJ, Griffiths RA (2005) The amphibian decline crisis: a watershed for conservation biology? Biol Conserv 125(3):271–285

Blanc CP, Frétey T (2000) Biogeographie des Amphibiens d'Afrique Centrale et d'Angola. Biogeographica 76(3):107–118

Bocage JVB (1866a) Lista dos reptis das possessões portuguesas d' Africa occidental que existem no Museu de Lisboa. Jornal de Sciências Mathemáticas, Physicas e Naturaes. Lisboa 1:37–56

Bocage JVB (1866b) Reptiles nouveaux ou peu connus recueillis dans les possessions portugaises de l'Afrique occidentale, qui se trouvent au Muséum de Lisbonne. Jornal de Sciencias Mathematicas, Physicas e Naturaes. Lisboa I(1):57–78

Bocage JVB (1867) Batraciens nouveaux de l'Afrique occidentale (Loanda et Benguella). Proc Zool Soc London 35:843–846

Bocage JVB (1873) Mélanges erpétologiques. Sur quelques Reptiles et Batraciens nouveux, rares ou peu connues de l'Afrique occidentale. Jornal de Sciências Mathemáticas, Physicas e Naturaes. Lisboa 4(1):209–227

Bocage JVB (1879a) Reptiles et batraciens nouveaux d' Angola. Jornal de Sciências Mathemáticas, Physicas e Naturaes. Lisboa 7(26):97–99

Bocage JVB (1879b) Subsidio para a fauna das possessões portuguezas d'África occidental. Jornal de Sciências Mathemáticas, Physicas e Naturaes. Lisboa (1) 7:85–95

Bocage JVB (1882) Reptiles rares ou nouveaux d'Angola. Jornal de Sciencias, Mathemáticas, Physicas e Naturaes. Lisboa (1) 8:299–304

Bocage JVB (1893) Diagnose de quelques nouvelles espéces de reptiles et batraciens d' Angola. Jornal de Sciências Mathemáticas, Physicas e Naturaes. Lisboa (2) 10:115–121

Bocage JVB (1895a) Herpétologie d'Angola et du Congo. Lisbonne, Imprimerie Nationale, 203 pp, 19 pls

Bocage JVB (1895b) Sur une espêce de Crapaud à ajouter à la faune herpétologique d'Angola. Jornal de Sciências Mathemáticas, Physicas e Naturaes. Lisboa (2) 4:51–53

Bocage JVB (1896a) Mamíferos, aves e réptis da Hanha, no sertão de Benguella. Jornal de Sciências Mathemáticas, Physicas e Naturaes. Lisboa (2) 14:105–114

Bocage JVB (1896b) Répteis de algumas possessões portuguesas de Africa que existem no Museu de Lisboa. Jornal de Sciências Mathemáticas, Physicas e Naturaes 2:65–104

Bocage JVB (1897a) Mamíferos, aves e reptis da Hanha, no sertão de Benguella. Segunda lista. Jornal de Sciências Mathemáticas, Physicas e Naturaes. Lisboa (2):207–211

Bocage JVB (1897b) Mamíferos, réptis e batrachios d'África de que existem exemplares típicos no Museu de Lisboa. Jornal de Sciências Mathemáticas, Physicas e Naturaes. Lisboa (2) 4:187–206

Boulenger GA (1882) Catalogue of the Batrachia Salientia s. Ecaudata in the collection of the British museum, 2nd edn. Taylor and Francis, London

Boulenger GA (1905) A list of the batrachians and reptiles collected by Dr. W. J. Ansorge in Angola, with descriptions of new species. Ann Mag Nat Hist Ser 7 16(92):8–115

Boulenger GA (1907a) Descriptions of three new lizards and a new frog, discovered by Dr. W J. Ansorge in Angola. Ann Mag Nat Hist Ser 7 19:212–214

Boulenger GA (1907b) Description of a new frog discovered by Dr. W. J. Ansorge in Mossamedes, Angola. Ann Mag Nat Hist Ser 7 20:109

Boulenger GA (1919) On *Rana ornatissima*, Bocage, and *R. ruddi*, Blgr. Trans Royal Soc S Afr 8:33–37

Branch WR, Bauer AM (2005) The life and herpetological contributions of Andrew Smith. pp. 1-19 in *Smith, A. The Herpetological Contributions of Sir Andrew Smith.* Society for the Study of Amphibians and Reptiles, Villanova, PA. iv + 84 pp

Branch WR, Conradie WC (2015) Herpetofauna da região da Lagoa Carumbo (Herpetofauna of the Carumba Lagoon Area). In: Huntley BJ (ed), Relatório sobre a Expedição Avaliação rápida da Biodiversidade de região da Lagoa Carumbo, Lunda-Norte – Angola, República de Angola. Ministério do Ambiente, pp 194–209, 219p

Branch WR, Baptista N, Keates CW, et al (2019) Rediscovery, taxonomic status, and phylogenetic relationships of two rare and endemic snakes (Serpentes: Psammophinae) from the Angolan Escarpment. Zootaxa, in press

Brito D (2010) Overcoming the Linnean shortfall: data deficiency and biological survey priorities. Basic Appl Ecol 11(8):709–713

Broadley DG (1971) The reptiles and amphibians of Zambia. The Puku, Occ Pap Dept Game Fish Zambia 7:1–143

Brooks C (2012) Biodiversity survey of the upper Angolan Catchment of the Cubango-Okavango River Basin. USAid-Southern Africa. 151 pp

Brooks C (2013) Trip report: aquatic biodiversity survey of the lower Cuito and Cuando river systems in Angola. USAid-Southern Africa. 43 pp

Cei JM (1977) Chaves para uma identificação preliminar dos batráquios anuros da R. P. de Angola. Boletim da Sociedade Portuguesa de Ciências Naturais 17:5–26

Ceríaco LMP, Bauer AM, Blackburn et al (2014a) The herpetofauna of the Capanda Dam region, Malanje, Angola. Herpetol Rev 45(4):667–674

Ceríaco LMP, Blackburn DC, Marques MP et al (2014b) Catalogue of the amphibian and reptile type specimens of the Museu de História Natural da Universidade do Porto in Portugal, with some comments on problematic taxa. Alytes 31(1):13–36

Ceríaco LMP, Bauer AM, Heinicke MP et al (2016a) Geographical distributions: Ptychadenidae, *Ptychadena mapacha* Channing, 1993 – Mapacha ridged frog in Namibia. Afr Herp News 63:19–20

Ceríaco LMP, de Sá SAC, Bandeira SA et al (2016b) Herpetological survey of Iona National Park and Namibe regional natural park, with a synoptic list of the amphibians and reptiles of Namibe Province, Southwestern Angola. Proc Calif Acad Sci 63(2):15–61

Ceríaco LMP, Marques MP, Bandeira SA et al (2016c) Anfíbios e répteis do Parque Nacional da Cangandala. Instituto Nacional da Biodiversidade e Áreas de Conservação & Museu Nacional de História Natural e da Ciência, 96 pp

Ceríaco LMP, Marques MP, Bandeira S et al (2018a) A new earless species of Poyntonophrynus (Anura: Bufonidae) from the Serra da Neve Inselberg, Namibe Province, Angola. Zookeys 780:109–136

Ceríaco LMP, Marques MP, Bandeira S et al (2018b) Herpetological survey of Cangandala National Park, with a synoptic list of the amphibians and reptiles of Malanje Province, Central Angola. Herpetological Review 49(3):408–431

Channing A (1993) A new grass frog from Namibia. S Afr J Zool 28:142–145

Channing A (1999) Historical overview of amphibian systematics in Southern Africa. Trans Royal Soc S Afr 54(1):121–135

Channing A (2001) Amphibians of central and southern Africa. Cornell University Press, New York, 470 pp

Channing A, Baptista N (2013) *Amietia angolensis* and *A. fuscigula* (Anura: Pyxicephalidae) in southern Africa: a cold case reheated. Zootaxa 3640(4):501–520

Channing A, Griffin M (1993) An annotated checklist of the frogs of Namibia. Modoqua 18:101–116

Channing A, Vences M (1999) The advertisement call, breeding biology, description of the tadpole and taxonomic status of *Bufo dombensis*, a little-known dwarf toad from southern Africa. S Afr J Zool 34:74–79

Channing, A., and D. G. Broadley. 2002. A new snout-burrower from the Barotse Floodplain (Anura: Hemisotidae: Hemisus). Journal of Herpetology 36: 367–372.

Channing A, Rödel MO, Channing J (2012) Tadpoles of Africa: the biology and identification of all known tadpoles in sub-Saharan Africa. Edition Chimaira, Frankfurt, 401pp

Channing A, Hillers A, Lötters S et al (2013) Taxonomy of the super-cryptic Hyperolius nasutus group of long reed frogs of Africa (Anura: Hyperoliidae), with descriptions of six new species. Zootaxa 3620(3):301–350

Channing A, Dehling JM, Lötters S et al (2016) Species boundaries and taxonomy of the African river frogs (Amphibia: Pyxicephalidae: Amietia). Zootaxa 4155(1):1–76

Clark VR, Barker NP, Mucina L (2011) The great escarpment of southern Africa: a new frontier for biodiversity exploration. Biodivers Conserv 20(12):2543–2561

Coetzer W (2017) Occurrence records of southern African aquatic biodiversity. Version 1.10. The south African Institute for Aquatic Biodiversity. Occurrence dataset https://doi.org/10.15468/pv7vds. Accessed via GBIF.org

Conradie W, Bills R (2017) Wannabe Ranid: notes on the morphology and natural history of the Lemaire's toad (Bufonidae: *Sclerophrys lemairii*). Salamandra 53(3):439–444

Conradie W, Branch WR, Measey GJ et al (2012) A new species of Hyperolius Rapp, 1842 (Anura: Hyperoliidae) from the Serra da Chela mountains, South-Western Angola. Zootaxa 3269(1):1–17

Conradie W, Branch WR, Tolley KA (2013) Fifty Shades of Grey: giving colour to the poorly known Angolan Ashy reed frog (Hyperoliidae: *Hyperolius cinereus*), with the description of a new species. Zootaxa 2636(3):201–223

Conradie W, Bills R, Branch WR (2016) The herpetofauna of the Cubango, Cuito, and lower Cuando river catchments of South-Eastern Angola. Amphibian Reptile Conserv 10(2):6–36

Crawford-Cabral JC (1966) Some new data on Angolan Muridae. Zool Afr 2:193–203

Crawford-Cabral J, Mesquitela LM (1989) Índice toponímico de colheitas zoológicas em Angola. Instituto de Investigação Cientifica Tropical, Centro de Zoologia, Lisbon, 206

Cunningham M, Cherry MI (2004) Molecular systematics of African 20-chromosome toads (Anura: Bufonidae). Mol Phylogenet Evol 32(3):671–685

Curtis B, Roberts KS, Griffin M et al (1998) Species richness and conservation of Namibian freshwater macro-invertebrates, fish and amphibians. Biodivers Conserv 7(4):447–466

Davic RD, Welsh JHH (2004) On the ecological roles of salamanders. Annu Rev Ecol Evol Syst 35:405–434

Dehling JM, Sinsch U (2013a) Diversity of *Ptychadena* in Rwanda and taxonomic status of *P. chrysogaster* Laurent, 1954 (Amphibia, Anura, Ptychadenidae). Zoo Keys 356:69–102

Dehling JM, Sinsch U (2013b) Diversity of Ridged Frogs (Anura: Ptychadenidae: *Ptychadena* spp.) in wetlands of the upper Nile in Rwanda: morphological, bioacoustic, and molecular evidence. Zoologischer Anzeiger 253(2):143–157

Du Preez L, Carruthers V (2009) A complete guide to the frogs of Southern Africa. Struik Publishers, Cape Town, 488 pp

Du Preez L, Carruthers V (2017) Frogs of Southern Africa: a complete guide. Struik Publishers, Cape Town, 520 pp

Ernst R, Nienguesso ABT, Lautenschlaeger T et al (2014) Relicts of a forested past: southernmost distribution of the hairy frog genus *Trichobatrachus* Boulenger, 1900 (Anura: Arthroleptidae) in the Serra do Pingano region of Angola with comments on its taxonomic status. Zootaxa 3779(2):297–300

Ernst R, Schmitz A, Wagner P, Branquima MF et al (2015) A window to Central African forest history: distribution of the *Xenopus fraseri* subgroup south of the Congo Basin, including a first country record of *Xenopus andrei* from Angola. Salamandra 52(1):147–155

Ferreira JB (1897) Lista dos reptis e amphibios que fazem parte da última remessa de J. d'Anchieta. Jornal de Sciências Mathemáticas, Physicas e Naturaes 5(2):240–246

Ferreira JB (1900) Sobre alguns exemplares pertencentes à fauna do norte de Angola (Reptis, Batrachios, Aves e Mammiferos). Jornal de Sciências Mathemáticas, Physicas e Naturaes, Lisboa 2(6):48–54

Ferreira JB (1904) Reptis e amphibios de Angola da região ao norte do Quanza (Collecção Newton – 1903). Jornal de Sciências Mathemáticas, Physicas e Naturaes, Segunda Série 7(26):111–117

Ferreira JB (1906) Algumas espécies novas ou pouco conhecidas de amphibios e reptis de Angola (Collecção Newton – 1903). Jornal de Sciências Mathemáticas, Physicas e Naturaes, Segunda Série 7(26):159–171

Frétey T, Dewynter M, Blanc CP (2011) Amphibiens d'Afrique central et d'Angola. Clé de détermination ilustrée desamphibiens du Gabo et du Mbini/Illustrated identification key of the amphibians from Gabon and Mbini. Biotope, Mèze/Muséum national d'Histoire naturelle, Paris, 232 pp

Frost DR (2018) Amphibian species of the world: an online reference. Version 6.0. Electronic database accessible at http://research.amnh.org/herpetology/amphibia/index.html. American Museum of Natural History, New York

Frost DR, Grant T, Faivovich J et al (2006) The amphibian tree of life. Bull Am Mus Nat Hist 297:1–370

Furman BL, Bewick AJ, Harrison TL et al (2015) Pan-African phylogeography of a model organism, the African clawed frog 'Xenopus laevis'. Mol Ecol 24(4):909–925

Greenbaum E, Meece J, Reed KD et al (2014) Amphibian chytrid infections in non-forested habitats of Katanga, Democratic Republic of the Congo. Herpetol Rev 45:610–614

Günther ACLG (1865 '1864') Descriptions of new species of batrachians from West Africa. Proc Zool Soc London 3:479–482

Guttman SI (1967) Transferrin and hemoglobin polymorphism, hybridization and introgression in two African toads, *Bufo regularis* and *Bufo rangeri*. Comp Biochem Physiol 23(3):871–877

Haacke WD (1999) Geographical distribution: *Ptychadena mapacha* Channing, 1993 – Mapacha Grass Frog. Afr Herp News 30:35

Hall BP (1960) The faunistic importance of the scarp of Angola. Ibis 102(3):420–442

Heinicke MP, Ceríaco LM, Moore IM et al (2017) *Tomopterna damarensis* (Anura: Pixicephalidae) is broadly distributed in Namibia and Angola. Salamandra 53(3):461–465

Hellmich W (1957a) Die reptilienausbeute der Hamburgischen Angola Expedition. Mitteilungen aus dem Hamburgischen Zoologischen Museum und Institut 55:39–80

Hellmich W (1957b) Herpetologische Ergebnisse einer Forschungsreise in Angola. Veröffentlichungen der Zoologischen Staatssammlung München 5:1–92

Herrmann HW, Branch WR (2013) Fifty years of herpetological research in the Namib Desert and Namibia with an updated and annotated species checklist. J Arid Environ 93:94–115

Huntley BJ (1974) Outlines of wildlife conservation in Angola. J S Afr Wildl Manag Assoc 4:157–166

Huntley BJ (2019) Angola in outline: physiography, climate and patterns of biodiversity. In: Huntley BJ, Russo V, Lages F, Ferrand N (eds) Biodiversity of Angola. Science & conservation: a modern synthesis. Springer, Cham

IUCN Red List of Threatened Species Version 2017-2. www.iucnredlist.org

IUCN Red List of Threatened Species. Version 2017-3. www.iucnredlist.org

Jongsma CF, Barej MF, Barratt CD et al (2018) Diversity and biogeography of frogs in the genus Amnirana (Anura: Ranidae) across sub-Saharan Africa. Mol Phylogenet Evol 120:274–285

Jordan K (1936) Dr Karl Jordan's expedition to South-West Africa and Angola. Narrative. Novitates Zooligicae 40:17–62, 2 maps, 5 pls

Jürgens N, Strohbach B, Lages F et al (2018) Biodiversity observation – an overview of the current state and first results of biodiversity monitoring studies. In: Revermann R, Krewenka KM, Schmiedel U et al (eds) Climate change and adaptive land management in southern Africa – assessments, changes, challenges, and solutions, Biodiversity & ecology, vol 6, pp 382–396

Köhler J, Vieites DR, Bonett RM et al (2005) New amphibians and global conservation: a boost in species discoveries in a highly endangered vertebrate group. AIBS Bull 55(8):693–696

Lamotte M, Perret JL (1961) Les formes larvaires de quelques espèces de *Leptopelis: L. aubryi, L. viridis, L. anchietae, L. ocellatus et L. calcaratus*. Bulletin de l'Institute fondamental d'Afrique noire, Sér. A 23:855–885

Laurent RF (1950) Reptiles et Batraciens de la region de Dundo (Angola du Nord-Est). Publicações culturais da Companhia de Diamantes de Angola 6:126–136

Laurent RF (1954) Reptiles et Batraciens de la région de Dundo (Angola) (Deuxième Note). Publicações culturais da Companhia de Diamantes de Angola 23:35–84

Laurent RF (1964) Reptiles et Amphibiens de l'Angola (Troisième contribution). Publicações culturais da Companhia de Diamantes de Angola 67:11–165

Marques MP, Ceríaco LMP, Bauer AM et al (2014) Geographic distribution of amphibians & reptiles of Angola: towards an Atlas of the Angolan Herpetofauna. 12th conference of the Herpetological Association of Africa, Gobabeb, Namibia

Marques MP, Ceríaco LMP, Blackburn DC et al (2018) Diversity and distribution of the amphibians and terrestrial reptiles of Angola atlas of historical and bibliographic records (1840¬–2017). Proceedings of the California academy of sciences, Series 4, Volume 65, Supplement II: 1-501

Mertens R (1938a) Amphibien und Reptilien aus Angola, gesammelt von W. Schack. Senckenbergiana 20:425–443

Mertens R (1938b) Herpetologische Ergebnisse einer Reise nach Kamerun. Abhandlungen der Senckenbergischen Naturforschenden Gesellschaft, Frankfurt am Main 442:1–52

MHNG – Muséum d'histoire naturelle de la Ville de Genève (2017) Partial Amphibians Collection. Occurrence Dataset https://doi.org/10.15468/iftvxc. Accessed via GBIF.org

Minter LR, Netherlands EC, Du Preez LH (2017) Uncovering a hidden diversity: two new species of Breviceps (Anura: Brevicipitidae) from northern KwaZulu-Natal, South Africa. Zootaxa 4300:195–216

Monard A (1937) Contribuition à la Batrachologie d'Angola. Bulletin de la Société Neuchâteloise des Sciences Naturelles 62:1–59

Nagy ZT, Kusamba C, Collet M et al (2013) Notes on the herpetofauna of Western Bas-Congo, Democratic Republic of the Congo. Herpetology Notes 6:413–419
Ohler A (1996) Systematics, morphometrics and biogeography of the genus *Aubria* (Ranidae, Pyxicephalinae). Alytes 13:141–166
Parker HW (1936) Dr. Karl Jordan's expedition to South West Africa and Angola: herpetological collection. Novitates Zoologicae 40:115–146
Passmore NI (1972) Intergrading between members of the "regularis group" of toads in South Africa. J Zool 167(2):143–151
Perret JL (1975) Les sous-espèces *d'Hyperolius ocellatus* Günther (Amphibia, Salientia). Annales de la Faculté des Sciences du Cameroun 20:23–31
Perret JL (1976) Révision des amphibiens africains et principalement des types, conservés au Musée Bocage de Lisbonne. Arquivos do Museu Bocage, Segunda Série 6(2):15–34
Perret JL (1977) Les *Hylarana* (Amphibiens, Ranidés) du Cameroun. Rev Suisse Zool 84:841–868
Perret JL (1996) Sur un énigmatique batracien d'Angola. Societé Neuchâteloise des Sciences Naturelles 119:95–100
Peters WCH (1877) Übersicht der Amphibien aus Chinchoxo (Westafrika), welche von der Afrikanischen Gesellschaft dem Berliner zoologischen Museum übergeben sind. Monatsberichte der Königlichen Preussische Akademie des Wissenschaften zu Berlin 1877:611–621
Peters WCH (1882) Neue Batrachier (*Amblystoma Krausei, Nyctibatrachus sinensis, Bufo buchneri*). Sitzungsberichte der Gesellschaft Naturforschender Freunde zu Berlin 1882:145–148
Pickersgill M (2007) Frog search. Results of expeditions to Southern and Eastern Africa from 1993–1999. Frankfurt contributions to natural history 28. Edition Chimaira, Frankfurt
Pietersen DW, Pietersen EW, Conradie W (2017) Preliminary herpetological survey of Ngonye Falls and surrounding regions in southwestern Zambia. Amphibian Reptile Conserv 11(1) [Special Section]:24–43 (e148)
Poynton JC (1964) The Amphibia of Southern Africa: a faunal study. Ann. Natal Mus 17:1–334
Poynton JC (1970) Guide to the *Ptychadena* (Amphibia: Ranidae) of the southern third of Africa. Ann. Natal Mus 20(2):365–375
Poynton JC, Broadley DG (1985a) Amphibia Zambesiaca 1. Scolecomorphidae, Pipidae, Microhylidae, Hemisidae, Arthroleptidae. Ann. Natal Mus 26(2):503–553
Poynton JC, Broadley DG (1985b) Amphibia Zambesiaca 2. Ranidae. Ann. Natal Mus 27(1):115–181
Poynton JC, Broadley DG (1987) Amphibia Zambesiaca 3. Rhacophoridae and Hyperoliidae. Ann. Natal Mus 28(1):161–229
Poynton JC, Broadley DG (1988) Amphibia Zambesiaca, 4. Bufonidae. Ann. Natal Mus 29(2):447–490
Poynton JC, Broadley DG (1991) Amphibia Zambesiaca 5. Zoogeography. Ann. Natal Mus 32:221–277
Poynton JC, Haacke WD (1993) On a collection of amphibians from Angola, including a new species of Bufo Laurenti. Ann Transv Mus 36(2):9–16
Poynton JC, Loader SP, Conradie W et al (2016) Designation and description of a neotype of *Sclerophrys maculata* (Hallowell, 1854), and reinstatement of *S. pusilla* (Mertens, 1937) (Amphibia: Anura: Bufonidae). Zootaxa 4098(1):73–94
Regester KJ, Lips KR, Whiles MR (2006) Energy flow and subsidies associated with the complex life cycle of ambystomatid salamanders in ponds and adjacent forest in southern Illinois. Oecologia 147(2):303–314
Rochebrune AT (1885) Vertebratorum novorum vel minus cognitorum orae Africae occidentalis incolarum. Diagnoses (1). Bulletin de la Société Philomathique de Paris 7(9):86–99
Royal Belgian Institute of Natural Sciences (2017). RBINS DaRWIN. Occurrence Dataset https://doi.org/10.15468/qxy4mc. Accessed via GBIF.org
Royal Museum for Central Africa, Belgium (2017). RMCA-HERP. Occurrence Dataset https://doi.org/10.15468/0inmlf. Accessed via GBIF.org

Ruas C (1996) Contribuição para o conhecimento da fauna de batráquios de Angola. Garcia de Orta, Série de Zoologia 21(1):19–41

Ruas C (2002) Batráquios de Angola em colecção no Centro de Zoologia. Garcia de Orta, Série de Zoologia 24(1–2):139–154

SASSCAL ObservationNet (2017) http://www.sasscalobservationnet.org/ consulted on March 6th 2018

Scheinberg L, Fong J (2017) CAS herpetology (HERP). Version 33.10. California Academy of Sciences. Occurrence dataset https://doi.org/10.15468/bvoyqy. Accessed via GBIF.org

Schick S, Kielgast J, Rödder D et al (2010) New species of reed frog from the Congo basin with discussion of paraphyly in Cinnamon-belly reed frogs. Zootaxa 2501:23–36

Schiøtz A (1999) Treefrogs of Africa. Editions Chimaira, Frankfurt am Main, 350 pp

Schiøtz A, Van Daele P (2003) Notes on the treefrogs (Hyperoliidae) of North-Western Province, Zambia. Alytes 20:137–149

Schmidt KP (1936) The amphibians of the Pulitzer-Angola expedition. Ann Carnegie Mus 25:127–133

Steindachner F (1867) Reise der österreichischen Fregatte Novara um die Erde in den Jahren 1857, 1858, 1859 unter den Bafehlen des Commodore B. von Wüllerstorf-Urbair. Zologischer Theil. 1. Amphibien. Wien: K. K. Hof- und Staatsdruckerei. 70 p + 5pls

Tandy M, Keith R (1972) *Bufo* of Africa. In: Blair WF (ed) Evolution in the Genus *Bufo*. University of Texas Press, Austin/London, pp 119–170

Turner AA, Channing A (2017) Three new species of *Arthroleptella* Hewitt, 1926 (Anura: Pyxicephalidae) from the Cape Fold Mountains, South Africa. Afr J Herpetol 66:53–78

Uyeda JC, Drewes RC, Zimkus BM (2007) The California Academy of Sciences Gulf of Guinea expeditions (2001, 2006) VI. Proc Calif Acad Sci 58(13–22):367

Van Dijk DE (1966) Systematic and field keys to the families, genera and described species of southern African anuran tadpoles. Ann Natal Mus 18:231–286

Van Dijk DE (1971) A further contribution to the systematics of Southern African anuran tadpoles—the genus *Bufo*. Ann Natal Mus 21:71–76

Waddle JH (2006) Use of amphibians as ecosystem indicator species. PhD Thesis. University of Florida, Gainesville

Whiles MR, Lips KR, Pringle CM et al (2006) The effects of amphibian population declines on the structure and function of Neotropical stream ecosystems. Front Ecol Environ 4(1):27–34

Zimkus BM, Rödel MO, Hillers A (2010) Complex patterns of continental speciation: molecular phylogenetics and biogeography of sub-Saharan puddle frogs (*Phrynobatrachus*). Mol Phylogenet Evol 55(3):883–900

Zimkus BM, Lawson LP, Barej MF et al (2017) Leapfrogging into new territory: how Mascarene ridged frogs diversified across Africa and Madagascar to maintain their ecological niche. Mol Phylogenet Evol 106:254–269

Chapter 13
The Reptiles of Angola: History, Diversity, Endemism and Hotspots

William R. Branch, Pedro Vaz Pinto, Ninda Baptista, and Werner Conradie

Abstract This review summarises the current status of our knowledge of Angolan reptile diversity, and places it into a historical context of understanding and growth. It is compared and contrasted with known diversity in adjacent regions to allow insight into taxonomic status and biogeographic patterns. Over 67% of Angolan reptiles were described by the end of the nineteenth century. Studies stagnated during the twentieth century but have increased in the last decade. At least 278 reptiles are currently known, but numerous new discoveries have been made during recent surveys, and many novelties await description. Although lizard and snake diversity is currently almost equal, most new discoveries occur in lizards, particularly geckos and lacertids. Poorly known Angolan reptiles and others from adjacent regions that

W. R. Branch (deceased)
National Geographic Okavango Wilderness Project, Wild Bird Trust, Hogsback, South Africa

Department of Zoology, Nelson Mandela University, Port Elizabeth, South Africa

P. Vaz Pinto
Fundação Kissama, Luanda, Angola

CIBIO-InBIO, Centro de Investigação em Biodiversidade e Recursos Genéticos, Universidade do Porto, Campus de Vairão, Vairão, Portugal
e-mail: pedrovazpinto@gmail.com

N. Baptista
National Geographic Okavango Wilderness Project, Wild Bird Trust, Hogsback, South Africa

CIBIO-InBIO, Centro de Investigação em Biodiversidade e Recursos Genéticos, Universidade do Porto, Campus de Vairão, Vairão, Portugal

Instituto Superior de Ciências da Educação da Huíla, Rua Sarmento Rodrigues, Lubango, Angola
e-mail: nindabaptista@gmail.com

W. Conradie (✉)
National Geographic Okavango Wilderness Project, Wild Bird Trust, Hogsback, South Africa

School of Natural Resource Management, Nelson Mandela University, George, South Africa

Port Elizabeth Museum (Bayworld), Humewood, South Africa
e-mail: werner@bayworld.co.za

B. J. Huntley et al. (eds.), *Biodiversity of Angola*,
https://doi.org/10.1007/978-3-030-03083-4_13

may occur in the country are highlighted. Most endemic Angolan reptiles are lizards and are associated with the escarpment and southwest arid region. Identification of reptile diversity hotspots are resolving but require targeted surveys for their delimitation and to enable protection. These include the Kaokoveld Centre of Endemism, Angolan Escarpment and the Congo forests of the north. The fauna of Angola remains poorly known and under-appreciated, but it is already evident that it forms an important centre of African reptile diversity and endemism.

Keywords Herpetofauna · History · Priority studies · Reptilia · Review

Introduction

> Systematic biology is the backbone of biology in that it describes the taxa and their relationships which then serve as the objects of research. (Uetz and Stylianou 2018)

> The extensive territory of Angola is for herpetologists one of the least known parts of Africa. This is particularly unfortunate because there are indications that it may be one of the most interesting areas of the continent. (Gans 1976)

The need to classify things is a basic human need. Initially it was simply utilitarian, driven by the necessity for rural people to know what was edible, venomous, poisonous, or useful. As the world's diverse civilizations developed all were faced with the need to refine this knowledge, and as they migrated, came together, experienced new habitats and new life forms, the need for classification became essential. Only when new technological innovations in such things as sailing and weaponry allowed the reach of various nations to become global, did universal categorisation and classification really need to be become standardised. Driven by the Enlightenment and during the rise of critical thinking and the scientific revolution there began the first steps in developing a universally recognised system for classifying Life. The current classification system was initiated by Carl Linnaeus (1707–1778), a Swedish botanist, physician, and zoologist who formalised the modern system of naming organisms, now known as binomial nomenclature. During the 250+ years since the Linnaean revolution, the rules used to name this diversity have been refined and modified, and increasingly sophisticated methods have been developed to gain insight into the relationships of its components.

Since the equally important Darwinian Revolution and awareness of the evolutionary relatedness of Life, modern systematics places emphasis on revealing patterns of relationship among groups, often figuratively represented as trees or cladograms. The branches of these 'trees' are monophyletic when they include only the descendants of a common ancestor. All modern classifications comprise hypotheses represented as phylogenies of nested groups (clades) exhibiting monophyly. Biochemical adjuncts to traditional taxonomy have proliferated since the middle of the last century, but detailed multi-taxon genomic analysis, linked with increasingly sophisticated computer processing of sequence data, is a phenomenon of the twenty-first century. These recent technological advances have allowed objective assessments of hypotheses of phylogenetic relationships. It is important to emphasise that the assignment of any individual specimen, first to a species and then to any higher taxonomic group, tests hypotheses of relationships. The placement of a specimen at any level in the nomenclatural hierarchy, from species to phylum, must conform to the definitions of those groups. It should be stressed, moreover, that taxonomy is a dynamic discipline and that every assignment of a specimen to a species or higher taxonomic group is a hypothesis of relationship. It is always subject to revision in the light of new knowledge.

It has become increasingly obvious that species may result from different mechanisms and histories, and there is increasing use of evolutionary and phylogenetic species concepts to reflect hypotheses about the boundaries of past and present gene transfer within evolutionary lineages of Life's diversity. Many phylogenies based on molecular/genetic data conflict with historical ideas of relationships previously based solely or largely on morphological analysis. It is evident that morphology is often conservative (maintained by selective pressures) that may mask underlying cryptic genetic diversity. This awareness has led to the burgeoning description of new species, genera, and higher categories.

History: Early Research on Angolan Reptiles

Early studies on the Angolan herpetofauna have been summarised by Baptista et al. (2019) in this volume, also see the recently published Angola Reptile and Amphibian Atlas (Marques et al. 2018). Other recent summaries can be found in introductions to regional herpetofaunas, e.g. Ceríaco et al. (2014a, 2016a) and Conradie et al. (2016). To avoid duplication much will not be repeated here, where instead emphasis is placed on the main publications discussing Angolan reptiles during and after the colonial period, and particularly the periodic attempts to overview its diversity (Bocage 1895; Monard 1937).

José Vincente Barbosa du Bocage is rightly known as the 'Father of Angolan Herpetology'. He was, however, more than just a scientist and for much of his life held multiple positions in government, administration and science, often simultaneously. De Almeida (2011) reviews diverse aspects of his multifaceted life; Madruga (2013) discusses the development of his scientific career; and Gamito-Marques (2017) explains Bocage's role in the foundation of National Museum of Lisbon and his importance, via his contacts with collectors in various Portuguese colonies as well as other zoologists at major European museums, in developing the collections and status of the museum. A list of his scientific publications is available at TRIPLOV (2018). In his first attempt to review the known Angolan herpetofauna he listed 26 reptile species from the Congo and 57 reptiles and amphibians from Angola in the Museu de Lisboa collection (Bocage 1866). Part of this material came from José de Anchieta collected during a zoological expedition in 1864 to Rio Quilo, Cabinda, and the coast of Loango, and other material was collected by Bayão Pinheiro when a military commander in the Duque de Bragança (Calandula) district. After nearly 30 years of subsequent study, during which Bocage published at least a paper a year on the herpetofauna of Portugal's African colonies (see full list in TRIPLOV 2018), he again summarised the Angolan herpetofauna in his monographic *Herpetology d'Angola et du Congo* (Bocage 1895). In this review he more than trebled his previous summary, recording 143 reptile and 39 amphibian species from Angola. Of these Bocage himself described 40 reptile species that are still considered valid, and of which no less that 26 (65%) had been collected by José de Anchieta, seven of which are still named in his honour. Following Bocage's retirement, herpetological studies at Lisbon were continued, but with less intensity, by Bethencourt Ferreira (1897, 1900, 1903, 1904, 1906), who added a number of additional species to the country list but described only one new snake, *Typhlops bocagei* (Ferreira 1904), that was later placed in the synonymy of *Afrotyphlops lineolatus* (Jan, 1864).

George A Boulenger of the British Natural History Museum remains the most prolific herpetologist of all time, and described a remarkable 659 reptiles still recognised today – no less than 5% of the world's 13,000+ known reptiles (Uetz and Stylianou 2018). In 1905 Boulenger published a note on the amphibians and reptiles collected by WJ Ansorge during a prolonged visit (1903–1905) to Angola (Boulenger 1905). The material included three new species of frogs and a snake, of which two, *Rana ansorgii* (= *Ptychadena ansorgii*) and *Psammophis ansorgii,* were named in the collector's honour. Two years later Boulenger described another three lizards and a frog collected by Ansorge, including the gecko *Phyllodactylus ansorgii,* the skinks *Mabuia ansorgii* (*Trachylepis sulcata ansorgii*) and *Mabuia laevis* (*Trachylepis laevis*), and the frog *Rana bunoderma* (= *Ptychadena bunoderma*) (Boulenger 1907a), soon followed (Boulenger 1907b) by another new frog from 'Mossâmedes' (actually Catequero, Cunene), *Rana cryptotis* (= *Tomopterna cryptotis*). Boulenger (1915) also prepared a *List of the snakes of Belgian and Portuguese Congo, Northern Rhodesia and Angola* in which he documented 139 snakes for the region, of which only 57 were considered to occur in Angola. This was lower than Bocage's (1895) assessment, but by this time Boulenger had completed his monographic, three volume 'Catalogue' of the World's snakes (Boulenger 1893, 1894, 1896), in which he had synonymised many of Bocage's taxa or shown some to be synonyms. This included Angola's most iconic snake, *Vipera heraldica,* which Boulenger (1896) surprisingly and incorrectly synonymised with *Bitis peringueyi,* and Bogert (1940) continued the confusion. There it languished until revalidated by Mertens (1958).

The dramatic discovery of the Giant Sable in Angola early in the twentieth century led to numerous expeditions to collect them as hunting trophies or taxidermy specimens for European and American museum dioramas. Some expeditions collected additional fauna, although the herpetological collections of the Vernay Angola Expedition (VAE 1925) seem to have been plagued by poor documentation. Although various specimens were incorporated in diverse taxonomic reviews, no publication dedicated to the full herpetological results was produced. Bogert (1940) incorporated 202 VAE snakes into his review of African snakes, but 10 species and 42 (21%) specimens lacked detailed locality information and were simply listed as from 'Angola'. Three new snakes were described from Angolan material: Vernay's File Snake (*Mehelya vernayi,* = *Limaformosa vernayi* Broadley et al. 2018) from Hanha, and Cowles' Shield Cobra (*Aspidelaps lubricus cowlesi*) and the Western

Banded Spitting Cobra (*Naja nigricollis nigricincta*) from Munhino. The first two have subsequently been discovered at numerous localities in northern Namibia (Haacke 1981; Broadley and Baldwin 2006), but remain known in Angola only from the type or a few other specimens, respectively. Loveridge (1944) described two new geckos (*Afroedura bogerti* and *Pachydactylus scutatus angolensis*) on VAE material, and Stanley et al. (2016) discussed VAE *Cordylus* material labelled simply as 'Angola' and that they assigned morphologically to a new species, *Cordylus namakuiyus*, discovered in the Namibe region. The description of at least one other new species from old VAE material is also in preparation (*Ichnotropis* sp. Branch in prep.).

The main targets of the Pulitzer Angola Expedition (1930–1931) were birds and mammals, but Rudyerd Boulton, who had earlier accompanied the VAE, also collected reptiles and amphibians. Karl Schmidt (1933, 1936) documented the reptiles and amphibians, respectively, collected during the expedition from diverse sites in the centre and south of the country. The reptiles included two new species, but *Lygodactylus laurae* was quickly synonymised when Schmidt realised he had overlooked Bocage's (1896) earlier description of *L. angolensis*. His description of *Rhoptropus boultoni* not only honoured Rudyerd's contribution to the collection of Angolan reptiles, but was also the first record of this interesting diurnal rupicolous gecko genus for the country. Two other new subspecies were also described, of which *Pachydactylus bibronii pulizerae* was latter transferred to *Chondrodactylus* (Bauer and Lamb 2005), and has also been recently validated as a full species, *C. pulizerae*, that is mainly restricted to Angola but also extends into northern Namibia (Heinz 2011; Ceríaco et al. 2014a). The Angolan Savanna Monitor (*Varanus albigularis angolensis*) was described by Schmidt (1933) from 'Gaúca, Bihe' (= Zaúca River, Malanje; Crawford-Cabral and Mesquitela 1989). Although additional material had been collected the validity of the morphological diagnosis (small nuchal scales and larger body scales) has not been reassessed, and neither has its genetic affinities.

In 1933–1934 Karl Jordan, an entomologist, collected through northern Namibia and Angola and made well documented collections and published a detailed itinerary of his trip (Jordan 1936). Among these was an important herpetological collection, particularly from scarp forest habitats at Congulo and Quirimbo. These were studied by Parker (1936), who recorded diverse Congo Basin snakes previously unknown from Angola; e.g. *Philothamnus heterodermus, Thelotornis kirtlandii, Toxicodryas blandingii, T. pulverulenta, Pseudohaje goldii, Chamaelycus parkeri* and *Hormonotus modestus,* as well as the new wolf snake *Lycophidion ornatum.*

Swiss zoologist Albert Monard explored Angola during two extensive trips (July 1928–February 1929 and April 1932–October 1933) that resulted in extensive reviews of Angolan birds (Monard 1934), mammals (Monard 1935), reptiles (Monard 1937), and amphibians (Monard 1938). Ceríaco et al. (2016a, b) note that Monard was so inspired by Angolan biodiversity that he unsuccessfully championed for the development of a local Natural History Museum that he offered to direct and manage. His detailed reptile 'Contribution' (Monard 1937) was the first overview of Angolan reptiles subsequent to Bocage's (1895) monograph, and within it Monard presented taxonomic updates and also initiated the first attempts to generate a biogeographic overview of the herpetofauna (see section below). He noted that only 19 lizards, 10 snakes and a single terrapin had been added to the reptile fauna of Angola, and he even envisaged (incorrectly!) that most of the knowledge on the subject was complete, and he therefore concentrated on understanding biogeographic patterns. However, eight of the 19 additional lizards and three of the 10 snakes he included had already been synonymised (Boulenger 1915) or subsequently were. Certainly Monard seemed little interested in taxonomy and described relatively few novelties, only one of which may remain valid. The rejected taxa include: the worm lizards *Amphisbaenia ambuellensis* (= *Zygaspis quadrifrons*), *Monopeltis granti kuanyamarum* (= *Dalophia pistillum*), *M. devisi* (= *Monopeltis anchietae*), and *M. okavangensis* (= *M. anchietae*); the serpent plated lizard *Tetradactylus lundensis* (= *T. ellenbergeri*); and the skink *Mabuia striata angolensis*. The latter, however, remains problematic and under investigation (Conradie et al. 2016). Marques et al. (2018) provided a replacement name, *Trachylepis monardi* nom. nov., to stabilise taxonomy.

Throughout the early part of the twentieth century various other publications discussed small collections made by explorers (e.g. Angel 1921, 1923; Mertens 1938). All added new locality records within and for the country, and also described a number of new species (some no longer valid, e.g. *Psammophis rohani*, Angel 1921). Some reptile discoveries were especially serendipitous, e.g. the discovery of the new lizard species *Ichnotropis microlepidota* (Marx 1956) based on three specimens all found in the crop of a Dark Chanting Goshawk (*Melierax metabates*) collected during a bird survey of Mount Moco, that still awaits discovery in the wild!

In 1952–1954 the Hamburg Museum made an expedition to various locations in western Angola to collect mammals and herpetofauna, the most important being numerous additional snake records from forested habitats at Piri-Dembos (= Piri, Cuanza-Norte) (Hellmich 1957a, b). These confirmed, and sometimes added to Parker's (1936) records of snakes from the scarp forest isolates of Congulo and Quirimbo. These included (e.g.): *Philothamnus heterodermus, Thrasops flavigularis, Toxicodryas blandingii, T. pulverulenta, Gonionotophis poensis, Pseudohaje goldii, Atheris squamigera* and *Bitis nasicornis.* All are Congo Basin species and form an important biogeographic component of Angolan reptile diversity. Hellmich undertook a follow up expedition in 1954–1955, but his expedition suffered delays in obtaining permits and he missed the wet season activity and therefore shifted his survey to more open habitats in the southern provinces. There he undertook some of the first ecological studies on Angolan reptiles that were briefly discussed in a six-part series of popular articles on his Angolan travels (Hellmich 1954–1955). On his return he studied the reptile collections of the combined Hamburg expeditions (Hellmich and Schmelcher 1956; Hellmich 1957a, b), but the amphibians were only studied much later (Ceríaco et al. 2014b). As with Monard, Hellmich discovered relatively few taxonomic novelties, i.e. the lizards *Gerrhosaurus nigrolineatus ahlefeldti* (currently not considered valid) and *Agama agama mucosoensis* (now a full species; Wagner et al. 2012).

Between 1950 and 1960 the Belgian herpetologist Raymond F Laurent lived in Rwanda and Katanga (then Belgian colonies) and undertook detailed studies on varied herpetological groups in the Congo Basin, describing numerous new species and subspecies. During this period he studied the herpetological collections of the Museu do Dundo in northeast Angola, made by António de Barros Machado, the museum's director. Summaries of the museum's snake collections appeared first (Laurent 1950, 1954), followed by another report on Dundo material including Machado's extensive herpetological collection from the arid southwest region of the country (Laurent 1964). This was completed after Laurent had moved to the United States (1961) and the report described a number of new Angolan species, including Bogert's Speckled Western Burrowing Skink (*Typhlacontias bogerti*), two Namib Day geckos (*Rhoptropus taeniostictus* and *R. boultoni montanus*), and finally Hellmich's Wolf Snake (*Lycophidion hellmichi*), based (in part) on material collected during the Hamburg expeditions. It also included additional information of many previously poorly known species, as well as making ecological observations. It set a new standard for herpetological research in the area, but sadly was the final major Angolan herpetological work of the colonial period. Laurent did not study the historical collections in the Museu de Lisboa, and therefore did not re-assign Bocage's original material to his new taxa or identify significant new distribution records. This was unfortunate as he was one of the last herpetologists working on the Angolan herpetofauna before the disastrous fire that destroyed (1978) the collections that Bocage had studied, as well as much of his correspondence with collectors and fellow researchers. Manaças reported on collections of lizards (Manaças 1963), snakes (Manaças 1973), and venomous snakes (Manaças 1981) from Angola.

Bringing Knowledge of Angolan Reptiles into the Modern Era

Awareness of the interesting reptiles of the Angolan Namib region started incidentally following expeditions in the 1950s by the enthusiastic entomologist Charles Koch of the Transvaal Museum (TMP, now the Ditsong Natural History Museum, Pretoria, South Africa). Koch did much to inventory the amazing diversity of the tenebrionid beetle fauna of the Namib Desert, and much of his collecting involved walking at night in the desert with a pressure lamp. In addition to his numerous beetle discoveries Koch also collected many nocturnal and terrestrial reptiles, particularly geckos. These he gave to his colleague at the TMP, the Curator of Lower Vertebrates Vivian FitzSimons. Koch visited the northern Namib in Angola on four occasions (1951–1957), accompanied on the last trip by the Swedish zoologists Lundholm and Rudebeck. The herpetological collections during these trips were significant, and FitzSimons (1953, 1959) described a new genus of plated lizard *Angolosaurus* (now subsumed within *Gerrhosaurus*) as well two new species, *Pachydactylus caraculicus* and *Prosymna visseri*. Unreported, however, were many of Koch's numerous other reptile discoveries, including new records of the iconic Namib Web-footed Gecko (*Palmatogecko rangei,* now included in *Pachydactylus*) in 1951 and 1954, then unknown from the Angolan Namib region. Laurent (1964) described the new Angolan Namib Day Gecko *Rhoptropus taeniostictus* from Angola, although nine specimens had already been collected by Koch during his trips, but remained undescribed. Also unrecorded were nine specimens of *Pachydactylus scutatus angolensis* from Lungo, Lucira and São Nicolau, the first collected since the description of the species by Loveridge (1944), and also 13 specimens of *Chondrodactylus fitzsimonsi*, at the time known from only one Angolan specimen (Pico Azevedo, Schmidt 1933).

In 1964 Wulf Haacke, born in Namibia, became the then Transvaal Museum's herpetologist with a special interest in the arid western areas of southern Africa. In March–April 1971 he undertook his first trip to Angola, concentrating on the northern extension of the Namib Desert into southwestern Angola. A follow up trip in April–June 1974 targeted specific genera to confirm the northern range limits and taxonomic status of *Cordylus, Cordylosaurus, Gerrhosaurus, Pachydactylus, Afroedura* and *Rhoptropus*. Both trips were exceptionally successful resulting in over 2000 specimens, the largest herpetological collections ever assembled by one researcher in Angola. Although the amphibian collections made during these trips were reviewed by Poynton and Haacke (1993), the vast majority of the numerous new and rare reptiles contained in these collections were never formally published. Haacke's second trip in 1974 was designed in particular to collect new material for his proposed thesis and revision of *Rhoptropus*. Prior to this trip, and excluding Koch's undescribed material, less than 30 specimens of *Rhoptropus* were known from Angola (Bocage 1895; Parker 1936; Mertens 1938; Laurent 1964). At the end of Haacke's surveys the Transvaal Museum held 650 specimens of the genus, included nearly 250 specimens of *R. barnardi* from over 25 localities, nearly 50 specimens of *R. biporosus*, and seven of *R. afer.* At the time *R. barnardi* in Angola was known from very few specimens (Bocage 1895; Schmidt 1933; Parker 1936; Laurent 1964) and *R. biporosus* was unknown in Angola and restricted to northern Namibia. The status of *R. afer* in Angola was particularly confused. Bocage's (1895) knowledge of Namib Day geckos (*Rhoptropus*) seems to have been limited, and he considered specimens from diverse localities from coastal Moçâmedes and the interior to Capangombe to all be applicable to *R. afer* Peters 1869. However, he noted that his material had 6–8 preanal pores whilst *R. afer*, as Peters (1869) had correctly recorded, had none. Schmidt also recorded *R. barnardi* for the first time from Angola, noting numerous other specimens from the Moçâmedes region in the British and Zoologisches museums that agreed with *R. barnardi.* These may have been the source of Bocage's confusion. The few recent records of Angolan *R. afer* have all been restricted to the vicinity of the Cunene River mouth, and it is evident that Bocage's material from further north was not based on true *R. afer.*

Due to the limited access to Angola during the protracted civil war the TM expeditions to Angola were curtailed, and for the next 34 years studies on Angolan reptiles were based on museum material collected earlier. Details of some of the geckos collected by the TM expeditions were incorporated into Haacke's studies on the burrowing geckos of southern Africa, which included the first records for Angola (Haacke 1976a) for the *Palmatogecko rangei* and *Kaokogecko vanzyli* (both now included in

Pachydactylus), and by implication *Colopus* (= *Pachydactylus*) *wahlbergii*, known to Haacke (1976b) from three specimens (TM 38526—28) from the Angola-SWA border, 18 degrees E). The bizarre and iconic Plume-tailed Gecko *Afrogecko plumicaudus*, collected during Haacke's 1971 trip, was of problematic taxonomic affinities and not described until much later (Haacke 2008). Its true affinities, however, were finally resolved later when fresh material was available for genetic analysis and it was placed in the monotypic genus *Kolekanus* (Heinicke et al. 2014). The 64 burrowing skinks of the genus *Typhlacontias* collected during these expeditions also formed an integral part of Haacke's (1997) revision of the genus, and which led to the recognition of a long overlooked Namib species, *T. johnstonii*, previously confused (Bocage 1895; Monard 1937) with *T. punctatissimus*, and described from Porto Alexandre (=Tômbwa) at the northern limit of its range. A new Angolan species, *T. rudebecki*, was also described, based on a single specimen collected from São Nicolau during Koch's 1956 expedition. Laurent's (1964) species *T. bogerti* was treated as a northern subspecies of the Speckled Burrowing Skink, *T. punctatissimus bogerti* (Haacke 1997). A number of other new Angola records were discovered by Haacke during his trips. The rupicolous gecko *Pachydactylus oreophilus*, described from northern Namibia (McLachlan and Spence 1967), was known only from the types until Haacke discovered similar material from numerous localities in southwest Angola. No additional Angolan material was collected until the Huntley expedition in the region in 2009 (see below), when it was realised that Angolan material was not conspecific with Namibian *P. oreophilus*. The status of the Angolan material is currently being investigated (Branch et al. in prep.). Haacke also collected the first records of true *P. scutatus* from Angola, as well as additional *P. angolensis* (Branch et al. 2017). Finally, Broadley (1975) referred some small skinks collected by Haacke to *Trachylepis lacertiformis*, creating a zoogeographic enigma as the nominotypic population of this small skink is restricted to the lower Zambezi River valley. Fuller details of the Koch and Haacke collections and other recent material will be incorporated into a full review of the herpetofauna of the Angolan Namib region (Branch in prep.).

Legless burrowing worm lizards (Amphisbaenidae) are rarely found due to their ability to burrow deep underground. Carl Gans (1976) described three new species from Angola, including *Monopeltis luandae* based on fresh (1971) and historical (1892) material from the Luanda region, and *M. perplexus* on material from the Vernay-Angola expedition collected from 'Hanha or Capelongo' in 1925. Similarly, Gans (1976) re-assessed old and new material when describing *Dalophia angolensis* from Calombe near Vila Luso (= Luena), and reassigned specimens identified as *M. ellenbergeri*, and then as *M. granti transvaalensis* (Monard 1937), to *D. angolensis*. None of Gans' three new species had been rediscovered in the intervening 40+ years until recently, when *M. luandae* was rediscovered from near the type locality (Branch et al. 2018). In a companion paper that radically changed understanding of the taxonomy of worm lizards in the southern half of Africa, Broadley et al. (1976) revised the genera *Monopeltis* and *Dalophia*. The revision of Broadley et al. (1976) affected most of the early names applied to Angolan worm lizards. The first large worm lizard described from Angola was *Lepidostsernon* (*Phractogonus*) *anchietae* Bocage 1873, from 'Humbe, dans l'intérior de Mossamedes', later transferred to the genus *Monopeltis* (Boulenger 1885). Broadley et al. (1976) relegated Monard's (1937) species *M. okavangensis* and *M. devisi* to *M. anchietae,* which is now known to have a wider range in northern Namibia and adjacent Botswana. *Monopeltis vanderysti vilhenai*, described by Laurent (1954) from Dundo, Angola, was not recognised by Broadley et al. (1976) and returned to *M. vanderysti,* which is widely distributed in the Congo region. The Dundo specimen remains the only Angolan record of the species. *Monopeltis capensis* in Angola was first recorded by Bocage (1873), and later by Monard (1937). Although provisionally placed in the *M. capensis capensis* Group B (Broadley et al. 1976), with a wide range through the Kalahari region (Northern Cape, South Africa, through Botswana to southern Angola), it was later treated as a separate species, *M. infuscata* Broadley 1997a. Monard's (1937) species *Monopeltis granti kuanyamarum,* described from a single specimen from Mupanda, was transferred to *Dalophia pistillum* (Broadley et al. 1976). The only Angolan specimen of *Dalophia ellenbergeri* was collected whilst trench digging during hostilities at Cuito Cuanavale (Branch and McCartney 1992; Broadley 1997b). Gray (1865) described *Dalophia welwitschii* from Pungo Andongo, and this has not been rediscovered. It is the type species for the genus *Dalophia,* and Gans (2005) was obviously in error when treating it as *M. welwitschii* and yet still continuing to recognise *Dalophia.* A phylogeny of African amphisbaenids (Measey and Tolley 2013), albeit based on poor taxon sampling, recovered *Monopeltis* and *Dalophia* as monophyletic clades, supporting the use of *Dalophia welwitschii*.

Two species of round-headed worm lizards of the genus *Zygaspis* are now known to occur in southeast Angola, but the genus was unknown to Bocage from Angola, and the first record from the country was Monard's (1931) description of *Amphisbaena ambuellensis* from 'Chimporo' (= Tchimpolo). This was subsequently synonymised with *A. quadrifrons* by Loveridge (1941) with some misgivings, and subsequently transferred to *Zygaspis* by Alexander and Gans (1966). It remained known only from Monard's material for many years, but has recently been collected in southern Angola (Conradie et al. 2016, Baptista et al. in prep.), and the availability of Monard's *ambuellensis* for this material is being reassessed. More recently, Laurent (1964) recorded *A. q. capensis* from Alto Chicapa in northeast Angola, which was shown to be the new species *Zygaspis nigra* by Broadley and Gans (1969). This small black worm lizard is near endemic to eastern Angola, with records from adjacent Zambia (Kalobo, Broadley and Gans 1969; Ngonya Falls, Pietersen et al. 2017) and Namibia (Katima Mulilo, Broadley and Gans 1969). Recent material is known from the Okavango catchment (Conradie et al. 2016).

Following the cessation of hostilities, modern biodiversity surveys were initiated by Brian Huntley with the multinational SANBI/ISCED/UAN Angolan Biodiversity Assessment and Capacity Building Project (Huntley 2009). Surveys were undertaken by botanists and zoologists in various habitats between Lubango and the Cunene River, and 15 Angolan students were involved in fieldwork and training sessions. The immediate reptile highlights of the survey involved the discovery of two new species of the lacertid *Pedioplanis* (Conradie et al. 2012a), two specimens of the rare Shovel-snout Snake *Prosymna visseri* were collected at Espinheira, Iona National Park, only the 5th and 6th Angolan specimens since its description

(FitzSimons 1959); the 1st record of the Namib Wolf Snake (*Lycophidion namibianum*) from Angola, again at Espinheira; the southernmost records of the newly-described Plume-tailed Gecko (*Afrogecko plumicaudatus* Haacke 2008) that allowed its generic assignment to later be readjusted; and topotypic material of the rare chameleon (*Chamaeleo anchietae*) were collected around Estação Zootécnica. This chameleon has an unusual, disjunct distribution with scattered populations (treated as separate subspecies by Laurent 1951) from the Upemba region, Democratic Republic of the Congo (DRC) and Udzungwa Mountains, Tanzania. The status of these disjunct *C. anchietae* populations is currently under investigation (Branch et al. in prep.). A new species of reed frog, *Hyperolius chelaensis*, completed the new discoveries (Conradie et al. 2012b). Following the success of the 2009 survey, another expedition was organised in 2011 to Lagoa Carumbo, the second-largest Angolan freshwater lake situated in Lunda Norte Province (Huntley and Francisco 2015). The herpetological results were summarised by Branch and Conradie (2015). Significant herpetological discoveries included the discovery of at least two new species of frog, one described (*Hyperolius raymondi* Conradie et al. 2013), and the description of the other (*Amnirana* sp.) is in preparation (Jongsma et al. 2018), and also the first record for Angola (Branch and Conradie 2013) of the Banded Water Cobra (*Naja annulata*). Other reports include new insights into the distributions of venomous snakes, such as Jameson's Green Mamba, *Dendroaspis jamesoni* (Vaz Pinto and Branch 2015) and the Gaboon Adder, *Bitis gabonica* (Oliveira et al. 2016), as well as a recent summary of Angolan venomous snakes (Oliveira 2017).

The unique World Heritage site of the Okavango Delta is situated in Botswana, but depends on the Okavango drainage, which arises and is almost entirely contained within southeastern Angola. During the last six years a series of international collaborative surveys have been undertaken to explore this poorly known region of southeast Angola, and to understand the hydrology and biodiversity of the Okavango drainage. The first surveys were organised by the Okavango River Basin Water Commission (OKACOM), in accord with the Angolan National Action Plan for the Sustainable Management of the Cubango/Okavango River Basin (OKACOM 2011), and occurred in the lower catchments of the Cubango and Cuito rivers (Brooks 2012, 2013). More recent surveys (2015–2018) formed part of the ongoing National Geographic funded Okavango Wilderness Project (NGOWP 2018), which have intensely surveyed the source lakes of the major Okavango tributaries in an unexplored region where the headwaters of the Cuanza, Zambezi and Okavango basins meet. The herpetological results of the OKACOM surveys (2012–2013) and first phase of the NGOWP surveys were presented by Conradie et al. (2016), who also reviewed the region's herpetofauna. In total 67 reptiles species are now known from the region, comprising 38 snakes, 32 lizards, five chelonians, and a single crocodile (NGOWP 2018). Three reptiles new for Angola, including *Causus rasmusseni* (although the specific status of this taxon still requires genetic confirmation), *Acontias kgalagadi kgalagadi* and *Panaspis maculicollis* were discovered (Conradie and Bourquin 2013; Medina et al. 2016). The results of more recent surveys (2016–2018) were presented (Conradie et al. 2017) and fuller details are being prepared for publication, and online species lists are planned for public access.

Contemporaneous with the above surveys a number of other Angolan biodiversity initiatives began. In partnership, the Kimpa Vita University in Uíge and Dresden University, Germany, undertook herpetological surveys of Serra do Pingano, Uíge Province, discovering diverse tropical Congo Basin species (Ernst 2015), including two frogs previously unrecorded for Angola (Ernst et al. 2014, 2015). In addition the California Academy of Sciences in conjunction with the National Institute of Biodiversity and Conservation Areas and the Ministry of the Environment of Angola (MINAMB/INBAC) initiated an ongoing Atlas of Angolan amphibian and reptiles (Marques et al. 2014, 2018). Various areas have been surveyed and preliminary results published in the scientific and popular literature (Ceríaco et al. 2014a, b, 2016a, b, 2018b). As part of the Southern African Science Service Centre for Climate Change and Adaptive Land Management program (SASSCAL) the Instituto Superior de Ciências da Educação da Huíla (ISCED), Lubango, has been undertaking herpetofaunal monitoring at several areas in Huíla Province and elsewhere in Angola (Baptista et al. 2018, 2019), with emphasis on the escarpment. A herpetofaunal archive is also being developed at ISCED.

Angolan citizen science is in its infancy, but the FaceBook site *Angola Ambiente* is a public group where members post observations (https://www.facebook.com/groups/1045499302182009/). It is designed "to raise awareness of the fantastic fauna and flora of this magnificent country", and requests observations with detailed locality data. It includes irregular lists of sightings with locality details in support of mapping initiatives.

Checklist of Angolan Reptiles

How Many Species?

The first attempt to summarise the herpetofauna of Angola was undertaken by JV Barbosa du Bocage (1866), who listed 50 reptile species from Angola in the Museu de Lisboa collection, including 23 snakes, 21 lizards, four chelonians, and one crocodilian. After nearly 30 years of study he again summarised the Angolan herpetofauna in his monographic *Herpetology d'Angola et du Congo* (Bocage 1895) in which he listed 143 reptile and 39 amphibian species from the country. Of these Bocage had described 37 of the taxa (although not all are now recognised). During two trips to Angola (1928–1929 and 1932–1933) the Swiss collector Albert Monard made important collections of amphibians and reptiles. In his monograph (Monard 1937) he presented an updated checklist of Angolan reptiles, listing 169 reptile species in 10 families and 28 genera. There has been no subsequent updated and checklist of the country's reptiles, although Blanc and Fretey (2002) noted a total of 257 Angolan reptile species and published a breakdown of its composition. However, no species list was included and therefore it is impossible to assess the accuracy or the validity of the species included. In contrast, the online Reptile Database (Uetz et al. 2018, as on 14 October 2018) currently generates a list of 255 reptiles for Angola, but unfortunately although close to the existing count, it is inaccurate in numerous respects. Some species have been included that are unrecorded from the country but may occur in the country (e.g. a snakes from the Congo Basin *Calabaria reinhardti,* Namibian *Lygodactylus lawrencei,* etc.). Moreover, many other species are duplicated and listed under both their historical and current taxonomic assignments (e.g. *Agama hispida = A. aculeata, Chamaesaura macrolepis = Ch. miopropus, Cordylus cordylus = C. namakuiyus, C. vittifer = C. machadoi,* etc.) These inaccuracies have been discussed in the recent Atlas and being rectified (see Marques et al. 2018 Appendix Table A2).

Currently (as of mid-2018), there are 278 reptile species recorded from Angola, comprising 15 chelonian, three crocodilian, 132 lizard and 128 snake species. Table 13.1 details the historical growth of knowledge of reptile diversity in Angola based on summaries in Bocage (1866, 1895), Monard (1937), and this study. Table 13.2 summarises the number of genera, species, and endemic taxa in the major reptile groups in Angola. Updated checklists of the major reptile groups, including details of common and scientific names, historical scientific names used by Bocage (1895) for the current taxa, as well as their endemic and conservation status are summarised in: Appendix 1 – chelonians; Appendix 2 – lizards; and Appendix 3 – snakes. Contained within these checklists are 43 Angolan species named by Bocage, i.e. 15.5% of the current reptile diversity. This is less than the 25.9% (37 of 143) in Bocage's (1895) summary, but no other researcher has described more from Angola.

Table 13.1 Historical development of reptile diversity in Angola based on summaries in Bocage (1866, 1895), Monard (1931), and this study

	Bocage		Monard	This study
Group	1866	1895	1937	2018
Snakes	23	71	81	128
Lizards	21	59	78	132
Chelonians	4	8	9	15
Crocodilians	2	1	1	3
Total [a]	**50**	**139**	**169**	**278**

[a]This includes 'species' known at the time of Bocage and Monard, some of which may have later been relegated to synonymy (see discussion on Monard's list)

Table 13.2 Summary of the taxonomic diversity and endemicity of the reptiles of Angola

Group	Genera	Species	Endemic
Chelonians	11	15	0
Crocodilians	3	3	0
Lizards	40	132	27
Snakes	49	128	6
Total	**103**	**278**	**33**

Recent Discoveries

In the last decade, and resulting from the burgeoning scientific interest in Angola, biodiversity surveys have led to the description of numerous new species and the validation of the specific status of others. Perhaps the most exciting was the long delayed description of the beautiful and bizarre Plume-tailed Gecko (*Afrogecko pulumicaudus* Haacke 2008) from the Angolan Namib region. Other novelties included the description of the lacertids *Pedioplanis haackei* and *P. huntleyi* (Conradie et al. 2012a and the cordylid *Cordylus namakuiyus* (Stanley et al. 2016). Some subspecies were validated as full species, including the geckos *R. boultoni benguellensis* and *R. boultoni montanus* (Ceríaco et al. 2016a) and the snake *Psammophylax rhombeatus ocellatus* (Branch et al. 2019), as well as the revival of the skink *Trachylepis damarana* from the synonymy of *T. varia* (Weinell and Bauer 2018). Some species, e.g. *Philothamnus nitidus loveridgei*, however, have been shown to lack genetic support for recognition (Engelbrecht et al. 2018) and are now not recognised.

In addition, preliminary studies have revealed numerous problematic specimens and populations that demonstrate the existence of cryptic, previously synonymised species or unnamed taxa awaiting description in numerous genera. Geckos – *Hemidactylus, Rhoptropus* (Ceríaco et al. 2016a; Bauer and Kuhn 2017), *Afroedura* (Branch et al. 2017), and various *Pachydactylus* groups (Branch et al. 2017; Ceríaco et al. 2016a; Heinz 2011); within the lacertids *Nucras, Pedioplanis* and *Heliobolus* (Branch and Tolley 2017); and a skink in the *Trachylepis varia* complex (Weinell and Bauer 2018). The descriptions of at least a dozen new species in these genera are in preparation. Ceríaco et al. (2016b, 2018b) signalled the presence of a new skink (*T.* cf. *megalura*) from Cangandala National Park. Snake-eyed skinks (*Panaspis wahlbergii-maculilabris* complex) have been shown to include numerous cryptic species in southern and east Africa (Medina et al. 2016). Records of *P. wahlbergii* in Angola are therefore also likely to represent taxonomic novelties. It is also likely that genetic studies will further validate a number of other lizard taxa currently treated as subspecies, e.g. *Ichnotropis bivittata palida* and *Trachylepis bayoni huilensis,* as full species. Moreover, the rare gecko *Afrogecko ansorgi,* described by Boulenger (1907a), as *Phyllodactylus ansorgi* and not collected again for nearly 100 years, was recently rediscovered and a re-assessment of its generic status is in preparation. In addition, ongoing surveys of the Angolan Okavango Project continue to confirm new species records for Angola, including most recently *Pachydactylus wahlbergii* (G Neef pers. comm. July 2018), previously assumed to enter southern Angola based on material collected on the Angolan-Nambian border in the 1970s (Haacke 1976b). Cryptic diversity in snakes is also being unravelled, and African forest cobras have been shown to include five species (Wüster et al. 2018), of which two enter Angola, whilst Angolan house snakes (*Boaedon*) is expected to comprise at least eight species, with four potential new country records, i.e. *B. fuliginosus, B. radfordi, B. virgatus* and *B. mentalis* (the latter signalled as a valid species by Kelly et al. 2011, and will be formally revived for western arid populations from South Africa to southwest Angola), revival from synonymy of two

Bocage names for Angolan endemics, i.e. *B. angolensis* and *B. variegatus,* and two additional taxonomic novelties (Hallermann et al. personal communication).

Overview of Reptile Diversity

Chelonians

This ancient lineage of reptiles has relatively little global diversity and includes the greatest proportion of threatened reptiles, particularly in Asia. They are relatively poorly known in Angola, and their diversity is discussed below. The first sea turtle to swim in the early South Atlantic, after the separation of Africa and South America 90 Million years ago, was the extinct chelonian *Angolachelys mbaxi,* discovered in Angola in 2009 near the village Iembe, Bengo Province (Mateus et al. 2009). Five of the seven species of sea turtles in the world have been recorded from Angolan waters (Carr and Carr 1991), although only four occur regularly. They include (in decreasing order of abundance): Olive Ridley Sea Turtle (*Lepidochelys olivacea*), Leatherback Sea Turtle (*Dermochelys coriacea*), Green Sea Turtle (*Chelonia mydas*), Loggerhead Sea Turtle (*Caretta caretta*), and Hawksbill Sea Turtle (*Eretmochelys imbricata*). Although early surveys (2000–2006) indicated the absence of Hawksbill in Angolan waters (Weir et al. 2007), juveniles were recently recorded in the Soyo and Cabinda region (Morais 2008, 2016). These may still be vagrants in Angolan waters (TTWG 2017) as Hawksbills forage on coral reefs which are absent in Angola. Nesting in Angolan waters has only been confirmed for Green, Olive Ridley and Leatherback sea turtles, and occurs during September–March, peaking in November–December in the north and a month later in the south (Morais 2017). It is widespread for the Olive Ridley, but restricted mainly to the south for the Green Sea Turtle. The latter remains common in the Cunene River estuary where adults and juveniles feed and also escape the cold waters of the Benguela Current (Elwen and Braby 2015). The giant Leatherback nests primarily in the warmer north, with little activity south of Benguela. The Angolan population (approximately 1000 in 2005–2016, Morais 2016) forms the southern part of the major Gabon nesting grounds, where 6000–7000 females breed annually (Billes et al. 2006). Sea turtles face numerous threats, including by-catch and drowning in trawler fishing nets, the collection of nesting females and their eggs for food, and disturbance of the nest sites by beach activities, etc. (Morais et al. 2005; Morais 2008; Weir et al. 2007). The Projecto Kitabanga of Universidade Agostinho Neto (https://www.facebook.com/Kitabanga/) is involved in research and public education of Angolan sea turtles.

Only three land tortoises are recorded for Angola. The Leopard Tortoise (*Stigmochelys pardalis*) is restricted to the southwest, with all records occurring below the escarpment south of Benguela and along the Cunene valley. Bell's Hinge-back Tortoises (*Kinixys belliana*) is considered widespread in Angola (TTWG 2017), and *Kinixys* material from central and eastern Angola were confirmed as this species in a molecular phylogeny of the genus (Kindler et al. 2012). Although material from Capanda Dam has been referred to *Kinixys spekii* (Ceríaco et al. 2014a), this species is not currently considered to occur in the country (TTWG 2017), but is known from the Zambezian region of Namibia and the Ikelenge pedicle of northwest Zambia. The Forest Hinge-back Tortoise (*Kinixys erosa*) occupies moist forests of the Congo basin and West Africa, but only enters Angola in the extreme northeast (Dundo, Laurent 1964) and the Cabinda enclave (Bocage 1895).

The Nile Soft-shelled Terrapin (*Trionyx triunguis*) is restricted to the coastal region, entering the estuaries and lower stretches of the major rivers. It tolerates sea water and grows to over a metre in length. Populations in the eastern Mediterranean and lower Nile River are threatened, and its status in Angola is poorly known, but is known from the coastal region and with populations in the Cunene River mouth and extending some distance upstream in the Cuanza River. Aubry's Soft-shelled Terrapin (*Cycloderma aubryi*) was recorded once from Cabinda (Peters 1869), but there are no recent records. Trade in chelonians, particularly soft-shelled terrapins for food in Asia, has pushed many species to the brink of extinction (TTWG 2011), and involvement in Africa is confirmed by the discovery of a turtle butchery on Lake Malawi (Face of Malawi 2013) and the recent confiscation of a large *T. triunguis* in a shopping mall in Luanda (Arruda 2018). All other Angolan terrapins have hard shells and are restricted to fresh water ecosystems. They withdraw the head into the shell sideways and are represented by the Pelomedusidae in Africa, including the genera *Pelomedusa* and *Pelusios*. Although Bocage (1866) listed the forest species *Pelusios gabonensis* from Cabinda and Duque de Bragança in his first review of Angolan reptiles, the species was subsequently omitted (Bocage 1895). However, it was subsequently recorded from Dundo (Laurent 1964), and mapped to enter extreme northwest Angola south of the Congo River (TTWG 2017), but no documentation supporting this is presented. It was not recorded at Soyo (W Klein pers. comm.), although Western Hinged Terrapin (*Pelusios castaneus*) was common. The most widespread Angolan terrapins are *Pelusios nanus, P. bechuanicus* and *P. rhodesianus* in the extensive wetlands of eastern Angola (Conradie et al. 2016; TTWG 2017).

Crocodilians

Of the three crocodilians that occur in Angola, only the Nile crocodile (*Crocodylus niloticus*) is widespread, being absent only from the southwest although occurring in the lower Cunene River. The remaining two species are both denizens of the Congo Basin and have only a peripheral presence in Angola. The Sharp-snouted Crocodile (*Mecistops cataphractus*) in Angola had been discussed by Machado (1952), who noted an unusual early record from Lunda and others from Dundo, later confirmed by Laurent (1964). Recent studies (Shirley et al. 2014) have found significant molecular and morphological support for two divergent taxa in *Mecistops* – one distributed entirely in West Africa and the other in Central Africa. As the type locality is Senegal, West African populations would keep the name and Congo Basin and Angolan populations have been considered to represent an undescribed species (Shirley et al. 2014), and was subsequently described as a new species, *Mecistops leptorhynchus* (Shirley et al. 2018). The Dwarf Crocodile (*Osteolaemus tetraspis*) is known from nineteenth century of records from the Cabinda enclave (Bocage 1866; Peters 1877), but no confirmed records exist for the natural occurrence of the species south of the Congo River (Eaton 2010). Ceríaco et al. (2018a) discuss a problematic specimen collected in the lagoon at Luanda that they consider to be indicative of an unknown population in the Cuanza River drainage and also the first record of *O. osborni* for Angola. However, the specimen's identity was not confirmed by genetic monophyly, and its presence in Angola may also result from an escapee brought to Luanda for the bushmeat trade. As with *Mecistops* recent genetic studies indicate the existence of at least three species within the *Osteolaemus tetraspis* complex (Eaton et al. 2009), but the taxonomic identity of Cabinda and the putative Cuanza *Osteolaemus* populations require further study.

Squamates

Scaled reptiles (Squamata) form the major component of reptile diversity (Pincheira-Donoso et al. 2013), with over 10,000 species currently recognised, of which over 60% are lizards. Reflecting this, lizards are also the dominant component of the Angolan reptile diversity and are the group in which most recent discoveries have been made (see above).

The 132 species of Angolan lizards are currently contained in nine families, with skinks (Scincidae) containing the greatest diversity. This contrasts with Namibia (Herrmann and Branch 2013) and South Africa (Branch 2014) where geckos form the greatest component of lizard diversity and endemicity (Table 13.3). It is likely that the current dominance of skinks in Angola is an artefact of our present knowledge. Most skinks are diurnal and active and therefore more easily discovered. Cryptic diversity has already been identified in certain Angolan gecko genera (e.g. *Afroedura, Pachydactylus* and *Rhoptropus*, see above), and the discovery of these and others is predicted to also promote geckos to dominance in diversity and endemicity in the Angolan reptile diversity. The evolutionary centre for girdled lizards (Cordylidae) occurs in southern African (Stanley et al. 2016), but the family is relatively poorly represented in Angola. Although it is unlikely to reach the species or generic diversity of even Namibia, there are indications that the diversity of rupicolous *Cordylus* in the escarpment and central uplands is under-represented (e.g. Stanley et al. 2016), and that rediscoveries and further new species await discovery and description.

The families Agamidae, Chamaeleonidae, Gerrhosauridae and Varanidae all have limited diversity in Angola, as do the last two families throughout Africa. Blue-headed tree agamas have been revised (Wagner et al. 2018), with populations from northern Namibia, Angola and northwest Zambia now referred to the revived *Acanthocercus cyanocephalus*. However, it is evident that current species boundaries

Table 13.3 Comparison of diversity and endemicity of Angolan and South African Squamates (excluding Chelonia), by genera (Gen.), species (Spp.), subspecies (Sub.), and endemism (End.)

	Angola			South Africa		
Family	Gen.	Spp.	End.	Gen.	Sp.	End.
Lizards						
Gekkonidae	8	34	8	12	89	55
Agamidae	2	7	2	2	7	0
Chamaeleonidae	2	5	0	2	19	15
Gerrhosauridae	4	8	0	5	13	6
Cordylidae	2	5	2	10	53	38
Scincidae	12	45	6	7	62	32
Lacertidae	6	15	6	8	29	9
Amphisbaenidae	3	11	3	4	12	2
Varanidae	1	2	0	1	2	0
subtotal	**40**	**132**	**27** (**20.5%**)	**51**	**286**	**157** (**54.9%**)
Snakes						
Leptotyphlopidae	2	5	2	3	10	3
Typhlopidae	2	8	1	3	7	0
Pythonidae	1	3	0	1	2	0
Colubridae	14	31	0	9	16	0
Natricidae	2	4	0	1	2	0
Lamprophiidae[a]	15	39	2	17	42	3
Atractaspididae	6	11	0	6	16	2
Elapidae	5	14	0	6	18	1
Viperidae	3	13	1	2	14	4
subtotal	**50**	**128**	**6** (**4.7%**)	**48**	**127**	**13** (**10.2%**)
TOTAL	**95**	**260**	**33** (**12.7%**)	**109**	**413**	**170** (**41.2%**)
Angola/South Africa	**83%**	**63%**				

[a]Excludes additional *Boaedon* species (Hallerman et al. in prep.)

in *Agama* and *Acanthocercus* do not full reflect Angolan agamid diversity. The remaining families, Lacertidae and Amphisbaenidae, are relatively well represented in Angola, with worm lizard diversity in Angola (three genera, 11 species) second only to that in South Africa (12) for diversity in Africa. Most are associated with the sands of the Kalahari Basin, or in secondary deposition in the coastal zone of South Africa and southern Mozambique. The role of river capture and hydrological changes associated with nascent rifting on fossorial reptiles awaits fuller study. Lacertid diversity in Angola (13 species) is reduced relative to South Africa (29) and Namibia (24), but is known to be under-represented and recently described *Pedioplanis* species (Conradie et al. 2012a), and recently discovered cryptic diversity in other lacertid genera (Branch and Tolley 2017; Conradie et al. 2016) will increase species numbers in the family. A number of additional tropical lacertids may also enter the northern regions of Angola (see below).

There are several aspects of Angolan reptile diversity that stimulate interest. The first is the absence of an endemic radiation of chameleons within Angolan forest refugia. African countries with the highest chameleon diversity (Tilbury 2018), i.e. South Africa and Tanzania, have largely endemic radiations of chameleons (*Kinyongia* and *Rhampholeon* in Tanzania, *Bradypodion* in South Africa). All three genera are absent from Angola, where only *Chamaeleo* and *Trioceros* occurs. Greater knowledge of the history of forest habitats in Angola may give insight as to the absence of a forest chameleon radiation. Sandy habitats in arid southwest Angolan include a radiation of skinks of the genera *Sepsina* and *Typhlacontias* that have reduced limbs, serpentine locomotion and fossorial behaviour. The ranges of some species within these genera extend south into adjacent Namibia and Botswana. In arid habitats at the southern end of the Namib Desert, in the southern Dune Sea and adjacent Succulent Karoo biome these Angolan fossorial skink radiations are almost completely replaced by another suite of sepentiform skinks of the genera *Scelotes, Typhlosaurus* and *Acontias.* Only one species, *Typhlacontias brevipes,* of the Angolan radiation occurs in the northern parts of the southern Dune Sea. Increased knowledge of the history of aridification and dune movements of the Namib Desert may again give insight into these distributions.

That snake diversity in Angola is probably the most well known component of the reptile fauna is unsurprising. However, their distribution, particularly of forest-adapted species in the northern and scarp forest isolates, remains poorly-known. The taxonomic status of these isolated forest populations calls for genetic studies on their phylogenetic relationships to confirm their conspecificity with northern populations. The diversity and composition of snake families in Angola reflects that of Africa, with relatively low diversity in primitive groups such as scolecophidians (Typhlopidae and Leptotyphlopidae) and haenophidians (Pythonidae). Again, in Angola as in southern Africa the venomous families Elapidae and Viperidae have slightly greater species diversity, but with more tropical representatives (e.g. the elapids *Pseudohaje goldi, Naja annulata* and *N. melanoleuca*, and viperids *Causus lichtensteini, C. maculatus, Atheris squamigera* and *Bitis nasicornis*). The dominant African snake family is the Lamprophiidae, of which the Atractaspididae is closely related and sometimes treated as a subfamily. The group appears to have originated in Africa and subsequently radiated into Arabia and Asia, and the subfamilies Lamprophiinae, Prosyminae and Psammophinae form important radiations in Sub-Saharan Africa. Lamprophids thus form the dominant component of the Angolan snake fauna (39 species), but includes only two endemic psammophines. As with elapids and viperids a number of Congo Basin species enter the northern forests, including some currently known from very few Angolan specimens, e.g. *Lycodonomorphus subtaeniatus, Chamaelycus parkeri, Boaedon olivaceus, Bothrophthalmus lineatus*, etc. Perhaps the greatest difference between South Africa and Angola is reflected in the greater diversity of colubrids (Colubridae) in Angola (28 vs 14 species). These include numerous tropical Congo Basin snakes that enter the northern and scarp forests, and of particular interest are the rare Congo Basin species *Toxicodryas blandingii, T. pulverulenta, Rhamnophis aethiopissa, Philothamnus nitidus, Dasypelis palmarum,* etc. The family is considered of Asian origin and to have entered and subsequently radiated in Africa.

Species Recorded from Angola but Poorly Known

Some species are known from Angola from either a single or very few specimens and their presence and taxonomic status requires confirmation. This summary does not include wide ranging species that peripherally enter Angola, either from the Congo Basin (e.g. *Pelusios gabonensis, Feylinia grandisquamis, Hypoptophis wilsoni,* etc.), or from the southern Kalahari or Namib deserts (e.g. *Rhoptropus afer, Pachydactylus rangei, P. vanzyli, Chamaeleo namaquensis, Amblyodipsas ventrimaculata*, etc.).

Grass Lizard – *Chamaesaura anguina oligopholis* Laurent (1964). Described from Calonda, Lunda, but with no recent material. It may deserve specific status.

Angola Girdled Lizard – *Cordylus angolensis* (Bocage, 1895). Known only from the type description of a single male from Caconda, but a population that conforms to the species has recently been discovered (Vaz Pinto Unpublished Data).

Scaled Sandveld Lizard -*Nucras scalaris* Laurent 1964. Still known only from type series of four specimens from Alto Chicapa and Alto Chilo.

Dewitte's Five-toed Skink – *Leptosiaphos dewittei* (Loveridge, 1934). Recorded by Parker (1936, as *Lygosoma dewittei*) from Congulo. However, the only known Angolan specimen lacks the diagnostic compressed tail. The species occurs in the eastern Congo Basin, a considerable disjunction from Congulo.

African Shovel-nosed Snake – *Scaphiophis albopunctatus* Peters, 1870. Only once recorded from Angola (Laurent 1950, Muita River) in Guinea-Congo savannah habitat.

The only other know record is a juvenile specimen collected from Capaia, Lunde Norte (Branch and Conradie 2015).

Collared Snake-eater – *Polemon collaris* (Peters 1881). Recorded by Peters (1881, Cuango), Ferreira (1904, Golungo Alto) and Hellmich 1957a, b, Bella Vista, as *Miodon gabonensis*). Isolated populations of small fossorial snakes such as *Polemon* often include cryptic diversity (Portillo et al. 2018), and fresh material is required for taxonomic assessment.

Lined Water Snake – *Lycodonomorphus* (?) *subtaeniatus* Laurent 1954. Described from Keseki (DRC), with four paratypes from Dundo the only Angolan records. Greenbaum et al. (2015) transferred *L. s. upembae* to *Boaedon*. This is probably where *L. subtaeniatus* belongs but fresh material is required for genetic analysis.

Speckled Wolf Snake – *Lycophidion meleagre* Boulenger 1893. Described from Angola and known from Cabinda to Luanda, but Broadley (1996) also includes records from coastal Tanzania in the species' range, creating a biogeographic anomaly that requires genetic assessment.

Parker's Banded Snake – *Chamaelycus parkeri* (Angel, 1934). Parker's (1936) Congulo specimen (as *Oophilositum parkeri*) remains the only known Angolan material. Elsewhere the species is restricted to Kivu (DRC) and Congo Brazzaville, and confirmation of the specific status of the Congulo population is required.

Angolan Coral Snake – *Aspidelaps lubricus cowlesi* Bogert 1940. Described from Munhino (101 km east of Moçamedes, via railroad), and known from Angola from the type and one additional specimen (Branch 2018). Considered widespread in northern Namibia, but genetic monophyly between Angolan and Nambian populations is required for confirmation.

Angolan Garter Snake – *Elapsoidea semiannulata moebiusi* Werner, 1897. Listed by Broadley (2006) from northern Angola, but with no specific localities given. All Bocage localities (1866, 1895, 1897) were restricted to Bissau specimens. A southern subspecies is now treated as a valid species (*E. boulengeri*). The status of *E. s. moebiusi* requires a modern taxonomic assessment and also confirmation for Angola.

Angolan Dwarf Adder – *Bitis heraldica* (Bocage, 1889). Angola's most iconic snake for which no new material was collected for over 50 years has recently been rediscovered. It has a disjunct distribution in montane grasslands of the Angolan inland plateau, and the fresh material will allow its subgeneric relationships within *Bitis* to be assessed as well as its conservation status.

Species Likely to Occur in Angola but Currently Unconfirmed

A number of species are recorded in close proximity to the Angolan border and live in habitats contiguous with those in Angola, and are therefore likely to occur in the country. They include:

Lizards

Heenen's Dwarf Day Gecko – *Lygodactylus heeneni* De Witte, 1933. This small diurnal gecko was recorded from the Ikelenge Pedicle in northwest Zambia (Broadley 1991; Haagner et al. 2000) within 25 km of the Angolan border.

Long-tailed Worm Lizard – *Dalophia longicauda* (Werner, 1915). This fossorial species was described from northern Namibia and is known to extend through the Caprivi region to western Zimbabwe (Broadley et al. 1976; Gans 2005) and also to southwest Zambia (Pietersen et al. 2017). Populations occur to the east and west of the Okavango River and are expected to occur in southeast Angola.

Maurice's Worm Lizard – *Monopeltis mauricei* Parker, 1935. This fossorial species was described from central Botswana and is known to extend through the Kalahari to Katima Mulilo in the Caprivi region (Broadley et al. 1976; Gans 2005). Elevated to a full species by Broadley (2001).

West African Striped Lizard – *Poromera fordii* (Hallowell, 1857). An arboreal species recorded during a survey in the Bas-Congo region (Nagy et al. 2013) within 30 km of the Angolan border but currently unknown from Angola.

Fine-scaled forest lizard – *Adolfus africanus* (Boulenger, 1906). A terrestrial species recorded from the Ikelenge Pedicle in northwest Zambia (Broadley 1991), within 25 km of the Angolan border.

Snakes

Western Thread Snake – *Namibiana occidentalis* (FitzSimons, 1962) occurs in extreme Kaokoveld (Broadley and Broadley 1999) but has not yet been recorded from southern Angola. The single record of the Damara Thread Snake (*N. labialis* Sternfeld, 1908) from southern Angola demonstrates that these small snakes can cross the Cunene River.

Leptotyphlops sp. An unidentified thread snake was recorded during a survey of the Bas-Congo region (Nagy et al. 2013). Based on its forest habitat it is unlikely to be referable to any known Angolan species.

Slender Quill-snouted Snake – *Xenocalamus b. bicolor* (Günther 1868). Although Broadley (1971) records no Angolan material, the species occurs in the Caprivi area and adjacent western Zambia, and it is usually associated with Kalahari sands. It is therefore likely to occur in southeast Angola

Bark Snake – *Hemirhaggheris nototaenia* (Günther, 1864). This dwarf arboreal snake is recorded from the western Caprivi and Okavango region, and extends eastwards through Zambia to East Africa. Earlier records from southwest Angola (Bocage, 1895) were later referred to *H. viperina* (Broadley and Hughes 2000). It is a secretive snake and may still be found in the miombo woodlands of southeast Angola.

Cunene Racer – *Mopanveldophis zebrinus* (Broadley and Schätti 2000). This enigmatic colubrid snake remains known from only a handful of specimens. The type locality is the Cunene River at Ruacana, western Owamboland, Namibia (17° 25'S, 14° 10′E), and it appears restricted to the Mopaneveld of northern Namibia and can be expected to occur in similar habitat in southern Angola.

Endemism in Angolan Reptiles

Species that are fully endemic or near endemic to a country (i.e. those that have over 90% of known records included in that country), should be highlighted for national conservation monitoring as their protection depends completely on the national authorities. Only six species of snake are endemic to Angola, but no chelonians or crocodilians. Endemic snakes include two species of primitive thread snake, *Namibiana latifrons* and *N. rostrata,* that are the northern members of a small genus (five species) endemic to the western arid region of southern Africa (Adalsteinsson et al. 2009). Three rare snakes are also endemic to the high plateau region, including the psammophines *Psammophis ansorgi* and *Psammophylax ocellatus* (Branch et al. 2019), as well as the rare and iconic *Bitis heraldica*, which may be now of high conservation concern. During the Hamburg Expedition 10 specimens were collected from Bela Vista (Hellmich 1957a, b), but only one other specimen (Mount Moco) has been recorded in last 60 years (FM Gonçalves, photo 2010). Extensive clearing of natural habitat for agriculture, and increased fire risk in these montane grasslands may threaten the species.

Lizards contain the greatest number of endemic and near-endemic Angolan reptiles, particularly among cordylids (two endemic, 50%), lacertids (one near endemic, six endemic, 53.8%), rupicolous geckos (eight endemic, 23.5%), amphisbaenians (one near endemic, three endemic species; 36.4%), and diverse skinks (one near endemic, six endemic, 16.3%). *Agama planiceps schacki* is certainly a full species that is well-defined morphologically, but requires genetic assessment. It would also be endemic to Angola. None of these endemic lizards are currently considered of conservation concern. Only 12.7% of all Angolan reptiles are endemic as opposed to 41.2% of those in South Africa. This number increases to nearly 20.5% when only lizards are considered, but is still much less than the 54.9% of endemic lizards in South Africa (Table 13.3). However, the number of endemic species in the country has increased with the description of new Angolan taxa (e.g. *Kolekanus plumicaudus, Pedioplanis haackei, P. huntleyi* and *Cordylus namakuiyus*), and will increase further as new species in the genera *Nucras, Heliobolus, Pedioplanis, Afroedura, Rhoptropus, Pachydactylus, Trachylepis* and *Boaedon* discovered during recent surveys are described.

Reptile Hotspots

The existing global protected area network and conservation priorities are heavily biased towards amphibian, avian and mammal faunas (Roll et al. 2017). Reptiles, which represent a third of terrestrial vertebrate diversity, have been largely ignored, in part, because their diversity and distribution was not globally assessed until 2017. Both the global (Roll et al. 2017) and African (Lewin et al. 2016) assessments demonstrated that whilst the distribution patterns of species richness of all reptiles combined, as well as those of snakes, revealed similar patterns to those of the other three tetrapod classes, the patterns displayed by hotspots of total and endemic lizards and chelonian richness do not overlap significantly with those of other terrestrial tetrapods. A detailed analysis of reptile hotspots within Angola awaits fuller details of species diversity and distributions, both of which are still in their formative period. However, it is already evident that certain regions and their associated habitats and reptile faunas, particularly for endemic or near endemic species, present unique associations, some of which may be confirmed as regionally important reptile hotspots.

Kaokoveld Centre of Endemism

Lizard diversity in southern Africa, particularly in the western arid regions, is the highest in Africa, and the existence of similar habitat structure and diversity in southwest Angola indicates that this African lizard hotspot may also extend into Angola in association with arid and hyper-arid habitats. In association with desert habitats a number of characteristic Namib reptiles cross the Cunene River and just enter extreme southwest Angola, including: *Gerrhosaurus skoogi, Pachydactylus rangei, P. vanzyli, Chamaeleo namaquensis, Meroles anchietae, M. reticulata, Trachylepis puncutula,* and *Bitis caudalis.* Recent discoveries also suggest the existence of an endemic Angolan Namib reptile fauna, including the existing Angolan Namib endemics *Pedioplanis benguellensis, Typhlacontias rudebecki,* and *T. punctatissimus bogerti,* as well as a number newly described species in the region, e.g. *Kolekanus plumicaudus* (Haacke 2008), *Pedioplanis huntleyi, P. haackei* (Conradie et al. 2012a), and *Cordylus namakuiyus* (Stanley et al. 2016). Moreover, recent surveys in the region have revealed numerous examples of cryptic diversity in some lizard genera, where new species of *Afroedura, Pachydactylus, Nucras, Pedioplanis* endemic to the Angolan Namib region have been identified and await description.

At its northern and southern limits, the Namib Desert transforms into semi-arid, often succulent vegetation that may be loosely termed the 'Pro-Namib' region. In the south this forms the Succulent Karoo, a botanical hotspot of regional endemism and floral beauty (CEPF 2003). The Succulent Karoo has diverse and specialised reptile endemics (Branch 1994; Bauer and Branch 2003), and the region has been highlighted as a regional reptile hotspot, including numerous species of conservation concern (Branch 2014). As with the Succulent Karoo, the recognition of a unique reptile fauna in southwest Angola supports a corresponding northern 'Pro-Namib', in some ways analogous to the Succulent Karoo, and that has been identified as a distinctive phytogeographical region – the Kaokoveld Centre of Endemism, which extends as a narrow strip north of Namibe to Lucira and is characterised by a number of localised succulents (see Craven 2009 for fuller discussion).

Angolan Escarpment

Inland from the coastal arid herpetofauna is the Bié section (sensu Clark et al. 2011) of the Angolan Escarpment and adjacent high plateau. The southern African Great Escarpment (GE) forms a semi-continuous U-shaped mountain chain that runs for 5000 km from western Angola through Namibia and South Africa to the Zimbabwe-Mozambique border. Clark et al. (2011) noted that the GE hosts more than half of southern Africa's centres of plant endemism and is a repository of palaeo- and neo-endemics. It also has a rich endemic fauna and its fragmented sections serve as refugia and as episodic corridors for biological continuity. However, many sections of the Great Escarpment have been poorly studied, particularly in Angola where the Bié Escarpment summit and adjacent highlands is one of the most isolated sections of the Afromontane archipelago in Africa. With ca. 20 endemic bird species it forms the core of the Western Angola Endemic Bird Area. Other faunal groups have not been as extensively studied, but endemic reptiles associated with the Serra da Chela grasslands and wetlands include two endemic snakes (*Psammophylax ocellatus* and *Psammophis ansorgi*), the chameleon *Chamaeleo anchietae*, the serpentine skink *Eumecia anchietae,* the skink *Trachylepis bayoni huilensis,* the gecko *Rhoptropus montanus,* and the lacertid *Ichnotropis bivittata pallida.* A new reed frog, *Hyperolius chelaensis* was also recently discovered (Conradie et al. 2012b). In the adjacent highlands, including Mount Moco, at least two new species of the *Afroedura bogerti* complex have also been signalled (Branch et al. 2017).

Northern Congo Forests

The Congo Basin has numerous forest specialists, particularly snakes. Many of these are found in forests in Cabinda and along the northern border of Angola. These forests have only been incidentally surveyed, particularly the numerous important snake records listed in a series of papers based on the Museu do Dundo collections (Laurent 1950, 1954, 1964; Tys van den Audenaerde 1967). Among these collections are the only known Angolan records *Gonionotophis brussauxi Letheobia praeocularis*, *Xenocalamus bicolour machadoi*, *Hypoptophis wilsoni katangae, Grayia tholloni, Philothamnus nitidus, Bothrophthalmus lineatus, Boaedon olivaceus, Lycodonomorphus subtaeniatus, Prosymna ambigua brevis* and *Causus lichtensteini.* In addition, other Congo Basin reptiles only recorded from Dundo include the terrapin *Pelusios gabonensis,* the worm lizard *Monopeltis vanderysti,* and the skinks *Lepidothyris hinkeli joei* (as *Mochlus fernandi*, Laurent 1964) and *Feylinia grandisquamis* (as *F. elegans*, Laurent 1964). Parker (1936) presented the first survey of the central scarp forests of the Angolan Escarpment and recorded numerous Congo Basin snakes. For many species these remain their southern records, and they probably occur as disjunct, relictual populations. Some were subsequently recorded further north in forest habitats from Dundo or during the Hamburg Angola Expedition at Piri Dembos (see above). They include: *Toxicodryas blandingii, T. pulverulenta, Atractaspis reticulata heterochilus, Bitis nasicornis,* and *Pseudohaje goldii.* Others remain known only from Parkers' records: i.e. the skinks *Panaspis breviceps, Leptosiaophis dewittei,* and *Trachylepis affinis*; and the snakes *Lycophidion ornatum, Chamaelycus parkeri* (as *Oophilositum parkeri*) and *Hormonotus modestus.* The Congo Basin snake *Rhamnophis aethiopissa* is recorded in Angola only from Piri Dembos (Hellmich 1957a, b). The taxonomic status of all these isolates requires genetic confirmation as some may have undergone vicariant speciation. A phylogenetic assessment may give insight towards dating the separation of these forest isolates and understanding their biogeographic importance.

The forests of Cabinda form part of the Congo Basin and a number of reptiles occur there which have not been recorded in Angola south of the Congo River. Currently Cabinda remains the southern limit of the African Dwarf Crocodile (*Osteolaemus tetraspis*) and the Soft-shelled Terrapin (*Cycloderma*). The presence of two other reptiles recorded from Cabinda by Peters (1876, 1877), e.g. Owen's Horned Chameleon (*Triceros oweni*) and the skink *Euprepes perrotetii* (= *Trachylepis perrotetii*) are problematic. The latter is widespread in West Africa but not known even from Gabon. Peters (1877) recorded *Euprepes perrotetii* from

Chinchoxo, Cabinda, and in a supplement to the same article noted a specimen from Pungo Andongo, upon which he considered it to form part of the fauna of Angola. However, no subsequent records of this distinctive species have been recorded from Angola. Although it is possible that these specimens were confused with large fire skinks (*Lepidothrys* sp.), Wagner et al. (2009) reviewed the genus and noted no misidentifications among the material they examined. It is more likely that Peters' specimens were simply accompanied by incorrect locality data. Forest chameleons are difficult to locate unless specifically targeted during faunal surveys, and Owen's Horned Chameleon is known from Gabon. No recent collections of both these species confirm their presence in Cabinda. Research underlying the proposed Mayombe Transfrontier Reserve (MTR) to protect forests in Cabinda and adjacent countries has concentrated on the large mammals, particular the Great Apes, and no detailed herpetological surveys have been undertaken. Recent surveys of the forest herpetofauna of the Serra do Pingano Ecosystem, Uíge Province (Ernst 2015) concentrated on amphibians but did record an number of interesting reptiles, particularly the arboreal lacertid *Holaspis guentheri* and water snake *Grayia ornata*, the former being the second record for the country (Laurent 1964) and the latter one of the few records for the country (Branch 2018). The northern tropical forests of Angola are threatened by massive timber extraction, and desperately need to be scientifically surveyed before their associated herpetofauna is lost.

Future Directions for Reptile Research in Angola

The Continued Need for Further Field Surveys and Taxonomic Studies

The conservation status and threats for African reptiles were reviewed by Tolley et al. (2016), who noted the large discrepancy between taxonomic sampling and documentation between many countries. They presented a scatter-plot of measured reptile species richness relative to log-transformed country area from African countries. This illustrated the great contrast between known reptile diversity in well surveyed countries such as South Africa, Kenya and Tanzania, with that of the majority of Africa. Angola is the seventh largest African country and has both habitat and topographic diversity. Together these features should generate rich biological diversity, but this is not reflected in our current knowledge of Angolan reptile diversity. Branch (2016) presented species accumulation curves documenting the growth in taxonomic knowledge of Angolan and southern African reptiles, noting that there has been no decline in the rate of new species discovery in the subcontinent during the last 150 years. This is in marked contrast to the relative stagnation of taxonomic discovery in Angola since the early part of the twentieth century (see Fig. 13.1 and Table 13.1 for comparison). As noted earlier, despite Angola and South Africa being of comparable size and habitat diversity there is a difference of over 150 species of

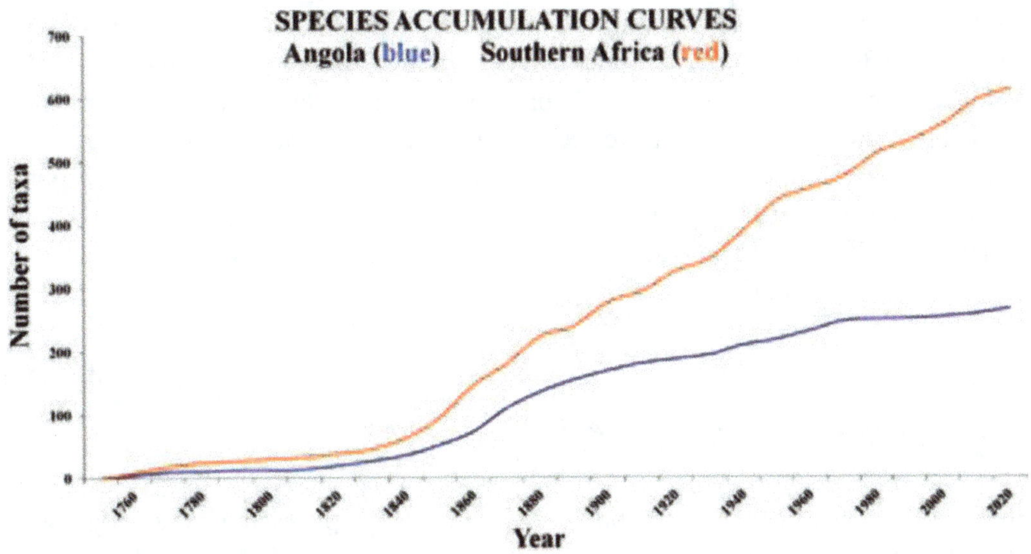

Fig. 13.1 Species accumulation curves for reptile discoveries in Angola (blue) and southern Africa (red) showing the relative stagnation of Angolan reptile species descriptions during the twentieth century. By the end of the nineteenth century 67.4% of Angolan reptiles had already been described, in contrast to less than half (47.8%) of those in southern Africa

lizards between the two countries (Angola 132, South Africa 286). This contrast is even higher in terms of endemicity, where only 27 (20.5%) of 132 Angolan lizards are endemic in contrast to 157 (54.9%) of 286 South African lizards. It would appear that perhaps as many as 75+ new lizard species await discovery in Angola, and that many of these will be endemic. Branch (2014) noted that endemicity in South African lizards was particularly evident in rupicolous forms (many geckos, cordylids and skinks) associated with rocky outcrops. Rock exposures may form an archipelago of 'sky islands' on which isolation inhibits gene flow and thus leads to speciation. It is the lizard families containing large numbers of rupicolous species, i.e. Gekkonidae, Cordylidae and Scincidae, already show the greatest levels of endemicity among Angolan reptiles, and in which recent surveys have already identified numerous cryptic taxa (Stanley et al. 2016; Branch et al. 2017).

Field Surveys of Potential Biodiversity 'Hotspots'

Many African protected areas underperform in their stated conservation goals (Lindsey et al. 2014; Bowker et al. 2017). It is now generally accepted that modern national and internationally co-ordinated networks of protected area should be designed to cover important biodiversity hotspots and also protect habitats essential for the maintenance of ecosystem services such as water flow and quality, nutrient transfer, etc. (NPAES 2010). Such a revised Angolan network was proposed many years ago (Huntley 1974), and initial biodiversity surveys to gain insight into biodiversity in potential sites were undertaken (Huntley 2009; Huntley and Francisco 2015). Recent studies have shown that protected reserves designed to protect mammals, birds and amphibians are effective in protecting snakes, but fare badly in protecting African lizard diversity (Lewin et al. 2016; Roll et al. 2017). Future systematic biodiversity surveys should be directed to unique habitats and landforms in undersampled regions. Some of the interesting species recorded on recent field surveys are illustrated in Fig. 13.2.

Biogeography of Angolan Reptiles

Huntley (2019) in the introduction to this volume has presented a biogeographic outline, summarising various aspects of climate, geology and vegetation, etc., that characterise Angola. He noted the complexity of the Angolan landscape, where seven of the nine African biomes are represented in Angola as well as the second largest representation of ecoregion diversity in Africa. Monard (1937) and Hellmich (1957a, b) made preliminary attempts to assess biogeographic patterns among Angolan reptiles. However, they did little more than look for coarse habitat associations within the Ethiopian region. These attempts were constrained by lack of knowledge of the true reptile diversity in the region and, more importantly, by the ignorance of reptile distributions as large tracts of the country were still unexplored. Moreover, recent studies indicate that reptile distributions, particularly those of lizards, are more influenced by substrate specificity and isolation than by vegetation type (Bauer and Lamb 2005; Roll et al. 2017). Recent biogeographic studies lay greater emphasis on evolutionary relationships within the group studied, and explore correlations between genetic divergence (as a proxy for time) and known dates of major events in landscape evolution. This approach searches for historical barriers to, or corridors for migration and gene flow. These may be generated, for example, by climatic changes associated with Ice Age cycles and the resultant contraction and expansion of forest and savanna, changes in historic coastlines and/or island connectivity, as well as the development of an ‘arid’ corridor at an Ice Age maxima, etc. The biological consequences of nascent rifting on river capture and other hydrological consequences on palaeolakes and wetlands have also been explored (Cotterill and De Wit 2011). However, the application of such approaches requires more detailed knowledge of reptile distribution within Angola, as well as the availability of genetic material and adequate taxon sampling within a chosen group. These will allow historic climatic and landform events to be meaningful correlated with speciation and radiation within groups for testing phylogeographic hypotheses. Such studies depend on meaningful progress in the topics discussed earlier in this section. Advances in all these areas are required to fully understand and conserve the diversity and evolution of Angolan reptiles.

Fig. 13.2 Angolan reptiles. Top to bottom, left to right. Bogert's Flat Gecko (*Afroedura* cf. *bogerti*), Omahua Lodge, Namibe; Angolan Namib Day Gecko (*Rhoptropus taeniostictus*), Chapéu Armado, Namibe; Ansorge's Leaf-toed Gecko (*Afrogecko ansorgii*), Meve, Benguela; Anchieta's Chameleon (*Chamaeleo anchietae*), Humpata, Huíla; Kaokoveld Girdled Lizard (*Cordylus namakuiyus*), Rio Makonga, Namibe; Bayon's Legless Skink (*Sepsina bayoni*), Quiçama National Park, Luanda; Water Cobra (*Naja annulata*), Lagoa Carumbo, Lunde Norte; Angolan Skaapstekker (*Psammophylax ocellatus*), Humpata, Huíla

Acknowledgements This review results from extensive collaboration by the authors and the Editor's of this volume during recent studies on the Angolan herpetofauna. It has been both synergistic and rewarding. Funding for research in Angola has been supported by: South Africa's National Research Foundation (2009–2017, WRB), National Geographic Society (Explorer Grant 2011, WRB; NGOWP and Wild Bird Trust 2015–2018, all authors), Fundação para a Ciência e Tecnologia (contract SFRH/PD/BD/140810/2018, NB). We are all particularly indebted to Fernandas Lages (ISCED), Brian Huntley (South Africa), and John Hilton (Wild Bird Trust) for their support with the documentation, logistics and permitting required for successful fieldwork in Angola.

Appendices

Appendix 1

Checklist of Angolan Chelonians and Crocodilians. C: Cabinda; Status: CITES (I, II = Appendix 1 or 2), IUCN Conservation Status[1]. Species listed under ORDER|**Family**

Common Name	Scientific Name	C	Bocage (1895)	Status
CHELONIA | **Chelonidae**				
Loggerhead Sea Turtle	*Caretta caretta* (Linnaeus, 1758)		*Thalassochelys caretta*	I, VU
Green Sea Turtle	*Chelonia mydas* (Linnaeus, 1758)	Y	*Chelonia mydas*	I, EN
Olive Ridley Sea Turtle	*Lepidochelys olivacea* (Eschscholtz, 1829)	Y		I, VU
Hawksbill Sea Turtle	*Eretmochelys imbricata* (Linnaeus, 1766)			I, CR
CHELONIA | **Dermochelyidae**				
Leatherback Sea Turtle	*Dermochelys coriacea* (Vandelli, 1761)	Y		I, VU
CHELONIA | **Testudinidae**				
Bell's Hinge-back Tortoise	*Kinixys belliana* (Gray, 1831)		*Cinixys belliana*	II
Forest Hinge-back Tortoise	*Kinixys erosa* (Schweigger, 1812)	Y	*Cinixys erosa*	II, EN[a]
Leopard Tortoise	*Stigmochelys pardalis* (Bell, 1828)		*Testudo pardalis*	II
CHELONIA | **Pelomedusidae**				
Southern Marsh Terrapin	*Pelomedusa subrufa* (Bonnaterre, 1789)		*Pelomedusa galeata*	
Okavango Hinged Terrapin	*Pelusios bechuanicus* (FitzSimons, 1932)			
Gabon Hinged Terrapin	*Pelusios gabonensis* (Duméril, 1856)			

(continued)

Common Name	Scientific Name	C	Bocage (1895)	Status
Dwarf Hinged Terrapin	*Pelusios nanus* (Laurent, 1956)			
Variable Hinged Terrapin	*Pelusios rhodesianus* (Hewitt, 1927)		*Sternothaerus sinuatus*	
Western Hinged Terrapin	*Pelusios castaneus* (Schweigger, 1812)	Y	*Sternothaerus Derbianus*	
CHELONIA | **Trionycidae**				
Nile Soft-shelled Terrapin	*Trionyx triunguis* (Forskål, 1775)	Y	*Trionyx triunguis*	II, VU[a]
Aubrey's Flap-shelled Terrapin	*Cycloderma aubryi* (Dumeri,, 1856)	Y	*Cycloderma Aubryi*	II, VU[a]
CROCODYLIA | **Crocodylidae**				
Nile Crocodile	*Crocodylus niloticus* (Laurenti, 1768)		*Crocodilus vulgaris*	II
Central African slender-snouted Crocodile	*Mecistops leptorhynchus* (Shirley et al. 2018)		*Crocodylus cataphractus*	I, DD
African Dwarf Crocodile	*Osteolaemus tetraspis* (Cope, 1861)	Y	*Ostelolaemus tetraspis*	I, EN

[a]Turtle Working Group 2017, Draft Red List
[b]IUCN Conservation Status categories. *CR* Critically Endangered, *EN* Endangered, *VU* Vulnerable, *DD* Data Deficient

Appendix 2

Checklist of Angolan Lizards. C: Cabinda; Obs: Observations (E:endemic; NE: near-endemic). Species listed under ORDER|**Family**|Subfamily

Common name	Scientific name	C	Bocage (1895)	Obs
SAURIA | **Agamidae**				
Angolan Tree Agama	*Acanthocercus cyanocephalus* (Falk, 1925)		*Stellio angolensis* Bocage 1866 is a nomen nudum *Stellio atricollis*	
Ground Agama	*Agama aculeata* (Merrem, 1820)		*Agama armata*	
Anchieta's Agama	*Agama anchietae* (Bocage, 1896)			
Congo Agama	*Agama congica* (Peters, 1877)	Y	*A. colonorum*	
Mucoso Agama	*Agama mucosoensis* (Hellmich, 1957)			E
Namib Rock Agama	*Agama planiceps planiceps* (Peters, 1862)		*Agama planiceps*	
Schack's Rock Agama	*Agama p. schacki* (Mertens, 1938)			E

(continued)

Common name	Scientific name	C	Bocage (1895)	Obs
Sauria | **Amphisbaenidae**				
Angola Blunt-tailed Worm Lizard	*Dalophia angolensis* (Gans, 1976)			NE
Ellenberger's Blunt-tailed Worm Lizard	*Dalophia ellenbergeri* (Angel, 1920)			
Zambezi Blunt-tailed Worm Lizard	*Dalophia pistillum* (Boettger, 1895)			
Welwitch's Blunt-tailed Worm Lizard	*Dalophia welwitschii* (Gray, 1865)		*Monopeltis Welwitschii*	E
Anchieta's Spade-snouted Worm Lizard	*Monopeltis anchietae* (Bocage, 1873)		*Monopeltis anchietae*	
Infuscate Spade-snouted Worm Lizard	*Monopeltis infuscata* (Broadley, 1997)		*Monopeltis capensis*	
Luanda Spade-snouted Worm Lizard	*Monopeltis luandae* (Gans, 1976)			E
Confusing Spade-snouted Worm Lizard	*Monopeltis perplexus* (Gans, 1976)			E
Vanderyst's Spade-snouted Worm Lizard	*Monopeltis vanderysti* (De Witte, 1922)			
Balck Round-headed Worm Lizard	*Zygaspis nigra* (Broadley and Gans, 1969)			
Kalahari Round-headed Worm Lizard	*Zygaspis quadrifrons* (Peters, 1862)			
Sauria | **Chamaeleonidae**				
Angolan Chameleon	*Chamaeleo anchietae* (Bocage, 1872)		*Chamaeleo anchietae*	
Flap-necked Chameleon	*Chamaeleo dilepis* (Leach, 1819)	Y	*Chamaeleon dilepis & C. quilensis*	
Etienne's Chameleon	*Chamaeleo gracilis etiennei* (Schmidt, 1919)	Y	*Chamaeleon gracilis*	
Namaqua Chameleon	*Chamaeleo namaquensis* (Smith, 1831)		*Chamaeleon namaquensis*	
Owen's Three-horned Chameleon	*Trioceros oweni* (Gray, 1831)			
Sauria | **Cordylidae**				
Northern Grass Lizard	*Chamaesaura miopropus* (Boulenger, 1895)		*Chamaesaura macrolepis*	
Angola Grass Lizard	*Chamaesaura anguina oligopholis* (Laurent, 1964)			E
Angola Girdled Lizard	*Cordylus angolensis* (Bocage, 1895)		*Zonurus cordylus*	E
Machado's Girdled Lizard	*Cordylus machadoi* (Laurent, 1964)			
Kaokoveld Girdled Lizard	*Cordylus namakuiyus* (Stanley et al, 2016)			E

(continued)

Common name	Scientific name	C	Bocage (1895)	Obs
SAURIA I **Gekkonidae**				
Bogert's Flat Gecko	*Afroedura bogerti* (Loveridge, 1944)			
Ansorge's Leaf-toed Gecko	*Afrogecko ansorgii* (Boulenger, 1907)			E
Button-scaled Gecko	*Chondrodactylus fitzsimonsi* (Loveridge, 1947)			
Pulitzer's Gecko	*Chondrodactylus pulitzerae* (Schmidt, 1933)		*Pachydactylus Bibronii* (part)	
Fisher's Gecko	*Chondrodactylus laevigatus* (Fischer, 1888)		*Pachydactylus Bibronii* (part)	
Pulitzer's Gecko	*Chondrodactylus pulitzerae* (Smith, 1933)			
Bayão's House Gecko	*Hemidactylus bayonii* (Bocage, 1893)		*Hemidactylus bayonii*	E
Benguela House Gecko	*Hemidactylus benguellensis* (Bocage, 1893)		*Hemidactylus benguellensis*	E
Western House Gecko	*Hemidactylus brooki angulatus* (Hallowell 1852)			
Long-headed House Gecko	*Hemidactylus longicephalus* (Bocage, 1873)		*Hemidactylus bocagii*	
Tropical House Gecko	*Hemidactylus mabouia* (Moreau De Jonnès, 1818)	Y	*Hemidactylus mabouia* & *H. benguellensis*	
Forest House Gecko	*Hemidactylus murecius* (Peters, 1870)		*Hemidactylus murecius*	
Plume-tailed Geco	*Kolekanos plumicaudus* (Haacke, 2008)			E
Angola Dwarf Day Gecko	*Lygodactylus angolensis* (Bocage, 1896)			
Bradfield's Dwarf Day Gecko	*Lygodactylus bradfieldi* (Hewitt, 1932)			
Cape Dwarf Day Gecko	*Lygodactylus capensis* (Smith, 1849)		*Lygodacylus capensis*	
Hewitt's Punctate Gecko	*Pachydactylus amoenoides* (Hewitt, 1935)		*Pachydactylus ocellatus*	
Angola Thick-toed Gecko	*Pachydactylus angolensis* (Loveridge, 1944)			E
Caricul Thick-toed Gecko	*Pachydactylus caraculicus* (FitzSimons, 1959)			
Kaokoveld Thick-toed Gecko	*Pachydactylus* cf. *oreophilus* (McLachlan and Spence, 1976)			

(continued)

Common name	Scientific name	C	Bocage (1895)	Obs
Punctate Thick-toed Gecko	*Pachydactylus punctatus* (Peters, 1854)			
Web-footed Gecko	*Pachydactylus rangei* (Andersson, 1908)			
Scherz's Thick-toed Gecko	*Pachydactylus scherzi* (Mertens, 1954)			
Rough-scaled Thick-toed Gecko	*Pachydactylus* cf. *rugosus* (Smith, 1849)			NR
Large-scaled Thick-toed Gecko	*Pachydactylus scutatus* (Hewitt, 1927)			
Kalahari Ground Gecko	*Pachydactylus wahlbergii* (Peters, 1869)			
Van Zyl's Web-footed Gecko	*Pachydactylus vanzyli* (Steyn and Haacke, 1966)			
Common Namib Day Gecko	*Rhoptropus afer* (Peters, 1869)			
Barnard's Namib Day Gecko	*Rhoptropus barnardi* (Hewitt, 1926)		*Rhoptropus afer* ?	
Benguella Namib Day Gecko	*Rhoptropus benguellensis* (Mertens 1938)			E
Two-pored Namib Day Gecko	*Rhoptropus biporosus* (FitzSimons, 1957)			
Boulton's Namib Day Gecko	*Rhoptropus boultoni* (Schmidt, 1933)			
Montane Namib Day Gecko	*Rhoptropus montanus* (Laurent, 1964)			E
Angolan Namib Day Gecko	*Rhoptropus taeniostictus* (Laurent, 1964)			E
SAURIA | **Gerrhosauridae**				
Dwarf Plated Lizard	*Cordylosaurus subtessellatus* (Smith, 1844)		*Cordylosaurus trivittatus*	
Kalahari Plated Lizard	*Gerrhosaurus auritus* (Boettger, 1887)			
Laurent's Plated Lizard	*Gerrhosaurus bulsi* (Laurent, 1954)			
Keeled Plated Lizard	*Gerrhosaurus multilineatus* (Bocage, 1866)			
Black-lined Plated Lizard	*Gerrhosaurus nigrolineatus* (Hallowell, 1857)	Y	*Gerrhosaurus nigrolineatus*	
Desert Plated Lizard	*Gerrhosaurus skoogi* (Andersson, 1916)			
Western Giant Plated Lizard	*Matobosaurus maltzahni* (De Grys, 1938)		*Gerrhosaurus validus*	
Ellenberger's Snake Lizard	*Tetradactylus ellenbergeri* (Angel, 1922)		*Caitia africana* Gray	

(continued)

Common name	Scientific name	C	Bocage (1895)	Obs
SAURIA | **Lacertidae**				
Bushveld Lizard	*Heliobolus lugubris* (Smith, 1838)		*Eremias lugubris*	
Northern Blue-tailed Tree Lizard	*Holaspis guentheri* (Gray, 1863)			
Western Rough-scaled Lizard	*Ichnotropis b. bivittata* (Bocage, 1866)		*Ichnotropis capensis*	
Pale Rough-scaled Lizard	*Ichnotropis b. pallida* (Laurent, 1964)			E
Cape Rough-scaled Lizard	*Ichnotropis c. capensis* (Smith, 1838)			
Overlaete's Rough-scaled Lizard	*Ichnotropis c. overlaeti* (Witte and Laurent 1942)			
Small-scaled Rough-scaled Lizard	*Ichnotropis microlepidota* (Marx, 1956)			E
Shovel-snouted Lizard	*Meroles anchietae* (Bocage, 1867)		*Pachyrhynchus Anchietae*	
Reticulate Desert Lizard	*Meroles reticulatus* (Bocage, 1867)		*Scaptira reticulata*	
Rough-scaled Desert Lizard	*Meroles squamulosus* (Peters, 1854)			
Laurent's Sandveld Lizard	*Nucras scalaris* (Laurent, 1964)			E
Western Sandveld Lizard	*Nucras* aff. *tesselata* (Smith, 1838)			NE
Benguella Sand Lizard	*Pedioplanis benguellensis* (Bocage, 1867)		*Eremias namaquensis*	E
Haacke's Sand Lizard	*Pedioplanis haackei* Conradie et al. 2012			E
Huntley's Sand Lizard	*Pedioplanis huntleyi* (Conradie et al. 2012)			E
SAURIA | **Scincidae** | Acontinae				
Japp's Burrowing Skink	*Acontias jappi* (Broadley, 1968)			
Kalahari Burrowing Skink	*Acontias kgalagadi kgalagadi* (Lamb et al., 2010)			
Western Burrowing Skink	*Acontias occidentalis* (FitzSimons, 1941)			
SAURIA | **Scincidae** | Eugongylinae				
Shorted-headed Snake-eyed Skink	*Panaspis breviceps* (Peters, 1873)			
Cabinda Snake-eyed Skink	*Panaspis cabindae* (Bocage, 1866)	Y	*Ablepharus cabindae*	
Speckle-lipped Snake-eyed Skink	*Panaspis maculicollis* (Jacobsen and Broadley, 2000)			

(continued)

Common name	Scientific name	C	Bocage (1895)	Obs
Angolan Snake-eyed Skink	*Panaspis* aff. *wahlbergii* complex		*Ablepharus wahlbergii*	
De Witte's Leaf-litter Skink	*Leptosiaphos dewittei* (Loveridge, 1934)			
SAURIA I **Scincidae** I Lygosominae				
Hinkel's Red-sided Skink	*Lepidothyris hinkeli* (Wagner et al., 2009)			
Sundevall's Writhing Skink	*Mochlus sundevalli* (Smith, 1849)		*Lygosoma Sundevallii*	
SAURIA I **Scincidae** I Mabuyinae				
Anchieta's Snake Skink	*Eumecia anchietae anchietae* (Bocage, 1870)		*Lygosoma Anchietae*	
Lunda Western Snake Skink	*Eumecia a. major* (Laurent, 1964)			E
Iven's Water Skink	*Lubuya ivensii* (Bocage, 1879)		*Lygosoma Ivensii*	
Wedge Snouted Skink	*Trachylepis acutilabris* (Peters, 1862)	Y	*Mabuia acutilabris*	
Senegal Skink	*Trachylepis affinis* (Gray, 1838)	Y	*Mabuia Raddonii* (not in Angola)	
Monard's Skink	*Trachylepis monardi* (Marques et al. 2018)			E
Bayão's Skink	*Trachylepis b. bayoni* (Bocage, 1872)		*Mabuia Bayonii*	
Huila Skink	*Trachylepis b. huilensis* (Laurent, 1964)			E
Ovambo Stree Skink	*Trachylepis binotata* (Bocage, 1867)		*Mabuia bionotata*	
Bocage's Skink	*Trachylepis bocagii* (Boulenger, 1887)		*Mabuia Petersi*	
Chimba Skink	*Trachylepis chimbana* (Boulenger, 1887)		*Mabuia chimbana*	
Damara Skink	*Trachylepis damarana* (Peters, 1870)		*Mabuia varia* (part)	
Hoesch's Skink	*Trachylepis hoeschi* (Mertens, 1954)			
Bronze Rock Skink	*Trachylepis* cf. *lacertiformis* (Peters, 1854)			
Angolan Blue-tailed Skink	*Trachylepis laevis* (Boulenger, 1907)			
Speckled-lipped Skink	*Trachylepis maculilabris* (Gray, 1845)	Y	*Mabuia maculilabris*	
Grass Skink	*Trachylepis* cf. *megalura* (Peters, 1878)			
Western Three Striped Skink	*Trachylepis occidentalis* (Peters, 1867)		*Mabuia occidentalis*	

(continued)

Common name	Scientific name	C	Bocage (1895)	Obs
Speckled Skink	*Trachylepis punctulata* (Bocage, 1872)		*Mabuia punctulata*	
Kalahari Tree Skink	*Trachylepis spilogaster* (Peters, 1882)			
Striped Skink	*Trachylepis striata* (Peters, 1844)		*Mabuia striata*	
Ansorge's Rock Skink	*Trachylepis sulcata ansorgii* (Boulenger, 1907)		*Mabuia sulcata*	
Angolan Variable Skink	*Trachylepis* cf. *albopunctata* (Bocage, 1867)		*Mabuia varia* (part)	
Wahlberg's Skink	*Trachylepis wahlbergi* (Peters, 1869)			
Sauria | **Scincidae** | Scincinae				
Curror's giant burrowing Skink	*Feylinia currori* (Gray, 1845)	Y	*Feylinia Currori*	
Large-scaled burrowing Skink	*Feylinia grandisquamis* (Müller, 1910)			
Western Limbless Skink	*Melanoseps occidentalis* (Peters, 1877)			
Angolan burrowing Skink	*Sepsina angolensis* (Bocage, 1866)		*Sepsina angolensis*	
Bayão's Burrowing Skink	*Sepsina bayoni* (Bocage, 1866)	Y	*Sepsina Bayonii*	NE
Cope's Burrowing Skink	*Sepsina copei* (Bocage, 1873)		*Sepsina Copei*	E
Johnson's Western Burrowing Skink	*Typhlacontias johnsonii* (Andersson, 1916)			
Speckled Western Burrowing Skink	*Typhlacontias p. punctatissimus* (Bocage, 1873)		*Typhlacontias punctatissimus*	
Bogert's Western Burrowing Skink	*Typhlacontias p. bogerti* (Laurent, 1964)			E
Rohan's Western Burrowing Skink	*Typhlacontias rohani* (Angel, 1923)			
Rudebeck's Western Burrowing Skink	*Typhlacontias rudebecki* (Haacke, 1997)			E
Sauria | **Varanidae**				
Savanna Monitor	*Varanus albigularis angolensis* (Schmidt, 1933)		*Varanus albigularis*	
Water Monitor	*Varanus niloticus* (Linneaus, 1766)	Y	*Varanus niloticus*	

Appendix 3

Checklist of Angolan Snakes. C: Cabinda; Obs: Observations (E:endemic; NE: near-endemic; NR: new record for Angola; RC: requires confirmation). Species listed under ORDER | **Family** | Subfamily

Common name	Scientific name	C	Bocage (1895)	Obs
SCOLECOPHIDIA | **Leptotyphlopidae**				
Shaba Thread Snake	*Leptotyphlops kafubi* (Boulenger, 1919)			
Peter's Thread Snake	*Leptotyphlops scutifrons* (Peters, 1854)		*Stenosoma scutifrons*	
Damara Thread Snake	*Namibiana labialis* (Sternfeld, 1908)			
Benguela Thread Snake	*Namibiana latifrons* (Sternfeld, 1908)			E
Angolan Beaked Thread Snake	*Namibiana rostrata* (Bocage, 1886)		*Stenosoma rostratum*	E
SCOLECOPHIDIA | **Typhlopidae**				
Angolan Blind Snake	*Afrotyphlops angolensis* (Bocage, 1866)		*Typhlops punctatus*	
Angolan Giant Blind Snake	*Afrotyphlops anomalus* (Bocage, 1873)		*Typhlops anomalus* & *T. anchietae*	
Blotched Blian Snake	*Afrotyphlops congestus* (Duméril and Bibron, 1844)	Y		
Lined Blind Snake	*Afrotyphlops lineolatus* (Jan, 1864)	Y	*Typhlops punctatus* var. *lineolatus* & *Typhlops boulengeri*	
Schmidt's Blind Snake	*Afrotyphlops schmidti* (Laurent, 1956)			
Schlegel's Blind Snake	*Afrotyphlops schlegelii* (Bianconi, 1847)		*Typhlops petersii*, *Typhlops humbo* & *Typhlops hottentotus*	
Giant Blind Snake	*Afrotyphlops mucruso* (Peters, 1854)		*Typhlops mucruso*	
Leopoldville Beaked Blind Snake	*Letheobia praeocularis* (Stejneger, 1894)			
HENOPHIDIA | **Pythonidae**				
Namib Dwarf Python	*Python anchietae* (Bocage, 1887)		*Python anchietae*	
Southern African Python	*Python natalensis* (Smith, 1840)		*Python natalensis*	
Northern African Python	*Python sebae* (Gmelin, 1789)	Y		

(continued)

Common name	Scientific name	C	Bocage (1895)	Obs
Henophidia | **Colubridae** | Colubrinae				
White-lipped Snake	*Crotaphopeltis hotamboeia* (Laurenti, 1768)	Y	*Crotaphopeltis rufescens*	
Barotse Water Snake	*Crotaphopeltis barotseensis* (Broadley, 1968)			NR
Confusing Egg-eater	*Dasylepis confusa* (Trape and Mané, 2006)			NR
Palm Egg-Eater	*Dasypeltis palmarum* (Leach, 1818)	Y	*Dasypeltis scabra* var. *palmarum*	
Rhombic Edd-Eater	*Dasypeltis scabra* (Linnaeus, 1758)		*Dasypeltis scabra*	
Shreve's Tree Snake	*Dipsadoboa shrevei* (Loveridge, 1932)			
Punctate Boomslang	*Dispholidus typus punctatus* (Laurent, 1955)		*Bucephalus capensis* (part)	
Green Boomslang	*Dispholidys t. viridis* (Smith, 1838)		*Bucephalus capensis* (part)	
Emerald Snake	*Hapsidophrys smaragdinus* (Schlegel, 1837)	Y	*Hapsidophrys smaragdina*	
Angolan Green Snake	*Philothamnus angolensis* (Bocage, 1882)	Y	*Philothamnus irregularis*	
Thirteen-scaled Green Snake	*Philothamnus carinatus* (Andersson, 1901)			
Striped Green Snake	*Philothamnus dorsalis* (Bocage, 1866)		*Philothamnus dorsalis*	
Emerald Green Snake	*Philothamnus heterodermus* (Hallowell, 1857)		*Philothamnus heterodermus*	
Slender Green Snake	*Philothamnus heterolepidotus* (Günther, 1863)		*Philothamnus heterolepidotus*	
Southeastern Green Snake	*Philothamnus hoplogaster* (Günther, 1863)			
Loveridge's Green Snake	*Philothamnus nitidus loveridgei* (Laurent, 1960)			
Ornate Green Snake	*Philothamnus ornatus* (Bocage, 1872)		*Philothamnus ornatus*	
Spotted Bush Snake	*Philothamnus semivariegatus* (Smith, 1840)		*Philothamnus semivariegatus*	
Large-eyed Green Treesnake	*Rhamnophis aethiopissa* (Günther, 1862)			
Hook-nosed Snake	*Scaphiophis albopunctatus* (Peters, 1870)		*Scaphiophis albopunctatus*	
Damara Tiger Snake	*Telescopus finkeldeyi* (Haacke, 2013)			
Western Tiger Snake	*Telescopus polystictus* (Mertens, 1954)		*Crotaphlopeltis semiannulatus*	
Oates' Vine Snake	*Thelotornis capensis oatesi* (Günther, 1881)		*Thelotornis kirtlandii*	

(continued)

Common name	Scientific name	C	Bocage (1895)	Obs
Forest Vine Snake	*Thelotornis kirtlandii* (Hallowell, 1844)		*Thelotornis kirtlandii*	
Yellow-throated Treesnake	*Thrasops flavigularis* (Hallowell, 1852)	Y	*Thrasops flavigularis*	
Jackson's Treesnake	*Thrasops jacksoni* (Günther, 1895)			
Blanding's Treesnake	*Toxicodryas blandingii* (Hallowell, 1844)	Y	*Dipsas Blandingii*	
Powdered Treesnake	*Toxicodryas pulverulenta* (Fischer, 1856)	Y	*Dipsas pulverulenta*	
Henophidia I **Colubridae** I Grayinae				
Ornate Water Snake	*Grayia ornata* (Bocage, 1866)	Y	*Grayia ornata*	
Smith's Water Snake	*Grayia smithii* (Leach, 1818)		*Grayia triangularis*	
Thollon's Water Snake	*Grayia tholloni* (Mocquard, 1897)			
Henophidia I **Natricidae**				
Bangweulu Swamp Snake	*Limnophis bangweolicus* (Mertens, 1936)			
Striped Swamp Snake	*Limnophis bicolor* (Günther, 1865)		*Helocops bicolour*	
Broadley's Marsh Snake	*Natriciteres bipostocularis* (Broadley, 1962)			
Olive Marsh Snake	*Natriciteres olivacea* (Peters, 1854)	Y	*Mizodon olivaceus*	
Henophidia I **Lamprophiidae** I Atractaspidinae				
Common Purple-glossed Snake	*Amblyodipsas polylepis* (Bocage, 1873)		*Calamelaps polylepis*	
Kalahari Purple-glossed Snake	*Amblyodipsas ventrimaculata* (Roux, 1907)			NR
Cape Centipede Eater	*Aparallactus capensis* (Smith, 1849)		*Uriechis capensis*	
Birbon's Burrowing Asp	*Atractaspis bibronii* (Smith, 1849)		*Atractaspis Bibronii*	
Congo Burrowing Asp	*Atractaspis congica* (Peters, 1877)	Y	*Atractaspis congica*	
Reticulate Burrowing Asp	*Atractaspis reticulata heterochilus* (Boulenger, 1901)			RC
Wilson's burrowing snake	*Hypoptophis wilsoni* (Boulenger, 1908)			
Collared Snake-Eater	*Polemon collaris* (Peters, 1881)		*Microsoma collare*	
Bi-coloured Quill-snouted Snake	*Xenocalamus bicolor machadoi* (Laurent, 1954)			
Elongate Quill-snouted Snake	*Xenocalamus mechowii mechowii* (Peters, 1881)			
Inorante Elongate Quill-snouted Snake	*Xenocalamus m. inorantus* (de Witte and Laurent, 1947)			

(continued)

Common name	Scientific name	C	Bocage (1895)	Obs
HENOPHIDIA I **Lamprophiidae** I Lamprophiinae				
Angolan House Snake	*Boaedon angolensis* (Bocage, 1895)		*Boodon lineatus* var. *angolensis,* Bocage, 1895	
Brown House Snake[a]	*Boaedon fuliginosus* (Boie, 1827)			
Olive House Snake	*Boaedon olivaceus* (Dumeril, 1856)	Y	*Boodon olivaceus*	
Red-Black Striped House Snake	*Bothrophthalmus lineatus* (Peters, 1863)		*Bothrophthalmus lineatus*	
Parker's Banded Snake	*Chamaelycus parkeri* (Angel, 1934)			
Mocquard's Dwarf File Snake	*Gonionotophis brusseauxi* (Mocquard, 1889)			
Yellow Forest Snake	*Hormonotus modestus* (Duméril, Bibron and Duméril, 1854)			
Western Forest File Snake	*Mehelya poensis* (Smith, 1849)			
Cape File Snake	*Limaformosa capensis* (Smith, 1847)		*Heterolepis Guirali ?*	
Savorgan's File Snake	*Limaformosa savorgani* (Moquard, 1887)	?		NR
Vernay's File Snake	*Limaformosa vernayi* (Bogert, 1940)			
White-bellied Water Snake	*Lycodonomorphus* (?) *subtaeniatus* (Laurent, 1954)			
Hellmich's Wolf Snake	*Lycophidion hellmichi* (Laurent, 1964)			
Flat Wolf Snake	*Lycophidion laterale* (Hallowell, 1857)		*Lycophidium laterale*	
Speckled Wolf Snake	*Lycophidion meleagre* (Boulenger, 1893)		*Lycophidium meleagris*	
Spotted Wolf Snake	*Lycophidion multimaculatum* (Boettger, 1888)	Y	*Lycophium capense*	
Namib Wolf Snake	*Lycophidion namibianum* (Broadley, 1991)			NR
Ornate Wolf Snake	*Lycophidion ornatum* (Parker, 1936)	?		
Viperine Rock Snake	*Hemirhagerrhis viperina* (Bocage, 1873)		Psammophylax nototaenia	
HENOPHIDIA I **Lamprophiidae** I Psammophinae				
Angolan Sand Snake	*Psammophis angolensis* (Bocage, 1872)		*Amphiophis angolensis*	
Ansorge's Sand Snake	*Psammophis ansorgii* (Boulenger, 1905)			E
Jalla's Sand Snake	*Psammophis jallae* (Peracca, 1896)			

(continued)

Common name	Scientific name	C	Bocage (1895)	Obs
Leopard Sand Snake	*Psammophis leopardinus* (Bocage, 1887)			
Namib Sand Snake	*Psammophis namibensis* (Broadley, 1975)			
Karoo Sand Snake	*Psammophis notostictus* (Peters, 1867)			
Mozambique Grass Snake	*Psammophis mossambicus* (Peters, 1882)	Y	*Psammophis sibilans* (Linnaeus, 1758)	
Strip-bellied Sand Snake	*Psammophis subtaeniatus* (Peters, 1882)			
Western Sand Snake	*Psammophis trigrammus* (Günther, 1865)			
Fork-marked Sand Snake	*Psammophis trinasalis* (Werner, 1902)			
Zambezi Sand Snake	*Psammophis zambiensis* (Hughes and Wade, 2000)			
Striped Beaked Skaapstekker	*Psammophylax acutus* (Günther, 1888)		*Rhageheris acutus*	
Huila Skaapstekker	*Psammophylax ocellatus* (Bocage, 1873)		*Psammophylax rhombeatus*	E
Striped Skaapstekker	*Psammophylax tritaeniatus* (Günther, 1868)		*Rhagerhis tritaeniata*	
HENOPHIDIA | **Lamprophiidae** | Prosymnidae				
Zambezi Shovel-snout Snake	*Prosymna ambigua* (Bocage, 1873)		*Prosymna ambigua*	
Angola Shovel-snout Snake	*Prosymna angolensis* (Boulenger, 1915)		*Prosymna frontalis*	
South-west Shovel-snout Snake	*Prosymna frontalis* (Peters, 1867)			
Visser's Shivel-snout Snake	*Prosymna visseri* (FitzSimons, 1959)			
HENOPHIDIA | **Lamprophiidae** | Pseudaspidae				
Mole Snake	*Pseudaspis cana* (Linnaeus, 1758)		*Pseudaspis cana*	
Western-keeled Snake	*Pythonodipsas carinata* (Günther, 1868)			
HENOPHIDIA | **Lamprophiidae** | Elapidae				
Cowles' Shield Cobra	*Aspidelaps lubricus cowlesi* (Bogert, 1940)			
Jameson's Mamba	*Dendroaspis jamesoni* (Traill, 1843)	?	*Dendroaspis neglectus*	
Black Mamba	*Dendroaspis polylepis* (Günther, 1864)	?	*Dendroaspis angusticeps*	
Günther's Garter Snake	*Elapsoidea guentherii* (Bocage, 1866)		*Elapsoidea Guentherii*	

(continued)

Common name	Scientific name	C	Bocage (1895)	Obs
Angolan Garter Snake	*Elapsoidea s. semiannulata* (Bocage, 1882)			
Western Garter Snake	*Elapsoidea s. moebiusi* (Werner, 1897)			
Anchiete's Cobra	*Naja* (*Ureaus*) *anchietae* (Bocage, 1879)		*Naja anchietae* & *Naja haje*	
Banded Water Cobra	*Naja* (*Boulengerina*) *annulata* (Peters, 1876)			
Central African Forest Cobra	*Naja* (*Boulengerina*) *melanoleauca* (Hallowell, 1857)			
Savanna Forest Cobra	*Naja* (*Boulengerina*) *subfulva* (Laurent, 1956)	?		
Mozambique Cobra	*Naja* (*Afronaja*) *mossambica* (Peters, 1854)			
Western Banded Spitting Cobra	*Naja* (*Afronaja*) *nigricincta* (Bogert, 1940)			
Black Spitting Cobra	*Naja* (*Afronaja*) *nigricollis* (Reinhardt, 1843)		*Naja nigricollis*	
Gold's Tree Cobra	*Pseudohaje goldii* (Boulenger, 1895)			
HENOPHIDIA | **Viperidae**				
Variable Bush Viper	*Atheris squamigera* (Hallowell, 1854)		*Atheris squamigera*	
Puff Adder	*Bitis arietans* (Merrem, 1820)		*Vipera arietans*	
Horned Adder	*Bitis caudalis* (Smith, 1839)		*Vipera caudalis*	
Gaboon Adder	*Bitis gabonica* Duméril, (Bibron and Duméril, 1854)	?	*Vipera rhinoceros*	
Angolan Adder	*Bitis heraldica* (Bocage, 1889)		*Vipera heraldica*	E
Rhinoceros Viper	*Bitis nasicornis* (Shaw, 1802)	?		
Peringuey's Adder	*Bitis peringueyi* (Boulenger, 1888)	?		
Two-lined Night Adder	*Causus bilineatus* (Boulenger, 1905)			
Lichtenstein's Night Adder	*Causus lichtensteini* (Jan, 1859)			
West African Night Adder	*Causus maculatus* (Hallowell, 1842)			
Rasmussen's Night Adder	*Causus rasmusseni* (Broadley, 2014)			
Angola Green Night Adder	*Causus resimus* (Peters, 1862)		*Causus resimus*	
Rhombic Night Adder	*Causus rhombeatus* (Lichtenstein, 1823)	Y	*Causus rhombeatus*	

[a]Don't include the additional *Boaedon* species (Hallerman et al. in prep.)

References

Adalsteinsson SA, Branch WR, Trapé S, Vitt LJ, Hedges SB (2009) Molecular phylogeny, classification, and biogeography of snakes of the family leptotyphlopidae (Reptilia, Squamata). Zootaxa 2244:1–50

Alexander AA, Gans C (1966) The pattern of dermal-vertebral correlation in snakes and amphisbaenians. Zool Med 41(11):171–190

Angel F (1921) Description d'un ophidien nouveau de l'Angola appartenant au genre *Psammophis*. Bull Soc Zool Fr Paris 46(8–10):116–118

Angel MF (1923) Reptiles. In: Rohan-Chabot (ed) Mission Rohan-Chabot, Angola et Rhodesia (1912–1914), Histoire Naturelle, Fascicule 1 (Mammifères – Oiseaux –Reptiles – Poissons), vol IV. Imprimerie Nationale, Paris, pp 157–169, 1 pl

Arruda M (2018) Confiscation by police of illegal soft-shell terrapin trade in Luanda. https://www.facebook.com/katimbala.ingles/videos/pcb.1576072472480736/1576072072480776/?type=3&theater

Baptista N, António T, Branch WR (2018) Amphibians and reptiles of the Tundavala region of the Angolan Escarpment. In: Revermann R, Krewenka KM, Schmiedel U et al (eds) Climate change and adaptive land management in southern Africa – assessments, changes, challenges, and solutions, Biodiversity & ecology, vol 6. Klaus Hess Publishers, Göttingen, pp 397–403

Baptista N, Conradie W, Vaz Pinto P et al (2019) The Amphibians of Angola: early studies and the current state of knowledge. In: Huntley BJ, Russo V, Lages F, Ferrand N (eds) Biodiversity of Angola. Science & conservation: a modern synthesis. Springer Nature, Cham

Bauer AM, Branch WR (2003) The herpetofauna of the Richterveld National Park and the adjacent northern Richtersveld, Northern Cape Province, Republic of South Africa. Herpetol Nat Hist 8(2):111–160

Bauer AM, Kuhn AL (2017) Historical climate change and the evolution of the Namib Day Geckos (Squamata: Gekkonidae: *Rhoptropus*) Oral Presentation (abst). Afr Herp News 66:9

Bauer AM, Lamb T (2005) Phylogenetic relationships of southern African geckos in the *Pachydactylus* group (Squamata: Gekkonidae). Afr J Herpetol 54(2):105–129

Billes A, Fretey J, Verhage B et al (2006) First evidence of leatherback movement from Africa to South America. Mar Turt Newsl 111:13–14

Blanc CP, Fretey J (2002) Analyse Zoogegraphique du peuplement reptilien de L'Afrique Centrale et de L'Angola. Biogeographica 78:49–75

Bocage JVB (1866) Lista dos reptis das possessões portuguesas d' Africa occidental que existem no Museu de Lisboa. J Sci Math Phys Nat Lisboa 1:37–56

Bocage JVB (1873) Reptiles nouveaux de l'interieur de Mossamedes. J Sci Math Phy Nat Lisboa 4:247–253

Bocage JVB (1895) Herpétologie d'Angola et du Congo. Lisbonne, Imprimerie Nationale, 203 pp, 19 pls

Bocage JVB (1896) Mammiferos, aves e reptis da Hanha, no sertào de Benguella. J Sci Math Phys Nat Lisboa 14(2):105–114

Bogert CM (1940) Herpetological results of the Vernay Angola Expedition. I. Snakes, including an arrangement of the African Colubridae. Bull Am Mus Nat Hist 77:1–107

Boulenger GA (1885) Catalogue of the lizards in the British Museum (Natural History). Volume II, Iguanidae, Xenosauridae, Zonuridae, Anguidae, Anniellidae, Helodermatidae, Varanidae, Xantusiidae, Teiidae, Amphisbaenidae. London: British Museum of (Natural History), London, xiv + 492 pp, 54 figs., 24 pls

Boulenger GA (1893) Catalogue of the snakes in the British Museum (Natural History). Volume I, containing the families Typhlopidae, Glauconiidae, Boidae, Ilysiidae, Uropeltidae, Xenopeltidae and Colubridae aglyphae, part. British Museum (Natural History), London, xiv + 448 pp, 26 figs., 28 pls

Boulenger GA (1894) Catalogue of the snakes in the British Museum (Natural History). Volume II, containing the conclusion of the Colubridae aglyphae. British Museum (Natural History), London, xii + 382 pp, 25 figs., 20 pls

Boulenger GA (1896) Catalogue of the snakes in the British Museum (Natural History). Volume III, containing the Colubridae (Opisthoglyphae and Proteroglyphae), Amblycephalidae, and Viperidae. British Museum (Natural History), London, xiv + 727 pp, 37 figs., 25 pls

Boulenger GA (1905) A list of the batrachians and reptiles collected by Dr. W. J. Ansorge in Angola, with descriptions of new species. Ann Mag Nat Hist Ser 7 16:105–115

Boulenger GA (1907a) Descriptions of three new lizards and a frog, discovered by Dr. W. J. Ansorge in Angola. Ann Mag Nat Hist Seventh Ser 19:212–214

Boulenger GA (1907b) Descriptions of a new frog discovered by Dr. W. J. Ansorge in Mossamedes, Angola. Ann Mag Nat Hist Seventh Ser 20:109

Boulenger GA (1915) A list of the snakes of the Belgian and Portuguese Congo, northern Rhodesia, and Angola. Proc Zool Soc London 1915:193–223

Bowker JN, De Vos A, Ament JM et al (2017) Effectiveness of Africa's tropical protected areas for maintaining forest cover. Conserv Biol 31(3):559–569

Branch WR (1994) Herpetofauna of the Sperrgebeit region of southern Namibia. Herpetol Nat Hist 2(1):1–11

Branch WR (2014) Reptiles of South Africa, Lesotho and Swaziland: conservation status, diversity, endemism, hotspots and threats. In: Bates MF, Branch WR, Bauer AM, Burger M, Marais J, Alexander GJ, de Villiers MS (eds) Atlas and Red Data Book of the Reptiles of South Africa, Lesotho and Swaziland, Suricata 1. South African National Biodiversity Institute, Pretoria, pp 22–50

Branch WR (2016) Preface, amphibian & reptile conservation special Angola-Africa issue. Amphib Rep Conserv 10(2):2. i-iii

Branch WR (2018) Snakes of Angola: an annotated checklist. Amphibian & Reptile Conservation 12(2) [General Section]: 41–82 (e159)

Branch WR, Conradie W (2013) *Naja annulata annulata* (Bucholtz & Peters, 1876). African Herp News 59:50–53

Branch WR, Conradie WC (2015) Vl Herpetofauna da regiáo da Lagoa Carumbo (Herpetofauna of the Carumba Lagoon Area). In: Huntley B & Francisco P (eds) *Relatório sobre a Expedição Avaliação rápida da Biodiversidade de regiáo da Lagoa Carumbo, Lunda-Norte – Angola*, pp 194–209. Republica de Angola Ministerio do Ambiente, 219p

Branch WR, McCartney CJ (1992) A report on a small collection of reptiles from southern Angola. J Herpetol Assoc Afr 41:1–3

Branch WR, Tolley KA (2017) Oral presentation (abst). New Lacertids from Angola. Afr Herp News 66:11

Branch WR, Haacke W, Vaz Pinto P et al (2017) Loveridge's Angolan geckos, *Afroedura karroica bogerti* and *Pachydactylus scutatus angolensis* (Sauria, Gekkonidae): new distribution records, comments on type localities and taxonomic status. Zoosyst Evol 93(1):157–166

Branch WR, Baptista N, Vaz Pinto P (2018) Angolan amphisbaenians: rediscovery of *Monopeltis luandae Gans* 1976, with comments on the type locality of *Monopeltis perplexus* Gans 1976 (Sauria: Amphisbaenidae). Herpetology Notes 11:603–606

Branch WR, Baptista N, Keates C et al (2019) Rediscovery, taxonomic status and phylogenetic relationships of two enigmatic Psammophine snakes (Serpentes: Psammophinae) from the southwestern Angola plateau. Zootaxa (in press)

Broadley DG (1971) A revision of the African snake genera *Amblyodipsas* and *Xenocalamus*. Occ Pap Natl Mus Rhod B4(33):629–697

Broadley DG (1975) A review of the *Mabuya lacertiformis* complex in southern Africa (Sauria: Scincidae). Arnoldia (Rhod) 7(18):1–16

Broadley DG (1991) The Herpetofauna of Northern Mwinilunga Distr., Northw. Zambia. Arnoldia Zimbabwe 9(37):519–538

Broadley DG (1996) A review of the tribe Atherini (Serpentes: Viperidae), with the descriptions of two new genera. Afr J Herpetol 45:40–48

Broadley DG (1997a) A review of the *Monopeltis capensis* complex in southern Africa (Reptilia: Amphisbaenidae). Afr J Herpetol 46(1):1–12

Broadley DG (1997b) Amphibaenia. *Dalophia ellenbergeri* (Angel, 1920). African Herp News 26:34–35

Broadley DG (2001) Geographical distribution. *Monopeltis sphenorhynchus*. Afr Herp News 32:23–24

Broadley DG, Baldwin AS (2006) Taxonomy, natural history, and zoogeography of the Southern African Shield Cobras, Genus *Aspidelaps* (Serpentes: Elapidae). Herpetol Nat Hist 9(2):163–176

Broadley DG, Broadley S (1999) A review of the African worm snakes from south of latitude 12°S (Serpentes: Leptotyphlopidae). Syntarsus 5:1–36

Broadley DG, Gans C (1969) A new species of *Zygaspis* (Amphisbaenia: Reptilia) from Zambia and Angola. Arnoldia (Rhod) 4(25):1–4

Broadley DG, Hughes B (2000) A revision of the African genus *Hemirhagerrhis* Boettger 1893 (Serpentes: Colubridae). Syntarsus 6:1–17

Broadley DG, Schätti B (2000) A new species of *Coluber* from northern Namibia (Reptilia: Serpentes). Modoqua 19(2):171–174

Broadley DG, Gans C, Visser J (1976) Studies on Amphisbaenians (6). The Genera *Monopeltis* and *Dalophia* in Southern Africa. Bull Am Mus Nat Hist 157(5):311–486

Broadley DG, Tolley KA, Conradie W et al (2018) A phylogeny and revision of the African File Snakes *Gonionotophis* Boulenger (Squamata: Lamprophiidae). Afr J Herpetol. https://doi.org/10.1080/21564574.2018.1423578

Brooks C (2012) Biodiversity survey of the upper Angolan Catchment of the Cubango-Okavango River Basin. USAid-Southern Africa, 151 p

Brooks C (2013) Trip report: aquatic biodiversity survey of the lower Cuito and Cuando river systems in Angola. USAid-Southern Africa. 43 p

Carr T, Carr P (1991) Surveys of the sea turtles of Angola. Biol Conserv 58(1):19–29

CEPF (2003) Ecosystem profile. The succulent Karoo hotspot, Namibia and South Africa. Critical ecosystem partnership fund. South African National Biodiversity Institute. https://www.sanbi.org/documents/ecosystem-profile-succulent-karoo-hotspot

Ceríaco LMP, Bauer AM, Blackburn DC et al (2014a) The herpetofauna of the Capanda Dam Region, Malanje, Angola. Herpetol Rev 45(4):667–674

Ceríaco LMP, Blackburn DC, Marques MP et al (2014b) Catalogue of the amphibian and reptile type specimens of the Museu de História Natural da Universidade do Porto in Portugal, with some comments on problematic taxa. Alytes 31(1):13–36

Ceríaco LMP, de Sá SAC, Bandeira S, Valério H et al (2016a) Herpetological survey of Iona National Park and Namibe Regional Natural Park, with a Synoptic list of the Amphibians and reptiles of Namibe Province, Southwestern Angola. Proc Calif Acad Sci 63(2):15–61

Ceríaco LMP, Marques MP, Bandeira SA (2016b) Anfíbios e Répteis do Parque Nacional da Cagandala. Publ. Instituto Nacional da Biodiversidade e Áreas de Conservação & Museu Nacional de História Natural e da Ciência, 96 p

Ceríaco LMP, de Sá S, Bauer AM (2018a) The genus *Osteolaemus* (Crocodylidae) in Angola and a new southernmost record for the genus. Herpetol Notes 11:337–341

Ceríaco LMP, Marques MP, Bandeira S et al (2018b) Herpetological survey of Cangandala National Park, with a synoptic list of the amphibians and reptiles of Malanje Province, Central Angola. Herpetol Rev 49(3):408–431

Clark VR, Barker NP, Mucina L (2011) The great escarpment of southern Africa: a new frontier for biodiversity exploration. Biodivers Conserv 20:2543–2561

Conradie W, Bourquin S (2013) Geographical Distributions: *Acontias kgalagadi kgalagadi* (Lamb, Biswas and Bauer, 2010). Afr Hep News 60:29–30

Conradie W, Branch WR, Measey GJ et al (2012a) Revised phylogeny of Sand lizards (*Pedioplanis*) and the description of two new species from south-western Angola. Afr J Herpetol 60(2):91–112

Conradie W, Branch WR, Measey JG et al (2012b) A new species of *Hyperolius Rapp*, 1842 (Anura: Hyperoliidae) from the Serra da Chela mountains, southwestern Angola. Zootaxa 3269:1–17

Conradie W, Branch WR, Tolley KA (2013) Fifty shades of grey: giving colour to the poorly known Angolan Ash reed frog (Hyperoliidae: *Hyperolius cinereus*), with the description of a new species. Zootaxa 3635(3):201–223

Conradie W, Bills R, and Branch WR (2016). The herpetofauna of the Cubango, Cuito, and lower Cuando river catchments of south-eastern Angola. Amphibian Reptile Conserv 10(2) [Special Section]:6–36

Conradie WC, Bills R, Baptista N et al (2017) Oral presentation (abst). Across river basins: Exploring the unknown southeastern Angola. African Herp News 66:14–15

Cotterill F, De Wit M (2011) Geoecodynamics and the Kalahari Epeirogeny: linking its genomic record, tree of life and palimpsest into a unified narrative of landscape evolution. S Afr J Geol 114:489–514

Craven P (2009) Phytogeographic study of the Kaokoveld Centre of Endemism. Unpublished Ph.D. thesis, University of Stellenbosch, 234 p

Crawford-Cabral J, Mesquitela LM (1989) *Índice toponímico de colheitas zoológicas em Angola.* Instituto de Investigação Cientifica Tropical. Centro de Zoologia, Lisbon 206 pp

De Almeida MAP (2011) José Vicente Barbosa du Bocage. In: Biographies of scientists and engineers, Centro Interuniversitário de História da Ciência e Tecnologia. http://ciuhct.org/pt/bocage-jose-vicente-barbosa-du

Eaton MJ (2010) Dwarf crocodile *Osteolaemus tetraspis*. In: Manolis SC, Stevenson C (eds) Crocodiles. Status survey and conservation action plan, 3rd edn. Crocodile Specialist Group, Darwin, pp 127–132

Eaton MJ, Martin A, Thorbjarnarson J, Amato G (2009) Species-level diversification of African dwarf crocodiles (genus *Osteolaemus*): a geographic and phylogenetic perspective. Mol Phylogenet Evol 50(3):496–506

Elwen S, Braby RJ (2015) Report on a turtle and cetacean assessment survey to the Kunene River mouth, northern Namibia – January 2014. Afr Sea Turtle Newsl 4:22–27

Engelbrecht HM, Branch WR, Greenbaum E et al (2018) Diversifying into the branches: species boundaries in African green and bush snakes, *Philothamnus* (Serpentes: Colubridae). Mol Phylo Evol. https://doi.org/10.1016/j.ympev.2018.10.023

Ernst R (2015) A rapid assessment of the herpetofauna of the Serra do Pingano ecosystem in Uíge Province, northern Angola. Unpubl. Report to Instituto Nacional da Biodiversidade e Áreas de Conservação, Ministério do Ambiente, República de Angola, 11 p

Ernst R, Nienguesso ABT, Lautenschläger T et al (2014) Relicts of a forested past: southernmost distribution of the hairy frog genus *Trichobatrachus* Boulenger, 1900 (Anura: Arthroleptidae) in the Serra do Pingano region of Angola with comments on its taxonomic status. Zootaxa 3779(2):297–300

Ernst R, Schmitz A, Wagner P et al (2015) A window to Central African forest history: distribution of the *Xenopus fraseri* subgroup south of the Congo Basin, including a first country record of *Xenopus andrei* from Angola. Salamandra 52(1):147–155

Face of Malawi (2013) Chinese 'managed' Turtle butchery discovered on Lake Malawi. http://www.faceofmalawi.com/2013/11/chinese-managed-turtle-butchery-discovered-on-lake-malawi/

Ferreira JB (1897) Lista dos reptis e amphibios que fazem parte da última remessa de J. d'Anchieta. J Sci Math Phys Nat Lisboa 5(2):240–246

Ferreira JB (1900) Sobre alguns exemplares pertencentes á fauna do norte de Angola. J Sci Math Phys Nat Lisboa 21:48–54

Ferreira JB (1903) Reptis de Angola da região norte do Quanza da collecção Pereira do Nascimento (1902). J Sci Math Phys Nat Lisboa Segunda Série 7(25):9–16

Ferreira JB (1904) Reptis e amphibios de Angola da região ao norte do Quanza (Collecção Newton – 1903). J Sci Math Phys Nat Lisboa Segunda Série 7(26):111–117

Ferreira JB (1906) Algumas espécies novas ou pouco conhecidas de amphibios e reptis de Angola (Collecção Newton – 1903). J Sci Math Phys Nat Lisboa Segunda Série 7(26):159–171

FitzSimons VFM (1953) A new genus of gerrhosaurid from southern Angola. Ann Transv Mus 22(2):215–217

FitzSimons VFM (1959) Some new reptiles from southern Africa and southern Angola. Ann Transv Mus 23(4):405–409

Gamito-Marques D (2017) A space of one's own: Barbosa du Bocage, the foundation of the National Museum of Lisbon, and the construction of a career in zoology (1851–1907). J Hist Biol. https://doi.org/10.1007/s10739-017-9487-6

Gans C (1976) Three new spade-snoted amphisbaenians from Angola (Amphisbaenia, reptilia). Am Mus Novit 2590:1–11

Gans C (2005) Checklist and bibliography of the Amphisbaenia of the World. Bull Am Mus Nat Hist 289(1):130

Gray JE (1865) A revision of the genera and species of amphisbaenians with the descriptions of some new species now in the collection of the British Museum. Proc Zool Soc London 1865:442–455

Greenbaum E, Portillo F, Jackson K et al (2015) A phylogeny of central African *Boaedon* (Serpentes: Lamprophiidae), with the description of a new cryptic species from the Albertine Rift. Afr J Herpetol 64(1):18–38

Haacke WD (1976a) The burrowing geckos of southern Africa, 2 (Reptilia: Gekkonidae). Ann Transv Mus 30(2):13–29

Haacke WD (1976b) The burrowing geckos of southern Africa, 3 (Reptilia: Gekkonidae). Ann Transv Mus 30(3):29–44

Haacke WD (1981) The file snakes of the genus *Mehelya* in Southern Africa with special reference to South West Africa/Namibia. Modoqua 12(4):217–224

Haacke WD (1997) Systematics and biogeography of the southern African scincine genus *Typhlacontias* (Reptilia: Scincidae). Bonner Zool Beiträge 47(1–2):139–163

Haacke WD (2008) A new leaf-toed gecko (Reptilia: Gekkonidae) from south-western Angola. Afr J Herpetol 57:85–92

Haacke WD (2013) Description of a new Tiger Snake (Colubridae, *Telescopus*) from south-western Africa. Zootaxa 3737(3):280–288

Haagner GV, Branch WR, Haagner AJF (2000) Notes on a collection of reptiles from Zambia and adjacent areas of the Democratic Republic of the Congo. Ann East Cape Mus 1:1–25

Heinicke MP, Daza JD, Greenbaum E et al (2014) Phylogeny, taxonomy and biogeography of a circum-Indian Ocean clade of leaf-toed geckos (Reptilia: Gekkota), with a description of two new genera. Syst Biodivers 12(1):23–42

Heinz HM (2011) Comparative phylogeography of two widespread geckos from the typically narrow-ranging Pachydactylus group in Southern Africa. Unpublished MSc thesis, Villanova University, Villanova, Pennsylvania, USA, vii + 107 pp

Hellmich W (1954–1955) Auf herpetologischer Forschungsfahrt in Angola (Portugeisisch Westafrika). Die Aquarien und Terrarien Zeitschrift. In six parts: 1954 – I, 7(11): 302–304; II, 7(12): 324–326; 1955 – III, 8(1): 23–26; IV, 8(2): 51–53; V, 8(3):78–81; VI, 8(4):103–107

Hellmich W (1957a) Die reptilienausbeute der Hamburgischen Angola Expedition. Mitteilungen aus dem Hamburgischen Zoologischen Museum und Institut 55:39–80

Hellmich W (1957b) Herpetologische Ergebnisse einer Forschungsreise in Angola. Veröffentlichungen der Zoologischen Staatssammlung München 5:1–92

Hellmich W, Schmelcher D (1956) Eine neue Rasse von *Gerrhosaurus nigrolineatus* Hallowell (Gerrhosauridae). Zool Anz 156(7/8):202–205

Herrmann H-W, Branch WR (2013) Fifty years of herpetological research in the Namib Desert and Namibia with an updated and annotated species checklist. J Arid Environ 93:94–115

Huntley BJ (1974) Outlines of wildlife conservation in Angola. J S Afr Wildl Manage Assoc 4:157–166

Huntley BJ (2009) SANBI/ISCED/UAN Angolan biodiversity assessment capacity building project. Report on Pilot project. Unpublished report to ministry of environment, Luanda, 97 pp, 27 figures

Huntley BJ (2019) Angola in outline: physiography, climate and patterns of biodiversity. In: Huntley BJ, Russo V, Lages F, Ferrand N (eds) Biodiversity of Angola. Science & conservation: a modern synthesis. Springer Nature, Cham

Huntley BJ, Francisco P (eds) (2015) Relatório sobre a Expedição Avaliação rápida da Biodiversidade de regiáo da Lagoa Carumbo, Lunda-Norte – Angola/Biodiversity Rapid Assessment of the Carumbo Lagoon Area, Lunda-Norte. Angola Ministerio do Ambiente, Luanda 219 pp

Jordan K (1936) Dr Karl Jordan's expedition to South-West Africa and Angola. Narrat Nov Zool 40:17–62, 2 maps, 5 pls

Jongsma GFM, Barej MF, Barratt CD, Burger M, Conradie W, Ernst R, Greenbaum E, Hirschfeld M, Leaché AD, Penner J, Portik DM, Zassi-Boulou A-G, Rödel M-O, Blackburn DC (2018) Diversity and biogeography of frogs in the genus *Amnirana* (Anura: Ranidae) across sub-Saharan Africa. Mol Phylogenet Evol 120:274–285

Kelly CMR, Branch WR, Broadley DG et al (2011) Molecular systematics of the African snake family Lamprophiidae Fitzinger, 1843 (Serpentes: Elapoidea), with particular focus on the genera *Lamprophis* Fitzinger 1843 and *Mehelya* Csiki 1903. Mol Phylogenet Evol 58:415–426

Kindler C, Branch WR, Hofmeyr MF et al (2012) Molecular phylogeny of African hinge-back tortoises (*Kinixys* Bell, 1827): implications for phylogeography and taxonomy (Testudines: Testudinidae). J Zool Syst Evol Res 50(3):192–201

Laurent RF (1950) Reptiles et Batraciens de la region de Dundo (Angola du Nord-Est). Publicações culturais da Companhia de Diamantes de Angola 10:7–17

Laurent RF (1954) Reptiles et Batraciens de la région de Dundo (Angola) (Deuxième Note). Publicações culturais da Companhia de Diamantes de Angola 23:35–84

Laurent RF (1964) Reptiles et Amphibiens de l'Angola (Troisième contribution). Publicações culturais da Companhia de Diamantes de Angola 67:11–165

Lewin A, Feldman A, Bauer AM et al (2016) Patterns of species richness, endemism and environmental gradients of African reptiles. J Biogeogr 43:2380–2390

Lindsey PA, Nyirenda VR, Barnes JI et al (2014) Underperformance of African protected area networks and the case for new conservation models: insights from Zambia. PLoS One 9(5):e94109

Loveridge A (1941) Revision of the African lizards of the family Amphisbaenidae. Bull Mus Comp Zool 87(5):353–451

Loveridge A (1944) New geckos of the genera *Afroedura*, new genus, and *Pachydactylus* from Angola. Am Mus Novit 1254:1–4

Machado AdB (1952) Generalidades acerca da Lunda e da sua exploraçao biolôgica. Publ. Cult. Comp. Diam. Angola, 12

Madruga CM (2013) José Vicente Barbosa du Bocage (1823–1907): a construção de uma persona científica. Unpublished M.Sc. thesis, Universidade de Lisboa

Manaças S (1963) Sáurios de Angola. Memórias da Junta de Investigações do Ultramar, Lisboa, 43, segunda série. Estudos Zool:223–240

Manaças S (1973) Alguns ofídeos de Angola. Memórias da Junta de Investigações do Ultramar, Lisboa, 58, segunda série. Estudos de Zoologia:187–200

Manaças S (1981) Ofídeos venenosos da guiné, S. Tomé, Angola e Moçambique. Garcia Orta Sér de Zool 10(1/2):13–46

Marques MP, Ceríaco LMP, Bauer AM, et al. (2014) Geographic distribution of Amphibians and reptiles of Angola: towards an Atlas of Angolan herpetofauna. Poster. 12th Herpetological association of Africa conference, Gobabeb, Namibia, 20–22 November, 2014

Marques MP, Ceríaco LMP, Blackburn DC, Bauer AM (2018) Diveristy and distribution of the amphibians and terrestrail reptiles of Angola atlas of historical and bibliographic records (1840–2017). Proceedings of the California academy of sciences, Series 4, Volume 65, Supplement II: 1–501

Marx H (1956) A new lacertid lizard from Angola. Fieldiana: Zoology 39:5–9

Mateus O, Jacobs L, Polcyn M (2009) The oldest African eucryptodiran turtle from the Cretaceous of Angola. Acta Palaeontol Pol 54(4):581–588

McLachlan GR, Spence JM (1967) A new species of *Pachydactylus* (*Pachydactylus oreophilus* sp. nov.) from Sesfontein, South West Africa. Cimbebasia (21):3–8

Measey J, Tolley KA (2013) A molecular phylogeny for sub-Saharan amphisbaenians. Afr J Herpetol 62(2):100–108

Medina MF, Bauer AM, Branch WR et al (2016) Molecular phylogeny of *Panaspis* and *Afroablepharus* skinks (Squamata: Scincidae) in the savannas of sub-Saharan Africa. Mol Phylogenet Evol 100:409–423

Mertens R (1938) Amphibien und Reptilien aus Angola, gesammelt von W. Schack. Senckenbergiana 20:425–443

Mertens R (1958) *Bitis heraldica*, eine oft verkannte Otter aus Angola. Senckenbergiana Biologica Frankfurt-am-Main 39(3–4):145–148, 4 figs

Monard A (1931) Mission scientifique Suisse dans l'Angola. Résultats scientifiques. Reptiles. Bull Soc Neuchâtel Sci Nat ser 2 40(55):89–111

Monard A (1934) Ornithologie de l'Angola. Arquivos do Museu Bocage 5:1–110

Monard A (1935) Contribution à la mammologie d'Angola et prodrome d'une faune d'Angola. Arquivos do Museu Bocage 6:1–314 44 fig

Monard A (1937) Contribution á l'herpétologie d'Angola. Arquivos do Museu Bocage 8:19–154

Monard A (1938) Contribution à la batrachologie d'Angola. Bull Soc Neuch Se Nat 62:5–59, 19 fig

Morais M (2008) Tartarugas Marinhas na Costa de Cabinda. Plano de conservação e gestão para a implementação do projecto de prospecção sísmica "on shore". Holisticos/Chevron. 67 p

Morais M (2016) Projecto Kitabanga – Conservação de tartarugas marinhas. Relatório final da temporada 2015/2016. Universidade Agostinho Neto/Faculdade de Ciências, Luanda

Morais M (2017) Projecto Kitabanga – Conservação de tartarugas marinhas. Relatório final da temporada 2016/2017. Universidade Agostinho Neto / Faculdade de Ciências, Luanda

Morais M, Torres MOF, Martins MJ (2005) Análise da Biodiversidade Marinha e Costeira e Identificação das Pressões de Origem Humana sobre os Ecossistemas Marinhos e Costeiros. Ministerio do Urbanismo e Ambiente, Luanda, 140 pp

Nagy ZT, Kusamba C, Collet M et al (2013) Notes on the herpetofauna of western Bas-Congo, democratic Republic of the Congo. Herpetology Notes 6:413–419

NGOWP (2018) National geographic Okavango wilderness project. Initial findings from exploration of the upper catchments of the Cuito, Cuanavale, and Cuando Rivers, May 2015–December 2016

NPAES (2010) National protected area expansion strategy for South Africa 2008. Priorities for expanding the protected area network for ecological sustainability and climate change adaptation. Government of South Africa, Pretoria

OKACOM (2011) National action plan for the sustainable management of the Cubango/Okavango River Basin, Angola – Draft 3. Available http://www.okacom.org/site-documents. Accessed 24 June 2016

Oliveira PRS d (2017) Sepentes em Angola. Uma visão toxinolŏgca e clínica dos envenenamentos. Glaciar, Lisbon, p 159

Oliveira PRS d, Rocha MT, Castro AG et al (2016) New records of Gaboon viper (*Bitis gabonica*) in Angola. Herpetol Bull 136:42–43

Parker HW (1936) Dr. Karl Jordan's expedition to South West Africa and Angola: herpetological collection. Nov Zool 40:115–146

Peters WCH (1869) Eine Mittheilung über neue Gattungen und Arten von Eidechsen. Monatsber Königl Preuss Akad Wissensch Berlin 1869:57–66

Peters WCH (1876) Über die von Hrn. Professor Dr. R. Buchholz in Westafrika gesammelten Amphibien. Auszug aus dem Monatsberich der Königl. Akademie der Wissenschafen zu Berlin:117–124

Peters WCH (1877) Übersicht der Amphibien aus Chinchoxo (Westafrika), welche von der Africanischen Gesellschaft dem Berliner zoologischen Museum übergeben sind. Monatsberichte der königlich Akademie der Wissenschaften zu Berlin 10:611–620

Peters WCH (1881) Zwei neue von Herrn Major von Mechow während seiner letzten Expedition nach West-Afrika entdeckte Schlangen und eine Übersicht der von ihm mitgebrachten herpetologischen Sammlung. Sitzungsberichte der Gesellschaft Naturforschender Freunde zu Berlin 9:147–150

Pietersen DW, Pietersen EW, Conradie W (2017) Preliminary herpetological survey of Ngonye falls and surrounding regions in southwestern Zambia. Amphibian Reptile Conserv 11(1) [Special Section]:24–43

Pincheira-Donoso D, Bauer AM, Meiri S, Uetz P (2013) Global taxonomic diversity of living reptiles. PLoS One 8(3):e59741

Portillo F, Branch WR, Conradie W et al (2018) Phylogeny and biogeography of the African burrowing snake subfamily Aparallactinae (Squamata: Lamprophiidae). Mol Phylogenet Evol 127:288–303

Poynton JC, Haacke WD (1993) On a collection of amphibians from Angola including a new species of *Bufo* Laurenti. Ann Transv Mus 36(2):9–16

Roll U, Feldman A, Novosolov M et al (2017) The global distribution of tetrapods reveals a need for targeted reptile conservation. Nat Ecol Conserv 1:1677–1682

Schmidt KP (1933) The reptiles of the Pulitzer-Angola expedition. Ann Carnegie Mus 22:1–15

Schmidt KP (1936) The amphibians of the Pulitzer-Angola expedition. Ann Carnegie Mus 25:127–133

Shirley MH, Vliet KA, Carr AN et al (2014) Rigorous approaches to species delimitation have significant implications for African crocodilian systematics and conservation. Proc R Soc B 281:20132483

Shirley MH, Carr AN, Nestler JH et al (2018) Systematic revision of the living African slender-snouted crocodiles (*Mecistops* Gray, 1844). Zootaxa 4504:151–193

Stanley EL, Ceríaco LMP, Bandeira S et al (2016) A review of *Cordylus machadoi* (Squamata: Cordylidae) in southwestern Angola, with the description of a new species from the Pro-Namib desert. Zootaxa 4061(3):201–226

Tilbury C (2018) Chameleons of Africa: an atlas, including the Chameleons of Europe, the Middle East and Asia. Edition Chimaira, Frankfurt M, 831 pp

Tolley KA, Alexander GJ, Branch WR et al (2016) Conservation status and threats for African reptiles. Biol Conserv 204:63–67

Trape J-F, Mediannikov O (2016) Cinq serpents nouveaux du genre *Boaedon* Duméril, Bibron & Duméril, 1854 (Serpentes : Lamprophiidae) en Afrique centrale. Bull Soc Herp France 159:61–111

TRIPLOV (2018) Publications José Vincente Barbosa du Bocage. http://www.triplov.com/biblos/bocage.htm

TTWG Turtle Taxonomy Working Group (2011) Turtle conservation Coalition. Turtles in trouble: the world's 25+ most endangered tortoises and freshwater turtles. Chelonian Research Foundation, Conservation International, Wildlife Conservation Society, and San Diego Zoo Global, 54 pp

TTWG Turtle Taxonomy Working Group (2017) Turtles of the world: annotated checklist and atlas of taxonomy, synonymy, distribution, and conservation status (8th Ed.). Chelonian Res Monogr 7:1–292

Tys van den Audenaerde DFE (1967) Les serpents des environs de Dundo (Angola) (Note complémentaire). Publicações culturais da Companhia de Diamantes de Angola 76:31–37

Uetz P, Freed P, Hošek J (eds) (2018) The reptile database. Available http://www.reptile-database.org. Accessed 11 Jan 2018

Vaz Pinto P, Branch WR (2015) Geographical distribution: *Dendroaspis jamesoni* (Thraill, 1843), Jameson's Mamba. African Herp News 62:45–47

Uetz P, Stylianou A (2018) The original descriptions of reptiles and their subspecies. Zootaxa 4375(2):257

Wagner P, Böhme W, Pauwels OSG, Schmitz A (2009) A review of the African red-flanked skinks of the *Lygosoma fernandi* (Burton, 1836) species group (Squamata: Scincidae) and the role of climate change in their speciation. Zootaxa 2050:1–30

Wagner P, Bauer AM, Wilms TM et al (2012) Miscellanea accrodontia: notes on nomenclature, taxonomy and distribution. Russ J Herpetol 19:177–189

Wagner P, Greenbaum E, Bauer AM et al (2018) Lifting the blue-headed veil – integrative taxonomy of the *Acanthocercus atricollis* species complex (Squamata: Agamidae). J Nat Hist. https://doi.org/10.1080/00222933.2018.1435833

Weinell JL, Bauer AM (2018) Systematics and phylogeography of the widely distributed African skink *Trachylepis varia* species complex. Mol Phylogenet Evol 120:103–117

Weir CR, Ron T, Morais M et al (2007) Nesting and at-sea distribution of marine turtles in Angola, West Africa, 2000–2006: occurrence, threats and conservation implications. Oryx 41(2):224–231

Wüster W, Chirio L, Trape J-F et al (2018) Integration of nuclear and mitochondrial gene sequences and morphology reveals unexpected diversity in the forest cobra (*Naja melanoleuca*) species complex in Central and West Africa (Serpentes: Elapidae). Zootaxa 4455:68–98

Chapter 14
The Avifauna of Angola: Richness, Endemism and Rarity

W. Richard J. Dean, Martim Melo, and Michael S. L. Mills

Abstract Angola has a rich history of ornithological exploration going back to the early 1800s. From the early-1970s to 2002, however, the civil war prevented access to many areas, and very little work on birds was done. From about the early 2000s information on birds in Angola has been gathered at an increasing rate, with new species being added to the list and a steady rise in publications on biogeography and biology of birds. With about 940 species, Angola has an impressive array of bird species, including c. 29 endemic species, and several species that are rare and poorly known. For the future, there are many areas of avian biology to attract studies, not only to gather more data on the rare and endemic species, but also local surveys of bird communities, the identification of major threats to the avifauna from landuse changes (concomitantly with suggestions for remedial action) and more. Understanding the role of birds in ecosystem processes, long term studies on the biology and breeding of individual species, and inferring the evolutionary history of the endemic species and of those species that occur in small isolated populations in Angola are all areas for future research. The future of ornithological research and conservation in Angola is dependent on it being carried out by Angolans – outreach,

W. R. J. Dean (✉)
DST-NRF Centre of Excellence at the FitzPatrick Institute, University of Cape Town, Rondebosch, South Africa
e-mail: lycium@telkomsa.net

M. Melo
DST-NRF Centre of Excellence at the FitzPatrick Institute, University of Cape Town, Rondebosch, South Africa

CIBIO-InBIO, Centro de Investigação em Biodiversidade e Recursos Genéticos, Laboratório Associado, Universidade do Porto, Vairão, Portugal

Instituto Superior de Ciências da Educação de Huíla, Lubango, Angola
e-mail: melo.martim@gmail.com

M. S. L. Mills
Instituto Superior de Ciências da Educação de Huíla, Lubango, Angola

A. P. Leventis Ornithological Research Institute, University of Jos, Jos, Plateau State, Nigeria
e-mail: birdsangola@gmail.com

B. J. Huntley et al. (eds.), *Biodiversity of Angola*,
https://doi.org/10.1007/978-3-030-03083-4_14

capacity building, and advanced training must all come together in order to find and train the motivated ornithologists that such a biodiversity-rich country deserves.

Keywords Afromontane forests · Angolan escarpment · Conservation · Endemic bird area · Ornithology · Rare birds

Early Ornithological History

The richness, endemism and rarity of the avifauna in Angola has attracted many ornithologists, with the early studies during the late 1800s and first few years of the 1900s being almost entirely simple collections of birds. A chronology of bird collections is given in Table 14.1. Publications and results of many of these collections have been well covered by several authors, including Traylor (1963), Pinto (1983) and Dean (2000). From the 1960s until the early 1970s extensive collections were made at a number of localities in Angola by the *Instituto de Investigação Científica de Angola* (IICA) (Fig. 14.1). Details of some of these collections, and records of special interest, were published in a series of papers by Pinto (see references) providing much needed data on the biogeography and habitats of birds. The bird specimen collection assembled by the IICA, and now held by the *Instituto Superior de Ciências da Educação* (ISCED) in Lubango was catalogued by Mills et al. (2010). This was revised by Fernanda Lages and colleagues in 2016, aided by the discovery of the field notebooks associated with the collections. This database, of what is probably the third largest bird collection in Africa, will be available soon through the Global Biodiversity Information Facility (GBIF) portal.

Although a war for independence in Angola had been going on since 1961, armed conflict escalated with the start of a civil war in 1975, immediately after Angola became independent from Portugal. The war went on to last almost three decades, pre-empting any significant field-based biological research. Many of the reports on the avifauna of Angola that were published during the 1960s, 1970s and 1980s were "desktop" studies of museum specimens, all using data that had been collected before 1974. Despite issues with security and hazards imposed by the localised patches of unexploded ordinances and the extensive use of landmines, some avian studies were done in this period. Two East German biologists, Dr. Rainer Günther and Dr. Alfred Feiler, based at the *Museum für Naturkunde der Humboldt-Universität zu Berlin* and the *Staatliches Museum für Tierkunde*, Dresden, respectively, were commissioned to survey biodiversity, including birds, in Angola (Günther and Feiler 1986a, b). The ICBP (International Council for Bird Preservation, now BirdLife International) attempted a project to gather data on the status of (*inter alia*) the threatened endemic bird species on the Angolan "Scarp". The subsequent report adds little to what was known about the avifauna of this area, but the report usefully outlines the major threats to the biodiversity in the southernmost patches of Guinea-Congolian forest (Hawkins 1993).

It was only after the war ended definitely in 2002 that ornithologists returned to the country with most expeditions targeting the regions that had been classified as Important Bird Areas (IBAs, Dean 2001) and, in particular, the core habitats of the

Table 14.1 A chronology of the collecting expeditions made in Angola, adapted from Traylor (1963) and Dean (2000), and reproduced, in part, by courtesy of the British Ornithologists' Union

1850–1892	J. Anchieta collected birds mainly in central Angola. The specimens, many of which were Types, were described and reported on by Bocage in a number of separate papers, summarised up to 1881 (Bocage 1877, 1881).
1858–1868	J. J. Monteiro lived in Angola and collected birds. His collections, together with those made by Charles Hamilton (a visitor), were reported on by Hartlaub and Monteiro (1860), Hartlaub (1865), Sharpe and Monteiro (1869) and Sharpe (1871).
1876–1877	A. Lucan and L. A. Petit collected birds in Cabinda. Some of their specimens were deposited in the Natural History Museum, Tring. The collection was reported on by Sharpe and Bouvier (1876a, b, 1877, 1878).
1880 and 1887	A. W. Eriksson collected in Cunene, between the Cunene and Cubango rivers. His specimens are in the *Älvsborgs Länsmuseum*, Vänersborg, Sweden (for details see Rudebeck 1955 and Lundevall and Ängermark 1989), and in the Zoological Museum, Uppsala, Sweden.
1884–1888	P. J. van der Kellen collected for the Nationaal Natuurhistorisch Museum, Leiden, in Namibe and Huíla (Büttikofer 1888, 1889a, b).
Early 1900s	Francisco Newton, a Portuguese naturalist, collected in southern Cuanza-Norte and along the coast (Seabra 1905–1907). Some of the material he collected is in the *Museu de História Natural – Zoologia*, Oporto (Seabra 1905a, b, 1906a, b, c, d, 1907).
1901	C. H. Pemberton collected along the Cunene River and in the area between the Cuanza River and Bailundo for the Rothschild's Museum at Tring, UK.
1903–1906	W. J. Ansorge collected extensively throughout western Angola for the Rothschild's Museum at Tring, UK.
1908–1909	W. J. Ansorge collected in Cuanza-Norte for the British Museum in London.
1910–1911	W. Lowe spent a few days collecting in December 1910 and March 1911 in the Luanda area (Bannerman 1912).
1912–1913	Some birds were collected in Cuando Cubango, Cunene and Huíla by the Mission Rohan-Chabot (Ménégaux and Berlioz 1923).
1920s and 1930s	R. Braun lived and studied birds mainly in Cuanza-Norte, northern Malanje and on the escarpment of Cuanza-Sul (Braun 1930, 1934; Sick 1934; Stresemann 1934, 1937).
1925	R. Boulton collected in Namibe, Huíla and Benguela for the American Museum of Natural History.
1926–1927	H. Lynes and B. B. Osmaston collected cisticolas on the Huambo highlands and along the Benguela coast.
1927	H. & C. Chapman collected on the central plateau for the American Museum of Natural History.
1928	P. Koester collected in the highlands of Huambo and southern Cuanza-Sul and sent the skins to O. Neumann. Some of these skins are now in the Museum of Comparative Zoology, Harvard University.
1928–1929, 1932–1933	A. Monard (1932, 1934) collected in eastern and southern Huíla and in Lunda-Norte. His specimens are in the Musée D'Histoire Naturelle, La-Chaux-de-Fonds, Switzerland.
1929–1930	The gray African expedition of the Academy of Natural Sciences, Philadelphia made two collections in Bié and southern Malanje (Bowen 1931, 1932).

(continued)

Table 14.1 (continued)

1930	L. Fenaroli collected in the northwest and on the north-central plateau (Moltoni 1932).
1930–1931	H. Lynes and J. Vincent collected cisticolas and other species on the plateau and from Dundo, Lunda-Norte, to Vila Luso (now Luena) in Moxico (Lynes and Sclater 1933, 1934).
1931	R. Boulton collected in central and southern Angola for the Carnegie Museum of Natural History (Boulton 1931).
1931–1934	Jean Bodaly made large collections in northern Bié and sent them to the Carnegie Museum and Chicago Natural History Museum (now Field Museum of Natural History).
1931–1934	H. K. Prior collected at Dondi in Huambo and sent skins to the Field Museum of Natural History.
1932–1933	The Phipps-Bradley Expedition made a collection on the plateau for the American Museum of Natural History.
1933–1934	H. Lynes and J. Vincent collected birds (mainly cisticolas) in Benguela, Huambo, southern Lunda-Sul and northern Lunda-Norte (Lynes 1938).
1944–1949	C. M. N. White (1950) collected in areas in the extreme east of Moxico.
1952	H. A. Beatty collected in the northwest and sent skins to the Field Museum of Natural History, Chicago.
1954	W. Serle visited coastal areas for only 1 week but published interesting data (Serle 1955).
1954–1955	G. Heinrich (1958a, b, c) collected extensively in the western half of Angola and sent some skins to the Field Museum of Natural History, Chicago, and the *Zoologisches Institut und Zoologisches Museum*, Hamburg (Meise 1958).
1956	G. Rudebeck collected for the Visser-Transvaal Museum Expedition in southwestern Angola, but only a few details have been published (Rudebeck 1958).
1957	B. P. Hall led an expedition to central and western Angola which resulted in two major publications on zoogeography and taxonomy (Hall 1960a, b).
1957	R. Boulton collected in northwestern and northeastern Angola, and sent skins to the Field Museum of Natural History, Chicago.
1957–1958	G. Heinrich collected in Cuanza-Norte, Malanje and Lunda-Norte for the Peabody Museum of Natural History, Yale University, and the Smithsonian Institution.
1958–1973	Staff at the *Instituto de Investigação Científica de Angola* (IICA) collected in Moxico, Bengo, Luanda, Malange, Cuanza-Sul, Bié, Benguela, Huíla, Namibe and Cuando-Cubango for the IICA collection at Lubango.
1972	W.R.J. Dean collected in Huíla, Malange and Cabinda for the Peabody Museum, New Haven (Dean 1974).
1972	M.E. Ferreira collected in Huíla for the *Zoologishes Forschungsinstitut und Museum Alexander Koenig*, Bonn.
1973	W.R.J. Dean collected in Huíla, Cuanza-Norte, Cuanza-Sul and Malange for the Peabody Museum, New Haven (Dean 1974).
1982–1983	R. Günther and A. Feiler collected in Luanda, Bengo, Uige, Cuanza-Norte and Lunda-Norte for the Staatliches Museum für Tierkunde, Dresden and the Museum für Naturkunde der Humboldt-Universität zu Berlin (Günther and Feiler 1986a, b)

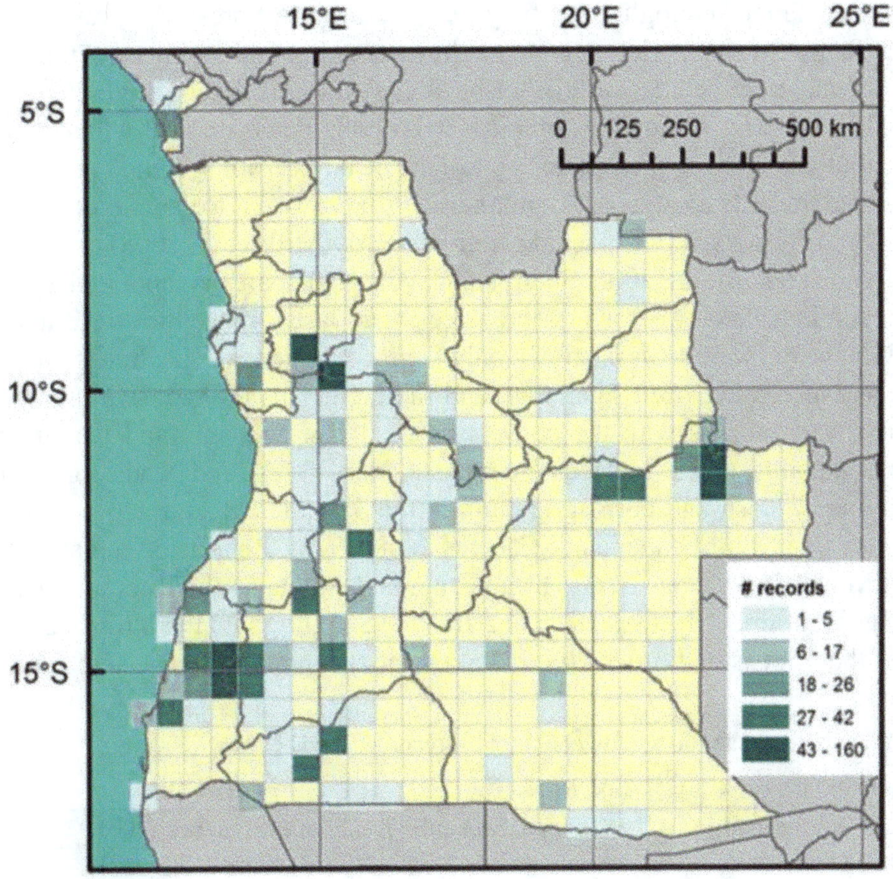

Fig. 14.1 Localities where bird specimens were collected by the *Instituto de Investigação Científica de Angola* (IICA) with duplicates held by the *Instituto de Investigação Científica Tropical* (IICT), shown as the number of specimens collected in each 30-minute square. (Figure extracted from Monteiro et al. (2014) and reproduced by permission of the authors)

endemic birds: the forests of the Western Escarpment and the highlands. Over the last 15 years the knowledge of avian diversity, distribution and biogeography for the country has steadily increased, with several species new to the Angola list being recorded from areas that had never been previously explored.

Recent History and the Exponential Increase in Information on Birds

One of the first "post war" studies was the publication of field notes on the Grey-striped Francolin *Francolinus griseostriatus* that provided some information on the biology of this endemic and rare species (Vaz Pinto 2002). Studies of individual species and surveys of areas of particular interest gained momentum during the early 2000s, with notes on the Gabela Akalat *Sheppardia gabela* (Mills et al. 2004), and more general reports on the conservation status and vocalisations of endemic and threatened bird species from the forests of the western escarpment ('Scarp forests') of Angola (Ryan et al. 2004; Mills 2010) and a survey of the birds in Cumbira

Forest, Gabela (Sekercioğlu and Riley 2005). Cumbira Forest, a representative of Central Scarp forests, is rich in endemic bird species, and has since been the focus of one of the most detailed studies of birds in Angola (Cáceres et al. 2015, 2016, 2017). Most recent publications on birds in Angola, however, deal with extensions of distributional range, vocalisations and lists of birds from specific areas, all providing data towards an atlas of Angolan birds (Table 14.2). Reports on the birds of particular areas, such as Cangandala National Park (Mills et al. 2008) and the Soyo area (Dean and le Maitre 2008; Stavrou and Mills 2013) are valuable in that there is a large gap in time between when the places were last surveyed for birds and now. The avifauna of many areas is known only from collections of specimens during the 1950s and there have been significant and rapid changes in landuse in many areas after the war, such as with the conversion of old-growth miombo woodland for charcoal making, or the replacement of secondary forest and shade-forest coffee plantations by slash-and-burn agriculture in the Scarp (Leite et al. 2018), the most important area of bird endemism (Cáceres et al. 2017). Reports on the avifauna of Mount Moco (Mills et al. 2011a, b), the Namba Mountains (Mills et al. 2013a, b), and Lagoa Carumbo (Mills and Dean 2013) have highlighted the bird species richness of these areas and, *inter alia*, noted threats to the local ecosystems and their

Table 14.2 New species on the Angolan bird list recorded since 1975

Common name	Species	References
Northern Royal Albatross	*Diomedea (epomorpha) sanfordi*	Lambert (2001)
Spectacled Petrel	*Procellaria conspicillata*	Lambert (2001)
Red-billed Tropicbird	*Phaethon aethereus*	Lambert (2001)
White-tailed Tropicbird	*Phaethon lepturus*	Lambert (2001)
Cape Vulture	*Gyps coprotheres*	Bamford et al. (2007)[a]
Lesser Spotted Eagle	*Aquila pomarina*	Meyburg et al. (2001)
Booted Eagle	*Hieraaetus pennatus*	Sinclair (1981)
Red-necked Falcon	*Falco chicquera*	Mills et al. (2016)
Pacific Golden Plover	*Pluvialis fulva*	Mills (2015)
Red Phalarope	*Phalaropus fulicarius*	Lambert (2001)
European Oystercatcher	*Haematopus ostralegus*	Simmons et al. 2009
Black-headed Gull	*Chroicocephalus ridibundus*	Lambert (2001)
Little Tern	*Sternula albifrons*	Lambert (2001)
Greater Crested Tern	*Thalasseus bergii*	Dean et al. (2002)
Lemon Dove	*Columba larvata*	Mills and Dowd (2007)
Yellow-throated Cuckoo	*Chrysococcyx flavigularis*	Mills et al. (2013a, b)
Pink-billed Lark	*Spizocorys conirostris*	Mills (2006)
Red-tailed Leaflove	*Phyllastrephus scandens*	Mills et al. (2013a, b)
Forest Swallow	*Petrochelidon fuliginosa*	Mills and Tebb (2015)
South African Cliff Swallow	*Petrochelidon spilodera*	Mills et al. (2013a, b)
Singing Cisticola	*Cisticola cantans*	Dean et al. (2003)
White-collared Oliveback	*Nesocharis ansorgei*	Mills and Vaz Pinto (2015)

The list does not include unconfirmed records
[a]Record based on satellite tracking

importance for conservation. Notes on rare and little known species, such as Brazza's Martin *Phedina brazzae* (Mills and Cohen 2007), Black-tailed Cisticola *Cisticola melanurus* (Mills et al. 2011a, b), Bocage's Sunbird *Nectarinia bocagii* (Mills 2013), and the Red-necked Falcon *Falco chicquera* (Mills et al. 2016), have provided some insights into the biology of these species. Breeding data for all species in Angola are few, but MSLM and co-workers (see references) have recently published several notes on the biology and first descriptions of nests and eggs and some useful notes on brood parasitism and nests of a number of species.

The creation of the Internet group *Angola Birders* by MSLM in 2012 has greatly facilitated and promoted the acquisition and sharing of data on bird distributions. This has led to a large number of records of the occurrence of species, some of which have cleared up distributional anomalies. For example, records of Black Bishop *Euplectes gierowii* by Pedro Vaz Pinto (2 June 2015), together with earlier records by MSLM, provide some evidence that the species is not as rare as previously thought (see Dean 2000). The *Angola Birders* Group has proved very useful, not only for new distribution records, but also for new breeding records. More importantly, it has generated interest in birds in many people, including diplomats and executives of companies now trading in Angola, who now spend their spare time "off the beaten track". Their contributions towards a national database (at present held by MSLM) of bird records are valuable. The facebook group "*Angola Ambiente*" has completely overtaken *Angola Birders* and is currently providing a lot of useful information.

One of the useful products from the national database of bird records is the bilingual annotated checklist published by Mills and Melo (2013), based on the catalogue of the specimen collection at Lubango (Mills et al. 2010) and sight records. This updates earlier lists, and includes some notes on unconfirmed and doubtful records of species unlikely to occur in Angola. These records vary from (probably) confusion with known species or sightings of species well outside their distribution range.

The Richness of the Angolan Avifauna

About 940 species of birds have been recorded in Angola (Mills and Melo 2013). This increases the number of species recorded for the country; a total of 915 species were listed by Dean (2000), and 12 additional species were added by Mills and Dean (2007), of which seven marine species, mostly pelagic, were recorded by Lambert (2001). Some specimens thought to be of Red-faced Cisticola proved to be those of an isolated population of Singing Cisticola *Cisticola cantans* (Dean et al. 2003), and the occurrence of the Lesser Spotted Eagle *Aquila pomarina* (Meyburg et al. 2001) and the Swift Tern *Sterna bergii* (Dean et al. 2002) have been verified. One additional species, Slaty Egret *Egretta vinaceigula*, has been added as a result of cataloguing the bird collection at Lubango (Mills et al. 2010). Observations made in poorly known areas of Angola by MSLM and co-workers have added additional species to the list, including the Yellow-throated Cuckoo *Chrysococcyx flavigularis*, [Red-tailed] Leaf-love *Phyllastrephus scandens* and South African Cliff Swallow *Petrochelidon spilodera* (Mills et al. 2013a, b). Developments in assessing species boundaries using molecular methods, have added another 16 species (Mills and Dean 2007). One example of this is the Common Fiscal *Lanius collaris* that has now been split into two species, the Northern Fiscal *L. humeralis* and Southern Fiscal *L. collaris* (Fuchs et al. 2011). There have also been some corrections to the list, such as specimens of the White-bellied Sunbird *Cinnyrus talatala* from Mount Moco and Mount Soque listed by Dean (2000) that were found to be miscatalogued Oustalet's Sunbirds *Cinnyrus oustaleti* (Mills and Dean 2007). There are also a number of avian taxa that require further investigation, either in the field or using molecular analyses to establish species boundaries (Mills and Dean 2007).

The list for Angola does not include 88 species for which there are sight records and that, in many cases, and for various reasons, are unlikely to occur in Angola (Mills and Melo 2013). Some species are simply misidentifications of similar, related species. Others, however, particularly migrant waders (shorebirds) from the Northern Hemisphere, may have been correctly identified, but require more records, photographs or specimens for verification. Unconfirmed species are dealt with by Mills and Dean (2007) and more fully by Mills and Melo (2013) and will not be listed here.

Endemism in the Angolan Avifauna

There are about 29 species of birds endemic to Angola, the number depending on the taxonomic authority that is followed (Table 14.3). Most occur in the forests of the Western Escarpment and in the last remnants of Afromontane forest of the highlands, the two core habitats of the Western Angola Endemic Bird Area (BirdLife International 2017). Two species, Red-backed Mousebird *Colius castanotus* and Bubbling Cisticola *Cisticola bulliens*, are widely distributed in Angola, including the Western Escarpment and associated coastal plains and in a range of woodlands and forest patches modified for the cultivation of coffee (Dean 2000).

The semi-evergreen humid forests of the Angolan Escarpment ('the Scarp Forests') are impoverished outliers of the Congolian rainforest (Huntley and Matos 1994). They have been the major speciation hotspot for birds in Angola by: (i) creating a barrier between arid-adapted species of the coastal plains and of the miombo woodlands of the plateau, (ii) creating a steep ecological gradient, and (iii) functioning as a refuge for moist forest specialists that were isolated here during the dry periods of the glacial cycles (Hall 1960a) – 75% of the endemic bird species are associated with this region.

The Afromontane forests of west-central Angola make the most isolated representatives of all the Afromontane centres of endemism, separated by >2000 km from other similar habitats. This isolation has allowed the development of plant and animal communities that are quite distinct from those of other montane centres. The

Table 14.3 Provisional list of bird species endemic or near endemic to Angola, with their IUCN Red List Category, and main area of occurrence

Scientific name	English name	I	S	M	O	N
Pternistis griseostriatus	Grey-striped Francolin	VU	•			
Pternistis swierstrai	Swierstra's Francolin	LC		•		
Tauraco erythrolophus	Red-crested Turaco	LC				
Colius castanotus	Red-backed Mousebird	LC	•			
Gymnobucco vernayi	Angola Naked-faced Barbet	LC	•			*
Lybius leucogaster	Angola White-headed Barbet	LC	•			*
Platysteira albifrons	White-fronted Wattle-eye	LC	•			
Prionops gabela	Gabela Helmetshrike	EN	•			
Malaconotus monteiri	Monteiro's Bushshrike	LC	•			1
Laniarius amboimensis	Gabela Bushshrike	EN	•			
Laniarius brauni	Braun's Bushshrike	EN	•			
Phyllastrephus viridiceps	Angola White-throated Greenbul	LC	•			*
Phyllastrephus fulviventris	Pale-olive Greenbul	LC	•			
Macrosphenus pulitzeri	Pulitzer's Longbill	LC	•			
Cisticola bulliens	Bubbling Cisticola	LC			•	
Cisticola bailunduensis	Huambo Cisticola	LC		•		*
Cisticola melanura	Black-tailed Cisticola	LC			•	
Sheppardia gabela	Gabela Akalat	LC	•			
Xenocopsychus ansorgei	Angola Cave-Chat	LC	•	•		2
Dioptrornis brunneus	Angola Slaty Flycatcher	LC	•			
Nectarinia bocagii	Bocage's Sunbird	LC			•	
Cinnyris ludovicensis	Ludwig's Double-collared Sunbird	LC	•	•		3
Ploceus temporalis	Bocage's Weaver	LC			•	
Euplectes aureus	Golden-backed Bishop	LC			•	4
Lagonosticta ansorgei	Ansorge's Firefinch	LC			•	
Coccopygia bocagei	Angolan Swee Waxbill	LC			•	
Estrilda thomensis	Cinderella Waxbill	LC			•	
Macronyx grimwoodi	Grimwood's Longclaw	LC			•	
Crithagra benguelensis	Benguela Seedeater	LC			•	

IUCN categories, *LC* Least Concern, *NT* Near Threatened, *VU* Vulnerable, *EN* Endangered, *S* Forests of the Western Escarpment ('Scarp forests'), *M* Afromontane forests, *O* Other habitats, *N* Notes

1. Recent records from Cameroon are considered doubtful (Mills 2010)
2. Near-endemic as a marginal population was recently found in Namibia (Swanepoel 2013)
3. Isolated populations in Malawi and Tanzania sometimes treated as sub-species are better treated as distinct species (Bowie et al. 2016)
4. Population on São Tomé Island was very likely introduced by humans as cage birds (Jones and Tye 2006)

*indicates recent species splits proposed by HBW and BirdLife International (2017) following the criteria in Tobias et al. (2010). In these cases only phenotypic data (morphology and song) was used and it would be useful to measure the levels of genetic differentiation from sister taxa total number of endemic bird species associated with these forests is small (Table 14.3), but many endemic subspecies are present (Mills et al. 2011a, b) and molecular studies are likely to support the treatment of several of these populations as distinct species. It is likely that because of their small size, the Afromontane forests of Angola were not included in the 'Afromontane archipelago' biome as defined by White (1978; cf. Fig. 1). Current research has uncovered a key role of these forests in the evolutionary history of the bird communities of the montane forests of Africa. Genetic data, together with the reconstruction of past climates and associated habitats, have shown that the small Angolan Afromontane forests were areas of high climatic stability throughout glacial cycles and constituted the link between the montane bird communities of East Africa and the Cameroon mountains (Vaz da Silva 2015). For species such as the African Hill Babbler *Sylvia [Pseudalcippe] abyssinica* and Bocage's Akalat *Sheppardia bocagei*, populations were isolated in the Angola mountains from very early on and are likely to constitute distinct species (Vaz da Silva 2015).

Apart from the Afromontane and Scarp forests, most other vegetation types and bird habitats are all part of much larger areas that extend into Angola from (i) the North: Guinea-Congolian forests; (ii) the East: miombo woodlands; and (iii) the South: Namib Desert. The avifauna of these biomes is endemic to the habitat type and thus not confined to Angola. An exception might be Bocage's Sunbird *Nectarinia bocagii* that is known only from Angola (Dean 2000; Mills 2013) and western DRC (Dowsett et al. 2008), Black-tailed Cisticola *Cisticola melanurus* (Irwin 1991; Mills et al. 2011a, b) and the White-headed Robin-Chat *Cossypha heinrichi* that show a similar distribution. There are no data on the relative abundance of these species in the DRC, but in Angola they are considered uncommon to locally common (Dean 2000; Mills and Melo 2013), and thus Angola is very likely to house most of the population, giving them near endemic status. On the other hand, the formerly endemic Angolan Cave Chat *Xenocopsychus ansorgei* is now treated as 'near-endemic' after the discovery of an isolated population in northern Namibia (Swanepoel 2013).

Commonness and Rarity

The relative abundance of birds in Angola is covered by Mills and Melo (2013). Most bird species that are widespread in Angola are, if not common, then frequently seen. About 170 species can be considered uncommon (134 species) or rare (35 species). The status of many of the uncommon and rare species is uncertain – some species are known from a few, or a single specimen, collected a long time ago and not subsequently recorded. Examples are the Congo Serpent Eagle *Dryotriorchis spectabilis*, collected in 1954 at Canzele, Cuanza-Norte, and another specimen collected at Mwaoka, Lunda-Norte, in 1964, and not seen since. The status of the Lemon-bellied Crombec *Sylvietta denti* of which a single specimen was collected at Dundo, Lunda-Norte, in 1958, and a second specimen sound-recorded at Lago

Carumbo (Mills and Dean 2013) is uncertain. Similarly, the status of the Long-tailed Hawk *Urotriorchis macrourus* is not known. A specimen was collected at Cacongo (Lândana) in Cabinda by L. Petit, probably in 1876, and not recorded since, despite an extensive collecting trip to Cabinda by the IICA in 1969 (Pinto 1972). Species that have restricted ranges with small populations in Angola, are generally not rare, and may be locally common within their particular habitat.

Anomalies in Bird Distribution Ranges and Recent Findings

A few species are known from isolated communities within certain areas, with the nearest conspecifics many kilometres away. These patterns could be real, or they could be the result of the geographical bias in surveys and collecting. Most collectors favoured the western half of Angola. With the exception of Lunda-Norte, and parts of Moxico, the coverage by collectors across the east-west gradient was poor (e.g., Fig. 3 in Monteiro et al. 2014). For a few species we can be certain that the gap in the distribution between western Angola and western Zambia is real, and is likely to be the result of relict mountain chains that no longer exist.

The recent exploration of places such as Lagoa Carumbo in Lunda-Norte has provided much new information on distributions thought to be disjunct (Mills and Dean 2013). Only 67 species had been collected in the Lagoa Carumbo area during the 1950s by Heinrich (1958a, b, c). Field surveys by MSLM in 2011 recorded 175 species, with 21 species that had been collected by Heinrich not seen. The data on the species seen at Lagoa Carumbo included new records for the area, extensions of ranges and two new records for Angola.

Some remarkable recent finds have been made. The presence of the White-collared Oliveback *Nesocharis ansorgei* in Angola was unknown and not even suggested until populations were found in 2011, 2012, and 2013 at Quibaxi and Quitexe, Cuanza-Norte, and at Uíge (Mills and Vaz Pinto 2015). Before this the nearest known populations were > 1500 km away in eastern DRC.

Ecotourism in Angola: Birding

Ecotourism is becoming a significant means of raising funds for the protection of sites of high biodiversity value. With recent changes in the entry requirements for visitors, improvements in road and hotel infrastructure, and its high biodiversity, Angola stands to attract a large number of visitors for ecotourism purposes. Most of the key bird watching sites are unprotected, thus making income from tourism even more important (Cáceres 2011). For visiting birders the endemic and near-endemic birds (Table 14.3) are a major drawcard, but the country also holds a variety of specials summarised in Mills (2018), including Finsch's Francolin *Scleroptila finschii*, Anchieta's Barbet *Stactolaema anchietae*, Angola Batis *Batis minulla*,

Yellow-throated Nicator *Nicator vireo*, Angola Lark *Mirafra angolensis*, Brazza's Martin *Phedina brazzae*, Black-and-rufous Swallow *Hirundo nigrorufa*, Sharp-tailed Starling *Lamprotornis acuticaudus*, White-headed Robin-Chat *Cossypha heinrichi*, Forest Scrub Robin *Erythropygia leucosticta*, Oustalet's Sunbird *Cinnyris oustaleti*, Black-chinned Weaver *Ploceus nigrimentus*, Dusky Twinspot *Euschistospiza cinereovinacea* and Black-faced Canary *Crithagra capistrata*, all arguably seen more easily in Angola than any other country. Table 14.4 lists the key sites for visiting birders, with main habitats and most sought-after birds – some of which are depicted on Fig. 14.2. Most 'Namibian specials' are also easily found in Angola.

Where to from Here? Future Directions for Ornithological Research in Angola

As noted in the introduction, it is clear that there have been major advances in the knowledge of avian species diversity and distribution and in the relative abundance of species during the last 15 years. This information is crucial to identify potential conservation areas, although more local surveys of bird communities are needed to paint a complete picture. We can run algorithms on the species distribution and abundance data to identify precisely where conservation areas should be, and the inclusion of other parameters such as endemism and/or threat levels can be used to

Table 14.4 Key sites for birdwatching in Angola

Site name	Habitat	Key birds
Northern Escarpment	Congo forest	Braun's Bushshrike, Congo basin birds
Calandula Falls area	Gallery forest, miombo	White-headed Robin-Chat, Anchieta's Barbet, Bannerman's Sunbird
Quiçama NP	Gallery forest, thickets	Grey-striped Francolin, White-fronted Wattle-eye, Monteiro's Bushshrike, Gabela Helmetshrike, Bubbling Cisticola, Red-backed Mousebird
Cumbira Forest	Forest	Gabela Akalat, Gabela Bushshrike, Pulitzer's Longbill, Red-crested Turaco, Hartert's Camaroptera, Black-faced Canary, Forest Scrub Robin
Mount Moco	Montane forest, grassland, miombo	Swierstra's Francolin, Finsch's Francolin, Ludwig's Double-collared Sunbird, Bocage's Sunbird, Black-and-rufous Swallow, Dusky Twinspot, Angola Lark
Benguela area	Arid bushveld	Hartlaub's Francolin, White-tailed Shrike, Bare-cheeked Babbler
Tundavala	Montane forest, grassland, rocks	Angola Cave Chat, Swierstra's Francolin, Angola Swee Waxbill, Angola Slaty Flycatcher, Oustalet's Sunbird, Ludwig's Double-collared Sunbird
Lubango-Namibe	Arid bushveld, desert	Cinderella Waxbill, Benguela Long-billed Lark, Rüppell's Korhaan

Fig. 14.2 Some special birds of Angola. Top to bottom, left to right: Red-crested Turaco, the endemic national bird of Angola. (Photo: Lars Petersson); Anchieta's Barbet, a sought-after species with a range extending to the DRC and Zambia but best seen in Angola. (Photo: Maans Booysen); Braun's Bushshrike, an endemic restricted to the forests of the northern escarpment. (Photo: Fiona Tweedie); Monteiro's Bushshrike, a difficult-to-see endemic associated primarily with the central escarpment. (Photo: Tasso Leventis); Gabela Helmetshrike, an endemic that occurs primarily at the base of the central escarpment, as in Quiçama NP. (Photo: Tasso Leventis); Bocage's Sunbird is only present in the highlands of Angola and the southwest of the DRC. (Photo: Alexandre Vaz)

prioritise conservation efforts. Detailed data on patterns of bird diversity and conservation threats have been obtained for the core habitats of the only Endemic Bird Area of Angola: the Afromontane forests at Mount Moco (Mills et al. 2011a, b) and at the Namba Mountains (Mills et al. 2013a, b); and the Angolan Scarp forests, in particular for the central scarp forests where most endemism is concentrated (Mills 2010), with special emphasis in Cumbira Forest (Cáceres et al. 2015, 2016, 2017).

So far, and understandably, almost all the research on the birds of Angola has been on species diversity and distribution patterns, and not processes. Avian diversity surveys – that would feed a permanently updated atlas for the breeding birds of Angola – should continue, as many areas remain poorly explored or not visited at all for decades. In parallel with such exploration efforts, research on ecological and evolutionary processes must be promoted, as this will provide the information that ultimately is essential for guiding conservation efforts.

Very few studies have been carried out on the biology of individual species, no long-term studies of breeding have been done, and the nests and eggs of many species have yet to be discovered (e.g., Mills and Vaz 2011). Seed dispersal and frugivory by birds in Angola is another field that needs investigation, particularly now where so much habitat is being destroyed for slash-and-burn cultivation and charcoal making. Birds can play a key role in the rehabilitation of damaged areas. Seeds regurgitated by birds often germinate below roost sites, and the seedlings can be collected and planted out. Rehabilitation initiatives are already underway, albeit at a small scale, on Mount Moco and Cumbira. The Mount Moco reforestation project has been running since 2010, with the community-run nursery holding over 1400 saplings grown from locally collected seeds, and with almost 950 trees planted back in the wild (MSLM, unpublished). The Cumbira project is still at its early steps, with the creation of a pilot-nursery (Aimy Cáceres & Ninda Baptista, unpublished).

Research on the evolutionary history of the endemic species and subspecies of Angola is likely to provide novel insights into bird diversification in Africa and on the uniqueness of the Angolan avifauna (see Endemism section above). The use of molecular tools will clarify the taxonomic status of species with small and isolated populations in Angola such as the Orange Ground Thrush *Geokichla gurneyi* restricted to Mount Namba or Margaret's Batis *Batis margaritae* present only on Mount Moco and Namba, and separated from its nearest conspecific in western Zambia by about 800 km. Moreover, the two subspecies occupy rather different habitats – in Angola Margaret's Batis is present in patches of Afromontane forest, whereas in Zambia the species is present in dissimilar evergreen *Cryptosepalum* forest. It is highly likely that the two forms constitute well-separated evolutionary lineages and could be considered different species. This situation, with one "subspecies" present in western Angola and the nearest other "subspecies" present 800–900 km to the east in Zambia or the Katanga area is repeated in many of the western Angolan avian taxa, and raises many questions about whether the isolated populations are two recently diverged forms of one species, or two species. Similarly, the identification of the complex of swamp-dwelling weavers along the eastern border of Angola, western Zambia and Katanga is still something of a mystery. The question has been addressed by several authors (Louette and Benson 1982; Louette 1984; Dean 1996) but remains unsettled. Molecular tools may be required to clarify the situation.

The distinctive endemic subspecies of Horus Swift (*Apus horus fuscobrunneus*) is known from a single series of specimens taken on the coastal plain of Namibe and has not been recorded in Angola since the early 1970s. Likewise, the endemic subspecies of White-headed Barbet (*Lybius leucocephalus leucogaster*), which was fairly common around the southern escarpment, was only rediscovered in 2017, in Tundavala, after almost 40 years of being undetected (Baptista and Mills 2018). Both taxa have been proposed as endemic species, so finding extant populations in the field is a high priority.

Future ornithological research will only succeed and grow with greater local input. There is a great need to stimulate more interest within Angola for the study of birds, both by engaging students more directly and producing relevant educational materials for local students. To these ends, joint Portuguese-English language books have already been produced on *The Common Birds of Luanda* (Mills and Melo 2015) and *The Special Birds of Angola* (Mills 2018), to raise interest and awareness. A basic handbook on ornithology relevant to Angola and written in Portuguese would be a welcome addition. There is also a need for field courses to provide training to Angolan students, working together with Angolan universities. Most importantly, finding ways to encourage the interest of local students in field studies is greatly needed.

Appendix 14.1 Publications Post 1975 Not Cited in the Text

Beel C (1992) Species new to the Angolan list. *Zambian Ornithological Society Newsletter* 22(1):2

Bowen PStJ (1979) Some notes on Margaret's Batis (*Batis margaritae*) in Zambia. *Bulletin of the Zambian Ornithological Society* 11:1–10

Bowen PStJ (1983) The Black-collared Bulbul *Neolestes torquatus* in Mwinilunga District and the first Zambian breeding record. *Bulletin of the Zambian Ornithological Society* 13–15:7–14

Bowen PStJ, Colebrook-Robjent JFR (1984) The nest and eggs of the Black-and-rufous Swallow *Hirundo nigrorufa*. *Bulletin of the British Ornithologists' Club* 104:146–147

Braine S (1990) Records of birds of the Cunene River estuary. *Lanioturdus* 25:38–44

Brooke RK (1981a) The Feral Pigeon - a 'new' bird for the South African list. *Bokmakierie* 33:37–40

Brooke RK (1981b) The seabirds of the Moçâmedes Province, Angola. *Gerfaut* 71:209–225

Collar NJ (1998) Monotypy of *Francolinus griseostriatus*. *Bulletin of the British Ornithologists' Club* 118:124–126

Dean WRJ (1976) Breeding records of *Crex egregia*, *Myrmecocichla nigra* and *Cichladusa ruficauda* from Angola. *Bulletin of the British Ornithologists' Club* 96:48–49

Dean WRJ (1988) The avifauna of Angolan miombo woodlands. *Tauraco* 1:99–104

Dean WRJ (2001) The distribution of vultures in Angola. *Vulture News* 45:20–25

Dean WRJ (2006) Age structure of a Palm-nut Vulture *Gypohierax angolensis* population. *Vulture News* 55:8–9

Dean WRJ (2007) Type specimens of birds (Aves) in the Transvaal Museum collection. *Annals of the Transvaal Museum* 44:67–121

Dean WRJ, Milton SJ (2005) Stomach contents of birds (Aves) in The Natural History Museum, Tring, U.K., collected in southern Africa, northern Mozambique and Angola. *Durban Museum Novitates* 30:15–23

Dean WRJ, Milton SJ (2007) Some additional breeding records for birds in Angola. *Ostrich* 78:645–648

Dean WRJ, Vernon CJ (1988) Notes on the White-winged Babbling Starling *Neocichla gutturalis* in Angola. *Ostrich* 59:39–40

Dean WRJ, Sandwith M, Milton SJ (2006) The bird collections of C. J. Andersson in southern Africa, 1850–1867. *Archives of Natural History* 33:159–171

Dean WRJ, Walters MP, Dowsett RJ (2003) Records of birds breeding collected by Dr WJ Ansorge in Angola and Gabon. *Bulletin of the British Ornithologists' Club* 123:239–250

Dean WRJ, Franke U, Joseph G, et al. (2012) Type specimens in the bird collection at Lubango, Angola. *Bulletin of the British Ornithologists' Club* 132:41–45

Dean WRJ, Franke U, Joseph G, et al. (2014) Further breeding records for birds (Aves) in Angola. *Durban Museum Novitates* 36:1–36

Dean WRJ, Huntley MA, Huntley BJ, et al. (1988) Notes on some birds of Angola. *Durban Museum Novitates* 14:43–92

Lambert K (2006) Seabirds sighted in the waters off Angola, 1966–1988. *Marine Ornithology* 34:77–80

Louette M (2002) Relationship of the Red-thighed Sparrowhawk *Accipiter erythropus* and the African Little Sparrowhawk *A. minullus*. *Bulletin of the British Ornithologists' Club* 122:218–222

Mendelsohn JM, Haraes L (2018) Aerial census of Cape Cormorants and Cape Fur Seals at Baía dos Tigres, Angola. *Namibian Journal of Environment* 2A:1–6

Meyburg B-U, Mendelsohn JM, Ellis DH, et al. (1995) Year-round movements of a Wahlberg's Eagle *Aquila wahlbergi* tracked by satellite. *Ostrich* 66:35–140

Mills MSL (2007a) Swierstra's Francolin *Francolinus swierstrai*: a bibliography and summary of specimens. *Bulletin of the African Bird Club* 14:175–180

Mills MSL (2007b) Vocalisations of Angolan Birds. Vol. 1. CD. Birds Angola & Birding Africa, Cape Town

Mills MSL (2009) Vocalisations of Angolan birds: new descriptions and other notes. *Bulletin of the African Bird Club* 16:150–166

Mills MSL (2014a) Dusky Twinspot *Euschistospiza cinereovinacea*, a new host species for indigobirds *Vidua*. *Bulletin of the African Bird Club* 21:193–199

Mills MSL (2014b) Observations of the rarely seen aerial display of Short-winged Cisticola *Cisticola brachypterus*. *Bulletin of the African Bird Club* 21:200–201
Mills MSL, Oschadleus HD (2013). Black-chinned Weaver *Ploceus nigrimentus* in Angola, and its nest. *Bulletin of the African Bird Club* 20:60–66
Mills MSL, Vaz Pinto P, Haber S (2012). Grey-striped Francolin *Pternistis griseostriatus*: specimens, distribution and morphometrics. *Bulletin of the African Bird Club* 19:172–177
Mills MSL, Melo M, Borrow N, et al (2011) The Endangered Braun's Bushshrike *Laniarius brauni*: a summary. *Bulletin of the African Bird Club* 18:175–181
Morant PD (Compiler). (1996) Environmental Study of the Kunene River Mouth. CSIR Report EMAS-C96023. CSIR, Stellenbosch
Oschadleus HD, Mills MSL, Monadjem A (2014) Roadside colony densities of weavers in southern Angola. *Lanioturdus* 47:17–20
Paterson J, Boorman M, Glendenning J, et al. (2009) Vagrants, range extensions and interesting bird records for Skeleton Coast Park Namibia and southern Angola. *Lanioturdus* 42:4–10
Ripley SD, Bond GM (1979) A third set of additions to the avifauna of Angola. *Bulletin of the British Ornithologists' Club* 99:140–142
Ryan PG, Cooper J, Stutterheim CJ (1984) Waders (Charadrii) and other coastal birds of the Skeleton Coast, South West Africa. *Madoqua* 14:71-78
Simmons RE (2010) First breeding records for Damara Terns and density of other shorebirds along Angola's Namib Desert coast. *Ostrich* 81:19—23
Simmons RE, Braby R, Braby SJ (1993) Ecological studies of the Cunene River mouth: avifauna, herpetofauna, water quality, flow rates, geomorphology and implications of the Epupa Dam. *Madoqua* 18:163-180
Simmons RE, Sakko A, Paterson J, et al. (2010) Birds and conservation significance of the Namib Desert's least known coastal wetlands: Baia and Ilha dos Tigres, Angola. *African Journal of Marine Science* 28:713-717
Sinclair I (2007) First record of Bob-tailed Weaver *Brachycope anomala* for Angola. *Bulletin of the African Bird Club* 14:78-78
Sinclair I, Chamberlain D, Chamberlain M, et al. (2007) Observations of three little-known bird species in northern Angola. *Bulletin of the African Bird Club* 14:55-56
Sinclair I, Spottiswoode CN, Cohen C, et al. (2004) Birding western Angola. *Bulletin of the African Bird Club* 16:211-212
Steinheimer FD, Dean WRJ (2007) Avian type specimens and their type localities from Otto Schütt's and Friedrich von Mechow's Angolan collections in the Museum für Naturkunde of the Humboldt-University of Berlin. *Zootaxa* 1387:1-25
Stjernstedt R, Aspinwall DR (1979) The nest and eggs of the Bar-winged Weaver *Ploceus angolensis*. *Bulletin of the British Ornithologists' Club* 99:138-140
Tye A (1992) A new subspecies of *Cisticola bulliens* from northern Angola. *Bulletin of the British Ornithologists' Club* 112:55-56

References

Bamford AJ, Diekmann M, Monadjem A et al (2007) Ranging behaviour of Cape Vultures *Gyps coprotheres* from an endangered population in Namibia. Bird Conserv Int 17:331–339

Bannerman DA (1912) On a collection of birds made by Mr Willoughby P. Lowe on the West Coast of Africa and outlying islands; with field notes by the collectors. Ibis 54:219–229

Baptista NL, Mills MSL (2018) Angola White-headed Barbet *Lybius [leucocephalus] leucogaster* rediscovered. Bull Afr Bird Club 25:225–229

BirdLife International (2017) Endemic bird areas factsheet: Western Angola. Downloaded from http://www.birdlife.org on 19/04/2017

Bocage JVB du (1877) Ornithologie d'Angola. Part 1. Imprimerie Nationale, Lisbonne (Lisbon), pp 1–256

Bocage JVB du (1881) Ornithologie d'Angola. Part 2. Imprimerie Nationale, Lisbonne (Lisbon), pp 257–576

Boulton R (1931) New species and subspecies of African birds. Ann Carnegie Museum 21:43–56

Bowen WW (1931) Angolan birds collected during the Gray African expedition – 1929. Proc Acad Natl Sci Phila 83:263–299

Bowen WW (1932) Angolan birds collected during the second Gray African expedition – 1930. Proc Acad Natl Sci Phila 84:281–289

Bowie RCK, Fjeldsa J, Kiure J et al (2016) A new member of the greater double-collared sunbird complex (Passeriformes: Nectariniidae) from the Eastern Arc Mountains of Africa. Zootaxa 4175:23–42

Braun R (1930) Beitrage zur Biologie der Vögel von Angola. J Ornithol 78:47–49

Braun R (1934) Biologische Notizen über einige Vögel Nord-Angolas. J Ornithol 82:553–560

Büttikofer J (1888) On birds from the Congo and South Western Africa. Notes Leyden Mus 10:209–244

Büttikofer J (1889a) On a new collection of birds from South Western Africa. Notes Leyden Mus 11:65–79

Büttikofer J (1889b) Third list of birds from South Western Africa. Notes Leyden Mus 11:193–200

Cáceres A (2011) Implementation of eco-tourism as a conservation tool to save the last remnants of Afromontane Forest of Mount Moco, Angola. MSc thesis. University of Porto, Porto

Cáceres A, Melo M, Barlow J et al (2015) Threatened birds of the Angolan Central Escarpment: distribution and response to habitat change at Kumbira Forest. Oryx 49:727–734

Cáceres A, Melo M, Barlow J et al (2016) Radio telemetry reveals key data for the conservation of *Sheppardia gabela* (Rand, 1957) in the Angolan Escarpment Forest. Afr J Ecol 54:317–327

Cáceres A, Melo M, Barlow J et al (2017) Drivers of bird diversity in an understudied African centre of endemism: the Angolan Escarpment Forest. Bird Conserv Int 27:256–268

Dean WRJ (1974) Breeding and distributional notes on some birds of Angola. Durban Mus Novit 10:109–125

Dean WRJ (1996) The distribution of the Masked Weaver *Ploceus velatus* in Angola. Bull Br Ornithol Club 116:254–256

Dean WRJ (2000) The birds of Angola: an annotated checklist. BOU Checklist No. 18. British Ornithologists' Union, Tring

Dean WRJ (2001) Angola. In: Evans MI (ed) Fishpool LDC. Important Bird Areas in Africa and Associated Islands – Priority Sites for Conservation. BirdLife International, Cambridge & Pisces Publications, Newbury

Dean WRJ, Le Maitre DC (2008) The birds of the Soyo area, Northwest Angola. Malimbus 30:1–18

Dean WRJ, Dowsett RJ, Sakko A et al (2002) New records and amendments to the birds of Angola. Bull Br Ornithol Club 122:180–184

Dean WRJ, Irwin MPS, Pearson DJ (2003) An isolated population of Singing Cisticola, *Cisticola cantans*, in Angola. Ostrich 74:231–232

Dowsett RJ, Aspinwall DR, Dowsett-Lemaire F (2008) The birds of Zambia. An atlas and handbook. Tauraco Press and Aves, Liège

Fuchs J, Crowe TM, Bowie RCK (2011) Phylogeography of the fiscal shrike (*Lanius collaris*): a novel pattern of genetic structure across the arid zones and savannas of Africa. J Biogeogr 38:2210–2222

Günther R, Feiler A (1986a) Zur phänologie, ökologie und morphologie angolanischer Vögel (Aves). Teil I: Non-Passeriformes. Faunistische Abhandlungen aus dem Staatlichen Museum für Tierkunde in Dresden 13:189–227

Günther R, Feiler A (1986b) Zur phänologie, ökologie und morphologie angolanischer Vögel (Aves). Teil II: Passeriformes. *Faunistische Abhandlungen aus dem Staatlichen Museum für* Tierkunde in Dresden 14:1–29

Hall BP (1960a) The faunistic importance of the scarp of Angola. Ibis 102:420–442

Hall BP (1960b) The ecology and taxonomy of some Angolan birds. Bull Brit Mus (Nat Hist) Zool 6:367–463

Hartlaub G (1865) Descriptions of seven new species of birds discovered by Mr J. J. Monteiro in the Province of Benguela, Angola, West Africa. Proc Zool Soc London 33:86–88

Hartlaub G, Monteiro JJ (1860) On some birds collected in Angola. Proc Zool Soc London 28:109–112

Hawkins F (1993) An integrated biodiversity conservation project under development: the ICBP Angola Scarp Project. Proceedings of the VIII Pan-African Ornithological Congress: 279–284. Kigali, Rwanda, 1992. Koninklijk Museum voor Midden-Afrika, Tervuren

HBW and BirdLife International (2017) Handbook of the Birds of the World and BirdLife International digital checklist of the birds of the world. Version 2. Available at: http://datazone.birdlife.org

Heinrich G (1958a) Zur Verbreitung und Lebensweise der Vögel von Angola. J Ornithol 99:121–141

Heinrich G (1958b) Zur Verbreitung und Lebensweise der Vögel von Angola. Systematischer Teil I (Galli – Muscicapidae). J Ornithol 99:322–362

Heinrich G (1958c) Zur Verbreitung und Lebensweise der Vögel von Angola. Systematischer Teil III (Hirundinidae – Fringillidae). Journal für Ornitholgie 99:399–421

Huntley BJ, Matos EM (1994) Botanical diversity and its conservation in Angola. Strelitzia 1:53–74

Irwin MPS (1991) The specific characters of the Slender-tailed Cisticola *Cisticola melanura* (Cabanis). Bull Br Ornithol Club 111:228–236

Jones PJ, Tye A (2006) The Birds of São Tomé and Príncipe, with Annobón: Islands of the Gulf of Guinea. BOU Checklist No. 22. British Ornithologists' Union & British Ornithologists' Club, Oxford

Lambert K (2001) Sightings of new and rarely reported seabirds in Southern African waters. Mar Ornithol 29:115–118

Leite A, Cáceres A, Melo M, Mills MSL, Monteiro AT (2018) Reducing Emissions from Deforestation and forest Degradation (REDD+) in Angola: insights from the Scarp Forest conservation hotspot. Land Degrad Dev 29:4291–4300. https://doi.org/10.1002/ldr.3178

Louette M (1984) The identity of swamp-dwelling weavers in North-East Angola. Bull Br Ornithol Club 104:22–24

Louette M, Benson CW (1982) Swamp-dwelling weavers of the *Ploceus velatus/vitellinus* complex, with the description of a new species. Bull Br Ornithol Club 102:24–31

Lundevall C-F, Ängermark W (1989) Fåglar från Namibia. Axel W. Erikssons fågelsamling från Sydvästafrika på Vänersborg Museum. Älvsborgs Länsmuseum, Vänersborg

Lynes H (1938) Contribution to the ornithology of the Southern Congo Basin. Revue de Zoologie et Botanique Africaines 31:3–128

Lynes H, Sclater WL (1933) Lynes-Vincent tour in Central and West Africa in 1930–1931. Part I. Ibis 75:694–729

Lynes H, Sclater WL (1934) Lynes-Vincent tour in Central and West Africa in 1930–1931. Part II. Ibis 76:1–51

Meise W (1958) Über neue Hühner-, Specht- und Singvögelrassen von Angola. *Abhandlungen des Naturwissenschaftlichen Vereins in Hamburg,* N.F. 2:63–83

Ménégaux A, Berlioz J (1923) Oiseaux. In: Mission Rohan-Chabot: Angola et Rhodesia (1912–1914). Tome IV: Histoire Naturelle. Fascicle 1: Mammifères (anatomie comparée, embryologie). Oiseaux. Reptiles. Poissons. Imprimerie Nationale, Paris, pp 107–155

Meyburg B-U, Ellis DH, Meyburg C et al (2001) Satellite tracking of two lesser spotted eagles, *Aquila pomarina*, migrating from Namibia. Ostrich 72:35–40

Mills MSL (2006) First record of Pink-billed Lark *Spizocorys conirostris* for Angola. Bull Afr Bird Club 13:212

Mills MSL (2010) Angola's central scarp forests: patterns of bird diversity and conservation threats. Biodivers Conserv 19:1883–1903

Mills MSL (2013) Little-known African bird: Bocage's Sunbird *Nectarinia bocagii*—an Angolan near-endemic. Bull Afr Bird Club 20:80–88

Mills MSL (2015) First record of Pacific Golden Plover *Pluvialis fulva* for Angola. Bull Afr Bird Club 22:223–224

Mills MSL (2018) The Special Birds of Angola/As Aves Especiais de Angola. Go-away-birding, Cape Town & Kissama Foundation, Luanda

Mills MSL, Cohen C (2007) Brazza's Martin *Phedina brazzae*: new information on range and vocalisations. Ostrich 78:51–54

Mills MSL, Dean WRJ (2007) Notes on Angolan birds: new country records, range extensions and taxonomic questions. Ostrich 78:55–63

Mills MSL, Dean WRJ (2013) The avifauna of the Lagoa Carumbo area, Northeast Angola. Malimbus 35:77–92

Mills MSL, Dowd AD (2007) First records of Lemon Dove *Aplopelia larvata* for Angola. Bull Afr Bird Club 14:77–78

Mills MSL, Melo M (2013) The checklist of the birds of Angola/A Lista das Aves de Angola. Associação Angolana para Aves e Natureza & Birds Angola, Luanda

Mills MSL, Melo M (2015) As Aves Comuns de Luanda/The Common Birds of Luanda. Associação Aves e Natureza Angola, Luanda

Mills MSL, Tebb G (2015) First record of forest swallow *Petrochelidon fuliginosa* for Angola. Bull Afr Bird Club 22:221–222

Mills MSL, Vaz A (2011) The nest and eggs of Margaret's Batis *Batis margaritae*. Bull Br Ornithol Club 131:208–210

Mills MSL, Vaz Pinto P (2015) An overlooked population of White-collared Oliveback *Nesocharis ansorgei*, in Angola. Bull Afr Bird Club 22:64–67

Mills MSL, Cohen C, Spottiswoode C (2004) Little-known African bird: Gabela Akalat, Angola's long-neglected *Gabelatrix*. Bull Afr Bird Club 11:149–151

Mills MSL, Vaz Pinto P, Dean WRJ (2008) The avifauna of Cangandala National Park, Angola. Bull Afr Bird Club 15:113–116

Mills MSL, Franke U, Joseph G et al (2010) Cataloguing the Lubango Bird Skin Collection: towards an atlas of Angolan bird distributions. Bull Afr Bird Club 17:43–53

Mills MSL, Melo M, Vaz A (2011a) Black-tailed Cisticola *Cisticola melanurus* in eastern Angola: behavioural notes and the first photographs and sound recordings. Bull Afr Bird Club 18:193–198

Mills MSL, Olmos F, Melo M et al (2011b) Mount Moco: its importance to the conservation of Swierstra's Francolin *Pternistis swierstrai* and the Afromontane avifauna of Angola. Bird Conserv Int 21:119–133

Mills MSL, Melo M, Vaz A (2013a) The Namba mountains: new hope for Afromontane forest birds in Angola. Bird Conserv Int 23:159–167

Mills MSL, Vaz Pinto P, Palmerim JM (2013b) First records for Angola of Yellow-throated Cuckoo *Chrysococcyx flavigularis*, South African Cliff Swallow *Petrochelidon spilodera* and Red-tailed Leaflove *Phyllastrephus scandens*. Bull Afr Bird Club 20:200–204

Mills MSL, Bennett B, Baptista N et al (2016) Red-necked Falcon *Falco chicquera* in Angola. Bull Afr Bird Club 23:89–90

Moltoni E (1932) Uccelli d'Angola raccolti da L. Fenaroli durante la spedizione 1930 Baragioli-Durini. Atti della Società Italiana di Scienze Naturali e del Museo Civico di Storia Naturale in Milano 71:169–178
Monard A (1932) Matériaux de la mission scientifique suisse en Angola. Oiseaux. Bull Soc Neuchâteloise Sci Nat 56:301–355
Monard A (1934) Ornithologie de l'Angola. Arquivos do Museu Bocage 5:1–110
Monteiro M, Reino L, Beja P et al (2014) The collection and database of birds of Angola hosted at IICT (Instituto de Investigação Científica Tropical), Lisboa, Portugal. ZooKeys 387:89–99
Pinto AA da R (1972) Contribuição para o estudo da avifauna do Distrito de Cabinda (Angola). *Memórias e Trabalhos do Instituto de* Investigação Científica de Angola 10:1–103
Pinto AA da R (1983) Ornitologia de Angola, vol 1. Instituto de Investigacão Cientifica Tropical, Lisboa
Rudebeck G (1955) Aves I. S Afr Anim Life 2:426–576
Rudebeck G (1958) A new race of the Bunting *Fringillaria capensis* (L.) from Angola. Bull Br Ornithol Club 78:129–132
Ryan PG, Sinclair I, Cohen C et al (2004) The conservation status and vocalizations of threatened birds from the scarp forests of the Western Angola Endemic Bird Area. Bird Conserv Int 14:247–260
Seabra A (1905a) Aves de Angola da exploração de F. Newton. J Sci Math Phys Nat 7(26):118–128
Seabra A (1905b) Mammiferos e aves da exploração de F. Newton em Angola. J Sci Math Phys Nat 7(26):103–110
Seabra A (1906a) Aves da exploração de Fr. Newton em Angola – Subsidios para o conhecimento da destribuição geographica das aves d'Africa occidental. Ann Sci Nat 10:153–159
Seabra A (1906b) Aves de Porto Alexandre. J Sci Math Phys Nat 7(27):143–148
Seabra A (1906c) Nota sobre a existencia de "*Diomedia imutabilis*" nas costas occidentaes de Africa. J Sci Math Phys Nat 7(27):141–142
Seabra A (1906d) Ribeirinhas e palmípedes das margens do Rio Cunene. Ann Sci Nat 10:83–90
Seabra A (1907) Sur quelques oiseaux d'Angola envoyés par Francisco Newton. Contribution à l'étude de la distribution géographique des oiseaux de l'Afrique occidentale. Bull Soc Portugaise Sci Naturelles 1:41–45
Sekercioğlu CH, Riley A (2005) A brief survey of the birds in Kumbira Forest, Gabela, Angola. Ostrich 76:111–117
Serle W (1955) The bird life of the Angolan littoral. Ibis 97:425–431
Sharpe RB (1871) On the birds of Angola.–Part III. Proc Zool Soc London 39:130–135
Sharpe RB, Bouvier A (1876a) Catalogue d'une collection recueille à Lândana et Chinchoxo (Congo), par M. Louis Petit, pendant les mois de janvier février, mars et avril 1876. Bull Soc Zool Fr 1:36–53
Sharpe RB, Bouvier A (1876b) Sur les collections recueilles dans la région du Congo par MM. le Dr A. Lucan et L. Petit, depuis le mois de mai jusqu'en septembre. Bull Soc Zool Fr 1:300–314
Sharpe RB, Bouvier A (1877) Nouvelle liste d'oiseaux recueillis dans la région du Congo par MM. le Dr A. Lucan et L. Petit, de Septembre 1876 à Septembre 1877. Bull Soc Zool Fr 2:470–481
Sharpe RB, Bouvier A (1878) Nouvelle liste d'oiseaux recueillis dans la région du Congo par MM. le Dr A. Lucan et L. Petit, de Septembre 1876 à Septembre 1877. Bull Soc Zool Fr 3:73–80
Sharpe RB, Monteiro JJ (1869) On the birds of Angola.–Part I. Proc Zool Soc London 37:563–571
Sick H (1934) Ueber einige Vogelbälge aus Nord-Angola, gesammelt von Herrn R. Braun. Ornithol Monatsberichte 42:167–172
Simmons RE, Mills MSL, Dean WRJ (2009) Oystercatcher *Haematopus* records from Angola. Bull Afr Bird Club 16:211–212
Sinclair JC (1981) First sight records of the Booted Eagle in Angola. Ostrich 52:57
Stavrou C, Mills MSL (2013) Observations of birds of the Soyo area, Northwest Angola. Malimbus 35:27–36
Stresemann E (1934) *Apalis rufogularis brauni* subsp. nov. Ornith Monatsber 62:156–157

Stresemann E (1937) Weitere Vogelbälge aus Nord-Angola, gesammelt von Herrn R. Braun. Ornithol Monatsberichte 45:51–53

Swanepoel W (2013) Rock star. Angola cave chat: a new species for Namibia. Afr Birdlife 1:30–32

Tobias JA, Seddon N, Spottiswoode CN et al (2010) Quantitative criteria for species delimitation. Ibis 152:724–746

Traylor MA (1963) Check-list of Angolan Birds. Publicações Culturais 6. Companhia de Diamantes de Angola (DIAMANG), Lisboa

Vaz da Silva B (2015) Evolutionary history of the birds of the Angolan Highlands – the missing piece to understand biogeography of the Afromontane Forests. MSc thesis. University of Porto, Porto

Vaz Pinto P (2002) Field notes on the Grey-striped Francolin (*Francolinus griseostriatus*) in w Angola. Newsletter of the Partridge, Quail and Francolin Specialist Group 17:3–5

White CMN (1950) Some records from Eastern Angola. Bull Br Ornithol Club 70:35

White F (1978) The Afromontane region. In: Werger MJA (ed) Biogeography and ecology of Southern Africa. Springer, Dordrecht, pp 463–513

Chapter 15
The Mammals of Angola

Pedro Beja, Pedro Vaz Pinto, Luís Veríssimo, Elena Bersacola, Ezequiel Fabiano, Jorge M. Palmeirim, Ara Monadjem, Pedro Monterroso, Magdalena S. Svensson, and Peter John Taylor

Abstract Scientific investigations on the mammals of Angola started over 150 years ago, but information remains scarce and scattered, with only one recent published account. Here we provide a synthesis of the mammals of Angola based on a thorough survey of primary and grey literature, as well as recent unpublished records. We present a short history of mammal research, and provide brief information on each species known to occur in the country. Particular attention is given to endemic and near endemic species. We also provide a zoogeographic outline and information on the conservation of Angolan mammals. We found confirmed records for 291 native species, most of which from the orders Rodentia (85), Chiroptera (73), Carnivora (39), and Cetartiodactyla (33). There is a large number of endemic and near endemic species, most of which are rodents or bats. The large diversity of species is favoured by the wide

P. Beja (✉)
CIBIO-InBIO, Centro de Investigação em Biodiversidade e Recursos Genéticos, Universidade do Porto, Vairão, Portugal

CEABN-InBio, Centro de Ecologia Aplicada "Professor Baeta Neves", Instituto Superior de Agronomia, Universidade de Lisboa, Lisboa, Portugal
e-mail: pbeja@cibio.up.pt

P. Vaz Pinto
Fundação Kissama, Luanda, Angola

CIBIO-InBIO, Centro de Investigação em Biodiversidade e Recursos Genéticos, Universidade do Porto, Campus de Vairão, Vairão, Portugal
e-mail: pedrovazpinto@gmail.com

L. Veríssimo
Fundação Kissama, Luanda, Angola
e-mail: lmnverissimo@gmail.com

E. Bersacola · M. S. Svensson
Nocturnal Primate Research Group, Faculty of Humanities and Social Sciences, Oxford Brookes University, Oxford, UK
e-mail: hellenbers@gmail.com; m.svensson@brookes.ac.uk

E. Fabiano
Department of Wildlife Management and Ecotourism, Katima Mulilo Campus, Faculty of Agriculture and Natural Resources, University of Namibia, Katima Mulilo, Namibia
e-mail: fabianoezekiel@gmail.com

B. J. Huntley et al. (eds.), *Biodiversity of Angola*,
https://doi.org/10.1007/978-3-030-03083-4_15

range of habitats with contrasting environmental conditions, while endemism tends to be associated with unique physiographic settings such as the Angolan Escarpment. The mammal fauna of Angola includes 2 Critically Endangered, 2 Endangered, 11 Vulnerable, and 14 Near-Threatened species at the global scale. There are also 12 data deficient species, most of which are endemics or near endemics to the country.

Keywords Africa · Angolan escarpment · Conservation · Endemism · History of mammalogy · Threatened species · Zoogeography

Introduction

The mammals of Africa, particularly the great apes, large herbivores, and carnivores are among the most iconic wild species in the world, catching the imagination of scientists and the general public alike (Monsarrat and Kerley 2018). These species provided the motivation in the late nineteenth and early twentieth century for some of the first efforts in wildlife conservation and sustainable use, initially with the establishment of game reserves and later with the creation of National Parks and other protected areas (Adams 2013). Today, over one hundred years later, the interest in these charismatic species has increased even further, attracting ever larger numbers of visitors each year from around the world to protected areas in Africa, and thus representing important sources of economic revenue in some African countries. This interest has also been fuelled by appreciation that many mammalian species have critical influences on the structure and functioning of African natural ecosystems (Keesing and Young 2014; Malhi et al. 2016), and that they may provide important services such as biological pest control in human-dominated landscapes (Kunz et al. 2011; Sirami et al. 2013; Taylor et al. 2018a). At the same time, however, African mammals have become involved in some of the most challenging and

J. M. Palmeirim
Departamento de Biologia Animal, Faculdade de Ciências, cE3c – Centre for Ecology, Evolution and Environmental Changes, Universidade de Lisboa, Lisboa, Portugal
e-mail: jmpalmeirim@fc.ul.pt

A. Monadjem
Department of Biological Sciences, University of Swaziland, Kwaluseni, Swaziland

Mammal Research Institute, Department of Zoology and Entomology, University of Pretoria, Pretoria, South Africa
e-mail: aramonadjem@gmail.com

P. Monterroso
CIBIO-InBIO, Centro de Investigação em Biodiversidade e Recursos Genéticos, Universidade do Porto, Vairão, Portugal
e-mail: pmonterroso@cibio.up.pt

P. J. Taylor
School of Mathematical & Natural Sciences, University of Venda, Thohoyandou, South Africa
e-mail: peter.taylor.univen@gmail.com

controversial conservation problems in the world, due in particular to the rapid growth of human populations, agricultural and pastoralism expansion and the associated loss of natural habitats (Laurance et al. 2014; Searchinger et al. 2015), deforestation (Hansen et al. 2013), conflicts due to crop raiding (Hoare 2015; Seiler and Robbins 2016) and predation on people and livestock (Loveridge et al. 2017; McNutt et al. 2017), and poaching for bushmeat (Wilkie et al. 2016; van Velden et al. 2018) and international trade (Biggs et al. 2013; Wasser et al. 2015; Cerling et al. 2016). Conservation of African mammals is thus at a crossroads, with a combination of multiple threats and opportunities, demanding a good understanding of species diversity and ecological requirements, and how they interact with humans in the context of complex and ever changing social-ecological systems.

In Angola, mammals have long been the focus of research and conservation efforts. Like elsewhere in Africa, albeit later than in some other countries, mammals provided the main motivation for the creation of the first Angolan Game Reserves and National Parks (NPs) in the 1930s, which were mostly located in areas with particularly important populations of large herbivores (Huntley et al. 2019). Scientific research started as early as the mid-nineteenth century, with collectors and zoologists describing the mammalian diversity of the country, including many species new to science. Research continued over the years and until the present, but it was plagued by long periods of interruption, particularly during the civil war of 1975–2002, making Angola one of the least known African countries in terms of its mammalian fauna. During this period of turmoil there were very few mammalogical studies (but see, e.g., Anstey 1991, 1993), but efforts to assess the status of the most charismatic and highly endangered species resumed soon after the situation improved in the early years of the twenty-first century (Morais et al. 2006a, b; Veríssimo 2008; Chase and Griffin 2011; Carmignani 2015; Chase and Schlossberg 2016; Fabiano et al. 2017; Overton et al. 2017; Vaz Pinto 2018; NGOWP 2018; Schlossberg et al. 2018). Scientific interest in Angolan mammals is slowly mounting again, with recent studies reporting the discovery of new species to science (Carleton et al. 2015; Svensson et al. 2017), describing important aspects of species distribution and ecology (Bersacola et al. 2015; Svensson 2017), and even using cutting-edge tools for answering complex questions related to species biogeography, phylogeography and evolution (Rodrigues et al. 2015; Vaz Pinto 2018). This renewed interest is timely, as Angola is currently striving to expand, reorganise and improve the management of its system of protected areas, in which there will once again be a strong focus on the conservation and sustainable use of mammal populations. This endeavour needs to be solidly rooted in scientific information, profiting from data that has been collected for over 150 years in the country, and promoting new studies that will help designing cost-effective conservation and management strategies.

This chapter provides a synthesis of what is known at present about the mammals of Angola. We have considered all mammalian species except cetaceans, which are treated in Weir (2019). Regarding the pinnipeds, we have only considered the Brown Fur Seal (*Arctocephalus pusillus*), which is the sole species of this group breeding in Angola. We start by presenting a short history of mammal research in the country, beginning with the studies of the pioneer Portuguese naturalist José Vicente Barbosa du Bocage, and finishing with the present-day efforts to resume mammological research and to clarify the status of many species that have virtually vanished during and in the years following the civil war. We then present a brief description of the mammalian species recorded in Angola, which accompanies the checklist presented in Appendix. Poorly known endemics and near endemics for which Angola may be particularly relevant at the global scale are highlighted, but special attention is also given to iconic species of high conservation concern, though the charismatic Giant Sable Antelope is dealt with in more detail by Vaz Pinto (2018, 2019). The next section provides an overview on the biogeography of the mammalian fauna in Angola, based primarily on the study of Linder et al. (2012) for sub-Saharan Africa, and the study of Rodrigues et al. (2015) dealing specifically with the mammals of Angola. Finally, we provide a summary of the conservation status of Angolan mammals, largely based on the global assessments by the IUCN (IUCN 2018). We also make a brief assessment of threats and conservation opportunities for mammals, but leave the details to Huntley et al. (2019), which deals specifically with the challenges of biodiversity conservation in the country.

History of Mammal Collecting in Angola

The first truly scientific studies on the vertebrate fauna of Angola, which included the classification and characterisation of several species of mammals, date back to the end of the nineteenth century. They are mainly due to José Vicente Barbosa du Bocage (1869, 1878, 1889a, b, 1890, 1897, 1902), professor of zoology at the Polytechnic School of Lisbon, to whom the explorer José Alberto de Oliveira Anchieta regularly sent specimens he collected in various parts of western Angola ('Sertão de Loanda'; 'Sertão de Benguella', 'Sertão de Mossâmedes'), and three other scientists of the time, who exchanged correspondence and opinions with Bocage. The latter included the German WCH Peters, who published on Angolan mammals based on the observations of the botanist Frederich MJ Welwitsch (Peters 1865) and the collections made by the German Expedition to Loango-Cabinda (Peters 1879), and the British WL Sclater and JE Gray, who studied the specimens sent to the British Museum of Natural History (Gray 1868, 1869) by geologist Joachim José Monteiro, who lived in Angola at the time. Other collectors, especially at the end of the nineteenth century and the first decade of the twentieth century, also sent specimens to the Polytechnic School's Zoological Museum. Initially founded on the collections studied by Bocage, this museum later came to be known as the Bocage Museum which officially constituted the Zoology Section of the National Museum of Natural History. Unfortunately, the collections deposited in it were lost in their entirety, due to a fire that broke out in 1978. Further material was also sent to other Portuguese museums and universities, such as the specimens offered to the museum of the University of Coimbra, by Lieutenant Colonel Teodoro da Cruz, and much later studied by A.A. THEMIDO (THEMIDO, 1931, 1946).

Besides museums in Portugal and the British Museum, other institutions receiving material from Angola at that time included: the Berlin Museum, which included material obtained by the German expeditions to Loango and the northeast of Angola, as well as the Kunene-Sambesi Expedition where the zoologist A Sokolowski collected mammals (Sokolowski 1903); the Leyden Museum, where P Van Der Kellen collected in southern Angola, and his material was studied by FA Jentink (1887, 1893, 1900, 1901); and the Tring Museum, which, along with the British Museum, received material collected by J Ansorge in various parts of Angola, largely to be studied by Thomas Oldfield (Thomas 1892, 1900, 1926, Thomas and Wroughton 1905). In 1916, Thomas would leave his name linked to the classification of *Hippotragus niger variani*, the Giant Sable Antelope (Thomas 1916), whose discovery and dispatch of specimens to the British Museum was due to the chief engineer of the Benguela Railway HF Varian. The amassing of ever larger collections of mammals from Angola and deposited at the British Museum at this time led to further publications of catalogues and other papers on Angola's mammalian fauna (Lydekker 1899, 1903, 1904; Lydekker and Blaine 1913–1916; Blaine 1922, 1925)

The 1920s and 1930s witnessed a resurgence of zoological holdings collected in Angola. Two Swiss missions by Albert Monard, curator of the Museum of Natural History of La Chaux-de-Fonds, published important contributions to the mammalian fauna of Angola (Monard 1930, 1931, 1933, 1935). Various American expeditions also carried out work in Angola during this period, including the Vernay Angola Expedition, organised by Arthur Vernay in 1925 to obtain material for the American Museum of Natural History, New York; the Gray African Expeditions, led by Prentiss Gray, who in 1929 obtained material, including specimens of *Hippotragus niger variani*, for the Philadelphia Academy of Natural Sciences; the Pulitzer Angola Expedition (1930–1931), organised by the Carnegie Museum and directed by Rudyard Boulton, who, despite being an ornithologist, collected mammalian material; and the Phipps-Bradley Expedition in 1932–1933, organised by John H Phipps, whose material was donated to the American Museum of Natural History. It was especially the material in these museums which served as the basis for the classic work of Hill and Carter (1941), *The Mammals of Angola, Africa*, published in 1941, as well as other papers (Hill 1941). Other minor expeditions included the Karl Jordan Expedition in 1934 whose material was deposited at the Tring Museum, and reported by St. Leger (1936); and the Percy Sladen and Kaffrarian Museum Expedition in 1934, organised by the Kaffrarian Museum and directed by Capt. GC Shortridge. Shortridge collected mainly in Namibia and was limited in Angola to the banks of the Cunene River. The increase in mammalian specimens collected in Angola and held in museums around the world up to this time allowed for the description of new subspecies by Hinton (1921), Matschie (1900, 1906), Zukowsky (1964) and Zukowsky and Haltenorth (1957).

Following World War II, Portuguese participation in zoological surveys became predominant in Angola. In fact, a board of overseas research was founded in Portugal as a branch office of the Ministry of the Colonies – the *Junta das Missões Geográficas e de Investigações Coloniais*. From the mid-1950s, the then-established Overseas Research Board, based in Lisbon, later becoming the Institute of Tropical Scientific Research (IICT), was the official institution in Portugal to oversee scientific missions to the Overseas Provinces at the time and, in fact, between 1957 and 1959, a zoological mission to Angola was conducted, directed by F Frade, the materials of which were deposited at the then IICT Zoology Center. This researcher, later director of the Zoology Center, was a prolific contributor of scientific papers on Angolan mammals on topics including anatomy, taxonomy and conservation (Frade 1933, 1936, 1955, 1956, 1958, 1959a, b, 1960, 1963; Frade and Sieiro 1960). Nonetheless, the majority of these scientific initiatives were mainly from institutions that were effectively based in Angola, which, during the 1950s and 1960s, promoted zoological explorations and collections in Angola.

Of the greatest importance was the Laboratory of Biology at the Dundo Museum, in Lunda-Norte, in the extreme northeast of Angola. This museum had two sections, one for ethnographic and the other for biological studies. Directed by António de Barros Machado, it became world-renowned for its invaluable collections, as well as for its prestigious magazine, Cultural Publications of the Diamond Company of Angola. Barros Machado, in spite of his specialisation as an entomologist, made an important contribution to the mammalogy of Angola (Machado 1952, 1968, 1969). RW Hayman of the British Museum studied the mammal material housed in the Dundo Museum (Hayman 1951, 1963).

The other Angolan institution of importance to mammalogy was the former Institute of Scientific Research of Angola (IICA), specifically its Sections of Ornithology and Mammalogy, based in Lubango, Huíla. The first section was directed by AA Rosa Pinto and the second by J Crawford-Cabral. As a result of several years of fieldwork and the collaborative work of various personnel, including collectors and taxidermists, it was possible to organise, in both these Sections, an excellent repository of zoological material from Angola. Both Sections still remain in Lubango, where they are currently part of the Higher Institute of Sciences and Education (*Instituto Superior de Ciências da Educação* – ISCED). The study of the material of the Mammalogy Section has been partly published mainly by Crawford-Cabral in an extensive number of articles, initially in the Bulletin of the Institute of Scientific Research of Angola and, more recently, in the Zoology Series of the magazine *Garcia de Orta*, and elsewhere (Crawford-Cabral 1961, 1966a,b, 1967, 1968, 1969a, b, 1970a, b, 1971, 1982, 1986, 1987, 1992, 1996, 1997, 1998; Crawford-Cabral and Fernandes 2001; Crawford-Cabral and Simões 1987, 1988; Crawford-Cabral and Veríssimo 2005).

However, the interest of foreign countries in the Angolan fauna had not diminished. During the 1950s and mid-1960s important collections were made by the German explorer Gerd Heinrich, mostly deposited in the Field Museum of Natural History, Chicago; Werner Trense, who undertook a collecting expedition in Angola between 1952–1954, which were deposited at the Hamburg Institute and Zoological Museum, and studied by him (Trense 1959); and, a decade later, another expedition from this last museum, which included the museum's anatomist H Oboussier, whose collections in Angola were related with her studies on the hypophysis of antelopes (Oboussier 1962, 1963, 1964, 1965, 1966, 1972, 1976; Oboussier and Von Tyszka 1964).

In the late 1960s and until the mid-1970s the interest of South African zoologists in Angolan mammalogy was also felt. In 1969, the State Museum of Namibia organised an expedition to southwestern Angola (mainly within the Namibe Provice), under the direction of its director, CG Coetzee, which was repeated in 1974; and, in June and July of the same year, the University of Cape Town and the Wildlife Society, undertook an expedition to the same regions (Broom et al. 1974). Worthy of reference, in this period just before independence, are the scientists who performed field work in Angola, such as Richard Estes, with his studies on the Giant Sable (Estes and Estes 1974) and the ecologist and conservationist Brian Huntley (1972a, b, 1973a, b, c, d, e, f, 1974).

Following the independence of Angola in 1975, the political situation deteriorated rapidly and soon after a civil war raged on until 2002. During this period very little was added to the knowledge of the Angolan mammalian fauna. However one should highlight the contributions of Alfred Feiler, assistant to AG Marques at the University Agostinho Neto, in Luanda, who undertook studies on mammal fauna (Feiler 1986, 1989, 1990); as well as by a short mammal survey conducted in some conservation areas (Juste and Carballo 1992); and a rapid assessment of the environmental conditions and fauna in some of the protected areas conducted by Huntley and Matos (1992).

With the end of the civil war, in 2002, the return of field work conditions and initiatives was severely hampered by the unknown status of the war legacy such as land mine fields, and the overall disruption of infrastructure and government institutions. However, the first aerial survey for large mammals in Iona NP was conducted in 2003 by a joint initiative between the government of Angola and the Namibia Ministry of Environment (Kolberg and Kilian 2003). At this same time, a concerted effort was ongoing to assess the status of the Giant Sable Antelope. This later culminated in the establishment of the Giant Sable Project with the assistance of the Kissama Foundation, which has since been in the forefront of the protection and recovery of this species Vaz Pinto (2019). The first complete historical review of the distribution of the ungulate fauna of Angola was published in 2005 (Crawford-Cabral and Verissimo 2005). Further wildlife monitoring initiatives have been developed in the southeast of the country. The first aerial surveys conducted in Cuando Cubango province were undertaken by the organisation *Elephants Without Borders* to assess the status of elephant populations, within the Luiana Partial Reserve, in 2004, 2005 and 2006, and extended in 2015 (Chase and Griffin 2011, Chase and Schlossberg 2016; Schlossberg et al. 2018). In 2007, the first systematic ground mammal survey was developed in the former Mucusso Game Reserve (Veríssimo 2008), in an effort to assist the Angolan Ministry of Environment to review the protected areas status of southeast Angola. Recent and ongoing initiatives, including a large carnivores assessment developed by the organisation *Panthera* in Cuando Cubango (see Funston et al. 2017); Huntley et al. 2019), as well as other initiatives of mammal surveys in Mupa, Bicuar and Iona NPs (Overton et al. 2017; Fabiano et al. 2017), and elsewhere (INBAC 2016), will continue to improve the knowledge of the unique mammalian fauna of Angola, and hopefully, its long term recovery and conservation. Despite these recent efforts, only a single recent publication has provided a checklist of mammals of Angola (Taylor et al. 2018c).

The Mammal Fauna

In this section we provide an overview of the mammalian fauna of Angola, giving at least a brief comment on each species recorded until now, all of which are presented in the checklist of Appendix. We also refer to some species that have never been collected in the country, but that occur very close to the border in neighbouring countries and thus are likely to occur in Angola. We have also reviewed cases of species that were once judged to occur in Angola, usually based on old records, but that have been probably misidentified and thus are no longer considered in the checklist. The section is based on a wide range of sources, including for example previous reviews focusing specifically on Angolan mammals (e.g., Hill and Carter 1941; Crawford-Cabral 1998; Crawford-Cabral and Simões 1987,1988; Crawford-Cabral and Veríssimo 2005), monographs on the mammals of Africa (e.g., Happold 2013; Happold and Happold 2013; Monadjem et al. 2010a, 2015), data from museums and historical observations available through GBIF (e.g. Bohm and Jonsson 2017; Conroy 2018; Grant and Ferguson 2018; MNHN 2018; MHNG 2018; Rodrigues et al. 2018; Taylor et al. 2018c), and unpublished data from the co-authors, among others. These sources reflect a highly uneven survey effort across Angola, as illustrated by the distribution of records in the GBIF database, and so it is likely that new mammal species for Angola are still to be discovered, particularly in less explored regions (Fig. 15.1).

In this review the higher taxonomy (i.e., family level and above) follows Kingdon et al. (2013), and the taxa are presented in alphabetical order, following the hierarchy of orders and families. The taxonomy at species and infraspecific levels is based largely on that adopted by the Red List of the IUCN (IUCN 2018), which in turn mostly follows the *3rd edition of Mammal Species of the World – A Taxonomic and Geographic Reference* (Wilson and Reeder 2005). This option was chosen because this is a generally recognised taxonomy, and because information on global conservation status is available for each of these species. In a few cases we have not followed this taxonomy, mainly when there were recent splitting of taxa treated as conspecific by the IUCN. Although they are not treated systematically, we have provided information on some particular subspecies, mainly in cases of type localities or restricted ranges in Angola, distinctive morphologies or ecologies, high conservation value, or that may warrant species status upon taxonomic revision.

Afrosoricida (Otter-shrews, Golden Moles)

The two species of Afrosoricida recorded in Angola are Congo Golden Mole (*Huetia leucorhina*) and Giant Otter-shrew (*Potamogale velox*). Little has been published on the Congo Golden Mole in the country, and the species is only known from a handful of records from northern Angola where it seems to occur in mosaics of grassland and moist forests (Hayman 1963; Crawford-Cabral and Veríssimo, unpublished

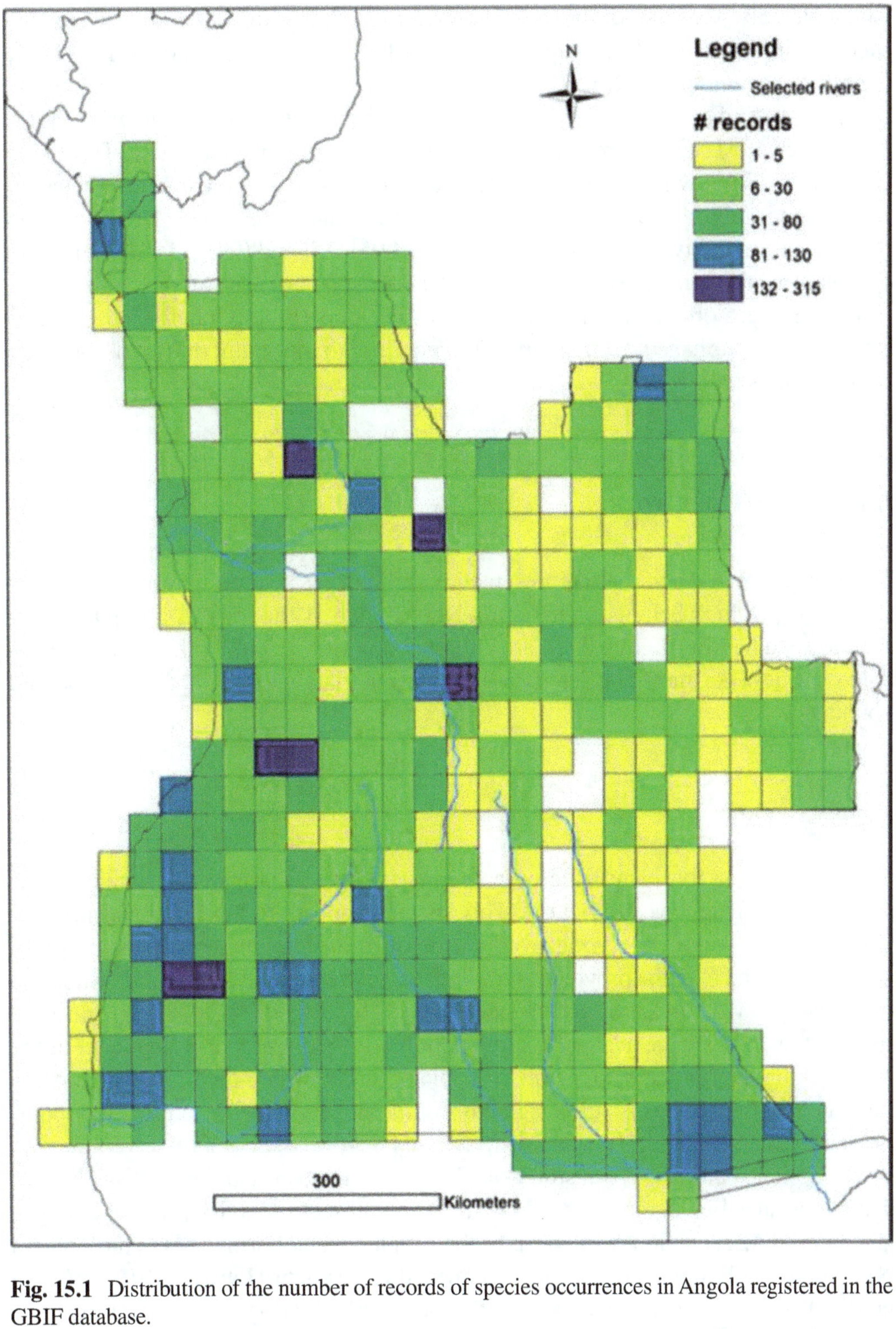

Fig. 15.1 Distribution of the number of records of species occurrences in Angola registered in the GBIF database.

data). Notably, the first record of Golden Mole in Angola was collected by Von Mechow in the Cuango river and initially attributed to the Hottentot Golden Mole (*Amblysomus hottentotus*), as *Chrysochloris albirostrus*, but was later assumed to be a mistake and has since been provisionally synonymised with *H. leucorhina* (Hill and Carter 1941; Crawford-Cabral and Veríssimo, unpublished data). The Giant Otter-shrew is known from relatively few records mostly dating from the nineteenth century and studied by Barbosa du Bocage (du Bocage 1865, 1882, 1890), or from the first half of the twentieth century (Seabra 1905; Hill and Carter 1941). This species was mainly found in small forest-lined streams in northern Angola, but a couple of records were obtained in the provinces of Bié and Huíla (Crawford-Cabral and Veríssimo, unpublished data), likely corresponding to the southernmost distribution of the species in Angola.

Carnivora (Carnivores)

The Carnivora in Angola are represented by at least 38 species within seven families, most of which belong to the family Herpestidae. Although this group is one of the most studied in Angola, there are still uncertainties regarding the occurrence of some species. For most species there is very little data on current distribution and abundance.

Family Canidae

There are at least five canid species in Angola, the most charismatic of which is the African Wild Dog (*Lycaon pictus*). This species appears to have once been widespread across Angola occurring from northeast in the province of Lunda-Norte to southwest in the Namibe province and southeast in the Cunene and Cuando Cubango province (Crawford-Cabral and Simões 1988; Huntley 1974). Although no estimates of abundance are available, some populations were probably abundant in the south, along the border with Namibia (Fabiano, unpublished data). Recent surveys indicate that the species is still resident in southern Angola, with confirmed populations at Bicuar, Luengue-Luiana and Mavinga NPs (Veríssimo 2008; Overton et al. 2017; Fabiano et al. 2017; Funston et al. 2017; Monterroso et al., unpublished data). There may also be other populations in the Angolan section of the greater Kavango-Zambezi (KAZA) region, westwards to the Mupa NP and northwards to the Cameia NP, where wild dogs were recently confirmed as resident (INBAC 2016; Fabiano et al. 2017). Preliminary surveys recently estimated wild dog densities at 0.65 individuals/100 km^2, which is comparable to other populations in southern Africa (Overton et al. 2017). Based on spoor counts and camera trapping, Overton et al. (2017) estimated a resident population size of 40–50 individuals in Bicuar NP, distributed through numerous small packs. In the same area, a camera trapping study by Fabiano et al. (2017) suggested a minimum population size of ca. 60 individuals in 10 different packs (2–38 individuals each), and confirmed reproduction in one wild dog pack. At the Luengue-Luiana and Mavinga NPs, Funston et al. (2017) estimated densities of 0.7 individuals/100 km^2 and a population size of 599 ± 260 individuals using camera trapping and spoor tracking.

The four species of small canids occurring in Angola are Bat-eared Fox (*Otocyon megalotis*), Cape Fox (*Vulpes chama*), Black-backed Jackal (*Canis mesomelas*), and Side-stripped Jackal (*Canis adustus*). Bat-eared Fox range based on few historical records seems to be confined to the Kaokoveld Desert, Namibian Savanna Woodlands and Angolan Mopane Woodlands, in southern Angola (Crawford-Cabral and Simões 1987). Recent surveys have confirmed its presence at Iona, Bicuar, Mupa and Luengue-Luiana and Mavinga NPs (Veríssimo 2008; Fabiano et al. 2017; Overton et al. 2017; Funston et al. 2017). The few historical records of Cape Fox also suggest that it is confined to the Kaokoveld Desert and Namibian Savanna Woodlands in southwest Angola (Crawford-Cabral and Simões 1987). Recent surveys indicate that the species still persist at Iona and Bicuar (Overton et al. 2017; Fabiano et al., unpublished data), but it was not detected at Mupa (Overton et al. 2017). The species is probably absent from the southeast, where it was not detected in surveys carried out at Luengue-Luiana and Mavinga NPs (Veríssimo 2008; Overton et al. 2017; Funston et al. 2017).

The historical records of the Black-backed Jackal suggest a distribution mostly restricted to the arid coastal belt within the Kaokoveld Desert, Namibian Savanna Woodlands and Angolan Mopane Woodlands (Crawford-Cabral and Simões 1987), but has been recorded recently as far north as the outskirts of Luanda and above 2000 m at the Humpata plateau (Vaz Pinto, unpublished data). Contemporary records indicate its presence in Iona NP (Fabiano et al., unpublished data), but also in Bicuar, Mupa, Luengue-Luiana and Mavinga NPs, which are dominated by the Zambezian Baikiaea Woodlands and Angolan Mopane Woodlands (Fabiano et al. 2017; Overton et al. 2017; Funston et al. 2017). The Side-stripped Jackal appears to have had a wider historical range, ranging from a few records in the Southern Congolian Forest-Savanna Mosaic in northeast Angola (Lunda-Norte province), and more records falling within the Angolan Miombo Woodlands from central northwest to central-southwest and highlands (Crawford-Cabral and Simões 1987). Recent surveys confirmed its presence in the Luengue-Luiana and Mavinga NPs (Veríssimo 2008; Funston et al. 2017), and in Cangandala NP (Vaz Pinto, unpublished data). Surprisingly, it has also been recorded recently in dry coastal savanna in Quiçama NP (Groom et al. 2018). It may also occur in Bicuar and Mupa NPs, though it was probably overlooked in recent surveys more directed to endangered species (Overton et al. 2017; Fabiano et al. 2017).

Family Felidae

The Felidae are represented in Angola by at least seven species, including iconic and threatened species such as African Lion (*Panthera leo*), Leopard (*Panthera pardus*), and Cheetah (*Acinonyx jubatus*). Historically, lions were probably widespread, inhabiting the Western Congolian Forest-Savanna Mosaic in the northeast, the Miombo Woodlands in central Angola, and the Savanna, Mopane and Baikiaea Woodlands in the south along the border with Namibia (Crawford-Cabral and Simões 1988; Veríssimo 2008; Huntley 1973c, 1974). The lion subspecies occurring in the country are poorly known, but recent phylogeographic studies suggest that Angola may represent a contact zone between the Central (*P. l. leo*) and Southern African (*P. l. melanochaita*) lineages (Barnett et al. 2014), and thus the genetic composition of Angolan lions could help elucidating the evolutionary history of this species in the African continent. Recent surveys indicate that lions still exist in the Luengue-Luiana and Mavinga NPs, while two recent records indicate their possible presence in the surroundings of the Cameia NP (iNaturalist.org 2018a, b). Funston et al. (2017) estimated the lion population of the Luengue-Luiana and Mavinga NPs to be about 10–30 individuals. They concluded that the very low biomass of preferred prey species was the main limiting factor for lions locally, as observed elsewhere (Bauer et al. 2015; Lindsey et al. 2017; Wolf and Ripple 2016). Recent surveys have failed to detect the species in Bicuar, Mupa and Quiçama NPs (Overton et al. 2017; Fabiano et al. 2017; Groom et al. 2018; Monterroso et al., unpublished data). However, park rangers and farmers in the vicinity of Bicuar NP have reported recent sightings of vagrant animals, suggesting that they may still occur in low numbers in the region (Fabiano, unpublished data). Recent observations of lone individuals have been obtained in Luando Reserve (Vaz Pinto, unpublished data). Other surveys, mostly relying on desktop surveys and interviews, indicate that lions might still occur in Cameia NP, and in the former Coutada do Mucusso (Veríssimo 2008; Purchase et al. 2007a, b), but these are unlikely to hold resident populations.

Cheetah historically appear to have occurred across Angola, inhabiting various habitats/ecoregions such as the Miombo Woodlands in the northeast and south-central Angola, the Angolan Scarp Savanna and Woodlands in the northwest, and in the Namibian Savanna and Mopane Woodlands as well as Zambezian Baikiaea Woodlands along the border with Namibia (Crawford-Cabral and Simões 1988; Veríssimo 2008). The subspecies represented in the country is the Southern African Cheetah (*A. j. jubatus*) (Kitchener et al. 2017). The current status of the Cheetah in Angola is poorly known, but it still occurs in some NPs (Funston et al. 2017; Kolberg and Kilian 2003; Purchase et al. 2007a, b; Fabiano, unpublished data; Álvaro Batista, personal communication). Funston et al. (2017) estimated that Cheetah occupy roughly 8% of the Luengue-Luiana and Mavinga NPs, occurring at a density of 0.2 individuals/100 km^2. Fabiano et al. (unpublished data) also using spoor counts estimated that cheetahs occupy approximately 28% of Iona's plains, at a density of 0.61 individuals/100 km^2 (0.17–1.98). This population is considered resident given the frequent report of sightings, including that of females with cubs (Bruce Bennett and Álvaro Batista, personal communication). The Iona cheetah population appears to be genetically similar to the Namibian counterpart based on a limited scat sample size ($n = 22$) genotyped at 8 loci (Fabiano et al., unpublished data). Other occasional sightings suggest cheetahs may be present in Cameia NP. Cheetahs were not detected in Bicuar and Mupa NPs (Overton et al. 2017; Fabiano et al. 2017), and

Overton et al. (2017) even suggests that they have been absent from the region for at least a decade. Recent camera-trapping detected the species in western Cuando Cubango (Stefan van Wyk, personal communication) and southern Moxico (NGOWP 2018).

African Leopard historically had a wide distribution across Angola, ranging from the northwest and the northeast, through central Angola to the southern border with Namibia. However, most historical records were from National Parks. The subspecies occurring in Angola is the African Leopard (*Panthera pardus pardus*) (Kitchener et al. 2017). The distribution range appears not to have reduced since the 1970s, as they still appear to be thriving throughout most of the country, including NPs and their surrounding areas (INBAC 2016). Using spoor counts Overton et al. (2017) estimated a density of 4.37–14.7 leopards/100 km^2 in the Bicuar NP. Camera-trapping also detected leopards in Mupa NP, though apparently at lower densities than in Bicuar (Overton et al. 2017). Based on spoor counts, Funston et al. (2017) found leopards to be widely distributed throughout Luengue-Luiana and Mavinga NPs, with an estimated population size of 518 ± 190 individuals. Their camera trapping efforts along the Cuando, Luiana and Luengue rivers allowed the detection of 120 different individuals, and estimated a density of 1.5 ± 0.14 leopards/100 km^2. Leopards are also found in the Iona NP (INBAC 2016), at a density of 1.02 (0.41–2.39) leopards/100 km^2 (Fabiano et al., unpublished data), as well as in Luando reserve, and Quiçama and Cangandala NPs (INBAC 2016; Groom et al. 2018; Fabiano, unpublished data; Vaz Pinto, unpublished data).

The other Felidae known to occur in Angola are Caracal (*Caracal caracal*), African Golden Cat (*Caracal aurata*), African Wildcat (*Felis silvestris*) and Serval (*Leptailurus serval*). Historical records indicate the presence of Caracal in the arid zone of southwestern Angola and the Miombo Woodlands of Cunene (Crawford-Cabral and Simões 1987). Recent surveys have confirmed its presence at Iona NP (Fabiano et al., unpublished data) and in Bicuar and Mupa NPs (Fabiano et al. 2017; Overton et al. 2017), and in the Luengue-Luiana and Mavinga NPs (Veríssimo 2008; Funston et al. 2017). The occurrence in Angola of the closely related African Golden Cat (*Caracal aurata*) was only confirmed very recently from an animal photographed on a local bushmeat market in northwestern Angola (Errol de Beer, personal communication). This cat is endemic to the forests of Equatorial Africa, particularly associated to areas of riverine forests with low human disturbance, and even penetrating savanna regions (Sunquist and Sunquist 2009; Bahaa-el-din et al. 2015). From the two recognised subspecies of Golden Cat, the one probably occurring in Angola is *C. a. aurata* (Sunquist and Sunquist 2009; Bahaa-el-din et al. 2015; Kitchener et al. 2017). African Wild Cat was historically widespread, occurring across most of the country (Crawford-Cabral and Simões 1987; Crawford-Cabral and Veríssimo, unpublished data). The species was associated with Miombo Woodlands, the Scarp Savanna and Woodlands, and the Kaokoveld Desert. Recent surveys have confirmed its presence in Bicuar and Mupa NPs (Fabiano et al. 2017; Overton et al. 2017), Quiçama NP (Groom et al. 2018), and in the Luengue-Luiana and Mavinga NP (Veríssimo 2008; Funston et al. 2017). Serval has a similar historical widespread distribution, occurring particularly across the western part of the country (Crawford-Cabral and Simões 1987). Within its range, Serval was associated with Miombo Woodlands, the Scarp Savanna and Woodlands, and the Kaokoveld Desert (Crawford-Cabral and Simões 1987). Two historical records were retrieved from northeastern Angola. Recent surveys have confirmed the presence of Serval in Bicuar and Mupa NPs (Fabiano et al. 2017; Overton et al. 2017), Quiçama NP (Groom et al. 2018), as well as in Mucusso area in Cuando Cubango (Veríssimo 2008). Additionally, it regularly surfaces as bushmeat near Luanda (Vaz Pinto, unpublished data). Black-Footed Cat (*Felis nigripes*) may occur marginally in southern Angola, but there seems to be no confirmed records (Sliwa 2013).

Family Herpestidae

The Family Herpestidae in Angola is represented by at least 12 species. The most well-known and charismatic is certainly the Meerkat (*Suricata suricatta*), a social mongoose widespread in the western parts of southern Africa (Jordan and Do Linh San 2015). Historical records suggest that in Angola it is confined to the arid southwest, including the Iona NP (Crawford-Cabral and Simões 1987), corresponding to the northwest edge of the species' range. Recent surveys suggest that meerkats still occur in this area (Fabiano et al., unpublished data; Monterroso et al., unpublished data). Crawford-Cabral (1971), proposed that the population occuring in Angola is a distinct subspecies, *Suricata suricatta iona*.

Another interesting group of species is that including the *Herpestes* mongooses, for which there are considerable taxonomic uncertainties (Taylor and Goldman 1993; Crawford-Cabral 1996; Gilchrist et al. 2009; Rapson et al. 2012; Veron et al. 2018). The three species known to occur in Angola are Kaokoveld Slender Mongoose (*Herpestes flavescens*), Egyptian Mongoose (*H. ichneumon*) and Common Slender Mongoose (*H. sanguineus*) (Crawford-Cabral 1996). Kaokoveld Slender Mongoose was described by Barbosa du Bocage from specimens collected in Angola, and is endemic to southwestern Angola and northwestern Namibia. However, there are marked variations between two subpopulations, which have been assigned to different subspecies or even species (Rathbun and Cowley 2008; Rapson et al. 2012). Individuals with tan or yellowish pelage confined to southwestern Angola are assigned to the subspecies *H. f. flavescens* (or *H. flavescens sensu stricto*), while those with very dark pelage and with a distinctive rufus tinge that occur in northwestern and northcentral Namibia have been assigned to *H. f. nigrata* (or *H. nigratus*) (Crawford-Cabral 1996; Tromp 2011; Taylor 2013). Very little is known about this species, particularly in Angola, where most assumptions on their distribution derive from habitat-interpreted satellite imagery (Rapson and Rathbun 2015). In contrast to the previous species, the Egyptian Mongoose and the Common Slender Mongoose are thought to be widespread (Crawford-Cabral and Simões 1987).

A group of five species of Herpestidae are thought to have wide distributions in Angola, though their current range and abundances are poorly known. Possibly the most widespread of these are the White-tailed Mongoose (*Ichneumia albicauda*), the Banded Mongoose (*Mungos mungo*) and the Dwarf Mongoose (*Helogale par-*

vula), as historical records report their presence throughout the country (Crawford-Cabral and Simões 1987; Trombone 2016; Figueira 2017; Grant and Ferguson 2018; Rodrigues et al. 2018). The White-tailed Mongoose has recently been confirmed in Cameia and Cangandala NPs, and the Banded Mongoose appears to be abundant in Luando Reserve and Cangandala NP (Vaz Pinto, unpublished data). The Marsh Mongoose (*Atilax paludinosus*) is associated with riparian habitats, such as swamps and streambed areas, though occurring also in estuarine and marine habitats (Gilchrist et al. 2009). Historical records of these species have mainly been from western Angola (Trombone 2016; Rodrigues et al. 2018), though it may be more widepread and has recently been confirmed in Cangandala NP (Vaz Pinto, unpublished data). The Selous Mongoose (*Paracynictis selousi*) has the northwestern edge of its range in Angola, with historical records from southern provinces (Crawford-Cabral and Simões 1987; Trombone 2016; Conroy 2018; Grant and Ferguson 2018; Rodrigues et al. 2018). Recent surveys confirmed its presence at Luengue-Luiana and Mavinga NPs (Funston et al. 2017) and Bicuar NP (Overton et al. 2017), albeit at low densities.

Finally, another three Herpestidae have very restricted or probably underestimated ranges in Angola. Ansorge's Cusimanse (*Crossarchus ansorgei*) may be more widespread than usually believed because it has a relatively wide distribution in rainforests of neighbouring regions of DRC (Angelici and Do Linh San 2015). Although in Angola it was known from a single specimen collected in 1908 north of the Cuanza River (Crawford-Cabral and Simões 1987), recent records suggest it may extend its distribution along the escarpment to Cuanza-Sul (Michael Mills, personal communication). The population in Angola is assigned to the endemic subspecies *C. a. ansorgei*. Black-legged Mongoose (*Bdeogale nigripes*) seems to be restricted in Angola to the enclave of Cabinda (Crawford-Cabral and Simões 1987). The Yellow Mongoose (*Cynictis penicillata*) only occurs in a narrow fringe in the extreme southwest of Angola (Crawford-Cabral and Simões 1987), though it is widespread elsewhere in southern Africa. We are unaware of recent records of any of these species in Angola.

Family Hyaenidae

There are three species of the family Hyaenidae recorded in Angola (Crawford-Cabral and Simões 1988). Historical and contemporary records indicate that the Brown Hyaena (*Parahyaena brunnea*) is largely confined to the arid parts of southwestern Angola, in Kaokoveld Desert and Namibian Savanna Woodlands (Huntley 1974; Fabiano et al., unpublished data). This range encompasses the Skeleton Coast Transfrontier Park between Angola and Namibia. Recent surveys indicate that it is widespread in the Iona NP (Fabiano et al., unpublished data), but it was not detected in Luengue-Luiana NP, despite its presence in the nearby Bwabwata NP of Namibia (Funston et al. 2017). Spotted hyaenas (*Crocuta crocuta*) historically were widely distributed across Angola with main populations occurring in Zambezian Baikiaea Woodlands in the south of the country, though it also appeared to have been widespread in the west. The Spotted Hyaena is one of the few large carnivores species that shows no evidence of recent population declines throughout its African range (Bohm and Höner 2015), although the situation may not necessarily be identical in Angola. Recently, populations were estimated at 10.8–18.0 individuals/100 km^2 in Bicuar NP (Overton et al. 2017), 1.4 individuals/100 km^2 in Mupa NP (Overton et al. 2017), and 0.9 individuals/100 km^2 in Luengue-Luiana and Mavinga NPs (Funston et al. 2017). The presence of Spotted Hyaena has not been confirmed in surveys of various protected areas, and they appear to have been extirpated from Luando Reserve and Cangandala NP (Vaz Pinto, unpublished data). In Quiçama NP an active den was known to be present on the Cuanza floodplain as recently as 2013 (Vaz Pinto, unpublished data), but a recent large mammal survey could not find evidence of the species (Groom et al. 2018). Overall, the species is expected to remain widely distributed in Angola (INBAC 2016). Aardwolf (*Proteles cristata*) is the least known of the three Hyaenidae of Angola, occurring only in the south of the country (Crawford-Cabral and Veríssimo, unpublished data). Recent records from direct observations and roadkills suggest the species to be relatively common along the arid coastal plain to as far north as Benguela, and on the highlands of Humpata plateau (Vaz Pinto, unpublished data).

Family Mustelidae

The Mustelidae are represented in Angola by Honey Badger (*Mellivora capensis*), Striped Polecat (*Ictonyx striatus*), African Striped Weasel (*Poecilogale albinucha*), and three species of otters (Crawford-Cabral and Simões 1987). Historical data on Honey Badger suggest that it was more frequent in the southwest and east of the country, within the Angolan Miombo Woodlands and the Zambezian Baikiaea Woodlands (Crawford-Cabral and Simões 1987), but also occurring in the provinces of Malanje and Moxico (Trombone 2016; MHNG 2018; Rodrigues et al. 2018). Recent surveys have confirmed its presence at Iona, Bicuar, Mupa, Quiçama, Cangandala, Luengue-Luiana and Mavinga NPs (Veríssimo 2008; Fabiano et al. 2017; Fabiano, unpublished data; Overton et al. 2017; Funston et al. 2017; Groom et al. 2018; Monterroso et al., unpublished data; Vaz Pinto, unpublished data).

The Striped Polecat is a generalist that occurs in most habitats except dense rainforests. Historically this species was recorded throughout the country (Trombone 2016; Figueira 2017; Grant and Ferguson 2018; Rodrigues et al. 2018), and is expected to maintain a wide distribution. It was detected in the recent surveys carried out in NPs of southern Angola (Veríssimo 2008; Funston et al. 2017; Monterroso et al., unpublished data). The African Striped Weasel is also a widespread habitat generalist, though it is often overlooked due to its secretive habits. Probably because of this it has a relatively small number of records in Angola, though widely distributed throughout the country (Crawford-Cabral and Simões 1987)

Little is known about the three otter species occurring in Angola, with no systematic surveys available to describe their current status (Crawford-Cabral and Simões 1987). The Congo Clawless Otter (*Aonyx congicus*) is associated with the rainforests of the Congo Basin (Jacques et al. 2015a), and so in Angola should be restricted to Cabinda and Lunda-Norte. This species is sometimes treated as conspecifc of the African Clawless Otter (*Aonyx capensis*) (Wozencraft 2005), which in spite of the paucity of historical records is thought to have a wide distribution in the east and south of Angola (Veríssimo 2008; Jacques et al. 2015b), and has recently been recorded at the Humpata plateau (Vaz Pinto, unpublished data). The Spotted-necked Otter (*Hydrictis maculicollis*) is likely to be the most widespread otter in freshwater systems throughout Angola (Reed-Smith et al. 2015), though the historical records for the country are relatively few and scattered.

Family Nandiniidae

The African Palm Civet (*Nandinia bionotata*) is the sole representative of the family Nandiniidae (Crawford-Cabral and Simões 1987). Historical records of the species have been mainly made in the provinces of Uíge, Cuanza-Norte, and Lunda-Norte, suggesting its occurrence throughout the north of Angola. There have been no recent published records of this species in Angola, though it probably is still widespread within its former known range in the northern half of the country.

Family Otariidae

The Brown Fur Seal (*Artocephalus pusillus*) is the only pinniped breeding in Angola, with several large colonies in the island of Tigres (Meÿer 2007). This is the northern limit of the species distribution, which extends along the coast of Namibia to Algoa Bay in South Africa (Hofmeyr 2015). Other species occur occasionally along the coast of Angola, with records of for instance Sub-Antarctic Fur Seal (*Arctocephalus tropicalis*) (Carr et al. 1985) and South-Atlantic Elephant Seal (*Mirounga leonina*) (França 1967).

Family Viverridae

The Viverridae in Angola include the African Civet (*Civettictis civetta*), and three species of the genus *Genetta*. The civet was reported to occur in northern (Bengo, Cuanza-Norte and Malanje), central-west (Benguela), and southwest Angola (Namibe and Huíla province) (Crawford-Cabral and Simões 1987), mainly associated with the Western Congolian Forest-Savanna Mosaic, Angolan Scarp Savanna and Woodlands, Angolan Miombo Woodlands. Recent surveys have confirmed its presence at the Mupa, Quiçama, Cangandala, Mavinga and Luengue-Luiana NPs (Veríssimo 2008; Overton et al. 2017; Funston et al. 2017; Groom et al. 2018; Vaz Pinto, unpublished data). The three genets currently recognised in Angola are the Small Spotted Genet (*Genetta genetta*), the Large Spotted Genet (*Genetta maculata*), and the Miombo or Angola Genet (*Genetta angolensis*) (Crawford-Cabral and Simões 1987). Small Spotted Genet was identified as *G. g. felina* by Crawford-Cabral and Simões (1987). Historical records suggest that the species occurred predominantly in southwestern Angola within the Kaokoveld Desert (Namibe) and the Angolan Mopane and Miombo Woodlands (Huíla and Cunene) (Crawford-Cabral and Simões 1987). Recent surveys have confirmed its presence in southeast Angola in Luengue-Luiana and Mavinga NPs (Funston et al. 2017). Small Spotted Genets may still occur in Bicuar and Mupa NPs, and their surrounding areas, as spotted genets (*Genetta* spp.) have been detected on cameras and roadkills, but not identified to the species level (Overton et al. 2017; Fabiano et al. 2017; Monterroso et al., unpublished data; Vaz Pinto, unpublished data). The taxonomy of the Large Spotted Genet is still to be resolved (Angelici et al. 2016), and it appears to be a 'superspecies' comprising several valid species. Large Spotted Genet was recorded as *G. m. rubiginosa* by Crawford-Cabral and Simões (1987). It appears to have had widespread distribution occurring across the western part of the country (Crawford-Cabral and Simões 1987). Recent surveys have confirmed its presence in southeast Angola in the Luengue-Luiana and Mavinga NPs (Funston et al. 2017), and in Quiçama NP (Groom et al. 2018). The Miombo Genet is considered a near-endemic of the miombo ecoregion (Timberlake and Chidumayo 2011). According to Gaubert et al. (2016), Miombo Genets' westernmost distribution range should be restricted to central Angola's miombo woodlands, as it was mainly present in central, south-west and south-central Angola (Crawford-Cabral and Simões 1987; Trombone 2016; Bohm and Jonsson 2017; Rodrigues et al. 2018). However, Huntley and Francisco (2015) suggest that the species could be widespread across Lunda-Norte province, suggesting its distribution in Angola may be underestimated. Recent surveys have confirmed its presence in Cangandala and Bicuar NPs (Overton et al. 2017; Vaz Pinto, unpublished data).

Cetartiodactyla (Pigs, Hippopotamuses, Chevrotain, Giraffes, Deer, Bovids)

The Cetartiodactyla includes 33 native species of 5 families in Angola. Most species belong to the Bovidae, which is represented by at least 27 species (Crawford-Cabral and Veríssimo 2005). Although this is one of the best-known animal groups in the country, there are many uncertainties regarding the current status and distribution of most species. Given the large number of species, information on this group is provided per family, while the bovidae are presented per tribe.

Family Bovidae

Tribe Aepycerotini

The tribe Aepycerotini comprises only one genus and one species, the Impala (*Aepyceros melampus*). Although up to six subspecies of impala have been listed, their validity was problematic and limits poorly defined (Ansell 1972; Fritz and

Bourgarel 2013). Most often only two races are recognised, the Common Impala (*A. m. melampus*) and the Black-faced Impala (*A. m. petersi*), which is also supported by molecular data (Lorenzen et al. 2006). The Black-faced Impala was described as a new species from a specimen collected at Humbe, Cunene Province (du Bocage 1879), and some authors maintained a dual-species classification (Shortridge 1934; Groves and Grubb 2011). Both taxa naturally occurred in Angola, in two disjunct and well demarcated populations (Crawford-Cabral and Veríssimo 2005). The Common Impala occurs in the southeast of the country between the Cubango and the Cuando rivers, with most historical records obtained along the former (Sokolowski 1903; Wilhelm 1933; Hill and Carter 1941; Huntley 1973c; Crawford-Cabral and Veríssimo 2005). The Common Impala was given as extirpated along the Cubango River by Veríssimo (2008), but subsequently relatively small numbers have been confirmed in the Luengue-Luiana NP (Chase and Schlossberg 2016; Funston et al. 2017). The Black-faced Impala is endemic to Kaokoland in northwestern Namibia and southwestern Angola (Fritz and Bourgarel 2013), but the bulk of its distribution used to be in Angola where it extended as far north as Benguela and mostly west of the Cunene River (Hill and Carter 1941; Swart 1967; Crawford-Cabral and Veríssimo 2005). Before 1975 the Black-faced Impala was represented in protected areas such as Iona, Bicuar and Mupa NPs (Huntley 1972c, 1973c; Crawford-Cabral and Veríssimo 2005). Currently it is likely on the verge of extinction in Angola, as recent surveys have failed to record it in Iona, Bicuar and Mupa NPs (Kolberg and Kilian 2003; Overton et al. 2017; van der Westhuizen et al. 2017). Very small numbers may still linger in eastern Iona (Álvaro Baptista, personal communication), and one single specimen was observed in 2016 near Serra da Neve in northern Namibe province (Vaz Pinto, unpublished data).

Tribe Alcelaphini

The Alcelaphini are represented in Angola by the genera *Damaliscus*, *Alcephalus* and *Connochaetes*. The earlier references mentioning the presence of Tsessebe (*Damaliscus lunatus*) in Angola are scattered (Sokolowski 1903; Monard 1935; Hill and Carter 1941; Varian 1953), but at some point the species was likely to be relatively common and widely distributed across southeastern Angola to the east of the Cunene river, and along the eastern border as far north as the upper Zambezi drainage (Newton da Silva 1970; Huntley 1973c; Crawford-Cabral and Veríssimo 2005). It was once considered common in the plains of Cameia NP and in the areas now included in Luengue-Luiana NP (Crawford-Cabral and Veríssimo 2005). Very little is known in terms of current status of the species in Angola, but it appears to have been much reduced in numbers. Veríssimo (2008) suggested their persistence in northern Luengue and Luiana based only on witness reports, but an extensive aerial survey did not record any Tsessebe in Luengue-Luiana and Mavinga NPs (Chase and Schlossberg 2016). However, the species was confirmed in the area through camera-trapping (Funston et al. 2017).

The taxonomy of the hartebeest, genus *Alcephalus*, is controversial, with some authors considering it a monospecific genus with up to eight subspecies (Flagstad et al. 2001; Gosling and Capellini 2013), while others recognise different species (e.g. Ansell 1972; Groves and Grubb 2011). Two morphologically very distinct taxa occur in Angola in disjunct populations, Red Hartebeest (*A. b. caama*) and Lichtenstein's Hartebeest (*A. b. lichtensteini*), which are often treated as full species (e.g. Huntley 1973c; Crawford-Cabral and Veríssimo 2005). Red Hartebeest used to be widely distributed in southern Angola in the region between the Cunene and Cubango rivers (Sokolowski 1903; Monard 1935; Hill and Carter 1941; Huntley 1973c; Crawford-Cabral and Veríssimo 2005). Red Hartebeest could be found in only one protected area, Mupa NP (Huntley 1973c; Crawford-Cabral and Veríssimo 2005), and by the 1970s the Angolan population was already much reduced and endangered (Huntley 1973c), and was feared extinct in the 1990s (Huntley and Matos 1992). Recent surveys found no evidence of Red Hartebeest in Mupa (Overton et al. 2017) and this taxon is possibly currently extinct in Angola. Lichtenstein's Hartebeest was found in eastern Angola where it was generally uncommon to rare (Huntley 1973c; Crawford-Cabral and Veríssimo 2005). Most records were obtained in Lundas and Moxico (Machado 1969), although some earlier authors reported the species in Luiana along the Cuando River, even if not fully supported (Statham 1926; Crawford-Cabral and Veríssimo 2005). Lichtenstein's Hartebeest was once present in Cameia NP, but their current status in Angola is unknown and likely extinct.

Blue Wildebeest (*Connochaetes taurinus taurinus*) used to be widespread across southern and eastern Angola, in Huíla, Cunene, Moxico and Cuando Cubango provinces (Hill and Carter 1941; Newton da Silva 1970; Crawford-Cabral and Veríssimo 2005). It was once common in the protected areas of Bicuar and Cameia NPs, and the region currently ascribed to Mavinga and to Luengue-Luiana NPs (Huntley 1973c; Crawford-Cabral and Veríssimo 2005). Wildebeest numbers in Angola must have plummeted during the war (Huntley and Matos 1992), while recent surveys and anecdotal records suggest that the species is likely extinct in the western areas of their Angolan range, such as Bicuar NP (Overton et al. 2017). Nevertheless small numbers have recently been recorded in the southeastern corner, both from camera-trapping (Funston et al. 2017) and aerial counts (Chase and Schlossberg 2016).

Tribe Antilopini

Springbok (*Antidorcas marsupialis*) is the only species of gazelle *sensu lato* occurring in Angola and southern Africa. Based on specimens collected south of Benguela, and comparing these with Springbok from South Africa, Blaine (1922) claimed enough differences to justify the description of a new species, Angolan Springbok (*A. angolensis*). It was subsequently considered as one of three subspecies of springbok, *A. m. angolensis*, and extending into Namibia (Ansell 1972; Hill and Carter 1941). However, the distinction among geographical boundaries and intergradation has led to questioning the validity of these races (Skinner 2013). In Angola,

Springbok is strongly associated with the arid coastal belt, and present in the protected areas of Chimalavera, Namibe and Iona (Huntley 1973c, 1974; Crawford-Cabral and Veríssimo 2005). Nevertheless, a few old records are also known from the region of Naulila (Monard 1935; Galvão and Montês 1943–1945; Crawford-Cabral and Veríssimo 2005). Currently it is still present along the coastal plain south of Benguela, albeit in much reduced numbers (Vaz Pinto, unpublished data). Recent aerial surveys in Iona NP have allowed population estimations, suggesting a decreasing trend with an estimated 21% reduction in total numbers between 2003 and 2017 (Kolberg and Kilian 2003; van der Westhuizen et al. 2017).

Tribe Bovini

Buffalo (*Syncerus caffer*) is the sole representative of the Bovini in Africa. Nevertheless its taxonomy remains controversial due to a marked geographical variation and the existence of intermediate forms, which has led to the recognition of several species, subspecies or variants (Grubb 1972; Prins and Sinclair 2013). In Angola at least two forms are known, the typical Cape Buffalo (*S. c. caffer*) of larger body size, dark coloration and large hook-shaped horns, and the Forest Buffalo (*S. c. nanus*) of smaller body size, reddish colour and smaller backward-pointing horns. Notwithstanding, quite a lot of variation has been observed particularly among Angolan forest buffalos, as specimens from northern Angola tend to be larger and darker than those in Cabinda (Crawford-Cabral and Veríssimo 2005). Even though Matschie (1906) has described an additional subspecies (*S. c. mayi*) based on a specimen from Luanda, it was subsequently synonymised with *S. c. nanus*. Forest Buffalo used to have a wide distribution across Cabinda and northern Angola, including in protected areas such as Quiçama NP and Luando Strict Reserve (Huntley 1973c; Crawford-Cabral and Veríssimo 2005). Cape Buffalo was mostly present in the southeast, even though an additional small population was known from Bicuar NP (Huntley 1973c) and a few other isolated pockets in Benguela, Cuanza-Sul and Lunda provinces (Crawford-Cabral and Veríssimo 2005). The buffalo populations were severely reduced during the war (Huntley and Matos 1992), but very small numbers of Forest Buffalo are still present in Quiçama NP (Groom et al. 2018), and Cangandala NP and Luando Reserve (Vaz Pinto, unpublished data). Cape Buffalo is not uncommon in Mucusso region and Luengue-Luiana NP (Veríssimo 2008; Chase and Schlossberg 2016; Funston et al. 2017). On the other hand, smaller and isolated populations of Cape Buffalo may have been extirpated, such as that which used to occur in Bicuar NP (Overton et al. 2017).

Tribe Cephalophini

The tribe Cephalophini is represented by three genera and six species in Angola. The genus *Cephalophus* corresponds to typical forest duikers, of which four species are known for Angola: White-bellied Duiker (*C. leucogaster*), Bay Duiker (*C.

dorsalis), Black-fronted Duiker (*C. nigrifrons*), and Yellow-backed Duiker (*C. silvicultor*). The White-bellied Duiker was reported from Cabinda based on a witness account obtained in the 1970s (Brian Huntley, personal communication), and subsequently added to the Angolan list (Crawford-Cabral and Veríssimo 2005). The species is known to occur in Maiombe forest across the border in Congo (Malbrant and Maclatchy 1949; East 1999), but further evidence of its presence in Angola is still lacking. Both Bay Duiker and Black-fronted Duiker have been recorded from moist forest habitats in the northern half of the country, including Cabinda (Huntley 1973c; Huntley and Matos 1992; Crawford-Cabral and Veríssimo 2005). Although no recent studies are available, the regular presence of both species in bushmeat markets in northwestern Angola (Vaz Pinto, unpublished data) suggest they may still be relatively common in spite of the poaching pressure. Yellow-backed Duiker is the largest of duikers and is also widely, yet discontinuously, distributed across the northern half of the country and including Cabinda (Machado 1969; Huntley 1973c; Crawford-Cabral and Veríssimo 2005). Unlike the previous species, Yellow-backed Duiker is less dependent on moist forest habitats, being mostly an ecotone species (Kingdon and Lahm 2013). In spite of the scarcity of records, it appears well adapted to the riverine forests and thickets of central Angola, venturing into nearby well-developed miombo woodlands (Vaz Pinto and Veríssimo 2016) and present even on the Angolan highlands (Statham 1922; Hill and Carter 1941), possibly in transition to Afromontane patches. Recent observations suggest the species to be relatively common in Luando Strict Reserve, and venturing into the upper catchments of the Okavango and Zambezi drainages (Vaz Pinto and Veríssimo 2016; NGOWP 2018). Angolan Yellow-backed Duikers are assigned to the subspecies *C. s. ruficrista*, of which the type locality given is Luanda (du Bocage 1869).

Blue Duiker (*Philantomba monticola*) is especially common along the escarpment and in various types of forests and thickets along the coastal plain north of 15° latitude, and including Cabinda (Crawford-Cabral and Veríssimo 2005). On the plateau it is present north of 13° latitude but it is here less common, patchily distributed, and associated with riverine forests (Crawford-Cabral and Veríssimo 2005; Vaz Pinto, unpublished data). Blue Duiker is still abundant in Quiçama NP (Groom et al. 2018), while present in Cangandala NP (Vaz Pinto unpublished data), and has recently been photographed on the upper catchments of the Cuito River (NGOWP 2018). Three subspecies have been tentatively ascribed to Blue Duiker from Angola, but their validity remains unclear (Ansell 1972; Crawford-Cabral and Veríssimo 2005; Hart and Kingdon 2013). Under this classification, the population in Cabinda is assigned to *P. m. congicus*, and those on the plateau to *P. m. defriesi*, while the blue duikers from the escarpment and western Angola correspond to an endemic race, *P. m. anchietae* (Ansell 1972; Crawford-Cabral and Veríssimo 2005; Hart and Kingdon 2013).

Grey or Common Duiker (*Sylvicapra grimmia*) is likely the most widespread and common of all Angolan antelopes (Statham 1922; Crawford-Cabral and Veríssimo 2005). It probably still occurs in all Angolan protected areas, except Iona and Maiombe NPs, and throughout the country (Crawford-Cabral and Veríssimo 2005; Veríssimo 2008; Funston et al. 2017; Overton et al. 2017; Groom et al. 2018; NGOWP 2018; Vaz Pinto, unpublished data). Although many subspecies of Grey Duiker have been suggested, the continuous distribution of the species in sub-

Saharan Africa and the existence of local intergrading variants prevents clear delineation of boundaries (Ansell 1972; Wilson 2013). In Angola most are assigned to *S. g. splendidula*, and the race is thought to intergrade with *S. g. steinhardti* in southwestern Angola (Hill and Carter 1941; Crawford-Cabral and Veríssimo 2005; Wilson 2013).

Tribe Hippotragini

The tribe Hippotragini contains seven extant species, of which three can be found in Angola, including two representatives of the genus *Hippotragus* and one of *Oryx*. The Roan Antelope (*Hippotragus equinus*) is the most common and widely distributed large antelope in Angola, historically being absent only from Cabinda and the arid southwest (Huntley 1973c; Newton da Silva 1970; Crawford-Cabral and Veríssimo 2005). It used to be present in all existing protected areas except Iona NP and Namibe Partial Reserve (Crawford-Cabral and Veríssimo 2005), and was once considered abundant in Quiçama and Bicuar NPs (Huntley 1973c; Huntley and Matos 1992). As a result of the civil war the species has been extirpated from Quiçama NP (Huntley and Matos 1992; Groom et al. 2018; Vaz Pinto, unpublished data), but it remains relatively common in Bicuar NP (Overton et al. 2017) and Luando Strict Nature Reserve (Vaz Pinto, unpublished data), while small numbers still linger in Mupa (Overton et al. 2017) and Cangandala NPs (Vaz Pinto et al. 2016). In addition the species has also been confirmed recently in various surveys conducted across central and eastern Angola (Veríssimo 2008; Chase and Schlossberg 2016; Funston et al. 2017; NGOWP 2018), and likely remains widespread even if in reduced numbers throughout most of the country except on the coastal plain. Roan intraspecific taxonomy is still unresolved, but the Angolan race is usually ascribed to the Zambezian region subspecies *H. e. cottoni* (Ansell 1972; Chardonnet and Crosmary 2013; Vaz Pinto 2018).

Sable Antelope (*Hippotragus niger*) had a highly fragmented distribution in Angola, with three disjunct populations corresponding to three different subspecies (Crawford-Cabral and Veríssimo 2005; Estes 2013; Vaz Pinto 2018, 2019). Giant Sable (*H. n. variani*) is an endemic and critically endangered taxon, confined to the Cuanza drainage, and being the most famous Angolan mammal is dealt with in a separate dedicated chapter (see Vaz Pinto 2019). The occurrence of Kirk's Sable (*H. n. kirkii*) in eastern Angola was confirmed by a few scattered records from Cazombo, east of the Zambezi River, and in the Lundas, on the western banks of the Cassai River (Huntley 1973c; Crawford-Cabral and Veríssimo 2005). No records for this taxon have been obtained in Angola for over 40 years and we are unaware of any witness reports, thus suggesting the possibility of local extinction. The typical race *H. n. niger* is known from southeastern Angola to the east of the Cuito river (Hill and Carter 1941; Huntley 1973c; Crawford-Cabral and Veríssimo 2005; Vaz Pinto 2018), a region that broadly corresponds to the newly proclaimed Mavinga and Luengue-Luiana NPs. Recent surveys have confirmed typical sable to be still relatively common in the region and clearly outnumbering congeneric roan antelope (Veríssimo 2008; Chase and Schlossberg 2016; Funston et al. 2017), and very recently a dispersing male has been recorded as far north as southern Moxico (Kerllen Costa, personal communication)

Gemsbok (*Oryx gazelle*) in Angola is mostly associated with the Namib desert in the southwestern corner, but its distribution used to extend along the semi-arid coastal plain as far north as near Benguela (Blaine 1922; Statham 1922; Hill and Carter 1941; Crawford-Cabral and Veríssimo 2005), and in Cunene province at least as far inland as Cuamato and Chimporo (Monard 1935; Crawford-Cabral and Veríssimo 2005). Specimens from southwestern Angola have led to the description of a local endemic subspecies of gemsbok *O. g. blainei* based on facial mask differences (Blaine 1922; Hill and Carter 1941; Newton da Silva 1970). However the species is currently assumed to be monotypic (Knight 2013). Gemsbok may have been extirpated from most of its Angolan range in the second half of the twentieth century (Newton da Silva 1970; Crawford-Cabral and Veríssimo 2005), while remaining abundant in Iona NP (Huntley 1973c). The numbers were then likely much reduced during the civil war (Huntley and Matos 1992), and although they may still be relatively common in Iona, recent surveys suggest the general trend has remained negative (Kolberg and Kilian 2003; van der Westhuizen et al. 2017).

Tribe Madoquini

The sole Madoquini in Angola is Kirk's Dik-dik (*Madoqua kirkii*). The species was first collected in Angola by Anchieta in 1878, and specimens obtained near Lobito led to the description of a new subspecies, *M. k. variani*, which was later synonymised with *M. k. damarensis* (Drake-Brockman 1909, 1930; Hill and Carter 1941; Newton da Silva 1970; Kingswood and Kumamoto 1997). Angolan dik-diks are part of a southwest African population that extends well into Namibia, ascribed to the subspecies *M. k. damarensis*, although the huge geographical gap that separates these from the populations in the horn of Africa added by some morphological characters and genetic evidence, suggest they should be best treated as full species (Kumamoto et al. 1994; Zhang and Ryder 1995; Brotherton 2013). In Angola dik-diks are associated with the semi-arid environments and particularly with Mopane (*Colophospermum mopane*) woodlands, but also extending into the southern plateau west of the Cunene River (Crawford-Cabral and Veríssimo 2005). The species is well represented in protected areas such as Chimalavera and Namibe Reserves, and Iona and Bicuar NPs (Huntley 1973c; Crawford-Cabral and Veríssimo 2005; Vaz Pinto, unpublished data).

Tribe Oreotragini

Klipspringer (*Oreotragus oreotragus*) is the only representative of the Oreotragini, but its taxonomy is one of the most hotly debated within African bovids, both in terms of the relationships with other clades and among various populations. Ansell (1972) recognised 11 subspecies, while Groves and Grubb (2011) distinguished up to 20 taxa and elevated them to full species, but the latter still lacks molecular support and the monospecific classification is still the most widely accepted. Angolan klipspringers have been attributed to the subspecies *O. o. tyleri* described from a specimen obtained in Equimina, Benguela (Hinton 1921). They likely form part of a metapopulation that extends into Namibia (Crawford-Cabral and Veríssimo 2005; Roberts 2013), where a second subspecies *O. o. cunenensis* (Zukowsky 1924) described near Ruacana falls has been synonymised with the former (Hill and Carter 1941; Ansell 1972). Klipspringer occurs in Angola in rocky mountainous habitats, particularly in southern Angola and along the escarpment, with the northernmost population present at Pungo Andongo (Crawford-Cabral and Veríssimo 2005), and the easternmost population was reported from the region of Cassinga (Monard 1935; Newton da Silva 1970; Crawford-Cabral and Veríssimo 2005). The species used to be relatively common in Iona NP, and Chimalavera and Namibe Reserves (Huntley 1973c; Juste and Carballo 1992; Crawford-Cabral and Veríssimo 2005), and in spite of absence of recent data it is still often observed in southwestern Angola and along the southern escarpment (Vaz Pinto, unpublished data).

Tribe Ourebiini

This tribe is monospecific, comprising only the Oribi (*Ourebia ourebi*), which has a wide distribution in Africa and across southcentral and eastern Angola (Crawford-Cabral and Veríssimo 2005; Brashares and Arcese 2013). Up to 13 subspecies of Oribi have been described, but their validity remains problematic (Brashares and Arcese 2013). Two subspecies were described from specimens collected in Angola, namely *O. o. rutila* from Luando Reserve (Statham 1922), and *O. o. leucopus* (Monard 1930). The latter was subsequently synonymised with the former, and considered to extend into Caprivi, Botswana and west Zambia (Ansell 1972; Crawford-Cabral and Veríssimo 2005; Brashares and Arcese 2013). In Angola the species occurs in open savanna habitats above 1000 m, and used to be present in protected areas such as Luando Reserve, Bicuar, Mupa and Cameia NPs, and across south-eastern Angola (Huntley 1973c; Crawford-Cabral and Veríssimo 2005). Two references reporting the presence of oribi on the southern coastal plain (Statham 1922; Fenykovi 1953) are dubious and may result from misidentification of Steenbok (Crawford-Cabral and Veríssimo 2005). Recent surveys have failed to record the species in Mupa and Bicuar (Overton et al. 2017) and in Mucusso (Veríssimo 2008). Anecdotal evidence suggestes its presence in the upper catchments of the Okavango (NGOWP 2018), and they were recorded by camera traps in Luengue-Luiana NP (Funston et al. 2017). Small numbers are still present and have been recently observed and photographed in Luando Reserve and Cameia NP (Vaz Pinto, unpublished data).

Tribe Raphicerini

Only one species of Raphicerini has been confirmed in Angola, the Steenbok (*Raphicerus campestris*), occurring south of 12° latitude, being most common in semi-arid habitats in the coastal plain but also present inland from Huíla to Cuando Cubango provinces (Crawford-Cabral and Veríssimo 2005). The species used to be common and is still present in Iona, Bicuar and Mupa NPs, and in the recently designated Luengue-Luiana and Mavinga NPs (Huntley 1973c; Veríssimo 2008; Funston et al. 2017; Overton et al. 2017). A remarkable record was one specimen collected south of Namibe at Lagoa S. João do Sul, with very long hooves, suggesting an isolated population and local adaptation to muddy terrain (Simões and Crawford-Cabral 1988; Crawford-Cabral and Veríssimo 2005). Recent surveys have extended the species distribution northwards into the upper catchment of the Okavango (NGOWP 2018), adding to records from the Cuito River source and from near Cuemba (Vaz Pinto, unpublished data). Interestingly, witness accounts reported unusual behaviour displayed by the latter Steenbok, also suggesting isolation and local adaptation (Vaz Pinto, unpublished data). Several subspecies have been proposed and Ansell (1972) recognised eight races, but these remain unclear and often only two are accepted, with the nominate subspecies *R. c. campestris* being assigned to all populations in southern Africa (du Toit 2013). A congeneric species, Sharpe's Grysbok (*Raphicerus sharpei*), has never been recorded in Angola but it may well be present in the regions of Cazombo or Luiana, as it is known to occur in western Zambia and east Caprivi, very close to the Angolan border (Ansell 1972; Hoffman and Wilson 2013).

Tribe Reduncini

The tribe is represented in Angola by the genera *Redunca* and *Kobus*, comprising four species in total. Southern Reedbuck (*Redunca arudinum*) had a wide, albeit discontinuous distribution associated with grassy patches near drainage lines, and was present throughout the country except in Cabinda and the arid southwest (Newton da Silva 1970; Huntley 1973c; Crawford-Cabral and Veríssimo 2005). The species was especially common in protected areas such as Quiçama, Cangandala, Cameia and Bicuar NPs, Luando Reserve, and the southeast regions (Huntley 1973c, 1974; Crawford-Cabral and Veríssimo 2005). Although a recent camera-trap survey has failed to record the species in Bicuar and Mupa (Overton et al. 2017), they may still be present, and have been recorded on surveys conducted in Mucusso, Luengue-Luiana and Mavinga (Veríssimo 2008; Chase and Schlossberg 2016; Funston et al. 2017). Even if in much reduced numbers, Reedbuck is also still found in Quiçama NP (Groom et al. 2018), and in Cangandala NP and Luando Strict Reserve (Vaz Pinto, unpublished data). The intraspecific taxonomy of Reedbuck is still unresolved, but often two subspecies are recognised and separated by the Zambezi River (Ansell 1972; Kingdon and Hoffmann 2013), with the Angolan populations corresponding to the typical race *R. a. arundinum*.

The genus *Kobus* comprises the remaining three species of Reduncini present in Angola. Puku (*Kobus vardonii*), is a relatively rare antelope that used to have its westernmost populations in Angola (Jenkins 2013). The species was mostly recorded in northeastern Angola, and a lot of what is known is due to the studies of Machado (1969). Puku used to occur, albeit in low numbers, in Luando Strict

Reserve (Statham 1922; Huntley 1973c), possibly corresponding to an isolated and westernmost subpopulation, but appears now to be absent (Vaz Pinto, unpublished data). An old record reported to have been obtained in Huíla province (du Bocage 1902), is generally dismissed as mistaken (Crawford-Cabral and Veríssimo 2005), and the southernmost record was obtained by Wilhelm Trense in Luiana (Crawford-Cabral and Veríssimo 2005). The species has not been recorded in Angola in over 40 years and it is thus possibly extinct. Southern Lechwe (*Kobus leche*) used to be widely distributed along river drainages in central and eastern Angola, only marginally overlapping with Puku in Luando Reserve and possibly Luiana (Crawford-Cabral and Veríssimo 2005). Angolan Lechwe was suggested by Sokolowski (1903) to constitute a separate species *Adenota* (=*Kobus*) *amboellensis*, but it was subsequently synonymised with the typical race also known as Red Lechwe *K. l. lechwe*. Red Lechwe seems to have been extirpated from a large portion of their former range in Angola, when only a relic population survived in Luando Reserve (Vaz Pinto, unpublished data), but larger populations have recently been recorded in the upper catchments of the Okavango (NGOWP 2018), and in the Luengue-Luiana and Mavinga NPs (Veríssimo 2008; Chase and Schlossberg 2016; Funston et al. 2017).

Waterbuck (*Kobus ellipsiprymnus*) comprises two very distinct subspecies that often are considered to warrant full species status, and both have been recorded in Angola. Common Waterbuck (*K. e. ellipsiprymnus*) is very localised in the country, and only known to occur along the lower Cuando River in Luengue-Luiana NP (Huntley 1973c; Crawford-Cabral and Veríssimo 2005). Current status is unknown, as various surveys failed to record the species (Veríssimo 2008; Chase and Schlossberg 2016), though a recent survey reported finding carcasses (Funston et al. 2017). The other taxon present in Angola is currently ascribed to Defassa Waterbuck (*K. e. defassa*), as previous attempts to recognise geographical variants attributed to Angolan waterbuck such as *K. e. penricei* (Hill and Carter 1941; Ansell 1972) have been abandoned in favour of synonymy with *K. e. defassa*. Defassa Waterbuck used to be widely distributed in Angola but were usually uncommon (Huntley 1973c; Crawford-Cabral and Veríssimo 2005), and in protected areas were only represented in Cangandala, Luando and Bicuar NPs (Statham 1922; Huntley 1973c). A recent survey failed to record the species in Bicuar NP where it is feared extinct (Overton et al. 2017). Nevertheless small numbers are still present across northeasten Angola, including Cangandala and Luando (Vaz Pinto, unpublished data).

Tribe Tragelaphini

Four species of Tragelaphini are known from Angola (Crawford-Cabral and Veríssimo 2005), all currently lumped within the genus *Tragelaphus*: Kudu (*T. strepsiceros*), Bushbuck (*T. scriptus*), Sitatunga (*T. spekii*) and Common Eland (*T. oryx*). Kudus used to be widely distributed in southern Angola, and along the semi-arid coastal plain (Huntley 1973c; Crawford-Cabral and Veríssimo 2005). Although reported previously (Huntley 1972c, 1973c), recent surveys have not recorded Kudu in Iona NP (Kolberg and Kilian 2003; van der Westhuizen et al. 2017), but they are still relatively common in Bicuar and Mupa NPs (Overton et al. 2017) and small numbers have been confirmed in the southeastern Cuando Cubango (Veríssimo 2008; Funston et al. 2017). The distinction among Kudu subspecies has remained dubious due to variability of morphological characteristics and intergradation (Owen-Smith 2013).

Bushbuck in Angola have been ascribed to the subspecies *T. scriptus ornatus* (Crawford-Cabral and Veríssimo 2005), but the species taxonomy is highly problematic. Genetic studies have revealed two highly divergent and non-monophyletic mitochondrial clades which provided support for two species: *Tragelaphus scriptus* and *T. sylvaticus* (Moodley and Bruford 2007; Moodley et al. 2009). This however still lacks confirmation with nuclear and morphological data (Hassanin et al. 2012). Angolan animals could be important to disentangle the phylogenetic relationships within Bushbuck, as they are relatively common and widespread across central and northern Angola, where the boundary between both clades may be found. Bushbuck has remained common in northern Angola, including in protected areas such as Quiçama, Cangandala and Luando (Huntley 1973c; Groom et al. 2018; Vaz Pinto, unpublished data), and features prominently in bushmeat markets (Bersacola et al. 2014; Groom et al. 2018; Vaz Pinto, unpublished data). It was generally uncommon in the southern regions, and recent surveys have failed to detect the species in Mupa and Bicuar NPs (Overton et al. 2017). Small numbers were recorded in the Luengue-Luiana region (Veríssimo 2008; Funston et al. 2017).

Sitatunga had a relatively wide distribution in Angola, following the main river systems but excluding the coastal plain (Huntley 1973c; Crawford-Cabral and Veríssimo 2005). The species is still present in protected areas such as Cangandala and Luando Reserve (Vaz Pinto, unpublished data) or in Luengue-Luiana (Veríssimo 2008; Funston et al. 2017). Additional records have been obtained from northern Malanje and Lunda-Norte (Huntley and Francisco 2015; Vaz Pinto, unpublished data). Traditionally two subspecies of Sitatunga have been recognised in Angola, *T. s. gratus* in northern and central Angola, and *T. s. selousi* within the Zambezi and Okavango drainages (Crawford-Cabral and Veríssimo 2005).

Common Eland used to be relatively widespread in Angola, except in the forest biomes of the northwest (Crawford-Cabral and Veríssimo 2005). It was once abundant in Quiçama NP and in the hunting concessions of southeastern Angola (Newton da Silva 1970; Huntley 1973c; Crawford-Cabral and Veríssimo 2005), while present in low numbers in Iona, Bicuar, Mupa and Cangandala NPs and Luando Reserve (Huntley 1973c; Crawford-Cabral and Veríssimo 2005). The species is now extinct in Quiçama NP (Groom et al. 2018) and Cangandala NP (Vaz Pinto, unpublished data), and has not been recorded in mammal surveys conducted in Iona and Mupa NPs (Kolberg and Kilian 2003; Overton et al. 2017; van der Westhuizen et al. 2017). A small population may still linger in Luando (Vaz Pinto, unpublished data) and Bicuar NP (Overton et al. 2017), and relatively larger numbers in Luengue-Luiana NP (Veríssimo 2008; Funston et al. 2017). Angolan Eland have often been attributed to the subspecies *T. oryx livingstoni* (Crawford-Cabral and Veríssimo 2005; Thouless 2013), but intergrading with the nominate race *T. o. oryx* in southern Angola (Ansell 1972; Crawford-Cabral and Veríssimo 2005; Thouless 2013). It has been argued that Eland of the semi-arid biomes along the coastal plain could be ascribed to the nominate race, while the remaining could be ascribed to *T. o. livingstoni* (Crawford-Cabral and Veríssimo 2005).

Family Giraffidae

Angolan giraffes used to be ascribed to *Giraffa camelopardalis angolensis* (Lydekker 1904; Ciofolo and Le Pendu 2013), although recent molecular studies have proposed a four species classification, under which they would correspond to *G. giraffa angolensis* (Fennessy et al. 2016). Two disjunct populations were known in Angola, one in Cunene and southern Huíla provinces, and the other in Cuando Cubango east of Cuito River (Newton da Silva 1970; Crawford-Cabral and Veríssimo 2005). Once the symbol of Mupa NP, in the early 1970s the giraffe was already on the brink of extinction (Huntley 1973c), and was presumed extinct in Angola by the 1990s (Huntley and Matos 1992; Juste and Carballo 1992). Recent surveys are consistent with the species extinction in Mupa (Overton et al. 2017), but records obtained from spoor and from aerial surveys have demonstrated the persistence of pocket populations in the areas corresponding to the current Luengue-Luiana NP (Veríssimo 2008; Chase and Schlossberg 2016).

Family Hippopotamidae

The sole hippopotamus species in Angola is the Common Hippopotamus (*Hippopotamus amphibius*). This species is relatively widespread, and used to be found across most large rivers and drainage basins in the country (Newton da Silva 1970; Crawford-Cabral and Veríssimo 2005). The hippopotamus in Angola have been ascribed to *H. a. constrictus*, although subspecific taxonomy remains controversial for this species and is often ignored (Klingel 2013). Although once common, the hippopotamus has suffered from human direct persecution. In the 1970s it had become rare in the country and reduced to small pockets in larger rivers, and this situation may have deteriorated since (Huntley 1973c; Huntley and Matos 1992; Juste and Carballo 1992). Although no study has addressed specifically the hippopotamus populations in Angola, recent general surveys have reported its presence in isolated pockets along rivers such as Cuanza, Queve and Luando (Vaz Pinto, unpublished data), and in the middle to lower sections of various rivers within the Okavango and Zambezi basins (Veríssimo 2008; Chase and Schlossberg 2016; Funston et al. 2017; NGOWP 2018). Small populations are still present in Quiçama NP and Luando Strict Reserve (Groom et al. 2018; Vaz Pinto, unpublished data), but based on recent surveys they appear to have disappeared from the southern Cunene system, namely within Iona, Mupa and Bicuar NPs (Kolberg and Kilian 2003; Overton et al. 2017; van der Westhuizen et al. 2017).

Family Suidae

Three species of wild suids occur in Angola, but no studies have yet focused on Angolan pigs. The Bushpig (*Potamochoerus larvatus*) is widespread, and possibly only naturally absent from Cabinda and the arid southwest (Crawford-Cabral and Veríssimo 2005). Data on Bushpig distribution have been reported by several authors (e.g. Statham 1922; Monard 1935; Hill and Carter 1941; Newton da Silva 1970; Huntley 1973c), but mostly refer to scattered and localised records. The species has recently been recorded in various surveys in protected areas (Veríssimo 2008; Chase and Schlossberg 2016; Funston et al. 2017; Overton et al. 2017; Groom et al. 2018; NGOWP 2018; Vaz Pinto, unpublished data). Two subspecies have been recognised in the past in Angola, *P. l. johnstoni* and *P. l. cottoni*, but the species displays a large amount of individual variation and both have been subsequently synonymised with *P. l. koiropotamus* (Grubb 1993; Crawford-Cabral and Veríssimo 2005; Seydack 2013). Formerly treated as conspecific with bushpigs, the Red River Hog (*Potomochoerus porcus*) in Angola is confirmed only in Cabinda, and little is known about its possible occurrence in Lunda's provinces or possible intergradation with the previous species (Crawford-Cabral and Veríssimo 2005; Leus and Vercammen 2013; Seydack 2013). Common Warthog (*Phacochoerus africanus*), were relatively widespread in Angola, and used to be locally common, even if irregularly distributed (Crawford-Cabral and Veríssimo 2005; Huntley 1973c). Recently, it has been regularly recorded in various protected areas (e.g. Funston et al. 2017; Overton et al. 2017; Groom et al. 2018; NGOWP 2018; Vaz Pinto, unpublished data). Angolan warthogs may correspond to the subspecies *P. a. sundervalli* (Cumming 2013), but geographic variation and transition among different forms remains unresolved. Some early authors (Statham 1922; Monard 1935; Varian 1953) have reported the existence of a giant pig in central Angola, based on local witness accounts, which they tentatively ascribed to the giant forest hog (*Hylochoerus meinertzhageni*), but these reports are generally dismissed, and were likely based on tales or misidentification of bushpigs (Newton da Silva 1970; Crawford-Cabral and Veríssimo 2005).

Family Tragulidae

Water Chevrotain (*Hyemoschus aquaticus*) was first confirmed for Angola in Cabinda (Huntley 1973c), although it was also reported to occur in Lunda-Norte along the Cassai River (Crawford-Cabral and Veríssimo 2005). More recent witness accounts suggest the occurrence of the species in forest streams and nearby flooded areas in Uíge Province (Michael Mills, personal communication). It is possible that the Water Chevrotain has a wider distribution across northern Angola in forest habitats, but the elusive nature of the species and lack of studies prevent any conclusions at this stage.

Chiroptera (Bats)

With 73 species recorded, bats (Chiroptera) are the second most speciose order of mammals in Angola after the rodents. This represents about one third of the bat species known to occur in mainland Africa (Happold and Happold 2013), and almost two thirds of the species reported for the southern Africa region (Monadjem et al. 2010a). Angola is therefore a particulary rich country for bats in southern Africa, compared with other species-rich countries such as Mozambique (67 species), Zambia (65), Malawi (62) and Zimbabwe (62) (Monadjem et al. 2010b). Considering that Angola is one of the least known southern African countries with respect to bats, more species are likely to be found as new surveys are carried out throughout the country. For instance, three of the 73 species reported here have been found recently during the course of just a few weeks of fieldwork (Taylor et al. 2018c). Even new (and possibly endemic) species may be found, particularly on the western escarpment region and Afromontane forests that are high in bird endemism, and in species with restricted and fragmented distributions (Mills et al. 2011, 2013).

The inventory of Angolan bats started in the nineteenth century (e.g. Peters 1870; du Bocage 1889a, 1898; Seabra 1898a, b). A substantial number of records was added later by various authors including Thomas (1904), Hill and Carter (1941), Sanborn (1950), Hayman (1963) and finally by Crawford-Cabral (1986), who also reviewed all the information then available. Little research on bats has been done in Angola in the past few decades, and to our knowledge no studies of bat ecology have been carried out in the country. There are important taxonomic issues to be resolved in most families, and the existing records for the majority of species are old and scarce, so the descriptions of bat ranges in the following sections should be considered as provisional. Also, no recent bat surveys have been conducted in any of the remaining Afromontane forests and adjacent grasslands along the escarpment, and this must remain a critical zone for future surveys.

Family Emballonuridae

There are three species of Emballonuridae in Angola (Monadjem et al. 2010a; Happold and Happold 2013). The Mauritian Tomb Bat (*Taphozous mauritianus*) occurs widely in a variety of habitats and often roosts in buildings inhabited by people, but in Angola it is only known from scattered records. African Sheath-tailed Bat (*Coleura afra*) is a rare species in southern Africa, which in Angola is known from just two records in the coastal area near Benguela, but this species may have been overlooked. Giant Pouched Bat (*Saccolaimus peli*) is a species mainly found in the Rainforest Belt Zone of Africa, with an isolated record in eastern Angola, but it may have been overlooked in much of the northern region of the country.

Family Hipposideridae

There are five species of Hipposideridae recorded in Angola (Monadjem et al. 2010a; Happold and Happold 2013). One of these is the African Trident Bat (*Triaenops afer*), which is now included in the separate family Rhinonycteridae (Foley et al. 2015). This is a tree-roosting species with a widespread, albeit patchy distribution, with the isolated Congolese population encompassing Cabinda and the extreme northwest of the country corresponding to the subspecies *T. a. majusculus* Allen et Brosset, 1968, though this is not widely recognised (Benda and Vallo 2009). The Sundevall's Leaf-nosed Bat (*Hipposideros caffer*) is also wide ranging, but represents a complex of species and requires urgent taxonomic revision (Vallo et al. 2008). Most existing records are from western Angola, where the populations presumably belong to the form *H. caffer angolensis* Seabra, 1898 (which is currently placed in the species *H. ruber* - see below), but the species also seems to be present in the extreme east of the country, representing an edge of a much larger range in Central and East Africa (Kock et al. 2008). The Giant Leaf-nosed Bat (*Macronycteris gigas*) is mostly a lowland forest and moist savanna species, which in Angola occurs in Cabinda and penetrates southwards along the northern section of the escarpment (Monadjem et al. 2010a). The type locality of this species is in Benguela, in semi-arid savanna that is atypical regarding its habitat requirements. This species is difficult to separate from Striped Leaf-nosed Bat (*M. vittatus*), making it problematic to ascertain its actual distribution. Noack's Leaf-nosed Bat (*H. ruber*) has similar habitat affinities but may be widespread in the country. However, due to previous confusion of this species with *H. caffer* means that the distribution of *H. ruber* remains mostly unknown in Angola, which itself represents a complex of multiple species (Monadjem et al. 2013a). Finally, Striped Leaf-nosed Bat is a species forming very large cave roosting colonies, which is judged to occur mainly in southern Angola.

Family Miniopteridae

The Natal Clinging Bat (*Miniopterus natalensis*) is the only Miniopteridae judged at present to occur in Angola (Monadjem et al. 2010a; Happold and Happold 2013). However, this is a species distributed mostly in East and southern Africa, with a handful of records in western and southern Angola (e.g., Grant and Ferguson 2018; MHNG 2018). The population of Namibia and Angola appears to be largely isolated from eastern populations, but they do not appear to be phylogenetically distinct (Monadjem et al. 2013b). Some specimens collected in 1954 in the Huíla province have been identified as Black Clinging Bat (*Miniopterus fraterculus*), but this is a great distance from the known distribution of the species in the eastern parts of South Africa, suggesting that they are probably misidentifications (Monadjem et al. 2013b). The Greater Long-fingered Bat (*Miniopterus inflatus*) may occur in Angola and has probably been overlooked there (Monadjem et al. 2010a).

Family Molossidae

The Molossidae is represented by ten species in Angola. Many of these are known from only a few and scattered records in the country (Monadjem et al. 2010a), making it difficult to recognise distribution patterns. The White-bellied Free-tailed Bat (*Mops niveiventer*) seems to be one of the most widespread species. It is mostly associated with mature miombo woodland and occurs widely in Central Africa. The similar Angolan Free-tailed Bat (*M. condylurus*) is widespread in sub-Saharan Africa, but its distribution in Angola is poorly known. Some Molossidae seem to have rather isolated populations in Angola (Monadjem et al. 2010a), which may prove to be phylogenetically distinctive. Ansorge's Free-tailed Bat (*Chaerephon ansorgei*) is a species of dry woodland savanna, with a restricted population in western Angola that is isolated from the remaining known range of the species in the eastern side of southern Africa. Little Free-tailed Bat (*C. pumilus*) occurs widely in sub-Saharan Africa and is widespread and abundant in eastern parts of southern Africa, but in the west seems to be restricted to an isolated population in northwestern Angola and neighbouring DRC. Pale Free-tailed Bat (*C. chapini*) is sparsely distributed in southern Africa, occurring mainly in northern Botswana and northeastern Zimbabwe and Zambia, with records in northern Namibia, western Angola and DRC representing an isolated population.

The other Molossidae occur marginally or have only scattered records in Angola, which may reflect environmental constraints but may also be due to poor survey efforts. Roberts's Flat-headed Bat (*Sauromys petrophilus*) is an arid zone specialist for which there is an old record from Moçamedes (Crawford-Cabral 1986). Egyptian Free-tailed Bat (*Tadarida aegyptiaca*) is widespread in southern Africa, but it occurs only marginally in the south of Angola. The range of Midas Free-tailed Bat (*M. midas*) also seems to reach the south of Angola, where it was tentatively detected accoustically in a recent survey (Taylor et al. 2018c). In southern Africa the Nigerian Free-tailed Bat (*C. nigeriae*) occurs in northwestern Namibia, northern Botswana, Zimbabwe, Zambia and marginally in Angola and the DRC, though modelling of the environmental niche suggests that the species may have been overlooked in the country (Monadjem et al. 2010a). The Large-eared Free-tailed Bat (*Otomops martiensseni*) has a localised distribution in southern Africa, with scattered records from Angola, Zimbabwe, Zambia, Malawi and the DRC.

Family Nycteridae

All but one of the six Nycteridae species recorded in Angola have distributions centered in Central and/or Western Africa and the country is at the southern edge of their distribution (Monadjem et al. 2010a; Happold and Happold 2013). Dwarf Slit-faced Bat (*Nycteris nana*), Intermediate Slit-faced Bat (*N. intermedia*), and Bate's Slit-faced Bat (*N. arge*) are mostly associated with lowland rainforests with very few records in Angola, where they are probably restricted to the northern regions. Large-eared Slit-faced Bat (*N. macrotis*) is thought to prefer savannas, and although the known records are all in northern Angola its range in Zambia and Botswana suggests that it is also present further south. Hairy Slit-faced Bat (*N. hispida*) is a species widespread in sub-Saharan Africa that uses a variety of habitats, which occurs throughout central and northern Angola. Egyptian Slit-faced Bat (*Nycteris thebaica*) is a savanna species with wide habitat tolerance and very widespread in southern Africa, though most records in Angola were obtained in the southwest (Monadjem et al. 2010a).

Family Pteropodidae

A total of 15 species of pteropid fruit bats has been reported in Angola (Monadjem et al. 2010a; Happold and Happold 2013), although the actual number of species present is likely to be greater than this. Further work may reveal the presence of additional species, particularly in the poorly surveyed Maiombe rainforest and savanna-forest mosaics of Cabinda. It is interesting to note that several of the currently recognised pteropid species have a type locality in the country. This is the case in the Angolan Soft-furred Fruit Bat (*Myonycteris angolensis*), Anchieta's Fruit Bat (*Plerotes anchietae*), Dobson's Epauletted Fruit Bat (*Epomops dobsonii*), Lesser Angolan Epauletted Fruit Bat (*Epomophorus grandis*), Hayman's Dwarf Epauletted Fruit Bat (*Micropteropus intermedius*), and Angolan Epauletted Fruit Bat (*E. angolensis*). The latter species is a near endemic, present only in Angola and northern Namibia, while Hayman's Dwarf and Lesser Angolan Epauletted Fruit Bats are only known from Angola and Congo.

The information available on the range of pteropids in Angola is insufficient to make a definitive identification of distribution patterns, but some can be provisionally suggested. Overall, species diversity declines from north to south, because most species of pteropids are dependent on well-wooded habitats and abundant fruit resources. In fact, a few rainforest species that have the centre of their range in Congo are known only in Cabinda and in some of the large pockets of moist forests in the northern provinces. Species displaying this pattern include the Woermann's Long-tongued Fruit Bat (*Megaloglossus woermanni*), Franquet's Epauletted Fruit Bat (*E. franqueti*), Angolan Soft-furred Fruit Bat, Hammer-headed Fruit Bat (*Hypsignathus monstrosus*), and Little Collared Fruit Bat (*Myonycteris torquata*). Existing data suggest that at least some of these species extend their range southward along the narrow band of forests of the Angolan Escarpment. Other species associated with moist tropical forests and savannas occur more widely, extending their range into northern Angola, as it is the case with Little Collared Fruit Bat and Peter's Dwarf Epauletted Fruit Bat (*Micropteropus pusillus*).

The most abundant pteropid in the species-poor south of Angola seems to be the Angolan Epauletted Fruit Bat. It has a broad latitudinal range, although all the records are in the western half of the country. In the drier regions it may be mostly dependent on riverine woodlands. Peters's Epauletted Fruit Bat (*E. crypturus*) has been captured in one locality in the south of Angola but its range in the neighbouring countries suggests that it may have a broad distribution in the south and east of

Angola, where only limited surveys have been conducted (Crawford-Cabral 1986). Wahlberg's Epauletted Fruit Bat (*E. wahlbergi*) and Straw-coloured Fruit Bat (*Eidolon helvum*) have not been recorded in the drier areas of southern Angola, but are present in the rest of the country. The latter occurs in Namibia and is known to make long migrations tracking fruiting patterns, so it may also occur in much of southern Angola, albeit only seasonally. Two species are only known from the central part of the country and may be associated with its highlands. This is the case of Dobson's Epauletted Fruit Bat and Anchieta's Fruit Bat, a rare or localised species. Finally, Egyptian Rousette (*Rousettus aegyptiacus*) is so far only known in the northwest of the country, which may be explained by the need to satisfy its requirements for abundant fruit resources and caves (Crawford-Cabral 1986).

Family Rhinolophidae

There are five species of Rhinolophidae currently recognised to occur in Angola (Monadjem et al. 2010a; Happold and Happold 2013). Rüppell's Horseshoe Bat (*Rhinolophus fumigatus*) is a savanna species with separate populations in the west and east of southern Africa, with the western population occupying southwest Angola and central and northern Namibia. Lander's Horseshoe Bat (*R. landeri sensu lato*) was thought until recently to have an isolated population in Angola. However, recent taxonomic work recognised populations from southern Africa as a separate species, Peters's Horseshoe Bat (*R. lobatus*), though no material from Angola was examined (Taylor et al. 2018b). Given the proximity to populations analysed it may be provisionally accepted that Peters's Horseshoe Bat is the species occurring in Angola, though it is possible that the country hosts a different species, described by Seabra (1898b) as Angola's Horseshoe Bat (*R. angolensis*) (Monadjem et al. 2010a; Taylor et al. 2018b). Damara Horseshoe Bat (*R. damarensis*) was previously treated as a subspecies of Darling's Horseshoe Bat (*R. darlingi*) (Monadjem et al. 2010a), but has been given full species status in the IUCN Red List (Monadjem et al. 2017). It is a species from arid habitats that is restricted to southwest Angola and western Namibia. Dent's Horseshoe Bat (*Rhinolophus denti*) is only known from the Ruacaná falls, on the Angola-Namibia border (Crawford-Cabral 1986), though it is widespread in Namibia and western Botswana. Eloquent Horseshoe Bat (*Rhinolophus eloquens*) has been collected from Jau, Huíla Province (Grant and Ferguson 2018), a location that is over 2000 km away from the closest records in the eastern DRC and Rwanda. These specimens from the American Museum of Natural History are worth re-examining as they may refer to a new species within the *R. eloquens*/*R. hildebrandtii* group. Further surveys are likely to record more species of this genus that are known to occur in the neighbouring countries close to the border with Angola.

Family Vespertilionidae

There are 28 species of Vespertilionidae recorded in Angola. This is a poorly known group in the country, with many species represented by just one or a few scattered and old records (Monadjem et al. 2010a). This is for instance the case of Beatrix's Butterfly Bat (*Glauconycteris beatrix*), Lesser Woolly Bat (*Kerivoula lanosa*), Dobson's Pipistrelle (*Neoromicia grandidieri*), White-bellied House Bat (*Scotophilus leucogaster*), Green House Bat (*S. viridis*), and Moloney's Flat-headed Bat (*Mimetillus moloneyi*), which are known in Angola from just one or two records. Seemingly, Silvered Woolly Bat (*K. argentata*) and Dusky Pipistrelle (*Pipistrellus hesperidus*) have been tentatively recorded at single locations based on acoustic identifications during recent surveys (Taylor et al. 2018c). Even species that are likely to be widespread and abundant in the region, and that are known to roost in houses in relatively large numbers, such as the Cape Serotine (*Neoromicia capensis*), have been scarcely recorded in the country possibly due to under-sampling.

Despite the paucity of studies, it is noteworthy that five currently recognised Vespertilionidae species have their type locality in Angola. This is the case of Anchieta's Pipistrelle (*Hypsugo anchietae*), which was collected in western Angola (Cahata) in atypical environmental conditions compared to the main range of the species, though it may be widespread in the southeast. Angolan Long-eared Bat (*Laephotis angolensis*) is an endemic to southern Africa that is known from just four specimens collected at one locality in DRC and two in central Angola, including the type locality (Tyihumbwe). Bocage's Mouse-eared Bat (*Myotis bocagii*) is widespread in tropical Africa, with just a few records in northern Angola, including the type locality. Welwitsch's Mouse-eared Bat (*M. welwitschii*) was described from northern Angola, although the great majority of the species' records are in East Africa. Angolan Hairy Bat (*Cistugo seabrae*) was described from specimens collected in Moçamedes. It is endemic to southern Africa, where it occurs from the extreme southwest of Angola through western Namibia to the extreme northwest of South Africa. Finally, it is worth mentioning Machado's Butterfly Bat (*Glauconycteris machadoi*), a species described from eastern Angola that is not retained here because it has been suggested that it may correspond to a colour phase of *G. variegata*, although this issue needs clarification (Crawford-Cabral 1986).

For several Vespertilionidae Angola is at the edge of their distribution, and so they occur marginally in the country, which may justify at least partly the scarcity of records. This is the case of species associated with rainforests and other forests of tropical Africa, which are largely restricted to Cabinda and northern Angola. Besides the above mentioned Bocage's Mouse-eared Bat, the species showing this pattern are Common Butterfly Bat (*G. argentata*), Beatrix's Butterfly Bat, White-winged Serotine (*Neoromicia tenuipinnis*), and Broad-headed Pipistrelle (*Hypsugo crassulus*). Other species have their core range in eastern and southern Africa and have isolate records in Angola, including Lesser Woolly Bat, Green House Bat, Damara Woolly Bat, and Dusky Pipistrelle. It is possible that these patterns are shaped to at least some extent by under-sampling in Angola. Finally, Long-tailed Serotine (*Eptesicus hottentotus*) is only known from southwest Angola, but it has a much wider, albeit sparse distribution in southern Africa (Monadjem et al. 2010a).

Most Vespertilionidae from woodlands and savannas have also been scarcely recorded in Angola, though climatic niche modelling suggests that many savanna species may occur more widely in the south of the country (Monadjem et al. 2010a). In contrast, only Thomas's Flat-headed Bat (*M. thomasi*) has been associated with the moist miombo belt of southcentral Africa, which occupies a large part of central Angola. Banana Bat (*Neoromicia nana*) is associated with well-wooded habitats such as riparian vegetation and forest patches, and it has been recorded in western and central Angola. Savanna and open woodland species, many of which are associated with aquatic and riparian habitats, include Variegated Butterfly Bat (*G. variegata*), Botswana Long-eared Bat (*Laephotis botswanae*), Zulu Serotine (*Neoromicia zuluensis*), Schlieffen's Twilight Bat (*Nycticeinops schlieffeni*), Rüppell's Pipistrelle (*Pipistrellus rueppellii*), Rusty Pipistrelle (*P. rusticus*), Thomas's House Bat (*Scotoecus hindei*), Yellow-bellied House Bat (*Scotophilus dinganii*), and White-bellied House Bat. The inclusion of Thomas's House Bat is tentative, because the taxonomy of this genus is still unresolved and confusion with White-bellied Lesser House Bat (*S. albigula*) is possible (Monadjem et al. 2010a). Specimens of Yellow-bellied House Bat from northern Angola, the southern DRC and northern Zambia were collected in environmental conditions considered unsuitable, and may represent a distinct species.

Erinaceopmorpha (Hedgehogs)

The Southern African Hedgehog (*Atelerix frontalis*) is the sole representative of the order Erinaceopmorpha in Angola, with a distribution restricted to the southwest of the country, in the Namibe and Huíla provinces, which extends into central Namibia (Cassola 2016b). This population is disjunct from another occurring in Zimbabwe, Botswana and South Africa (Cassola 2016b).

Hyracoidea (Hyraxes)

Four species of Hyracoidea are known from Angola. The Angolan Bush Hyrax (*Heterohyrax brucei bocagei*) is widely distributed in rocky habitats across the western half of the country, and is mostly a plateau species absent below 500 m of altitude (Gray 1869; du Bocage 1889b; Crawford-Cabral and Veríssimo 2005). Nevertheless, the Bush Hyrax may be sympatric just below the southern escarpment with the Kaokoveld Rock Hyrax (*Procavia capensis welwitschii*), the latter being mostly associated with rock outcrops in the southwest arid coastal plain (du Bocage 1889b; Hill and Carter 1941; Crawford-Cabral and Veríssimo 2005). In addition, two species of Tree Hyrax, the Southern Tree Hyrax (*Dendrohyrax arboreus* cf *braueri*) and the Western Tree Hyrax (*Dendrohyrax dorsalis nigricans*) are present in forest habitats, the former associated with gallery forests and nearby miombo in eastern Angola (Machado 1969; Crawford-Cabral and Veríssimo 2005). The

Western Tree Hyrax is only known from one confirmed record obtained in gallery forest in Cabinda (Peters 1879; Crawford-Cabral and Veríssimo 2005), but recent observations have suggested that the species' may be present on the central escarpment region (Vaz Pinto, unpublished data). The Angolan Bush Hyrax is likely an endemic taxon and the Kaokoveld Rock Hyrax has a global distribution restricted to the Namib Desert between coastal Namibia and Angola, but to date no study has specifically addressed Angolan hyraxes.

Lagomorpha (Hares)

A recent review recognises only two species of lagomorphs in Angola (Smith et al. 2018), but taxonomic uncertainties, controversial identifications and lack of recent studies in the country, have blurred the matter and it is likely that at least three species occur. The African Savanna Hare (*Lepus victoriae*) is the most widespread species, occurring throughout the country except the arid southwest and the forested northwest (Smith et al. 2018). There are also records of additional *Lepus* species, but at least some of these may refer to misidentifications and thus need further consideration. Genest-Villard (1969) recorded the presence of two species of hares based on specimens collected by Crawford-Cabral, with Cape Hare (*Lepus capensis*) occuring in semi-desertic areas and *L. crawshayi* (= *L. victoriae*) in less arid areas. Actually, hares collected in Moçâmedes were initially classified as a distinct species, *L. salai* Jentink 1880, and then ascribed to a subspecies of Cape Hare (*L. c. salai*) occurring in subdesertic areas of southwestern Angola and western Namibia (Petter and Genest 1965). In the mammal collection of IICA, currently housed in ISCED-Huíla, various skins from coastal Namibe were identified as *L. capensis* (Crawford-Cabral and Veríssimo, unpublished data), and recent field observations suggest the Cape Hare to be common in Iona NP and along the southwestern arid coastal plain into Benguela province (Vaz Pinto, unpublished data). A recent record of Cape Scrub Hare (*Lepus saxatilis*) in the Malanje province by Moraes and Putzke (2014) is probably a misidentification, as the species is a narrow endemic to the Cape province of South Africa (Smith et al. 2018). Jameson's Red Rock Hare (*Pronolagus randensis*) occurs in southern Angola, corresponding to the northern tip of a larger range extending into central Namibia, which is disjunct from another population with its core distribution in Zimbabwe, eastern Botswana and northeast South Africa (Smith et al. 2018). Interestingly, the Rock Hare in Angola appears to be present in two subpopulations segregated by altitude. One subpopulation is present in altitude on top of the escarpment, often above 2000 m in Huíla Province, and has been tentatively ascribed by Crawford-Cabral to (*P. r. waterbergensis*) (Crawford-Cabral and Veríssimo, unpublished data). The other subpopulation is found on rocky outcrops and inselbergs on the coastal plain, and distinct coloration patterns led the same author to suggest a new taxon (*P. r. moçamedensis*) (Crawford-Cabral and Veríssimo, unpublished data). No subspecies of Red Rock Hare are currently recognised (Happold 2013), but the Angolan populations and their distribution pattern remain unresolved. There have also been claims suggesting the occurrence of Bunyoro Rabbit (*Poelagus marjorita*) in Angola (Petter 1972), but these were subsequently challenged and dismissed in recent publications, which consider the species to be restricted to relatively small disjunct populations in the Central African Republic, South Sudan, and Uganda (Happold and Wendelen 2006; Happold 2013; Smith et al. 2018). Nevertheless, a couple of museum specimens collected in 1941 on the Angolan Escarpment near Gabela and attributed to this species (RBINS 2017) are intriguing and require further verification, particularly as the region is known as an endemism hotspot (Hall 1960; Happold and Wendelen 2006; Clark et al. 2011; Svensson et al. 2017).

Macroscelidea (Sengis)

Three species of Macroscelideae have been collected in Angola (Crawford-Cabral and Veríssimo, unpublished data). Short-snouted Sengi (*Elephantulus brachyrhynchus*) is reportedly distributed across the plateau, associated with grasslands and thickets in miombo woodlands (Monard 1935; Hill and Carter 1941; Crawford-Cabral and Veríssimo, unpublished data). The species is possibly represented by three subspecies in Angola, *E. b. brachyrhynchus*, *E. b. brachyurus* and *E. b. schinzi* (Crawford-Cabral and Veríssimo, unpublished data), but a clinal variation has also been suggested and Perrin (2013) recognises no subspecies. Bushveld Sengi (*Elephantulus intufi*) is present in the semi-arid savannas of the southwest and often associated with mopane woodlands, with the subspecies *E. i. mossamedensis* occurring on the coastal plain, while *E. i. alexandri* is found inland (Hill and Carter 1941; Crawford-Cabral and Veríssimo, unpublished data). Four-toed Sengi (*Petrodromus tetradactylus tordayi*) has been collected from gallery forests in Lunda-Norte, and it is likely that the nominate race *P. t. tetradactylus* also occurs in eastern Moxico (Hayman 1963; Crawford-Cabral and Veríssimo, unpublished data; Rathbun 2013). As no studies have focused on Angolan sengis, and considering the known range of sengis in neighbouring countries, it is possible that future surveys may add new species to the country list, such as the Western Rock Sengi (*Elephantulus rupestris*) in the arid southwest region.

Perissodactyla (Rhinoceros, Zebras)

The Order Perissodactyla is represented in Angola by Black Rhinoceros (*Diceros bicornis*), and two species of zebra. Although White Rhinoceros (*Ceratotherium simum*) has occasionally been included on the Angolan mammal lists, this is not supported by any hard data. Claims referring White Rhinoceros to having been present in the southeastern regions were not based in any collected specimens, but rather on poor and indirect evidence or old unsubstantiated witness reports (Shortridge

1934; Newton da Silva 1970; Crawford-Cabral and Veríssimo 2005). Most authors agree that even if they had once occurred, which is possible but remains unproven, by the mid-twentieth century they had long been extirtpated (Hill and Carter 1941; Newton da Silva 1970; Crawford-Cabral and Veríssimo 2005).

Family Equidae

The zebras native to Angola are Hartmann's Mountain Zebra (*Equus zebra hartmannae*) and Plains Zebra (*Equus quagga*). A zebra collected in the nineteenth century by Penrice 70 km north of Moçamedes led Oldfield Thomas to describe a new species of mountain zebra (Thomas 1900) as *E. penricei*, but these were later revaluated and synonymised with Hartmann's Zebra (Hill and Carter 1941; Crawford-Cabral and Veríssimo 2005). In Angola, Hartmann's Zebras were found in the arid southwest, mostly in Namibe Province but its distribution may have once extended north to southern Benguela (Shortridge 1934; Crawford-Cabral and Veríssimo 2005). The stronghold of the species in Angola was in Iona NP where a healthy population was present in the 1970s (Huntley 1973c), although 20 years later they were on the verge of extinction and their status remained unaltered until the end of the war (Huntley and Matos 1992; Novellie et al. 2002; Crawford-Cabral and Veríssimo 2005; Penzhorn 2013). Nevertheless a few herds survived the war in Iona NP, and aerial surveys in the park have reported an increase in estimated numbers from 263 to 434 individuals between 2003 and 2016 (Kolberg and Kilian 2003; van der Westhuizen et al. 2017). On the other hand, recent observations and circumstantial evidence strongly suggest that some herds in Iona NP may have been hybridizing with feral donkeys, as various individuals with intermediate phenotypes have been photographed in recent years (Vaz Pinto, unpublished data). Genetic confirmation of the hybridisation is lacking and the extent of the phenomenon remains unknown.

Angolan Plains Zebras have been tentatively ascribed to various subspecies, however recent molecular studies failed to distinguish among traditional subspecies (Lorenzen et al. 2008). Plains Zebra used to be relatively common and widespread in Angola across the southern half of the country and possibly also present in eastern Moxico (Newton da Silva 1970; Crawford-Cabral and Veríssimo 2005). A sixteenth century report has even suggested that zebras in those days may have extended along the coast to Ambriz, but the claim remains controversial (Crawford-Cabral and Veríssimo 2005). In the 1970s they were still numerous at least in protected areas such as Iona NP and Bicuar NP, and in the Cuando Cubango province (Huntley 1973c, 1974; Crawford-Cabral and Veríssimo 2005). It is likely that Plains Zebra were more affected by war-time poaching than most other large ungulates, and already by 1992 they were feared extinct in Angola (Huntley and Matos 1992). Some surveys after the end of the war have failed to record the species in southern Angola (Kolberg and Kilian 2003; Veríssimo 2008; van der Westhuizen et al. 2017), but recent reports suggest that a few animals may still linger in Bicuar NP (Overton et al. 2017). Small numbers were confirmed in general surveys conducted in Luengue-Luiana NP (Chase and Schlossberg 2016; Funston et al. 2017).

Family Rhinocerotidae

Black Rhinoceros were known in Angola from two disjunct populations probably corresponding to different subspecies, although the distinctions among black rhino subspecies remain controversial (Crawford-Cabral and Veríssimo 2005; Rookmaaker 2005; Emslie and Adcock 2013). A population of arid-adapted rhinos occurring in Angola to the west of the Cubango (Okavango) river, has been ascribed to *D. b. minor* (Ansell 1972; Crawford-Cabral and Veríssimo 2005), but it is now generally recognised to represent instead the former northern limit for the typical race *D. b. bicornis* (Emslie and Brooks 1999; Emslie and Adcock 2013). The other population used to extend to the east of the Cuito River in southeastern Angola, and was once considered as *D. b. chobiensis* (Ansell 1972; Crawford-Cabral and Veríssimo 2005). However, this putative race has been more often synonymised with *D. b. minor* (Emslie and Brooks 1999; Rookmaaker 2005; Emslie and Adcock 2013).

No research has specifically focused on Angolan rhinos, and the existing knowledge is based in the few specimens collected during early expeditions, scattered reports from trophy hunters and the work by ecologists in the 1970s (Hill and Carter 1941; Newton da Silva 1970; Huntley 1973c, 1974; Crawford-Cabral and Veríssimo 2005). Black Rhinoceros were likely always scarce in numbers throughout historical times (Huntley 1973c, 1974; Crawford-Cabral and Veríssimo 2005), and in the 1970s they were estimated at around 30 in Iona NP with small populations in southern Cuando Cubango (Huntley 1973c). The situation deteriorated fast during the armed conflict that followed independence, and by 1992 they were already gone or on the verge of extinction (Huntley and Matos 1992; Crawford-Cabral and Veríssimo 2005). By the turn of the millennium rhinos were considered extinct in Angola (Emslie and Brooks 1999), and have remained since, in spite of occasional unconfirmed sightings that suggest the possibility of a few scattered individuals surviving in remote locations. Recent general surveys in regions where they used to occur have consistently failed to record the species (e.g. Veríssimo 2008; Chase and Schlossberg 2016; Funston et al. 2017; Overton et al. 2017; NGOWP 2018).

Pholidota (Pangolins)

Three species of pangolins occur in Angola (Crawford-Cabral and Veríssimo, unpublished data), but no specific studies have been conducted regarding the status or ecology of these species within the country. White-bellied Pangolin (*Phataginus tricuspis*) has been recorded, and collected (Hill and Carter 1941; Trense 1959) in the provinces of Cabinda, Lunda-Norte, Zaire and Cuanza-Norte, and although assumed overall scarce, this species is expected to be widely distributed across northern Angola and the upper plateau, and likely extending south along the escarpment forests (Crawford-Cabral and Veríssimo, unpublished data). The species has also been recorded recently in Cangandala NP, Malanje Province (Vaz Pinto, unpublished data), and it has been found in Angolan bushmeat markets (Svensson et al. 2014b). Temminck's Ground Pangolin (*Smutsia temminckii*) is the most common species of pangolin in Angola, with a wide distribution in the central and southern areas of the country. It has been recorded in the provinces of Cuanza-Sul, Benguela, Bié, Huíla and adjacent areas of Namibe, and Cuando Cubango (Crawford-Cabral and Veríssimo, unpublished data), although the only known material available in natural history collections is housed in the AMNH (Hill and Carter 1941) and ISCED (Crawford-Cabral and Veríssimo, unpublished data). Giant Ground Pangolin (*Smutsia gigantea*) is restricted to the forests of Cabinda, and its occurrence there has been known since the mid-1970s (Huntley 1973e). More recently, the species has been re-confirmed in the Maiombe region of the enclave (Ron 2005).

Primates (Monkeys, Apes, Pottos, Galagos)

Angola has a great diversity of primates, including up to 15 diurnal and possibly 8 nocturnal species. The vast majority of Angola's primate species are found in the rainforests and riverine forest-woodland mosaics in the north, including Cabinda, within the biodiversity-rich Guinea-Congolian biome (Huntley 1973e; Kuedikuenda and Xavier 2009; IUCN 2018). The montane forests/dry woodland mosaics of the Angolan Escarpment along the coast, which connect to the rainforests in the north, are also rich in primates, including at least four diurnal and four nocturnal species (Bersacola et al. 2015).

Family Cercopithecidae

The Cercopithecidae is the most speciose group of primates occurring in Angola, including two baboons, one mangabey, six guenons, two talapoin monkeys, and one colobus. The two baboons are Chacma Baboon (*Papio ursinus*), ranging in the arid southwestern regions (Benguela, Namibe, Huíla and Cunene Provinces), and Kinda Baboon (*P. kindae*), occupying the central and northeast of the country (Machado 1969). Chacma Baboon is adapted to woodland, sub-desert, savanna and montane habitats, whereas the Kinda Baboon tends to be present in miombo woodland, deciduous and semi-deciduous forests, savanna, gallery and riverine forests habitats (Kingdon 2016; Rowe and Myers 2016). Up-to-date information on baboon distribution in Angola is not available. Black Crested Mangabey (*Lophocebus aterrimus*) inhabit the tropical forests south of the River Congo. The subspecies occurring in Angola, the Southern Black Crested Mangabey (*L. a. opdenboschi*) was recorded in the gallery forests of Lunda-Norte (Machado 1969) and just on the eastern side of the Cuango River in the Lunda-Norte province (Hart et al. 2008). Little is known about the current status of this subspecies (Hart et al. 2008; Rowe and Myers 2016).

Red-tailed Monkey (*Cercopithecus ascanius*) is represented by three subspecies in Angola, namely Black-nosed Red-tailed Monkey (*Cercopithecus a. atrinasus*), Black-cheeked Red-tailed Monkey (*C. a. ascanius*), and Katanga Red-tailed Monkey (*C. a. katangae*), all ranging in the north of the country (Machado 1969;

Oates et al. 2008a, b; Rowe and Myers 2016). The existence of Black-nosed Red-tailed Monkey in Angola is based on only nine individuals collected in the 1960s (Machado 1969; Sarmiento et al. 2001; Oates et al. 2008b). Red-tailed Monkeys are typically found in a wide range of habitats, including rainforests, swamp, riverine and montane forests, and deciduous and semi-deciduous forests (Sarmiento et al. 2001; Rowe and Myers 2016). All three subspecies of Red-tailed Monkey have been known to hybridize among each other in Angola (Machado 1969; Detwiler et al. 2005). Blue Monkey (*Cercopithecus mitis*) occurs in two disjunct populations in Angola, corresponding to two subspecies. The Pluto Monkey (*Cercopithecus mitis mitis*) is endemic to Angola and ranges along the Angolan Escarpment areas in the west part of the country (Machado and Crawford-Cabral 1999; Kingdon 2008a; Lawes et al. 2013). In 2013, this was the second most commonly occurring species in a bushmeat survey (Bersacola et al. 2014). The Rump-spotted Blue Monkey (*C. mitis opisthostictus*), ranges in a small part of eastern Angola, corresponding to the edge of a much wider distribution in East Africa (Kingdon 2008b). Black-footed Crowned Monkey (*Cercopithecus pogonias nigripes*) supposedly occurs in the Cabinda Province, likely occurring in the Maiombe NP. The species uses high vegetation strata, mainly occurring in primary and lowland tropical forests, as well as savanna, gallery forests, mature secondary forests and montane forests (Zinner et al. 2013). Moustached Monkey (*Cercopithecus cephus*) was common in Cabinda (Machado 1969; Huntley 1973e). The species is mainly found in lowland tropical rainforest, but also in secondary habitats (Gautier-Hion et al. 1999). Machado (1969) also recorded Putty-nosed Monkey (*C. nictitans*) in Cabinda, a species found in lowland and montane tropical moist forests, as well as gallery and secondary forests (Oates and Groves 2008). De Brazza's Monkey (*C. neglectus*) was common in Lunda-Norte (Machado 1969). The species is mainly found in riverine forest habitats, in lowland and submontane semi-deciduous or tropical moist forest, as well as in swamp forest (Struhsaker et al. 2008). The Malbrouck Monkey (*Chlorocebus cynosuros*) occurs throughout Angola, in various habitats and elevation gradients (Huntley 1973c; Sarmiento 2013). This species is mainly present in open woodland, savanna and forest-grassland mosaic, and tends to occur close to water sources (Butynski 2008; Sarmiento 2013). However, it is also able to occupy both rural and urban environments (Butynski 2008).

The two Talapoin Monkeys of Angola are Southern Talapoin Monkey (*Miopithecus talapoin*) and Northern Talapoin Monkey (*M. ogouensis*). The first species occurs along the Angolan Escarpment, including Quiçama, Cumbira Forest, up to the Congo River (Machado 1969; Gautier-Hion 2013a; Groom et al. 2018). The second species is present in Cabinda (Gautier-Hion 2013b). The Southern Talapoin Monkey is one of Africa's least studied primates, but it is assumed to be ecologically similar to the Northern Talapoin Monkey, therefore preferring dense forest environments, such as riverine forest (Machado 1969). Sclater's Angolan Colobus (*Colobus angolensis angolensis*) was known to occur in the northeast of Angola (Malanje, Lunda-Norte; Machado 1969), where it was recorded in 2009 near Lóvua (Pedro Vaz Pinto, unpublished data) and in riverine forests in the Lagoa Carumbo area in 2011 (Huntley and Francisco 2015).

Family Galagidae

The Galagidae are represented by six species in Angola. Four of these species have been confirmed in recent surveys, namely Thick-tailed Greater Galago (*Otolemur crassicaudatus*), Southern Lesser Galago (*Galago moholi*), Demidoff's Dwarf Galago (*Galagoides demidoff*), and importantly, a new, recently-described Dwarf Galago believed to be endemic to Angola, the Angolan Dwarf Galago (*Galagoides kumbirensis*) (Bersacola et al. 2015; Svensson et al. 2017). The Angolan Dwarf Galago was named after Cumbira Forest where it was first observed (Svensson et al. 2017), but it appears to be adapted to a wide range of habitats (Bersacola et al. 2015). The geographical distribution of this new species is not yet established, though it might range as far as the Congo River in DRC (Svensson et al. 2017).

Thick-Tailed Greater Galago occurs throughout Angola, except the extreme south (Bearder 2008). It is typically associated with open woodland and savanna habitats as well as in forest edges and thickets, using mid to high strata (Bearder et al. 2003; Bearder and Svoboda 2013). Its occurrence was confirmed in semi-arid savanna environments and Cumbira Forest (Cuanza-Sul), and in miombo woodlands in Malanje Province (Bersacola et al. 2015). Southern Lesser Galago is widespread in the miombo woodlands of Angola (Huntley 1973c; Bersacola et al. 2015). It is known to use all strata in open woodland, savanna, forest edges and other semi-arid habitats (Bearder et al. 2003; Bearder et al. 2008; Pullen and Bearder 2013). Demidoff's Dwarf Galago is known to occur across central to northeastern parts of Angola, including Cuanza-Sul, Cuanza-Norte, Malanje, Lunda-Sul and Lunda-Norte Provinces (Machado 1969; Svensson 2017). The occurrence of this species in semi-arid savanna zones of the Angolan Escarpment extended the species' range c. 190 km further southwest (Svensson 2017). This is the smallest of all the galagos and is typically associated with forest habitats, including deciduous and semi-deciduous forests, evergreen and gallery forests, mainly in the edge and understory habitats (Bearder et al. 2003; Ambrose and Butynski 2013). Their relative abundance in Angola was correlated with undergrowth density, canopy cover and tree density (Bersacola et al. 2015).

Two additional species have been reported from Angola, but their current status is unknown. Thomas's Dwarf Galago (*Galagoides thomasi*) is believed to range in the northern parts of Angola, but this still needs confirmation (Bersacola et al. 2015). Bersacola et al. (2015) proposed that competitive exclusion between the Thomas's and the Angolan Dwarf Galagos could explain why the former species was not observed in the Angolan Escarpment forests. The Southern Needle-clawed Galago (*Euoticus elegantulus*) was listed as possibly occurring in Cabinda (Huntley 1973e). The species is known to occur in both primary and secondary forests at low- to medium-altitude, including in deciduous and semi-deciduous, evergreen and littoral forests (Ambrose and Butynski 2013).

Family Hominidae

The two species of the family Hominidae in Angola are the Western Lowland Gorilla (*Gorilla gorilla gorilla*) and the Central Chimpanzee (*Pan troglodytes troglodytes*), which are confined to the Cabinda Province (Maisels et al. 2016a, b). Both species were known to be present in the area currently included in the Maiombe NP, a c. 2000 km^2 area consisting of mainly tropical forest ecosystems which are part of the Guinea-Congolian biome. The landscape in Cabinda is characterised by semi-deciduous tropical forests in the northeast (including Maiombe), agroforest mosaics largely covering the south, as well as mangrove and flooded swamp forests along the coast. Western Lowland Gorilla occur in different types of forest environments (Robbins et al. 2004; Tutin and Fernandez 1984). Across their range chimpanzees occupy a great variety of habitats, from tropical rainforests to semi-arid savanna environments (Boesch and Boesch-Achermann 2000; Pruetz 2006). Chimpanzees show high socioecological flexibilities in human-dominated environment (Hockings et al. 2012; McLennan 2013; Bessa et al. 2015). Considering the high socioecological flexibility of chimpanzees across their range, this species' range in Cabinda is likely to include human-dominated areas in the south of the province. Despite the urgency for surveys highlighted previously (Tutin et al. 2005), population estimates for the two great apes in Cabinda remain unavailable. In the province, chimpanzees are likely targeted for the commercial bushmeat trade (Ron and Golan 2010), but the scale of this trade is poorly known. Huntley (2017) records that gorilla and chimpanzee were by tradition not included among bushmeat species in Cabinda in 1973, while Bersacola et al. (2014) reports on chimpanzees occuring in the pet trade in Angola that were believed to originate from Cabinda. Future studies to assess the distribution and population status of the two great apes in Cabinda, including in-depth investigation on the human-great ape interactions in this region should be considered a priority.

Family Lorisidae

Milne-Edwards's Potto (*Perodicticus edwardsi*) and Golden Angwantibos (*Arctocebus aureus*) are the two species of the family Lorisidae reported for Angola (Huntley 1973e; Bersacola et al. 2015; Svensson 2017). Milne-Edwards's Potto is the largest of the potto species, generally occurring in both primary and secondary forests, from low altitude to montane forests (Butynski and de Jong 2007; Pimley 2009; Oates 2011; Pimley and Bearder 2013). The species was reported from Cabinda by Huntley (1973e), while recent surveys confirmed its occurrence in the tall, sub-humid forests of the Cuanza-Norte Province, and for the first time it was observed in the agro-forest mosaics in Cumbira Forest (Cuanza-Sul), extending the previously known geographical range c. 320 km further south (Bersacola et al. 2015). The Golden Angwantibos was reported from Cabinda by Huntley (1973e), but no information has been obtained thereafter.

Proboscidae (Elephants)

African elephants (genus *Loxodonta*) have traditionally been considered monotypic and to comprise two subspecies, but recent genetic studies, albeit confirming hybridisation along a contact zone, have also provided compelling evidence to sustain the validity of those two forms as full species (Grubb et al. 2000; Roca et al. 2001, 2015; Palkopoulou et al. 2018), namely Savanna Elephant (*L. africana*) and Forest Elephant (*L. cyclotis*). Interestingly, the earliest morphological studies suggesting that both species should be recognised result from the efforts of Frade (1933, 1936, 1955) and were based on the analyses of specimens collected in different parts of Angola. Although the studies by Frade reflect the occurrence of both species, since then no material from Angola has been critically evaluated and records were often assumed by default to refer to Savanna Elephant. On the other hand, some authors, have distinguished among various 'types' including potential subspecies of Savanna Elephants (Hill and Carter 1941), and some hunters even suggested the occurrence in the north of a dwarf elephant (Crawford-Cabral and Veríssimo 2005), but these likely correspond to geographical variants of either species.

Forest Elephant was once likely common to abundant in moist forested habitats across northwestern Angola, and including Cabinda, and a large number of records from hunters suggest their former presence throughout the provinces of Zaire, Uíge and Cabinda (Crawford-Cabral and Veríssimo 2005; Crawford-Cabral and Veríssimo, unpublished data). Soon after the end of the civil war, Forest Elephants were still not uncommon in the Maiombe rainforest of Cabinda (Heffernan 2005) but more recent accounts are lacking. There remains little doubt that their numbers have plummeted in recent decades as a result of hunting and habitat destruction, and Forest Elephant may have been extirpated from large parts of its former range in Angola. Surprisingly, a few herds of what appear to be Forest Elephant still seem to linger in forest blocks in Cuanza-Norte and Bengo Provinces (Vaz Pinto, unpublished data). If confirmed, this pocket of Forest Elephant in Cuanza-Norte and Bengo may consist of an isolated, southernmost population of the species, which would much increase its conservation importance. The possibility of a hybrid zone in northern Angola where Savanna Elephant also used to occur (Crawford-Cabral and Veríssimo 2005) cannot be excluded, but this hypothesis remains untested. Unfortunately, and paradoxically considering the large interest and funding channelled to research and conservation of elephants in general, very little is known about the Angolan Forest Elephant, including its taxonomic status and relationships with other populations, ecology, and numbers.

Savanna Elephant used to have a wide distribution in Angola, including along the coastal plain and in the east and northeast, but the core was in the southern half of the country. An extrapolation of figures based on aerial counts suggested the population in the southeast province of Cuando Cubango to reach up to 23,000 animals (Hall-Martin and Pienaar 1992), but the ensuing civil war prevented reliable counts. Quantification of elephant populations for the whole country was also attempted in the 1990s but based on little ground data, and yet suggesting a steep reduction in numbers from 50,000 to less than 10,000 (Anstey 1991, 1993). As a result of the armed conflict, the Savanna Elephant populations were most affected and may have disappeared completely from extensive regions, particularly along the coastal plain, while in other regions they may have survived in much reduced pockets. Following the end of the war, they may have recolonised extensive regions of the southeast, as migratory routes have been reopened, allowing the dispersal of Savanna Elephants coming from neighbouring countries, particularly Botswana (Chase and Griffin 2011), although more recent evidence suggest negative trends as a result of increased poaching pressure and human encroachment (Chase and Schlossberg 2016; Funston et al. 2017; Schlossberg et al. 2018). A small contingent of Savanna Elephant were introduced into Quiçama NP in 2000 and 2001, and these have since increased from 32 to about 90 (Carmignani 2015), but a small number of the original population may also have survived in the park (Groom et al. 2018). Much of what is currently known on the distribution and status of elephants in Angola is summarised in the IUCN African Elephant Status Report (Thouless et al. 2016).

Rodentia (Mole Rats, Mice, Dormice, Rats, Voles, Squirrels)

The rodents are a vast group with at least 85 species currently recognised to occur in Angola (Monadjem et al. 2015; Taylor, unpublished data Taylor et al. (2018c). Rodents show a high degree of endemism in Angola, with at least 13 endemic or near-endemic species. However, there are considerable uncertainties regarding the taxonomy of African rodents, and in the future it is likely that many species will be split after the development of detailed taxonomic and genetic studies, thus increasing the number of endemics or near endemics (Monadjem et al. 2015). For instance, while Taylor (2016) treats the African Marsh Rat (*Dasymys incomtus*) as a single widespread species, this may indeed be a complex of several similar species, some of which may have restricted distributions (Monadjem et al. 2015). Continued field surveys are also likely to increase the rodent list, as several species have been recorded on or close to the borders of Angola and will probably be shown to occur there in the future.

Family Anomaluridae

There are only two species of Anomaluridae in Angola, both of which are known from just a few records (Monadjem et al. 2015). Lord Derby's Scaly-tailed Squirrel (*Anomalurus derbianus*) occurs in the northern half of Angola. It is a nocturnal and arboreal species, which is mostly associated with the forest zone of tropical Africa, though it also occurs in miombo woodlands. Beecroft's Scaly-tailed Squirrel (*A. beecrofti*) is also a species from the forests of tropical Africa, which in Angola occurs mainly in Cabinda, though there are also records south of the Congo River (Happold 2013).

Family Bathyergidae

There are two species of Bathyergidae recorded in Angola (Monadjem et al. 2015). Bocage's Mole Rat (*Fukomys bocagei*) is near endemic, occurring widely in the west of Angola and extending narrowly into northern Namibia (Faulkes et al. 2016). Mechow's Mole Rat (*F. mechowi*) is a highly adaptable species often found in villages and croplands, and which occurs in central and northeast Angola. The population in western Angola seems to be separated by a large gap from those in eastern Angola, Zambia and DRC, but this may reflect the paucity of surveys. Damara Mole Rat (*F damarensis*) is known in Angola from only two specimens collected in 1964 in the southeast (Orrell and Hollowell 2018), but it has not been recorded thereafter. The presence of this species thus needs confirmation, though this is likely because it is known from neighbouring areas in Namibia and Zambia. Old records refer to the presence of African Mole Rat (*C. hottentotus* or *C. h. bocagei*) in Angola (e.g., Conroy 2018, MNHN 2018), but these were more likely Bocage's Mole Rats.

Family Gliridae

There are five Gliridae in Angola (Monadjem et al. 2015). Angolan African Dormouse (*Graphiurus angolensis*) is a near endemic savanna species, which is restricted to Angola and to a small disjunct area in western Zambia. Kellen's Dormouse (*G. kelleni*) has a range in central Angola that extends into western Zambia, but that is disjunct from other populations across the savanna zone of sub-Saharan Africa. Monard's Dormouse (*G. monardi*) is a species associated with miombo woodland, with a small distribution in northeastern Angola and northwestern Zambia. Stone Dormouse (*G. rupicola*) is a rupicolous species with its core distribution in a narrow belt along the escarpment of Namibia, and marginally into southwest Angola and South Africa. Lorrain Dormouse (*G. lorraineus*) is a forest species with a known distribution in Angola restricted to the northeast, though the species occurs in Zambia close to the border of Angola.

Family Hystricidae

Cape Porcupine (*Hystrix africaeaustralis*) is the sole Hystricidae occurring in Angola (Monadjem et al. 2015). It is an eclectic species that occurs in most habitats except dense forests, and that it is widespread throughout the country.

Family Muridae

The Muridae include 48 native species recorded in Angola, of which eight are endemic (Monadjem et al. 2015). Endemics are mainly associated with the central plateau, such as Thomas's Rock Rat (*Aethomys thomasi*), Angolan Marsh Rat

(*Dasymys nudipes*), and Angolan Vlei Rat (*Otomys anchietae*), or the western highlands, such as Angolan Multimammate Mouse (*Myomyscus angolensis*) and Cuanza Vlei Rat (*O. cuanzensis*). Coetzee's Praomys (*Praomys coetzeei*) is a species recently described that is known from just a few specimens collected in northern Angola (van der Straeten 2008). Angolan Hylomyscus (*Hylomyscus carillus*) is only known from Angola, but may also occur in neighbouring DRC. Heinrich's Hylomyscus (*H. heinrichorum*) was very recently described from specimens collected in 1954 at Mount Moco and Mount Soque (Carleton et al. 2015). Another seven species are near endemics, occurring in Angola and neighbouring countries. Bocage's Rock Rat (*A. bocagei*), Griselda's Single-striped Mouse (*Lemniscomys griselda*), Angolan Brush-furred Rat (*Lophuromys angolensis*), Callewaert's Mouse (*Mus callewaerti*), and Bell Groove-toothed Swamp Rat (*Pelomys campanae*) are restricted to Angola and DRC. Shortridge's Mastomys (*Mastomys shortridgei*) is known from just a few scattered localities in Angola, the Caprivi Strip (Namibia) and the extreme northwestern region of Botswana. Cabral's Marsh Rat (*D. cabrali*) and Setzer's Mouse (*M. setzeri*) are endemic to a narrow area in southeastern Angola, northwestern Botswana and northeastern Namibia. The latter species was only recently recorded in the Okavango source lakes region of Angola (Taylor et al. 2018c).

The Muridae species with the widest distribution in Angola are those able to thrive in association with agricultural fields and homesteads, including for instance the Natal multimammate mouse (*Mastomys natalensis*). There are also widespread species associated with the woodlands and grasslands of the central plateau, though some of these are associated with moister tropical conditions and have their distributions biased towards the north and/or the west, while others are more associated with drier savanna habitats and have their distribution biased towards the south and east. Overall, this is a large group of species including Marsh Rat (*D.* cf *incomtus*, *sensu* Monadjem et al. 2015), Savanna Gerbil (*Gerbilliscus validus*), Bushveld Gerbil (*G. leucogaster*), Woodland Thicket Rat (*Grammomys dolichurus*), Pygmy Mouse (*M. minutoides*), Gray-bellied Pygmy Mouse (*M. triton*), Thomas's Pygmy Mouse (*M. sorella*), Creek Groove-toothed Swamp Rat (*P. fallax*), Angoni Vlei Rat (*O. angoniensis*), Mesic Four-striped Grass Rat (*Rhabdomys dilectus*), and Hildegarde's Broad-headed Mouse (*Zelotomys hildegardeae*). Some of these species, however, are known from just a few scattered records, including for instance Thomas's Pygmy Mouse and Hildegarde's Broad-headed Mouse.

Some rodent species have relatively restricted distributions in Angola, because they are associated with habitats represented only marginally in the country. This is the case of species associated with rainforests and other moist tropical habitats, which occur mainly in Cabinda and/or relatively small areas in the north of the country, including African Wading Rat (*Colomys goslingi*), Shining Thicket Rat (*G. poensis*), Typical Striped Grass Mouse (*L. striatus*), Dollman's Brush-furred Rat (*L. rita*), Jackson's Soft-furred Mouse (*P. jacksoni*) and Big-eared Swamp Rat (*Malacomys longipes*). Some species largely restricted to the north may penetrate southwards along the Angolan Escarpment, as it is the case of Rufous-nosed Rat (*Oenomys hypoxanthus*). In contrast, species associated with deserts, arid and semi-arid habitats occur mainly in the southwest of Angola, including Cape Short-eared Gerbil (*Desmodillus auricularis*), Hairy-footed Gerbil (*G. paeba*), Setzer's Gerbil (*G. setzeri*), Black-tailed Tree Rat (*Thallomys nigricauda*), and Striped Mouse (*Rhabdomys bechuanae*). Other species with marginal distributions in Angola due possibly to environmental or biogeographic constraints include Kaiser's Rock Rat (*A. kaiseri*), Nyika Rock Rat (*A. nyikae*) and Least Groove-toothed Swamp Rat (*P. minor*), in the northeast, and Red Rock Rat (*A. chrysophilus*), Highveld Gerbil (*G. brantsii*), and Woosnam's Broad-headed Mouse (*Zelotomys woosnami*), in the south. The latter species was only confirmed in recent surveys (Taylor et al. 2018c). Records of Desert Pygmy Mouse (*Mus indutus*) in southeast Angola need to be confirmed through molecular data. Namaqua Rock Rat (*Micaelamys namaquensis*) is a species widespread in southern Africa, which penetrates northwards through western Angola.

Besides native species, the rodents of Angola also include three non-native invasives, namely House Mouse (*Mus musculus*), a widespread commensal species, Brown Rat (*Rattus norvegicus*), mainly occurring in coastal cities, and Black Rat (*Rattus rattus*), widespread throughout the country. It is likely that future surveys will increase the list of native murids in the country, including tropical species that are known to occur close to the border of Cabinda and the north of Angola, such as Congo Forest Mouse (*D. ferrugineus*), Ansorge's Brush-furred Rat (*L. ansorgei*), Peter's Striped Mouse (*Hybomys univittatus*), Beaded Wood Mouse (*H. aeta*), Ansell's Wood Mouse (*H. anselli*), African Groove-toothed Rat (*Mylomys dybowskii*), Petter's Praomys (*Praomys petteri*), and Target Rat (*Stochomys longicaudatus*).

Family Nesomyidae

The Nesomyidae include 15 species recorded in Angola, of which four are endemic or near endemic to the country (Monadjem et al. 2015). Angolan Gray African Climbing Mouse (*Dendromus leucostomus*) is known only from its type locality (Caluquembe) in the highlands of Angola, but some authors treat it as conspecific with the Gray African Climbing Mouse (*D. melanotis*). Vernay's Climbing Mouse (*D. vernayi*) is only known at present from a series of specimens collected near Chitau in the central Angolan highlands. Bocage's Fat Mouse (*Steatomys bocagei*) is restricted to northern Angola and neighbouring regions of the DRC. Shortridge's Rock Mouse (*Petromyscus shortridgei*) is known only from a few scattered localities in northwestern Namibia and southwestern Angola. Another two species have largely isolated populations in Angola. Nyika Climbing Mouse (*D. nyikae*) occurs along the Angolan Escarpment, well separated from other patchy populations in Central and East Africa. Tiny Fat Mouse (*S. parvus*) occurs widely in East Africa and the northern savannas of southern Africa, but the population in southwestern Angola seems to be largely isolated.

Six Nesomyidae are associated with savanna woodlands and/or grasslands throughout their range, and are widespread in Angola. This includes Northern Giant

Pouched Rat (*Cricetomys ansorgei*), Chestnut Climbing Mouse (*D. mystacalis*), Gray African Climbing Mouse (*D. melanotis*), Southern African Pouched Mouse (*Saccostomus campestris*), Fat Mouse (*S. pratensis*), and Kreb's Fat Mouse (*S. krebsii*). In contrast to these species, the Forest Giant Pouched Rat (*Cricetomys emini*) is associated with tropical rainforests, and in Angola its presence is vouched in Cabinda by Musser and Carleton (2005) and Monadjem et al. (2015), but the species does not appear to occur as widely in northern Angola as indicated in Happold (2013). Gerbil Mouse (*Malacothrix typica*) and Pygmy Rock Mouse (*P. collinus*) are restricted to arid environments in the southwest of the country.

Banana Climbing Mouse (*D. messorius*) was recorded by three specimens from the Field Museum from Dundo in the extreme northeast Angola collected by Barros Machado in 1948 (Grant and Ferguson 2018), but there are no known records close to this (Monadjem et al. 2015). Previous comments on these specimens by Hayman (1963) suggest they may be African Climbing Mouse (*D. mystacalis*) (Taylor unpublished data). The monotypic *Dendroprionomys*, Velvet Climbing Mouse (*D. rousselotti*), is known only from the type locality Brazzaville which is close to Cabinda and may be shown to occur there.

Family Pedetidae

Spring Hare (*Pedetes capensis*) is the sole Pedetidae in Angola (Monadjem et al. 2015). It occurs throughout the country, except in the arid southwest, and in the moist and forested areas of the north.

Family Petromuridae

Dassie Rat (*Petromus typicus*) is the single species of Petromuridae. It has been recorded in a small area in southwest Angola, with its range expanding southwards through Namibia and into the Northern Cape Province of South Africa (Monadjem et al. 2015; Cassola 2016a). The species is confined to the western escarpment and adjoining mountainous areas, as well as inselbergs.

Family Sciuridae

There are nine species of Sciuridae in Angola (Monadjem et al. 2015). Rope Squirrels of the *Funisciurus* genus typically have distributions towards the north of the country, probably due to their association with moist tropical forests. Congo Rope Squirrel (*Funisciurus congicus*) is the most widespread species, ranging widely in the Congo basin, south through the western provinces of Angola and into northwestern Namibia. Ribboned Rope Squirrel (*F. lemniscatus*) occurs widely in the Lower Guinea Forest zone of Central Africa but in Angola has only been recorded in Cabinda, while Fire-footed Rope Squirrel (*F. pyrropus*) is widely distributed in the forest zone of tropical Africa but in Angola is restricted to Cabinda and the northwest. Lunda Rope Squirrel (*F. bayonii*) is globally restricted to northern and northeastern Angola and neighbouring areas of DRC. This species is associated with moist savanna mosaics, sandy woodlands, and low to medium elevation moist forests (Thorington et al. 2012). Thomas's Rope Squirrel (*F. anerythrus*) has been recorded near the border of Cabinda and could possibly occur there, but this is yet to be confirmed.

Gambian Sun Squirrel (*Heliosciurus gambianus*) is widely distributed in moister savannas of tropical Africa, occurring in central and northeastern Angola. African Giant Squirrel (*Protoxerus stangeri*) is the largest squirrel in Africa, having a wide distribution across the rainforest belt of tropical Africa. In Angola it occurs in the northwest, penetrating to the south along the escarpment, corresponding to the endemic subspecies *P.* s. *loandae* (Happold 2013). Red-legged Sun Squirrel (*Heliosciurus rufobrachium*) is not known to occur south of the Congo River but a doubtful record from 'Raca Camele, north of Quionlungo' was attributed to a specimen from Yale Peabody Museum identified by A Heinrich. Boehm's Bush Squirrel (*Paraxerus boehmi*) has been reported for Angola based on old records from Cabinda (Wendelen and Noé 2017) and Benguela (MNHN 2018), quite far from the core of species distribution in tropical forests of Central Africa. Two other species occur in the south of the country, with Damara Ground Squirrel (*Xerus princeps*) occurring in a small area in southwest Angola, corresponding to the northern tip of a larger distribution in western Namibia, and Smith's Bush Squirrel (*Paraxerus cepapi*) occurring in southern savannas.

Family Thryonomyidae

The family Thryonomyidae is represented in Angola by the Greater Cane Rat (*Thryonomys swinderianus*), which is a habitat generalist and occurs throughout the country (Monadjem et al. 2015). There are also three old records of Lesser Cane Rat (*T. gregorianus*), but they are dubious because the species is difficult to distinguish from Greater Cane Rat and the closest known records of this species are from central DRC and western Zambia (Happold 2013; Monadjem et al. 2015). These records refer to three specimens from Mount Moco collected by GH Heinrich in 1954 (Grant and Ferguson 2018)

Sirenia (African Manatee)

The African Manatee (*Trichechus senegalensis*) occurs in Angola at the southern limit of its global distribution (Powell 1996), where it is associated with mangroves along the lower sections of large rivers in northern Angola. The species has been confirmed from Cabinda and various rivers between the Congo and the Cuanza (Hatt 1934; Crawford-Cabral and Veríssimo 2005; Morais et al. 2006a, b; Dodman et al. 2008; Collins et al. 2011). There seems to be some uncertainty regarding its current distribution. The southernmost records have often been suggested to be the Longa or the Queve rivers (Crawford-Cabral and Veríssimo 2005), but recent surveys found no evidence of their existence south of the Cuanza (Morais et al. 2006a). It has also been suggested that they might occur throughout the Angolan coast, including in coastal lagoon systems such as Mussulo and as far south as the Cunene River (Powell 1996). However, these claims lack supporting data and the habitat present is not adequate, and so they these reports should therefore be treated with caution (Dodman et al. 2008).

Soricomorpha (Shrews)

There are 15 species of Soricomorpha thus far recognised to occur in Angola, all from the genus *Crocidura* and *Suncus* (Hill and Carter 1941; Hayman 1963, Crawford-Cabral and Veríssimo unpublished data). For many of these species, however, there are only a few old records and their occurrence in Angola needs to be confirmed. Greater Gray-brown Musk Shrew (*C. luna*) is known to occur in the northeast of Angola, although until now only validated by a single specimen from Lunda-Norte (Hayman, 1963). Reddish-gray Musk Shrew (*C. cyanea*) has a very restricted range in southwestern Angola, representing the northwestern tip of a much wider distribution in Namibia, South Africa, and elsewhere in the eastern part of southern Africa (Baxter et al. 2016). Roosvelt's Shrew (*C. roosevelti*) is a species occurring in moist savanna around the Congo Basin forest block, which in Angola seems to be restricted to the northeast (Hutterer and Peterhans 2016). Records of two additional species, Lesser Gray-brown Musk Shrew (*C. silacea*) and Dent's Shrew (*C. denti*), are either doubtful or in need of further re-identification because their known range is nowhere near Angola (Happold and Happold 2013). The only potentially valid Angolan record for the Lesser Gray-brown Musk Shrew is a specimen in alcohol in the Museum of Dundo (Lunda-Norte), identified by Heim de Balsac and quoted by Hayman (1963). However, this species was not considered in the checklist, because the location of this single record is many hundreds of kilometers from other known records, and difficulties in identification at the time may signify misidentification. Dent's Shrew record was an undated record from the Natural History Museum with no recorded locality. The endangered Ansell's Shrew (*C. ansellorum*) is known only from two locations in gallery forests of northwestern Zambia, close to the Angolan border where the species may also occur (Kennerley 2016).

Two of the *Crocidura* species recorded in Angola are endemics with restricted distributions. The Heather Shrew (*C. erica*) is a poorly known species found in Western Angola (Gerrie and Kennerley 2016), with records collected in the provinces of Cuanza-Norte, Malanje, Huambo, Benguela, Huíla (Crawford-Cabral and Veríssimo, unpublished data), while Blackish White-toothed Shrew (*C. nigricans*) occurs in the southwest (Crawford-Cabral 1987; Hutterer 2016), particularly in localities along the mountainous western belt of the Angolan plateau (Huambo, Benguela, Huíla, Cunene) (Crawford-Cabral and Veríssimo, unpublished data). The only *Crocidura* shrew that is thought to be widespread throughout Angola is Oliver's Shrew (*C. olivieri*) (Crawford-Cabral and Veríssimo, unpublished data; Cassola 2016c) which represents a species complex and is in urgent need of revision. However, Tumultuous Shrew (*C. turba*), African Black Shrew (*C. nigrofusca*) and Swamp Musk Shrew (*C. mariquensis*) are also widespread in some regions of the north and along the Angolan Escarpment. Small-footed Shrew (*C. parvipes*) is known to occur in the provinces of Bié and Huíla (Hill and Carter 1941; Crawford-Cabral and Veríssimo, unpublished data). Other species in Angola have their distributions associated with their much wider ranges in Africa, including Lesser Red Musk Shrew (*C. hirta*) recorded from Lunda-Norte, Lunda-Sul, Cuanza-Sul, Huambo and Huíla, and may occur everywhere on the Angolan highlands, being locally rather common, and likely representing the subspecies *C. hirta luimbalensis* (Crawford-Cabral and Veríssimo, unpublished data). In the south (Huíla and Cuando Cubango) there were two specimens identified as Desert Lesser Red Shrew (*C. deserti*), which may be a subspecies of Lesser Red Musk Shrew (Cassola 2016d). Bicolored Musk Shrew (*C. fruscomurina*) has been recorded from Bengo/Luanda, Cuanza-Norte, Cuanza-Sul, Malanje, Lunda-Norte, Huambo, Huíla, Namibe and Cuando Cubango, and is thus thought to occur throughout the country (Crawford-Cabral and Veríssimo, unpublished data).

Both Greater Dwarf Shrew (*S. lixus*) and Climbing Shrew (*S. megalura*) are known to occur in Angola (Happold and Happold 2013; Crawford-Cabral and Veríssimo, unpublished data). Greater Dwarf Shrew is known to occur in the east, with records in Lunda-Norte (Heim de Balzac and Meester 1977). Climbing Shrew has only been reported in Angola from Cuanza-Sul and Lunda-Norte provinces, yet very scarcely, but it may occur throughout most of northern Angola. Recent field collections in the Okavango source lakes area in 2016 added an additional species for Angola, the Lesser Dwarf Shrew (*S. varilla*) (Taylor et al. 2018c). The species has a sparse distribution and was previously known from southeast DRC so its occurrence in central Angola is not surprising.

Tubulidentata (Aardvark)

The Aardvark (*Orycteropus afer*) is the only Tubulidentata occurring in Angola, where it seems to be widespread, although it is known from relatively few and scattered records due to its cryptic nature (Hill and Carter 1941; Crawford-Cabral and Veríssimo 2005). No studies have focused on this species, and although it is often hunted for bushmeat, it is likely not threatened due to its widespread distribution and elusive habits. The species has also been frequently reported in general mammal surveys conducted in protected areas in southern Angola (Veríssimo 2008; Funston et al. 2017; Overton et al. 2017; NGOWP 2018).

Zoogeographic Outline

A quantitative regionalisation of Africa based on plants and vertebrates was carried out by Linder et al. (2012), providing information on the biogeographic position of Angola in the context of the African continent. In the analysis based on mammals, this study located most of the country in a wide Zambezian band across south-central Africa, crossing the continent from the Atlantic coast of Angola to the Indian Ocean coasts of Tanzania and Mozambique. This band was bordered to the north by the Guinean-Congolian region, corresponding to the tropical moist forests of the Congo Basin and West Africa, which encompasses the enclave of Cabinda and a narrow fringe in north and northeast Angola. To the south, the Zambezian region was bordered by a Southern African Region, which extends into a narrow strip in southern Angola. This rather coarse regionalisation was refined in analyses using a dataset combining all plants and vertebrates. This analysis recognised a broad southern transition zone (the Shaba subregion) between the Congolian and the Zambezian regions, which forms an arc from the Angolan Atlantic coast to the southern Ugandan uplands, and that expands southwards in Angola along the coastal escarpment. Also, it split the Southern African Region in several subregions, including the small biogeographic unit of Southwest Angola. In this new analysis the Zambezian zone was bordered in the south by the Kalahari subregion.

Early efforts to undertake a zoogeographic analysis of Angola based on the mammalian fauna were made by Crawford-Cabral (1982, 1997) and Feiler (1990). More recently, the theme was revisited by Rodrigues et al. (2015), aiming to refine the broad scale analysis of Linder et al. (2012) and to understand the environmental determinants of biogeographic patterns (Fig. 15.2). The quantitative regionalisation developed by Rodrigues et al. (2015) focused solely on Angola (excluding Cabinda) and used data on ungulates (Cetartiodactyla, Perissodactyla, Hyracoidea), carnivores and small mammals available in the literature (Crawford-Cabral and Simões 1987, 1988; Crawford-Cabral 1998; Crawford-Cabral and Veríssimo 2005). The study retrieved four main biogeographical units, which were particularly clear in the analysis focusing on ungulates (Fig. 15.2): one northern region (Zaire-Lunda-Cuanza), one central region (Central Plateau) and two regions in the south (Namibe and Cunene-Cuando Cubango). This biogeographical pattern was strongly affected by environmental factors, reflecting the dominant climate gradients in this region of Africa (Le Houérou 2009), and the associated variation in soil and vegetation types. The regions identified also closely matched the strong north–south gradient in closed canopy forest cover (Hansen et al. 2013) with a progressive southwards transition to savannas (Murphy and Bowman 2012).

In the north, the Zaire-Lunda-Cuanza Region largely matched the Angolan portion of the Shaba Region (Linder et al. 2012), and was mainly characterised by indicator species that have their core range within the Congo Basin forests and reach their southern limit in Angola, such as Forest Buffalo, Yellow-backed Duiker, Blue Duiker, Black-fronted Duiker, and Bay Duiker. The Central Plateau corresponds roughly to the Zambezian region of Linder et al. (2012) and encompasses to

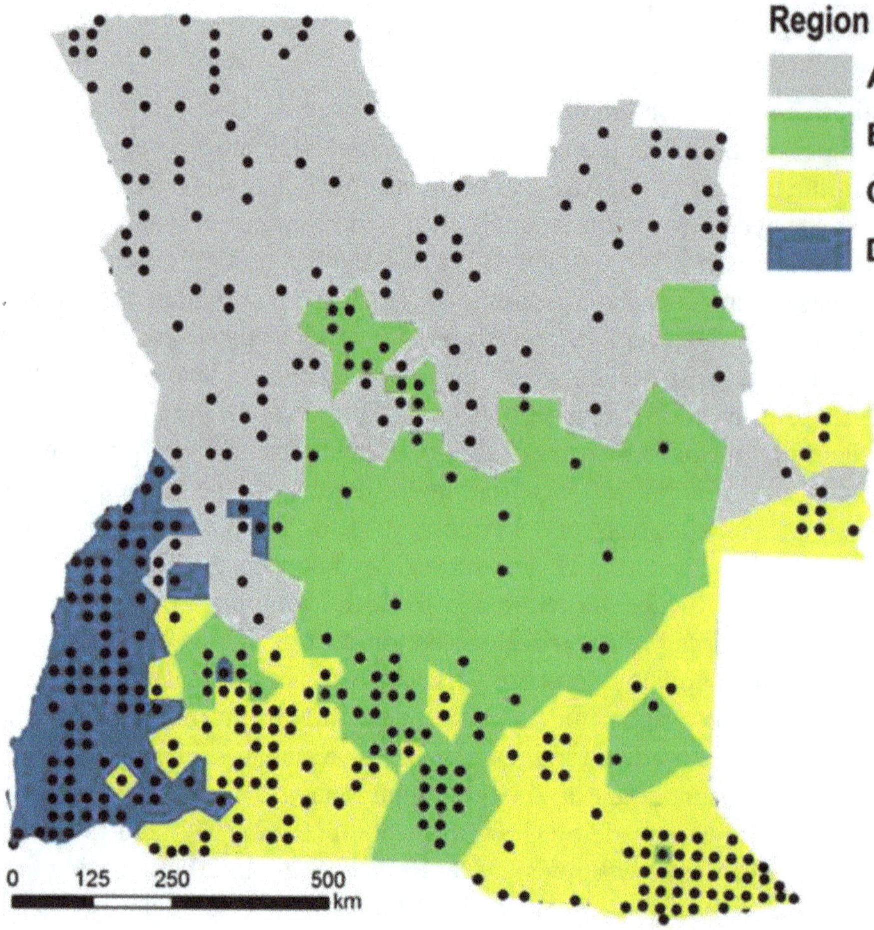

Fig. 15.2 Zoogeographic regions of Angola identified with a cluster analysis of quarter degree grid cells (approximately 25 × 25 km) characterised using the occurrences of ungulate species (Cetartiodactyla, Perissodactyla, Hyracoidea). Analysis used information documented by Crawford-Cabral and Veríssimo (2005), corresponding to data collected mainly in 1930–1980. As there was no information in many areas, grid cells were converted into a network of Thiessen polygons, each containing the centre of a single grid cell with occurrence records for five or more species (black dots). Colours are used to differentiate regions; *A* – Zaire-Lunda-Cuanza; *B* – Central Plateau; *C* – Cunene-Cuando Cubango; *D* – Namibe. (Redrawn from Rodrigues et al. (2015) and published under license by John Wiley and Sons)

a large extent the WWF Ecoregion of the Angolan miombo woodlands (Olson et al. 2001). It intergrades to some extent with the Zaire-Lunda-Cuanza region, suggesting that there is a north–south gradient in mammalian assemblages, rather than two well-defined regions. Indicator species of this region were Oribi, Roan Antelope, Eland, Common Warthog, Bush Duiker, and the local endemic Giant Sable Antelope.

In the south, the Cunene-Cuando Cubango region was clearly identified in analysis, corresponding to a savanna band running along the southern border of Angola with Namibia, eastward of the Namibe region and continuing farther north along the border with Zambia. The region corresponds to the northern limit of the Kalahari subregion of Linder et al. (2012), though extending further north into Angola than previously recognised. It also corresponds to the Zambezian Baikiaea woodlands and the Angolan mopane woodlands defined by WWF (Olson et al. 2001). The indicators of this region are species such as Wildebeest, Giraffe, Sable Antelope, Black Rhino, Common Tsessebe, Buffalo, Hartebeest and Impala, which are widespread in savannas of southern and eastern Africa, but that have restricted distributions elsewhere in Angola. Finally, the Namibe region matches closely the south-western Angola region of Linder et al. (2012), and the WWF Ecoregion of the Kaokoveld desert (Olson et al. 2001), representing the northern part of the vast Namib Desert, and the Namibian savanna woodlands. The main indicators of this region were species such as Rock Hyrax, Yellow-spotted Rock Hyrax, Black-faced Impala, Kirk's Dik-dik, Klipspringer, Springbok, Gemsbok and Mountain Zebra.

Although this biogeographic regionalisation emerged from the analysis of occurrence data for ungulates, it is likely that similar patterns would be obtained for other groups, though a quantitative analysis was not possible due to the scarcity of information. However, the recent reviews on African bats and rodents by (Monadjem et al. 2010a, 2015), Schoeman et al. (2013) and Cooper-Bohannon et al. (2016), clearly suggest marked north-south distribution gradients, with species from Guinean and Congolian tropical forests penetrating southwards in northern Angola, species from arid areas occurring in the southwest, and savanna species occurring in a southern band. The same seems to happen with other groups such as primates.

As described in 'The Mammal Fauna' section above, Angola has a number of endemic and near endemic species, which are particularly numerous in the case of rodents. The number of endemics recognised has been growing in recent years, mainly due to the use of molecular techniques to understand phylogenetic relationships among taxa, new field surveys, and the re-examination of specimens collected several decades ago (Carleton et al. 2015; Svensson et al. 2017). In addition, there are several endemic and near endemic subspecies, though their taxonomic status is often uncertain. Finally, there are in Angola some isolated or otherwise disjunct populations, often far from the main distribution range of the corresponding species, which in the future may prove to warrant taxonomic recognition (Monadjem et al. 2010a, 2015). Despite this richness, however, there has been no systematic attempt to identify the regions or habitats where such endemisms occur in Angola, or on the phylogeographic processes that have driven their divergence from sister taxa. Nevertheless, analogies with other taxonomic groups suggest that in Angola there are well-defined regions that are centres of endemism, and thus merit further surveys and conservation attention. These are mostly concentrated along the Angolan Escarpment and Afromontane forests of western Angola, which are known to have high prevalence of plant, invertebrate and vertebrate endemisms (Hall 1960; Figueiredo et al. 2009; Clark et al. 2011; Mills et al. 2011, 2013), including mammals (Carleton et al. 2015).

Conservation

The global conservation status of 95.2% of the 290 native mammal species known to occurr in Angola has been evaluated by IUCN (2018), of which there are 2 Critically Endangered, 2 Endangered, 11 Vulnerable, 14 Near-Threatened, and 12 Data Deficient species (Appendix). The order Carnivora has the largest number of threatened species (5), but there are also threatened species in the orders Pholidota (3), Cetartiodactyla (2), Perissodactyla (2), Primates (2) and Sirenia (1). The Critically Endangered species are Black Rhinoceros, which is likely to be currently extinct in the country, and Western Gorilla, which seems to be restricted to a very small area in the Maiombe forests of Cabinda. There is also one Critically Endangered subspecies, the Giant Sable Antelope, though its parental species is considered Not Threatened. The Endangered species are the African Wild Dog, which in Angola maintains some populations that may be relevant for species conservation at the global scale (Veríssimo 2008; Overton et al. 2017; Fabiano et al. 2017; Funston et al. 2017; Monterroso et al., unpublished data), and Chimpanzee, which like the Western Gorilla is restricted to small areas in the forests of Cabinda.

The Vulnerable category includes a diverse array of species, including Cheetah, African Golden Cat, Lion, Leopard, Giraffe, Hippopotamus, Mountain Zebra, White-bellied Pangolin, Giant Ground Pangolin, Temminck's Ground Pangolin and African Manatee. Although classified as Vulnerable by IUCN, the Elephant was not considered in this group because we assumed the scientifically well-established division in two separate species, Forest and Savanna Elephants, which have hitherto not been evaluated by IUCN. Given the precipituous decline of Forest Elephant due to poaching and its very low intrinsic growth rate, the species may qualify to the Endangered or even Critically Endangered status (e.g., Cerling et al. 2016; Poulsen et al. 2017; Turkalo et al. 2017). Black-faced Impala is a subspecies considered Vulnerable, though its parental species is classified as Not Threatened. Regarding Near Threatened species, these include four carnivores (Brown Hyena, African Clawless Otter, Congo Clawless Otter, Spotted-necked Otter), six Cetartiodactyla (Bay Duiker, White-bellied Duiker, Yellow-backed Duiker, Waterbuck, Southern Lechwe, Puku), three bats (Striped leaf-nosed Bat, Large-eared Free-tailed Bat, Angolan Epauletted Fruit Bat), and one primate (Black Crested Mangabey).

There is very little information on the current status in Angola of most threatened and near-threatened species, but many of them are feared extinct or at the brink of extinction. This is mainly the case for large carnivores and herbivores, which were heavily hunted during and in the years following the civil war, and for which virtually no information on distribution and numbers has been collected for over three decades (Huntley 2017). A few recent surveys confirm this situation, showing that many species once common in Angola only persist at present in remote areas, usually having small and fragmented populations (Veríssimo 2008; Overton et al. 2017; Fabiano et al. 2017; Funston et al. 2017; Monterroso et al., unpublished data). This is illustrated for instance by the critical condition of the iconic Giant Sable Antelope, which has declined to very small numbers over the past decades (e.g., Vaz Pinto et al. 2016 and Vaz Pinto 2019). Another iconic species, Savanna Elephant, also appeared to have declined precipitously, even after the end of the civil war (Milliken et al. 2006; Chase and Griffin 2011; Schlossberg et al. 2018), and the Forest Elephant may be on the verge of extinction in the country. Poaching and habitat destruction are likely the main threats, and a major cause of concern is the illegal trade of ivory, channelled through the capital Luanda, and which is fuelling most of the poaching activities (Milliken et al. 2006; Svensson et al. 2014b). African Manatee is another globally vulnerable species that is likely on the verge of extinction in Angola, resulting from unsustainable hunting associated with the bushmeat trade (Morais et al. 2006a, b; Collins et al. 2011). Information is even scarcer for smaller and less charismatic species, and it is noteworthy that many of the data deficient mammal taxa listed by IUCN for Angola are endemics or near-endemics to the country. These include for instance Lesser Angolan Epauletted Fruit Bat, Hayman's Dwarf Epauletted Fruit Bat, Angolan Long-eared Bat, Black-nosed Red-tailed Monkey, Pluto Monkey, Southern Black Crested Mangabey, Angolan African Dormouse, Monard's Dormouse, Angolan Shaggy Rat, Vernay's Climbing Mouse, Lunda Rope Squirrel, and Heather Shrew, all of which have rather restricted distribution and may qualify to a threatened or near-threatened category upon further investigation.

The problems affecting the conservation of biodiversity in Angola and some of their potential solutions are discussed at length in Huntley et al. (2019). The same considerations broadly apply to the mammal species. It should be noted, however, that mammals will inevitably be at the forefront of biodiversity conservation in Angola, as this group includes some of the most threatened species at the global scale, as well as some of the species that have suffered the most from decades of persecution and management neglect. Furthermore, the key threats to many of the most endangered species are likely to continue or even intensify in the country (Huntley 2017), due to a detrimental combination of factors such as commercial and illegal wildlife trade, bushmeat harvesting, and habitat destruction through deforestation, agricultural expansion, and infrastructure development. At the same time, however, there are encouraging signs for biodiversity conservation in Angola, many of which involving efforts to preserve endangered mammals. Despite multiple problems, it has been possible to secure the critically endangered populations of the Giant Sable (e.g., Vaz Pinto et al. 2016 and Vaz Pinto 2019), while new surveys have revealed the potential of recovery of other iconic species such as Lion and Wild Dog, among others (Veríssimo 2008; Overton et al. 2017; Fabiano et al. 2017; Funston et al. 2017; Monterroso et al., unpublished data). For at least some of these species, Angola may play an important role for conservation efforts, by securing relevant populations and thus reducing risks at the global scale (e.g., Riggio et al. 2013). Many of these species have persisted in areas that are now protected by national legislation, and it is expected that current conservation efforts, albeit yet modest, will help them to recover in range and numbers. It is now necessary to move forward, increasing the conservation efforts for these threatened species, at the same time that new surveys are carried out to obtain a more complete appreciation of the diversity and conservation needs of the overall mammal fauna of Angola (Fig. 15.3).

Fig. 15.3 Angolan mammals. *1* Pack of African Wild Dogs (*Lycaon pictus*) in Luando Strict Nature Reserve; *2* Herd of Forest Buffalo (*Syncerus caffer nanus*) in Quiçama National Park; *3* Yellow-backed Duiker (*Cephalophus silvicultor ruficrista*) in Luando Strict Nature Reserve; *4* Black Rhinoceros (*Diceros bicornis bicornis*) in Iona National Park; *5* Angolan Bush Hyrax (*Heterohyrax brucei bocagei*) at Serra da Neve; *6* White-bellied Pangolin (*Phataginus tricuspis*) in Cangandala National Park; *7* Angolan Dwarf Galago (*Galagoides kumbirensis*) in Cumbira Forest; *8* Pluto Monkey (*Cercopithecus mitis mitis*) in Quiçama National park. (Photo Credits: 1, 4 – Brian J. Huntley, 2 – Merle Huntley: 1970s personal archive; 3, 5, 6, 8 – Pedro Vaz Pinto; 7 – Elena Bersacola)

Appendix

Checklist of the native mammals of Angola. The table provides all species with occurrence confirmed in the country (e.g., collected specimens, photos, reliable sightings), as well as all subspecies (in grey font) confirmed in the country and that are evaluated separately in the IUCN Red List (IUCN 2018). The ORDER, **Family**, *Latin name* (author, date), English name, IUCN conservation status (**CS**), and key references (**Ref**) confirming species presence are provided for each taxa. Taxa are arranged by alphabetic order of Order, Family and Latin name

Species	English name	CS[a]	Ref.[b]
AFROSORICIDA	**Chrysochloridae**		
Huetia leucorhina (Huet, 1885)	Congo Golden Mole	DD	21
AFROSORICIDA	**Tenrecidae**		
Potamogale velox (Du Chaillu, 1860)	Giant Otter Shrew	LC	3-5
CARNIVORA	**Canidae**		
Canis adustus (Sundevall, 1847)	Side-striped Jackal	LC	12
Canis mesomelas (Schreber, 1775)	Black-backed Jackal	LC	12
Lycaon pictus (Temminck, 1820)	African Wild Dog	EN	13
Otocyon megalotis (Desmarest, 1822)	Bat-eared Fox	LC	12
Vulpes chama (A. Smith, 1833)	Cape Fox	LC	12
CARNIVORA	**Felidae**		
Acinonyx jubatus (Schreber, 1775)	Cheetah	VU	13
Caracal aurata (Temminck, 1827)	African Golden Cat	VU	16
Caracal caracal (Schreber, 1776)	Caracal	LC	12
Felis silvestris (Schreber, 1777)	Wild Cat	LC	12
Leptailurus serval (Schreber, 1776)	Serval	LC	12
Panthera leo (Linnaeus, 1758)	Lion	VU	13
Panthera pardus (Linnaeus, 1758)	Leopard	VU	13
CARNIVORA	**Herpestidae**		
Atilax paludinosus (G.[Baron] Cuvier, 1829)	Marsh Mongoose	LC	12
Bdeogale nigripes (Pucheran, 1855)	Black-legged Mongoose	LC	12
Crossarchus ansorgei (Thomas, 1910)	Ansorge's Cusimanse	LC	12
Cynictis penicillata (G.[Baron] Cuvier, 1829)	Yellow Mongoose	LC	12
Helogale parvula (Sundevall, 1847)	Common Dwarf Mongoose	LC	12
Herpestes flavescens (Bocage, 1889)	Kaokoveld Slender Mongoose	LC	10
Herpestes ichneumon (Linnaeus, 1758)	Egyptian Mongoose	LC	12
Herpestes sanguineus (Rüppell, 1835)	Common Slender Mongoose	LC	10
Ichneumia albicauda (G.[Baron] Cuvier, 1829)	White-tailed Mongoose	LC	12
Mungos mungo (Gmelin, 1788)	Banded Mongoose	LC	12
Paracynictis selousi (de Winton, 1896)	Selous's Mongoose	LC	12
Suricata suricatta (Schreber, 1776)	Meerkat	LC	12
CARNIVORA	**Hyaenidae**		
Crocuta crocuta (Erxleben, 1777)	Spotted Hyaena	LC	13

(continued)

Species	English name	CS[a]	Ref.[b]
Parahyaena brunnea (Thunberg, 1820)	Brown Hyaena	NT	13
Proteles cristata (Sparrman, 1783)	Aardwolf	LC	13
CARNIVORA	**Mustelidae**		
Aonyx capensis (Schinz, 1821)	African Clawless Otter	NT	12
Aonyx congicus (Lönnberg, 1910)	Congo Clawless Otter	NT	12
Hydrictis maculicollis (Lichtenstein, 1835)	Spotted-necked Otter	NT	12
Ictonyx striatus (Perry, 1810)	Striped Polecat	LC	12
Mellivora capensis (Schreber, 1776)	Honey Badger	LC	12
Poecilogale albinucha (Gray, 1864)	African Striped Weasel	LC	12
CARNIVORA	**Nandiniidae**		
Nandinia binotata (Gray, 1830)	African Palm Civet	LC	12
CARNIVORA	**Otariidae**		
Arctocephalus pusillus	Brown Fur Seal	LC	36
CARNIVORA	**Viverridae**		
Civettictis civetta (Schreber, 1776)	African Civet	LC	12
Genetta angolensis Bocage, 1882	Miombo Genet	LC	12
Genetta genetta (Linnaeus, 1758)	Common Genet	LC	12
Genetta maculata (Gray, 1830)	Large-spotted Genet	LC	12
CETARTIODACTYLA	**Bovidae**		
Aepyceros melampus (Lichtenstein, 1812)	Common Impala	LC	14
Aepyceros melampus ssp. *melampus* (Lichtenstein, 1812)	Common Impala	LC	14
Aepyceros melampus ssp. *petersi* (Bocage, 1879)	Black-faced Impala	VU	14
Alcelaphus buselaphus (Pallas, 1766)	Hartebeest	LC	14
Alcelaphus buselaphus ssp. *lichtensteinii* (Peers, 1849)	Lichtenstein's Hartebeest	LC	14
Alcelaphus buselaphus ssp. *caama* (É. Geoffroy Saint-Hilaire, 1803)	Red Hartebeest	LC	14
Antidorcas marsupialis (Zimmermann, 1780)	Springbok	LC	14
Cephalophus dorsalis (Gray, 1846)	Bay Duiker	NT	14
Cephalophus leucogaster (Gray, 1873)	White-bellied Duiker	NT	14
Cephalophus nigrifrons (Gray, 1871)	Black-fronted Duiker	LC	14
Cephalophus silvicultor (Afzelius, 1815)	Yellow-backed Duiker	NT	14
Connochaetes taurinus (Burchell, 1823)	Common Wildebeest	LC	14
Damaliscus lunatus (Burchell, 1823)	Topi	LC	14
Hippotragus equinus (É. Geoffroy Saint-Hilaire, 1803)	Roan Antelope	LC	14
Hippotragus niger (Harris, 1838)	Sable Antelope	LC	14
Hippotragus niger ssp. *variani* (Thomas, 1916)	Giant Sable Antelope	CR	14
Kobus ellipsiprymnus (Ogilbyi, 1833)	Waterbuck	NT	14
Kobus ellipsiprymnus ssp. *defassa* (Ruppell, 1835)	Defassa Waterbuck	NT	14
Kobus ellipsiprymnus ssp. *ellipsiprymnus* (Ogilbyi, 1833)	Common Waterbuck	LC	14
Kobus leche (Gray, 1850)	Southern Lechwe	NT	14
Kobus leche leche (Gray, 1850)	Red Lechwe	NT	14

(continued)

Species	English name	CS[a]	Ref.[b]
Kobus vardonii (Livingstone, 1857)	Puku	NT	14
Madoqua kirkii (Günther, 1880)	Kirk's Dik-dik	LC	14
Oreotragus oreotragus (Zimmermann, 1783)	Klipspringer	LC	14
Oryx gazella (Linnaeus, 1758)	Gemsbok	LC	14
Ourebia ourebi (Zimmermann, 1783)	Oribi	LC	14
Philantomba monticola (Thunberg, 1789)	Blue Duiker	LC	14
Raphicerus campestris (Thunberg, 1811)	Steenbok	LC	14
clinginRedunca arundinum (Boddaert, 1785)	Southern Reedbuck	LC	14
Sylvicapra grimmia (Linnaeus, 1758)	Common Duiker	LC	14
Syncerus caffer (Sparrman, 1779)	Forest Buffalo	LC	14
Tragelaphus oryx (Pallas, 1766)	Common Eland	LC	14
Tragelaphus scriptus (Pallas, 1766)	Bushbuck	LC	14
Tragelaphus spekii (Speke, 1863)	Sitatunga	LC	14
Tragelaphus strepsiceros (Pallas, 1766)	Greater Kudu	LC	14
CETARTIODACTYLA	**Giraffidae**		
Giraffa camelopardalis (Linnaeus, 1758)	Giraffe	VU	14
CETARTIODACTYLA	**Hippopotamidae**		
Hippopotamus amphibius (Linnaeus, 1758)	Hippopotamus	VU	14
CETARTIODACTYLA	**Suidae**		
Phacochoerus africanus (Gmelin, 1788)	Common Warthog	LC	14
Potamochoerus larvatus (F. Cuvier, 1822)	Bushpig	LC	14
Potamochoerus porcus (Linnaeus, 1758)	Red River Hog	LC	14
CETARTIODACTYLA	**Tragulidae**		
Hyemoschus aquaticus (Ogilby, 1841)	Water Chevrotain	LC	14
CHIROPTERA	**Emballonuridae**		
Coleura afra (Peters, 1852)	African Sheath-tailed Bat	LC	8, 29
Saccolaimus peli (Temminck 1853)	Pel's Pouched Bat	LC	8, 29
Taphozous mauritianus (E. Geoffroy, 1818)	Mauritian Tomb Bat	LC	8, 29
CHIROPTERA	**Hipposideridae**		
Hipposideros caffer (Sundevall, 1846)	Sundevall's Leaf-nosed Bat	LC	8, 29
Hipposideros vittatus (Peters, 1852)	Striped Leaf-nosed Bat	NT	8, 29
Hipposideros ruber (Noack, 1893)	Noack's leaf-nosed Bat	LC	8, 29
Macronycteris gigas (Wagner, 1845)	Giant Leaf-nosed Bat	LC	8, 29
Triaenops afer (Peters, 1877)	African Trident Bat	LC	20
CHIROPTERA	**Minipteridae**		
Miniopterus natalensis (A. Smith 1833)	Natal Long-fingered Bat	LC	8,29
CHIROPTERA	**Molossidae**		
Chaerephon ansorgei (Thomas, 1913)	Ansorge's Free-tailed Bat	LC	8, 29
Chaerephon chapini (J.A. Allen, 1917)	Pale Free-tailed Bat	LC	8, 29
Chaerephon nigeriae (Thomas, 1913)	Nigerian Free-tailed Bat	LC	8, 29
Chaerephon pumilus (Cretzschmar, 1826)	Little Free-tailed Bat	LC	8, 29
Mops condylurus (A. Smith, 1833)	Angolan Free-tailed Bat	LC	8, 29
Mops midas (Sundevall, 1843)	Midas Free-tailed Bat	LC	8, 29
Mops niveiventer (Cabrera and Ruxton, 1926)	White-bellied Free-tailed Bat	LC	8, 29

(continued)

Species	English name	CS[a]	Ref.[b]
Otomops martiensseni (Matschie, 1897)	Large-eared Free-tailed Bat	NT	8, 29
Sauromys petrophilus (Roberts, 1917)	Roberts's Flat-headed Bat	LC	8
Tadarida aegyptiaca (E. Geoffroy St.-Hilaire, 1818)	Egyptian Free-tailed Bat	LC	8, 29
CHIROPTERA	**Nycteridae**		
Nycteris arge (Thomas, 1903)	Bate's Slit-faced Bat	LC	8, 29
Nycteris hispida (Schreber, 1775)	Hairy Slit-faced Bat	LC	8, 29
Nycteris intermedia (Aellen, 1959)	Intermediate Slit-faced Bat	LC	8, 29
Nycteris macrotis (Dobson, 1876)	Large-eared Slit-faced Bat	LC	8, 29
Nycteris nana (K. Andersen, 1912)	Dwarf Slit-faced Bat	LC	8, 29
Nycteris thebaica (E. Geoffroy St.-Hilaire, 1818)	Egyptian Slit-faced Bat	LC	8, 29
CHIROPTERA	**Pteropodidae**		
Eidolon helvum (Kerr, 1792)	Straw-coloured Fruit Bat	LC	8, 29
Epomophorus angolensis (Gray, 1870)	Angolan Epauletted Fruit Bat	NT	8, 29
Epomophorus crypturus (Peters, 1852)	Peters's Epauletted Fruit Bat	LC	8, 29
Epomophorus grandis (Sanborn, 1950)	Lesser Angolan Epauletted Fruit Bat	DD	8, 29
Epomophorus wahlbergi (Sundevall, 1846)	Wahlberg's Epauletted Fruit Bat	LC	8, 29
Epomops dobsoni (Bocage, 1889)	Dobson's Epauletted Fruit Bat	LC	8, 29
Epomops franqueti (Tomes, 1860)	Franquet's Epauletted Fruit Bat	LC	8, 29
Hypsignathus monstrosus (H. Allen, 1862)	Hammer-headed Fruit Bat	LC	8, 29
Megaloglossus woermanni (Pagenstecher, 1885)	Woermann's Long-tongued Fruit Bat	LC	8, 29
Micropteropus intermedius (Hayman, 1963)	Hayman's Dwarf Epauletted Fruit Bat	DD	8, 29
Micropteropus pusillus (Peters, 1868)	Peters's Dwarf Epauletted Fruit Bat	LC	8, 29
Myonycteris angolensis (Bocage, 1898)	Angolan Soft-furred Fruit Bat	LC	8, 29
Myonycteris torquata (Dobson, 1878)	Little Collared Fruit Bat	LC	8, 29
Plerotes anchietae (Seabra, 1900)	Anchieta's Fruit Bat	DD	8, 29
Rousettus aegyptiacus (E. Geoffroy St.-Hilaire, 1810)	Egyptian Rousette	LC	8, 29
CHIROPTERA	**Rhinolophidae**		
Rhinolophus damarensis (Roberts, 1946)	Damara Horseshoe Bat	LC	1
Rhinolophus denti (Thomas, 1904)	Dent's Horseshoe Bat	LC	8
Rhinolophus eloquens (K. Andersen, 1905)	Eloquent Horseshoe Bat	LC	19
Rhinolophus fumigatus (Rüppell, 1842)	Rüppell's Horseshoe Bat	LC	8, 29
Rhinolophus lobatus (Peters, 1852)	Peters's Horseshoe Bat	NE	33
CHIROPTERA	**Vespertilionidae**		
Cistugo seabrai (Thomas, 1912)	Angolan Hairy Bat	LC	8, 29
Eptesicus hottentotus (A. Smith, 1833)	Long-tailed Serotine	LC	8, 29
Glauconycteris argentata (Dobson, 1875)	Common Butterfly Bat	LC	8, 29

(continued)

Species	English name	CS[a]	Ref.[b]
Glauconycteris beatrix (Thomas, 1901)	Beatrix's Butterfly Bat	LC	8, 29
Glauconycteris variegata (Tomes, 1861)	Variegated Butterfly Bat	LC	8, 29
Hypsugo anchietae (Seabra, 1900)	Anchieta's Pipistrelle	LC	8, 29
Hypsugo crassulus (Thomas, 1904)	Broad-headed Pipistrelle	LC	20
Kerivoula argentata (Tomes, 1861)	Damara Woolly Bat	LC	34
Kerivoula lanosa (A. Smith, 1847)	Lesser Woolly Bat	LC	29
Laephotis angolensis (Monard, 1935)	Angolan Long-eared Bat	DD	8, 29
Laephotis botswanae (Setzer, 1971)	Botswana Long-eared Bat	LC	26
Mimetillus moloneyi (Thomas, 1891)	Moloney's Flat-headed Bat	LC	20
Mimetillus thomasi (Hinton, 1920)	Thomas's Flat-headed Bat	NE	29
Myotis bocagii (Peters, 1870)	Bocage's Mouse-eared Bat	LC	8, 29
Myotis welwitschii (Gray, 1866)	Welwitsch's Mouse-eared Bat	LC	8, 29
Neoromicia capensis (A. Smith, 1829)	Cape Serotine	LC	8, 29
Neoromicia grandidieri (Dobson, 1876)	Dobson's Pipistrelle	DD	29
Neoromicia nana (Peters, 1852)	Banana Bat	LC	8, 29
Neoromicia tenuipinnis (Peters, 1872)	White-winged Serotine	LC	8, 29
Neoromicia zuluensis (Roberts, 1924)	Zulu Serotine	LC	29
Nycticeinops schlieffeni (Peters, 1859)	Schlieffen's Bat	LC	8, 29
Pipistrellus hesperidus (Temminck, 1840)	Dusky Pipistrelle	LC	20
Pipistrellus rueppellii (J. Fischer, 1829)	Rüppell's Pipistrelle	LC	8, 29
Pipistrellus rusticus (Tomes, 1861)	Rusty Pipistrelle	LC	34
Scotoecus hindei (Thomas, 1901)	Thomas's House Bat	NE	29
Scotophilus dinganii (A. Smith, 1833)	Yellow-bellied House Bat	LC	8, 29
Scotophilus leucogaster (Cretzschmar, 1826)	White-bellied House Bat	LC	29
Scotophilus viridis (Peters, 1852)	Green House Bat	LC	8, 29
ERINACEOMORPHA	**Erinaceidae**		
Atelerix frontalis (A. Smith, 1831)	Southern African Hedgehog	LC	7
HYRACOIDEA	**Procaviidae**		
Dendrohyrax arboreus (A. Smith, 1827)	Southern Tree Hyrax	LC	14
Dendrohyrax dorsalis (Fraser, 1855)	Western Tree Hyrax	LC	14
Heterohyrax brucei (Gray, 1868)	Bush Hyrax	LC	14
Procavia capensis (Pallas, 1766)	Kaokoveld Rock Dassie	LC	14
LAGOMORPHA	**Leporidae**		
Lepus capensis (Linnaeus, 1758)	Cape Hare	LC	18
Lepus victoriae (Thomas, 1893)	African Savanna Hare	LC	31
Pronolagus randensis (Jameson, 1907)	Jameson's Red Rock Hare	LC	31
MACROSCELIDEA I	**Macroscelididae**		
Elephantulus brachyrhynchus (A. Smith, 1836)	Short-snouted Elephant-shrew	LC	15
Elephantulus intufi (A. Smith, 1836)	Bushveld Elephant-shrew	LC	15
Petrodromus tetradactylus (Peters, 1846)	Four-toed Elephant-shrew	LC	15
PERISSODACTYLA	**Equidae**		
Equus quagga (Boddaert, 1785)	Plains Zebra	LC	14
Equus zebra (Linnaeus, 1758)	Mountain Zebra	VU	14

(continued)

Species	English name	CS[a]	Ref.[b]
Equus zebra ssp. *hartmannae* (Matschie, 1898)	Hartmann's Mountain Zebra	VU	14
PERISSODACTYLA	**Rhinocerotidae**		
Diceros bicornis (Linnaeus, 1758)	Black Rhino	CR	14
Diceros bicornis ssp. *bicornis* (Linnaeus, 1758)	South-western Black Rhino	VU	14
Diceros bicornis ssp. *minor* (Drummond, 1876)	Southern-central Black Rhino	CR	14
PHOLIDOTA	**Manidae**		
Phataginus tricuspis (Rafinesque, 1821)	White-bellied Pangolin	VU	22
Smutsia gigantea (Illiger, 1815)	Giant Ground Pangolin	VU	22
Smutsia temminckii (Smuts, 1832)	Temminck's Ground Pangolin	VU	25
PRIMATES	**Cercopithecidae**		
Cercopithecus ascanius (Audebert, 1799)	Red-tailed Monkey	LC	27
Cercopithecus ascanius ssp. *atrinasus* (Machado, 1965)	Black-nosed Red-tailed Monkey	DD	27
Cercopithecus cephus (Linnaeus, 1758)	Moustached Monkey	LC	27
Cercopithecus mitis (Wolf, 1822)	Blue Monkey	LC	28
Cercopithecus mitis ssp. *mitis* (Wolf, 1822)	Pluto Monkey	DD	28
Cercopithecus mitis ssp. *opisthostictus* (Sclater, 1894)	Rump-spotted Blue Monkey	LC	28
Cercopithecus pogonias (Bennett, 1833)	Crowned Monkey	NE	?
Cercopithecus pogonias ssp. *nigripes* (Du Chaillu, 1860)	Black-footed Crowned Monkey	LC	?
Cercopithecus neglectus (Schlegel, 1876)	De Brazza's Monkey	LC	27
Cercopithecus nictitans (Linnaeus, 1766)	Putty-nosed Monkey	LC	27
Chlorocebus cynosuros (Scopoli, 1786)	Malbrouck Monkey	LC	23
Colobus angolensis (P. Sclater, 1860)	Angola Colobus	LC	27
Colobus angolensis ssp. *angolensis* (P. Sclater, 1860)	Sclater's Angolan Colobus	LC	27
Lophocebus aterrimus (Oudemans, 1890)	Black Crested Mangabey	NT	27
Lophocebus terrimus ssp. *opdenboschi* (Schouteden, 1944)	Southern Black Crested Mangabey	DD	27
Miopithecus ogouensis (Kingdon, 1997)	Northern Talapoin Monkey	LC	17
Miopithecus talapoin (Schreber, 1774)	Southern Talapoin Monkey	LC	27
Papio kindae (Lönnberg, 1919)	Kinda Baboon	LC	27
Papio ursinus (Kerr, 1792)	Chacma Baboon	LC	27
Papio ursinus ssp. *urșinus* (Kerr, 1792)	Southern Chacma Babbon	LC	27
PRIMATES	**Galagidae**		
Euoticus elegantulus (Le Conte, 1857)	Southern Needle-clawed Galago	LC	24
Galago moholi (A. Smith, 1836)	Southern Lesser Galago	LC	2
Galagoides demidoff (G. Fischer, 1806)	Demidoff's Dwarf Galago	LC	27
Galagoides kumbirensis (Svensson et al. 2017)	Angolan Dwarf Galago	NE	32
Galagoides thomasi (Elliot, 1907)	Thomas's Dwarf Galago	LC	2
Otolemur crassicaudatus (É. Geoffroy Saint-Hilaire, 1812)	Garnett's Greater Galago	LC	2

(continued)

Species	English name	CS[a]	Ref.[b]
PRIMATES	**Hominidae**		
Gorilla gorilla (Savage, 1847)	Western Gorilla	CR	24
Gorilla gorilla ssp. *gorilla* (Savage, 1847)	Western Lowland Gorilla	CR	24
Pan troglodytes (Blumenbach, 1799)	Chimpanzee	EN	24
Pan troglodytes ssp. *troglodytes* (Blumenbach, 1799)	Central Chimpanzee	EN	24
PRIMATES	**Lorisidae**		
Arctocebus aureus de (Winton, 1902)	Golden Potto	LC	2
Perodicticus edwardsi (Bouvier, 1879)	Milne-Edwards's Potto	LC	2
PROBOSCIDEA	**Elephantidae**		
Loxodonta africana (Blumenbach, 1797)	Savanna Elephant	NE	14
Loxodonta cyclotis (Matschie, 1900)	Forest elephant	NE	14
RODENTIA	**Anomaluridae**		
Anomalurus beecrofti (Fraser, 1853)	Beecroft's Scaly-tailed Squirrel	LC	30
Anomalurus derbianus (Gray, 1842)	Lord Derby's Scaly-tailed Squirrel	LC	30
RODENTIA	**Bathyergidae**		
Fukomys bocagei (de Winton, 1897)	Bocage's Mole Rat	LC	30
Fukomys damarensis (Ogilby, 1838)	Damara Mole Rat	LC	30
RODENTIA	**Gliridae**		
Graphiurus angolensis (de Winton, 1897)	Angolan African Dormouse	DD	30
Graphiurus kelleni (Reuvens, 1890)	Kellen's Dormouse	LC	30
Graphiurus lorraineus (Dollman, 1910)	Lorrain Dormouse	LC	30
Graphiurus monardi (St. Leger, 1936)	Monard's Dormouse	DD	30
Graphiurus rupicola (Thomas & Hinton, 1925)	Stone Dormouse	LC	30
RODENTIA	**Hystricidae**		
Hystrix africaeaustralis (Peters, 1852)	Cape Porcupine	LC	30
RODENTIA	**Muridae**		
Aethomys bocagei (Thomas, 1904)	Bocage's Rock Rat	LC	11, 30
Aethomys chrysophilus (de Winton, 1897)	Red Rock Rat	LC	11, 30
Aethomys kaiseri (Noack, 1887)	Kaiser's Rock Rat	LC	11, 30
Aethomys nyikae (Thomas, 1897)	Nyika Rock Rat	LC	30
Aethomys thomasi (de Winton, 1897)	Thomas's Rock Rat	LC	11, 30
Colomys goslingi (Thomas & Wroughton, 1907)	African Wading Rat	LC	11, 30
Dasymys cabrali (Verheyen et al. 2003)	Cabral's Marsh Rat	NE	30
Dasymys cf incomtus	African Marsh Rat	LC	30
Dasymys nudipes (Peters, 1870)	Angolan Shaggy Rat	DD	11, 30

(continued)

Species	English name	CS[a]	Ref.[b]
Desmodillus auricularis (A. Smith, 1834)	Cape Short-eared Gerbil	LC	11, 30
Gerbilliscus brantsii (A. Smith, 1836)	Highveld Gerbil	LC	11, 30
Gerbilliscus leucogaster (Peters, 1852)	Bushveld Gerbil	LC	11, 30
Gerbilliscus paeba (A. Smith, 1836)	Hairy-footed Gerbil	LC	11, 30
Gerbilliscus setzeri (Schlitter, 1973)	Setzer's Hairy-footed Gerbil	LC	11, 30
Gerbilliscus validus (Bocage, 1890)	Savanna Gerbil	LC	11, 30
Grammomys dolichurus (Smuts, 1832)	Woodland Thicket Rat	LC	11, 30
Grammomys poensis (Eisentraut, 1965)	Shining Thicket Rat	NE	11, 30
Hylomyscus carillus (Thomas, 1904)	Angolan Wood Mouse	LC	11, 30
Hylomyscus heinrichorum (Carleton et al. 2015)	Heirich's Hylomyscus	NE	6
Lemniscomys griselda (Thomas, 1904)	Griselda's Striped Grass Mouse	LC	11
Lemniscomys striatus (Linnaeus, 1758)	Typical Striped Grass Mouse	LC	11, 30
Lophuromys angolensis (Verheyen et al. 2000)	Angolan's Brush-furred Rat	NE	30
Lophuromys rita (Dollman, 1910)	Dollman's Brush-furred Rat	NE	30
Malacomys longipes (Milne-Edwards, 1877)	Big-eared Swamp Rat	LC	11, 30
Mastomys natalensis (Smith, 1834)	Natal Multimammate Mouse	LC	11, 30
Mastomys shortridgei (St. Leger, 1933)	Shortridge's Multimammate Mouse	LC	11, 30
Micaelamys namaquensis (A. Smith, 1834)	Namaqua Rock Rat	LC	11,30
Mus callewaerti (Thomas, 1925)	Callewaert's Mouse	LC	11,30
Mus indutus (Thomas, 1910)	Desert Pygmy Mouse	LC	30
Mus minutoides (Smith, 1834)	Pygmy Mouse	LC	30
Mus setzeri (Petter, 1978)	Setzer's Mouse	LC	30
Mus sorella (Thomas, 1909)	Thoma's Mouse	LC	30
Mus triton (Thomas, 1909)	Gray-bellied Pygmy Mouse	LC	11, 30
Myomyscus angolensis (Bocage, 1890)	Angolan Multimammate Mouse	LC	11, 30
Oenomys hypoxanthus (Pucheran, 1855)	Rufous-nosed Rat	LC	11, 30
Otomys anchietae (Bocage, 1882)	Angolan Vlei Rat	LC	11, 30
Otomys angoniensis (Wroughton, 1906)	Angoni Vlei Rat	LC	11, 30

(continued)

Species	English name	CS[a]	Ref.[b]
Otomys cuanzensis (Hill & Carter, 1937)	Kuanza Vlei Rat	LC	30
Pelomys campanae (Huet, 1888)	Bell Groove-toothed Swamp Rat	LC	11, 30
Pelomys fallax (Peters, 1852)	Creek Groove-toothed Swamp Rat	LC	11, 30
Pelomys minor (Cabrera & Ruxton, 1926)	Least Groove-toothed Swamp Rat	LC	11, 30
Praomys coetzeei (Van der Straeten, 2008)	Coetzee Praomys	NE	30, 35
Praomys jacksoni (de Winton, 1897)	Jackson's Soft-furred Mouse	LC	11, 30
Rhabdomys bechuanae (Thomas, 1893)	Thoma's Four-striped Grass Mouse	NE	30
Rhabdomys dilectus (de Winton, 1897)	Mesic Four-striped Grass Rat	NE	11, 30
Thallomys nigricauda (Thomas, 1882)	Black-tailed Tree Rat	LC	11, 30
Zelotomys hildegardeae (Thomas, 1902)	Hildegarde's Broad-headed Mouse	LC	11, 30
Zelotomys woosnami (Schwann, 1906)	Woosnam's Broad-headed Mouse	LC	34
Rodentia	**Nesomydae**		
Cricetomys ansorgei (Thomas, 1904)	Southern Giant Pouched Rat	LC	30
Cricetomys emini (Wroughton, 1910)	Forest Giant Pouched Rat	LC	30
Dendromus leucostomus (Monard, 1933)	Gray African Climbing Mouse	LC	30
Dendromus melanotis (A. Smith, 1834)	Gray African Climbing Mouse	LC	30
Dendromus mystacalis (Heuglin, 1863)	Chestnut Climbing Mouse	LC	30
Dendromus nyikae (Wroughton, 1909)	Nyika Climbing Mouse	LC	30
Dendromus vernayi (Hill & Carter, 1937)	Vernay's Climbing Mouse	DD	30
Malacothrix typica (A. Smith, 1834)	Gerbil Mouse	LC	30
Petromyscus collinus (Thomas & Hinton, 1925)	Pygmy Rock Mouse	LC	30
Petromyscus shortridgei (Thomas, 1926)	Shortridge's Rock Mouse	LC	30
Saccostomus campestris (Peters, 1846)	Southern African Pouched Mouse	LC	30
Steatomys bocagei (Thomas, 1892)	Bocage's Fat Mouse	LC	30
Steatomys krebsii (Peters, 1852)	Kreb's Fat Mouse	LC	30
Steatomys parvus (Rhoads, 1896)	Tiny Fat Mouse	LC	30
Steatomys pratensis (Peters, 1846)	Fat Mouse	LC	30
Rodentia	**Pedetidae**		
Pedetes capensis (Forster, 1778)	Spring Hare	LC	30
Rodentia	**Petromuridae**		
Petromus typicus (A. Smith, 1831)	Dassie Rat	LC	30
Rodentia	**Sciuridae**		
Funisciurus bayonii (Bocage, 1890)	Lunda Rope Squirrel	DD	30

(continued)

Species	English name	CS[a]	Ref.[b]
Funisciurus congicus (Kuhl, 1820)	Congo Rope Squirrel	LC	30
Funisciurus lemniscatus (Le Conte, 1857)	Ribboned Rope Squirrel	LC	30
Funisciurus pyrropus (F. Cuvier, 1833)	Fire-footed Rope Squirrel	LC	30
Heliosciurus gambianus (Ogilby, 1835)	Gambian Sun Squirrel	LC	30
Paraxerus boehmi (Reichenow, 1886)	Boehm's Bush Squirrel	LC	30
Paraxerus cepapi (A. Smith, 1836)	Smith's Bush Squirrel	LC	30
Protoxerus stangeri (Waterhouse, 1842)	African Giant Squirrel	LC	30
Xerus princeps (Thomas, 1929)	Damara Ground Squirrel	LC	30
RODENTIA	**Thryonomydae**		
Thryonomys swinderianus (Temminck, 1827)	Greater Cane Rat	LC	30
SIRENIA	**Trichechidae**		
Trichechus senegalensis (Link, 1795)	African Manatee	VU	14
SORICOMORPHA	**Soricidae**		
Crocidura cyanea (Duvernoy, 1838)	Reddish-gray Musk Shrew	LC	25
Crocidura erica (Dollman, 1915)	Heather Shrew	DD	15
Crocidura fuscomurina (Heuglin, 1865)	Bicolored Musk Shrew	LC	15
Crocidura hirta (Peters, 1852)	Lesser Red Musk Shrew	LC	15
Crocidura luna (Dollman, 1910)	Greater Gray-brown Musk Shrew	LC	21
Crocidura mariquensis (A. Smith, 1844)	Swamp Musk Shrew	LC	15
Crocidura nigricans (Bocage, 1889)	Blackish White-toothed Shrew	LC	9
Crocidura nigrofusca (Matschie, 1895)	African Black Shrew	LC	15
Crocidura olivieri (Lesson, 1827)	African giant shrew	LC	15
Crocidura parvipes (Osgood, 1910)	Small-footed Shrew	LC	15
Crocidura roosevelti (Heller, 1910)	Roosvelt's Shrew	LC	25
Crocidura turba Dollman, 1910	Turbo Shrew	LC	15
Suncus lixus (Thomas, 1898)	Greater Dwarf Shrew	LC	15
Suncus megalura (Jentink, 1888)	Climbing Shrew	LC	15
Suncus varilla (Thomas, 1895)	Lesser Dwarf Shrew	LC	34
TUBULIDENTATA	**Orycteropodidae**		
Orycteropus afer (Pallas, 1766)	Aardvark	LC	14

[a]IUCN Conservation Status categories. *CR* critically endangered, *EN* endangered, *VU* vulnerable, *NT* near threatened, *LC* least concern, *DD* data deficient, *NE* not evaluated

[b]References – 1. Monadjem et al. (2017); 2. Bersacola et al. (2015); 3. du Bocage (1865); 4. du Bocage (1882); 5. du Bocage (1890); 6. Carleton et al. (2015); 7. Cassola (2016b); 8. Crawford-Cabral (1986); 9. Crawford-Cabral (1987); 10. Crawford-Cabral (1996); 11. Crawford-Cabral (1998); 12. Crawford-Cabral and Simões (1987); 13: Crawford-Cabral and Simões (1988); 14. Crawford-Cabral and Veríssimo (2005); 15. Crawford-Cabral and Veríssimo, unpublished data; 16: Errol de Beer, unpublished data; 17. Gautier-Hion (2013b); 18. Genest-Villard (1969); 19. Grant and Ferguson (2018); 20. Happold and Happold (2013); 21. Hayman (1963); 22. Hill and Carter (1941); 23. Huntley (1973c); 24. Huntley (1973e); 25. IUCN RedList Map; 26. Monadjem et al. (2010a); 27. Machado (1969); 28. Machado and Crawford-Cabral (1999); 29. Monadjem et al. (2010a); 30. Monadjem et al. (2015); 31. Smith et al. (2018); 32. Svensson et al. (2017); 33. Taylor et al. (2018c); 34. Taylor (in press); 35. Van der Straeten (2008); 36. Meÿer (2007)

References

Adams WB (2013) Against extinction: the story of conservation. Routledge, London

Ambrose L, Butynski TM (2013) Galagoides demidovii Demidoff's Dwarf Galago. In: Butynski TM, Kingdon J, Kalina J (eds) The mammals of Africa. Vol II (Primates). Bloomsbury Publishing, London, pp 459, 556 pp–461

Angelici FM, Do Linh San E (2015) *Crossarchus ansorgei*. The IUCN Red List of threatened species 2015:e.T41594A45205422. Downloaded on 10 April 2018

Angelici FM, Gaubert P, Do Linh San E (2016) *Genetta maculata*. The IUCN Red List of threatened species 2016:e.T41699A45218948. Downloaded on 12 April 2018

Ansell WFH (1972) Part 15. Order Artiodactyla. In: Meester JA, Setzer HW (eds) The mammals of Africa: an identification manual. Smithsonian Institution Press, Washington, DC, pp 1, 432 pp–93

Anstey S (1991) Plano de Conservação do Elefante para Angola. Unpublished report. Ministério da Agricultura. Instituto de Desenvolvimento Florestal, Luanda

Anstey S (1993) Angola: elephants, people and conservation – a preliminary assessment of the status and conservation of elephants in Angola. Unpublished report. IUCN Regional Office for Southern Africa, Harare

Bahaa-el-din L, Mills D, Hunter L, Henschel P (2015) *Caracal aurata*. The IUCN Red List of threatened species 2015:e.T18306A50663128. Downloaded on 12 April 2018

Barnett R, Yamaguchi N, Shapiro B et al (2014) Revealing the maternal demographic history of *Panthera leo* using ancient DNA and a spatially explicit genealogical analysis. BMC Evolut Biol 14:70

Bauer H, Chapron G, Nowell K et al (2015) Lion (*Panthera leo*) populations are declining rapidly across Africa, except in intensively managed areas. Proc Natl Acad Sci U S A 112:14894–14899

Baxter R, Hutterer R, Griffin M, Howell K (2016) *Crocidura cyanea* (errata version published in 2017). The IUCN Red List of threatened species 2016:e.T40625A115176043. Downloaded on 10 May 2018

Bearder S (2008) *Otolemur crassicaudatus*. The IUCN Red List of threatened species 2008:e. T15643A4943752. Downloaded on 20 March 2018

Bearder SK, Svoboda NS (2013) *Otolemur crassicaudatus* Large-eared Greater Galago. In: Butynski TM, Kingdon J, Kalina J (eds) The mammals of Africa. Vol II (Primates). Bloomsbury Publishing, London, pp 409, 556 pp–413

Bearder SK, Ambrose L, Harcourt C et al (2003) Species-typical patterns of infant contact, sleeping site use and social cohesion among nocturnal primates in Africa. Folia Primatologica 74:337–254

Bearder S, Butynski TM, Hoffmann M (2008) *Galago moholi*. The IUCN Red List of threatened species 2008:e.T8788A12932349. Downloaded on 20 March 2018

Benda P, Vallo P (2009) Taxonomic revision of the genus *Triaenops* (Chiroptera: Hipposideridae) with description of a new species from southern Arabia and definitions of a new genus and tribe. Folia Zoologica 58(Monograph 1):1–45

Bersacola E, Svensson MS, Bearder SK et al (2014) Hunted in Angola: surveying the bushmeat trade. SWARA 2014(January–March):31–36

Bersacola E, Svensson MS, Bearder SK (2015) Niche partitioning and environmental factors affecting abundance of strepsirrhines in Angola. Am J Primatol 77:1179–1192

Bessa J, Sousa C, Hockings KJ (2015) Feeding ecology of chimpanzees (*Pan troglodytes verus*) inhabiting a forest-mangrove-savanna-agricultural matrix at Caiquene-Cadique, Cantanhez National Park, Guinea-Bissau. Am J Primatol 77:651–665

Biggs D, Courchamp F, Martin R et al (2013) Legal trade of Africa's rhino horns. Science 339:1038–1039

Blaine G (1922) Notes on the Zebras and some Antelopes of Angola. J Zool 92:317–339

Blaine G (1925) New subspecies of *Connochaetes taurinus*. Ann Mag Nat Hist 15(Series 9):129–130

Boesch C, Boesch-Achermann H (2000) The Chimpanzees of the Taï Forest: behavioural ecology and evolution. Oxford University Press, New York, 328 pp

Bohm T, Höner OR (2015) *Crocuta crocuta*. The IUCN Red List of threatened species 2015:e. T5674A45194782. Downloaded on 11 April 2018

Bohm C, Jonsson C (2017) Vertebrates of the Gothenburg Natural History Museum (GNM). Version 4.2. Gothenburg Natural History Museum, Gothenburg Accessed via GBIF.org on 24 Feb 2018

Brashares JS, Arcese P (2013) *Ourebia ourebi* Oribi. In: Kingdon J, Hoffmann M (eds) The mammals of Africa. Vol VI (Hippopotamuses, pigs, deer, giraffe and bovids). Bloomsbury Publishing, London, pp 406, 680 pp–413

Broom J, Milton S, Davis C et al (1974) Expedition to South-Western Angola June/July 1974. Unpublished report. University of Cape Town/Wild Life Society, Cape Town

Brotherton PNM (2013) *Madoqua (kirkii)* Kirk's Dik-Dik Species Group. In: Kingdon J, Hoffmann M (eds) The mammals of Africa. Vol VI (Hippopotamuses, pigs, deer, giraffe and bovids). Bloomsbury Publishing, London, pp 327, 680 pp–333

Butynski TM (2008) *Chlorocebus cynosuros*. The IUCN Red List of threatened species 2008:e. T136291A4270290. Downloaded on 21 March 2018

Butynski TM, De Jong YA (2007) Distribution of the potto *Perodicticus potto* (Primates: Lorisidae) in Eastern Africa, with a description of a new subspecies from Mount Kenya. J East Afr Nat Hist 96:113–147

Carleton MD, Banasiak RA, Stanley WT (2015) A new species of the rodent genus *Hylomyscus* from Angola, with a distributional summary of the *H. anselli* species group (Muridae: Murinae: Praomyini). Zootaxa 4040:101–128

Carmignani E (2015) Elephant assessment and report on the status on Quissama National Park's Special Conservation Area October – November 2015. Unpublished report. Instituto Nacional de Biodiversidade e Áreas de Conservação, Luanda

Carr T, Carr N, David JHM (1985) A record of the sub-Antarctic fur seal Arctocephalus tropicalis in Angola. Afr Zool 20:77

Cassola F (2016a) *Petromus typicus*. The IUCN Red List of threatened species 2016:e. T16776A22240649. Downloaded on 19 April 2018

Cassola F (2016b) *Atelerix frontalis* (errata version published in 2017). The IUCN Red List of threatened species 2016:e.T2274A115061260. Downloaded on 19 April 2018

Cassola F (2016c) *Crocidura olivieri* (errata version published in 2017). The IUCN Red List of threatened species 2016:e.T41348A115180235. Downloaded on 01 May 2018

Cassola F (2016d) *Crocidura hirta* (errata version published in 2017). The IUCN Red List of threatened species 2016:e.T41323A115178068. Downloaded on 10 May 2018

Cerling TE, Barnette JE, Chesson LA et al (2016) Radiocarbon dating of seized ivory confirms rapid decline in African elephant populations and provides insight into illegal trade. Proc Natl Acad Sci U S A 113:13330–13335

Chardonnet P, Crosmary W (2013) *Hippotragus equinus* Roan Antelope. In: Kingdon J, Hoffmann M (eds) The mammals of Africa. Vol VI (Hippopotamuses, pigs, deer, giraffe and bovids). Bloomsbury Publishing, London, pp 548, 680 pp–556

Chase MJ, Griffin CR (2011) Elephants of south-east Angola in war and peace: their decline, re-colonization and recent status. Afr J Ecol 49:353–361

Chase MJ, Schlossberg S (2016) Dry-season fixed-wing aerial survey of elephants and other large mammals in Southeast Angola. Unpublished report. Elephants Without Borders, Kasane, Botswana

Ciofolo I, Le Pendu Y (2013) *Giraffa camelopardalis* Giraffe. In: Kingdon J, Hoffmann M (eds) The mammals of Africa. Vol VI (Hippopotamuses, pigs, deer, giraffe and bovids). Bloomsbury Publishing, London, pp 98, 680 pp–110

Clark VR, Barker NP, Mucina L (2011) The Great Escarpment of southern Africa: a new frontier for biodiversity exploration. Biodiver Conserv 20:2543

Collins T, Keith L, Rosembaum H (2011) Inshore and Congo River Marine Mammals. Final Report. Unpublished report, Ocean Giants Program. Wildlife Conservation Society, New York

Conroy C (2018) MVZ mammal collection (Arctos). Version 35.13. Museum of Vertebrate Zoology, Berkeley. Accessed via GBIF.org on 2018-04-19

Cooper-Bohannon R, Rebelo H, Jones G et al (2016) Predicting bat distributions and diversity hotspots in southern Africa. Hystrix 27:38–48

Crawford-Cabral J (1961) Considerações em torno de *Equus quagga intermedia* Taborda Morais. Boletim do Instituto de Angola 15:77–79

Crawford-Cabral J (1966a) Some new data on Angolan Muridae. Zoologica Africana 2:193–203

Crawford-Cabral J (1966b) Quatro formas de mamíferos novas para Angola. Boletim do Instituto de Investigação Científica de Angola 3:137–148

Crawford-Cabral J (1967) Mamíferos da Reserva do Luando. Boletim do Instituto de Investigação Científica de Angola 4:33–44

Crawford-Cabral J (1968) Notas sobre a variação geográfica da pelagem de alguns carnívoros. Boletim do Instituto de Investigação Científica de Angola 5:15–122

Crawford-Cabral J (1969a) As Genetas de Angola. Boletim do Instituto de Investigação Científica de Angola 6:25–26

Crawford-Cabral J (1969b) A study of the Giant Sable (*Hippotragus niger variani*). News Bulletin of the Zoological Society of Southern Africa 10:1–7 [Erratum: ibid 10: 32]

Crawford-Cabral J (1970a) Alguns aspectos da ecologia da Palanca real. Boletim do Instituto de Investigação Científica de Angola 7:7–42

Crawford-Cabral J (1970b) As Genetas da África Central (República do Zaire, Ruanda e Burundi). Boletim do Instituto de Investigação Científica de Angola 7:3–23

Crawford-Cabral J (1971) A Suricata do Iona, subspécie nova. Boletim do Instituto de Investigação Científica de Angola 8:65–83

Crawford-Cabral J (1982) Esboço zoogeográfico de Angola em ordem à fauna de mamíferos terrestres. Unpublished report

Crawford-Cabral J (1986) A list of Angolan Chiroptera with notes on their distribution. Boletim do Instituto de Investigação Científica de Angola 13:7–48

Crawford-Cabral J (1987) The taxonomic status of *Crocidura nigricans* Bocage, 1889 (Mammalia, Insectivora). Garcia de Orta, Série de Zoologia 14:3–12

Crawford-Cabral J (1992) Parapatry as a secondary event. Garcia de Orta, Série de Zoologia 19:1–6

Crawford-Cabral J (1996) The species of *Galerella* (Mammalia: Carnivora: Herpestinae) occurring in the southwestern corner of Angola. Garcia de Orta, Série de Zoologia, Lisboa 21:7–17

Crawford-Cabral J (1997) A zoogeographical division of Western Angola (Africa), based on the distribution of Muroidea (Rodentia). In: Ulrich H (ed) Tropical biodiversity and systematics: Proceedings of the international symposim on biodiversity and systematics in tropical ecosystems, Bonn, 1994. Zoologisches Forschungsinstitut und Museum Alexander Koenig, Bonn, pp 221–227

Crawford-Cabral J (1998) The Angola rodents of the superfamily Muroidea. An account on their distribution. Estudos, Ensaios e Documentos do Instituto de Investigação Científica Tropical 161:1–222

Crawford-Cabral J, Fernandes CA (2001) The Rusty-spotted Genets as a group with three species in Southern Africa (Carnivora: Viverridae). In: Denys C, Granjon L, Poulet A (eds) African small mammals/Petits Mammifères Africains. IRD Éditions, Paris, pp 65, 570 pp–80

Crawford-Cabral J, Simões AP (1987) Distributional data and notes on Angolan carnivores (Mammalia: Carnivora) I – small and medium-sized species. Garcia de Orta, Série de Zoologia 14:3–27

Crawford-Cabral J, Simões AP (1988) Distributional data and notes on Angolan carnivores (Mammalia: Carnivora) II – larger species. Garcia de Orta, Série de Zoologia 15:9–20

Crawford-Cabral J, Veríssimo LN (2005) The ungulate fauna of Angola: systematic list, distribution maps, database report. Estudos, Ensaios e Documentos do Instituto de Investigação Científica Tropical 163:1–277

Cumming DHM (2013) *Phacochoerus africanus* Common Warthog. In: Kingdon J, Hoffmann M (eds) The mammals of Africa. Vol VI (Hippopotamuses, pigs, deer, giraffe and bovids). Bloomsbury Publishing, London, pp 54, 680 pp–60

Detwiler KM, Burrell AS, Jolly CJ (2005) Conservation implications of hybridization in African cercopithecine monkeys. Int J Primatol 26:661–684

Dodman T, Diop NMD, Sarr K (2008) Conservation strategy for the West African Manatee. UNEP and Wetlands International Africa, Nairobi/Dakar, 128 pp

Drake-Brockman RE (1909) VI. – On a new species and a new subspecies of the genus Madoqua and a new subspecies of the genus *Rhynchotragus*. J Nat Hist 4:48–51

Drake-Brockman RE (1930) 4. A review of the Antelopes of the Genera Madoqua and *Rhynchotragus*. J Zool 100:51–57

du Bocage JVB (1865) 3. Sur quelques Mammifères rares ou peu connus d'Afrique occidentale qui se trouvent au Muséum de Lisbonne. Proc Zool Soc Lond *1865*:401–402

du Bocage JVB (1869) Sur une espèce de '*Cephalophus*' à taille plus forte, d'Afrique occidentale, qui parait identique au '*C. longiceps*' Gray. Jornal de Sciências Mathemáticas, Physicas e Naturaes 2:220–222

du Bocage JVB (1878) Liste des Antilopes d'Angola. Proc Zool Soc Lond 1878:741–745

du Bocage JVB (1879) Subsídios para a Fauna das possessões portuguesas d'Africa occidental. Jornal de Sciências Mathemáticas, Physicas e Naturaes 7:85–96

du Bocage JVB (1882) Liste des mammifères envoyés de Caconda «Angola» par M. D'Anchieta. Jornal de Sciências Mathemáticas, Physicas e Naturaes 9:25–29

du Bocage JVB (1889a) Chiroptères africains nouveaux, rares ou peu connus. Jornal de Sciências Mathemáticas, Physicas e Naturaes, Segunda Série 1:1–7

du Bocage JVB (1889b) Les Damans d'Angola. Jornal de Sciências Mathemáticas, Physicas e Naturaes, Segunda Série 1:186–196

du Bocage JVB (1890) Mammifères d'Angola et du Congo (Suite). Jornal de Sciências Mathemáticas, Physicas e Naturaes, Segunda Série 1:8–32

du Bocage JVB (1897) Mammiferos, Reptis e Batrachios d'Africa de que existem exemplares typicos no Museu de Lisboa. Jornal de Sciências Mathemáticas, Physicas e Naturaes, Segunda Série 4:187–206

du Bocage JVB (1898) Sur une nouvelle espèce de *Cynonycteris* d'Angola. Jornal de Sciências Mathemáticas, Physicas e Naturaes, Segunda Série 5:133–139

du Bocage JVB (1902) Les Antilopes d'Angola. Jornal de Sciências Mathemáticas, Physicas e Naturaes, Segunda Série 4:234–242

Du Toit JT (2013) *Raphicerus campestris* Steenbok. In: Kingdon J, Hoffmann M (eds) The mammals of Africa. Vol VI (Hippopotamuses, pigs, deer, giraffe and bovids). Bloomsbury Publishing, London, pp 311, 680 pp–314

East R (1999) African Antelope Database 1998. Occasional Paper of the IUCN Species Survival Commission No. 21. IUCN, Gland, 434 pp

Emslie RH, Adcock K (2013) *Diceros bicornis* Black Rhinoceros. In: Kingdon J, Hoffmann M (eds) The mammals of Africa. Vol V (Carnivores, Pangolins, Equids and Rhinoceros). Bloomsbury Publishing, London, pp 455, 544 pp–466

Emslie R, Brooks M (1999) African Rhino: status survey and conservation action plan. IUCN, Gland, 92 pp

Estes RD (2013) *Hippotragus niger* Sable Antelope. In: Kingdon J, Hoffmann M (eds) The mammals of Africa. Vol VI (Hippopotamuses, pigs, deer, giraffe and bovids). Bloomsbury Publishing, London, pp 556, 680 pp–565

Estes RD, Estes RK (1974) The biology and conservation of the giant sable antelope, *Hippotragus niger variani* Thomas, 1916. Proc Acad Nat Sci U S A 126:73–104

Fabiano EC, Álvares F, Kosmas S et al (2017) The conservation status of the Endangered African Wild Dogs in Angola: an historical and contemporary perspective. Final progress report (Project No.162513063). Unpublished report. University of Namibia, Katima Mulilo

Faulkes C, Maree S, Griffin M (2016) *Fukomys bocagei*. The IUCN Red List of threatened species 2016:e.T5752A22184407. Downloaded on 19 April 2018

Feiler A (1986) Zur faunistik and biometrie angolanisher Fledermause. Zoologische Abhandlungen, Staatliches Museum für Tierkunde Dresden 42:65–77

Feiler A (1989) Individuelle variation bei Buschböcken (*Tragelaphus scriptus*) aus Angola (Mammalia, Artiodactyla: Bovidae). Zoologische Abhandlungen, Staatliches Museum für Tierkunde Dresden 44:149–154

Feiler A (1990) Distribution of mammals in Angola and notes on biogeography. In: Peters G, Hutterer R (eds) Vertebrates in the tropics: Proceedings of the international symposium on vertebrate biogeography and systematics in the tropics, Bonn, June 5–8, 1989. Alexander Koenig Zoological Research Institute and Zoological Museum, Bonn, pp 221–236

Fennessy J, Bidon T, Reuss F et al (2016) Multi-locus analyses reveal four giraffe species instead of one. Curr Biol 26:2543–2549

Fenykovi J (1953) Angola, en el visor del rifle y de la cámara. Cayrel Ediciones, Madrid, 255 pp

Figueira R (2017) IICT Colecção Zoológica. Version 4.2. Instituto de Investigação Científica Tropical, Lisbon. Accessed via GBIF.org on 23 Feb 2018

Figueiredo E, Smith GF, César J (2009) The flora of Angola: first record of diversity and endemism. Taxon 58:233–236

Flagstad Ø, Syversten PO, Stenseth NC et al (2001) Environmental change and rates of evolution: the phylogeographic pattern within the hartebeest complex as related to climatic variation. Proc R Soc Lond B 268:667–677

Foley NM, Thong VD, Soisook P et al (2015) How and why overcome the impediments to resolution: lessons from rhinolophid and hipposiderid bats. Mol Biol Evolut 32:313–333

Frade F (1933) Eléphants d'Angola. Bulletin de la Société Portugaise des Sciences Naturelles 11:319–333

Frade F (1936) Distribution géographique des eléphants d'Afrique. Compte Rendu du XII Congres International de Zoologie, Lisbonne 1935:1191–1202

Frade F (1955) Ordre des Proboscidiens (Proboscidea Illiger, 1811). In: Grassé PP (ed) Traité de Zoologie. Anatomie, Systématique, Biologie, Tome XVII, 1° fascicule. Masson et Cie, Paris, pp 715–783

Frade F (1956) Reservas naturais de Angola – I (alguns mamíferos da Reserva da Quiçama). Anais da Junta de Investigações do Ultramar 11:228–245

Frade F (1958) Mesures adoptées pour la protection de l'hippotrague géant en Angola. Mammalia 22:476–477

Frade F (1959a) Medidas para a protecção da Palanca gigante de Angola (*Hippotragus niger variani* Thomas). Memórias da Junta de Investigações do Ultramar (Série 2) 8:11–18

Frade F (1959b) Breve notícia a propósito da Reserva da Quiçama. Garcia de Orta 4:215–223

Frade F (1960) Os animais na etnologia ultramarina. Estudos, Ensaios e Documentos da Junta de Investigações do Ultramar 84:211–240

Frade F (1963) Linhas gerais da distribuição dos Vertebrados em Angola. Memórias da Junta de Investigações do Ultramar 43:241–257

Frade F, Sieiro DM (1960) Palanca preta gigante de Angola. Garcia de Orta 8:21–38

França P (1967) Sur la présence d'*Arctocephalus pusillus* (Schreber) (Otariidae) et de *Mirounga leonina* (Linne) (Phocidae) au sud de l'Angola. Mammalia 31:50–54

Fritz H, Bourgarel M (2013) *Aepyceros melampus* Impala. In: Kingdon J, Hoffmann M (eds) The mammals of Africa. Vol VI (Hippopotamuses, pigs, deer, giraffe and bovids). Bloomsbury Publishing, London, pp 480, 680 pp–487

Funston P, Henschel P, Petracca L et al (2017) The distribution and status of lions and other large carnivores in Luengue-Luiana and Mavinga National Parks, Angola. KAZA TFCA Secretariat, Kasane

Galvão HCF, Montês A (1943–1945) A Caça no Império Português, 2 vols. Editorial Primeiro de Janeiro, Porto, 639 pp

Gaubert P, Fischer C, Hausser Y et al (2016) *Genetta angolensis*. The IUCN Red List of threatened species 2016:e.T41696A45218468. Downloaded on 10 March 2018

Gautier-Hion A (2013a) *Miopithecus talapoin* Southern talapoin monkey. In: Butynski TM, Kingdon J, Kalina J (eds) The mammals of Africa. Vol II (Primates). Bloomsbury Publishing, London, pp 252, 556 pp–253

Gautier-Hion A (2013b) *Miopithecus ogouensis* – Northern talapoin monkey. In: Butynski TM, Kingdon J, Kalina J (eds) The mammals of Africa. Vol II (Primates). Bloomsbury Publishing, London, pp 253, 556 pp–256

Gautier-Hion A, Colyn M, Gautier J-P (1999) Histoire Naturelle des Primates d'Afrique Centrale. ECOFAC, Libreville, 162 pp

Genest-Villard H (1969) Particularités des lièvres du Sud-Ouest de l'Angola. Mammalia 33:124–132

Gerrie R, Kennerley R (2016) *Crocidura erica* (errata version published in 2017). The IUCN Red List of threatened species 2016:e.T5626A115078377. Downloaded on 01 May 2018

Gilchrist JS, Jennings AP, Veron G, Cavallini P (2009) Family Herpestidae. In: Wilson DE, Mittermeier RA (eds) Handbook of the mammals of the world, Vol I (Carnivores). Lynx Edicions, Barcelona, pp 262, 728 pp–328

Gosling LM, Capellini I (2013) *Alcephalus buselaphus* Hartebeest. In: Kingdon J, Hoffmann M (eds) The mammals of Africa. Vol VI (Hippopotamuses, pigs, deer, giraffe and bovids). Bloomsbury Publishing, London, pp 511, 680 pp–526

Grant S, Ferguson A (2018) Field Museum of Natural History (Zoology) Mammal Collection. Version 9.3. Field Museum, Chicago. Accessed via GBIF.org on 19 Apr 2018

Gray JE (1868) VI. – Revision of the species of hyrax, founded on the specimens in the British Museum. Ann Mag Nat Hist Ser 4 1:35–51

Gray JE (1869) New species of hyrax. Ann Mag Nat Hist Ser 4 3:242–243

Groom R, Elizalde S, Elizalde D, de Sá S, Alexandre G (2018) Large and medium sized terrestrial mammals survey in Quiçama National Park. Preliminary results report based on species presence and bushmeat. Instituto Nacional de Áreas de Conservação, Luanda, 17 pp

Groves C, Grubb P (2011) Ungulate taxonomy. Johns Hopkins University Press, Baltimore, 336 pp

Grubb P (1972) Variation and incipient speciation in the African buffalo. Zeitschrift für Säugetierkunde 37:121–144

Grubb P (1993) The afrotropical suids (*Phacochoerus*, *Hylochoerus* and *Potamochoerus*). In: Oliver WLR (ed) Status survey and conservation plan – pigs, peccaries and hippos. IUCN, Gland, pp 66, 202 pp–75

Grubb P, Groves CP, Dudley JP, Shoshani J (2000) Living African elephants belong to two species: *Loxodonta africana* (Blumenbach 1797) and *Loxodonta cyclotis* (Matschie 1900). Elephant 2:1–4

Hall BP (1960) The faunistic importance of the scarp of Angola. Ibis 102:420–442

Hall-Martin A, Pienaar D (1992) A note on the elephants of Southeast Angola. Unpublished report. African Elephant and Rhino Specialist Group, Nairobi

Hansen MC, Potapov PV, Moore R et al (2013) High-resolution global maps of 21st-century forest cover change. Science 342:850–853

Happold DCD (2013) The mammals of Africa. Vol III (Rodents, hares and rabbits). Bloomsbury Publishing, London, 789 pp

Happold M, Happold DCD (2013) The mammals of Africa. Vol IV (Hedgehogs, shrews and bats). Bloomsbury Publishing, London, 800 pp

Happold DCD, Wendelen W (2006) The distribution of *Poelagus marjorita* (Lagomorpha: Leporidae) in central Africa. Mammalian Biology-Zeitschrift für Säugetierkunde 71:377–383

Hart JA, Kingdon J (2013) *Philatomba monticola* Blue Duiker. In: Kingdon J, Hoffmann M (eds) The mammals of Africa. Vol VI (Hippopotamuses, pigs, deer, giraffe and bovids). Bloomsbury Publishing, London, pp 228, 680 pp–234

Hart J, Groves CP, Ehardt C (2008) *Lophocebus aterrimus* ssp. *opdenboschi*. The IUCN Red List of threatened species 2008:e.T12311A3334719. Downloaded on 21 March 2018

Hassanin A, Delsuc F, Ropiquet A et al (2012) Pattern and timing of diversification of Cetartiodactyla (Mammalia, Laurasiatheria), as revealed by a comprehensive analysis of mitochondrial genomes. Comptes Rendus Biologies 335:32–50

Hatt RT (1934) A manatee collected by the American Museum Congo Expedition: with observations on the recent manatees. Bull Am Mus Nat Hist 66:553–566

Hayman RW (1951) Notes on some Angolan Mammals. Publicações Culturais da Companhia de Diamantes de Angola 11:33–35

Hayman RW (1963) Mammals from Angola, mainly from the Lunda District. Publicações Culturais da Companhia de Diamantes de Angola 66:81–139

Heffernan J (2005) Elephants of Cabinda; Mission report, Angola, April 2005. Unpublished report. Fauna & Flora International & United Nations Development Programme in co-operation with the Deptartment of Urban Affairs & Environment, Cabinda

Heim de Balzac H, Meester J (1977) Part 1. Order Insectívora. In: Meester JA, Setzer HW (eds) The mammals of Africa: an identification manual. Smithsonian Institution Press, Washington, DC

Hill JE (1941) A collection of mammals from Dondi, Angola. J Mammal 22:81–85

Hill JE, Carter TD (1941) The mammals of Angola, Africa. Bull Am Mus Nat Hist 78:1–211

Hinton MAC (1921) Klipspringers of Rhodesia, Angola and Northern Nigeria. Ann Mag Nat Hist Ser 9 8:129–133

Hoare R (2015) Lessons from 20 years of human–elephant conflict mitigation in Africa. Hum Dimens Wildl 20:289–295

Hockings KJ, Anderson JR, Matsuzawa T (2012) Socioecological adaptations by chimpanzees, *Pan troglodytes verus*, inhabiting an anthropogenically impacted habitat. Anim Behav 83:801–810

Hoffmann M, Wilson V (2013) *Raphicerus sharpei* Sharpe's Grysbok. In: Kingdon J, Hoffmann M (eds) The mammals of Africa. Vol VI (Hippopotamuses, pigs, deer, giraffe and bovids). Bloomsbury Publishing, London, pp 308, 680 pp–310

Hofmeyr GJG (2015) *Arctocephalus pusillus*. The IUCN Red List of threatened species 2015: e.T2060A45224212. Downloaded on 10 May 2018

Huntley BJ (1972a) Relatório do Ecólogo Sobre a Ocupação do Parque Nacional da Quiçama pela Pecuária da Barra do Cuanza. Unpublished report. Direcção Provincial dos Serviços de Veterinária, Luanda, 11pp

Huntley BJ (1972b) Plano para o Futuro da Palanca Real em Angola. Unpublished report. Direcção Provincial dos Serviços de Veterinária, Luanda

Huntley BJ (1972c) Report on visit to Iona National Park. 24 February to 25 March 1972. Unpublished report. Direcção Provincial dos Serviços de Veterinária, Luanda, 13 pp

Huntley BJ (1973a) Aspectos Gerais da Conservação do Bravio em Angola. Relatório N° 15. Unpublished report. Direcção Provincial dos Serviços de Veterinária, Luanda, 30 pp

Huntley BJ (1973b) Reordenamento da População Humana no Parque Nacional da Quiçama. Relatório N° 19. Unpublished report. Direcção Provincial dos Serviços de Veterinária, Luanda, 10 pp

Huntley BJ (1973c) Distribuição e Situação da Grande Fauna Selvagem de Angola com Referência Especial às Espécies Raras e em Perigo de Extinção – Primeiro Relatório Sobre o Estado Actual. Relatório N°. 21. Unpublished report. Direcção Provincial dos Serviços de Veterinária, Luanda, 37 pp

Huntley BJ (1973d) Parque Nacional do Iona: Administração, Maneio, Investigação e Turismo. Relatório N°. 23. Unpublished report. Direcção Provincial dos Serviços de Veterinária, Luanda, 37 pp

Huntley BJ (1973e) Proposta para a Criação de uma Reserva Natural Integral na Floresta do Maiombe, Cabinda. Unpublished report. Direcção Provincial dos Serviços de Veterinária, Luanda, Angola, 9 pp

Huntley BJ (1973f) Distribution of larger mammals of Angola according to vegetation types of Dr. Grandvaux Barbosa's Map. Unpublished report

Huntley BJ (1974) Outlines of wildlife conservation in Angola. S Afr J Wildl Res 4:157–166

Huntley BJ (2017) Wildlife at war in Angola. The rise and fall of an African Eden. Protea Book House, Pretoria, 432 pp

Huntley BJ, Francisco P (eds) (2015) Avaliação Rápida da Biodiversidade de Região da Lagoa Carumbo, Lunda-Norte – Angola/Rapid Biodiversity Assessment of the Carumbo Lagoon Area, Lunda-Norte – Angola. Ministério do Ambiente, Luanda, 219 pp

Huntley BJ, Matos E (1992) Angola environment status Quo Assessment report. Unpublished report. IUCN Regional Office for Southern Africa, Harare, 255 pp

Huntley BJ, Beja P, Vaz Pinto P et al (2019) Biodiversity conservation: history, protected areas and hotspots. In: Huntley BJ, Russo V, Lages F, Ferrand N (eds) Biodiversity of Angola. Science & conservation: a modern synthesis. Springer, Berlin

Hutterer R (2016) *Crocidura nigricans*. The IUCN Red List of threatened species 2016:e. T41345A22310112. Downloaded on 01 May 2018

Hutterer R, Peterhans JK (2016) *Crocidura roosevelti* (errata version published in 2017). The IUCN Red List of threatened species 2016: e.T41355A115181119. Downloaded on 10 May 2018

iNaturalist.org (2018a) iNaturalist research-grade observations. Occurrence dataset https://doi.org/10.15468/ab3s5x. Accessed via GBIF.org on 27 June 2018. https://www.gbif.org/occurrence/1802602131

iNaturalist.org (2018b) iNaturalist research-grade observations. Occurrence dataset https://doi.org/10.15468/ab3s5x. Accessed via GBIF.org on 27 June 2018. https://www.gbif.org/occurrence/1135207782

INBAC – Instituto Nacional da Biodiversidade e Áreas de Conservação (2016) Plano de Acção Nacional de Conservação da Chita e Mabeco em Angola. Unpublished report. Ministério do Ambiente, República de Angola, Luanda, 30pp

IUCN (2018) The IUCN Red List of threatened species v. 2017.3. IUCN, Gland Accessed 20 Mar 2018. www.redlist.org

Jacques H, Reed-Smith J, Davenport C, Somers MJ (2015a) *Aonyx congicus*. The IUCN Red List of threatened species 2015:e.T1794A14164772. Downloaded on 13 April 2018

Jacques H, Reed-Smith J, Somers MJ (2015b) *Aonyx capensis*. The IUCN Red List of threatened species 2015:e.T1793A21938767. Downloaded on 13 April 2018

Jenkins R (2013) *Kobus vardonii* Puku. In: Kingdon J, Hoffmann M (eds) The mammals of Africa. Vol VI (Hippopotamuses, pigs, deer, giraffe and bovids). Bloomsbury Publishing, London, pp 445, 680 pp–449

Jentink FA (1887) On mammals from Mossamedes. Notes Leyden Mus 9:171–180

Jentink FA (1893) On some mammals from Cahama. Notes Leyden Mus 15:262–265

Jentink FA (1900) The species of the antelope – genus *Pediotragus*. Notes Leyden Mus 22:33–43

Jentink FA (1901) The antelopes in the Leyden Museum. Notes Leyden Mus 23:17–31

Jordan NR, Do Linh San E (2015) *Suricata suricatta*. The IUCN Red List of threatened species 2015:e.T41624A45209377. Downloaded on 30 May 2018

Juste J, Carballo C (1992) Proyecto Para la Definición y Justificación de la Recuperación y Conservación de los Parques Nacionales de Quissama, Bikuar, Mupa, Iona y la Reserva Parcial de Namibe (Angola). Asociacíon Amigos del Coto de Doñana, Sevilla, 79 pp

Keesing F, Young TP (2014) Cascading consequences of the loss of large mammals in an African Savanna. BioScience 64:487–495

Kennerley R (2016) *Crocidura ansellorum* (errata version published in 2017). The IUCN Red List of threatened species 2016:e.T5558A115073943. Downloaded on 10 May 2018

Kingdon J (2008a) *Cercopithecus mitis* ssp. *mitis*. The IUCN Red List of threatened species 2008: e.T136943A4351535. Downloaded on 20 March 2018

Kingdon J (2008b) *Cercopithecus mitis* ssp. *opisthostictus*. The IUCN Red List of threatened species 2008:e.T136850A4346858. Downloaded on 20 March 2018

Kingdon J (2016) *Papio kindae*. The IUCN Red List of threatened species 2016:e. T136848A92251482. Downloaded on 20 March 2018

Kingdon J, Hoffmann M (2013) *Redunca arudinum* Southern Reedbuck. In: Kingdon J, Hoffmann M (eds) The mammals of Africa. Vol VI (Hippopotamuses, pigs, deer, giraffe and bovids). Bloomsbury Publishing, London, pp 426, 680 pp–431

Kingdon J, Lahm SA (2013) *Cephalophus silvicvultor* Yellow-backed Duiker. In: Kingdon J, Hoffmann M (eds) The mammals of Africa. Vol VI (Hippopotamuses, pigs, deer, giraffe and bovids). Bloomsbury Publishing, London, pp 288, 680 pp–293

Kingdon J, Happold D, Butynski T et al (2013) The mammals of Africa. Vols 1–6. Bloomsbury Publishing, London

Kingswood SC, Kumamoto AT (1997) *Madoqua kirkii*. Mamm Species 569:1–10

Kitchener AC, Breitenmoser-Würsten C, Eizirik E et al (2017) A revised taxonomy of the Felidae. The final report of the Cat Classification Task Force of the IUCN/SSC Cat Specialist Group. Cat News Spec Issue 11:1–80

Klingel H (2013) *Hippopotamus amphibius* Common Hippopotamus. In: Kingdon J, Hoffmann M (eds) The mammals of Africa. Vol VI (Hippopotamuses, pigs, deer, giraffe and bovids). Bloomsbury Publishing, London, pp 68, 680 pp–78

Knight M (2013) *Oryx gazella* Gemsbok (Southern Oryx). In: Kingdon J, Hoffmann M (eds) The mammals of Africa. Vol VI (Hippopotamuses, pigs, deer, giraffe and bovids). Bloomsbury Publishing, London, pp 572, 680 pp–576

Kock D, Amr Z, Mickleburgh S et al (2008) *Hipposideros caffer*. The IUCN Red List of threatened species 2008:e.T10115A3166805. Downloaded on 24 April 2018

Kolberg H, Kilian W (2003) Report on an aerial survey of Iona National Park, Angola, 6 to 14 June 2003. Directorate of Scientific Services, Ministry of Environment and Tourism, Windhoek, 22 pp

Kuedikuenda S, Xavier NG (2009) Framework report on Angola's Biodiversity. Unpublished report. Republic of Angola, Ministry of Environment, Luanda, 60 pp

Kumamoto AT, Kingswood SC, Hugo W (1994) Chromosomal divergence in allopatric populations of Kirk's dik-dik, *Madoqua kirki* (artiodactyla, Bovidae). J Mammal 75:357–364

Kunz TH, Braun de Torrez E, Bauer D et al (2011) Ecosystem services provided by bats. Ann N Y Acad Sci 1223:1–38

Laurance WF, Sayer J, Cassman KG (2014) Agricultural expansion and its impacts on tropical nature. Trends Ecol Evolut 29:107–116

Lawes MJ, Cords M, Lehn C (2013) *Cercopithecus mitis* Gentle monkeys. In: Butynski TM, Kingdon J, Kalina J (eds) The mammals of Africa. Vol II (Primates). Bloomsbury Publishing, London, pp 354, 556 pp–362

Le Houérou HN (2009) Bioclimatology and biogeography of Africa. Springer, Heidelberg, 240 pp

Leus K, Vercammen P (2013) *Potamochoerus porcus* Red River Hog. In: Kingdon J, Hoffmann M (eds) The mammals of Africa. Vol VI (Hippopotamuses, pigs, deer, giraffe and bovids). Bloomsbury Publishing, London, pp 37, 680 pp–40

Linder HP, de Klerk HM, Born J et al (2012) The partitioning of Africa: statistically defined biogeographical regions in sub-Saharan Africa. J Biogeogr 39:1189–1205

Lindsey PA, Petracca LS, Funston PJ et al (2017) The performance of African protected areas for lions and their prey. Biol Conserv 209:137–149

Lorenzen ED, Arctander P, Siegismund HR (2006) Regional genetic structuring and evolutionary history of the impala *Aepyceros melampus*. J Hered 97:119–132

Lorenzen ED, Arctander P, Siegismund HR (2008) High variation and very low differentiation in wide ranging plains zebra (*Equus quagga*): insights from mtDNA and microsatellites. Mol Ecol 17:2812–2824

Loveridge AJ, Kuiper T, Parry RH et al (2017) Bells, bomas and beefsteak: complex patterns of human-predator conflict at the wildlife-agropastoral interface in Zimbabwe. PeerJ 5:e2898

Lydekker R (1899) The great and small game of Africa. Rowland Ward, London, 642 pp

Lydekker R (1903) Hutchinson's animal life, vol 2. Hutchinson, London

Lydekker R (1904) On the subspecies of *Giraffa camelopardalis*. Proc Zool Soc Lond 1904:202–227

Lydekker R, Blaine G (1913–1916) Catalogue of Ungulate mammals in the British Museum (Natural History), 5 vols. British Museum (Natural History), London

Machado AB (1952) Generalidades acerca da Lunda e da sua exploração biológica. Publicações Culturais da Companhia de Diamantes de Angola 12:1–107

Machado AB (1968) A exploração biológica da Lunda. Memórias da Academia de Ciências de Lisboa 12:35–71

Machado AB (1969) Mamíferos de Angola ainda não citados ou pouco conhecidos. Publicações Culturais da Companhia de Diamantes de Angola 46:93–232

Machado AB, Crawford-Cabral J (1999) As subespécies de *Cercopithecus mitis* Wolf, 1922 (Primates, Cercopithecidae) existentes em Angola. Garcia de Orta, Série de Zoologia 23:99–117

Maisels F, Strindberg S, Breuer T et al (2016a) *Gorilla gorilla* ssp. *gorilla* (errata version published in 2016). The IUCN Red List of threatened species 2016:e.T9406A102328866. Downloaded on 21 March 2018

Maisels F, Strindberg S, Greer D et al (2016b) *Pan troglodytes* ssp. *troglodytes* (errata version published in 2016). The IUCN Red List of threatened species 2016:e.T15936A102332276. Downloaded on 21 March 2018

Malbrant R, Maclatchy A (1949) Faune de l'Équateur Africain Français. Tome 2: Mammifères. Lechavalier, Paris, 324 pp

Malhi Y, Doughty CE, Galetti M et al (2016) Megafauna and ecosystem function from the Pleistocene to the Anthropocene. Proc Natl Acad Sci U S A 113:838–846

Matschie P (1900) Über *Equus penricei*. Sitzungsberichte der Gesellschaft Naturforschender Freunde zu Berlin 1900:231

Matschie P (1906) Einige noch nicht beachrieben des Arten africanischen *Büffels*. Sitzungsberichte der Gesellschaft Naturforschender Freunde zu Berlin 7:161–179

McLennan MR (2013) Diet and feeding ecology of chimpanzees (*Pan troglodytes*) in Bulindi, Uganda: foraging strategies at the forest–farm interface. Int J Primatol 34:585–614

McNutt JW, Stein AB, McNutt LB, Jordan NR (2017) Living on the edge: characteristics of human–wildlife conflict in a traditional livestock community in Botswana. Wildl Res 44:546–557

Meÿer MA (2007) The first aerial survey of Cape Fur Seal numbers at Baia dos Tigres, southern Angola. In: Kirkman SP (ed) Final report of the BCLME (Benguela current large marine ecosystem). Project on top predators as biological indicators of ecosystem change in the BCLME. Avian Demography Unit, Cape Town, pp 307–308, 382 pp

MHNG (2018). Mammals housed at MHNG, Geneva. Muséum d'Histoire Naturelle de la Ville de Genève. Accessed via GBIF.org on 2018-04-23

Milliken T, Pole A, Huongo A (2006) No peace for elephants: unregulated domestic ivory markets in Angola and Mozambique. TRAFFIC online report series no. 11. Traffic East/Southern Africa, Harare, 46 pp

Mills MS, Olmos F, Melo M, Dean WRJ (2011) Mount Moco: its importance to the conservation of Swierstra's Francolin *Pternistis swierstrai* and the Afromontane avifauna of Angola. Bird Conserv Int 21:119–133

Mills MS, Melo M, Vaz A (2013) The Namba mountains: new hope for Afromontane forest birds in Angola. Bird Conserv Int 23:159–167

MNHN (2018) The mammals collection (ZM) of the Muséum National d'Histoire Naturelle. Version 43.58. Museum National d'Histoire Naturelle, Paris Accessed via GBIF.org on 19 Apr 2018

Monadjem A, Taylor PJ, Cotterill W, Schoeman MC (2010a) Bats of southern and Central Africa: a biogeographic and taxonomic synthesis. Wits University Press, Johannesburg, p 596

Monadjem A, Schoeman MC, Reside A et al (2010b) A recent inventory of the bats of Mozambique with documentation of seven new species to the country. Acta Chiropterologica 12:371–391

Monadjem A, Richards L, Taylor PJ, Denys C, Dower A, Stoffberg S (2013a) Diversity of Hipposideridae in the Mount Nimba massif, West Africa, and the taxonomic status of *Hipposideros lamottei*. Acta Chiropterologica 15:341–352

Monadjem A, Goodman SM, Stanley WT et al (2013b) A cryptic new species of *Miniopterus* from south-eastern Africa based on molecular and morphological characters. Zootaxa 3746:123–142

Monadjem A, Taylor PJ, Denys C, Cotterill FP (2015) Rodents of Sub-Saharan Africa: a biogeographic and taxonomic synthesis. de Gruyter, Berlin, 1102 pp

Monadjem A, Jacobs D, Taylor P, Cohen L, MacEwan K, Richards LR, Sethusa T (2017) Rhinolophus damarensis. The IUCN Red List of threatened species 2017: e.T67369846A67369914. Downloaded on 25 April 2018

Monard A (1930) Mission Scientifique Suisse dans l'Angola. Résultats scientifiques. Mammifères. Part I: Ongulés. Bulletin de la Société Neuchâteloise des Sciences Naturelles 54:73–102

Monard A (1931) Mission Scientifique Suisse dans l'Angola. Résultats scientifiques. Mammifères. Part I: Carnivores. Bulletin de la Société Neuchâteloise des Sciences Naturelles 55:51–71

Monard A (1933) Mission Scientifique Suisse dans l'Angola. Résultats scientifiques. Mammifères. Part IV: Ongulés (Suite). Bulletin de la Société Neuchâteloise des Sciences Naturelles 57:45–66

Monard A (1935) Contribution à la mammologie d'Angola et prodrome d'une faune d'Angola. Arquivos do Museu Bocage 6:1–103

Monsarrat S, Kerley GI (2018) Charismatic species of the past: biases in reporting of large mammals in historical written sources. Biol Conserv 223:68–75

Moodley Y, Bruford MW (2007) Molecular biogeography: towards an integrated framework for conserving pan-African biodiversity. PLoS One 2:e454

Moodley Y, Bruford MW, Bleidorn C et al (2009) Analysis of mitochondrial DNA data reveals non-monophyly in the bushbuck (*Tragelaphus scriptus*) complex. Mammalian Biology-Zeitschrift für Säugetierkunde 74:418–422

Moraes J, Putzke J (2013) Ocorrência de Lepus saxatilis F. Cuvier, 1823 na província de Malanje, norte de Angola. Cad Pesqui 25:40–43

Morais M, Velasco L, Carvalho E (2006a) Avaliação da Condição e Distribuição do Manatim Africano (Trichechus senegalensis) ao longo do Rio Cuanza. Unpublished report. Universidade Agostinho Neto & Ministério do Urbanismo e Ambiente, Luanda

Morais M, Torres MOF, Martins MJ (2006b) Biodiversidade Marinha e Costeira em Angola. Projecto de Estratégia e Plano de Acção Nacionais para a Biodiversidade (NBSAP). Ministério do Urbanismo e Ambiente, Luanda, Angola

Murphy BP, Bowman DMJS (2012) What controls the distribution of tropical forest and savanna? Ecol Lett 15:748–758

Musser GG, Carleton MD (2005) Superfamily Muroidea. In: Wilson DE, Reeder DM (eds) Mammal species of the world: a taxonomic and geographic reference. The Johns Hopkins University Press, Baltimore, pp 894–1531

Newton da Silva S (1970) A Grande Fauna Selvagem de Angola. Direcção Provincial dos Serviços de Veterinária, Luanda, 151 pp

NGOWP (2018) National Geographic Okavango Wilderness Project. Initial Findings from Exploration of the Upper Catchments of the Cuito, Cuanavale and Cuando Rivers in Central and South-Eastern Angola (May 2015 to December 2016). National Geographic Okavango Wilderness Project, 352 pp

Novellie P, Lindeque M, Lindeque P et al (2002) Status and Action Plan for the Mountain Zebra (*Equus zebra*). In: Moehlman PDR (ed) Equids: zebras, asses, and horses. Status Survey and Conservation Action Plan. IUCN, Gland, pp 28, 190 pp–42

Oates JF (2011) Primates of West Africa: a field guide and natural history. Conservation International, Arlington, 556 pp

Oates JF, Groves CP (2008) *Cercopithecus nictitans*. The IUCN Red List of threatened species 2008:e.T4224A10682370. Downloaded on 02 April 2018

Oates JF, Hart J, Groves CP et al (2008a) *Cercopithecus ascanius*. The IUCN Red List of threatened species 2008. Downloaded on 30 May 2018

Oates JF, Hart J, Groves CP et al (2008b) *Cercopithecus ascanius* ssp. *atrinasus*. The IUCN Red List of threatened species 2008:e.T136869A4347775. Downloaded on 20 March 2018

Oboussier H (1962) Zur Kenntnis des Kaffernbüffels (*Syncerus caffer* Sparrman, 1779). Hirn und Hypophyse. Ergebnisse einer Forshungsreisen nach Süd-Angola. Boletim do Instituto de Investigação Científica de Angola 1:39–47

Oboussier H (1963) Die Pferdeantilope (*Hippotragus equinus cottoni* Dollman and Burlace, 1928). Ergebnisse der Forschungsreisen nach Sud-Angola. Zeitschrift fuer Morphologie und Oekologie der Tiere 52:688–713

Oboussier H (1964) Ein ungewöhlnliches Warzenschwein (*Phacochoerus aethiopicus shortridgei* St. Leger, 1932). Säugetierkundliche Mitteilungen 12:94–97

Oboussier H (1965) Zur Kenntnis der Schwarzfersenantilope (Impala) *Aepyceros melampus* unter besonderer Berücksichtigung des Grosshirnfurchenbildes und der Hypophyse. Ergebnisse der Forschungsreisen nach Süd-Angola und Ostafrika. Zeitschrift fuer Morphologie und Oekologie der Tiere 54:531–550

Oboussier H (1966) Das Grosshirnfurchenbild als Merkmal der Evolution. Untersuchungen an Boviden II (Subfamilien Cephalophinae und Antilopinae nach Simpson 1945). Mitteilungen aus dem Hamburgischen Zoologischen Museum und Institut 63:159–182

Oboussier H (1972) Morphologische und quantitative Neocortexuntersuchungen bei Boviden, ein Beitrag zur Phylogenie dieser Familie. II. Formen geringen Körpergewichts (3 kg – 25 kg) aus den Subfamilien Cephalophinae und Antilopinae. Mitteilungen aus dem Hamburgischen Zoologischen Museum und Institut 68:231–269

Oboussier H (1976) Zur Kenntnis der Moorantilopen (Mammalia, Bovidae, Reduncini). Mitteilungen aus dem Hamburgischen Zoologischen Museum und Institut 73:281–294

Oboussier H, Von Tyszka H (1964) Beiträge zür Kenntnis der Reduncini (Hippotraginae – Bovidae) Süd-Angolas Hirnfurchenbild und Hypophyse. Ergebnisse der Forschungsreisen von Prof. Dr. H. Oboussier nach Angola 1959 und 1961. Zeitschrift fuer Morphologie und Oekologie der Tiere 53:362–386

Olson DM, Dinerstein E, Wikramanayake ED et al (2001) Terrestrial ecoregions of the world: a new map of life on Earth. BioScience 51:93–938

Orrell T, Hollowell T (2018) NMNH extant specimen records. Version 1.16. National Museum of Natural History, Smithsonian Institution, Washington, DC Accessed via GBIF.org on 19 Apr 2018

Overton J, Fernandes S, Elizalde D et al (2017) A large mammal Survey of Bicuar and Mupa National Parks, Angola. Unpublished report. Instituto Nacional da Biodiversidade e Áreas de Conservação, Luanda

Owen-Smith N (2013) *Tragelaphus strepsiceros* Greater Kudu. In: Kingdon J, Hoffmann M (eds) The mammals of Africa. Vol VI (Hippopotamuses, pigs, deer, giraffe and bovids). Bloomsbury Publishing, London, pp 152, 680 pp–159

Palkopoulou E, Lipson M, Mallick S et al (2018) A comprehensive genomic history of extinct and living elephants. Proc Natl Acad Sci U S A 115:E2566–E2574

Penzhorn B (2013) *Equus zebra* Mountain Zebra. In: Kingdon J, Hoffmann M (eds) The mammals of Africa. Vol V (Carnivores, pangolins, equids and rhinoceros). Bloomsbury Publishing, London, pp 438, 544 pp–443

Perrin M (2013) *Elephantulus brachyrhynchus* Short-snouted Sengi. In: Kingdon J, Happold D, Hoffmann M et al (eds) The mammals of Africa. Vol I (Introductory chapters and afrotheria). Bloomsbury Publishing, London, pp 263, 351 pp–265

Peters WC (1865) Note on the Mammalia observed by Dr. Welwitsch in Angola. Proc Zool Soc Lond 1865:400–401

Peters WC (1870) Lista dos Mammiferos das possessões portuguesas d'Africa occidental e diagnose d'algumas especies novas. Jornal de Sciências Mathemáticas, Physicas e Naturaes, Primeira Série 1:123–127

Peters WC (1879) Eine Neue Art der Säugethiergattung Hyrax (*H. nigricans*) aus Chinchoxo und über eine neue Eidechse, Platysaurus torquatus, aus Mossambique. Sitzungsberichte der Gesellschaft Naturforschender Freunde zu Berlin 1879:10–11

Petter F (1972) Part 5. Lagomorpha. In: Meester J, Setzer HW (eds) The mammals of Africa: an identification manual. Smithsonian Institution Press, Washington, DC, pp 1–7

Petter F, Genest H (1965) Variation morphologique et répartition géographique de *Lepus capensis* dans le Sud-Ouest Africain. Mammalia 27:238–255

Pimley ER (2009) A survey of nocturnal primates (Strepsirrhini: Galaginae, Perodictinae) in southern Nigeria. Afr J Ecol 47:784–787

Pimley ER, Bearder SK (2013) *Perodicticus potto* – Potto. In: Butynski TM, Kingdon J, Kalina J (eds) The mammals of Africa. Vol II (Primates). Bloomsbury Publishing, London, pp 393, 556 pp–398

Poulsen JR, Koerner SE, Moore S et al (2017) Poaching empties critical Central African wilderness of forest elephants. Curr Biol 27:R134–R135

Powell JA (1996) The distribution and biology of the West African manatee (*Trichechus senegalensis* Link, 1795). Unpublished report. United Nations Environment Programme, Regional Seas Programme, Oceans and Coastal Areas, Nairobi, 68 pp

Prins HHJ, Sinclair ARE (2013) *Syncerus caffer* African Buffalo. In: Kingdon J, Hoffmann M (eds) The mammals of Africa. Vol VI (Hippopotamuses, pigs, deer, giraffe and bovids). Bloomsbury Publishing, London, pp 125, 680 pp–136

Pruetz JD (2006) Feeding ecology of savanna chimpanzees (*Pan troglodytes verus*) at Fongoli, Senegal. In: Hohmann G, Robbins MM, Boesch C (eds) Feeding ecology in apes and other primates. Cambridge University Press, Cambridge, pp 161, 540 pp–182

Pullen S, Bearder SK (2013) *Galago moholi* Southern Lesser Bushbaby. In: Butynski TM, Kingdon J, Kalina J (eds) The mammals of Africa. Vol II (Primates). Bloomsbury Publishing, London, pp 430, 556 pp–433

Purchase GK, Marker L, Marnewick K et al (2007a) Regional assessment of the status, distribution and conservation needs of cheetahs in southern Africa. Cat News 3:44–46

Purchase GK, Mateke C, Purchase D (2007b) A review of the status and distribution of carnivores, and levels of human carnivore conflict, in the protected areas and surrounds of the Zambezi Basin. Unpublished report. The Zambezi Society, Bulawayo, 79 pp

Rapson S, Rathbun GB (2015) *Herpestes flavescens*. The IUCN Red List of threatened species 2015:e.T41599A45205933. Downloaded on 20 April 2018

Rapson SA, Goldizen AW, Seddon J (2012) Species boundaries and hybridization between the black mongoose (*Galerella nigrata*) and the slender mongoose (*Galerella sanguinea*). Mol Phylogenet Evolut 65(3):831–839

Rathbun GB (2013) *Petrodromus tetradactylus* Four-toed Sengi. In: Kingdon J, Happold D, Hoffmann M et al (eds) The mammals of Africa. Vol I (Introductory chapters and afrotheria). Bloomsbury Publishing, London, pp 279, 351 pp–281

Rathbun GB, Cowley TE (2008) Behavioural ecology of the black mongoose (*Galerella nigrata*) in Namibia. Mammalian Biology-Zeitschrift für Säugetierkunde 73:444–450

RBINS (2017) RBINS DaRWIN. Royal Belgian Institute of Natural Sciences, Brussels Accessed via GBIF.org on 19 Apr 2018

Reed-Smith J, Jacques H, Somers MJ (2015) *Hydrictis maculicollis*. The IUCN Red List of threatened species 2015:e.T12420A21936042. Downloaded on 13 April 2018

Riggio J, Jacobson A, Dollar L et al (2013) The size of savannah Africa: a lion's (*Panthera leo*) view. Biodiver Conserv 22:17–35

Robbins MM, Bermejo M, Cipolletta C et al (2004) Social structure and life-history patterns in western gorillas (*Gorilla gorilla gorilla*). Am J Primatol 64:145–159

Roberts SC (2013) *Oreotragus oreotragus* Klipspringer. In: Kingdon J, Hoffmann M (eds) The mammals of Africa. Vol VI (Hippopotamuses, pigs, deer, giraffe and bovids). Bloomsbury Publishing, London, pp 470, 680 pp–476

Roca AL, Georgiadis N, Pecon-Slattery J et al (2001) Genetic evidence for two species of elephant in Africa. Science 293:1473–1477

Roca AL, Ishida Y, Brandt AL et al (2015) Elephant natural history: a genomic perspective. Annu Rev Anim Biosci 3:139–167

Rodrigues P, Figueira R, Vaz Pinto P et al (2015) A biogeographical regionalization of Angolan mammals. Mamm Rev 45:103–116

Rodrigues P, Figueira R, Beja P (2018) Bibliographic records of Angola mammals. Version 2.2. Instituto de Investigação Científica Tropical, Lisboa Accessed via GBIF.org on 24 Apr 2018

Ron T (2005) The Maiombe forest in Cabinda: conservation efforts, 2000–2004. Gorilla J 30:18–21

Ron T, Golan T (2010) Angolan Rendezvous: man and nature in the shadow of war. 30 Degrees South Publishers, Kwa-Zulu Natal, 272 pp

Rookmaaker LC (2005) The black rhino needs a taxonomic revision for sound conservation. Int Zoo News 52:280–282

Rowe N, Myers M (2016) All the world's primates. Pogonias Press, Charlestown, 777 pp

Sanborn CC (1950) Chiroptera from Dundo, Lunda, northeastern Angola. Publicações Culturais da Companhia de Diamantes de Angola 10:51–62

Sarmiento EE (2013) *Chlorocebus cynosures* Malbrouck monkey. In: Butynski TM, Kingdon J, Kalina J (eds) The mammals of Africa. Vol II (Primates). Bloomsbury Publishing, London, pp 284, 556 pp–286

Sarmiento EE, Stiner EO, Brooks EGE (2001) Red-tail monkey *Cercopithecus ascanius* distinguishing characters and distribution. Afr Primate 5:18–24

Schlossberg S, Chase MJ, Griffin CR (2018) Poaching and human encroachment reverse recovery of African savannah elephants in south-east Angola despite 14 years of peace. PLoS One 13:e0193469

Schoeman MC, Cotterill FPDW, Taylor PJ et al (2013) Using potential distributions to explore environmental correlates of bat species richness in southern Africa: Effects of model selection and taxonomy. Curr Zool 59:279–293

Seabra A (1898a) Sobre a determinação dos géneros da família Pteropodidae fundada nos caracteres extrahidos da fórma, disposição e numero das pregas do paladar e lista das especies d'esta familia, existentes nas collecções do Museu de Lisboa. Jornal de Sciências Mathemáticas, Physicas e Naturaes, Segunda Série 5:163–171

Seabra A (1898b) Sobre um caracter importante para a determinação dos generos e especies dos 'Microchiropteros' e lista das espécies d'este grupo existentes nas collecções do Museu Nacional. Jornal de Sciências Mathemáticas, Physicas e Naturaes, Segunda Série 5:247–258

Seabra AF (1905) Mammiferos e aves da exploração de F. Newton em Angola. Jornal de Sciências Mathemáticas, Physicas e Naturaes, Segunda Série 7:103–110

Searchinger TD, Estes L, Thornton K et al (2015) High carbon and biodiversity costs from converting Africa's wet savannahs to cropland. Nat Clim Chang 5:481–486

Seiler N, Robbins MM (2016) Factors influencing ranging on community land and crop raiding by mountain gorillas. Anim Conserv 19:176–188

Seydack AHW (2013) *Potamochoerus larvatus* Bushpig. In: Kingdon J, Hoffmann M (eds) The mammals of Africa. Vol VI (Hippopotamuses, pigs, deer, giraffe and bovids). Bloomsbury Publishing, London, pp 32, 680 pp–36

Shortridge GC (1934) The mammals of South West Africa, 2 vols. Heinemann, London

Simões AP, Crawford-Cabral J (1988) Notice on a large hoofed steenbok, *Raphicerus campestris* (Mammalia: Artiodactyla), from Angola. Garcia de Orta, Série de Zoologia 15:1–8

Sirami C, Jacobs DS, Cumming GS (2013) Artificial wetlands and surrounding habitats provide important foraging habitat for bats in agricultural landscapes in the Western Cape, South Africa. Biol Conserv 164:30–38

Skinner JD (2013) *Antidorcas marsupialis* Springbok. In: Kingdon J, Hoffmann M (eds) The mammals of Africa. Vol VI (Hippopotamuses, pigs, deer, giraffe and bovids). Bloomsbury Publishing, London, pp 398, 680 pp–403

Sliwa A (2013) *Felis nigripes* Black-footed Cat. In: Kingdon J, Hoffmann M (eds) The mammals of Africa. Vol V (Carnivores, pangolins, equids and rhinoceros). Bloomsbury Publishing, London, pp 203, 544 pp–206

Smith AT, Johnston CH, Alves PC et al (2018) Lagomorphs: pikas, rabbits, and hares of the world. John Hopkins University Press, Baltimore, p 280

Sokolowski A (1903) Die Antilopenarten der von der Kunene- Sambesi Expedition durchzogenen Gebiete auf Grund der von der Expedition mitgebrachten Gehörne. In: Baum H (ed) Kunene-Zambesi expedition. Verlag des Kolonial-Wirtschaftlichen Komitees, Berlin, pp 517–593

St. Leger J (1936) Dr. Karl Jordan's Expedition to South-West Africa and Angola: Mammals. Novitates Zoologicae 40:75–81

Statham JCB (1922) Through Angola: a coming colony. W. Blackwood & Sons, London, 388 pp

Statham JCB (1926) With my wife across Africa by Canoe and Caravan. Simpkin Marshall Hamilton Kent & Co Ltd, London, 323 pp

Struhsaker T, Oates JF, Hart J et al (2008) *Cercopithecus neglectus*. The IUCN Red List of threatened species 2008:e.T4223A10680717. Downloaded on 02 April 2018

Sunquist M, Sunquist F (2009) Family Felidae (Cats). In: Wilson DE, Mittermeier RA (eds) Handbook of the mammals of the world, Vol I (Carnivores). Lynx Edicíons, Barcelona, pp 54, 728 pp–168

Svensson MS (2017) Conservation and ecology of nocturnal primates: night monkeys, galagos, pottos and Angwantibos as case studies. PhD thesis, Oxford Brookes University, Oxford

Svensson MS, Bersacola E, Bearder SK et al (2014a) Open sale of elephant ivory in Luanda, Angola. Oryx 48:13

Svensson MS, Bersacola E, Bearder SK (2014b) Pangolins in Angolan bushmeat markets. IUCN/SSC Pangolins Specialist Group Newsletter. www.pangolinsg.org/news

Svensson MS, Bersacola E, Mills MS et al (2017) A giant among dwarfs: a new species of galago (Primates: Galagidae) from Angola. Am J Phys Anthropol 163:30–43

Swart PS (1967) New data on the black-faced impala (*Aepyceros melampus petersi* Bocage). Cimbebasia 20:3–18

Taylor ME (2013) *Herpestes flavescens* Kaokoveld Slender Mongoose (Angolan Slender Mongoose). In: Kingdon J, Hoffmann M (eds) The mammals of Africa. Vol V (Carnivores, pangolins, equids and rhinoceroses). Bloomsbury Publishing, London, pp 304, 680 pp–306

Taylor PJ (2016) *Dasymys incomtus* (errata version published in 2017). The IUCN Red List of threatened species 2016:e.T6269A115080446. Downloaded on 23 April 2018

Taylor ME, Goldman CA (1993) The taxonomic status of the African mongooses, *Herpestes sanguineus*, *H. nigratus*, *H. pulverulentus* and *H. ochraceus* (Carnivora: Viverridae). Mammalia 57:375–391

Taylor PJ, Grass I, Alberts AJ et al (2018a) Economic value of bat predation services – a review and new estimates from macadamia orchards. Ecosyst Serv 30:372–381

Taylor PJ, Macdonald A, Goodman SM et al (2018b) Integrative taxonomy resolves three new cryptic species of small southern African horseshoe bats (*Rhinolophus*). Zool J Linn Soc https://doi.org/10.1093/zoolinnean/zly024

Taylor PJ, Neef G, Keith M et al (2018c) Tapping into technology and the biodiversity informatics revolution: updated terrestrial mammal list of Angola, with new records of mammals from the Okavango Basin. ZooKeys 2018:51–88

Themido AA (1931) Catalogue des ongulés et siréniens existants dans les collections du Muséum Zoologique de Coimbra. Memórias e Estudos do Museu Zoológico da Universidade de Coimbra, Série *1*(49):5–22

Themido AA (1946) Mamíferos das colónias portuguesas (catálogo das colecções do Museu Zoológico de Coimbra). Memórias e Estudos do Museu Zoológico da Universidade de Coimbra, Série 1(174):1–52

Thomas O (1892) On the species of the Hyracoidea. Proc Zool Soc Lond 1892:50–76

Thomas O (1900) On *Equus penricei*, a representative of the Mountain Zebra (*Equus zebra*, L.) discovered by Mr. W. Penrice in Angola. Ann Mag Nat Hist Ser 7 6:456–466

Thomas O (1904) On the mammals from northern Angola collected by Dr. W. J. Ansorge. Ann Mag Nat Hist Ser 7 13:405–421

Thomas O (1916) A new Sable Antelope from Angola. Proc Zool Soc Lond 1916:298–301

Thomas O (1926) On Mammals from Ovamboland and the Cunene River, obtained during Capt. Shortridge's third Percy Sladen and Kaffrarian Museum Expedition into South-West Africa. Proc Zool Soc Lond 1926:285–312

Thomas O, Wroughton RC (1905) On a second collection of Mammals obtained by Dr. W. J. Ansorge in Angola. Ann Mag Nat Hist Ser 7 16:169–178

Thorington RW Jr, Koprowski JL, Steele MA et al (2012) Squirrels of the world. The John Hopkins University Press, Baltimore, p 472

Thouless C (2013) *Tragelaphus oryx* Common Eland. In: Kingdon J, Hoffmann M (eds) The mammals of Africa. Vol VI (Hippopotamuses, pigs, deer, giraffe and bovids). Bloomsbury Publishing, London, pp 191-198, 680 pp

Thouless C, Dublin HT, Blanc J et al (2016) African elephant status report 2016. Occasional paper of the IUCN species survival commission no. 60. IUCN, Gland, 37 pp

Timberlake J, Chidumayo E (2011) Miombo ecoregion vision report. Occasional publication in biodiversity no. 20. Biodiversity Foundation for Africa, Famona, Bulawayo, 76 pp

Trense W (1959) Die Saugetiere Angolas, ihre Beziehungen zucinander, zu den benachbarten Gebieten und ihre Geschichte (Nach den Ergebnissen der Hamburgischen Angola-Expedition, 1952–1954). Unpublished report. Centro de Zoologia, Instituto de Investigação Científica Tropical, Lisboa, 211 pp

Trombone T (2016) AMNH mammal collections. American Museum of Natural History, New York Accessed via GBIF.org on 23 Feb 2018

Tromp S (2011) The effects of past major climatic fluctuations on the genetic structures of fauna endemic to Namibia's Granite Inselbergs. PhD thesis, The University of Queensland, Brisbane

Turkalo AK, Wrege PH, Wittemyer G (2017) Slow intrinsic growth rate in forest elephants indicates recovery from poaching will require decades. J Appl Ecol 54:153–159

Tutin CE, Fernandez M (1984) Nationwide census of gorilla (*Gorilla g. gorilla*) and chimpanzee (*Pan t. troglodytes*) populations in Gabon. Am J Primatol 6:313–336

Tutin C, Stokes E, Boesch C et al (2005) Regional action plan for the conservation of Chimpanzees and Ggorillas in Western Equatorial Africa. IUCN/SSC Primate Specialist Group, Conservation International, Washington, DC

Vallo P, Guillén-Servent A, Benda P et al (2008) Variation of mitochondrial DNA in the Hipposideros caffer complex (Chiroptera: Hipposideridae) and its taxonomic implications. Acta Chiropterologica 10:193–206

van der Straeten E (2008) Notes on the *Praomys* of Angola with the description of a new species (Mammalia: Rodentia: Muridae). Stuttgarter Beiträge zur Naturkunde A, Neue Serie 1:123–131

van der Westhuizen J, Thomas J, Haraes L et al (2017) An aerial photographic wildlife survey of the Iona National Park, Angola, November 2016 to February 2017. Unpublished report. Ministry of Environment of Angola, Luanda, 31 pp

van Velden J, Wilson K, Biggs D (2018) The evidence for the bushmeat crisis in African savannas: a systematic quantitative literature review. Biol Conserv 221:345–356

Varian HF (1953) Some African milestones. George Ronald, Oxford, 272 pp

Vaz Pinto P (2018) Evolutionary history of the critically endangered giant sable antelope (Hippotragus niger variani). Insights into its phylogeography, population genetics, demography and conservation. PhD thesis, University of Porto, Porto

Vaz Pinto P (2019) The giant sable antelope: Angola's national icon. In: Huntley BJ, Russo V, Lages F, Ferrand N (eds) Biodiversity of Angola. Science & conservation: a modern synthesis. Springer, Cham

Vaz Pinto P, Veríssimo L (2016) Yellow-backed duiker in miombo woodland in Angola. Gnusletter 35:13

Vaz Pinto P, Beja P, Ferrand N et al (2016) Hybridization following population collapse in a critically endangered antelope. Sci Rep 6:18788

Veríssimo LN (2008) Mucusso reserve. Larger mammals assessment. Preliminary report. Unpublished report. United States Agency International Development, Washington, DC, 56 pp

Veron G, Patou M-L, Jennings AP (2018) Systematics and evolution of the mongooses (Herpestidae, Carnivora). In: Do Linh San E, Sato JJ, Belant JL et al (eds) Small carnivores: evolution, ecology, behaviour and conservation. Wiley-Blackwell, Oxford

Wasser SK, Brown L, Mailand C et al (2015) Genetic assignment of large seizures of elephant ivory reveals Africa's major poaching hotspots. Science 349:84–87

Weir CR (2019) The Cetaceans (Whales and Dolphins) of Angola. In: Huntley BJ, Russo V, Lages F, Ferrand N (eds) Biodiversity of Angola. Science & conservation: a modern synthesis. Springer Nature, Cham

Wendelen W, Noé N (2017) African Rodentia. Belgian Biodiversity Platform, Brussels Accessed via GBIF.org on 19 Apr 2018

Wilhelm JH (1933) Das Wild des Okawangogebietes und des Caprivizipfels. J South-West Afr Sci Soc 6:51–74

Wilkie DS, Wieland M, Boulet H et al (2016) Eating and conserving bushmeat in Africa. Afr J Ecol 54:402–414

Wilson VJ (2013) *Sylvicapra grimmia* Common Duiker. In: Kingdon J, Hoffmann M (eds) The mammals of Africa. Vol VI (Hippopotamuses, pigs, deer, giraffe and bovids). Bloomsbury Publishing, London, pp 235, 680 pp–243

Wilson DE, Reeder DM (2005) Mammal species of the world: a taxonomic and geographic reference, 3rd edn. Johns Hopkins University Press, Baltimore, 2000 pp

Wolf C, Ripple WJ (2016) Prey depletion as a threat to the world's large carnivores. R Soc Open Sci 3:160252

Wozencraft WC (2005) Order Carnivora. In: Wilson DE, Reeder DM (eds) Mammal species of the world: a taxonomic and geographic reference, 3rd edn. Johns Hopkins University Press, Baltimore, pp 532, 2000 pp–628

Zhang YP, Ryder OA (1995) Different rates of mitochondrial DNA sequence evolution in Kirk's Dik-dik (*Madoqua-kirkii*) populations. Mol Phylogenet Evol 4:291–297

Zinner D, Fickenscher GH, Roos C (2013) Cercopithecidae (old world monkeys). In: Mittermeier RA, Rylands AB, Wilson DE (eds) Handbook of the mammals of the World. Vol III (Primates). Lynx Edicíons, Barcelona, pp 550, 952 pp–753

Zukowsky L (1924) Beitrag zur Kenntnis der Säugetiere der nördlichen Teile Deutsch-Südwestafrikas unter besonderer Berücksichtigung des Großwildes. Archiv für Naturgeschichte, Abteilung 1:29–164

Zukowsky L (1964) Die Systematic der Gattung *Diceros* Gray, 1821. Zoologische Garten 30:1–178

Zukowsky L, Haltenorth T (1957) Das Erdferkel (*Orycteropus afer*) aus Angola, eine eigene Unterart? Säugetierkundliche Mitteilungen 5:24–126

Chapter 16
The Cetaceans (Whales and Dolphins) of Angola

Caroline R. Weir

Abstract The history of whale and dolphin (cetacean) research in Angolan waters is scant. Prior to the 2000s it primarily consisted of information from historical (1700s to the 1920s) and modern (1920s–1970s) whaling catches, from which baleen whales and the sperm whale were confirmed. Very few species were added to Angola's cetacean checklist between the whaling era and the 2000s. However, observations since 2003 have confirmed Angola as a range state for at least 28 species, comprising seven baleen whales, two sperm whale species, at least two beaked whales, and at least 17 delphinids. There is potential for approximately seven more species to be identified in the region based on their known worldwide distributions. Angola has one of the most diverse cetacean faunas in Africa, and indeed worldwide, due to its varied seabed topography and transitional ocean climate which supports both (sub)tropical species and those associated with the Benguela Current. While no cetacean species are truly endemic to Angola, the country is one of few confirmed range states for the Critically Endangered Atlantic humpback dolphin and the Benguela-endemic Heaviside's dolphin. Those species, together with endangered baleen whales and breeding populations of sperm and humpback whales, are highlighted as conservation priorities.

Keywords Benguela current · Checklist · Conservation · Endangered species · Endemism · Whaling

Introduction

The occurrence of cetaceans along the west coast of Africa in the eastern tropical Atlantic (ETA) is poorly-studied, due to factors including remoteness, the history of political unrest in many countries, deficiencies in funding and logistical support

---

C. R. Weir (✉)
Ketos Ecology, Kingsbridge, Devon, UK
e-mail: caroline.weir@ketosecology.co.uk

B. J. Huntley et al. (eds.), *Biodiversity of Angola*,
https://doi.org/10.1007/978-3-030-03083-4_16

(especially for marine work requiring boats), and a lack of training programmes to support local marine scientists (Jefferson et al. 1997; Weir 2010a, 2011a,b). Located at the southern limit of the ETA, Angola is expected to support a diverse cetacean community due to its varied marine environment. This chapter provides the history of Angolan cetacean research, reviews cetacean biodiversity and identifies priorities for future research and conservation options.

Methods

Study Area

Angolan waters are defined as marine habitat from the coast to the 200 nautical mile seaward limit of the Exclusive Economic Zone (EEZ), which is located in oceanic habitat over 4000 m deep (Fig. 16.1). They extend from the southern border with Namibia (17°15′S) northwards to the border with the Republic of Congo in Cabinda (5°02′S), but excluding the Democratic Republic of the Congo (DRC) EEZ which divides Angola from the exclave of Cabinda. Some maritime areas in the northernmost EEZ are the subject of disputed ownership with neighbouring countries (Fig. 16.1), but are included here in the non-political context of assessing cetacean occurrence.

Weir (2011a) described the oceanography of the Angolan EEZ as habitat for cetaceans. The Angolan continental shelf is widest in the north, extending to 80 km from the coast off Soyo where it is intersected by the deep Congo Canyon at the mouth of the Congo River. In the southern part of the country, the shelf is narrow and depth increases strongly, bringing deep waters (>1000 m) to within 15 km of the coast in places. The region is predominantly tropical, with warm (>24 °C) nutrient-poor water flowing southward from the Gulf of Guinea as the Angola Current. However, the Benguela Current influences the southern area, bringing nutrient-rich cold water northwards from Namibia. The two currents converge at latitudes of between 14° and 16°S (depending on season) to form the Angola–Benguela Front (Fig. 16.1).

Data

Published (and some available unpublished) papers and reports were reviewed for information on Angolan cetaceans (see Weir 2011a). Whaling catch statistics were acquired from the International Whaling Commission (IWC). Since 2003, Marine Mammal Observers (MMOs), sometimes supported by Passive Acoustic Monitoring (PAM), have been used during seismic surveys by the oil and gas industry to mitigate the potential impacts of airgun sound on cetaceans (Weir 2008). With the exception of published subsets, MMO data are not publicly available and are therefore not included here.

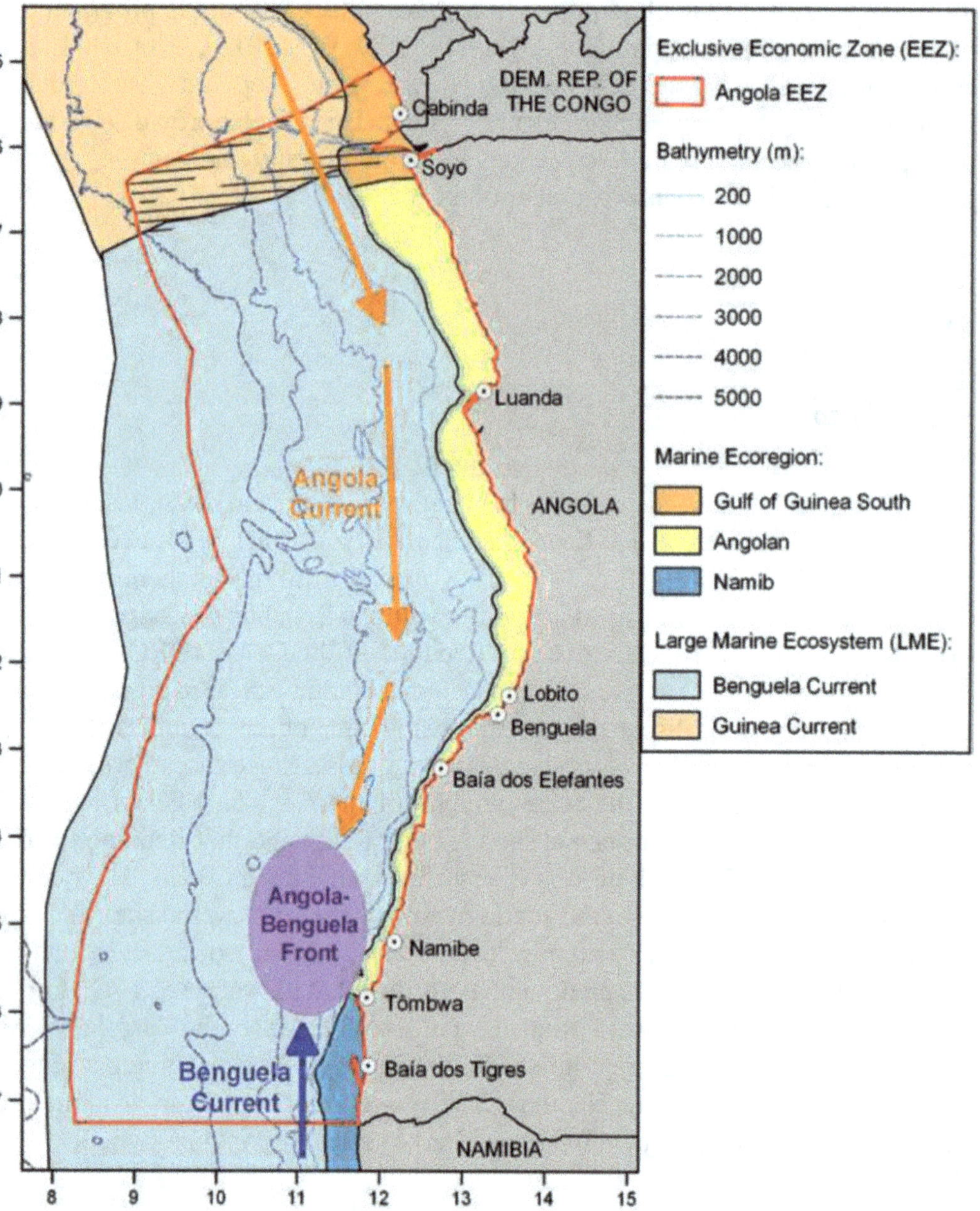

Fig. 16.1 Angolan waters showing the places and major current systems mentioned in this chapter. Hatched areas show some areas of disputed ownership of the Exclusive Economic Zone (EEZ) with neighbouring countries

Species Identification

Cetaceans are often seen briefly and only partially by an observer, and there are morphological similarities between many species in the ETA region (e.g. within *Stenella* dolphins, beaked whales and *Balaenoptera* whales) that causes confusion. High potential for species misidentification exists, even for established cetacean observers and trained MMOs (many of whom lack previous field experience with the particular species occurring off Angola). Published records therefore require careful evaluation (e.g. Best 2001; Fertl et al. 2003; Weir et al. 2014), particularly records originating prior to the 2000s, after which knowledge of key identification features increased markedly with the advent of digital photography, modern field guides and genetic work. Consequently, some Angolan records were not considered sufficiently well-supported for inclusion (e.g., Brown 1959, Mörzer Bruyns 1971, Tormosov et al. 1980).

History of Cetacean Research in Angola

Angola's Whaling Era

Whaling has been practiced since prehistoric times, and whaling data provides the earliest information available on the species identification, distribution, migrations and population status of whale stocks around the world. Whaling also generated much of the best-available information on the life histories, morphology and diet of large whales. Consequently, the whaling era is still considered a prime source of scientific data on the larger baleen whales and the sperm whale (*Physeter macrocephalus*).

It was not until the 1700s that American pelagic whalers first visited the west coast of Africa in search of the relatively slow-moving and oil-rich sperm whales and southern right whales (*Eubalaena australis*). They reached the coast of Angola by 1770 (Best 1981), and catches from this period onward provide the earliest documentation of whale species in Angola. *The distribution of certain whales as shown by logbook records of American whale ships*, published by Charles Haskins Townsend in 1935, included the capture locations of over 50,000 whales taken during American pelagic whaling between 1761 and 1920, including three species from Angolan waters (sperm whales, southern right whales and humpback whales, *Megaptera novaeangliae*: Fig. 16.2). Similar and expanded analyses of whaling logbook catch datasets including Angolan waters have also been published by other authors (e.g., Richards 2009; Smith et al. 2012).

Whaling changed drastically from the mid-1800s with the development of exploding harpoon guns, modern steam-driven whaling boats ('catcher boats'), cannon-fired bow-mounted harpoons and the technique of inflating dead whales with air to keep them afloat (Harmer 1928; Mackintosh 1965; Tønnessen and Johnsen 1982). Species that had previously been inaccessible to whalers, especially the *Balaenoptera* whales that were fast-swimming and sank after death, could now be harvested, and were either towed to shore stations or processed at factory vessels moored in coastal bays. Shore-based whaling stations were established in several African countries during the early 1900s (Tønnessen and Johnsen 1982; Best 1994). Summarised statistics on whale catches worldwide since 1900 (together with some incomplete information on catches taken in the late 1800s) are maintained by the IWC (Allison 2016a). There is also a catch database for individual captures that

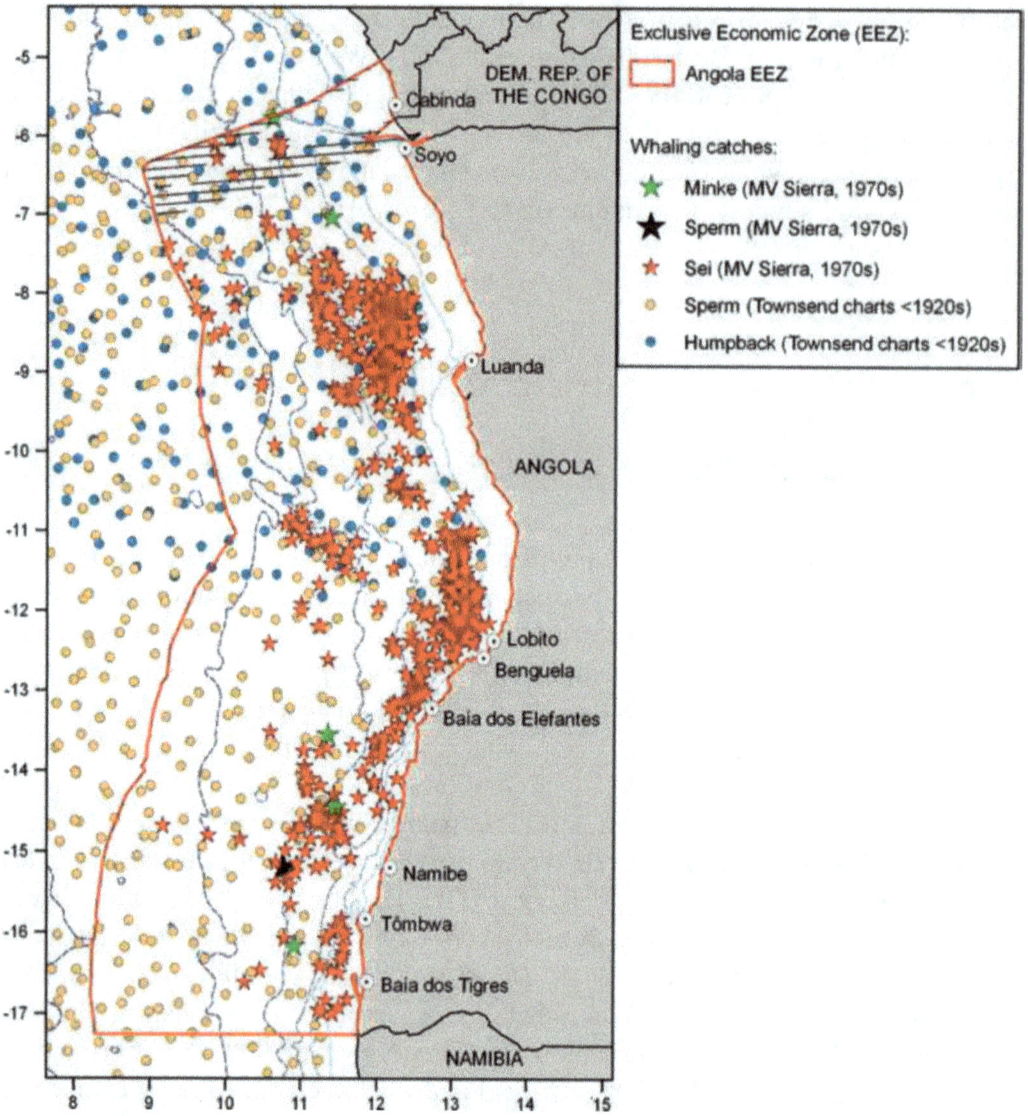

Fig. 16.2 Distribution of whale catch positions in the Angola EEZ. *MV Sierra* catches from the IWC database (Allison 2016b). Digitised Townsend (1935) charts are available from https://canada.wcs.org/wild-places/global-conservation/townsend-whaling-charts.aspx

contains date, length, sex, foetus details, stomach contents and location (when those are available; Allison 2016b). These databases are continually updated (Allison and Smith 2004), and consequently the total species catches reported by various sources has altered over time (e.g. Best 1994; Figueiredo and Weir 2014; this chapter). Catches of whales in Angola since 1900 are presented in Table 16.1.

The first modern coastal whaling operation in Angola was established at Tômbwa (formerly Porto Alexandre), with the moored Norwegian factory ship *Ambra* taking around 237 whales in 1909 (number revised by the IWC from 270 whales in earlier sources; Figueiredo 1960; Tønnessen and Johnsen 1982; Best 1994). The *Ambra* returned to Tômbwa in 1910 and took 650 whales, with a second operation (shore station and a Portuguese catcher vessel) commencing at Moçâmedes and taking

Table 16.1 Estimated whale catches in Angola from the International Whaling Commission databases (Allison 2016a, b)

Year	Locations	Blue	Fin	Sperm	Humpback	Sei/ Bryde's	Right	Minke	Total
Shore-based operations									
1909	Tômbwa	1	0	0	236	0	0	0	237
1910	Tômbwa, Moçâmedes	2	1	0	718	0	0	0	721
1911	Tômbwa, Lobito, Baía dos Elefantes, Baía dos Tigres, Moçâmedes	2	2	0	2281	4	0	0	2289
1912	Tômbwa, Baía dos Elefantes, Baía dos Tigres, Moçâmedes	0	0	18	3417	0	0	0	3435
1913	Tômbwa, Baía dos Elefantes, Baía dos Tigres, Moçâmedes	121	38	39	2419	700	1	0	3318
1914	Tômbwa, Baía dos Elefantes, Baía dos Tigres, Moçâmedes	542	200	138	596	102	0	0	1578
1915	Tômbwa, Baía dos Elefantes, Moçâmedes	360	260	79	201	79	0	0	979
1916	Baía dos Elefantes, Moçâmedes	118	85	26	65	26	0	0	320
1923	Pelagic floating factory just outside of territorial waters	168	26	17	2	0	0	0	213
1924	Baía dos Elefantes	75	17	17	47	274	0	0	430
1925	Baía dos Elefantes	134	42	27	17	68	0	0	288
1926	Moçâmedes	303	40	14	6	33	0	0	396
1927	Moçâmedes	186	73	3	3	305	0	0	570
1928	Moçâmedes	58	32	141	37	246	0	0	514
Total shore landings		*2070*	*816*	*1837*	*10,045*	*519*	*1*	*0*	*15,288*
Pelagic operations									
1934	Angola EEZ	1	21	44	7	10	0	0	83
1936	Angola EEZ	0	1	2	17	6	0	0	26
1956	Angola EEZ	0	0	20	0	0	0	0	20
1971	Angola EEZ	0	0	44	0	234	0	0	278
1972	Angola EEZ	0	0	0	0	10	0	1	11
1973	Angola EEZ	0	0	0	0	228	0	2	230
1974	Angola EEZ	0	0	0	0	221	0	2	223
1975	Angola EEZ	0	0	42	0	100	0	0	142
1976	Angola EEZ	0	0	48	0	0	0	0	48
Total pelagic catches		*1*	*22*	*200*	*24*	*809*	*0*	*5*	*1061*

around 70 whales (Mackintosh 1942; Best 1994; Allison 2016a, b). Operations increased during 1911, with the issuing of five licenses to Norwegian floating factory ships (based at Tômbwa, Lobito, Baía dos Elefantes and Baía dos Tigres) at the end of 1910, and the continuation of the Portuguese operation at Moçâmedes (Figueiredo 1960, Allison 2016a, b). Whaling in Angola boomed between 1911 and 1914, capturing over 10,000 animals (mostly humpbacks: Table 16.1). However, the catch in 1914 was half that of 1912 and 1913, and a collapse in whale stocks was suggested (Figueiredo 1960). The combination of declining whale stocks and the occurrence of the First World War meant that no whales were caught off Angola between 1917 and 1922 (Best 1994).

Whaling was re-established off Angola in 1923, with a Norwegian floating factory ship operating just outside of territorial waters, and coastal operations resuming at Baía dos Elefantes and Moçâmedes between 1924 and 1928. This second period did not yield sufficient captures to be profitable (Table 16.1), and marked the end of coastal whaling from Angolan shore stations (Figueiredo 1960; Tønnessen and Johnsen 1982).

The 1920s saw the development of new ocean-going factory ships (fitted with a stern slipway and a flensing station to process whales) that could operate for long periods with a fleet of smaller catcher vessels and allowed whaling to move into offshore waters. Between 1934 and 1937 the Norwegian factory ships *Pioner*, *Haugar* and *Norskhavet* operated in the ETA including Angola. Catches in Angolan waters over this period included one blue (*Balaenoptera musculus*), 22 fin (*B. physalus*), 24 humpback, 16 sei/Bryde's (*B. borealis/B. edeni*), and 46 sperm whales (Allison 2016b). In later decades factory ships opportunistically took sperm whales encountered while transiting through Angolan waters. For example, the *Olympic Challenger* caught 20 in March 1956, the *Peder Huse* took 41 in early 1971, and the *Sovetskaya Ukraina* took 90 in 1975 and 1976 (Mikhalev et al. 1981a; Allison 2016b).

Most recently, the combined catcher/factory vessel MV '*Run/Sierra*' operated year-round between South Africa and the Gulf of Guinea during the 1970s. The IWC database includes 801 whales taken by the vessel in the Angolan EEZ between 1971 and 1975, comprising five minke whales, three sperm whales and 793 'sei whales' (Fig. 16.2; Allison 2016a,b). However, the *Run/Sierra* 'sei whale' catches are now considered to predominantly comprise Bryde's whales (Tønnessen and Johnsen 1982; Best 1996, 2001).

The composition of whaling catch data altered over time as each species declined to levels where protection in the Southern Hemisphere was introduced by the IWC, beginning with the southern right whale in the 1930s, continuing with the blue and humpback whales in the 1960s, fin and sei whales in the 1970s, and finally with the worldwide ban on the exploitation of all whale species under the 1986 moratorium. Consequently, the whaling era in Angola was ended in the 1970s by the protection of most Southern Hemisphere whale stocks.

Opportunistic Sightings and Specimen Records

Weir (2011a) recognised a 'stranding and specimen era' of cetacean research in the ETA (1950s–1970s), during which new information emerged on the taxonomy, morphometry and distribution of many small cetaceans (see Cadenat 1959, Jefferson et al. 1997). However, the majority of this work was carried out by French scientists in Mauritania, Senegal and Côte d'Ivoire, and the only information emerging from Angola during this period appears to be the 1972 paper of Bree and Purves, which included a single skull from Angola in an evaluation of the *Delphinus* genus. Some opportunistic sightings in Angolan waters by the Dutch sea captain Mörzer Bruyns were also published (Mörzer Bruyns 1968, 1971), although the species identification for many of his records cannot be confirmed. Effort has been made to locate cetacean specimens that may have been captured off Angola during this period and preserved by naturalists in Lisbon museum collections. However, it appears that no cetaceans from Angola are present in Portuguese collections (Cornelis Hazevoet, pers. comm.). The dearth of papers from Angola in this period was also noted in the compilation of African cetacean research by Elwen et al. (2011).

During the 1980s and 1990s a few publications from the wider Atlantic region included opportunistic at-sea sightings (species identifications unsupported) from Angolan waters, for example Tormosov et al. (1980), Mikhalev et al. (1981b) and Wilson et al. (1987). In 1997 Jefferson et al. published a review of dolphin and porpoise records off West Africa, but their study area (to 6°S) included only the exclave of Cabinda and not the rest of Angola. The only 'Angolan' cetacean records located by Jefferson et al. (1997) were common dolphins (*Delphinus* sp.) reported by Simmons (1968). However, careful reading of Simmons (1968) indicates that the observations were actually recorded off Cape Palmas in Liberia rather than Angola.

Targeted At-Sea Cetacean Surveys

Although instability related to the Angolan civil war from 1975 to 2002 is known to have interrupted field studies of terrestrial fauna (other chapters, this volume), dedicated cetacean research had still not yet developed prior to the outbreak of war. In fact, the first dedicated field study of cetaceans in Angolan waters began during the final period of the war in September 1998, when the Whale Unit of the Mammal Research Institute in South Africa was invited to northern Angola (6°52′S) by an oil company to conduct a preliminary investigation into large numbers of humpback whales reported in the area. This initial field study was successful in acquiring acoustic and behavioural data, photographing whale tail flukes for photo-identification and acquiring 13 genetic samples via biopsy sampling (Best et al. 1999). Although the authors recommended that a full survey programme should be initiated to assess the distribution, abundance and status of humpback whales in Angolan waters using aerial surveys and small boat work, such work never developed.

During the early 2000s, some cetacean data were collected concurrently with pelagic fish abundance assessments in Angolan waters as part of an agreement between the Norwegian Institute of Marine Research (IMR) and the Angolan *Instituto de Investigação Pesqueira e Marinha* (INIP). The IMR research vessel *Dr Fridtjof Nansen* surveyed a series of transects across the continental shelf in the Angolan EEZ. These investigations were carried out in cooperation with the Benguela Current Large Marine Ecosystem (BCLME) research programme. Cruise reports outlining the fish stock results are available from the IMR website, and cetacean observations are included for surveys (all between July and September) in 2003 (Krakstad et al. 2003), in 2004 (Axelsen et al. 2004), in 2005 (Axelsen et al. 2005, Roux et al. 2007), and in 2015 (Michalsen et al. 2015).

The Best et al. (1999) study was the first of several suites of cetacean survey work in Angola to be associated with, and funded by, the burgeoning oil and gas industry. From 2003 many oil companies began to use MMOs during their seismic surveys in Angolan waters, leading to a sudden increase in the potential for biologists to use geophysical survey vessels as 'platforms of opportunity' to collect data on cetacean occurrence. This was a landmark development in the documentation of Angola's cetacean biodiversity, since many seismic surveys covered deep, oceanic waters that had previously been inaccessible to cetacean scientists. A resulting surge of information on Angolan cetacean occurrence was published from 2006 to 2014 including: (1) the documentation of species records for Angola (Weir 2006a, b, c, Weir et al. 2008, 2010, 2014); (2) evaluations of seasonal relative abundance and spatial distribution (Weir 2007, 2011a, b); (3) examinations of morphology and taxonomy (Weir and Coles 2007, Weir et al. 2014); (4) assessment of habitat preferences (Weir et al. 2012); and (5) studies of behaviour (Weir 2008). Weir (2010a) also published a comprehensive review of cetacean records in the Angola to Gulf of Guinea region, which together with her fieldwork on oceanic cetaceans and Atlantic humpback dolphins was published as the first doctoral thesis focused on Angolan cetaceans (Weir 2011a).

Between 2008 and 2009, some marine mammal survey work was also carried out in association with the construction of a Liquefied Natural Gas (LNG) terminal at the mouth of the Congo River at Soyo, including the use of Marine Autonomous Recording Units (MARUs) between March and December 2008 at two locations along the edge of the Congo Canyon (6°S). The MARUs recorded singing humpback whales between June and early December (Cerchio et al. 2014), and blue whale calls on one date in October (Cerchio et al. 2010).

The year 2008 saw the onset of independent (non-industry) cetacean field research, when Weir (2009, 2011a) visited Namibe Province in southern Angola during two seasons to conduct an ecological study of the Atlantic humpback dolphin. That work provided the first comprehensive assessment of an Atlantic humpback dolphin population, collecting information on abundance (via photo-identification), distribution, movements, seasonality and behaviour (including vocal behaviour: Weir 2010b). The study also produced information on several other cetacean species in coastal waters (Weir 2010c).

Cetacean Species Recorded in Angola

A checklist of Angolan cetacean species is provided in Table 16.2 and some images of the most frequently-recorded species are shown in Fig. 16.3. The SMM (2018) currently recognises 89 species of cetacean worldwide, of which 28 (including unidentified beaked whales of the *Mesoplodon* genus, and only accounting for a single species of common dolphin) have been confirmed to occur in Angola to date. At least seven further species might potentially be added to Angola's fauna in the future.

Baleen Whales

Southern right whale—The majority of Angolan records are from Baía dos Tigres (17°S), which was the northernmost ground for southern right whale catches in the 1700s and 1800s; over 30 were taken there in 1801 (Best 1981; Richards 2009). Catches occurred predominantly in June and July (and thus likely represent a winter breeding presence: Best 1981). The northernmost record in Angola is at approximately 6°S to the southwest of the Congo River mouth (Townsend 1935), but may be atypical. An animal taken off Tômbwa in 1913 is the only record in the 1900s (Table 16.1; Allison 2016a). Best (1990) reported that a catch of 17 right whales at Baía dos Elefantes during 1925 was probably erroneous and actually related to Bryde's whales.

Blue whale—A comprehensive review of blue whale records in Angolan waters was provided by Figueiredo and Weir (2014). Over 2000 blue whales were captured off Angola between 1909 and 1928 (Table 16.1; Allison 2016a), and all were landed at stations in the southern half of Angola (south of 13°S). A single animal was also taken close to Baía dos Tigres in 1934 (Figueiredo and Weir 2014). Several blue whale calls were recorded on an acoustic device off the Congo River mouth (6°S) in October 2008 (Cerchio et al. 2010). Four photographically-verified sightings of blue whales were recently reported from deep waters (>1000 m) off central Angola, at latitudes between 11 and 12°30′S (Figueiredo and Weir 2014). The presence of calves in whaling catches and one sighting indicates the potential use of Angolan waters as a calving or nursery ground (Figueiredo and Weir 2014).

Fin whale—Primarily documented from whaling catches, with over 800 animals captured off Angola between 1910 and 1928, and an additional 22 taken by pelagic whalers between 1934 and 1936 (Table 16.1; Allison 2016a, b). Four sightings were reported off Angola between 2003 and 2006 (Weir 2007); however, two of those were downgraded after subsequent evaluation (Weir 2011a, b). The two remaining sightings occurred in deep-water (>1500 m) during winter (August).

Sei and Bryde's whales—It is considered that the majority of reported 'sei whale' catches in the ETA were misidentifications and more likely comprised Bryde's whales (Harmer 1928, Ruud 1952, Best 1994, 1996, 2001). An estimated total of 1837 sei/Bryde's whales were landed at Angolan shore stations between

Table 16.2 Cetacean species confirmed in Angola. IUCN (November 2018) conservation status: DD (Data Deficient); LC (Least Concern); NT (Near Threatened); VU (Vulnerable); EN (Endangered); CR (Critically Endangered)

English name	Portuguese name	Scientific name	IUCN
Southern right whale	Baleia-franca-austral	*Eubalaena australis*	LC
Blue whale	Baleia-azul	*Balaenoptera musculus*	EN
Fin whale	Baleia-comum	*Balaenoptera physalus*	VU
Sei whale	Baleia-sardinheira	*Balaenoptera borealis*	EN
Bryde's whale	Baleia-de-Bryde	*Balaenoptera brydei*/*B. edeni*	LC
Antarctic minke whale	Baleia-anã-antártica	*Balaenoptera bonaerensis*	NT
Humpback whale	Baleia-de-bossa	*Megaptera novaeangliae*	LC
Sperm whale	Cachalote	*Physeter macrocephalus*	VU
Dwarf sperm whale	Cachalote-anão	*Kogia sima*	DD
Cuvier's beaked whale	Zífio; Baleia-de-bico-de-Cuvier	*Ziphius cavirostris*	LC
Mesoplodon sp.	Baleia-de-bico	*Mesoplodon* sp.	DD
Killer whale	Orca	*Orcinus orca*	DD
Short-finned pilot whale	Baleia-piloto-tropical	*Globicephala macrorhynchus*	LC
False killer whale	Falsa-orca	*Pseudorca crassidens*	NT
Melon-headed whale	Cabeça-de-melão	*Peponocephala electra*	LC
Atlantic humpback dolphin	Golfinho-de-bossa do Atlântico; Golfinho-corcunda do Atlântico	*Sousa teuszii*	CR
Rough-toothed dolphin	Caldeirão; Golfinho-de-bico-comprido	*Steno bredanensis*	LC
Dusky dolphin	Golfinho-cinzento	*Lagenorhynchus obscurus*	DD
Risso's dolphin	Grampo; Golfinho-de-Risso	*Grampus griseus*	LC
Common bottlenose dolphin	Roaz-corvineiro; Golfinho-roaz	*Tursiops truncatus*	LC
Pantropical spotted dolphin	Golfinho-malhado-pantropical	*Stenella attenuata*	LC
Atlantic spotted dolphin	Golfinho-malhado do Atlântico; Golfinho-pintado	*Stenella frontalis*	LC
Spinner dolphin	Golfinho-fiandeiro-de-bico-comprido; Golfinho-fiandeiro	*Stenella longirostris*	LC
Clymene dolphin	Golfinho-fiandeiro-de-bico-curto	*Stenella clymene*	LC
Striped dolphin	Golfinho-riscado	*Stenella coeruleoalba*	LC
Common dolphin sp.	Golfinho-comum	*Delphinus* sp.	LC/DD
Fraser's dolphin	Golfinho-de-Fraser; Golfinho-do-Bornéu	*Lagenodelphis hosei*	LC
Heaviside's dolphin	Golfinho-de-Heaviside	*Cephalorhynchus heavisidii*	NT

Fig. 16.3 Photographs of the 10 most frequently-recorded cetacean species in Angolan waters (>55 records; Weir 2011a, b): (**a**) Bryde's whale; (**b**) humpback whale; (**c**) sperm whale; (**d**) short-finned pilot whale; (**e**) Atlantic humpback dolphin; (**f**) Risso's dolphin; (**g**) bottlenose dolphin; (**h**) Atlantic spotted dolphin; (**i**) striped dolphin; and (**j**) common dolphin. All photographs taken in Angolan waters by the author

1911 and 1928, with a further 809 animals taken by pelagic whalers between 1934 and 1975 (Table 16.1; Harmer 1928; Best 1994; Allison 2016a, b). The majority of pelagic catches were included in the comprehensive assessment of the distribution, migration and diet of ETA Bryde's whales by Best (1996, 2001). Only one sighting of sei whales has been reported for Angola; two animals observed in deep water southwest of Soyo during August 2004 (Weir 2007). In contrast, 63 sightings of Bryde's whales were recorded off northern Angola, mostly in oceanic waters of 1000 to 3000 m depth (Weir 2007, 2011a, b). Bryde's whales have also been confirmed in central and southern Angola, from the *Nansen* surveys (Axelsen et al. 2004, 2005), during coastal dolphin surveys off Namibe Province (Weir 2010c), north of Baía dos Tigres (Dyer 2007), and off Tômbwa and Lobito (Olsen 1913). Best (1996, 2001) described a seasonal migration of the offshore Bryde's whale population in and out of Angolan waters. However, sightings have been reported year-round (Weir 2007, 2010c, 2011a, b), although seasonal fluctuations occur. For example, Weir (2010c) only recorded Bryde's whales during the summer in coastal Namibe Province, while most sightings from northern Angola are in winter and spring (August and September; Weir 2011a, b).

Minke whale—While there are unspecific mentions of minke whales off Angola in several sources (e.g. Mörzer Bruyns 1971; Stewart and Leatherwood 1985), the number of verified records is very low. The vessel *Run/Sierra* caught five Antarctic minke whales at latitudes of 5°S to 16°S (Allison 2016b). An Antarctic minke whale stranded at the Coroca River mouth near Tômbwa (15°45′S) during March 1970 (photograph held in the Museu do Mar, Cascais, Portugal; Peter Best pers. comm.), also confirming this species in Angolan waters (Best 2007).

Humpback whale—Townsend (1935) noted that the region between the equator and 12°S produced the highest nineteenth century humpback catches on the west coast of Africa, particularly between June and October. Angolan whaling catches from 1909 to 1928 included over 10,000 humpbacks, with a strong peak between 1911 and 1913 (Table 16.1; Best 1994; Allison 2016a). No new information on humpback whales emerged until the 1998 field study off northern Angola by Best et al. (1999), which recorded many surface-active groups, cow-calf pairs and singing males, and led those authors to conclude that the area was (or was very close to) a breeding ground. Acoustic monitoring off northern Angola (6°S) during 2008 recorded humpback whale singing activity, which was also considered indicative of breeding behaviour (Cerchio et al. 2014). Numerous sightings of humpback whales have been recorded during sighting surveys, including in southern Angola (Axelsen et al. 2004; Dyer 2007; Weir 2010c), central regions (Krakstad et al. 2003; Axelsen et al. 2005; Roux et al. 2007; Michalsen et al. 2015), and the northern areas off Soyo and Cabinda (Weir 2007, 2011a, b). The highest densities occur over the shelf, but sightings also occur far offshore (to at least 4000 m depth: Weir 2011b). Strong seasonality is evident in Angolan waters, with all captures, sightings and acoustic records occurring between May and January, and with a strong peak between July and October (Weir 2011a, b; Cerchio et al. 2014). The humpback whales using Angolan waters originate from Southern Hemisphere IWC stock B (Rosenbaum et al. 2009), and migrate between breeding areas in the ETA and summer Antarctic feeding grounds.

Synopsis—Both whaling data and sightings surveys indicate that humpback and Bryde's whales are the most numerous baleen whale species in the region, with the remaining species either naturally less common or still to recover from whaling exploitation. The timing of catches (Allison 2016a, b), and observations from year-round sighting surveys (Weir 2011a, b), indicate that most baleen whales exhibit strong seasonality in Angolan waters, occurring during the austral winter and spring (June to October) which corresponds with the breeding period of Southern Hemisphere whale stocks. There is evidence for breeding in Angolan waters of at least humpback whales and blue whales. Many humpback whales may also use Angolan waters as a migratory corridor to reach well-established calving grounds off Gabon and in the Gulf of Guinea (Rosenbaum et al. 2009). The Bryde's whale is one of few baleen whale species that inhabit warm waters year-round (Best 2001), and its seasonal movements in Angolan waters more likely relate to prey availability. Although there are no confirmed records to date in Angola, three additional baleen whale species may be recorded in the future including two documented elsewhere from warm Atlantic Ocean waters (Common minke whale *Balaenoptera acutorostrata* and Omura's whale *B. omuraii*) and one cool water species that has been recorded further south off northern Namibia (19°28′S; pygmy right whale *Caperea marginata*; Leeney et al. 2013) and could extend into the Benguela-influenced waters of southern Angola.

Sperm Whales

Sperm whale—The whaling charts of Townsend (1935) reveal numerous sperm whale captures on the 'Coast of Africa' whaling ground (3–23°S), including the entire coast of Angola. Over 500 sperm whales were landed at Angolan shore stations between 1912 and 1928, with an additional 200 taken by pelagic fleets from the 1930s to the 1970s (Table 16.1; Harmer 1928; Mikhalev et al. 1981a; Best 1994; Allison 2016a, b). Sighting surveys in Angolan waters found that the sperm whale was one of the most frequently-recorded cetacean species (Weir 2011a, b). Sightings were distributed exclusively in deep waters from 800 to 3800 m and usually comprised singletons or nursery schools of ≤20 animals, although loose aggregations of up to 65 animals have been observed (Weir 2011a, b). Sperm whales are present in Angolan waters year-round, but there may be fine-scale spatio-temporal fluctuations in their occurrence and an overall preference for warmer waters where sea surface temperatures (SSTs) exceed 23 °C (Weir et al. 2012).

Dwarf sperm whale—Twenty-six sightings of this species were reported by Weir (2011a, b) from Angolan waters, comprising small groups of one to three animals seen in deep waters in the 1000–2000 m range. The closely-related pygmy sperm whale (*Kogia breviceps*) has not been confirmed off Angola to date, but may be expected to occur based on its worldwide distribution (Caldwell and Caldwell 1989).

Beaked Whales

Of the 22 currently-recognised beaked whale species (SMM 2018), only the Cuvier's beaked whale (*Ziphius cavirostris*) has been positively-confirmed in Angolan waters to date, with four sightings in slope waters of 847–2040 m depth (Weir 2006a, 2011a, b). Eleven additional sightings of unidentified beaked whales (including *Mesoplodon* species) are documented off Angola in deep waters exceeding 730 m (Weir 2006a, 2011a, b). Mörzer Bruyns (1968) also observed three unidentified *Mesoplodon* whales off Angola in July 1966. There is one record of a stranded adult male Gervais' beaked whale (*Mesoplodon europaeus*) from the mouth of the Cunene River (on the Angola–Namibia border) in 1997. Although considered a Namibian record (Griffin and Coetzee 2005), this stranding is highly-supportive of an occurrence in Angolan waters. The warm Atlantic distribution of Blainville's beaked whale (*M. densirostris;* MacLeod et al. 2006) is also indicative of a likely occurrence off Angola.

Delphinids

Killer whale—Records in Angola include observations south of Moçâmedes during July 1966 (Mörzer Bruyns 1971), from a pelagic whaler (Mikhalev et al. 1981b), and from the *Nansen* surveys (Axelsen et al. 2005). Weir et al. (2010) provided information on 18 sightings from Angolan waters between 1991 and 2008. An additional two sightings were reported in 2009 (Weir 2011a, b). Sightings have comprised 1 to 12 animals observed at latitudes of 5°S to 12°S and in water depths ranging from very shallow coastal waters to well over 2000 m. In January 2005, a group of five killer whales was seen attacking sperm whales off northern Angola (Weir et al. 2010).

Short-finned pilot whale—All pilot whales observed in northern Angola to date have been conclusively identified as the short-finned pilot whale (*Globicephala macrorhynchus*). However, it is likely that long-finned pilot whales (*G. melas*) also occur in the Benguela Current-influenced areas and will be confirmed in the future. Pilot whales were the third most frequently-observed species in Angolan waters (perhaps partly because they are easy to identify at distance), with 125 sightings reported by Weir (2011a, b). Over 94% of sightings consisted of ≤50 animals, and all records were located over the slope or in oceanic waters (400–4000 m depth). This species was also reported by Krakstad et al. (2003), Axelsen et al. (2004, 2005) and Dyer (2007).

False killer whale—Thirteen sightings of false killer whales were reported in oceanic habitat (1400–2600 m depth) off northern Angola, comprising groups of 2–50 animals (Weir 2011a, b).

Melon-headed whale—Four sightings of melon-headed whales have been reported in oceanic waters (>1300 m depth) off the northern half of Angola (Weir 2011a, b). Three of the schools were large, comprising 100–300 animals.

Atlantic humpback dolphin—First documented in Angola from a photograph taken near Tômbwa in 2004 (Van Waerebeek et al. 2004). The 'numerous reports' cited in Van Waerebeek et al. (2004) from opportunistic observers in northern Angola and Cabinda have not been upheld by subsequent scientific fieldwork in those areas (Weir 2009, 2011a; Weir and Collins 2015), and are considered likely misidentifications. Dedicated photo-identification surveys in Namibe Province in January and June/July 2008 revealed a very small population of 10 humpback dolphins that inhabit nearshore (<1.4 km) waters along a small 40 km stretch of coast year-round, and use the area for both feeding and calving (Weir 2009, 2010c). Published information on the whistles of this species represents one of few cetacean acoustic studies in Angola to date (Weir 2010b).

Rough-toothed dolphin—Weir (2006b) reported three sightings of rough-toothed dolphins from Angolan waters in 2004 and 2005, while Weir (2011a, b) added an additional 15 sightings up until 2009. All records were seaward of the shelf (700–2200 m), and usually comprised ≤60 animals although several larger groups were observed. An interesting account of an interaction between rough-toothed dolphins and a sport fishing tournament off Luanda was described by Weir and Nicolson (2014), with dolphins taking bait from the fishing lines of several vessels.

Dusky dolphin—Two were photographed off Lobito (12°22′S: Kramer 1961; Findlay et al. 1992; Best and Meÿer 2009). A group of 40 was reported by Axelsen et al. (2004) at 16°48′S off Baía dos Tigres, while four schools of 6–40 animals were recorded during August 2005 south of 16°06′S (Axelsen et al. 2005; Roux et al. 2007). Dyer (2007) observed a group of six at 15°40′S just north of Tômbwa. Dusky dolphins inhabit cool Benguela Current-influenced waters along the west coast of Africa, and are likely limited to southern Angola.

Risso's dolphin—A total of 75 Angolan sightings was described in Weir (2011a, b), and included in the global review of Jefferson et al. (2013). Sightings occurred in slope and oceanic habitat from 900 to 2500 m depth. Group size was generally ≤10 animals, but some larger groups of 35–75 animals were recorded.

Bottlenose dolphin—Fifty-six sightings were reported in Angolan waters by Weir (2011a, b), occurring in water depths varying from 10 m by the coast to 3700 m in oceanic areas. Group size in Angola is typically small at 15 or fewer animals, and in oceanic regions they frequently form mixed-species associations with pilot whales (Weir 2011a, b). They have also been regularly reported during the *Nansen* surveys, including mixed groups with pilot whales (Krakstad et al. 2003, Axelsen et al. 2004, 2005). Weir (2010c) reported 24 sightings (1–50 animals) in the coastal waters between Tômbwa and Moçâmedes in 2008, with more frequent sightings during the winter.

Pantropical spotted dolphin—Weir (2011a, b) reported four sightings from Angola in slope and oceanic habitat (≥820 m depth) north of 8°40'S. The groups ranged from 50 to 200 animals.

Atlantic spotted dolphin—A total of 101 sightings was recorded by Weir (2011b), making it the most commonly recorded species of the *Stenella* genus in Angola. Water depth ranged from 800 to 3000 m, and group sizes were 1–500 animals.

Spinner dolphin—A single sighting exists for Angola, comprising three animals in 1000 m depth off northern Angola in 2004 (Weir 2007, 2011a, b). There have been 11 additional sightings of animals identified as either spinner or Clymene dolphins, but too distant to confirm (Weir 2011a).

Clymene dolphin—The first record for Angola was reported by Weir (2006c). A comprehensive review of Clymene dolphins in the ETA was conducted by Weir et al. (2014) and included 16 records for Angola from 6°S off the Congo River to 14°S. Clymene dolphins in Angola were sighted in water depths ranging from 466 to 2362 m, and in groups of 12 to 1000 animals (Weir et al. 2014).

Striped dolphin—Two sightings were reported by Wilson et al. (1987; No's 40082 and 40083) at 13°59′S and 09°15′S off central Angola in October 1974. A total of 66 sightings were reported from the northern half of Angola by Weir (2011a, b), occurring in slope and oceanic waters from 800 to 2700 m depth.

Common dolphin—The taxonomic status of *Delphinus* dolphins worldwide remains unresolved (Cunha et al. 2015). A few Angolan common dolphin skulls have been included in morphological analyses of the *Delphinus* genus (Bree and Purves 1972), identifying 'short-beaked' and 'long-beaked' forms (Van Waerebeek 1997). However, these may be morphotypes of a single species (Cunha et al. 2015). The external appearance of Angolan animals appears intermediate between short-beaked (*D. delphis*) and long-beaked (*D. capensis*) common dolphins (Weir and Coles 2007; Weir 2011a), and until their taxonomy is better clarified then they are referred to simply as 'common dolphin'. The surveys by Weir (2011a, b) reported 62 sightings of common dolphins off Angola, including in shelf, slope and oceanic habitat (to 2600 m depth), and in group sizes of up to 500 animals. Sightings have been reported as far south as Moçâmedes (15°20′S: Axelsen et al. 2004). Weir et al. (2012) identified a preference for cooler SSTs (≥22.1 °C) in Angola, suggesting the species is associated with areas of upwelling.

Fraser's dolphin—The occurrence of Fraser's dolphins off Angola was first described by Weir et al. (2008) from two sightings recorded in 2007 and 2008. An additional record was added by Weir (2011a, b). All sightings have occurred at latitudes of around 07°30'S off northern Angola, and in deep waters exceeding 1300 m.

Heaviside's dolphin—Two animals were caught by a trawler approximately 12 km north of the Cunene River mouth near the Angola-Namibia border (17°09'S: Findlay et al. 1992; Peter Best pers. comm.). Another was caught in a fishing net off the Cunene River mouth just south of Angola during January 1982 (Windhoek Museum specimen WM 11708; Peter Best pers. comm.), supporting an occurrence in southern Angolan waters. Two Heaviside's dolphin sightings were recorded during the *Nansen* 2004 surveys, in water depths of 20–120 m and at latitudes south of 16°48′S between Baía dos Tigres and the Namibian border (Axelsen et al. 2004, Best 2007). This species appears to inhabit water temperatures ≤15 °C (Best and Abernethy 1994), and is likely restricted to Benguela-influenced regions in the far south of Angola (Best 2007).

Synopsis—At least 17 delphinid species have been confirmed in Angolan waters (assuming only one species of *Delphinus*). Most are likely to occur year-round, although there may be seasonal fluctuations in the distributions of some species depending on the extent of the Benguela Current influence. This applies particularly to dusky dolphins and Heaviside's dolphins, which reach the northern limits of their African distribution range in the southern part of Angola. Sighting surveys indicated that some delphinid species were relatively more common than others off Angola, with Atlantic spotted and common dolphins being frequently-sighted, while pantropical spotted and spinner dolphins were far less common. The relative frequency of dolphin species likely relates to (at least) water temperature, water depth and productivity, with some niche partitioning evident (Weir et al. 2012). Despite large amounts of survey effort in suitable habitat, there are no published (verified) sightings of the pygmy killer whale (*Feresa attenuata*) to date in Angola. This species is likely to be added to the Angola cetacean list in the future, along with the long-finned pilot whale.

Endemism

As highly mobile oceanic predators, none of the reported cetacean species are endemic to Angolan waters. However, four species are endemic to the Atlantic Ocean, including the Atlantic spotted dolphin, Clymene dolphin, Atlantic humpback dolphin and the Heaviside's dolphin. The latter two species have restricted geographic ranges, with the Atlantic humpback dolphin occurring only in nearshore waters of the ETA (Weir and Collins 2015), and the Heaviside's dolphin occupying cool shelf waters of the Benguela Current system (Best and Abernethy 1994). Consequently, Angolan waters are of particular relevance for those species in terms of their very limited global range.

Cetacean Biodiversity and the Marine Environment

The occurrence of cetacean species is strongly related to seabed topography (i.e. depth, slope) and oceanographic variables such as SST, turbidity, salinity and chlorophyll (e.g. Davis et al. 2002; Hamazaki 2002). Consequently, cetacean biodiversity in Angola varies according to habitat (Weir et al. 2012).

Large marine ecosystems (LMEs) have been recognised worldwide based on ecological criteria including bathymetry, hydrography, productivity, and trophically-dependent populations, with the majority of the Angolan EEZ situated within the Benguela Current LME (Fig. 16.1; Sherman 2014). The Angola Front at 5°S forms the northern limit of the Benguela Current LME, and the waters off Cabinda therefore fall into the tropical Guinea Current LME. A biogeographic system to classify marine regions was also developed by Spalding et al. (2007) for coastal waters. In this system, the majority of the Angolan EEZ is situated in the Angolan ecoregion of the Gulf of Guinea province in the Tropical Atlantic realm (Fig. 16.1). However, the northernmost area (north of 6°30′S) falls into the more tropical Gulf of Guinea South ecoregion, while the area south of 15°45′S is recognised as an entirely different biogeographic region located in the Namib ecoregion of the Benguela province in the Temperate Southern Africa realm (Spalding et al. 2007). Consequently, both the LME (Sherman 2014) and marine ecoregion (Spalding et al. 2007) approaches support transition zones within the Angolan EEZ between tropical and temperate (Benguela-influenced) biomes.

Cetacean species in Angola can be broadly classified into communities, based on their occurrence in shelf (less than 200 m depth) versus oceanic (greater than 200 m depth) waters and on their distribution according to marine ecoregion (which broadly corresponds with water temperatures). Using this method, three distinct communities are apparent, with the most diverse comprising the warm water species found in oceanic waters (Fig. 16.4). A second community inhabits cool shelf waters in the south of the study area, while the Atlantic humpback dolphin occupies a unique niche being found only in warm waters on the shelf. There are also six spe-

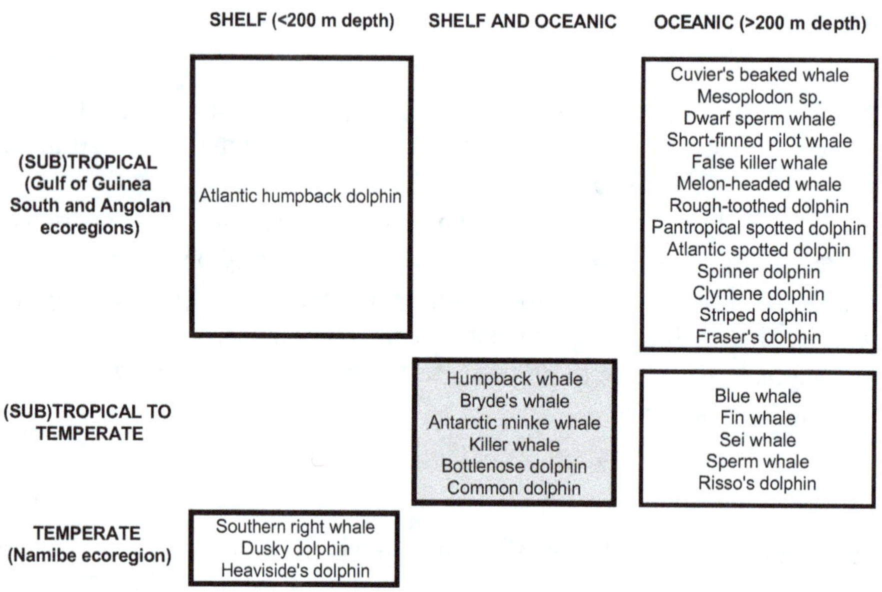

Fig. 16.4 Classification of Angolan cetacean communities. Some species have wider ecological niches than shown here; for example, blue, fin and sei whales are found in shelf waters in some geographic regions, while right whales and dusky dolphins may also be oceanic. However, the information is based solely on documented occurrence in Angola to date. The species in the grey box are those with the most cosmopolitan distributions. The Risso's dolphin is included as a temperate species due to additional sightings of this species during survey work offshore of Lobito (Weir unpublished data)

cies that might be expected to occur throughout the temperature range, primarily comprising the migrating baleen whales and several very cosmopolitan species (e.g. killer whales, bottlenose dolphin and common dolphin) that occupy wide habitat ranges (Fig. 16.4). This system is a useful starting point for considering the underlying drivers of cetacean biodiversity off Angola, and further research into species distribution and environmental parameters should narrow down the habitat preferences for some species in the future. The seasonally variable and transitional oceanographic environment off Angola explains the high cetacean biodiversity recorded relative to most other (solely tropical) ETA countries (Weir 2010a, 2011a).

The association of particular cetacean communities with oceanographic biomes means that species diversity in central and southern Angola will fluctuate on a seasonal basis. The Angola-Benguela Front exhibits spatio-temporal variation over the year as the Benguela Current strengthens and weakens, and Weir (2011a) showed corresponding seasonal SST variations of over 7 °C along the Angolan coast. Consequently, species with preferences for cold or tropical waters may shift in distribution northwards or southwards in response to seasonal changes in oceanography.

Environmental parameters also influence the relative abundance of different species in Angolan waters. For example, in the genus *Stenella* the prevalence of Atlantic spotted dolphins, striped dolphins and Clymene dolphins off Angola in comparison with very few sightings of pantropical and spinner dolphins, may be the result of the productive Benguela-influence. Pantropical spotted and spinner dolphins are more characteristic of tropical oligotrophic waters (Au and Perryman 1985), and are replaced in more productive, slightly cooler areas by the other members of the genus.

The specific use of Angolan waters by some cetacean species also relates to environmental conditions. For example, Cabinda is located in the tropical Gulf of Guinea LME in the far north of Angola, and has consistently warmer SSTs during the winter than further south. This may explain why humpback whale calving and singing behaviour (i.e. breeding activity) has only been confirmed to date in that region of Angola (Best et al. 1999; Cerchio et al. 2014).

Conservation

There are few published accounts of the conservation issues facing cetaceans in Angolan waters, but identified threats in other ETA regions include directed takes (i.e. for human food as 'marine bushmeat'), bycatch in fishing gear, entanglement, prey reduction due to over-fishing, habitat loss and degradation (including noise disturbance and pollution), vessel strikes, marine ecotourism and live captures for display in aquaria (review by Weir and Pierce 2013).

In 1986, the International Whaling Commission's moratorium effectively ended commercial whaling in Angolan waters, but there is also evidence for the capture of small cetaceans. Brito and Vieira (2009) found reports of catches of '*toninhas*' (unidentified dolphins) in Angola between 1940 and 1954 in the national fishing books kept in the National Institute of Statistics in Lisbon, with an average of 20 dolphins landed annually. Those authors considered it likely that bow-riding dolphins were purposefully harpooned by hand for their meat (Brito and Vieira 2009).

There are no specific published records of cetacean fisheries bycatch in Angolan waters, but bycatch affects small cetaceans worldwide and its absence in the literature can be considered a lack of reporting rather than a lack of occurrence in Angola. Weir et al. (2011) reported high numbers of artisanal gillnets deployed in nearshore waters in Namibe Province, and identified them as a major threat to coastal dolphins in the area. Weir and Nicolson (2014) described the potential for bycatch of dolphins during depredation of recreational and commercial fisheries.

Several studies have reported the potential for seismic survey operations to disturb cetaceans in Angola, including spatial avoidance (Weir 2008) and reductions in singing by humpback whales (Cerchio et al. 2014).

The lack of population size information and the absence of quantitative data on impacts on Angolan cetaceans make it impossible to currently assess status and conservation threats. However, the small population of humpback dolphins identified in Namibe Province is clearly of high conservation concern (Weir 2009; Weir et al. 2011), especially given the recent upgrading of the species to Critically Endangered by the IUCN (2018).

Research in Angola: What Next?

Cetacean research in Angola is still in its infancy. Although the species checklist is more complete for Angola than many other ETA countries (Weir 2010a, 2011a), this relates predominantly to MMO data collected during offshore seismic surveys. MMO data can provide information on 'presence', species composition and group sizes, but cannot provide robust 'absence' data due to the unknown potentially-adverse affects of airgun sound on species occurrence and the fact that sightings often remain unidentified to species level due to the lack of ability to approach animals.

Most survey effort and records of cetaceans to date have originated from the (sub)tropical waters between Luanda and Cabinda, where the oil and gas industry is most active (Weir 2011a, b). With the exception of several short periods of effort (e.g. Axelsen et al. 2004, 2005; Roux et al. 2007; Weir 2009, 2010c), the waters in the southern half of Angola have not been surveyed for cetaceans. Consequently, establishing the year-round species composition, distribution and abundance of cetaceans in the Benguela-influenced region south of Lobito should be a priority for future research, especially since whaling captures and recent sightings indicate that region may be most important for large endangered whales (e.g. blue whale; Figueiredo and Weir 2014).

Information on population sizes, population structure (via genetic sampling), spatio-temporal distribution, movements and diet are required for all cetacean species in Angolan waters. Critical to this is the development of comprehensive ongoing training programmes for local biologists in species identification and techniques such as photo-identification, genetic sampling, necropsying of dead animals and passive acoustic monitoring. In particular the field identification of cetacean species takes significant training and field experience, and building this capacity within Angola will be fundamental to the success of long-term population monitoring.

The collection of quantitative data to assess threats is also highlighted as a research priority, and could be achieved through a trained bycatch observer programme for fishing communities, and monitoring at artisanal and commercial landing sites.

Species priorities for Angolan research include Endangered large whale species (Table 16.2) and the Critically Endangered Atlantic humpback dolphin. A decade has passed since Weir's (2009) study of humpback dolphins in Namibe Province, and the current status of the species in nearshore waters along the entire coast requires urgent assessment if it is to be conserved in future decades (Weir et al. 2011). Additionally, Angolan waters are potentially of global importance for breeding populations of sperm whale (Weir 2011a, b), and the waters off Cabinda appear to comprise a calving area for humpback whales (Best et al. 1999, Weir 2011a, b, Cerchio et al. 2014). A systematic research programme would be valuable for informing the management of both of those species.

References

Allison C (2016a) IWC summary catch database Version 6.1; Date: 18 July 2016. Available from the International Whaling Commission, Cambridge, UK

Allison C (2016b) IWC individual catch database Version 6.1; Date: 18 July 2016. Available from the International Whaling Commission, Cambridge, UK

Allison C, Smith TD (2004) Progress on the construction of a comprehensive database of twentieth century whaling catches. Paper SC/56/O27 presented to the International Whaling Commission

Au DWK, Perryman WL (1985) Dolphin habitats in the eastern tropical Pacific. Fish Bull 83:623–643

Axelsen BE, Lutuba-Nsilulu H, Zaera D, et al. (2004) Surveys of the fish resources of Angola. Survey of the pelagic resources 28 July–27 August 2004. Cruise Reports Dr Fridtjof Nansen. Available at: http://hdl.handle.net/11250/107204

Axelsen BE, Luyeye N, Zaera D et al. (2005) Surveys of the fish resources of Angola. Survey of the pelagic resources 16 July–24 August 2005. Cruise Reports Dr Fridtjof Nansen. Available at: http://hdl.handle.net/11250/107244

Best PB (1981) The status of right whales (*Eubalaena glacialis*) off South Africa, 1969–1979. Investigational Report of the Sea Fisheries Institute, South Africa 123:1–44

Best PB (1990) The 1925 catch of right whales off Angola. Rep Int Whaling Commission 40:381–382

Best PB (1994) A review of the catch statistics for modern whaling in southern Africa, 1908-1930. Reports of the International Whaling Commission 44:467–485

Best PB (1996) Evidence of migration by Bryde's whales from the offshore population in the Southeast Atlantic. Rep Int Whaling Commission 46:315–322

Best PB (2001) Distribution and population separation of Bryde's whale *Balaenoptera edeni* off southern Africa. Mar Ecol Progr Ser 220:277–289

Best PB (2007) Whales and dolphins of the southern African Subregion. Cambridge University Press, Cape Town

Best PB, Abernethy RB (1994) Heaviside's dolphin *Cephalorhynchus heavisidii* (Gray, 1828). In: Ridgway SH, Harrison R (eds) Handbook of marine mammals, volume 5, the first book of dolphins. Academic, San Diego, pp 289–310

Best PB, Meÿer MA (2009) Neglected but not forgotten: southern Africa's dusky dolphins. In: Würsig B, Würsig M (eds) The dusky dolphin: master acrobats off different shores. Elsevier Science & Technology, Oxford, pp 291–312

Best PB, Reeb D, Morais M, et al. (1999) A preliminary investigation of humpback whales off northern Angola. Paper SC/51/CAWS33 presented to the IWC Scientific Committee, 12 pp

Bree PJH v, Purves PE (1972) Remarks on the validity of *Delphinus bairdii* (Cetacea, Delphinidae). J Mammal 53:372–374

Brito C, Vieira N (2009) Captures of "toninhas" in Angola during the 20th century. Paper SC/61/SM18 presented to the Scientific Committee of the International Whaling Commission, Madeira, Portugal, 2009

Brown SG (1959) Whales observed in the Atlantic Ocean: notes on their distribution. Norsk Hvalfangst-Tidende 48:289–308

Cadenat J (1959) Rapport sur les petits cétacés ouest-africains: résultats des recherches entreprises sur ces animaux juqu'au mois de mars 1959. Bulletin de l'Institut Français D'Afrique Noire Série A, Sciences Naturelles 21:1367–1409

Caldwell DK, Caldwell MC (1989) Pygmy sperm whale *Kogia breviceps* (de Blainville, 1838): Dwarf sperm whale *Kogia sima* Owen, 1866. In: Ridgway SH, Harrison R (eds) Handbook of marine mammals, volume 4, river dolphins and the larger toothed whales. Academic, San Diego, pp 235–260

Cerchio S, Collins T, Mashburn S et al. (2010) Acoustic evidence of blue whales and other baleen whale vocalizations off northern Angola. Paper SC/62/SH13 presented to the Scientific Committee of the International Whaling Commission, Agadir, June 2010. 8 pp. (Available from the IWC Secretariat, Cambridge, UK)

Cerchio S, Strindberg S, Collins T et al (2014) Seismic surveys negatively affect humpback whale singing activity off northern Angola. PLoS One 9(3):e86464

Cunha HA, de Castro RL, Secchi ER et al (2015) Molecular and morphological differentiation of common dolphins (*Delphinus* sp.) in the Southwestern Atlantic: testing the two species hypothesis in sympatry. PLoS One 10(11):e0140251

Davis RW, Ortega-Ortiz JG, Ribic CA et al (2002) Cetacean habitat in the northern oceanic Gulf of Mexico. Deep-Sea Res I Oceanogr Res Pap 49:121–142

Dyer BM (2007) Report on top-predator survey of southern Angola including Ilha dos Tigres, 20–29 November 2005. In Kirkman SP (ed) Final Report of the BCLME (Benguela Current Large Marine Ecosystem) Project on Top Predators as Biological Indicators of Ecosystem in the BCLME. BCLME Project LMR/EAF/03/02. 381 pp. Available at: http://www.adu.uct.ac.za/adu/projects/sea-shore-birds/communication/report-bclmetpp

Elwen SH, Findlay KP, Kiszka J et al (2011) Cetacean research in the southern African subregion: a review of previous studies and current knowledge. Afr J Mar Sci 33:469–493

Fertl D, Jefferson TA, Moreno IB et al (2003) Distribution of the Clymene dolphin *Stenella clymene*. Mamm Rev 33:253–271

Figueiredo JM (1960) Pescarias de baleia nas províncias africanas Portuguesas. Boletim da Pesca 66:29–37

Figueiredo I, Weir CR (2014) Blue whales (*Balaenoptera musculus*) off Angola: recent sightings and evaluation of whaling data. Afr J Mar Sci 36(2):269–278

Findlay KP, Best PB, Ross GJB et al (1992) The distribution of small odontocete cetaceans off the coasts of South Africa and Namibia. S Afr J Mar Sci 12:237–270

Griffin M, Coetzee CG (2005) Annotated checklist and provisional national conservation status of Namibian mammals. Technical Reports of Scientific Services No. 4. Directorate of Scientific Services, Ministry of Environment and Tourism, Windhoek, 207 pp

Hamazaki T (2002) Spatiotemporal prediction models of cetacean habitats in the mid-western North Atlantic Ocean (from Cape Hatteras, North Carolina, U.S.A. to Nova Scotia, Canada). Mar Mamm Sci 18:920–939

Harmer SF (1928) History of whaling. Proc Linnaean Soc Lond 140:51–95

IUCN (2018) The IUCN Red List of Threatened Species. Version 2017–3. www.iucnredlist.org. Downloaded on 20 January 2018

Jefferson TA, Curry BE, Leatherwood S et al (1997) Dolphins and porpoises of West Africa: a review of records (Cetacea: Delphinidae, Phocoenidae). Mammalia 61:87–108

Jefferson TA, Weir CR, Anderson RC et al (2013) Global distribution of Risso's dolphin *Grampus griseus*: a review and critical evaluation. Mamm Rev 44:56–68

Krakstad J-O, Vaz-Velho F, Axelsen BE et al. (2003) Surveys of the fish resources of Angola. Survey of the pelagic resources 20 July–19 August 2003 (Including observations of marine seabirds and mammals). Cruise Reports Dr Fridtjof Nansen. Available at: http://hdl.handle.net/11250/107254

Kramer MO (1961) Dolphins have the laugh on us … as far as speed goes. South African Yachting News 1961:28–30

Leeney RH, Post K, Best PB et al (2013) Pygmy right whale *Caperea marginata* records from Namibia. Afr J Mar Sci 35(1):133–139

Mackintosh NA (1942) The southern stocks of whalebone whales. Discovery Rep 22:197–300

Mackintosh NA (1965) The stocks of whales. Fishing News (Books) Ltd, London

Macleod CD, Perrin WF, Pitman RL et al (2006) Known and inferred distributions of beaked whale species (Ziphiidae: Cetacea). J Cetacean Res Manag 7(3):271–286

Michalsen K, Alvheim OB, Zaera-Perez D et al. (2015) Surveys of the fish resources of Angola. Cruise Report N0. 8/2015. Surveys of the pelagic fish resources of Angola, 15 August–13 September 2015. Cruise reports Dr. Fridtjof Nansen. EAF—N2015/8. Available at: http://hdl.handle.net/11250/2374573

Mikhalev JA, Savusin VP, Kishiyan NA et al (1981a) To the problem of the feeding of sperm whales from the Southern Hemisphere. Rep Int Whaling Commission 31:737–745

Mikhalev YA, Ivashin MV, Savusin VP et al (1981b) The distribution and biology of killer whales in the Southern Hemisphere. Rep Int Whaling Commission 31:551–565

Mörzer Bruyns WFJ (1968) Sight records of cetacea belonging to the genus *Mesoplodon* Gervais, 1850. Zeitschrift für Säugetierkunde 33:106–107

Mörzer Bruyns WFJ (1971) Field guide of whales and dolphins. CA Mees, Amsterdam

Olsen Ø (1913) On the external characters and biology of Bryde's whale (*Balaenoptera brydei*), a new rorqual from the coast of South Africa. Proc Zool Soc Lond 1913:1073–1090

Richards R (2009) Past and present distributions of southern right whales (*Eubalaena australis*). N Z J Zool 36(4):447–459

Rosenbaum HC, Pomilla C, Mendez M et al (2009) Population structure of humpback whales from their breeding grounds in the South Atlantic and Indian oceans. PLoS One 4(10):e7318

Roux J-P, Dundee BL, da Silva J (2007) Seabirds and marine mammals distributions and patterns of abundance. Chapter 41 in Kirkman, S.P. (ed.) Final report of the BCLME (Benguela Current Large Marine Ecosystem) Project on Top Predators as Biological Indicators of Ecosystem in the BCLME. BCLME Project LMR/EAF/03/02. 381 pp. Available at: http://www.adu.uct.ac.za/adu/projects/sea-shore-birds/communication/report-bclmetpp

Ruud JT (1952) Catches of Bryde-whale off French Equatorial Africa. Norsk Hvalfangst-Tidende 12:662–663

Sherman K (2014) Toward ecosystem-based management (EBM) of the world's large marine ecosystems during climate change. Environ Dev 11:43–66

Simmons DC (1968) Purse seining off Africa's west coast. Commer Fish Rev 30:21–22

Smith TD, Reeves RR, Josephson EA et al (2012) Spatial and seasonal distribution of American whaling and whales in the age of sail. PLoS One 7(4):e34905

SMM (2018) Committee on Taxonomy. List of marine mammal species and subspecies, updated 2017. Society for Marine Mammalogy: www.marinemammalscience.org, consulted on 20 January 2018

Spalding MD, Fox HE, Allen GR et al (2007) Marine ecoregions of the world: a bioregionalization of coastal and shelf areas. Bioscience 57:573–583

Stewart BS, Leatherwood S (1985) Minke whale *Balaenoptera acutorostrata* Lacépède, 1804. In: Ridgway SH, Harrison R (eds) Handbook of marine mammals, volume 3, the sirenians and baleen whales. Academic, San Diego, pp 91–136

Tønnessen JN, Johnsen AO (1982) The history of modern whaling. University of California Press, Berkeley/Los Angeles

Tormosov DD, Budylenko GA, Sazhinov EG (1980) Biocenoological aspects in the investigations of sea mammals. Paper SC/32/02 presented to the IWC Scientific Committee, July 1980, 9 pp

Townsend CH (1935) The distribution of certain whales as shown by logbook records of American whaleships. Zoologica 19:3–50

Van Waerebeek K (1997) Long-beaked and short-beaked common dolphins sympatric off central-West Africa. Paper SC/49/SM46 presented to the IWC Scientific Committee, October 1997, 4 pp

Van Waerebeek K, Barnett L, Camara A et al (2004) Distribution, status, and biology of the Atlantic humpback dolphin, *Sousa teuszii* (Kükenthal, 1892). Aquat Mamm 30:56–83

Weir CR (2006a) Sightings of beaked whales (Cetacea: Ziphiidae) including first confirmed Cuvier's beaked whales *Ziphius cavirostris* from Angola. Afr J Mar Sci 28:173–175

Weir CR (2006b) Sightings of rough-toothed dolphins (*Steno bredanensis*) off Angola and Gabon, South-east Atlantic Ocean. Abstracts of the 20th Annual Conference of the European Cetacean Society, Gdynia, Poland, 2–7 April 2006

Weir CR (2006c) First confirmed records of Clymene dolphin *Stenella clymene* (Gray, 1850) from Angola and Congo, South-East Atlantic Ocean. Afr Zool 41:297–300

Weir CR (2007) Occurrence and distribution of cetaceans off northern Angola, 2004/05. J Cetacean Res Manag 9:225–239

Weir CR (2008) Overt responses of humpback whales (*Megaptera novaeangliae*), sperm whales (*Physeter macrocephalus*), and Atlantic spotted dolphins (*Stenella frontalis*) to seismic exploration off Angola. Aquat Mamm 34:71–83

Weir CR (2009) Distribution, behaviour and photo-identification of Atlantic humpback dolphins (*Sousa teuszii*) off flamingos, Angola. Afr J Mar Sci 31:319–331

Weir CR (2010a) A review of cetacean occurrence in West African waters from the Gulf of Guinea to Angola. Mamm Rev 40:2–39

Weir CR (2010b) First description of Atlantic humpback dolphin (*Sousa teuszii*) whistles, recorded off Angola. Bioacoustics 19:211–224

Weir CR (2010c) Cetaceans observed in the coastal waters of Namibe Province, Angola, during summer and winter 2008. Mar Biodivers Rec 3:e27

Weir CR (2011a) Ecology and conservation of Cetaceans in the waters between Angola and the Gulf of Guinea, with focus on the Atlantic Humpback Dolphin (Sousa teuszii). PhD thesis, University of Aberdeen, Aberdeen

Weir CR (2011b) Distribution and seasonality of cetaceans in tropical waters between Angola and the Gulf of Guinea. Afr J Mar Sci 33:1–15

Weir CR, Coles P (2007) Morphology of common dolphins (*Delphinus* spp.) photographed off Angola. *Abstracts of the 17th Biennial Conference of the Society for Marine Mammalogy*, Cape Town, South Africa, 29 November–3 December 2007. Society for Marine Mammalogy, San Diego

Weir CR, Collins T (2015) A review of the geographical distribution and habitat of the Atlantic humpback dolphin (*Sousa teuszii*). Adv Mar Biol 72:79–117

Weir CR, Nicolson I (2014) Depredation of a sport fishing tournament by rough-toothed dolphins (*Steno bredanensis*) off Angola. Aquat Mamm 40(3):297–304
Weir CR, Pierce GJ (2013) A review of the human activities impacting cetaceans in the eastern tropical Atlantic. Mammal Rev 43(4):258–274
Weir CR, Debrah J, Ofori-Danson PK et al (2008) Records of Fraser's dolphin *Lagenodelphis hosei* Fraser, 1956 from the Gulf of Guinea and Angola. Afr J Mar Sci 30:241–246
Weir CR, Collins T, Carvalho I et al (2010) Killer whales (*Orcinus orca*) in Angolan and Gulf of Guinea waters, tropical West Africa. J Mar Biol Assoc U K 90:1601–1611
Weir CR, Van Waerebeek K, Jefferson TA et al (2011) West Africa's Atlantic humpback dolphin (*Sousa teuszii*): endemic, enigmatic and soon endangered? Afr Zool 46:1–17
Weir CR, MacLeod CD, Pierce GJ (2012) Habitat preferences and evidence for niche partitioning amongst cetaceans in the waters between Gabon and Angola, eastern tropical Atlantic. J Mar Biol Assoc U K 92:1735–1749
Weir CR, Coles P, Ferguson A et al (2014) Clymene dolphins (*Stenella clymene*) in the eastern tropical Atlantic: distribution, group size, and pigmentation pattern. J Mammal 95(6):1289–1298
Wilson CE, Perrin WF, Gilpatrick JW et al (1987) Summary of worldwide locality records of the striped dolphin, '*Stenella coeruleoalba*'. NOAA-TM-NMFS-SWFC-90. 72 pp

Chapter 17
The Giant Sable Antelope: Angola's National Icon

Pedro Vaz Pinto

Abstract The giant sable antelope *Hippotragus niger variani* is the most widely recognised representative of Angolan biodiversity, owing to its endemic status, rarity and physical attributes. One of the last large mammals to be described in Africa, it is confined to the upper Cuanza basin, in central Angola. Studies on the biology of giant sable were mostly conducted in the 1970s, but ongoing efforts using modern tools such as DNA analyses, GPS tracking, camera trapping and satellite imagery are improving our knowledge. Past explanations for the extent of the isolation and relationships with other sable populations have been controversial. Molecular studies have only recently made significant contributions to interpret the evolutionary history of giant sable. Although much pursued by hunters during the first half of the twentieth century, the conservation needs of giant sable were recognised early on, with the proclamation of two protected areas and the setting in place of strict regulations. Park management and efficient protection was enforced in the 1960s, but these protected areas were abandoned soon after the country's independence, leading to population crashes and interspecific hybridization, which left the subspecies on the verge of extinction. The giant sable is currently the main focus of a conservation programme supervised by the Angolan Government that is successfully promoting its recovery.

Keywords Cangandala · Conservation · Cuanza · Evolutionary history · Extinction · Hybridisation · Luando · Population collapse · Trap cameras

P. Vaz Pinto (✉)
Fundação Kissama, Luanda, Angola

CIBIO-InBIO, Centro de Investigação em Biodiversidade e Recursos Genéticos, Universidade do Porto, Campus de Vairão, Vairão, Portugal
e-mail: pedrovazpinto@gmail.com

B. J. Huntley et al. (eds.), *Biodiversity of Angola*,
https://doi.org/10.1007/978-3-030-03083-4_17

Introduction and Background

Having captured the imagination of naturalists and the general public for over one hundred years, the giant sable antelope *Hippotragus niger variani* is the undisputed icon of Angola's natural heritage (Fig. 17.1). Its cultural significance extends from local totemic status among resident communities (where it is known as 'kolo' or 'sumbakaloko') to global recognition as an antelope symbol and flagship for conservation. Soon after its discovery, the giant sable was elevated to a high pedestal among the hunting community as one of the most sought after trophy prizes, and fuelled the lust of big game hunters from all over the world. In Angola, the giant sable was the first animal to receive full legal protection and was soon embraced as an icon during colonial rule. Since Angola's independence in 1975, its status has been reinforced. The importance of the current unanimous recognition of the giant sable as a national symbol should not be underestimated, constituting a key factor uniting Angola's people regardless of their different ethnic groups, religious beliefs or political ideologies, and thus contributing to social cohesion and national pride.

Fig. 17.1 A splendid Giant Sable Antelope bull. (Photo: P. Vaz Pinto)

Observations on the biology of giant sable were reported soon after its scientific description by Thomas (1916), resulting from the many expeditions undertaken to collect material for museums in Europe and the United States of America. Of particular importance were very detailed morphological observations, with taxonomic implications and accurate ecological descriptions, made by Gilbert Blaine (1922). In addition, behavioural observations were made by trophy hunters (e.g. Statham 1922; Gray 1930, 1933; Powell-Cotton 1932; Curtis 1933), while concerns expressed by the subspecies' discoverer himself led to timely conservation interventions (Varian 1953). For a long time these reports would remain as the most reliable sources of knowledge on the taxonomy and biology of giant sable, although several other publications addressed its conservation status (Harper 1945; Heim 1954; Newton da Silva 1958). In the late 1950s and during the following decade, the first systematic efforts focusing on giant sable addressed its biology and related conservation issues, and were undertaken by Portuguese researchers working for the *Centro de Zoologia da Junta de Investigação do Ultramar* or *Instituto de Investigação Científica de Angola* (e.g. Frade 1958, 1967; Frade and Sieiro 1960; Sieiro 1962; Crawford-Cabral 1965, 1966, 1969, 1970). Later, Huntley (1972) made important contributions on conservation planning for giant sable while adding ecological insights. Published in 1972, the book *A Palanca Real* (Silva 1972) provided a comprehensive and appealing photographic compilation, including a few ecological and behavioural observations. The researcher who became inextricably linked with the giant sable, the famous biologist Richard Estes, spent one full year studying the giant sable in Luando Nature Strict Reserve between 1969 and 1970. Estes is still the most relevant contributor to the current knowledge on the biology of the species (e.g. Estes and Estes 1970, 1972, 1974). Another invaluable source is the book *A Certain Curve of Horn* (Walker 2004), in which the author gives a detailed and vivid account of the various explorations, studies and conservation initiatives around the giant sable, a history spanning over 100 years. In recent years, molecular studies have addressed giant sable looking into phylogenetic relationships, genetic diversity or hybridisation (e.g. Pitra et al. 2006; Jansen van Vuuren et al. 2010; Vaz Pinto et al. 2015, 2016). Currently, comprehensive ongoing research is being developed on the giant sable and addressing a wide range of topics, including evolutionary history, biology and conservation (Vaz Pinto 2018).

Scientific Discovery

The giant sable was not discovered and described until the twentieth century, but two intriguing and obscure previous records are worth mentioning. The oldest proven giant sable material is a single horn, named 'The Florence Horn' after the museum in Italy where it was deposited in 1873 (Walker 2004; Vaz Pinto 2018). Nothing is known about the provenance details of the Florence Horn. Although recognised early on as extraordinary, and suspected later to have been obtained from the Angolan relic population (Thomas 1916; Walker 2004), only recently has a molecular study provided convincing evidence of it being a giant sable horn (Vaz Pinto 2018). The Florence Horn may, however, have been preceded by a skull collected by the famous Austrian botanist Friedrich Welwitsch in Angola between 1853 and 1861, and classified by Bocage as *Hippotragus niger* (Bocage 1878, 1890; Thomas 1916). As with so many other priceless biological material from Angola, this specimen was tragically lost in the fire that destroyed the Museu Bocage in 1978, and it was never possible to attribute it to subspecies, even though the reported length of the horns suggest that it was a giant sable. The specimen was stated as having been collected by Welwitsch inland from Moçâmedes (Bocage 1878, 1890; Thomas 1916; Hill and Carter 1941). The exact collecting locality remains in doubt, but it is known that Welwitsch collected in the Malanje region (Crawford-Cabral and Mesquitela 1989), which could have given him access to giant sable.

The discovery and scientific description of this taxon had to wait another half-century, and followed the efforts of the chief engineer overseeing the construction of the Benguela Railway, the British citizen Frank Varian (Varian 1953; Walker 2004). The first mention of a sable with extraordinary horns from the Cuanza district was made by Varian in 1909 and simply based on one photograph and witness accounts, but it was met with disbelief in Europe (Varian 1953; Walker 2004). The first material was secured in 1911, but it was only in 1916 that skulls and skins were shipped by Varian to the British Museum and led to the formal description of *Hippotragus niger variani* (Thomas 1916), thus honouring the discoverer.

Description

The giant sable is a large, compact and muscular antelope carrying massive scimitar-shaped horns. The original description was based on a giant sable skull with horns that measured 57 inches in length along the curvature and 11 inches in circumference at the base (Thomas 1916). Horn length is a distinctive trademark for giant sable as they are usually above 50 and often above 60 inches, while bulls from all other sable antelope populations only very rarely surpass 50 inches (Halse 1998; Vaz Pinto 2018).

The size of horns alone made the type specimen stand out, but equally striking was the darker face, in which unlike other sable, the ante-orbital white spots are not connected by a white streak to the sides of the muzzle (Thomas 1916). These features proved consistent as more specimens were analysed, but other peculiarities were revealed to be distinctive, as described in detail by Gilbert Blaine (1922). Very clear differences have been found in the skulls, both structurally and from measurements, and these helped to sustain a claim for elevating giant sable to specific status (Blaine 1922). Giant sable have longer and narrower forefaces and less prominent forehead compared to typical sables, and relatively small ears in adult males, while the neck in mature bulls is short, massive, oval in section and wedge shaped (Blaine 1922). Other published measurements on both skulls (Groves and Grubb 2011) and teeth (Klein 1974) have relied on very small sample sizes, and therefore added little to Blaine's observations.

The combination of skull and body build, crowned by the massive arched horns, must have greatly contributed to the 'giant' epithet attributed to this taxon, but as no specimen was ever weighted and few were measured (Blaine 1922; Harper 1945; Estes 1982, 2013), doubts remain regarding the size of giant sable in comparison with other sable antelope populations. Giant sable, particularly bulls, have often been considered as much larger in size and weight than other sable (Blaine 1922; Statham 1922; Harper 1945), but this view was not shared by Estes (1982, 2013). Nevertheless, recent observations made on immobilised animals suggest that mature bulls may be at least heavier than other sable subspecies (Vaz Pinto Unpublished Data).

Sexual dimorphism is pronounced in sable, but appears to be more extreme in giant sable (Fig. 17.2), possibly as a by-product of their sedentary nature and the mesic environmental in which they occur (Estes 2013; Vaz Pinto 2018). The distinct characteristics observed in males and female giant sable is obvious in body size, horn length, and pelage colouration. Various physical and colouration features were described in great detail by Blaine (1922). In short, mature giant sable bulls are glossy black with white bellies, white facial markings, reddish hocks and yellowish back to the ears (Blaine 1922; Frade and Sieiro 1960; Estes 2013). In females, the black colour of bulls is replaced by brown tones which have been described as differently as bright golden chestnut (Blaine 1922), light to dark brown-chocolate (Silva 1972) or deep chestnut to blackish brown (Groves and Grubb 2011). Female colouration has been used to assist differentiation among sable subspecies (Groves and Grubb 2011), but pelage colouration of giant sable females is variable irrespective of age (Vaz Pinto Unpublished Data).

Taxonomy

While examining the Angolan material, Oldfield Thomas, the mammal curator of the British Museum, was so impressed by some peculiarities and obvious differences when comparing Varian's material with ordinary sable that he considered the possibility of awarding specific status to the giant sable (Thomas 1916; Blaine 1922; Walker 2004). Subsequently, and having assembled additional material, Gilbert Blaine (1922) found it justifiable to recognise giant sable as a full species on its own *Hippotragus variani*, a view that was followed by some authors (e.g. Harper 1945). Nevertheless, Blaine's claim eventually fell into disuse and the giant sable became widely accepted as one of four to five subspecies of sable antelope (e.g. Hill and Carter 1941; Ansell 1972; Groves and Grubb 2011; Estes 2013).

With the advent of molecular tools, the taxonomic placement of giant sable could be revisited. The first molecular studies to include giant sable samples were based on small mitochondrial fragments and reported paraphyly for giant sable with respect to sable antelope across the southern parts of their range. These findings cast doubt on the recognition of *H. n. variani* as a valid subspecies (Mathee and Robinson

Fig. 17.2 Sexual dimorphism in antelopes is unique to the Sable Antelope but particularly pronounced in the more sedentary Giant Sable of Angola, where mature bulls are jet black (bottom) and cows have a rich cinnamon pelage. (Photos: P Vaz Pinto)

1999; Pitra et al. 2006). The results, in combination with the fact that sable collected in western Zambia often present similar facial markings, were further interpreted as blurring the differences between giant and western Zambian sable antelope (Wessels 2007). However, clear genetic differences exist between these western Zambian sable and giant sable, confirming that the giant sable represents a distinct mitochondrial evolutionary lineage (Jansen van Vuuren et al. 2010).

Much progress has recently been achieved allowing the clarification of the taxonomic status of giant sable, following more ambitious molecular efforts that included the use of mitogenomics and nuclear markers (Vaz Pinto 2018). The full sequencing of mitochondria on a large dataset pooled from across the full range of sable distribution, concluded that giant sable constitutes a reciprocally monophyletic group, one of six geographically discrete sable clusters (Vaz Pinto 2018). Confirming previous suggestions (Pitra et al. 2006; Jansen van Vuuren et al. 2010), the giant sable maternal lineage proved to be more closely related to some west Tanzanian lineages, than to those found elsewhere in southern Africa (Vaz Pinto 2018). A population approach resorting to 57 autosomal microsatellites applied to an even more comprehensive dataset showed giant sable as being the most divergent of five clearly identified and geographically coherent populations (Fig. 17.3, Vaz Pinto 2018). The molecular results combining mitogenomics and nuclear markers

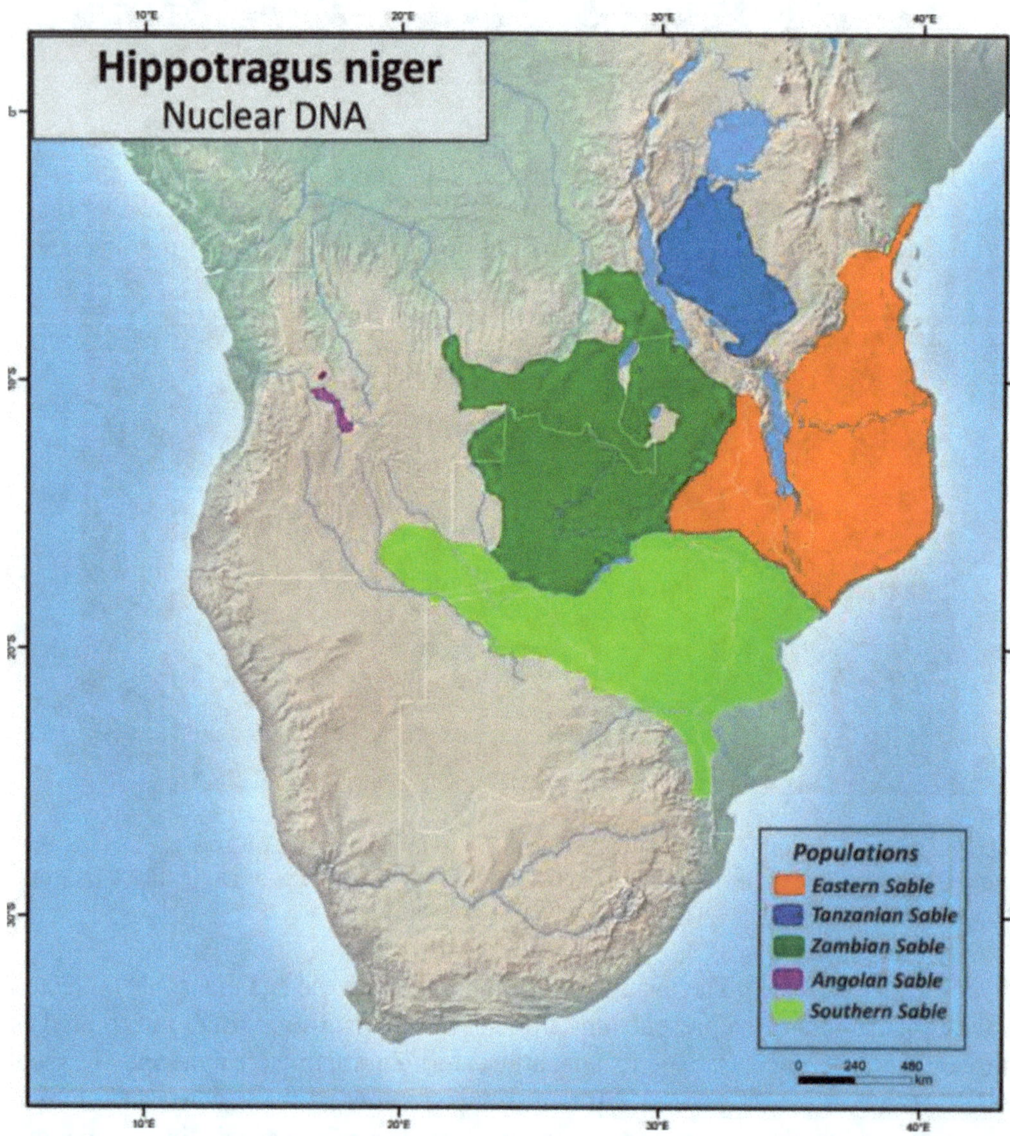

Fig. 17.3 Sable Antelope populations (and sub-species) as determined through nuclear DNA analysis (Vaz Pinto 2018). Eastern Sable *Hippotragus niger roosevelti*; Tanzanian Sable (possibly *H. n.* subsp. nov); Zambian Sable *H. n. kirki*; Angolan Sable *H. n. variani*; Southern Sable *H. n. niger*

provide compelling evidence to support the uniqueness of giant sable, and are therefore concordant with its unequivocal recognition as a separate endemic taxon (Vaz Pinto 2018).

Evolutionary History

The origin of giant sable has remained intriguing and subject to different interpretations, and only recently by resorting to modern molecular tools, a more robustly sustained and coherent picture is emerging. Attempting to explain the rarity of giant sable and its oddly restricted distribution, Huntley (1972) argued it not to be a recent artefact driven by anthropogenic causes, but rather a result of very specific habitat requirements that forced this population to a long-standing confinement in central Angola. Other authors have suggested alternative views such as the possibility of *H. n. kirkii* from western Zambian having had at some point a continuous distribution into the Angolan plateau before receding and a subpopulation becoming isolated in the Cuanza Basin (Crawford-Cabral and Veríssimo 2005; Wessels 2007).

The first molecular studies to include giant sable, based on small mitochondrial fragments and limited sampling (Mathee and Robinson 1999; Pitra et al. 2006) found *H. n. variani* samples to cluster within other sables, and results suggested recent connectivity with other southern African populations. The suggestion that giant sable may share a closer ancestry with some (but not all of) west Tanzanian lineages rather than with the geographically closer west Zambian populations (Pitra et al. 2006; Jansen van Vuuren et al. 2010) was difficult to interpret and explain. It was speculated that central Angola and southern Tanzania could have been colonised from a common source population (Pitra et al. 2006). Furthermore, a tentative date for this split was set at approximately 200,000 years ago (Pitra et al. 2006). In an effort to disentangle this phylogenetic riddle, Groves and Grubb (2011) suggested that *H. n. variani* could have migrated from Angola to southern Tanzania and hybridised with *H. n. roosevelti*, but this scenario is difficult to justify and, if anything, would complicate things even further.

A more recent molecular study based on mitogenomics revealed the phylogeographic patterns that may have shaped the evolutionary history of sable antelope, influenced by a complex interplay between Pleistocene climatic oscillations and geomorphological features (Vaz Pinto 2018). The study also provided credible time estimations for time splitting events. It was hypothesised that during the dry climatic period of the Elster Glaciation, corresponding to the Marine Isotopic Stage (MIS) 6 and estimated at approximately between 130,000 and 186,000 years ago, an ancestral sable lineage evolved in Central Congo and separated from two other lineages in eastern and southern Africa (Vaz Pinto 2018). When the climate became warmer and receding savannas were replaced by forests, at around 120,000 years ago, the Central Congo lineage would have split into an Angolan lineage and another that ended up in the rift region (Vaz Pinto 2018). Subsequently, the Angolan lineage may have since evolved confined to the Cuanza river basin and adapted to particular habitat conditions (Vaz Pinto 2018). The population approach using autosomal nuclear markers confirmed the uniqueness of the giant sable population, with low levels of gene flow for a long time, and as a result being genetically very distinctive (Vaz Pinto 2018). In addition it was possible to genetically differentiate the populations from Luando and Cangandala, even though the genetic signal may have been enhanced by an extreme recent bottleneck affecting the latter (Vaz Pinto 2018). It seems possible that the giant sable evolved in isolation in central Angola since the beginning of the late Pleistocene until the present day, and exposed to relatively rare male-mediated gene flow events with neighbouring sable populations (Vaz Pinto 2018).

Habitat and Ecology

As with other sable races, the giant sable is a specialist of miombo, a type of woodland and mesic savanna that occurs on poor dystrophic soils dominated by trees from the genus *Brachystegia*, *Julbernardia* and *Isoberlinia* (Estes 2013, Fig. 17.4). Giant sable are ecotone species, showing a preferential use of the edge between woodland and grassland (Estes and Estes 1974; Estes 2013). A peculiar feature in the gently undulating giant sable region are vast termite-infested open savannas covered by fire-prone geoxyle vegetation, and known by the local name of *anharas* (Barbosa 1970; Estes and Estes 1974; Zigelski et al. 2019). A mosaic of woodland and *anharas* seems to constitute the prime habitat for giant sable herds both in Luando Strict Nature Reserve (LSNR) and Cangandala National Park (CNP) (Estes and Estes 1974). It has also been suggested that the presence of these *anharas*, characterised by unique vegetation types and influenced by poor soils and unique local climatic conditions, may explain the current pattern of giant sable distribution (Vaz Pinto 2018). Giant sable are also water-dependant, and the availability of water, in streams or water holes, during the dry season, is a key component determining habitat value and can become a limiting factor affecting their distribution patterns (Estes and Estes 1974).

Giant sable are herbivores and mostly selective grazers, with preference for perennial grasses such as *Brachiaria*, *Digitaria*, *Panicum* or *Setaria* spp., and typically biting off only the tender outer portion of the plants (Estes and Estes 1974). Although predominantly grazers, they also browse frequently, in particular the locally abundant shrub species *Diplorhynchus condylocarpon* which seems to be favoured by giant sable all year round (Estes and Estes 1974; Vaz Pinto Unpublished Data), to the point of being referred by locals as '*palanca* bush' (Statham 1922). Also browsed often are the tree *Julbernardia paniculata*, and the dwarf shrubs *Mucana stans*, *Cryptosepalum maraviense* and *Dolichus sp* (Statham 1922; Crawford-Cabral 1970; Silva 1972; Estes and Estes 1974; Vaz Pinto 2018). In addition it has been found that giant sable resort often to geophagy, eating soil excavated on some ancient termite mounds, a habit likely evolved as a consequence of very low nutrient soils (Estes and Estes 1974; Baptista et al. 2013).

Fig. 17.4 Aerial view of the preferred habitat of the Giant Sable Antelope: a mosaic of grassland and miombo woodland – the latter pictured in the bottom photograph. (Photos: P Vaz Pinto)

In common with most other social antelopes, giant sable are gregarious and are structured according to three social classes: the breeding or nursery herds, bachelor groups, and territorial bulls (Estes and Estes 1974; Estes 2013). The matriarchal herd, composed of breeding cows, calves and young, constitute the main unit (Estes and Estes 1974; Estes 2013). Numbers and composition of giant sable breeding herds change seasonally and sometimes even daily, and different average figures have been obtained by various authors, usually ranging from eight to 24 animals (Blaine 1922; Crawford-Cabral 1966, 1970; Estes and Estes 1974; Vaz Pinto Unpublished Data). The nursery herds are sedentary and tend to perpetuate their home ranges across generations (Estes and Estes 1974; Estes 2013). Young males will be tolerated within the herd until around 3 years of age, before dispersing where after they either initiate solitary life or temporarily join other males to form bachelor groups (Estes and Estes 1974). Around their sixth year of age, bulls become territorial, and will demarcate their own territory by scrapping, defecating and bush-thrashing with their horns (Estes and Estes 1974). Dominant bulls typically display aggressive behaviour towards intruders, exerting domination by physical intimidation and chasing, and only exceptionally will the confrontation lead to a serious fight (Estes and Estes 1974).

Giant sable are seasonal breeders. The rutting coincides with miombo springtime and has been stated to start around late August (Estes and Estes 1974), although recent observations suggest the mating season to peak in September and October (Vaz Pinto 2018). The gestation likely follows what has been found for other sable populations, estimated at 8.5–9 months (Wilson and Hirst 1977). The calving season for giant sable matches the onset of the dry season, peaking during a 2-month period between May and July (Estes and Estes 1972), but off-season calving is not unusual (Estes and Estes 1974; Vaz Pinto Unpublished Data). As the calving season approaches, the breeding herds tend to break and the most heavily pregnant cows become isolated (Estes and Estes 1974). Giant sable are 'hiders', meaning that females will calve alone and hide their calves, attending them at irregular intervals for several days or weeks, before re-joining the herd with their offspring, and forming crèches with other calves of similar age (Estes and Estes 1974; Estes 2013).

Giant sable herds have been found to make use of home ranges of varying size, with one of two giant sable herds monitored for 1 year in Luando Strict Nature Reserve staying within 12 km^2 while the other moved a distance of 15 km to use different areas in the dry and rainy seasons, raising the size of the annual home range of one of those groups to 40 km^2 (Pedrosa 1971; Estes and Estes 1974). The same authors found home range size to be affected by food availability and seasonality (Estes and Estes 1974). Herds will tend to break up after the onset of rains, and move from the *anharas* into the woodlands (Estes and Estes 1974). During the wettest periods, sables will avoid waterlogged areas such as floodplains, and spend most of the time on high ground within the woodland (Crawford-Cabral 1970; Estes and Estes 1974). Nevertheless, recent data obtained with GPS transmitters, have found relatively little seasonal variation, and yet a huge contrast in home range size, ranging from 14 to 110 km^2 measured by Minimum Convex Polygon (MCP) (Vaz Pinto 2018). The daily movements of herds is reported to be modest, typically varying from one to two km (Estes and Estes 1974), and this finding is consistent with GPS data (Vaz Pinto Unpublished Data). In general, herd movement patterns can be summarised as concentrations in open areas during the dry season, followed by group partition as the rain starts and confinement of smaller stable groups in wooded parts of their range, and then increased movements towards the end of the rains and further group fragmentation prior to the calving season (Estes and Estes 1974). Different herds will not overlap in home ranges and are frequently separated by several km of seemingly suitable habitat (Estes and Estes 1974). Sable bulls have been reported to hold relatively small territories separated from neighbours by 1 to 3 km apart, while spending most of their time within 3–4 km^2, and are able to expand their area to at least 10–12 km^2 when accompanying breeding herds (Estes and Estes 1974). However, preliminary data obtained from GPS monitoring of several giant sable bulls over a period of 5 years suggest a very different spatial use, as bulls tend to move a lot more than previously thought, and with overlapping territories ranging above 200 km^2 measured as MCP's (Vaz Pinto 2018).

Historical Distribution and Abundance

Soon after the description of *H. n. variani*, it was assumed that they only occurred within the Cuanza watershed, and especially confined to the lowlands between the upper Cuanza and its tributary the Luando (Blaine 1922; Statham 1922; Hill and Carter 1941; Varian 1953). This region, also known as the 'land between two rivers' (Walker 2004), is a narrow strip of land stretching across 200 km oriented NW – SE, and spanning up to 60 km at its widest. Most of the hunters and naturalists exploring the region in search of giant sable entered the sable lands from the Benguela Railroad, heading northwards and thus focusing on the southern region within the species' distribution (Walker 2004). Nevertheless, one of the earliest giant sable explorations, by Col. Statham, entered from the north, and traversed across the land between the two rivers on foot, and was able to confirm the presence of giant sable up to the confluence of the Luando and Cuanza rivers (Statham 1922). Effectively limited by the two massive rivers to the north, to the south giant sable seemed to be contained by three features: swamps, an inland escarpment, and a desolate territory to the east crossed by the railway and known as the 'Hungry Country' (Varian 1953; Walker 2004). Within the 'land between two rivers', sable appear to prefer the Luando and Lingoio sub-basins while avoiding the remaining Cuanza drainage (Blaine 1922; Statham 1922; Vaz Pinto Unpublished Data). The land between two rivers was set aside as a conservation area in 1938, first as the Giant Sable Reserve, and since 1955 as Luando Strict Nature Reserve (Huntley 1971), and covering approximately 828,000 hectares.

The possible existence of a giant sable population north of the Luando River, near the village of Cangandala, was first suggested by Statham (1922). This was based on trophies acquired by Statham from a Portuguese settler. However, Statham seems to have been deliberately misled by local tribal chiefs who denied that giant sable occurred elsewhere than to the south of the Luando River (Statham 1922; Walker 2004). More than 30 years would pass before the existence of a second, albeit much smaller, sable population north of the Luando was eventually confirmed near Cangandala between the Cuque and Cuije rivers (Frade 1958), and a mere 50 km south of Malanje town. As result of this find, Cangandala Strict Nature Reserve was proclaimed in 1963, and elevated to a National Park status in 1970. The occurrence of giant sable outside the boundaries of Luando Reserve and Cangandala National Park was often claimed but never proven conclusively. Witness accounts mentioned sable to be present between the Cuanza River and its western tributary the Cutato (Blaine 1922), but by the 1970's their persistence in the region was at best doubtful (Estes and Estes 1974; Huntley Unpublished Data). Other unconfirmed records were reported from the 'Hungry Country' and from the areas that lie in between Cangandala National Park and Luando Strict Nature Reserve (Estes and Estes 1974; Crawford-Cabral and Veríssimo 2005; Huntley Unpublished Data). Two of the most geographically extreme giant sable records, falling outside the Cuanza basin, were from Quibala in Cuanza-Sul, and from Baixa de Cassanje in Lundas, but only males were reported (Estes and Estes 1974; Huntley Unpublished Data). A third case was a trophy obtained near Lupire in Cuando Cubango in 1966 and described by the collector as a very old lonely bull in poor condition (Francisco Sousa Machado pers. comm.). The trophy would have become the world record if accepted as a typical sable (Halse 1998; Wessels 2007), but a recent molecular study has established it as giant sable (Vaz Pinto 2018). Other reports of giant sables from as far from Luando as Katanga in the Democratic Republic of Congo or western Zambia (Schouteden 1947; Wessels 2007) are likely erroneous and lack molecular support (Ansell 1972; Jansen van Vuuren et al. 2010; Vaz Pinto 2018).

As bulls may disperse far and unpredictably, it is fair to conclude that the historical distribution of giant sable has remained well contained within the Cuanza Basin, and likely centred in Luando Reserve (Estes and Estes 1974), with relatively small population pockets reaching Cangandala National Park and surrounding areas.

The first estimations of giant sable populations were based on limited knowledge of the species distribution and lacked reliable quantifiable data, suggesting the total numbers to reach a few hundred individuals. Based on guestimates provided by Portuguese naturalists, numbers were set at less than 750–800 (Harper 1945) or at around 700 (Heim 1954). By partially surveying approximately one third of Luando Reserve, 159 giant sable were counted (Frade and Sieiro 1960; Newton da Silva 1970), leading Fernando Frade to suggest that the total number could be even less than 500 (Frade 1958, 1967). As conservation measures increased and biologists renewed their interest in the giant sable, subsequent efforts raised the estimations to range between 1500 and 2500 animals (Crawford-Cabral 1970; Huntley 1972, 1973). Nevertheless, and as pointed out by Richard Estes, these estimations should be taken with care as they were not based on actual counts (Estes and Estes 1972, 1974). In the early 1970s a few aerial counts were attempted by plane and helicopter, but the results did not contribute much for determining the population size (Estes and Estes 1970, 1974, Pedrosa 1971). With the bulk of giant sable present in Luando Reserve, the population in Cangandala NP was estimated to range between 100 and 150 animals (Crawford-Cabral 1970; Huntley 1973; João Serôdio

Unpublished Data). The existing estimations suggest therefore an average giant sable density of 0.0025 sable/ha.

Collapse, Rediscovery, Hybridisation

The civil war that ravaged Angola following the country's independence in 1975 had a dramatic impact on the populations of giant sable. As result of the war, management was abandoned in protected areas and infrastructure destroyed, while accounts reported on the widespread killing of giant sable (Walker 2004). A brief visit made in 1982 by Richard Estes, then chairman of the IUCN Antelope Specialist Group, was able to confirm various herds of giant sable that were still being monitored by a park warden in Cangandala (Estes 1982). But by then the Luando Reserve had been taken over by UNITA (*União Nacional para a Independência Total de Angola*) military and could not be surveyed by Estes, and the reports were worrisome (Walker 2004). Soon after Estes' visit the warden was forced to flee Cangandala as UNITA took control of the park and in Luando the rangers that stayed behind were killed. When a short peace allowed a national biodiversity assessment to be carried out in 1992, the situation was stated as mostly unknown in spite of recent sightings (Huntley and Matos 1992), and as the armed conflict resumed and grew in intensity many questioned if the survival of giant sable was even a realistic possibility (Walker 2004). Efforts to find the species were put on hold, and could only be implemented when peace and security were restored in the bush (Walker 2004).

In November 2001, a few months before the end of the war, the Kissama Foundation, a local biodiversity Non-Governmental Organisation, made a flight over Cangandala National Park on a large military helicopter without results (Walker 2004). A subsequent and more ambitious effort was carried out in August 2002, soon after the war ended, by the Kissama Foundation and again with strong support from the Angolan military. On this occasion, Cangandala was subject to a one-day survey by a large party on foot, and a couple of flights on a military helicopter were done over Luando Reserve (Walker 2004). No animals were seen from the air, but brief observations were reported in Cangandala of what could have been giant sable, although they could not be substantiated (Walker 2004). Ground surveys were then implemented in Cangandala by the *Centro de Estudos e Investigação Científica* (CEIC) of the Catholic University of Angola starting in 2003, and in the following year surveys extended into Luando Reserve where an aerial survey was also performed. No giant sable were seen, but *Hippotragus* dung was collected in both protected areas. In 2004, a monitoring programme with trap cameras was also initiated. Finally, in early 2005, it was possible to prove the survival of giant sable when DNA extracted from dung samples revealed that the mitochondria was typical of *H. n. variani*, and trap camera photographs showed sable herds in Cangandala (Pitra et al. 2006).

Although giant sable had proved resilient enough to persist in Cangandala National Park, the situation proved to be much worse than anticipated. As the trap camera records increased and additional ground observations were made in Cangandala, an alarming scenario emerged: only one herd of giant sable was present, no sable bulls could be found, a roan bull was seen escorting the sable herd, and based on morphological features alone, calves and young animals appeared to be hybrids (Vaz Pinto 2007). The possibility of interspecific hybridisation was subsequently confirmed with modern molecular tools (Vaz Pinto et al. 2016). The occurrence and extent of this phenomenon was unexpected, and allowed for a quite unprecedented study among mammals, as it was possible to document in detail the underlying mechanisms that led to such an extreme outcome (Vaz Pinto et al. 2016). By 2009, the last sable herd in Cangandala included nine old pure females and nine hybrids (Vaz Pinto et al. 2016). In spite of being naturally sympatric and with deep independent evolutionary histories, roan and sable were not only able to hybridise but also produced at least two confirmed second generation hybrids (Vaz Pinto et al. 2016).

In Luando, a combination of ground surveys, trap camera monitoring and four aerial surveys with helicopter between 2009 and 2016, suggested that as result of the war, giant sable had been extirpated from approximately 75% of their historical range, but a few herds did manage to survive in the remaining area, and may have totalled around 100 animals at the end of the war in 2002 (Vaz Pinto 2018).

The extreme population crashes that affected the giant sable populations, were further unravelled by an extensive genetic study using mitogenomes and nuclear DNA, applied on a very extensive dataset that included dozens of modern samples and pre-war material obtained from natural history collections in museums around the world (Vaz Pinto 2018). The population collapse in Cangandala, caused a severe loss of heterozygosity that ranks among the lowest ever recorded in mammals (Vaz Pinto 2018). In Luando, the heterozygosity was only moderately reduced as result of the bottleneck, but the loss of mitochondrial diversity was extreme, when 11 haplotypes were found in samples dating from the early twentieth century and a single haplotype became fixed in the extant population (Vaz Pinto 2018). Such a pattern is consistent with a scenario in which at least two subpopulations coexisted in Luando with gene flow maintained by male dispersal, before the more ancient nucleus that used to function as repository of maternal diversity became extinct during the war (Vaz Pinto 2018).

Conservation

Being an endemic population, restricted to the north-central Angolan plateau, and extremely rare, the giant sable has always been a species of conservation concern, firstly listed in the 1933 London Convention for the Preservation of Fauna and Flora, as a Class A species worthy of absolute formal protection (Walker 2004). It also featured as Endangered in the IUCN Red List of Threatened Species since its creation in 1964, with the status revised to Critically Endangered in 1996 and maintained since then (IUCN 2017). The giant sable is also included in Appendix I of the Convention on International Trade of Endangered Species (CITES) since its creation in 1975.

The earliest known giant sable conservation policies, in the form of regional hunting bans, were adopted by the District Governor of Moxico in 1913 just prior to the formal species description, and by the high commissioner for the colony of Angola, General Norton de Matos in 1922 (Varian 1953; Walker 2004). On both occasions, these bans intended to counteract excessive hunting of giant sable, and must be credited to the vision and perseverance of Frank Varian, without whom the species may not have lasted long (Walker 2004). This however, did not prevent many trophy hunters from travelling to Angola to shoot giant sable in the 1920s and 1930s (Walker 2004). Following the listing as a Class A protected species by the London Convention of 1933, giant sable hunting and the possession of trophies was prohibited in 1934 (Heim 1954).

Nevertheless, these regulations were poorly enforced, and many specimens were still being collected by foreign hunters, Portuguese traders and local communities (Harper 1945). In 1938 the 'Land between Two Rivers' was demarcated as a hunting reserve, and named *Reserva da Caça do Luando*, thus laying the ground for the implementation of future integrated conservation policies. Recommendations were made by the International Union for the Conservation of Nature and Natural Resources (IUCN) for the creation of a National Park (Heim 1954), and in response the colonial government upgraded the category of Luando from a hunting reserve to a strict nature reserve in 1955 (Huntley 1971). In the same year, a new law regulating hunting and nature conservation in Angola underlined the giant sable as priority species worthy of full protection, and regulations published in 1957 criminalised the killing of giant sable, setting the fine at 100,000 escudos (Frade and Sieiro 1960), which would be equivalent to more than US $10,000 at current value. Although giant sable conservation finally started to be taken seriously and hunting had been almost terminated, one last trophy hunting expedition was authorised in 1961 following negotiations at the highest level (Agundis 1965; Vaz Pinto 2018). Conservation measures for the giant sable were discussed in the late 1950s (Newton da Silva 1958; Frade 1958; Frade and Sieiro 1960), while for the first time the need to protect the small pocket population found near the town of Cangandala was addressed (Frade 1958). The recognition of this second population would lead to the creation of Cangandala Strict Nature Reserve in 1963, and then elevated to a National park in 1970 (Huntley 1971).

By the early 1970s the giant sable was relatively well protected and benefited from conservation management practices in both Cangandala and Luando (Pedrosa 1971; Huntley 1973; Estes and Estes 1974). This was achieved in spite of relatively moderate budgets and small management teams, comprising two senior rangers and seven assistant-rangers in Luando, and one senior ranger with four assistant rangers in Cangandala (Huntley 1971). As the direct harvest of giant sable was much reduced and virtually eliminated with increased enforcement of protection and management, other threats gained in perceived relevance, particularly habitat deterioration due to slash and burn agriculture (Crawford-Cabral 1970; Huntley 1972; Estes and Estes 1974). Luando Reserve had a human resident population estimated at 18,000 in 1971 in addition to 800 people living inside Cangandala National Park, mostly resorting to tree-cutting to plant manioc, and thus negatively affecting the natural miombo woodland (Huntley 1971; Pedrosa 1971; Estes and Estes 1974). The possible relocation of human populations, conversion of agricultural practices, and even translocation of giant sable to safe havens was then suggested (Huntley 1972, 1973; Estes and Estes 1974). Nevertheless, such conservation concerns would soon become irrelevant.

During the Angolan civil war the situation deteriorated rapidly, and if conservation measures were still in place in Cangandala until 1982 (Estes 1982), soon after this all management and protection ceased across the giant sable areas. Conservation initiatives could only be reinstated in recent years, following the arrival of peace and the location of the last surviving population pockets (Vaz Pinto 2018). In the absence of formal management in Luando and Cangandala, the Giant Sable Conservation Project was launched in 2003 by the Catholic University of Angola, and since 2010 has been led by the Kissama Foundation. The Giant Sable Project has been assisting the Angolan Government in the implementation of conservation practices and management in both protected areas. Members of resident communities were trained and appointed as conservation agents, and some have already been transferred to the park management as government rangers. In Cangandala National Park, the Giant Sable Project has rehabilitated park infrastructure, deployed equipment and built fences (Vaz Pinto 2018). Extraordinary conservation measures were adopted in 2009 to tackle the hybridisation crisis in Cangandala National Park, leading to translocations, sterilisation of hybrids and the constitution of a breeding nucleus (Vaz Pinto et al. 2016).

The most critical issue that has affected the giant sable in recent years, causing the populations to crash and risking compromising their recovery, is widespread uncontrolled poaching driven by the bushmeat trade (Vaz Pinto 2018). Although giant sables do not appear to be specifically targeted by poachers, they are still shot at, but the most pervasive and negative impact is the large scale use of snares and foot traps that are having a huge toll, affecting mostly young females and immature sables (Vaz Pinto 2018).

The Giant Sable Project has installed a network of trap cameras covering a good part of the reserves, which has allowed regular monitoring and the identification of individuals, and was instrumental in detecting and documenting the hybridisation phenomenon in Cangandala (Vaz Pinto et al. 2016; Vaz Pinto 2018). Between 2009 and 2016, a total of 74 giant sable and nine hybrids were darted and marked, 65 of them released with tracking collars, which included 32 GPS collars, allowing much increased surveillance power and the gathering of knowledge on the biology of the species (Vaz Pinto 2018).

The Angolan Government has been boosting law enforcement measures, and in 2016 the fine for killing one giant sable was increased to a value equivalent to roughly US$100,000, although no one has been prosecuted for such an offense in recent decades. Currently, the conservation of giant sable is understandably focused on anti-poaching, relying on an increase of surveillance, close monitoring of animals and strengthening park management. By 2018, the giant sable populations had recovered to around 70 animals in Cangandala National Park and an estimated 150 in Luando.

The Way Forward

The unquestionable importance of securing the future of a critically endangered taxon, which also happens to be a flagship species and a national icon, must frame current and future activities. In addition to the obvious need to implement more effective law enforcement measures and enhance park management through infrastructure rehabilitation and staff recruitment and training, some specific issues deserving consideration include erecting new fences and making relocations to recover parts of the historical giant sable distribution.

As tools ultimately benefitting giant sable conservation and management, several lines of research can be developed, either as completely new approaches, or by building on previous work. A study of the food preferences of giant sable is a crucial topic which can now be addressed in detail with modern molecular tools, complemented with remote tracking of movements. Studies on the use of other local resources may also prove critical, such as water and natural salt licks. A better understanding of factors that affect the breeding success and calf mortality, is also needed. An epidemiological study is required, including a study on parasites potentially affecting the giant sable, and a monitoring programme should be implemented and extended to other species and domestic animals in the region. The impact of frequent dry-season fires, and how they reflect on vegetation and on the movements of giant sable should be assessed, and the use of fire as a management tool should be explored. The fact that many giant sable have been collared with GPS satellite transmitters and more will probably be in the future, opens up unique opportunities to develop research on the spatial use of herds and bulls, address territoriality, use of local resources, breeding and response to extrinsic factors. Existing molecular tools should continue to be employed to address individual identification, parameters of genetic diversity, with clear application on the management of existing populations. Comparisons with other subspecies and historical giant sable material, and development of more advance molecular tools such as with genomics, will also greatly improve our knowledge, and may assist future breeding efforts, and help preserve some critical and unique genetic features of this magnificent antelope.

References

Agundis TM (1965) El Llamado de la Montaña, Viajes de Cacería, Angola-Tanzania-Alaska. Rustica Editorial, Mexico

Ansell WFH (1972) Part 15: order Artiodactyla. In: Meester JAJ, Setzer HW (eds) The mammals of Africa: an identification manual. Smithsonian Institution Press, Washington, DC, pp 1–93

Baptista SL, Pinto PV, Freitas MDC et al (2013) Geophagy by African ungulates: the case of the critically endangered giant sable antelope of Angola (*Hippotragus niger variani*). Afr J Ecol 51(1):139–146

Barbosa LAG (1970) Carta Fitogeográfica de Angola. Instituto de Investigação Científica de Angola, Luanda

Blaine G (1922) Notes on the zebras and some antelopes of Angola. Proc Zool Soc London 92(2):317–339

Bocage JVB du (1890) Mammifères d'Angola et du Congo (Suite). Jornal de Sciencias, Mathemáticas, Physicas e Naturaes, Lisboa, Segunda Série 1(1):8–32

Crawford-Cabral J (1965) A palanca preta gigante, sua situação e medidas a adoptar. Luanda, Unpublished mimeograph

Crawford-Cabral J (1966) A palanca preta gigante, aditamentos e correcções ao relatório do ano anterior. Luanda, Unpublished mimeograph

Crawford-Cabral J (1969) A study of the giant sable. The zoological society of Southern Africa. News Bull 10(2):1–7

Crawford-Cabral J (1970) Alguns aspectos da ecologia da palanca real (Hippotragus niger variani Thomas). Boletim do Instituto de Investigação Científica de Angola 7:5–38

Crawford-Cabral J, Mesquitela LM (1989) Índice toponímico de colheitas zoológicas em Angola (Mammalia, Aves, Reptilia e Amphibia). Estudos, Ensaios e Documentos 151:1–206

Crawford-Cabral J, Veríssimo LN (2005) The ungulate fauna of Angola: systematic list, distribution maps, database report. Lisboa: Instituto de Investigação Científica Tropical, Lisboa

Curtis CP (1933) Giant sable antelope. In: Grinnell GB, Roosevelt K (eds) Hunting trails on three continents. Boone and Crocket Club, New York, pp 237–252

du Bocage JVB (1878) Liste des Antilopes d'Angola. Proc Zool Soc London 1878:741–745

Estes RD (1982) The giant sable and wildlife conservation in Angola. Report to IUCN/Species Survival Commission, Gland, Switzerland

Estes RD (2013) *Hippotragus niger* sable antelope. In: Kingdon J et al (eds) Mammals of Africa, vol 6. Bloomsbury, London, pp 556–565

Estes RD, Estes RK (1970) Preliminary report on the giant sable. Unpublished manuscript, p 22

Estes RD, Estes RK (1972) The giant sable antelope *Hippotragus niger variani*. Summary report and recommendations. Unpublished manuscript, p 55

Estes RD, Estes RK (1974) The biology and conservation of the giant sable antelope, *Hippotragus niger variani* Thomas, 1916. Proc Acad Natl Sci Phila 26:73–104

Frade F (1958) Mesure adoptées pour la protection de l'Hippotrague géant en Angola. Mammalia 22(3):476–477

Frade F (1967) Palanca Preta Gigante, Relíquia da Fauna de Angola. Unpublished mimeograph, p 3

Frade F, Sieiro D (1960) Palanca preta gigante de Angola. Garcia de Orta 8:21–38

Gray PN (1930) African game lands. The Sportsman 8(4):1–34

Gray PN (1933) Along the livingstone trail. In: Grinnell GB, Roosevelt K (eds) Hunting trails on three continents. Boone and Crocket Club, New York, pp 103–143

Groves C, Grubb P (2011) Ungulate taxonomy. Johns Hopkins University Press, Baltimore, p 317

Halse ARD (ed) (1998) Rowland Ward's records of big game, XXV edn. Rowland Ward Publications, Johannesburg

Harper F (1945) Extinct and vanishing mammals of the old world. American Committee for International Wildlife Protection, New York

Heim F (1954) Les Fossiles des Demain: treize mammifères menaces d'extinction. Étudiés par le "Service de Sauvegarde". Union Internationale pour la Protection de la Nature, p 112

Hill JE, Carter TD (1941) The mammals of Angola Africa. Bull Am Mus Nat Hist 78(1):1–211

Huntley BJ (1971) Guia Preliminar dos Parques e Reservas de Angola. Relatório 3. Repartição Técnica da Fauna, Direcção Provincial dos Serviços de Veterinária, Luanda, Mimeograph report, p 17

Huntley BJ (1972) Plano para o futuro da palanca real de Angola. Relatório 11. Repartição Técnica da Fauna, Direcção Provincial dos Serviços de Veterinária, Luanda, Mimeograph report, p 9

Huntley BJ (1973) Distribuição e situação da grande fauna selvagem da Angola com referência especial às espécies raras e em perigo de extinção – primeiro relatório sobre o estado actual. Relatório 21. Repartição Técnica da Fauna, Direcção Provincial dos Serviços de Veterinária, Luanda, Mimeograph report, p 14

Huntley BJ, Matos EM (1992) Biodiversity: Angolan environmental status quo assessment report. IUCN Regional Office for Southern Africa, Harare

IUCN SSC Antelope Specialist Group (2017) *Hippotragus niger* ssp. *variani*. The IUCN red list of threatened species 2017: e.T10169A50188611

Jansen van Vuuren BJ, Robinson TJ, Vaz Pinto P et al (2010) Western Zambian sable: are they a geographic extension of the giant sable antelope? S Afr J Wildl Res 40(1):35–42

Klein RG (1974) On the taxonomic status, distribution and ecology of the blue antelope, *Hippotragus leucophaeus* Pallas 1766. Ann S Afr Mus 65(4):99–143

Mathee CA, Robinson TJ (1999) Mitochondrial DNA population structure of roan and sable antelope: implications for the translocation and conservation of the species. Mol Ecol 8(2):227–238

Newton da Silva S (1958) A Caça e a Protecção da Fauna em Angola. Author's Edition, Lisboa, p 174

Newton da Silva S (1970) A Grande Fauna Selvagem de Angola. Direcção Provincial dos Serviços de Veterinária, Luanda

Pedrosa V (1971) Deslocação à Reserva Natural e Integral do Luando. Direcção dos Serviços de Veterinária, Luanda, Unpublished manuscipt, p 24

Pitra C, Vaz Pinto P, O'Keeffe BW et al (2006) DNA-led rediscovery of the giant sable antelope in Angola. Eur J Wildl Res 52(3):145–152

Powell-Cotton PHG (1932) Angola. In: Maydon HC (ed) Big game shooting in Africa. Seeley, London, p 445

Schouteden H (1947) De zoogdieren van Belgisch-Congo en van Ruanda-Urundi (Les mammifères du Congo Belge et du Ruanda-Urundi), III. Ungulata (2) Rodentia. Annales du Musée du Congo Belge Série 2 3(3):333–576

Sieiro D (1962) Novas observações acerca da palanca preta gigante *Ozanna grandicornis variani* (Thomas). Boletim do Instituto de Investigação Científica de Angola 1:49–57

Silva JA (1972) A palanca real – contribuição para o estudo bioecológico da palanca real (*Hippotragus niger variani*). Junta de Investigações do Ultramar, Lisboa

Statham JCB (1922) Through Angola: a coming colony. Blackwood & Sons, London

Thomas O (1916) A new sable antelope from Angola. Proc Zool Soc Lond 1916:298–301

Varian HF (1953) Some African milestones. Books of Rhodesia, Bulawayo, p 272

Vaz Pinto P (2007) Hybridization in giant sable. A conservation crisis in a critically endangered Angolan icon. IUCN/SSC Antelope Spec Group Gnuslett 26:47–58

Vaz Pinto P (2018) Evolution history of the critically endangered giant sable antelope (*Hippotragus niger variani*) – insights into its phylogeography, population genetics, demography and conservation. PhD thesis. University of Porto, Porto

Vaz Pinto P, Lopes S, Mourão S et al (2015) First estimates of genetic diversity for the highly endangered giant sable antelope using a set of 57 microsatellites. Eur J Wildl Res 61(2):313–317

Vaz Pinto P, Beja P, Ferrand N et al (2016) Hybridization following population collapse in a critically endangered antelope. Sci Rep 6:18788

Walker JF (2004) A certain curve of horn: the hundred-year quest for the giant sable antelope of Angola. Grove/Atlantic Inc, New York

Wessels J (2007) Western Zambian sable: a giant sable look-alike or the real thing. Game Hunt 13:32–36

Wilson DE, Hirst SM (1977) Ecology and factors limiting roan and sable antelope populations in South Africa. Wildl Monogr 54:3–111

Zigelski P, Gomes A, Finckh M (2019) Suffrutex dominated ecosystems in Angola. In: Huntley BJ, Russo V, Lages F, Ferrand N (eds) Biodiversity of Angola. Science & conservation: a modern synthesis. Springer Nature, Cham

Part V
Research and Conservation Opportunities

Chapter 18
Biodiversity Conservation: History, Protected Areas and Hotspots

Brian J. Huntley, Pedro Beja, Pedro Vaz Pinto, Vladimir Russo, Luís Veríssimo, and Miguel Morais

Abstract Angola is a large country of great physiographic, climatic and habitat diversity, with a corresponding richness in animal and plant species. Legally protected areas (National Parks and Game Reserves) were established from the 1930s and occupied 6% of the country's terrestrial area at the time of independence in 1975. As a consequence of an extended war, the Protected Areas were exposed to serious neglect, poaching and land invasions. Many habitats of biogeographic importance, and many rare and endemic species came under threat. The recently strengthened administration gives cause for optimism that a new era for biodiversity conservation is at hand. The Protected Areas system was greatly expanded in 2011, and increasing resources are being made available towards achieving management effectiveness.

B. J. Huntley (✉)
CIBIO-InBIO, Centro de Investigação em Biodiversidade e Recursos Genéticos, Universidade do Porto, Vairão, Portugal
e-mail: brianjhuntley@gmail.com

P. Beja
CIBIO-InBIO, Centro de Investigação em Biodiversidade e Recursos Genéticos, Universidade do Porto, Vairão, Portugal

CEABN-InBio, Centro de Ecologia Aplicada "Professor Baeta Neves", Instituto Superior de Agronomia, Universidade de Lisboa, Lisboa, Portugal
e-mail: pbeja@cibio.up.pt

P. Vaz Pinto
Fundação Kissama, Luanda, Angola

CIBIO-InBIO Centro de Investigação em Biodiversidade e Recursos Genéticos, Universidade do Porto, Campus de Vairão, Vairão, Portugal
e-mail: pedrovazpinto@gmail.com

V. Russo · L. Veríssimo
Fundação Kissama, Luanda, Angola
e-mail: vladyrusso@gmail.com; vladimir.russo@holisticos.co.ao; lmnverissimo@gmail.com

M. Morais
Faculdade de Ciências, Universidade Agostinho Neto, Luanda, Angola
e-mail: dikunji@yahoo.com.br

B. J. Huntley et al. (eds.), *Biodiversity of Angola*,
https://doi.org/10.1007/978-3-030-03083-4_18

Keywords Angola · Bushmeat · Ecoregions · Marine ecosystems · Protected areas · Threatened species · Wildlife trade

Introduction: Wildlife Conservation During the Colonial Era

In common with most colonial territories in Africa, conserving wildlife was not a general consideration in Angola until the twentieth century. However, the first expression of concern regarding wildlife numbers came much earlier, and from none other than the most famous zoological collector to work in Angola – José Anchieta. In correspondence with the great Barbosa du Bocage, Anchieta (1869, in Andrade 1985:87) noted that inland of Luanda "the big game, abundant until fifty years ago, has moved into the interior because of the increased population and general use of firearms." But worse was to come. In 1880, the 'Angola Boers' arrived in Humpata, having crossed the Kalahari on their fateful Thirstland Trek (Stassen 2016). The Boers' hunting depredations soon spread across the country. Professional hunters such as William Chapman (Stassen 2010) described the wealth of game in the southwest, and personally contributed to its depletion.

Globally, by the end of the nineteenth century many repentant hunters were becoming alarmed at the fate of the once abundant herds and mobilised action to address the problem. The first international agreement on nature conservation was the *Convention for the Preservation of Wild Animals, Birds and Fish in Africa*, otherwise known as the London Convention, held in London in 1900. Attended by 11 European powers, the Convention was not ratified by several countries, including Portugal, and was abandoned with the onset of World War I (Carruthers 2017). Interestingly, the convention was the brain-child of the German hunter/explorer/military officer Hermann von Wissman, who with Paul Pogge collected in Malange and the Lundas in the early1880s, before he crossed Africa on the first of two trans-continental expeditions.

By the early twentieth century the impact of Boer biltong hunters had become notorious, and Thomas Varian, who introduced the Giant Sable Antelope to science, convinced the Governor of Moxico in 1913, and the then Portuguese High Commissioner, Norton de Matos, to close the sable lands to hunting (Varian 1953). The fame of the Giant Sable drew numerous trophy-hunting and scientific expeditions to Angola through the 1920s and 1930s (Walker 2004) and the zoological collections they made contributed much to our knowledge of Angola's biodiversity.

The London Convention of 1900 was followed in 1933 by the *Convention Relative to the Preservation of Fauna and Flora in their Natural State* – also known as the London Convention. Whereas the 1900 convention focused on hunting regulations, the 1933 convention promoted the idea that each colonial power should establish national parks and reserves in their colonial territories, following the model of the Kruger National Park established by South Africa in 1926. The 1933 convention also required states to give special protection to an internationally selected list of species – a list that included Giant Sable Antelope and the enigmatic desert plant *Welwitschia mirabilis*. Interest in protecting Angola's fauna was rising, and hunters such as Henrique Galvão and Teodósio Cabral, administrators such as Norton da Matos and Abel Pratas, and scientists such as Fernando Frade and Luis Carrisso, immediately championed the National Park model. Portugal set about creating Angola's National Parks and Game Reserves even though it did not ratify the 1933 convention until 1950. The first of these protected areas was Iona, established as a Game Reserve by decree on 2 October 1937, followed by Cameia, Quiçama, Bicuar and Luando proclaimed on 16 April 1938. The first four of these game reserves were raised to National Park status in the 1950 and 1960s.

The good work of the 1930s and 1940s was reversed by the *caça livre* (free hunting) period of the early 1950s, when the wildlife populations of the cattle ranching region of the southwest were decimated because of concerns regarding stock diseases carried by wild species. The voices of reason were raised by a younger generation – Luis Carmo, Armando Malacriz and Newton da Silva (Newton da Silva 1952, 1970) and by 1955 Angola had a new and detailed legislative instrument, the *Decreto 40,040* (*Regulamento sobre a Protecção do Solo, Flora e Fauna*) which was only revoked in 2017. Wildlife conservation was given formal status as a public concern by the establishment of the *Conselho de Protecção da Natureza* (CPN) in 1965, chaired by the Governor General. The CPN played a pivotal role in the expulsion of cattle ranches from Quiçama in the 1970s (Huntley 2017). The increasing support of the Portuguese government for conservation came to a head when a major conference of its African territories was convened in Lubango in 1972. Titled *Reunião para o Estudo dos Problemas da Fauna Selvagem e Protecção da Natureza no Ultramar Português*, the meeting ran for 2 weeks and was attended by 73 delegates. It prepared 53 recommendations for action to improve the protection of nature throughout Angola, leading the government to double the budget of the department responsible for National Parks – the *Repartição Técnica da Fauna.*

Post-independence History of Conservation in Angola

Following the 'Carnation Revolution' of 25 April 1974 in Portugal, and soon after gaining independence, Angola entered a period of increasing difficulty and ultimately war, which only ended in February 2002. The impact of this period of violence and displacement on the wildlife and protected areas of Angola is described elsewhere (Walker 2004; Huntley 2017). During the war years, efforts to bring public support to the National Parks were made through convening annual *Semanas do Ambiente* (Environment Weeks) led by a small network of Angolans, most notably Carlos Pinto Nogueira, Serôdio d'Almeida and Vladimir Russo. Most of the protected areas were abandoned and the wildlife populations decimated during the early years of the war. In 1992 the International Union for the Conservation of Nature (IUCN) led an international study of the situation (IUCN 1992) that concluded:

> Since 1975, most, if not all populations of large mammals have been severely reduced, if not eliminated. Wholesale slaughter of elephant, rhino, eland, roan, oryx, springbok, zebra, bushbuck, reedbuck, lechwe and many other species occurred in all parks and reserves. It is possible that some nucleus herds still survive, sufficient to recover if given effective protection.

In an ironic twist of wildlife conservation practice, in 1995 the Kissama Foundation (KF) was established, led by a group of conservation-minded military generals. Wishing to support Quiçama National Park on its road to recovery, the KF raised funds to re-introduce species that had been severely reduced during the first decades of the war. Unfortunately the initiative, promoted as 'Operation Noah's Ark' in 2000 introduced many species never previously known to occur in Quiçama. Despite international concern regarding the introductions, the programme was continued and expanded in 2014 by the then Minister of Environment, as an essentially private effort to create a mixed collection of animals in the tiny encampment – ca. 1% of Quiçama –that formed the 'Special Protected Area'. Sadly, the remaining 99% of Quiçama has since been left to the ravages of the bushmeat trade and illegal land occupation. Species never previously recorded in Quiçama but introduced with Ministerial approval in 2000 and 2014 included Plains Zebra, Giraffe, Kudu, Nyala, Common Waterbuck, Blue Wildebeest, Red Hartebeest, Blesbok, Oryx, and Common Impala. Nyala and Blesbok have never been recorded in Angola, or within 2300 km of Quiçama. Only two of the species introduced, Savanna Elephant and Eland, were native to the park, but the poorly documented animals introduced by wildlife dealers were from different genepools to the original Quiçama populations.

During the early 2000s, international interest in Angola led to several initiatives, most notably those of the Global Environment Facility (GEF), to support conservation in the country. A fundamental step supported by the GEF was the development of a National Biodiversity Strategy and Action Plan (NBSAP), (GoA 2006) that gave direction to policy established in the Base Law for the Environment (GoA 1998). In support of the objectives of the NBSAP, a survey of the Cuando Cubango paved the way to the proclamation of the Luengue-Luiana National Park (Bergman and Verissimo 2008). The proclamation of the Maiombe National Park in Cabinda resulted from the Mayombe Transfrontier Conservation initiative. GEF funding was raised to help rehabilitate and expand the protected areas system of Angola, and this and other initiatives continue to support the government in its programme.

The Protected Areas System

Angola's protected areas system, proposed in 1936, with the first reserve established in 1937, expanded rapidly through to the 1970s, by which time 13 Protected Areas (PAs) had been established, totaling 75,267 km^2 or 6,0% of national territory. During the early 1970s, extensive surveys were undertaken to identify key biodiversity hotspots or other areas deserving inclusion in an expanded conservation network (Huntley 1974a, b, c, d, 2010). The objective was to increase the representation of Angola's vegetation types and faunal species diversity within the PA system. Unfortunately the interruption of war and the weakness of governance systems delayed the consideration and approval of the recommendations until 2011, when the *Conselho de Ministros* not only approved the proposals of 1974 but added

Table 18.1 Terrestrial protected areas of Angola

Name	Category	Date established	Area 1, km^2	Area 2, km^2
Iona	National Park	1937	15,150	15,196
Cameia	National Park	1938	14,450	14,688
Quiçama	National Park	1938	9960	9227
Mupa	National Park	1938	6600	6039
Bicuar	National Park	1938	7900	6748
Cangandala	National Park	1963	650	637
Mavinga	National Park	2011	Unknown	Unknown
Luengue-Luiana	National Park	2011	45,818	22,720
Maiombe	National Park	2011	1930	2074
Chimalavera	Regional Nature Park	1971	150	102
Luando	Integral Nature Reserve	1938	8280	9930
Ilhéu dos Pássaros	Integral Nature Reserve	1973	2	1.5
Búfalo	Partial Reserve	1974	400	405
Namibe	Partial Reserve	1957	4450	4642
Total Area, km^2			115,740	92,409.5

Two game reserves established in the 1930s – Ambriz, of 1125 km^2, and Milando, of 6150 km^2 – and since deproclaimed – are not included in this listing. Furthermore, the boundaries of Mavinga await clarification. Sources for Area – 1: GoA 2018; 2: Veríssimo Unpublished Data 2018

several new areas. In terms of Law 38/11 of 29 December 2011, the PA system increased to over 115,000 km^2 of national territory in one step, Table 18.1, Fig. 18.1. However, some debate continues regarding the definition of the boundaries of the individual PAs, with recent estimates by Verissimo (2018 Unpublished Data) providing new insights. While the area proclaimed as PAs was nearly doubled, the budget has remained on a very low plateau. Most National Parks still lack the most basic management capacity and effectiveness, despite the wealth of legislation promulgated since the Base Law for the Environment was approved in 1998.

The contradictions of global conservation policy, on one hand pressuring governments to reach a target of 17% of national territory under protected areas by 2020 (CBD 2010) and on the other, expecting the governments of developing countries to provide funds to effectively manage such PAs, is well illustrated by the situation in Angola. The drive to double the area under legislation has been accompanied by the neglect of iconic protected areas such as Quiçama, Iona and Luando. Since peace was achieved in 2002, the illegal occupation of the vulnerable Quiçama coastline by tourism lodges, fishing villages, oil exploration infrastructure, commercial enterprises and quarries, and by cattle ranches and commercial agricultural schemes, in addition to the rampant activities of bushmeat poachers and charcoal producers, has continued unabated. Iona, once a pristine desert environment, is now occupied by nomadic pastoralists who have invaded the heart of the park, supported by government sponsored water points which give permanence to the occupation. While some of the land invasions were a consequence of the war, most have occurred since the peace of 2002.

Fig. 18.1 Protected areas of Angola: • 1 Maiombe • 2 Quiçama • 3 Cangandala • 4 Cameia • 5 Iona • 6 Bicuar • 7 Mupa • 8 Luengue-Luiana • 9 Luando • 11 Chimalavera • 12 Búfalo • 13 Namibe. (Mavinga is not indicated on this map due to incomplete details regarding its boundaries in its gazettement)

The difficulties attending limited budgets, weak technical capacity and poorly trained human resources suggests that a triage approach should be considered to bring a focus to where the government's limited conservation resources should be targeted (Huntley 2017). Recent government policy has been to expand the PA estate, regardless of the management capacity of such 'paper parks'. Fortunately, despite the reverses of the past decades, each protected area still includes areas of sufficient dimension that can, with effective management, achieve significant biodiversity conservation goals. Since 2017, the new government leadership gives promise for a revitalised and energetic approach to conservation in Angola, as demonstrated in the recent Strategic Plan for the Conservation Areas System of Angola (GoA 2018).

Wildlife Populations

In contrast to most southern African countries, where reliable statistics of wildlife population dynamics have been recorded over many decades, the data sets for Angola are extremely sparse. Estimates made during the 1970s tended to be conservative, but indicated robust populations of Elephant (600), Forest Buffalo (6000), Eland (3000) and Roan Antelope (3000) in Quiçama (Huntley 1971). These species were extinct or nearly so in the park by 1992 (IUCN 1992). The populations of Giant Sable Antelope and Red Lechwe in Luando, each estimated at 2000 in 1972 (Huntley 1972), had dropped to less than 100 Giant Sable and with Lechwe on the verge of extinction by 2017 (Vaz Pinto 2018, 2019). Savanna Elephant, Blue Wildebeest and Eland, abundant in Bicuar in the 1970s, had fallen to low numbers by 2017 (Beja et al. 2019). Across Angola, wildlife populations declined precipitously after 1974, but remarkably, very small but tenacious surviving populations of most species, including top predators such as Lion, Leopard, Cheetah and Wild Dog, have held out (Beja et al. 2019). Of considerable conservation concern is the number of large mammal species for which no recently confirmed records are available, including Gorilla, Black Rhinoceros, Puku, Red Hartebeest and Lichtenstein's Hartebeest (Beja et al. 2019).

The wildlife population densities and biomasses of Angolan protected areas have never been comparable to those in eastern and southern Africa. While this might be a factor of hunting pressure, more fundamental ecological factors are at play. As demonstrated by Bell (1982), herbivore population density and biomass in Africa is related to rainfall and soil nutrients, and more directly to the ratio of soluble to structural carbohydrates in plant material available to herbivores. The vast area of Angola covered by miombo woodlands with low-nutrient grasses, shrubs and trees accounts for the notoriously low game populations of central Angola. Only in the more arid savannas of the southwest and southeast were relatively large populations of herbivores found in colonial times. The popular perception of vast populations of game across Angola in the nineteenth century is an illusion, certainly if compared with eastern Africa, as manifest by the much lower volumes of ivory exported from Angola relative to Kenya throughout the period (Walker 2009). The highest biomasses for ungulates during the 1970s, based on estimates from field surveys, were the western littoral grasslands of Quiçama, occupied by Eland, Roan and cattle, and the northern 'Baixa dos Elefantes' forests and floodplains of the Cuanza, occupied by Savanna Elephant, Forest Buffalo and Hippopotamus. Ungulate biomass in Cangandala, Luando and Bicuar, in dense miombo woodlands, was very low, as was that of Iona. The richest hunting areas (*Coutadas*) of the southeastern Cuando

Cubango (Mucusso and Luiana) possibly had biomasses approximating those of similar nutrient poor mixed miombo woodlands of eastern Africa, such as Selous Game Reserve (Huntley Unpublished Data).

Species Richness, Endemism, Threatened Species and Biodiversity Hotspots

The seminal paper on Systematic Conservation Planning by Margules and Pressey (2000) triggered the wide adoption of objective measures for the identification of biodiversity conservation priorities. The process has been effectively applied in southern Africa, where fine-scale spatial data on species distribution and status are available, such as that required by IUCN categories of threat (Raimondo et al. 2009), vegetation and habitat maps (Mucina and Rutherford 2006), and surveys of marine ecosystems and their dynamics (Kirkman et al. 2016; Holness et al. 2014). These rich data sets have been used to produce detailed national and regional biodiversity conservation management plans (Driver et al. 2012; Kirkman et al. 2016) which provide models for future work in Angola.

A preliminary survey of the conservation status of Angolan mammals (Huntley 1973) gave subjective estimates for 70 species, none of which were considered threatened with extinction but several, in particular Gorilla, Chimpanzee and Black Rhinoceros, were deemed vulnerable. A summary of recent assessments of rarity and threat in various taxonomic groups in Angola are presented in Table 18.2. More specific details of conservation status or threats are provided in the sources for each major taxonomic group referenced in Table 18.2.

In an early objective assessment of habitats, the areas of 32 vegetation units mapped by Barbosa (1970) were measured to evaluate the proportional representation of each unit in the protected areas system (Huntley 1974a). The results were then used to focus attention on under-represented types, taking into consideration faunal as well as floral distribution and status (Huntley 1974c). Of the 32 vegetation types described by Barbosa (1970), only 11 fell within protected areas in 1974. The disparity of protection afforded to representatives of the major biogeographic divisions was considerable. The Karoo-Namib, represented by Barbosa vegetation types 27, 28 and 29, which occupy 2.6% of the country's land surface, had 50.6% of its area conserved, while the Guineo-Congolian forest/savanna mosaic, comprising 25.7% of Angola's total area, and holding probably over 70% of its biodiversity, was not represented in any protected area. The small relict fragments of Afromontane forest, without doubt the most threatened of all ecosystems in Angola, and currently reduced to less than 1000 ha in area, were also without protection. Both the Afromontane forests (Humbert 1940; Hall and Moreau 1962) and the Angolan Escarpment Zone (Hall 1960) have long been regarded as key centres of avifaunal speciation and floristic importance. But both remain unmapped and unprotected.

Table 18.2 Species richness, endemism and threatened status for selected taxa

Group	Total species	Endemic species n°	Endemic species %	IUCN status	Source
Plants	6850 indigenous species	997	14.6	399 species have been formally assessed, of which: 36 threatened:	1–3
	230 naturalised species			32 vulnerable,	
				4 endangered,	
				49 threatened or near-threatened	
Butterflies & Skippers	792	57	7.2	Not evaluated	4
Dragonflies & Damselflies	260	16	6.1	1 vulnerable	5
				4 near threatened	
				16 data deficient	
				6 not evaluated	
Fishes	358	78	22	0	6
Amphibians	111	21	19.3	Not evaluated	7
Reptiles	278			Not evaluated	8
Birds	940	29	3.1	Not evaluated	9
Mammals	291	12	4.1	2 critically endangered,	10
				2 endangered	
				11 vulnerable	
				14 near threatened	
				12 data deficient	
				235 least concern	

1: Figuerido and Smith (2008), 2: Goyder and Gonçalves (2019), 3: IUCN (2018), 4: Mendes et al. (2019), 5: Kipping et al. (2019), 6: Skelton Unpublished Data, 7: Baptista et al. (2019), 8: Branch et al. (2019), Dean et al. (2019), 10: Beja et al. (2019)

The identification of sites of high biodiversity importance (in terms of endemism, species richness, and threat) – popularly termed biodiversity 'hotspots' (Myers 1988; Myers et al. 2000) – was the focus of the Angolan Protected Areas Expansion Strategy (Huntley 2010) submitted to the Minister of Environment and adopted, with additional recommendations in 2011 (GoA 2011, 2018). The sites recommended for future gazettement included examples of:

- *Guineo-Congolian Forest and Savanna*: (Maiombe – Cabinda; Serra Pingano – Uíge; Lagoa Carumbo – Lunda-Norte; Serra Mbango – Malange; Gabela forest – Cuanza-Sul; Cumbira forest – Cuanza-Sul;
- *Afromontane Forest and Grassland*: (Mount Namba – Cuanza-Sul; Mount Moco – Huambo; Serra da Neve – Namibe; Serra da Chela – Huíla Province);
- *Zambezian Flooded Grassland*: (Luiana – Cuando Cubango).

The inclusion of these proposals in the Angolan protected areas system would effectively address the asymmetry of ecosystem representation, with the number of

Barbosa vegetation units increasing from 11 to 23. To date, Maiombe forest and Luiana (with adjacent Luengue, and Mavinga) have been gazetted as additional National Parks.

Recent studies by the National Geographic Okavango Wilderness Project have identified further biodiversity hotspots in the upper reaches of the Cuando and Cubango drainages (NGOWP 2018). Field surveys in Huíla (Mendelsohn Unpublished Data) and Zaire (Vaz Pinto Unpublished Data) and Cuanza-Norte (Hines Unpublished Data) have identified sites of high biodiversity interest that are also deserving of further study and evaluation as future protected areas. As biodiversity surveys become more inclusive of Angola's less accessible areas, more sites of conservation merit will undoubtedly be added to the list of priorities.

Coastal and Marine Ecosystems

At the vast scale of marine environments, the recently concluded multinational programme of research in the Benguela Current Large Marine Ecosystem (BCLME) has provided very detailed assessments of the demersal fish biodiversity hotspots and the dynamics of the oceanic and climatic systems that influence this biodiversity (Kirkman et al. 2013, 2016; Kirkman and Nsingi 2019). These researchers found that hotspots of species richness were associated with greater water depths and cooler bottom temperatures. From consideration of the relevance of measured climate changes, they concluded that range shifts in species associated with warming temperatures could conceivably affect the spatio-temporal persistence of hotspots in the long term (Kirkman et al. 2013).

In a detailed analysis of the spatial characterisation of the BCLME, based on the physical driving forces, primary and secondary production, trophic structures and species richness, Kirkman et al. (2016) found four different sub-systems, of which two fall within Angolan waters. The first lies to the north of the Angola-Benguela Front and the second between the Angola-Benguela Front and Luderitz. Using the products of the BCLME projects, Holness et al. (2014) used Systematic Conservation Planning concepts and approaches to identify potential marine protected areas for the benthic and coastal ecosystems of Angola, Namibia and South Africa. A total of 248 distinct ecosystem types within the BCLME of these countries were mapped and classified according to Ecosystem Threat Assessments and Ecosystem Protection Level Assessments. In Angola, five ecosystem types were found to be both Critically Endangered and Not Protected, mainly situated in areas subject to intensive coastal development, in the oil and gas fields in the north, or in particular inshore areas subject to more intense fishing pressure. If the Endangered and Poorly Protected categories are also included, there are an additional 23 priority ecosystem types for protection in Angola. The BCLME studies (Kirkman et al. 2013, 2016; Holness et al. 2014) provide excellent models for the application of Systematic Conservation Planning meriting replication across the terrestrial ecosystems of Angola.

The coastal ecosystems of Angola are particularly vulnerable to human disturbance, both directly through over-exploitation of living resources and indirectly through urbanisation and industrialisation within coastal environments (Weir et al. 2007; Morais et al. 2005, 2008, 2016). The marine turtle species that depend on Angola's sandy beaches for nesting are particularly vulnerable. Despite these challenges, Angola remains a very important sea turtle conservation nation, with Olive Ridley (*Lepidochelys olivacea*), Leatherback (*Dermochelys coriacea*) and Green (*Chelonia mydas*) sea turtles nesting regularly during the summer (Morais 2016, 2017). Loggerhead (*Caretta caretta*) nest very sporadically, while Hawksbill (*Eretmochelys imbricata*) sea turtles are not known to nest on Angolan shores although juveniles have been recorded on the Soyo and Cabinda coast (Morais 2016). Recent studies estimate that between 33,000 and 102,000 Olive Ridley Sea Turtles made use of the Angolan coast to nest during the 2015/2016 summer, showing a decrease from between 38,000 and 110,000 estimated during the 2014/2015 season. These figures demonstrate that the coast of Angola is one of the most important nesting regions for this species in the Eastern Atlantic (Morais 2016; Kitabanga Project 2017). Leatherback Sea Turtle are much less abundant, with estimates of between 495 and 1320 animals nesting along the entire coast of Angola during the 2015/2016 breeding season (Morais 2016). Angola provides the southern extension of the Gabon nesting grounds, where 6000 to 7000 females nest annually (Billes et al. 2006). As such, Angola may be second in importance on the Eastern Atlantic coastline for the nesting of this species. Inadequate data are available to determine trends in Green Sea Turtle populations on the Angolan coast (Morais 2015, 2016).

Drivers of Species Loss

One of the immediate causes of population declines and species loss in vertebrates since 1975 has been hunting for bushmeat during the prolonged war, undertaken by rural communities faced with starvation, or by soldiers seeking to supplement very limited rations. Moreover, the illegal trade in wildlife products (ivory, rhino horn and teak) became significant during the war as the leaders of UNITA (*União Nacional para a Independencia Total de Angola*) sought funds to purchase arms (Breytenbach 2015). Luanda has long provided an open market for the illegal trade in wildlife products (Milliken et al. 2006; Svensson et al. 2014) and was described by Martin and Vigne (2014) as the biggest ivory market in Africa. The *Mercado do Artesanato* in Luanda has openly traded in ivory (mainly sourced from the DRC), leopard skins and other wildlife products, in the full knowledge of the Angolan authorities. Following international condemnation of the practice, trading in ivory has been banned in Angola since 2017. Despite the proclamation of two mega-parks in the Cuando Cubango in 2011, the poaching of elephants for ivory has increased in the parks and the elephant population is estimated to have decreased by 21% from 2005 to 2015 (Schlossberg et al. 2018). The inclusion of the area within the much-publicised Kavango-Zambezi Transfrontier Conservation Area, promoted as the largest TFCA on Earth (Peace Parks Foundation 2016) has yet to demonstrate conservation benefits.

A more pervasive impact than ivory poaching, recorded across the country despite the prohibition of hunting since the late 1970s, is the informal trade in bushmeat (Bersacola et al. 2014). During a survey in September 2013, travelling 1700 km along the Angolan Escarpment, Bersacola and colleagues stopped at 13 market places and counted 71 specimens of 15 prey species. The surveys were mostly in forested areas, where mammals have been more resilient to poaching pressures than in open savannas and woodlands. The most numerous species found were Blue Duikers (45%), Blue (Pluto) Monkeys (11%), Bush Hyraxes (10%) and Yellow-backed Duikers (8%). "For 25 fresh carcasses, the hunting technique was evident. A total of 84% of these fresh carcasses were hunted with shotguns, 16% were trapped using metal or string snares." The National Geographic Okavango Wilderness Project has described 'industrial-scale' bushmeat harvesting operations in many areas of the Cuando Cubango (NGOWP 2018).

While the illegal trade in wildlife products has been documented for animal species, a much larger trade in timber products has exploded over the past 5 years but without any measurement or monitoring. In an effort to stimulate alternative foreign income streams following the global collapse of oil prices, the then Angolan president signed decrees in 2016 that facilitated the rapid issue of concessions for timber extraction over much of the country. Chinese agents have mobilised the massive extraction of hardwoods from across Angola, accelerating the deforestation of vast areas, even in the previously near-pristine woodlands of Moxico and Cuando Cubango provinces (Mendelsohn 2019 Unpublished Data).

Land transformation, as described by Mendelsohn (2019), is perhaps the most potent of all drivers of biodiversity loss, but like the timber trade, its impacts on biodiversity have not been quantified at a species level. Mendelsohn and Mendelsohn (2018) draw attention to the transformation of rural to urban economies, and from subsistence to cash-based economies. The result has been the demand of the newly urbanised populations for cash to purchase goods and services previously provided by rural ecosystems. For rural dwellers, cash is now derived from the sale of bushmeat and charcoal, not of fruit and vegetables.

Another insidious driver of species loss is that of invasive alien species. The presence of potentially invasive alien fish species introduced for aquaculture has been reported for *Oreochromis mossambicus* in the Cuanza and *Oreochromis niloticus* in Cabinda, and recently in the upper Cubango (Skelton 2019). Invasive alien plants have already become established over extensive areas of western Angola. Rejmánek et al. (2017) conducted a rapid assessment of invasive plant species across 13 primary vegetation types (Barbosa 1970) in western Angola and recorded populations of 44 naturalised plant species, 19 of which are conclusively invasive (spreading far from introduction sites). They found that dense invasive populations of *Chromolaena odorata*, *Inga vera* and *Opuntia stricta* pose the greatest threats. *Opuntia stricta* has invaded large areas of the arid coastal plain northwards from Dombe Grande, and along the Chela escarpment. *Inga vera* is widespread in the moist 'coffee forests' of the central escarpment, while *Chromolaena odorata* is prevalent in the northern escarpment. These species have become major environmental and economic problems elsewhere in Africa and the lack of any control actions in Angola is cause for concern.

Science and Protected Area Management

The development during the nineteenth century of pragmatic wildlife management practices into a sophisticated conservation science is reflected in the histories of the protected areas systems of South Africa, Namibia and Tanzania (Carruthers 2017). Angola has had very limited investment in research in its national parks and reserves until the present decade. Notwithstanding limited resources, biologists from the *Instituto de Investigação Científica de Angola* (IICA) and the *Instituto de Investigação Agronómica de Angola* conducted important surveys of birds (Pinto 1983), mammals (Frade 1956, 1960; Crawford-Cabral 1970, 1971) and vegetation (Teixeira et al. 1967; Teixeira 1968; Barbosa 1970) in various parks during the 1960s and 1970s. Estes and Estes (1974) undertook detailed behavioural studies of Giant Sable in Luando in 1970/71. Huntley undertook general ecological surveys in the protected areas and across most of Angola (Huntley 1973, 1974d, 2017), while Dean (2000) studied the avifauna of Angola in the field and in the key collections of museums of Angola, Europe and the USA. But it was not until the present century that more detailed studies were initiated in the protected areas of Angola, such as the long-term studies of Giant Sable in Cangandala and Luando by Vaz Pinto (2018) and of sea turtles on the Angolan coast (Kitabanga Project 2017). Nevertheless, there have recently been important surveys of the remnant populations of large mammals in the protected areas of Angola (Beja et al. 2019), and on the threatened and endemic avifauna of the escarpment (Dean et al. 2019).

Despite these recent advances, the need for full-time biologists posted to and living in Angola's protected areas is urgent. This is a vacuum that should be filled by young Angolan researchers, with the guidance and support of mentors from across the globe, as the successful models of many other African countries have demonstrated. The modern tools of remote sensing, geographic information systems, immobilisation drugs, radio-tracking collars, trap-cameras, drones, genetic fingerprinting, and much more are readily available. The opportunities are endless and the difficult challenges of the past are being resolved each year as access to Angola improves and both international and national government support increases.

Key Priorities for Biodiversity Conservation

Priorities for the conservation of species within different taxonomic groups (plants, invertebrates, vertebrates) are summarised by Russo et al. 2019). Here we focus on generic issues of concern.

The effective management of protected areas is one of the key mechanisms that governments have available to achieve biodiversity conservation goals (CBD 2010). With over 10 million ha of Protected Areas (PAs) gazetted (GoA 2018), Angola has a considerable proportion of its terrestrial landscape under formal legislation. This provides the potential for a broad base to the PA estate, with many species and ecoregions represented in the system. However, many of the biodiversity hotspots identified in successive PA expansion strategies (Huntley 1973, 2010; GoA 2011, 2018) are yet to be accurately surveyed, described and gazetted. A first priority would be to ensure that legislative protection is given to Angola's most critically endangered biodiversity hotspots, such as the forests of the escarpment, the central highlands and the northern borders with the Democratic Republic of Congo.

As urgently important as legal protection is effective management of PAs. The existing network of extensive PAs such as Iona, Quiçama, Cangandala, Luando, Bicuar, and Luengue-Luiana lack adequate resources, and these need reinforcement through provision of personnel, training, equipment and operational budgets. The options of joint ventures with international conservation organisations and public/private partnerships such as those that have succeeded in Botswana, Mozambique, Namibia, Zambia and other southern African countries needs consideration. Field training of rangers and researchers with ongoing mentoring is a fundamental process for professional development. At a national scale, Angola has excellent conservation strategies (GOA 2006, 2018), and several parks already have pragmatic 'emergency' management plans (Huntley 1974b, 2003; Anderson and Morkel 2009). These need adaptation and implementation rather than repetition. For many PAs, a triage approach to zonation and investment is appropriate where land invasions, illegal infrastructure developments and other irreversible developments have taken place (Huntley 2017).

Concluding Remarks

The engagement of the public at large in conservation is a first priority for Angola's biodiversity agenda. The use of social media has already born unexpected results. The Facebook forum *Angola Ambiente* has over 1000 members and the posting of dragonfly photos on its page has led to 12 species being identified as new to science (Chris Hines, pers. comm.). The conservation of flagship species that attract public attention at national and international scales is also of the utmost importance. A well-publicised example is the conservation project that has successfully saved the Giant Sable Antelope in Cangandala and Luando protected areas, as described by Vaz Pinto (2019). Another noteworthy example is the Kitabanga Project of Agostinho Neto University, which has monitored sea turtle populations and undertaken conservation actions since 2013 (Projecto Kitabanga 2017). The project involves research and environmental education on sea turtles, with specific emphasis on protecting the nesting beaches. The Kitabanga Project provides an excellent model of a locally driven conservation research and education initiative and deserves replication in Angola. The National Geographic Okavango Wilderness Project (NGOWP 2018) has also brought wide attention to Angola's biodiversity, and stimulated young Angolans to join biodiversity exploration and research initiatives. These and other projects that will be developed in the future contribute effectively to leveraging conservation action in Angola, attracting funders and the public administration to initiatives with high visibility and meaningful impact. The conservation of Angola's remarkably rich biodiversity is first and last an Angolan responsibility, to be led to success by Angolans.

References

Anderson JL, Morkel PV (2009) Parque Nacional da Quiçama. Status quo, as ameaças e a necessidade de accões eficazes. Report to Ministry of Environment, Luanda, 24 pp

Andrade AA (1985) O Naturalista José de Anchieta. Instituto de Investigação Científica Tropical, Lisboa, 187 pp

Baptista N, Conradie W, Vaz Pinto P et al (2019) The amphibians of Angola: early studies and the current state of knowledge. In: Huntley BJ, Russo V, Lages F, Ferrand N (eds) Biodiversity of Angola. Science & conservation: A modern synthesis. Springer, Cham

Barbosa LAG (1970) Carta Fitogeográfica de Angola. Instituto de Investigação Científica de Angola, Luanda, 343 pp

Beja P, Vaz Pinto P, Veríssimo L et al (2019) The mammals of Angola. In: Huntley BJ, Russo V, Lages F, Ferrand N (eds) Biodiversity of Angola. Science & conservation: A modern synthesis. Springer, Cham

Bell RHV (1982) The effect of soil nutrient availability on community structure in African ecosystems. In: Huntley BJ, Walker BH (eds) Ecology of tropical savannas. Springer, Heidelberg, pp 193–216

Bergman B, Verissimo L (2008) Avaliação do Estatuto de Áreas Protegidas do Sudeste do Kuando Kubango,- Projecto Integrado de Gestão da Bacia Hidrográfica do rio Okavango. OKACOM-USAID, 48 pp

Bersacola E, Svensson M, Bearder S et al (2014) Hunted in Angola. Surveying the Bushmeat trade. SWARA, January–March 2014:31–36

Billes A, Fretey J, Verhage B et al (2006) First evidence of leatherback movement from Africa to South America. Mar Turt Newsl 111:13–14

Branch WR, Vaz Pinto P, Baptista N et al (2019) The reptiles of Angola: history, diversity, endemism and hotspots. In: Huntley BJ, Russo V, Lages F, Ferrand N (eds) Biodiversity of Angola. Science & conservation: A modern synthesis. Springer, Cham

Breytenbach J (2015) Eden's Exiles. Protea Boekhuis, Pretoria, 306 pp

Carruthers J (2017) National park science. A century of research in South Africa. Cambridge University Press, Cambridge, 512 pp

CBD (2010) Strategic plan for biodiversity 2011–2020 and the Aichi targets. Secretariat for the Convention on Biodiversity, Montreal

Crawford-Cabral JC (1970) Alguns aspectos da ecologia da Palanca real. Bol Instituto de Investigação Cientifica de Angola 7:7–42

Crawford-Cabral JC (1971) A Suricata do Iona, subspécie nova. Bol Instituto de Investigação Cientifica de Angola 8:65–83

Dean WRJ (2000) The birds of Angola. An annotated checklist. British Ornithologists Union, Tring, 433 pp

Dean WRJ, Melo M, Mills MSL (2019) The avifauna of Angola: richness, endemism and rarity. In: Huntley BJ, Russo V, Lages F, Ferrand N (eds) Biodiversity of Angola. Science & conservation: A modern synthesis. Springer, Cham

Driver A, Sink KJ, Nel JL et al (2012) National Biodiversity Assessment 2011: an assessment of South Africa's biodiversity and ecosystems. Synthesis Report. South African National Biodiversity Institute and Department of Environmental Affairs, Pretoria

Estes RD, Estes RK (1974) The biology and conservation of the Giant Sable Antelope *Hippotragus niger variani* Thomas, 1916. Proc Acad Natl Sci Phila 126(7):73–104

Figueiredo E, Smith GF (2008) Plants of Angola/Plantas de Angola. Strelitzia 22:1–279

Frade F (1956) Reservas naturais de Angola – I. Alguns mamíferos da Reserva da Quiçama. Anais Junta Invest Ultram 11(3):228–245

Frade F (1960) Os animais na etnologia ultramarina. Estudos, Ensaios e Documentos 84:211–240

GoA (Government of Angola) (1998) Lei de Bases do Ambiente. Ministry of Fisheries and Environment, Luanda

GoA (Government of Angola) (2006) National biodiversity strategy and action plan. Ministry of Urban Affairs and Environment, Luanda, 54 pp

GoA (Government of Angola) (2011) Plano Estratégico da Rede Nacional de Áreas de Conservação de Angola (PLENARCA). Ministry of Environment, Luanda

GoA (Government of Angola) (2018) Plano Estratégico para o Sistema de Áreas de Conservação de Angola (PESAC). Ministry of Environment, Luanda

Goyder DJ, Gonçalves FMP (2019) The flora of Angola: collectors, richness and endemism. In: Huntley BJ, Russo V, Lages F, Ferrand N (eds) Biodiversity of Angola. Science & conservation: A modern synthesis. Springer, Cham

Hall BP (1960) The faunistic importance of the scarp of Angola. Ibis 102:420–442

Hall BP, Moreau RE (1962) The rare birds of Africa. Bull Brit Mus Nat Hist (Zool) 8:315–381

Holness S, Kirkman S, Samaai T et al (2014) Spatial biodiversity assessment and spatial management, including marine protected areas. Final report for the Benguela current commission project BEH 09-01

Humbert H (1940) Zones e Étages de Végétation dans le Sud-Oest de l'Angola. *C.R. somm.* Scéanc Soc Biogèogra 17:47–57

Huntley BJ (1971) Preliminary guide to the National parks and reserves of Angola. Report 3. Repartição Tecnica da Fauna, Direcção Provincial dos Serviços de Veterinária, Luanda, Mimeograph report, 17 pp

Huntley BJ (1972) A Plan for the future of the Giant Sable of Angola. Report 11. Repartição Tecnica da Fauna, Direcção Provincial dos Serviços de Veterinária, Luanda, Mimeograph report, 9 pp

Huntley BJ (1973) Distribution and status of the larger mammals of Angola, with particular reference to rare and endangered species. Report 21. Repartição Tecnica da Fauna, Direcção Provincial dos Serviços de Veterinária, Luanda, Mimeograph report, 14 pp

Huntley BJ (1974a) Vegetation and Flora Conservation in Angola. Report 22. Repartição Tecnica da Fauna, Direcção Provincial dos Serviços de Veterinária, Luanda, Mimeograph report, 13 pp

Huntley BJ (1974b) Iona national park: administration, management, research and tourism. Report 23. Repartição Tecnica da Fauna, Direcção Provincial dos Serviços de Veterinária, Luanda, Mimeograph report, 19 pp, 5 Figs

Huntley BJ (1974c) Ecosystem conservation priorities in Angola. Report 28. Repartição Tecnica da Fauna, Direcção Provincial dos Serviços de Veterinária, Luanda, Mimeograph report, 22 pp

Huntley BJ (1974d) Outlines of wildlife conservation in Angola. J S Afr Wildl Manag Assoc 4:157–166

Huntley BJ (2003) Quiçama national park. Integrated conservation management plan. Report to Ministry of Urban Affairs and Environment, Luanda, 20 pp

Huntley BJ (2010) Estratégia de Expansão de Rede da Áreas Protegidas da Angola/proposals for an Angolan protected area expansion strategy (APAES). Unpublished report to the Ministry of Environment, Luanda, 28 pp, map

Huntley BJ (2017) Wildlife at war in Angola. The rise and fall of an African Eden. Protea Book House, Pretoria, 432 pp

IUCN (1992) Environment status quo assessment report. IUCN Regional Office for Southern Africa, Harare, 255 pp

IUCN (2018) The IUCN Red List of threatened species. Ver. 2017–3. http://www.iucnredlist.org

Kipping J, Clausnitzer V, Fernandes Elizalde SRF et al (2019) The dragonflies and damselflies of Angola. In: Huntley BJ, Russo V, Lages F, Ferrand N (eds) Biodiversity of Angola. Science & conservation: A modern synthesis. Springer, Cham

Kirkman SP, Nsingi KK (2019) Marine biodiversity of Angola: biogeography and conservation. In: Huntley BJ, Russo V, Lages F, Ferrand N (eds) Biodiversity of Angola. Science & conservation: A modern synthesis. Springer, Cham

Kirkman SP, Yemane D, Kathena J et al (2013) Identifying and characterizing demersal fish biodiversity hotspots in the Benguela current large marine ecosystem: relevance in the light of global changes. ICES J Mar Sci 70:943–954

Kirkman SP, Blamey L, Lamont T et al (2016) Spatial characterization of the Benguela ecosystem for ecosystem-based management. Afr J Mar Sci 38:7–22

Margules CR, Pressey RI (2000) Systematic conservation planning. Nature 405:243–253

Martin E, Vigne L (2014) Luanda – the largest illegal ivory market in Africa. Pachyderm 55:30–37

Mendelsohn JM (2019) Landscape changes in Angola. In: Huntley BJ, Russo V, Lages F, Ferrand N (eds) Biodiversity of Angola. Science & conservation: A modern synthesis. Springer, Cham

Mendelsohn JM, Mendelsohn S (2018) Sudoeste de Angola: um retrato da terra e da vida. South West Angola: a portrait of land and life. Raison, Windhoek

Mendes L, Bivar-De-Sousa A, Williams M (2019) The butterflies and skippers of Angola. In: Huntley BJ, Russo V, Lages F, Ferrand N (eds) Biodiversity of Angola. Science & conservation: A modern synthesis. Springer, Cham

Milliken T, Pole A, Huongo A (2006) No peace for elephants: unregulated domestic ivory markets in Angola and Mozambique. TRAFFIC online report series no. 11. Traffic East/Southern Africa, Harare, Zimbabwe, 46 pp

Morais M (2008) Tartarugas Marinhas na Costa de Cabinda. Plano de conservação e gestão para a implementação do projecto de prospecção sísmica "on shore". Holisticos/Chevron, 67p

Morais M (2015) Projecto Kitabanga – Conservação de tartarugas marinhas. Relatório final da temporada 2014/2015. Universidade Agostinho Neto/Faculdade de Ciências, Luanda

Morais M (2016) Projecto Kitabanga – Conservação de tartarugas marinhas. Relatório final da temporada 2015/2016. Universidade Agostinho Neto/Faculdade de Ciências, Luanda

Morais M (2017) Projecto Kitabanga – Conservação de tartarugas marinhas. Relatório final da temporada 2016/2017. Universidade Agostinho Neto/Faculdade de Ciências, Luanda

Morais M, Torres MOF, Martins MJ (2005) Análise da Biodiversidade Marinha e Costeira e Identificação das Pressões de Origem Humana sobre os Ecossistemas Marinhos e Costeiros. Ministerio do Urbanismo e Ambiente, Luanda, 140 pp

Mucina L, Rutherford MC (2006) The vegetation of South Africa, Lesotho and Swaziland. Strelitizia 19:1–807

Myers N (1988) Threatened biotas: 'hotspots' in tropical forests. Environmentalist 8:120

Myers N, Mittermeier RA, Mittermeier CG et al (2000) Biodiversity hotspots for conservation priorities. Nature 403:853–858

Newton da Silva S (1952) Wild life and its protection in Angola. Oryx 1:7

Newton da Silva S (1970) A Grande Fauna Selvagem de Angola. Serviços de Veterinária, Luanda, 151 pp

NGOWP (2018) National Geographic Okavango wilderness project. Initial findings from exploration of the upper catchments of the Cuito, Cuanavale and Cuando Rivers in Central and South-Eastern Angola (May 2015 to December 2016). National Geographic Okavango Wilderness Project, 352 pp

Peace Parks Foundation (2016) Annual review 2016. Peace Parks Foundation, Stellenbosch

Pinto AA d R (1983) Ornitologia de Angola, vol 1. Instituto de Investigação Científica Tropical, Lisbon, 696 pp

Projecto Kitabanga (2017) Projecto Kitabanga – Conservação de tartarugas marinhas – Folhetos 2017. Universidade Agostinho Neto/Faculdade de Ciências, Luanda

Raimondo D, von Staden L, Foden W et al (2009) Red list of South African plants 2009. Strelitzia 25:1–668

Rejmánek M, Huntley BJ, le Roux JJ et al (2017) A rapid survey of the invasive plant species in western Angola. Afr J Ecol 55:56–69

Russo V, Huntley BJ, Lages F, Ferrand N (2019) Conclusions: biodiversity research and conservation opportunities. In: Huntley BJ, Russo V, Lages F, Ferrand N (eds) Biodiversity of Angola. Science & Conservation: a modern synthesis. Springer, Cham

Schlossberg S, Chase MJ, Griffin CR (2018) Poaching and human encroachment reverse recovery of African savannah elephants in south-East Angola despite 14 years of peace. PLoS One 13(3):e0193469. https://doi.org/10.1371/journal.pone.0193469

Skelton PH (2019) The freshwater fishes of Angola. In: Huntley BJ, Russo V, Lages F, Ferrand N (eds) Biodiversity of Angola. Science & conservation: A modern synthesis. Springer, Cham

Stassen N (ed) (2010) Reminiscences concerning the life of William James Bushnell Chapman & An account of the entry of the Trek Boers into Angola and of their sojourn during the forty-eight years they struggled in that country under Portuguese rule. Protea Book House, Pretoria, 476 pp

Stassen N (2016) The Thirstland Trek, 1874–1881. Protea Book House, Pretoria, 73 pp

Svensson MS, Bersacola E, Bearder SK et al (2014) Open sale of elephant ivory in Luanda, Angola. Oryx 48:13–14

Teixeira JB (1968) Parque Nacional do Bicuar. Carta da vegetação (1a aproximação) e memória descritiva. Instituto de Investigação Agronómica de Angola, Nova Lisboa

Teixeira JB, Matos GC, Sousa JNB (1967) Parque Nacional da Quiçama. Carta da vegetação e memória descritiva. Instituto de Investigação Agronómica de Angola, Nova Lisboa

Varian HF (1953) Some African milestones. Books of Rhodesia, Bulawayo, 78 pp

Vaz Pinto P (2018) Evolutionary history of the critically endangered giant sable antelope (Hippotragus niger variani): insights into its phylogeography, population genetics, demography and conservation. PhD thesis. University of Porto, Porto

Vaz Pinto P (2019) The Giant Sable Antelope: Angola's national icon. In: Huntley BJ, Russo V, Lages F, Ferrand N (eds) Biodiversity of Angola. Science & conservation: A modern synthesis. Springer, Cham

Walker JF (2004) A certain curve of horn. The hundred-year quest for the Giant Sable Antelope of Angola. Grove/Atlantic Inc., New York, 514 pp

Walker JF (2009) Ivory's ghosts. The white gold of history and the fate of elephants. Atlantic Monthly Press, New York, 296 pp

Weir CR, Ron T, Morais M (2007) Nesting and at-sea distribution of marine turtles in Angola, West Africa, 2000–2006: occurrence, threats and conservation implications. Oryx 41(2):224–231

Chapter 19
Museum and Herbarium Collections for Biodiversity Research in Angola

Rui Figueira and Fernanda Lages

Abstract The importance of museum and herbarium collections is especially great in biodiverse countries such as Angola, an importance as great as the challenges facing the effective and sustained management of such facilities. The interface that Angola represents between tropical humid climates and semi-desert and desert regions creates conditions for diverse habitats with many rare and endemic species. Museum and herbarium collections are essential foundations for scientific studies, providing references for identifying the components of this diversity, as well as serving as repositories of material for future study. In this review we summarise the history and current status of museum and herbarium collections in Angola and of information on the specimens from Angola in foreign collections. Finally, we provide examples of the uses of museum and herbarium collections, as well as a road-map towards strengthening the role of collections in biodiversity knowledge generation.

Keywords Bioinformatics · Catalogue of life · Checklists · Conservation · Expeditions · GBIF · Natural history collections

---

R. Figueira (✉)
CIBIO-InBIO, Centro de Investigação em Biodiversidade e Recursos Genéticos, Universidade do Porto, Vairão, Portugal

CEABN-InBio, Centro de Ecologia Aplicada "Professor Baeta Neves", Instituto Superior de Agronomia, Universidade de Lisboa, Lisboa, Portugal
e-mail: ruifigueira@isa.ulisboa.pt

F. Lages
ISCED – Instituto Superior de Ciências da Educação da Huíla, Lubango, Angola
e-mail: f_lages@yahoo.com.br

B. J. Huntley et al. (eds.), *Biodiversity of Angola*,
https://doi.org/10.1007/978-3-030-03083-4_19

Introduction

Natural History Collections (NHCs) are the basic building blocks for the discovery and understanding of the diversity of life. Scientific names are currently available for about two million of the eight million species estimated to live on Earth (Mora et al. 2011). In museum and herbarium collections, researchers try to compile and organise the most complete representation of biological diversity. This is motivated not (only) because some people have a:

> metaphysical angst, perhaps because they cannot bear the idea of chaos being the one ruler of the universe, which is why (...) they attempt to impose some order on the world...

as surmised by the Nobel literature prize winner José Saramago in his book '*All the Names*' (Saramago 2000) but also by the need to have reference samples to identify species, know how and where they live, their biotic and abiotic interactions, their links to communities and ecosystems, and finally, all subjects that define natural history (Tewksbury et al. 2014).

These needs are met by specimens serving as vouchers of a species' occurrence, collected in a specific habitat and in certain circumstances of time, space, traits and sampling methods. Globally, it is estimated that biological collections contain three billion preserved specimens (Brooke 2000; Wheeler et al. 2012). Specimens maintained in biological collections include the material samples on which new species are described – the type specimens – but also additional specimens that represent the variety and variability that a biologist needs to recognise to become a good practitioner of species identification. Collections are essential for taxonomic and systematic research, but also for studies in ecology, evolution, biogeography, conservation, climate change effects, and other fields, as will be discussed later in this chapter.

Building an inventory of the biodiversity of Angola, as a national checklist, begins with an initial register of species present in biological collections in Angola and abroad. With fifteen ecoregions (Olson et al. 2001; Burgess et al. 2004; Huntley 2019), Angola is one of the most biodiverse countries of the world, so biological collections will or should reflect that diversity. The number of endemic species is recognised to be high in several groups, for example in birds (Mills and Melo 2013; Dean et al. 2019) and plants (Figueiredo et al. 2009a, b; Goyder and Gonçalves 2019). However, despite this richness, the Angolan Escarpment could not be recognised as one of the biodiversity hotspots of the world due to the lack of information on its species diversity (Myers et al. 2000). This might be a consequence of the under-representation of Angola's biological diversity in natural history collections. This situation results, in part at least, from a combination of factors like the restrictions placed on field explorations due to the war situation in the country over several decades, and to limited access to the country's natural history collections such as those of the *Museu do Dundo*.

It is possible, nevertheless, to redress this situation. The pressure on biological collections has been very high for the last two decades due to cuts in budgetary support. For example, in the USA, 100 herbaria have closed since 1997 (Deng 2015). Paradoxically, the use of such collections has increased at an exponential rate in the same period (Pyke and Ehrlich 2010; Lavoie 2013). This increase in use might be related to the fact that biodiversity informatics and cyber-infrastructure developments now contribute to compress time and space, facilitating virtual access to specimens, data and literature. A researcher of Angolan biodiversity, working in Angola or working abroad, can now access hundreds of thousands of biodiversity records online. Such databases include images of the specimens in collections hosted elsewhere in the world, and species descriptions in old or inaccessible journals. In parallel, molecular tools have recently seen great advances, with the arrival of next-generation sequencing tools that promise to overcome limitations of DNA fragmentation caused by certain preservation methods used in collections (Yeates et al. 2016). This will facilitate attaching genetic sequences to specimens, and support biodiversity field surveys to achieve faster results. An optimistic view is held by a group of experts who suggest that it is possible to describe ten million species in 50 years, virtually describing all species that currently remain to be discovered (Wheeler et al. 2012). The authors conclude that this goal might be reached by interdisciplinary partnerships using and developing cyber technologies.

This background creates a favourable environment for the development and the increase of the role of museum and herbarium collections to support the advance of biodiversity knowledge globally, but also in Angola. In this chapter, we will provide a short review of the current status and knowledge of Angolan biodiversity based on these collections, their importance for biodiversity research, and some indications on how biodiversity informatics and cyber-infrastructure could facilitate their use in biodiversity knowledge generation.

Museum and Herbarium Collections from Angola

Internet access to information about species from Angola represented in museum and herbarium collections around the world is now possible. This has been facilitated by global networks and infrastructure resulting from several initiatives, based on biodiversity informatics standards, protocols, tools, manuals and quality control procedures that, being interoperable, create a digital global biodiversity observatory. The most visible facility is probably the intergovernmental Global Biodiversity Information Facility (GBIF), created in 2001, through which more than one thousand institutions share, in 2018, over one billion records, including 145 million records based on specimens preserved in collections. These records are freely and openly accessible to all using an Internet browser, at www.gbif.org. However, we are still far from having all specimens from collections catalogued in databases. In Europe, for example, only 10% of NHC specimens are digitally catalogued. But these inventories provide a good basis for knowing what expeditions and studies have contributed specimens of the different biological groups to museum and herbarium collections.

Most of the world's' biodiversity is located in the tropics. But the NHCs of Europe and North America hold the largest collections of material from the tropics, not the NHCs of the countries of origin (Peterson et al. 2016). For historical reasons, many of the best collections of African countries are not located in the country of origin, but in the former colonising countries, or in other countries that conducted field expeditions in Africa. This is the case of Angola, where the highest representation of Angolan biodiversity in collections is in Portuguese or other European or North American NHCs. For this reason, the repatriation of information that can be achieved through GBIF, where data is mobilised and made available to the country of origin, represents an important asset to assist biodiversity research and conservation in Angola.

The representation in NHCs of Angola's biodiversity varies between taxonomic groups, regions and time periods, and depends on the history of expeditions and studies conducted in the country through the last 150 years. To provide an overview of the museum and herbarium specimens collected in Angola, we compiled a dataset with information from several sources. We used the full dataset available through GBIF as on May 25, 2018 (GBIF.org 2018), containing 149,701 records for all groups. This was merged with other sources for specific groups, information that is not yet published by GBIF. These datasets include the database of the bird collections of the Herbarium and Museum of Ornithology and Mammalogy of Lubango (Lages 2016, pers. com.) containing 34,471 records, and the Herbarium of the University of Coimbra (Santos and Sales 2018), with 7864 herbarium records. Both resources are soon to be published through GBIF. Finally, we accessed RAINBIO (Gilles et al. 2016) that holds 1884 herbarium records from Angola. In the aggregated dataset, the possible duplication of records between RAINBIO and GBIF was checked and removed, as well as a careful check of the information about province, which was standardised or completed whenever possible. The total aggregated dataset for this analysis contains 193,839 records, of which 158,185 records contain information about the province and 154,631 records with information about the sampling year (Table 19.1). These records are published by more than 200 institutions from 28 countries (Fig. 19.1), and should be considered a partial view of the complete holding of specimens from Angola in collections worldwide.

The oldest specimens from Angola known in museums date from the late seventeenth and beginning of the eighteenth centuries. These are of plants included in the herbarium collection of the Natural History Museum, London and include 36 specimens collected in the region of Luanda by Mason in 1669 (Romeiras 1999; Goyder and Gonçalves 2019), followed by samples collected by John Kirckwood in Cabinda. The first records available through GBIF for Angola are from 1758, of mussels from the Malacology collection of the Balley-Matthews National Shell Museum, in the USA, which includes 70 specimens collected up to the end of the eighteenth century.

The time profile of the specimens collected in Angola (Fig. 19.2) only shows three mussel records from before 1800, presented as small peaks. The first records of the nineteenth century also create a small peak in 1804, based on material collected during the *Viagens Philosophicas*. These expeditions were organised by

Table 19.1 Data records of specimens from Angola in collections available in online resources, as of May 2018

Source	Reference	Collection type	N° of records	N° of types
GBIF[a]	GBIF.org (2018)	Herbarium	85,360	8877
		Fungi	601	283
		Mammals	4641	39
		Birds	58,821	187
		Herpetological	7269	269
		Fish	9227	569
		Arthropods	11,480	1518
		Invertebrates	5425	657
		Microorganisms	220	10
		Not classified	1046	
		TOTAL	148,573	12,409
ISCED[b]	Lages (2016, pers. com.)	Bird	34,471	
Coimbra[c]	Santos and Sales (2018)	Herbarium	7864	634
RAINBIO	Gilles et al. (2016)	Herbarium	1884	Not available

[a]Global Biodiversity Information Facility
[b]Museum of Ornithology and Mammalogy, ISCED-Huíla, Lubango
[c]Herbarium of the University of Coimbra

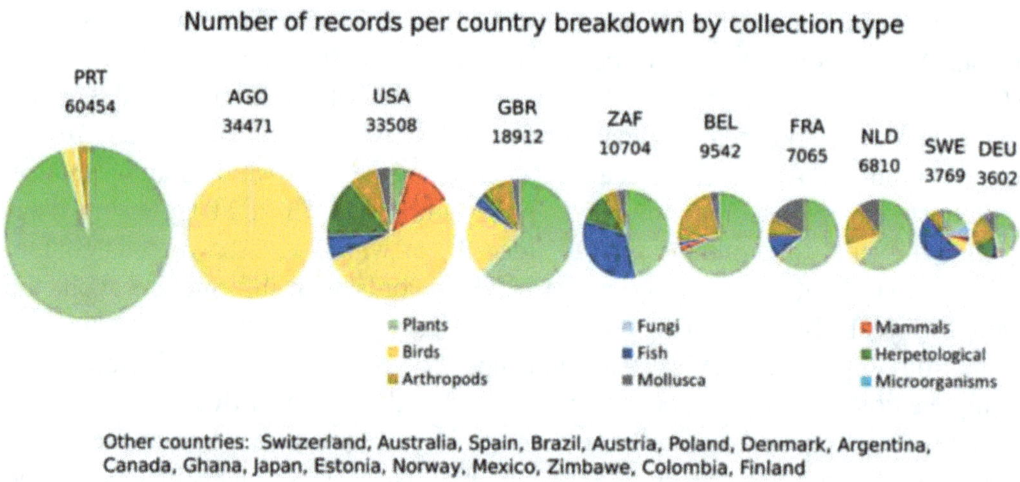

Fig. 19.1 Hosting country of specimens from Angola in collections available in online resources, as of May 2018. The size of the pie relates to the number of records published by each country, decreasing in logarithmic basis. *PRT* Portugal, *AGO* Angola, *USA* United States, *GRB* United Kingdom, *ZAF* South Africa, *BEL* Belgium, *FRA* France, *NLD* The Netherlands, *SWE* Sweden, *DEU* Germany

Portugal to explore the former Portuguese overseas territories of Brazil, Goa, Cape Verde, Mozambique and Angola. The naturalist Joaquim José da Silva was in charge of sampling plants and animals from Angola, and stayed in the country between 1783 and 1808. The materials collected were sent to Lisbon. However, during the

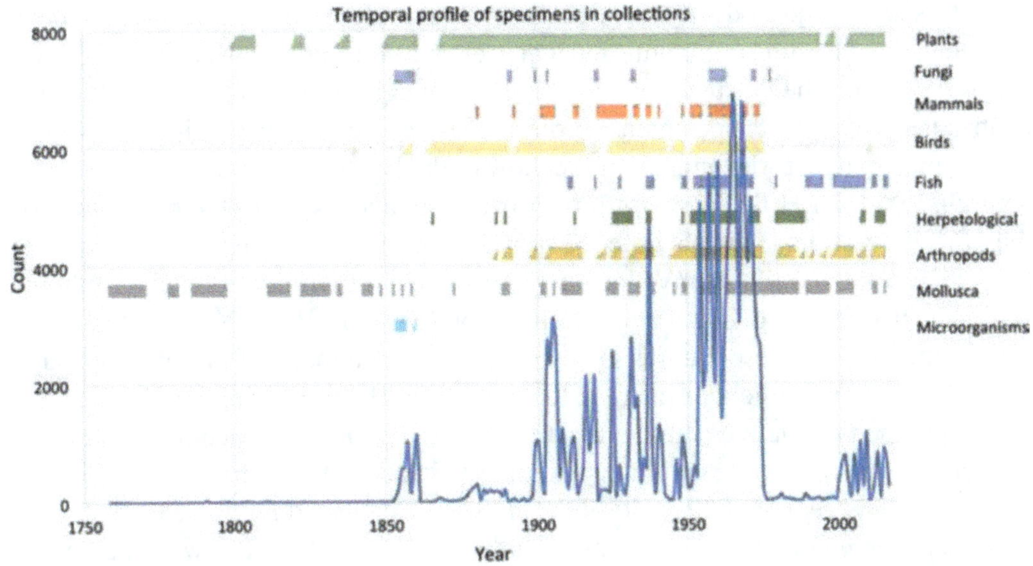

Fig. 19.2 The time profile of the specimens collected in Angola. The horizontal bars for each collection type, indicated on the right axis, indicates presence of specimens of that collection type in the corresponding period

French invasion of Portugal, these specimens were removed to the MNHN in Paris by Saint-Hillaire in 1808 (Barbosa du Bocage 1862 in Alves et al. 2014).

In the time profile (Fig. 19.2), the first significantly high value is visible in the beginning of the second half of the nineteenth century, when the Austrian botanist Friedrich Welwitsch was commissioned by the Portuguese government to explore the flora of Angola. In the expedition named *Iter Angolense*, he sampled more than 10,000 specimens (Albuquerque et al. 2009), between 1853 and 1860, of which more than 1000 were used to describe new species. Several sets of the collection were made by Welwitsch and distributed to several herbaria, but the most complete sets are located in Lisbon, at the LISU herbarium, and in the BM herbarium, in London. In total, more than 20,000 duplicates were sent to the major herbaria in Europe (Albuquerque and Correia 2010). Welwitsch sampled not only vascular plants but also cryptogams, including 350 lichen specimens with 50 type specimens, and also mammals. The most well-known and notable species he found is *Welwitschia mirabilis* from the Namib Desert, in southwest Angola. The genus was named in Welwitsch's honour by Sir Joseph Hooker, and is the most iconic plant species of Angola.

Other expeditions contributed to diversify, in terms of biological groups, the addition of specimens to collections. These collections are detailed in other chapters of this volume (Baptista et al. 2019; Beja et al. 2019; Branch et al. 2019; Dean et al. 2019; Kipping et al. 2019; Mendes et al. 2019; Skelton 2019). The institutions to which foreign collectors sent material are listed in Appendix. The Portuguese naturalist José Anchieta collected plants and animals of several groups (birds, reptiles and amphibians, mammals, fishes) between 1850 and 1897 (Albuquerque and Correia 2010; Mills et al. 2010; Alves et al. 2014; Ceríaco 2014). Other main collectors in that period are the German botanist Hugo Baum, with specimens deposited at the Berlin Herbarium, and which expedition was reviewed by Figueiredo et al. (2009a, b). The Portuguese explorers Hermenegildo Capelo and Roberto Ivens contributed with plants specimens to the LISU herbarium (University of Lisbon), the German botanist Alexandre von Mechow to the Berlin herbarium, the American naturalist William H Brown, with birds, mammals and fishes to the collections of the NMNH, Smithsonian Institute. Several bird collectors also promoted expeditions, like Axel W Eriksson (Vänersborg Museum, Sweden), the French ornithologists Albert Lucan and Louis Petit (NHM, London), and P van Kellen (Naturalis, The Netherlands). These naturalists also collected specimens of biological groups other than their main field of interest, like butterflies, bees, hemipterans, etc.

During the twentieth century, a crescendo of the number of specimens added to collections was observed continuously until 1957 and remained high until a drop in 1975. After this year, with the start of the civil war that lasted for 27 years, very few specimens were added to collections. Finally, after the year 2000, with the end of the civil war (in 2002), there is a recovery in the deposition of specimens into collections, but not to the levels observed before 1975. However, for the recent period we need to consider the time lags between the end of expeditions, the deposition of specimens in collections and the making of data accessible through databases. Currently, the average time gap between specimens being collected and identified is 21 years (Fontaine et al. 2012). Simultaneously, a change in sampling ethics and tight permit issuing control by national authorities might also explain lower sampling rates per expedition (Prathapan et al. 2018).

In the twentieth century, a series of large expeditions to Angola increased knowledge about the flora and fauna of the country. In terms of plants, the largest plant collections were those of the botanist John Gossweiler, sampling in all provinces of the country, with a total of 14,600 numbers, between years 1900 and 1950. The most complete set of this collector is deposited in the LISC herbarium, at the University of Lisbon, but many duplicate specimens were distributed to other herbaria, namely COI, BM, LISU, P, K, LUA. All herbarium acronyms are according to the Index Herbariorum (Thiers 2018). The second most prolific collector was JM Brito Teixeira, a follower of Gossweiler who collected about 13,000 numbers in all provinces between 1949 and 1969. Several botanical expeditions were organised from Portugal, either with a focus in botany, agronomy or forestry, while others were promoted by institutions newly created in Angola. The *Instituto de Investigação Científica de Angola* (IICA) was created in 1958, with herbarium and zoological collections based in Lubango. Another research institution was the *Instituto de Investigação Agronómica*, established in 1961, based in Huambo. Frequently, staff from institutions from both Portugal and Angola worked together in field expeditions, because, formally, the new institutions in Angola were dependencies of the equivalent Portuguese institutions. Therefore, duplicate samples were distributed to herbaria in Angola (LUBA or LUAI).

The main collectors were Luiz Carrisso (based in COI), Francisco de Ascensão Mendonça (LISC), Francisco de Sousa (LISC), Eduardo Mendes (LISC), Romeu Santos (LUBA), Óscar Azancot de Menezes, Carlos Henriques, Luís

Grandvaux-Barbosa (LISC) and the British botanist Arthur W Exell (BM). A considerable set of specimens without the individual collector being indicated were collected in the scope of the *Missão de Estudos Florestais a Angola* (MEFA), between 1957 and 1960. The last large plant collectors in the twentieth century sampled between 1970 and 1974 were António RF Raimundo, Gilberto Cardoso de Matos, Paul Bamps, Roger Dechamps and Eurico S. Martins. A detailed list of collectors, including the time range of the collections and provinces is available in Figueiredo and Smith (2008).

Concerning animal collections, the largest collections created or with the largest growth in the twentieth century are of birds. In this case, the ornithological collection of the Museum and Herbarium of Lubango is by far the largest and most representative of Angola, with circa 40,000 specimens, and probably one of the largest collections of birds based in Africa. The oldest specimens in the collection are from 1948, but the relevant sampling started in 1958, the year of the creation of IICA. The collection was established by António da Rosa Pinto, with many staff of IICA contributing significantly with specimens: at least 13 people each added more than 500 bird specimens. The collection contains specimens from throughout the country, but 75% of the specimens are from the western and southern provinces, with 25% from Huíla where the collection is based (Lubango).

Dean et al. (2019) presents details on the ornithological collectors of Angola. In the first half of the century, about 13,000 specimen records can be found through GBIF, the main publishing institutions being AMNH, NHMUK, CM, FMNH and GNM. The main collectors are WJ Ansorge, R Boulton, H Lynes, CH Pemberton and the main provinces with records are Cuanza-Norte, Bengo, Malanje, Benguela, Namibe, and Bié. Between 1950 and 1974, apart from IICA staff, the main collectors were Gerd H Heinrich and T Archer, which collected about 900 specimens held by YPM and USNM collections.

The history of mammal collecting in Angola is presented by Beja et al. (2019). In the twentieth century, the year 1925 presents an exceptionally high number of records of about 1400 specimens in mammal collections. This corresponds to the Arthur Vernay expedition to Angola, which specimens are in the AMNH collection. The following years with high values are 1932 and 1933, corresponding to the Phipps-Bradley Expedition, specimens also at the AMNH collection. In 1936, KH Prior sampled in Benguela, which specimens are located at the collection MVZ. In 1954 and 1955, GH Heinrich sampled in several provinces, which materials are held at FMNH. Considering the collections with the highest number of records available through GBIF, in general, the order with the highest number of specimens is Rodentia.

The number of records for mammals is in general low in the dataset compiled. A reason for this might be the incomplete status of digitisation of mammal collections in databases. This means that the values mentioned might not be representative of the total holdings of mammals from Angola. For example, the mammal collection from Lubango was not taken into account in this analysis, because the digitising process is still ongoing, in terms of data quality verification and import to a data management system. For the same reason, it is possible that in other collection types, the number of specimen records available online is also not representative of the true sampling effort for these collections. For example, in the dataset used in this chapter, no records are included from collections based in Portuguese institutions for fish or butterflies, although several zoological expeditions were organised by the *Instituto de Investigação Científica Tropical* (IICT) to Angola, since 1950. This is the case of ichthyological expeditions to lakes Cameia and Dilolo, by Fernando Frade and Teixeira Pinto, in 1958. Another example of a collection that still needs to be mobilised (although already studied) is the Lepidoptera order of the Entomological collection of IICT. This collection was extensively studied in the preparation of the book *Butterflies of Angola* (Mendes et al. 2013), with more than 15,000 specimens reviewed. However, the records of these specimens are not yet accessible.

In a related subject on collection accessibility, it should be noted that the *Instituto de Investigação Científica Tropical* (IICT) was integrated in the University of Lisbon in July 2015, as a special unit. This will not, however, change the possibility of accession to collections, except for the period while the collections are being moved. The new unit in the University shares the Director with the National Museum of Natural History and Science of the University of Lisbon, but all the zoological and herbarium collections of IICT will be retained as distinct collections. This is relevant for the study of the biodiversity of Angola because these collections are important, not only because they host many type specimens, but also because some are the most representative worldwide of the biodiversity of Angola. For example, the LISC herbarium of IICT has ca. 70,000 specimens, which is the largest worldwide for Angola, because it merged duplicate specimens from several expeditions, while in Angola these were hosted in separate herbaria (LUBA, LUAI, LUA).

The specimen-based atlas of the butterflies of Angola highlighted the importance of access to valuable but privately owned collections. In that example, four private collections were consulted.

The specimens collected in Angola are not evenly distributed across the country, as it is often observed in natural history collections (Lavoie 2013; see also Dean et al. 2019; Mendes et al. 2019; Beja et al. 2019). The bias is reflected both in the spatial coverage of the collections, as well as in the groups' representativeness across the country (Fig. 19.3). Some regions of Angola are clearly under-represented in collections, as is the case of the provinces of Zaire and Uíge, in the northwest, and most of the eastern provinces, including Lunda-Norte, Lunda-Sul, Moxico and Cuando Cubango. According to Crawford-Cabral (2010), there is a triangle that extends from Bié northeast to Lunda-Sul and southeast to the Cuando River, where there is a serious lack of knowledge about the fauna. This includes the interesting areas of the Upper Zambezi. The province of Huíla stands out has having almost twice the number of specimens compared with the second province, Namibe. This is possibly a result of the establishment, in Huíla, of the collections and research staff of IICA, which impact is also noted in its neighbouring provinces.

In most provinces, plant collections outnumber other taxa, but in four – Benguela, Cuanza-Sul, Malanje, Namibe – the number of birds in collections exceeds plants. The number of mammals in collections has some expression in Bié, Benguela, Cuanza-Sul, Huíla and Malanje. As for fish collections, these are more present for

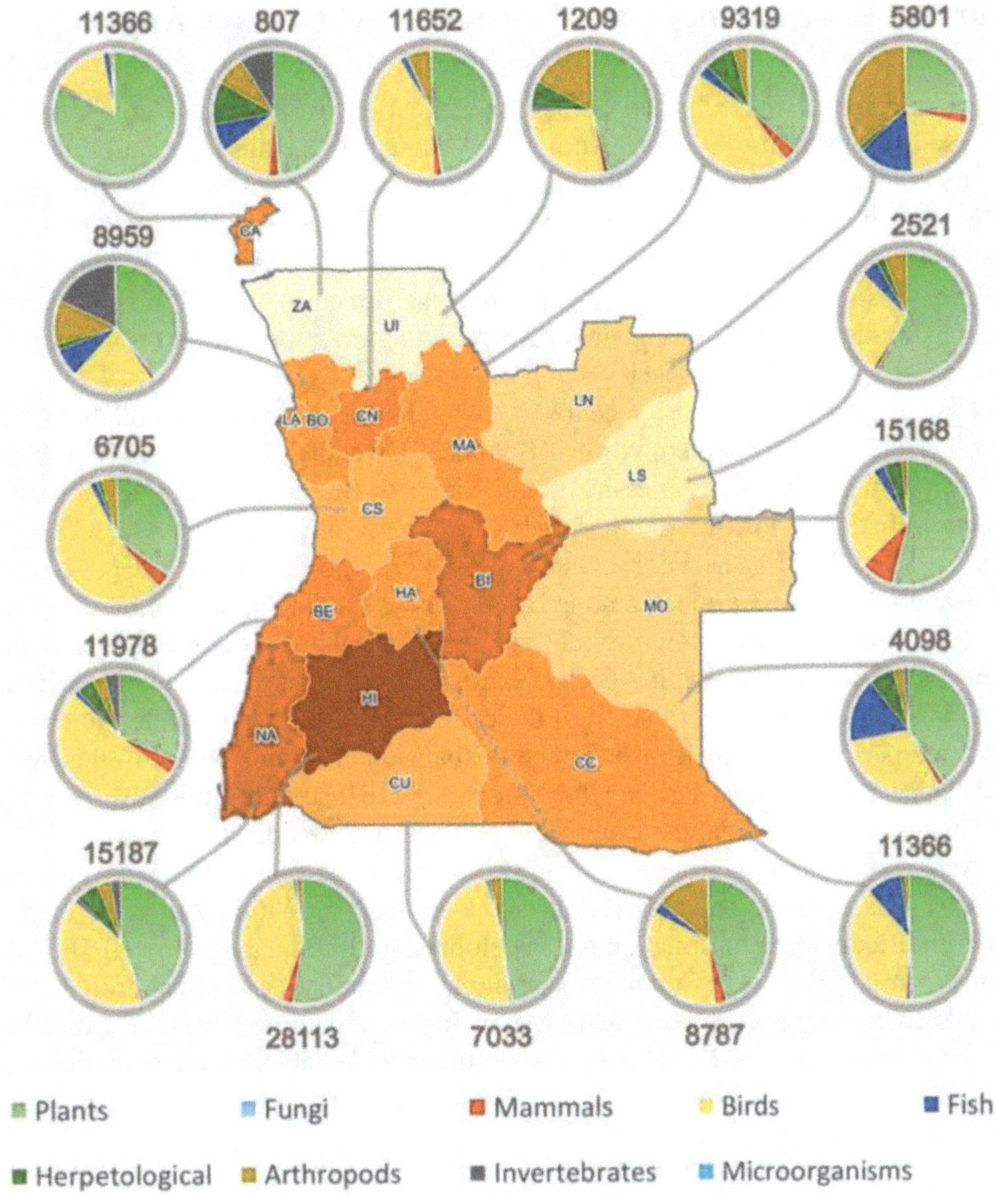

Fig. 19.3 Number of records per province, with darker colours corresponding to higher numbers. Each pie chart depicts the breakdown per collection type, and displays the number of records for the province. The records of Luanda and Bengo were aggregated in one chart. Province names: *Bo* Bengo, *BE* Benguela, *BI* Bié, *CA* Cabinda, *CC* Cuando Cubango, *CN* Cuanza-Norte, *CS* Cuanza-Sul, *Cu* Cunene, *HA* Huambo, *HI* Huíla, *LA* Luanda, *LN* Lunda-Norte, *LS* Lunda-Sul, *MA* Malanje, *MO* Moxico, *NA* Namibe, *UI* Uíge, *ZA* Zaire

the areas of Zambezi River, in Moxico, and in the Lundas. In this last region, we find an important representation of arthropods in collections, which might be a result of the activity of the *Museu do Dundo*, created in 1942. The museum included the Laboratory of Biology, where A Barros de Machado and E Luna de Carvalho established numerous international connections with specialists, with exchange of specimens with other collections.

Current Status of the Natural History Collections in Angola

The landscape of museum and herbarium collections in Angola is considerably diverse in terms of institutional governance. Although all the hosting institutions are public entities, they depend on different government ministries, which implies different priorities and funding programmes. As a result, it has been difficult to develop a common strategy for the development and use of NHCs in Angola. Currently, the different institutions have different capacities and dimensions. Most are still inactive or starting their activities, mainly by performing inventories, digitising and systematising information on species.

Research activities are also recent and have relied on international collaboration projects such as the Southern African Botanical Diversity Network (SABONET), the Angolan Biodiversity Assessment and Capacity Building Project, the African Plants Initiative (API), the Southern African Science Service Centre for Climate Change (SASSCAL), the Future Okavango (TFO), and the National Geographic Okavango Wilderness Project, among others. These funding opportunities allowed the support of the cooperation with relevant international institutions (Royal Botanic Gardens, Kew and the South African National Biodiversity Institute – SANBI), enabling the rehabilitation of some of the country's collections and the training of qualified personnel for their management, expansion and valorisation. In this context, since 2013, Angola has benefited from the GBIF training initiatives by the Portuguese node, for computerisation and publication of biodiversity data.

Another cooperation programme has also recently contributed to the advance of data mobilisation and capacity enhancement in Angola. Within the scope of the program Biodiversity Information for Development (BID), managed by GBIF with funds from the European Commission, Angola was granted a national project, led by SASSCAL, which started in 2016. Apart from the data mobilisation activities, some of which are based on collections, the project will enable Angola to participate in training workshops promoted by GBIF and other partners on biodiversity data publication, data quality, and data use.

Active Collections

In the scope of this chapter, we consider active collections those that are supporting or developing research activities, contributing to the increase of the value of the collection, by means of new additions of biological specimens, or valorisation through taxonomic revision, digitisation, and use by researchers on-site and online.

Herbarium LUBA, *Instituto Superior de Ciências de Educação da Huíla* (ISCED), Lubango

Founded in 1958 with the creation of the Instituto de Investigação Científica de Angola (IICA), the collection currently houses 15,902 plant specimens belonging to 202 families and 3520 species. The main collectors were G Barbosa, A de Menezes, R Santos, R Correia and JM Daniel. After independence, approximately half of the collection was transferred to Luanda, becoming part of the LUAI herbarium. Currently, the database of the collection is being prepared, and 200 specimens are available in high resolution through the Global Plants repository at http://plants.jstor.org.

Herbarium LUAI, *Universidade Agostinho Neto*, Luanda

The Luanda Herbarium incorporated part of the LUBA Herbarium, which was partially transferred to Luanda to be hosted at the National Centre for Scientific Research of the University of Agostinho Neto (Martins and Martins 2002). LUAI currently houses about 35,000 botanical specimens representing approximately 5000 species. The main collectors were A de Menezes, M Batalha, JM Daniel, M Lopes, R Santos, B Sousa and F Sousa. From 1995 until 2007, this herbarium housed 45,000 botanical samples of the LUA Herbarium. The herbarium currently has the digital infrastructure for databasing, but the pace of cataloguing has been slow.

Herbarium LUA, *Instituto de Investigação Agrária* (IIA), Ministry of Agriculture, Huambo

LUA was the first herbarium created in Angola, in Huambo Province, in 1941 (Martins and Martins 2002). Its collection includes about 40,000 specimens. The main contributors were G Barbosa, J Gossweiler, C de Matos, OA Leistner, EJ Mendes, FA Mendonça, R Monteiro and F Murta. There is a digitisation programme in preparation, within the scope of the National Project of the BID programme.

Bird and Mammal Collections, *Instituto Superior de Ciências de Educação da Huíla* (ISCED), Lubango

These collections were created as an IICA section installed in Lubango in the late 1950s. The first record dates from 1958, and until 1975 about 40,000 specimens were incorporated. The bird collection consists of 34,471 skins, as well as eggs, nests and embryos, distributed across 26 orders, 84 families and 305 genera. It is worth noting the contribution of António Rosa Pinto, representing 21% of the specimens of the collection sampled between 1958 and 1972 throughout the country (see map in

Dean et al. 2019). The collection of mammals consists of 4299 skins of 157 species distributed by 11 Orders, 56 Families and 103 Genera, and an unregistered number of skeletons and skulls. The temporal coverage is between 1960 and 1978, and the main collectors were J Crawford Cabral, AP Simões, C Simões and E Epalanga. The full bird collection and the group of chiroptera within mammals, with about 300 specimens, are on database, and will be published through GBIF.

Herpetological Collection, *Instituto Superior de Ciências de Educação da Huíla* (ISCED), Lubango

This is the first herpetological collection created after 1975, as an output of a study included in the SASSCAL project, under the responsibility of Ninda Baptista. It houses 1081 specimens of reptiles and amphibians (approx. 30–70% respectively), preserved in alcohol, as well tadpoles, eggs and tissue samples. Two collectors – Ninda Baptista and Pedro Vaz Pinto are important contributors to this collection.

Entomological Collection, *Instituto de Investigação Agrária* (IIA), Ministry of Agriculture, Huambo

Based on registry books, the collection contains the 44,884 specimens. There is also a digitisation programme for this collection, with the Odonata (1006 records) already digitised and published through GBIF (Cassinda et al. 2018), and with other orders to follow.

Snake Collection of the Research and Information for Drug and Toxicology Center (CIMETOX), Malanje

A collection of snakes was recently created in Malanje, at the Research and Information for Drug and Toxicology Center of the Medical Faculty, in Malanje (Oliveira et al. 2016). The number of specimens is not available at this stage.

Inactive Collections

Zoological Collection of Dundo, *Museu do Dundo*, Ministry of Culture, Dundo

Although the creation of the *Museu do Dundo* dates from 1942, the zoological collection began in 1936 (Machado 1952). This museum is best known for its valuable ethnographic collection, so at the time of its restoration at the beginning of this century, biological collections were not covered and some of them are in danger of deterioration. According to EC Afonso, curator of the Biological section in the 1980s, the Museum houses about 50,000 specimens of mammals, fishes, reptiles, amphibians and insects, the latter being the largest collection with about 30,000 specimens. Due to its value, we expect that in the future this collection can be studied and restored. The museum originally also had a herbarium (DIA), that no longer exists.

Museu Nacional de História Natural, Ministry of Culture, Luanda

The museum was created in 1938, and moved to the current location in 1956. The museum holds mounted specimens of mammals, fishes, birds, reptiles and insects. However, it was not possible to determine if there is an active collection, the number of specimens, and their value for scientific research. There is no insect collection. This museum is currently responsible for the management of *Museu do Dundo*.

Current and Potential Uses of Biological Collections

Museum and herbarium collections are examples where the whole is greater than the sum of the parts. Each specimen, as a voucher of a species found in nature, carries biological data (in its genes, tissues, traits, biochemistry) and metadata (on its label or attached field notes) providing contextual information on the location, date, habitat and ecology of that specimen. But from a set of specimens that make a collection, it is possible to do comparisons, grouping and separating them by their features, in what, ultimately, leads to the description or identification of a species. The possibility of making and analysing comparisons between specimens is fundamental to our developing the knowledge of what the species is, and therefore, our understanding of biodiversity. There is great value in having a collection. Because of this, we need to resist the fate predicted for collections in the previous quote from Saramago, that in the continuation of the text, says:

> … and for a short while they manage [to impose some order on the world], but only as long as they are there to defend their collection, because when the day comes when it must be dispersed, and that day always comes, either with their death or when the collector grows weary, everything goes back to its beginnings, everything returns to chaos.

We would lose too much in letting everything return to chaos, we cannot afford it.

Without a doubt, there is a cost in maintaining a collection. Several reports have been published about the termination of collections due to budget restrictions, either by closing doors, restricting or diverting staff to other tasks, or aggregating collections in large facilities (Gropp 2003; Deng 2015; Kemp 2015). For example, in Italy, by 2014, it was estimated that one-third of the biological specimens were lost through lack of preservation or bad practice (Nature Editorial 2014), and in the USA 100 herbaria have been closed since 1997 (Deng 2015). Discarding a collection brings with it the loss of investments in field expeditions, the costs of subsequent preservation of specimens across decades or centuries, and even the fact that many species may not be found anymore in the original collecting sites, due to the loss of habitats or restrictions in sampling because of conservation or ethical reasons. As an indicator, insurance companies in Norway value herbarium specimens at €21 each (Hannu Saarenmaa 2017, pers. com.). Here we will focus on the uses of collections and how they underpin scientific research, biodiversity conservation, food security, and other societal and economic benefits. Several reviews have discussed these uses, providing examples (Brooke 2000; Suarez and Tsutsui 2004; Tewksbury et al. 2014; Rocha et al. 2014). It is worth mentioning that, with the development of laboratory methods, technology and other tools still to be invented, there are potential applications of the collections in the future that we cannot foresee.

Preserving and Documenting Biodiversity

The most fundamental use of museum and herbarium collections is to support taxonomy and systematics, serving as references for species description, identification, and naming of species. One of the most important roles of collections is to preserve the physical specimens that served as samples for the formal scientific description of a new species for science. These specimens, often more than one, are called types. One of these specimens is usually designated as the holotype by the author of the species – the one chosen to be most representative of the characteristics of the species – but duplicates can also be mentioned in the publication of the species, and distributed to other collections. This distribution is important for security reasons, to ensure that if the holotype is lost because of an unfortunate event, other specimens that were used the initial description of the species are preserved. In 2017, a package with specimens sent by the National Museum of Natural History in Paris to the Queensland Herbarium in Australia, including some type specimens, was incinerated by Australian customs officers (Stokstad 2017).

Types are, therefore, special specimens, so their management is undertaken with extreme care. Digitisation programmes normally prioritise these specimens, to provide digital preservation and alternative access to the specimens via the internet. In the case of plant type specimens, the Global Plants Initiative framed this task, with support of the Andrew W Mellon Foundation. The repository Global Plants (http://plants.jstor.org) aggregates and provides access to more than two million high resolution images of types, including 3461 images of type specimens collected in Angola. The total number of types from Angola, from all groups, reported through GBIF and Global Plants is 6983 (Table 19.2).

Museum and herbarium collections serve not only to preserve types and other specimens already identified, but also organisms yet to be identified. In fact, many specimens remain unidentified for several years, either because there is no capacity for their immediate processing after being collected, or its identification represents taxonomic challenges, sometimes at the level of the description of a new species for

Table 19.2 Number of type specimens from Angola in NHC collections worldwide

Collection type	Holotypes	Other types	Total per collection
Plants	1236	2225	3461
Fungi	24	259	283
Mammals	25	14	39
Birds	108	79	187
Fish	82	487	569
Herpetological	40	229	269
Arthropods	300	1218	1518
Invertebrates	146	511	657
Microorganisms		10	10
Total	**1961**	**5022**	**6983**

The sources of the data are Global Plants (http://plants.jstor.org), for plant specimens, and GBIF (GBIF.org 2018), for other collection types

science. In plants, only 16% of newly collected species are described within the first 5 years after being collected, and approximately 25% of new species are described using specimens more than 50 years after their collection (Bebber et al. 2010), and in animals, the situation is likely to be similar (Kemp 2015).

Collections are the main source for documenting diversity not only between species but also within species. The majority of specimens in collections are not types, but regular specimens sampled at a certain date and location, by one or more collectors, and representing a species. These specimens and their associated information represent what we call primary biodiversity data, supporting different types of studies and applications. The set of specimens of a species, from one or more collections, allow identifying the range of natural variation of the several traits that are analyzed in the process of defining the species. Often, labels attached to the specimens, or registry books associated with the collections include information on traits that are recorded at the time of collection or when the specimen is added to the collection (e.g. size, weight, length, maturity stage, colour of the flower in plants, the presence of fruits, etc.). Additionally, the habitat, interactions with other species (e.g. parasite of, epiphyte on), its use by local populations in traditional medicine, food, and construction (mainly plants) is recorded.

Most taxonomic studies will require access to more than one collection, to allow a comprehensive analysis of the variability of the species of interest. Finding which collections have specimens important to the study might be demanding, but current digitisation projects underway in many collections do facilitate the task enormously. This is the case of some collections based in Angola at the Museum and Herbarium of Lubango, the entomological collections of *Instituto de Investigação Agronómica*, in Huambo, and the Herbarium of the *Centro de Botânica da Universidade Agostinho Neto*, in Luanda. The first two initiatives are currently preparing the publication of their databases through GBIF, for global and open access to data, which is possible even though Angola is not yet a member of the organisation. Through GBIF, researchers can have access to recorded information including most of the details included in the specimen's label, and if available, an image of the specimen, a sound recording or a video attached to the specimen record. There are currently approximately 49,000 specimens from Angola with media attachments.

Other mega-science digital platforms also support capacity development in taxonomic studies (Triebel et al. 2012). These include the Catalogue of Life, a global checklist for all groups, that integrates more than 168 international or group-specific checklists or taxonomic databases (Catalogue of Life 2018); the Biodiversity Heritage Library (2018) which provides digital access to legacy literature, including many publications with the original descriptions of the species, and automatic functions for searching scientific names; the Encyclopedia of Life (2018), also an integrative portal to information about species description, classification, multimedia, and distribution maps of; the Barcode of Life (Ratnasingham and Hebert 2007) which provides access to barcode sequence data; the IUCN Red List of Threatened Species (2018), that promotes global and regional assessments of species conservation status; and the previously mentioned Global Plants Initiative. All these initiatives agree on common biodiversity informatics data standards, protocols and tools that ensure the inter-operability across platforms under a common framework (Hobern et al. 2012). This means that when researchers and institutions contribute to or use one of these initiatives, they are reaching a global and transversal set of resources covering several biodiversity dimensions which, although global, provide detailed data applicable at the local or regional level.

These combinations of data are instrumental to prepare, for example, a national checklist, like the national vascular plant list (Figueiredo and Smith 2008, Figueiredo et al. 2009a, b). This task requires not only the compilation of information about species and their distribution, but also synonyms, and sometimes helping to solve taxonomic problems, species distribution ranges, and dealing with the scarcity of information. Internet access to these and similar platforms is fundamental and an important factor to reducing total costs of biodiversity research (Smith and Figueiredo 2010).

An increase of the role of collections in preserving reference material on biodiversity is likely to occur with the addition of new methods for biodiversity identification. Barcoding is one of the methods that determines DNA sequences that are species-specific (Gross 2012), and it can speed up the identification of new species. These sequences are stored in gene bank repositories, while the related physical specimens, the source of the sequences, are stored as vouchers in collections. Sometimes, specimens already represented in collections are found by barcoding to belong to different species, therefore new arrangements are needed within collections. Museums are currently adopting new workflows with duplicate specimen processing for traditional and molecular taxonomy (Gross 2012), in an articulated new way of producing natural history knowledge (Strasser 2011).

Detecting Changes in Species Distribution and in the Environment

Understanding species distribution is usually a multidimensional problem that involves information about species occurrences, climatic information, species migrations and the availability of resources such as food and water. Primary biodiversity data from collections is often the only data resource to document the presence of species, either because of loss of habitats due to change of land use or because of the local extinction of the species. This primary data, when combined with environmental data, can be used to model the species distribution by numerical tools that identify the environmental factors that are most closely associated with occurrence of the species. This is, in turn, converted to a species distribution model (SDM), expressed as a spatial map of probability of occurrence. Although subject to problems associated with sample bias (Beck et al. 2014; Gomes et al. 2018), these models can be an improved approach to gap analysis (Peterson and Kluza 2003), or help to plan sampling effort to sites with a high potential of occurrence but that has not yet been surveyed.

Distribution data from museums frequently supports assessments of the impact of climate change on species distribution. By comparing two models, one for the present distribution and another for a hypothesised future distribution, it is possible to identify changes, including the expansion or reduction of the area of distribution. The present distribution can be modeled from collection data, using a matching period for climate date. A projection for a future state can be performed, using the same distribution data, but with future climate scenarios data. Using this approach, Warren et al. (2013) found that half of the plant species and one-third of animal species can lose half of the suitable climate range by 2080. Another example with links to human health is provided by Capinha et al. (2014), using as target species the mosquito *Aedes aegypti*, the vector of dengue fever. These authors used collection data combined with other data sources to determine the macroclimatic conditions presently occupied by this mosquito, and the shifts in its distribution in the near future (2010–2039), based on models of possible climate scenarios.

In assessing biological invasions, the use of NHCs is also essential. The historical record of an alien species needs to be determined so that its native distribution is identified as well as its habitat and environmental requirements, life cycle, biotic and abiotic interactions. Frequently, data and information on these parameters are only available from museum or herbarium collections. These data will not only allow to assess the invasion risk of a species, but also to predict its spread to new regions, which can be done by projections of species distribution models, as for example, the Giant African Snail (Sarma et al. 2015) or a result of climate change in lantana (Taylor and Kumar 2014). Even if historical records were not available in collections to support a study, these are essential as repositories of new records of surveyed areas for future assessments (Rejmánek et al. 2017). Collections are also important for the assessment of impacts in areas of invasion, in the determination of species affected by the alien species.

Biodiversity Conservation

The IUCN Red List has become a standard with which to monitor a species' conservation status. Several of the criteria to determine the IUCN category of threat can be obtained from natural history collections, such as features of life history, biology, and geographical range. Williams and Crouch (2017) investigated whether herbarium records could suffice for accurate estimation of the plant geographical range in South Africa following IUCN Red List criteria. They concluded that results improve when information from national herbaria is complemented by local or regional herbarium datasets. For Cape Verde, herbarium data was also used in Red List assessments of the endemic flora (Romeiras et al. 2016). However, the role of collections in Red List assessments starts from the point of acertaining the correct identification of each specimen in a survey, and before the application of any criteria, based on a common taxonomy for the group of species under scrutiny (e.g. Grubb et al. 2003).

Another contribution of NHCs to biodiversity conservation is in reintroduction programmes. When a local population of one species is extinct or threatened, the reintroduction of new individuals can be done to increase population levels. However, the genetic profile of the local population should be determined, in order to ensure that new reintroduced individuals are the closest possible to the original population, and thus well adapted to the environmental conditions of the new location. If the local populations are extinct, museum or herbarium collections might be the only resource to determine the genetic profile of the original populations, if specimens are preserved in the NHCs from the original population. Collections are also instrumental in determining other aspects of translocation planning, including climate and habitat requirements (IUCN/SSC 2013).

A service provided by NHCs is also related to the trade of wild animals and plants, within the scope of the CITES. Frequently, specimens of wild species are seized by customs officials and the species identification is needed to check against the species lists in the annexes of the convention. NHC taxonomists are frequently asked to assist customs officials in identifying the species and the most likely source of these organisms.

Museum and herbarium collections can also be used to verify if the network of protected areas is effective in ensuring the protection of threatened species. Romeiras et al. (2014) used collection data to make a biogeographic analysis of 18 high-value timber trees from Angola. The authors concluded that these species could be grouped within four regions, which had little correspondence to currently recognised WWF ecoregions. They suggested that conservation plans based on WWF ecoregions might provide the inappropriate basis for the conservation of these trees, in which eight species were found to require high conservation priority because of their very restricted distribution in Angola.

Supporting Sustainable Food Production

As in the definition of biodiversity adopted by the Convention of Biological Diversity, the definition of agrobiodiversity by FAO highlights three levels of diversity: diversity of genetic resources (varieties, breeds); diversity of species used for food, fodder, fiber, fuel and pharmaceuticals; and the diversity of non-harvested species that support production and diversity in the wider environment that support agro-ecosystems (FAO 1999). Sustainable food production systems demand attention to all these dimensions, which reinforces the role of NHCs. For example, the conservation of crop wild relatives, which are wild species closely related to crops, is important to ensure sources of genetic diversity useful to develop more productive and resilient crops (Castañeda-Álvarez et al. 2016). A priority in the conservation of such wild relatives is, therefore, the importance of correct identification, based on herbarium data among other sources Castañeda-Álvarez et al. 2016).

Natural history collections are also of importance in many other aspects of agrosystems. This is the case of weed identification and the control of pests caused by insects and fungi. Collections provide the resources for the identification of these problem organisms, data for their first detection in a certain area, and information about life history and distribution needed to determine the potential areas of occurrence using bioclimatic and other modelling approaches. High biodiversity in agrosystems can in some cases contribute to increasing productivity, by promoting ecosystem services, e.g. through biological regulation of soil fertility (Duru et al. 2015), for which NHCs are important to guide actions to increase biodiversity in such systems, providing information on the original or potential species native to the region.

Two further types of collections in support of food and forestry are also associated with herbarium collections. One of these is germplasm or seed bank collections. These seed collections are fundamental to the preservation of plant species, by maintaining live and viable seeds for future use. There are more than 1300 seed banks worldwide (Rajasekharan 2015), both for crop or wild species. These include the Global Seed Vault, in Svalbard, Norway, which holds crop seeds for more than 5000 plant species, and the Millenium Seed Bank, at Kew and Wakehurst Place, United Kingdom, which holds seeds for 10% of the world wild species. In Angola, the Universidade Agostinho Neto hosts the only seed bank in the country as part of the center for plant genetic resources. The other type of collection is the xylarium, composed of wood samples, sometimes several pieces with different anatomical sections of the wood of the same species. These samples are used, sometimes supplemented by molecular genetics technologies, in identifying the products of illegal activities in the timber trade (Yu et al. 2017).

Connecting Biodiversity to Society Through Education

Biodiversity is present, but rarely noticed, in the everyday life of humans. Natural History Museums are important in bringing evidence of this to the population. Through displays and exhibitions, it is possible to explain or demonstrate, in simple terms, the value of biodiversity. In attractive displays, the direct value of biodiversity can be shown in food, medicine, fuel, fiber, rubber, oils and building materials, but also the indirect values through climate regulation, nutrient recycling, water and air purification, pollination, and cultural, religious and aesthetic aspects. People need to be informed by appropriate NHC displays in order to relate these values of biodiversity to their daily life. These topics can also be explored more deeply to explain biological concepts to students.

Specimens from collections facilitate the explanation of complex topics to visitors. For example, concepts as life stages, evolution, adaptation to the environment, species interaction and many others, are better explained using specimens as support. But these can also be used to provide insights to the science behind the scenes. The causes of diversity, how genes express into forms and colors, the roles of microscopic organisms that can move or destroy bodies a thousand times bigger, the work of a taxonomist, a geneticists or a bioinformatician in understanding phylogenies, the role of organisms as bioindicators of environmental changes, are all examples of what might interest the visitor to a NHC. All these approaches can be complemented by digital formats, through web pages that provide deeper coverage of the topics displayed in the exhibitions. Via the Internet, it is also possible to use virtual means to place species in their habitats and environments, or inform the visitor of the species in one's own neighbourhood. Other engagements with the public are also possible, turning the visitor into a collaborator. For example, crowdsourcing activities were implemented by some museums to make the databasing of specimens' labels, which is a time-consuming task in collection digitisation (Les Herbonautes (2018); Notes from Nature (2018); DIGIVOL (2018).

Another example is citizen science participation, through which platforms citizens can submit records, supported by images and other information, of species occurrence, normally referring a date and a location (through GPS coordinates) attached. Many projects of this type have emerged in recent years, the most visible of the global scope being eBird (2018) for bird observations and iNaturalist.org (2018) for any type of organism. Although subject to errors, these initiatives have the enormous merit to expand the network of voluntary observers and are improving their internal quality control mechanisms (using image analysis algorithms, for example), to suggest or correct identifications. Records with attached images thus become openly accessible to researchers, permitting validation of the records.

Museums can be windows to connect biodiversity science to society. Not only can they contribute to educating people on concepts of biodiversity conservation and sustainability but also to attracting new students and practitioners to biodiversity-related topics. They can also be vibrant regional poles for research and natural history activities, interlinked with a global community of scientists and naturalists through digital platforms. In some cases, their importance has also been recognised at an economic level: the Natural History Museum in London has free entry to visitors, a benefit that was earned after the demonstration of the economic benefit from attracting foreign tourists to London.

Roadmap for the Museum and Herbarium Collections of Angola

The goal of developing knowledge about the biodiversity of Angola should be intimately linked to strengthening the role of natural history collections as a reference of biodiversity resources. That connection should be bi-directional, first to ensure that natural history collections are used in studies about Angolan biodiversity, and second that specimens documenting new distributional information are included in collections for future reference. To ensure that the NHCs of Angola are prepared to play this role, we propose the following roadmap for the museum and herbarium collections of Angola.

Compile an Inventory of Collections from Angola

An inventory of the collections with specimens from Angola, either in Angolan institutions or abroad, is important to produce an index of the available resources and support a gap analysis of the biodiversity coverage of collections. This can be done by a metadata description of the holdings of such collections, mentioning the main taxonomic groups, time period, geographic area covered, main preservation methods, total (estimated) number of specimens and number of species in the database. An assessment of data needs (Asese and Schiwinger 2018) would provide elements for future prioritisation of data mobilisation activities.

Identify Taxonomic Expertise and Promote Networking

A network of experts is essential to support NHC activities, in order to avoid, for example, large time gaps between sampling and identification of specimens. These can be taxonomists working in collections, but increasingly, ecologists, molecular biologists and experts from other fields are performing taxonomic activities (Kemp 2015). This network of contacts should be developed to cover many biological groups and be strongly tied to cooperation activities in training, study programmes and projects focused on biodiversity. A route to promote the creation of this network is the effective participation of Angola in the leading international networks, as is the case of GBIF, or the Biodiversity Information Standards (TDWG) international community, which leads and promotes developments in biodiversity informatics worldwide.

Promote Data Repatriation Activities

Specimens from the country hosted in collections abroad contain important information that should be available for studies and biodiversity management in Angola. The repatriation of data can be promoted by a combination of initiatives that can facilitate and speed up the access to it. For example, in the late 1990s, Mexico used government funds to support visits of Mexican ornithologists to the largest bird collection in the United States and in Europe, to catalog in a database bird specimens from Mexico (Peterson et al. 2016). A similar approach could be done in the support of visiting students or researchers from Angola to institutions hosting Angolan collections. Furthermore, many collections have existing databasing projects, so coordination with such activities could facilitate the prioritisation of data mobilisation. The framework for these data mobilisation activities can be provided by the participation of countries and institutions in GBIF.

Include NHC Activities in University Curricula

In many graduate curricula in universities worldwide, a decrease in the importance of natural history has been experienced through the last quarter of the twentieth century. Fewer or no credits have been dedicated to traditional taxonomy, compared to subjects in ecology, cellular and molecular biology, evolution and biotechnology. However, natural history collections can now encompass these new methods, remaining central to the goal of understanding the world's biodiversity. We can see NHCs as vibrant facilities that merge specimen and biomolecular preservation, and biodiversity informatics infrastructure, being prepared to respond to societal challenges such as climate change, biodiversity loss and food security. Therefore, natural history can be attractive to teachers and graduate and postgraduate students in universities, provided that its activities can be properly compensated. One way of doing this is to reward data publication in career assessments of researchers and to provide proper recognition through traceable citations to the use of collections in scientific publications (Rouhan et al. 2017).

Align NHCs with National and International Agendas on Biodiversity

The 2030 Agenda for Sustainable Development approved by the United Nations includes several goals in which biodiversity and ecosystems take a central role. Goal 15 (Biodiversity, Forests, Desertification) is specifically targeted to halt biodiversity loss, but biodiversity is also relevant for other Goals, as the Goal 2 (Hunger and Food Security), Goal 12 (Sustainable Consumption and Production), Goal 13 (Climate Change) and Goal 14 (Oceans), if we consider ecosystem services or agro-biodiversity. Angola will have the opportunity to participate in this agenda with actions that fulfill international requirements, which in turn translate to national priorities. Institutions with NHCs in Angola should be prepared to respond to the needs that the implementation of this agenda requires, namely in providing the essential information and expertise to support projects and reporting.

Conclusions

Museum and herbarium collections are restoring their paramount role in the study of biodiversity, with the rapid developments seen recently in molecular biology and in biodiversity informatics. These new tools contribute to speed up and add layers of analysis to biodiversity assets represented in collections, not only for the materials sampled in current projects but also for specimens collected through the history of each collection. Many specimens have been kept hidden in collections for decades before they were discovered as new species for science. Collections thus represent an important asset by preserving the known (and unknown) biodiversity of a region or a country, especially if they combine with these new approaches of analysis and providing access to biodiversity information.

There is presently an under-representation of Angola's biodiversity in NHCs. The vastness of the country and the diversity of its ecoregions and habitats means that this is a demanding task, but essential to support biodiversity knowledge and conservation in the country. Obtaining a figure for the total number of specimens in collections from Angola worldwide is difficult. However, from data available through GBIF it is possible to obtain approximations. The current number of records available online through the facility is circa 150,000, which is in the same order of magnitude of other countries in southern Africa (except for South Africa, with 2,9 million and Democratic Republic of Congo, with 800,000). The situation is likely to improve in the near future, with the start of participation by Angolan institutions in GBIF (and the possible participation of the country), but the significance of these numbers will need to be translated into effective access after a fitness for use and a gap analysis of taxonomic and spatial coverage and biases of the data has been conducted.

There are three herbaria and four zoological collections based in Angola, but not all of them are currently active in the support of research or other biodiversity-related activities. The three herbaria are or have plans to create a database for their collections, and the hosting institution of two of them (ISCED and IIA) are already registered as publishers of biodiversity data through GBIF, indicating that these datasets will be openly available in the future. In terms of zoological collections, the collections in ISCED and IIA also are developing databasing activities on their collections, namely of birds and mammals, in the first case, and entomological, in the second, with perspectives of online publishing through GBIF very soon. Some other important collections remain, however, hidden or not easily accessible to researchers, such as the collections of *Museu do Dundo*, and the collections of the *Museu Nacional de História Natural*. Little information is available for the current situation and accessibility of these collections, although an extensive literature is available about the activity of *Museu do Dundo* in a publication issued by that institution in the 1950s and 1960s, with references to specimens in the collection (Machado 1995).

Online data availability is very important to attract national and international researchers and specialists to use the collections in the country. This is important to promote international collaboration and raise the capacity to use collections to improve the knowledge of Angolan biodiversity, on topics related to ecology, evolution, and conservation. These collaborations are also important to promote data mobilisation and quality improvement of collections based in Angola and abroad, which is now supported by a framework of international digital platforms. But collections in Angola need to be prepared to support new research activities in the field, considering that the biodiversity of parts of the country is still relatively unknown and in need of field surveys, as described in other chapters of this volume. Furthermore, to face big environmental challenges like the loss of biodiversity, climate change, and invasive alien species, it is urgently necessary to provide more information and knowledge about biodiversity, and collections are certainly the most accessible way to begin.

Natural History Collections are also important to link biodiversity to society. Many aspects of the importance of biodiversity to everyday life can be achieved through attractive displays that link the natural curiosity of humans with features of the structure and functioning of biodiversity, resulting in important impacts on the education and awareness of communities. Stimulating displays and activities can also contribute to attracting more young researchers to work in NHCs. Education is one of the most important roles of collections, in association with other uses for preservation, documentation and biodiversity conservation. Therefore, Natural History Collections represent strategic infrastructures for a country: reason enough to contradict the fate predicted by Saramago that all of them would return to chaos.

Appendix

Natural History Collections holding specimens from Angola

Acronym	Institution
AMNH	American Museum of Natural History (USA)
ARC	Agricultural Research Council, Plant Protection Research Institute (South Africa)
B	Botanic Garden and Botanical Museum Berlin-Dahlem (Germany)
BMSM	Bailey-Matthews National Shell Museum (USA)
BR	Botanic Garden Meise (Belgium)
CAS	California Academy of Sciences (USA)
CM	Carnegie Museums (USA)
COI	Herbarium of the Universidade de Coimbra
E	Royal Botanic Garden Edinburgh (United Kingdom)
FCEyN, UBA	ArOBIS Centro Nacional Patagónico (Argentina)
FishBase	FishBase
FMNH	Field Museum (USA)
GNM	Gothenburg Natural History Museum (Sweden)
Ifremer	French Research Institute for Exploitation of the Sea (France)
IICT	Instituto de Investigação Científica Tropical of the Universidade de Lisboa
ISCED	Instituto Superior de Ciências de Educação da Huíla (Angola)
K	Royal Botanic Gardens, Kew (United Kingdom)
KU	University of Kansas Biodiversity Institute (USA)
LEGON-GC	University of Ghana – Ghana Herbarium (Ghana)
MACN	Museo Argentino de Ciencias Naturales (Argentina)
MHNG	Muséum d'Histoire Naturelle de la Ville de Genève (Switzerland)
MNCN	Spanish National Museum of Natural Sciences (Spain)
MNHN	Museum National d'Histoire Naturelle (France)
MUHNAC	Museu Nacional de História Natural e da Ciência da Universidade de Lisboa (Portugal)
MVZ	Museum of Vertebrate Zoology (USA)
NHMUK	Natural History Museum (United Kingdom)
RBINS	Royal Belgian Institute of Natural Sciences (Belgium)
RMCA	Royal Museum for Central Africa (Belgium)
S	Swedish Museum of Natural History (Sweden)
SAIAB	South African Institute for Aquatic Biodiversity
SANBI	South African National Biodiversity Institute
SMF	Senckenberg (Germany)
SNSB-M	Staatliche Naturwissenschaftliche Sammlungen Bayerns (Germany)
TM	Ditsong National Museum of Natural History Collection (South Africa)
UPS	Museum of Evolution in Uppsala (Sweden)
USNM	National Museum of Natural History, Smithsonian Institution (USA)
VM	Vänersborg Museum (Sweden)

(continued)

Acronym	Institution
YPM	Yale University Peabody Museum (USA)
ZMB	Collection Crustacea, Senckenberg (Germany)
ZMUC	Zoological Museum, Natural History Museum of Denmark (Denmark)

References

Albuquerque S, Correia AI (2010) The Welwitsch collections – Iter Angolense (1853–1860) at LISU. In: Van Der Burgt X, Van Der Maesen J, Onana J-M (eds) Systematics and conservation of African plants. Royal Botanic Gardens, Kew, pp 787–790

Albuquerque S, Brummitt RK, Figueiredo E (2009) Typification of names based on the Angolan collections of Friedrich Welwitsch. Taxon 58:641–646

Alves MJ, Bastos-Silveira C, Cartaxana A et al (2014) As Coleções Zoológicas do Museu Nacional de História Natural e da Ciência. In: Alves MJ, Cartaxana A, Correia AM et al (eds) Professor Carlos Almaça (1934–2010) – Estado da Arte em Áreas Científicas do Seu Interesse. Museus da Universidade de Lisboa, Lisboa, pp 289–301

Asase A, Schwinger GO (2018) Assessment of biodiversity data holdings and user data needs for Ghana. Biodivers Inform 13:27–37

Baptista N, Conradie W, Vaz Pinto P et al (2019) The amphibians of Angola: early studies and the current state of knowledge. In: Huntley BJ, Russo V, Lages F, Ferrand N (eds) Biodiversity of Angola. Science & conservation: a modern synthesis. Springer Nature, Cham

Bebber DP, Carine MA, Wood JRI et al (2010) Herbaria are a major frontier for species discovery. Proc Natl Acad Sci 107:22169–22171

Beck J, Böller M, Erhardt A et al (2014) Spatial bias in the GBIF database and its effect on modeling species' geographic distributions. Eco Inform 19:10–15

Beja P, Vaz Pinto P, Veríssimo L et al (2019) The mammals of Angola. In: Huntley BJ, Russo V, Lages F, Ferrand N (eds) Biodiversity of Angola. Science & conservation: a modern synthesis. Springer Nature, Cham

Biodiversity Heritage Library (2018) URL https://www.biodiversitylibrary.org/. Accessed 12 June 2018

Branch WR, Vaz Pinto P, Baptista N et al (2019) The reptiles of Angola: history, diversity, endemism and hotspots. In: Huntley BJ, Russo V, Lages F, Ferrand N (eds) Biodiversity of Angola. Science & conservation: a modern synthesis. Springer Nature, Cham

Brooke ML (2000) Why museums matter. Trends Ecol Evol 15:136–137

Burgess N, Hales JD, Underwood E et al (2004) Terrestrial ecoregions of Africa and Madagascar – a conservation assessment. Island Press, Washington, DC, p 499

Cabral JC (2010) João Crawford Cabral (depoimento, 2009). IICT, Lisboa, 16 pp

Capinha C, Rocha J, Sousa CA (2014) Macroclimate determines the global range limit of Aedes aegypti. EcoHealth 11(3):420–428

Cassinda S, Fernandes Elizalde S, Bassimba D (2018) Colecção Entomológica IIA. Instituto de Investigação Agronómica – IIA. Occurrence dataset. https://doi.org/10.15468/bhqdhp. Accessed via GBIF.org on 18 June 2018

Castañeda-Álvarez NP, Khoury CK, Achicanoy HA et al (2016) Global conservation priorities for crop wild relatives. Nat Plants 2:16022

Catalogue of Life (2018) URL http://www.catalogueoflife.org/. Accessed 12 June 2018

Ceríaco LMP (2014) O "Arquivo Histórico Museu Bocage" e a História da História Natural em Portugal. In: Alves MJ, Cartaxana A, Correia AM et al (eds) Professor Carlos Almaça (1934–2010) – Estado da Arte em Áreas Científicas do Seu Interesse. Museus da Universidade de Lisboa, Lisboa, pp 329–358

Dean WRJ, Melo M, Mills MSL (2019) The avifauna of Angola: richness, endemism and rarity. In: Huntley BJ, Russo V, Lages F, Ferrand N (eds) Biodiversity of Angola. Science & conservation: a modern synthesis. Springer Nature, Cham
Deng B (2015) Plant collections left in the cold by cuts. Nature 523:16–16
DIGIVOL (2018) URL https://digivol.ala.org.au/. Accessed 15 June 2018
Duru M, Therond O, Martin G et al (2015) How to implement biodiversity-based agriculture to enhance ecosystem services: a review. Agron Sustain Dev 35:1259–1281
eBird (2018) URL https://ebird.org/home. Accessed 15 June 2018
Encyclopedia of Life (2018) URL http://eol.org/. Accessed 12 June 2018
FAO (1999) Agricultural biodiversity, multifunctional character of agriculture and land conference, background paper 1. FAO, Maastricht
Figueiredo E, Smith GF (2008) Plants of Angola/Plantas de Angola. Strelitzia 22:1–279
Figueiredo E, Smith GF, Cesar J (2009a) The flora of Angola: first record of diversity and endemism. Taxon 58:233–236
Figueiredo E, Soares M, Seibert G et al (2009b) The botany of the Cunene-Zambezi Expedition with notes on Hugo Baum (1867-1950). Bothalia 39:185–211
Fontaine B, Perrard A, Bouchet P (2012) 21 years of shelf life between discovery and description of new species. Curr Biol 22:R943–R944
GBIF.org (25 May 2018) GBIF occurrence download. https://doi.org/10.15468/dl.urk4kx
Gilles D, Zaiss R, Blach-Overgaard A et al (2016) RAINBIO: a mega-database of tropical African vascular plants distributions. PhytoKeys 74:1–18
Gomes VHF, Ijff SD, Raes N et al (2018) Species distribution modelling: contrasting presence-only models with plot abundance data. Sci Rep 8:1003
Goyder DJ, Gonçalves FMP (2019) The flora of Angola: collectors, richness and endemism. In: Huntley BJ, Russo V, Lages F, Ferrand N (eds) Biodiversity of Angola. Science & conservation: a modern synthesis. Springer Nature, Cham
Gropp RE (2003) Are university natural science collections going extinct? Bioscience 53:550–550
Gross M (2012) Barcoding biodiversity. Curr Biol 22:R73–R76
Grubb P, Butynski TM, Oates JF et al (2003) Assessment of the diversity of African primates. Int J Primatol 24:1301–1357
Hobern D, Apostolico A, Arnaud E, et al (2012) Global biodiversity informatics outlook: delivering biodiversity knowledge in the information age. Global Biodiversity Information Facility, Copenhagen. Available at: https://www.gbif.org/document/80859
Huntley BJ (2019) Angola in outline: physiography, climate and patterns of biodiversity. In: Huntley BJ, Russo V, Lages F, Ferrand N (eds) Biodiversity of Angola. Science & conservation: a modern synthesis. Springer Nature, Cham
iNaturalist.org (2018) URL https://www.inaturalist.org/. Accessed 15 June 2018
IUCN/SSC (2013) Guidelines for reintroductions and other conservation translocations. IUCN Species Survival Commission, Gland
Kemp C (2015) Museums: the endangered dead. Nature 518:292–294
Kipping J, Clausnitzer V, Fernandes Elizalde SRF et al (2019) The dragonflies and damselflies of Angola. In: Huntley BJ, Russo V, Lages F, Ferrand N (eds) Biodiversity of Angola. Science & conservation: a modern synthesis. Springer Nature, Cham
Lavoie C (2013) Biological collections in an ever changing world: herbaria as tools for biogeographical and environmental studies. Perspect Plant Ecol Evol Syst 15:68–76
Les herbonautes (2018) URL http://lesherbonautes.mnhn.fr/. Accessed 15 June 2018
Machado AB (1952) Generalidades acerca da Lunda e da sua exploração biológica. Companhia de Diamantes de Angola, Publicações Culturais 12:1–111
Machado AB (1995) Notícia sumária sobre a acção cultural da Companhia de Diamantes de Angola. In: Diamang – Estudo do património cultural da ex-Companhia dos Diamantes de Angola. Publicações do Centro de Estudos Africanos, Museu Antroplógico da Universidade de Coimbra, Coimbra, pp 11–24
Martins E, Martins T (2002) Herbários de Angola: que futuro? Garcia da Orta Sér Bot 16(1–2):1–4

Mendes LF, Bivar-de-Sousa A, Figueira R (2013) Butterflies of Angola. Lepidoptera. Papilionoidea. I. Hesperiidae, Papilionidae. IICT and CIBIO, Lisboa and Porto

Mendes L, Bivar-de-Sousa A, Williams M (2019) The butterflies and skippers of Angola. In: Huntley BJ, Russo V, Lages F, Ferrand N (eds) Biodiversity of Angola. Science & conservation: a modern synthesis. Springer Nature, Cham

Mills M, Melo M (2013) The checklist of the birds of Angola 2013. Associação Angolana para Aves e Natureza, Luanda, Angola

Mills M, Franke U, Joseph G, Miato F, Milton S, Monadjem A, Oschadleus D, Dean W (2010) Cataloguing the Lubango Bird Skin Collection: towards an atlas of Angolan bird distributions. Bull ABC 17:43–53

Mora C, Tittensor DP, Adl S et al (2011) How many species are there on Earth and in the Ocean? PLoS Biol 9:e1001127

Myers N, Mittermeier RA, Mittermeier CG, da Fonseca GAB, Kent J (2000) Biodiversity hotspots for conservation priorities. Nature 403:853–858

Nature Editorial (2014) Save the museums. Nature 515:311–312

Notes from Nature (2018) URL https://www.notesfromnature.org/. Accessed 15 June 2018

Oliveira PSD, Rocha MT, Castro AG et al (2016) New records of Gaboon viper (*Bitis gabonica*) in Angola. Herpetol Bull 136:42–43

Olson DM, Dinerstein E, Wikramanayake ED et al (2001) Terrestrial ecoregions of the world: a new map of life on earth. Bioscience 51:933–938

Peterson AT, Kluza DA (2003) New distributional modelling approaches for gap analysis. Anim Conserv 6:47–54

Peterson AT, Navarro-Sigüenza AG, Gordillo-Martínez A (2016) The development of ornithology in Mexico and the importance of access to scientific information. Arch Nat Hist 43:294–304

Prathapan KD, Pethiyagoda R, Bawa KS et al (2018) When the cure kills – CBD limits biodiversity research. Science 360:1405–1406

Pyke GH, Ehrlich PR (2010) Biological collections and ecological/environmental research: a review, some observations and a look to the future. Biol Rev 85:247–266

Rajasekharan PE (2015) Gene banking for ex situ conservation of plant genetic resources. In: Plant biology and biotechnology. Springer, New Delhi, pp 445–459

Ratnasingham S, Hebert PDN (2007) BOLD: the barcode of life data system (www.barcodinglife.org). Mol Ecol Notes 7:355–364

Rejmánek M, Huntley BJ, Roux JJL et al (2017) A rapid survey of the invasive plant species in western Angola. Afr J Ecol 55:56–69

Rocha LA, Aleixo A, Allen G, Almeda F et al (2014) Specimen collection: an essential tool. Science 344:814–815

Romeiras M (1999) Subsídio para o conhecimento dos colectores botânicos em Angola. Revista de Ciências Agrárias 22:73–83

Romeiras MM, Figueira R, Duarte MC et al (2014) Documenting biogeographical patterns of African timber species using herbarium records: a conservation perspective based on native trees from Angola. PLoS ONE 9:e103403

Romeiras MM, Catarino S, Gomes I et al (2016) IUCN Red List assessment of the Cape Verde endemic flora: towards a global strategy for plant conservation in Macaronesia. Bot J Linn Soc 180:413–425

Rouhan G, Dorr LJ, Gautier L et al (2017) The time has come for Natural History Collections to claim co-authorship of research articles. Taxon 66:1014–1016

Santos J, Sales F (eds) (2018) Catalogue of the herbarium of the University of Coimbra (COI). Department of Life Sciences, Faculty of Sciences and Technology of the University of Coimbra. http://coicatalogue.uc.pt. Accessed on 5 May 2018

Saramago J (2000) All the names. The Harvill Press, London

Sarma RR, Munsi M, Ananthram AN (2015) Effect of climate change on invasion risk of giant African snail (*Achatina fulica* Férussac, 1821: Achatinidae) in India. PLoS ONE 10:e0143724

Skelton PH (2019) The freshwater fishes of Angola. In: Huntley BJ, Russo V, Lages F, Ferrand N (eds) Biodiversity of Angola. Science & conservation: a modern synthesis. Springer Nature, Cham

Smith GF, Figueiredo E (2010) E-taxonomy: an affordable tool to fill the biodiversity knowledge gap. Biodivers Conserv 19:829–836

Stokstad E (2017) Botanists fear research slowdown after priceless specimens destroyed at Australian border. Science News 11-05-2017. https://doi.org/10.1126/science.aal1175. Accessed 13 June 2018

Strasser BJ (2011) The experimenter's museum: GenBank, natural history, and the moral economies of biomedicine. Isis 102:60–96

Suarez AV, Tsutsui ND (2004) The value of museum collections for research and society. Bioscience 54:66–74

Taylor S, Kumar L (2014) Impacts of climate change on invasive *Lantana camara* L. distribution in South Africa. Afr J Environ Sci Technol 8:391–400

Tewksbury JJ, Anderson JGT, Bakker JD et al (2014) Natural history's place in science and society. Bioscience 64:300–310

The IUCN Red List of Threatened Species (2018) URL http://www.iucnredlist.org/. Accessed 12 June 2018

Thiers B (2018) Index Herbariorum: a global directory of public herbaria and associated staff. New York Botanical Garden's Virtual Herbarium. http://sweetgum.nybg.org/science/ih/

Triebel D, Hagedorn G, Rambold G (2012) An appraisal of megascience platforms for biodiversity information. MycoKeys 5:45–63

Warren R, VanDerWal J, Price J et al (2013) Quantifying the benefit of early climate change mitigation in avoiding biodiversity loss. Nat Clim Chang 3:678–682

Wheeler QD, Knapp S, Stevenson DW et al (2012) Mapping the biosphere: exploring species to understand the origin, organization and sustainability of biodiversity. Syst Biodivers 10:1–20

Williams VL, Crouch NR (2017) Locating sufficient plant distribution data for accurate estimation of geographic range: the relative value of herbaria and other sources. S Afr J Bot 109:116–127

Yeates DK, Zwick A, Mikheyev AS (2016) Museums are biobanks: unlocking the genetic potential of the three billion specimens in the world's biological collections. Curr Opin Insect Sci 18:83–88

Yu M, Jiao L, Guo J et al (2017) DNA barcoding of vouchered xylarium wood specimens of nine endangered *Dalbergia* species. Planta 246:1165–1176

Chapter 20
Conclusions: Biodiversity Research and Conservation Opportunities

Vladimir Russo, Brian J. Huntley, Fernanda Lages, and Nuno Ferrand

Abstract Angola is a country full of opportunities. Few countries offer more exciting prospects for young scientists to discover and document the rich biodiversity, complex ecosystem processes and undescribed species of plants and animals that are to be found in its amazing diversity of landscapes and seascapes. The current expanding support of the Angolan government and of international partners is unprecedented, and the positive response from young students ensures the growth of a new generation of biodiversity researchers and conservation professionals. Based on a synthesis of biodiversity research and conservation activities of the past century, we outline opportunities, approaches and priorities for a strengthened collaborative research and conservation agenda.

Keywords Africa · Angola · Biological discovery · Checklists · Endemic species · Research priorities · Science collaboration · Socio-ecological systems · Threatened species

---

V. Russo (✉)
Fundação Kissama, Luanda, Angola
e-mail: vladyrusso@gmail.com; vladimir.russo@holisticos.co.ao

B. J. Huntley
CIBIO-InBIO, Centro de Investigação em Biodiversidade e Recursos Genéticos, Universidade do Porto, Vairão, Portugal
e-mail: brianjhuntley@gmail.com

F. Lages
ISCED – Instituto Superior de Ciências da Educação da Huíla, Lubango, Angola
e-mail: f_lages@yahoo.com.br

N. Ferrand
CIBIO-InBIO, Centro de Investigação em Biodiversidade e Recursos Genéticos, Laboratório Associado, Campus de Vairão, Universidade do Porto, Vairão, Portugal

Departamento de Biologia, Faculdade de Ciências, Universidade do Porto, Porto, Portugal

Department of Zoology, Auckland Park, University of Johannesburg, Johannesburg, South Africa
e-mail: nferrand@cibio.up.pt

B. J. Huntley et al. (eds.), *Biodiversity of Angola*,
https://doi.org/10.1007/978-3-030-03083-4_20

Context: Challenges and Opportunities

This book was conceived on the basis of three simplistic assumptions. First, that very little information is available on the biodiversity of Angola. The second assumption was that colonial governments had neglected, and post-independence authorities have been unsupportive of research on the fauna, flora and ecosystems of Angola. Thirdly, we assumed that existing biodiversity knowledge was mostly lost in dusty government archives or inaccessible scientific journals. In truth, the rich assemblage of information gathered together in this synthesis volume demonstrates the error of these assumptions. The perceived challenges of the past have become a mosaic of opportunities for the future.

The energy, knowledge and dedication of this book's 46 contributing authors has resulted in a comprehensive synopsis of the 'state of the science' on the evolution and diversity of Angola's landscapes, flora, vegetation, all vertebrates, two iconic invertebrate taxa, and key ecosystems in both marine and terrestrial environments. What is immediately obvious from each chapter is the wealth of Angola's natural heritage, and how fragile it is to anthropogenic impacts and the vicissitudes of climate change. The vulnerability of the remnant forests of Angolan Escarpment and Afromontane ecosystems to these pressures places a tremendous responsibility on Angolans to study and protect these fingerprints of the past. Angola's universities, scientific organisations and government research institutions can lead the way to strengthening our understanding of the evolution, structure and functioning of these and the many other special habitats that make Angola unique in Africa in terms of the diversity of biomes and ecoregions found within its borders.

Each chapter of this volume reveals research and conservation opportunities relevant to the environment or taxon under discussion and presents compelling arguments for greater levels of investment in both research and conservation. Some needs are very specific, such as the importance of biodiversity surveys, vegetation maps and socio-economic assessments of the country's many protected areas as a basis to achieving their effective management and the delivery of long-term benefits to society. Other priorities are more broadly based, aimed at developing tools for wide-scale natural resource-use planning, such as a new and detailed vegetation map for the entire country, building on the results of the current forest inventory, and for a national biodiversity data portal. Yet others focus on selected taxa that can help, through modern genomic studies, to explain the processes of speciation that have led to the richness of the country's flora and fauna. All have a common purpose – to effectively inform decisions that will ensure increasingly sustainable development for all Angolans and for humanity at large.

Towards a Biodiversity Conservation Research Strategy

Over 40 research topics are identified in the chapters of this volume. Such compilations of research opportunities need to be embraced within national strategies, matching needs with resources and priorities within a research agenda, while understanding the country's challenges. In recent decades, a first point of departure in the design of conservation science programmes has been the frameworks provided by international multilateral environmental conventions – most specifically the Convention on Biological Diversity to which Angola has been a signatory since 1998. The development of Angola's National Biodiversity Strategy and Action Plan (NBSAP) brought a logical structure to both policy and planning (GoA 2006, 2018). These strategies have been further developed through setting nationally relevant implementation goals such as those detailed in the Aichi Targets (CBD 2010). An early lesson learned in the implementation of such strategies was the fundamental importance of cooperative partnerships between multiple institutions. The global scarcity of taxonomists, for example, has meant that no single country has the capacity to study, understand and document all taxa and ecosystems. The strength of organisations such as IUCN, GBIF, IPBES, etc., is based on shared information and skills. The benefits of the recent surge of interest in Angola's biodiversity demonstrated by foreign universities, museums, non-governmental organisations and intergovernmental agreements have yet to be fully exploited. For this, a strategic approach, implemented opportunistically, is advantageous.

Building a programme of work for conservation science needs to be mobilised at several levels, often simultaneously, rather than sequentially. At a first level, biodiversity assessments are a priority. Biodiversity surveys, which provide both the building blocks of knowledge and unique training and capacity building opportunities, must maintain the positive momentum of the past decade. The preliminary checklists of species presented in this volume should be regularly updated and expanded to integrate these and other taxonomic groups through an electronic biodiversity data portal. Ideally, international institutions, in liaison with national hosts, should assist in the coordination and curation of checklists, atlases and field guidebooks such as those already available for the reptiles and amphibians of Cangandala National Park (Ceríaco et al. 2016), the 'special birds' of Angola (Mills 2018) and the atlas of Angolan reptiles and amphibians (Marques et al. 2018). Checklists and the natural history collections on which they are based need integration within international databases such as those of GBIF and the Catalogue of Life. A direct outcome of this component of research is the development of taxonomic skills and of para-taxonomists in Angola.

Second, the socio-ecological systems within which research and conservation play out should not be neglected. Biodiversity does not exist in a vacuum – it has a human face that governs the success or failure of interventions. Biodiversity scientists who ignore the human dimensions of research and conservation do so at their peril. This is especially true within and around formal protected areas, where studies are needed on the socio-economic drivers of change (from a subsistence to a cash-based economy in rural areas), and the impacts of the bushmeat trade, slash-and-burn agricultural practices, charcoal production and nomadic pastoralism, on biodiversity conservation across the country. Such studies should also draw on local indigenous knowledge in developing management plans and research projects. At a regional scale, the assessment of the role of ecosystem services for sustainable living in rural and urban environments provides information essential to long-term development planning, ideally guided by a new and detailed vegetation map of Angola. The interdependence of basic research, biodiversity assessments and applied studies is obvious.

Third, there is no questioning the paramount importance of strengthening the capacity of young Angolan researchers and of research institutions. These actions should be framed within a collaborative and mutually beneficial strategy. International collaboration is already assisting in this, but needs further support. Scholarships, internships and mentoring programmes are fundamental, but field experience is critical for young biodiversity researchers, and the promotion of bush camps for student training in protected areas and biodiversity hotspots is a highly effective mechanism for inspiring the new generation. The establishment of a virtual network of Angolan conservation professionals using social media such as the *Angola Ambiente* Facebook site is a highly effective initiative. Angolan institutions also need strengthening and closer cooperation across government departments, and the integration of their research results through a formal information portal, is a key opportunity.

Research Opportunities from Genes to Landscapes

While the above three pillars form the foundations of a nascent biodiversity conservation research agenda, with an emphasis on immediate and practical needs, there are many fascinating questions relating to the functioning of Angola's diverse ecosystems that challenge the minds of biologists as they explore the country. The research strategy should be reinforced by studies on the evolutionary and ecological processes that account for Angola's biodiversity. An understanding of the evolution of Angola's biota will be strengthened by the development of modern phylogenies of key taxonomic groups, as proposed in many to the chapters of this volume. An understanding of the evolution of species assemblages and of individual species is of great value to guiding conservation measures, as already demonstrated by the Angolan studies of Vaz da Silva (2015) and Vaz Pinto (2018). Advanced molecular technologies allow new insights into many patterns revealed by basic surveys and assessments. The population genetics and hybridisation of Angola's two elephant species needs urgent study before the last remnants of Forest Elephant are driven to extinction. Similarly, iconic plants such as the baobab *Adansonia digitata* that in Angola manifests as widely divergent phenotypes, from dwarf baobabs in Namibe to the obese giants of Cunene and the slender columns of Cuanza-Norte, merit studies on their genetic diversity, ecology and traditional uses. Angola's once vast populations of *Welwitschia mirabilis* await research on their population dynamics and potential resilience to overgrazing by cattle in Iona National Park. The bizarre patterns of 'fairy circles' of the Namib, 'fairy forests' of the Lundas and Moxico, of dwarf succulents along the desert margins of Benguela, and the ecological role of fog along the entire coast, are all ecological puzzles that need elucidation.

Many of these topics might at first sight appear of little more than academic interest, but every element of applied science and technology rests on the fundamentals of curiosity-driven enquiry. The baseline compilation of checklists leads to the identification of patterns of endemism and rarity, to be understood through phylogeographic studies across the country. From such studies, increasingly robust scenarios of evolutionary processes may be built. It is these phenomena of the patterns in nature, their ecological functioning and the interactions that drive large-scale environmental dynamics that will ultimately guide sustainable landuse management and inform responses to the impacts of climate change. It is at the level of landscapes and seascapes that the nation's economy and progress is built, and such wide-horizon visioning, underpinned by fundamental research, is needed for the sustainable development agendas of the twenty-first century.

At a landscape scale, an early priority should be to update the vegetation map of Angola to give a more balanced and objective delineation of the country's major vegetation units, for landuse planning and conservation purposes. This can best be achieved with the input of a multi-national team of workers, using modern remote sensing technologies for vegetation classification, mapping and monitoring. Vegetation classification and mapping skills take many years, even decades to develop, and the close collaboration, even leadership, of foreign experts would be valuable. An improved vegetation map will help expedite assessments of ecosystem conservation status, landuse potential, research priorities and opportunities, and help develop a predictive understanding of ecosystem structure and function.

At ecosystem scale, the importance of securing the effective management of protected areas, large and small, terrestrial and marine, is a *sine qua non* to the future of Angola's biodiversity. The biological and cultural importance of relatively small protected areas, such as Ilhéu dos Pássaros, and future protected areas such as Mount Moco, Namba, Cumbira, Tundavala and the rainforests of Cuanza-Norte, Uíge and Zaire, and of turtle nesting grounds along the coast, should not be overlooked during the pursuit of mega-parks that might excite the public and motivate politicians. Targeted studies of the existing protected areas and of the key biodiversity hotspots identified by the Ministry of Environment in successive strategies (GoA 2006, 2011, 2018) should be given priority, as these protected areas and hotspots most probably hold more than 80% of Angola's floral and faunal diversity in less than 15% of the country's land area. As training grounds for young biologists and conservation scientists, protected areas have no equal. Furthermore, identifying and rigorously protecting near-pristine zones within otherwise threatened protected areas such as Quiçama, Luando, Iona and Mupa, and of the country's extensive coastline, should be an integral part of any protected areas strategy. The marine environment is especially sensitive to the impacts of human activities and science-based marine spatial planning is essential if long-term conflicts between humans and the marine environment are to be avoided. The importance of a focus on the biodiversity hotspots and on the existing protected areas of Angola – the repositories of the country's natural wealth – is self-evident.

Conclusions

These outlines are of necessity simplistic and preliminary. Each of the chapters in this book identifies research questions that can be addressed by an emerging generation of Angolan biodiversity scientists and conservation professionals. The challenges are exciting and demanding – offering multiple opportunities for intellectual stimulation, knowledge generation and international collaboration. Angola is truly alive with research and conservation opportunities. The country is still blessed with vast areas of rich wilderness and unique habitats, and has the opportunity to stimulate scientists, conservationists and the general public to participate in programmes of research and effective biodiversity conservation management. As this synthesis volume demonstrates, the limits are boundless. *Carpe diem*!

References

CBD (2010) Strategic plan for biodiversity 2011–2020 and the Aichii targets. Secretariat of the Convention on Biological Diversity, Montreal

Ceríaco LMP, Marques MP, Bandeira SA et al (2016) Anfíbios e Répteis do Parque Nacional da Cangandala. Instituto Nacional da Biodiversidade e Áreas de Conservação/Museu Nacional de História Natural e da Ciência, Luanda/Lisboa, 96 pp

GoA (Government of Angola) (2006) National biodiversity strategy and action plan (2007–2012). Ministry of Urban Affairs and the Environment, Luanda, 55 pp

GoA (Government of Angola) (2011) Plano Estratégico da Rede Nacional de Áreas de Conservação de Angola (PLERNACA). Ministério do Ambiente, Luanda

GoA (Government of Angola) (2018) Plano Estratégico para o Sistema de Áreas de Conservação de Angola (PESAC). Ministério do Ambiente, Luanda

Marques MP, Ceríaco LMP, Blackburn DC, Bauer AM (2018) Diversity and distribution of the amphibians and terrestrial reptiles of Angola. Atlas of historical and bibliographic records (1840–2017). Proceedings of the California Academy of Sciences Series 4, 65:1–501

Mills MSL (2018) The special birds of Angola/As Aves Especiais de Angola. Go-Away-Birding, Cape Town

Vaz da Silva B (2015) Evolutionary history of the birds of the Angolan highlands – the missing piece to understand the biogeography of the Afromontane forests. MSc thesis. University of Porto, Porto

Vaz Pinto P (2018) Evolutionary history of the critically endangered giant sable antelope (Hippotragus niger variani): insights into its phylogeography, population genetics, demography and conservation. PhD thesis. University of Porto, Porto

www.ingramcontent.com/pod-product-compliance
Lightning Source LLC
LaVergne TN
LVHW080058220926
848484LV00007BA/1914